PRINCIPLES OF SEDIMENTOLOGY AND STRATIGRAPHY

SECOND EDITION

PRINCIPLES OF SEDIMENTOLOGY AND STRATIGRAPHY

Sam Boggs, Jr.
University of Oregon

Prentice Hall
Upper Saddle River, New Jersey 07458

Library of Congress Cataloging-in-Publication Data

Boggs, Sam.
    Principles of sedimentology and stratigraphy / Sam Boggs, Jr.—2nd ed.
       p.   cm.
    Includes bibliographical references and index.
    ISBN 0-02-311792-3
    1. Sedimentation and deposition.   2. Geology, Stratigraphic. I. Title.
QE571.B66   1995
552'.5—dc20
                                          94-35120
                                                CIP

Editor: Robert A. McConnin
Production Editor: Mary Harlan
Text Designer: Anne Flanagan
Cover Designer: Anne Flanagan
Production Buyer: Pamela D. Bennett
Illustrations: Academy ArtWorks, Inc.

This book was set in Melior by The Clarinda Company and was printed and bound by R. R. Donnelley & Sons Company. The cover was printed by Phoenix Color Corp.

 © 1995 by Prentice-Hall, Inc.
A Simon & Schuster Company
Upper Saddle River, New Jersey 07458

Printed in the United States of America

10 9 8 7 6

ISBN: 0-02-311792-3

Prentice-Hall International (UK) Limited, *London*
Prentice-Hall of Australia Pty. Limited, *Sydney*
Prentice-Hall of Canada, Inc., *Toronto*
Prentice-Hall Hispanoamericana, S. A., *Mexico*
Prentice-Hall of India Private Limited, *New Delhi*
Prentice-Hall of Japan, Inc., *Tokyo*
Simon & Schuster Asia Pte. Ltd., *Singapore*
Editora Prentice-Hall do Brasil, Ltda., *Rio de Janeiro*

To Sumiko, Barbara, Steve, Cindy, Ashley, Jamie, and Amy

Preface

New concepts and methods of studying sedimentary rocks and stratigraphic successions have arisen in the short time since publication of the first edition of *Principles of Sedimentology and Stratigraphy* in 1987. The second edition has been thoroughly revised to incorporate these new concepts, and substantial changes have been made in the arrangement and treatment of topics to make the book easier to use. The second edition remains a comprehensive, in-depth treatment of the fundamental principles of sedimentology and stratigraphy, which covers the processes that form sedimentary rocks, describes the important physical, chemical, biological, and stratigraphic characteristics of these rocks, and explores the various ways in which study of sedimentary rocks is used to interpret depositional environments, changes in ancient sea levels, and other aspects of Earth history. The book also discusses recent developments in the important fields of magnetostratigraphy, seismic stratigraphy, sequence stratigraphy, and isotope stratigraphy. It is suitable for a combined sedimentology–stratigraphy course or for separate sedimentology or stratigraphy courses; thus it is not necessary for students to purchase separate texts for these courses. The book is aimed primarily at undergraduate students, particularly upper-division students, but it is suitable also for graduate students. Professionals may likewise find the book useful to update their knowledge of recent developments in sedimentology and stratigraphy. Although thorough and rigorous, the book requires only modest preparation in geology, chemistry, and mathematics.

All chapters in the second edition of *Principles of Sedimentology and Stratigraphy* have been updated, including the addition of approximately 400 new references and 150 new figures and tables. Important additions include a section on soils and paleosols in chapter 2 and a new chapter on basin analysis to emphasize the practical applications of sedimentology and stratigraphy. Although the separate chapter on deposition of chemical/biochemical sedimentary rocks has been eliminated, pertinent material on this topic has been incorporated into other, appropriate parts of the book.

The separate chapter on diagenesis has also been deleted; discussion of diagenesis has been woven into the chapters dealing with the composition and characteristics of sedimentary rocks to make the discussion more meaningful. The book is profusely illustrated with figures and tables, which reduces the need for instructors to provide additional handouts. It is extensively referenced and up to date, thus encouraging students to consult recently published literature. References are grouped for ease of use at the end of the book, and suggestions for further reading are located at the end of each chapter.

The second edition is divided into six parts. Part 1 introduces the subject of sedimentology and stratigraphy and gives a brief overview of their importance in geologic study. Part 2 discusses the fundamental processes of weathering, soil formation and soils, and sediment transport and deposition. Part 3 describes the textures and structures of sedimentary rocks, and Part 4 deals with their composition, classification, origin, and diagenesis. Part 5 treats the interpretation of ancient sedimentary environments and the characteristics of sedimentary rocks formed in the major sedimentary environments. Part 6 covers the fundamental principles of stratigraphy (with separate chapters on lithostratigraphy, seismic stratigraphy, magnetostratigraphy, biostratigraphy, and chronostratigraphy) and basin analysis.

ACKNOWLEDGMENTS

I wish to thank the following individuals who reviewed the manuscript for the second edition and gave me their critical comments: Ian Evans, University of Houston; Douglas W. Haywick, University of South Alabama; Gary D. Johnson, University of South Dakota; James G. Ogg, Purdue University; and Roger C. Walker, McMaster University. Those who commented on portions of the text include David W. Andersen, San Jose State University; Roger J. Bain, University of Akron; Charles W. Byers, University of Wisconsin–Madison; James Clark, Calvin College; William J. Frazier, Columbus College; Lucy E. Harding, Middlebury College; Patricia Kelley, University of Mississippi; David T. King, Jr., Auburn University; Christopher T. Ledvina, Northeastern Illinois University; Robert Lowright, Susquehanna University; James McKee, University of Wisconsin–Oshkosh; Charlotte Mehrtens, University of Vermont; Barry Perlmutter, Jersey City State College; Anthony F. Randazzo, University of Florida; Robert Stanton, Texas A&M University; Donald J. Thompson, California University of Pennsylvania; and Michael Anthony Velbel, Michigan State University. Reviewers for the first edition include Charles W. Byers, University of Wisconsin–Madison; H. Edward Clifton, Conoco, Inc.; Gordon G. Goles, University of Oregon; Ralph E. Hunter, U.S. Geological Survey; Charlotte J. Mehrtens, University of Vermont; Jeffrey Mount, University of California–Davis; and William N. Orr, University of Oregon.

I wish also to thank my wife, Sumiko, for her unstinting support during preparation of the manuscript.

Contents

PART 5   SEDIMENTARY ENVIRONMENTS   287

PART **6** PRINCIPLES OF STRATIGRAPHY AND BASIN ANALYSIS   489

14 Lithostratigraphy   491

15 Seismic Stratigraphy   531

PART 1

INTRODUCTION

Chapter 1 Development and Application of Sedimentology and Stratigraphy

Delicate Arch, Arches National Park, near Moab, Utah; Entrada Formation (Jurassic). La Sal Mountains are in the background.

Sedimentary rocks cover roughly three-fourths of Earth's surface. Their sheer volume commands our attention; however, there are many reasons other than their abundance why geologists are interested in sedimentary rocks. All geologic study is aimed in one way or another at developing a better understanding of Earth history. All rocks, whether sedimentary, igneous, or metamorphic, contain clues to some aspect of this history, but sedimentary rocks are unique with regard to the information they provide. From the fossils, textures, and structures in sedimentary rocks, trained geologists can decipher clues that provide insight into past climates, oceanic environments and ecosystems, and even the configurations of ancient land systems and the locations and compositions of mountain systems long since vanished. Thus, study of sedimentary rocks forms the primary basis for the sciences of paleoclimatology, paleogeography, paleoecology, and paleoceanography. In addition, many sedimentary rocks have economic significance. Most of the world's oil and gas and all of its coal are contained in sedimentary rock successions. Uranium, gypsum, phosphorites, and many other economically valuable minerals occur also in these rocks.

The record of Earth history locked up in sedimentary rocks dates back to almost 4 billion years. It is the study of this reservoir of Earth history that constitutes the sciences of sedimentology and stratigraphy. **Sedimentology** is the scientific study of the classification, origin, and interpretation of sediments and sedimentary rocks. It is often difficult to draw a sharp distinction between sedimentology and **stratigraphy,** which is defined simply and broadly as the science of rock strata. In general, however, sedimentology is concerned with the physical (textures, structures, mineralogy), chemical, and biologic (fossils) properties of sedimentary rocks and the processes by which these properties are generated. It is these properties that provide much of the basis for interpreting paleoclimatology, paleogeography, and paleoecology. Stratigraphy, on the other hand, is concerned more with age relationships of strata, successions of beds, local and worldwide correlation of strata, and stratigraphic order and chronological arrangement of beds in the geologic column. Stratigraphy finds special applications in the study of plate reconstructions (plate tectonics) and in the unraveling of the intricate history of landward and seaward movements of ocean shorelines (transgressions and regressions) and rise and fall of sea level through time. Particularly exciting developments in stratigraphy have come about recently through application of the principles of seismology and paleomagnetism to stratigraphic problems.

This book provides an integrated view of sedimentology and stratigraphy. The first chapter offers a brief history of sedimentology and stratigraphy to give readers some insight into their development as sciences. Chapter 1 also describes some of the techniques and tools that geologists use to study sedimentary rocks, and it discusses in general terms some applications of sedimentologic and stratigraphic study to interpretation of Earth history and exploitation of economic resources. The remaining chapters are devoted to description and discussion of the processes that form sedimentary rocks; the physical, chemical, and biological properties of rocks that result from these processes; and interpretation of these properties and stratigraphic relationships in terms of Earth history.

1

Development and Application
of Sedimentology and Stratigraphy

1.1 HISTORICAL DEVELOPMENT OF SEDIMENTOLOGY AND STRATIGRAPHY AS SCIENCES

Introduction

Today, we think of sedimentology and stratigraphy as fully developed modern sciences in which sophisticated techniques and tools are used routinely to pursue and explore a host of advanced concepts and problems. The development of sedimentology and stratigraphy as modern sciences has not, however, come about quickly and easily. The study of sedimentary rocks in some form can be traced back at least to the sixteenth century. The gradual evolution of sedimentology and stratigraphy as sciences since this early beginning constitutes one of the more fascinating chapters in the overall history of the earth sciences.

The development of sedimentologic and stratigraphic study has been described by several workers, including Dunbar and Rogers (1957), Weller (1960), Krumbein and Sloss (1963), Pettijohn (1975), and Friedman and Sanders (1978). From these and other accounts we can place the beginning of sedimentology and stratigraphy at about A.D. 1500 with the observations of Leonardo da Vinci on the fossils in sedimentary rocks of the Italian Apennines. Da Vinci deduced that fossils are the remains of ancient organisms and concluded that the shells visible in the rocks belonged to animals that lived in a sea that once covered that area. Very little additional scientific study of sedimentary rocks appears to have taken place until about the middle of the seventeenth century, when Nicolas Steno began investigating the fossil-bearing strata around Rome. On the basis of his study, Steno made the first known attempt to place strata in some kind of positional order. In 1669 he postulated that in any sequence of flat-lying strata, the oldest layers are at the bottom and the youngest at the top, a concept referred to as the **principle of superposition.** He also proposed the principle of **original horizontality,** which states that beds are always deposited initially in a nearly horizontal position (Fig. 1.1), even though they may later be found dipping steeply. These principles are still considered fundamental to stratigraphy. About the same time that Steno made his studies in Italy, Robert Hooke in England initiated use of the microscope to study fossils. Hooke also apparently suggested the possibility of using fossils to make chronologic comparison of sedimentary rocks, although such comparisons were not actually attempted until much later.

From this rather modest beginning in the sixteenth and seventeenth centuries, study and understanding of sedimentary strata have continued to grow, although progress at times has been slow and somewhat erratic. A very generalized discussion of some of the important stages in the gradual evolution of sedimentology and stratigraphy into modern sciences is presented in the following section.

Organization of Sedimentary Rocks into Stratigraphic Successions

As interest in sedimentary rocks gradually increased into the eighteenth century, it became apparent to serious workers that systematic study of rock strata required organization of the strata into some kind of stratigraphic sequence. The most important examples of early attempts at stratigraphic organization are those made by Giovanni Arduino, an Italian professor and provincial director of mines, and Johann Gottlob Lehman, a German professor of mineralogy. Arduino (1714–1795) divided all rocks into four groups:

1. **primary** mountains composed of rocks containing metallic ores but devoid of fossils
2. **secondary** mountains consisting of stratified and well-lithified rocks containing fossils but without ore deposits
3. **tertiary** low mountains consisting of fossiliferous but unconsolidated gravels, sand, and clays with associated volcanic rocks
4. **alluvium** consisting of earth and rocky materials washed down from mountains and overlying the other kinds of rocks

FIGURE 1.1 These Tertiary sedimentary rocks, exposed in the Oregon Coast Range, were deposited by turbidity currents in the deep ocean in a nearly horizontal position. The rocks were folded slightly (arrow) during uplift above sea level but remain in a nearly horizontal position. Rock layers shown at the bottom of the photograph are older than layers at the top.

Here we see introduced the concept of stratigraphic ordering by relative age, as implied by the terms primary, secondary, and tertiary. Lehmann (1719–1767) recognized three classes of mountains very similar to those of Arduino:

1. primitive mountains composed of crystalline rocks devoid of fossils and unstratified or poorly bedded
2. layered mountains, or secondary mountains, consisting of well-bedded strata with fossils and containing material eroded from older rocks
3. mountains composed of loosely consolidated surficial sands and gravels called alluvium

Arduino and Lehmann did not know the actual ages of the strata in their groups and may have grouped together rocks of widely differing ages. Nonetheless, their efforts at organization were important steps in developing the concept of relative age as a basis for ordering stratigraphic successions. The Tertiary even survived as a name to become part of modern stratigraphic terminology. The term Quaternary, which is also used today for one of the geologic systems of Cenozoic rocks, is based on Arduino's fourth category of rocks (alluvium). It was introduced into the geologic literature by Desnoyers in France in 1829 as a specific term to parallel primary, secondary, and tertiary.

The Geologic Cycle and Uniformitarianism

The emergence of geology as a modern science began in the late eighteenth century with the work of James Hutton (1727–1797). Hutton, a Scottish physician and gentleman farmer, was the first worker to recognize and describe the cyclic behavior of earth processes and materials. He visualized tectonic uplift, erosion, sediment transport, and deposition as parts of a continuous cycle, repeated throughout geologic time. He wrote in 1788 that "the result, therefore, of our present enquiry is that we find no vestige of a beginning, no prospect of an end." This concept was labeled by later workers as the **geologic cycle.** On the basis of his observations of the cyclic behavior of geologic processes and his penetrating ideas concerning the significance of the rock record, Hutton has been credited by many subsequent workers with conceiving the **principle of uniformitarianism.** This principle, also sometimes called actualism, is commonly expressed to mean that the processes that shaped Earth throughout geologic time were the same as those observable today. It has often been stated simply as "The present is the key to the past."

Uniformitarianism became one of the guiding principles of geologic philosophy and has exerted great influence on geologic thinking since Hutton's time. Nonetheless, controversy exists regarding the exact meaning and usefulness of the term. Shea (1982) challenges the commonly held concept that the present is the key to the past and sets forth what he calls "twelve fallacies of uniformitarianism". He argues that the geologic literature is riddled with false and misleading statements about uniformitarianism and that uniformitarianism consists only of the scientific approach to the study of nature. That is, as scientists we must follow the rule of simplicity, the rule of choosing first the simplest hypothesis that fits the relevant observations and that also leads to least complexity in overall theory. Shea also suggests that Hutton was not the first to propose uniformitarianism in the sense of proposing scientifically reasonable interpretations of geologic phenomena. Nonetheless, Hutton's place in the history of geology is secure. He was the first scientist to present the ideas embodied in the concept of uniformitarianism in such a way that, to quote Bushman (1983), they "illuminated the rock record as it had never been illuminated before, and geologists were able to see and interpret things so effectively that geology underwent its greatest development and became established as an important field of science." For additional insight into uniformitarianism, see Hallam's (1989) *Great Geological Controversies,* Chapter 2.

Birth of Biostratigraphy and Stratigraphic Correlation

William Smith (1769–1839), an English surveyor and engineer, is commonly credited with initiating the science of biostratigraphy. In his work as a canal builder, Smith discovered that different layers of strata are characterized by unique assemblages of fossils. He displayed remarkable insight for his time by initiating the use of fossils for correlation of sedimentary strata from one area to another. He demonstrated the practical importance of the principle of superposition, first suggested by Steno and later expressed more fully by Hutton. Through his study of the relationship of fossils and rock strata, he also laid the foundation for development of the **law of faunal succession,** the formal statement of the principle that fossil organisms succeed each other in the stratigraphic record in an orderly, recognizable fashion. Smith was also one of the first persons to do useful geologic mapping. In recognition of his pioneer work in biostratigraphy, Smith is sometimes referred to as the "father of stratigraphy."

Smith's work was followed in 1842 by introduction of the concept of the biologic **stage.** A French paleontologist named Alcide d'Orbigny conceived stages as major subdivisions of strata, each systematically following the other and each containing a characteristic assemblage of fossils. D'Orbigny believed that his stages had worldwide extent and could be recognized everywhere. Following closely on the heels of the stage concept, a German geologist named Albert Oppel introduced in 1856 the concept of the biologic **zone** or **biozone.** Oppel visualized biozones as small-scale stratigraphic units that include all the strata deposited during the existence of specific fossil organisms. He based his zones on the overlapping stratigraphic ranges of these organisms. Stratigraphic range is the stratigraphic interval between the first appearance and the last appearance of a fossil species in the stratigraphic record. Using the overlapping ranges of species, he found that he could subdivide stages and delineate the boundaries between small-scale rock units on the basis of fossil content, irrespective of the lithology of the fossil-bearing beds. Furthermore, these zones could be correlated over wide distances.

The pioneer work of Smith, d'Orbigny, and Oppel laid the foundation for developing a standard worldwide stratigraphic column. The stratigraphic succession gradually unfolded with study of well-exposed sections of strata in different areas of Europe. By the early part of the twentieth century, a composite **standard stratigraphic column** covering the entire rock record had been erected, and major subdivisions, called systems, were named and defined. Relative ages were initially assigned to the systems on the basis of their fossil content. With the development of radiochronologic methods for estimating absolute ages of rocks, absolute ages were gradually assigned to the boundaries of the systems.

Development of Petrographic Microscopy

An English scientist named Henry Clifton Sorby initiated microscopic study of rocks around 1850 through his work on limestones. Sorby's work paved the way for development of the science of **petrography**—the branch of geology dealing with the description and systematic classification of rocks, of all types, especially by means of microscopic examination of thin slices of rock called **thin sections** (Fig. 1.2). Students of sedimentary rocks failed to follow up on the momentum begun by Sorby and allowed

FIGURE 1.2 Small fossil organisms and other calcium carbonate grains as they appear in a thin section of a limestone viewed with a petrographic microscope. The two prominent *Fusulinid* foraminifers (arrows) are about 1 mm in diameter.

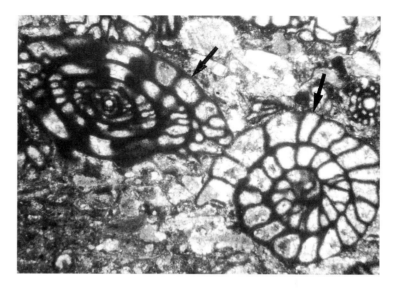

igneous geologists to largely develop the science of petrography; however, interest in sedimentary petrography resurged in the early part of the twentieth century. Today, thin-section petrography remains a standard and essential tool for study of sedimentary rocks.

Sedimentologic and Stratigraphic Syntheses

For a century following Sorby's introduction of thin-section microscopic techniques, progress in sedimentology and stratigraphy was relatively slow. New stratigraphic, paleontologic, and sedimentologic data were collected and analyzed during this period, and knowledge of sedimentary rock characteristics steadily grew. One searches in vain, however, for evidence of significant new discoveries or important new concepts that brought advances in understanding of sedimentology and stratigraphy of the magnitude of those made by Hutton and Smith. This period is distinguished particularly by the appearance in print of several classic sedimentologic and stratigraphic syntheses that brought together the cumulative knowledge of the time. Particularly noteworthy are Lyell's *Principles of Geology* published in 1833, Grabeau's *Principles of Stratigraphy* (1913), and Twenhofel's *Treatise on Sedimentation* (1926, 1932) and *Principles of Sedimentation* (1939, 1950). Lucien Cayeux's *Introduction à l'étude Pétrographique des roches sédimentaires* published in 1931 (in French) and *Les roches sédimentaires de France: Roches carbonatées* published in 1935 are also classic examples of outstanding syntheses produced during this period.

The Geologic Revolution—Seafloor Spreading and Global Plate Tectonics

The late 1950s and 1960s ushered in a new era in geologic study marked by significant increase in research activity and data gathering by all types of earth scientists. This renaissance in geology came about for many reasons, including greater availability of federal and private funds to support research and development of new tools and techniques for field and laboratory study; an accelerated pace in exploration, drilling, and research by energy companies; and expanded geophysical exploration, particularly in the ocean basins. This period of renewed and intensified research soon led to discoveries that brought about an almost quantum jump in understanding of Earth history and rapid advancement in ideas and concepts in every branch of the earth sciences, including sedimentology and stratigraphy.

Exploration of the ocean basins intensified in the 1950s using geophysical techniques involving magnetic, seismic, and gravity surveying. These geophysical surveys, along with other research efforts, brought rapid advances in knowledge and understanding of tectonic relationships and depositional settings in the ocean. They generated a massive body of data that led to significant new ideas about the tectonic evolution of continents and ocean basins. These fresh insights paved the way in the late 1950s and early 1960s for the birth of one of the most far-reaching concepts in geologic philosophy, **seafloor spreading** and **global plate tectonics.** This concept envisions Earth's crust as a rigid layer broken into several distinct segments or plates. These crustal plates move slowly about with respect to each other, sliding over a deeper, plastic layer beneath the crust. They spread apart along mid-ocean ridges where new crustal rocks are generated by volcanism. Plates move together or converge in major deep-sea trenches, where one plate may thrust beneath another to form a **subduction zone** (Fig. 1.3). The implications of seafloor spreading and plate tectonics have revolutionized all branches of the earth sciences and have dramatically changed many previously held ideas about the thickness and age of oceanic rocks, the tectonic setting in which sediments accumulate, and the processes by which sedimentary rocks become tectonically emplaced.

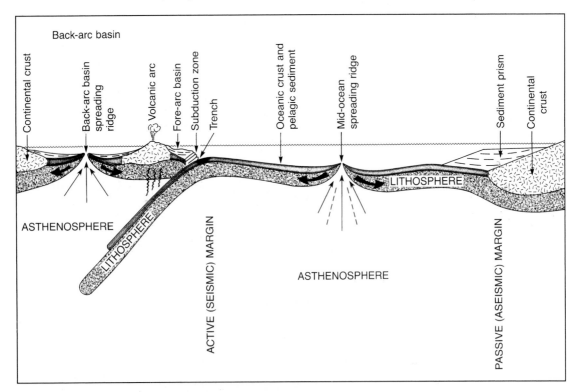

FIGURE 1.3   Idealized, schematic illustration of the principles of plate tectonics.

One of the most significant research spinoffs spawned by the plate tectonics revolution was initiation of a drilling program to recover cores of sedimentary and volcanic rock from the deep ocean floor. This program, commonly referred to as the Deep Sea Drilling Project (DSDP), was initiated in 1963 by the Joint Oceanographic Institutes Deep Earth Sampling Program (JOIDES), a group of U.S. planning institutions. The Deep Sea Drilling Project was financed mainly by the U.S. National Science Foundation until 1975; subsequently, it received financial and planning support from a number of other countries. The National Science Foundation commissioned the construction of a research drilling vessel called the *Glomar Challenger,* launched in 1968. The *Glomar Challenger* was capable of taking incremental cores up to 1000 m in cumulative length from a single hole while floating above the ocean floor in water as much as 6000 m deep. Prior to development of this capability to do rotary drilling and coring in deep water, the longest cores that could be taken in deep water by piston coring were about 18 m.

Under the management of Scripps Institution of Oceanography, San Diego, California, and with international cooperation, the *Glomar Challenger* drilled almost 600 holes and recovered cores having a cumulative length exceeding 56 km (35 mi). It steamed more than 483,000 km (300,000 mi), a distance greater than the distance from Earth to the moon (Warme, Douglas, and Winterer, 1981).

The *Glomar Challenger* was decommissioned in 1983 and a new research vessel, the *JOIDES Resolution* (Fig. 1.4) was leased to continue drilling operations. Manage-

FIGURE 1.4 The deep ocean drilling ship *JOIDES Resolution,* operated by Texas A & M University under the Ocean Drilling Program. The *Resolution* is 470 ft (143 m) long with a drilling tower approximately 200 ft (61 m) high. (Photograph courtesy of Ocean Drilling Program, Texas A & M University.)

ment of the new research vessel shifted to Texas A & M University, College Station, Texas in 1984. The name of the program was changed to the Ocean Drilling Program (ODP), and the first ODP cruises began in 1985. ODP has continued drilling, with an international group of scientists, in all the major basins of the world. International participants in the program (early 1994) include France, Germany, Japan, the United Kingdom, Canada, Australia, and the European Science Foundation (Belgium, Denmark, Finland, Iceland, Italy, Greece, The Netherlands, Norway, Spain, Sweden, Switzerland, and Turkey). As of January 1994, the *JOIDES Resolution* has drilled 773 holes at 311 sites and recovered approximately 88 km (55 mi) of new core; the deepest hole drilled extended to a depth of 2000 m below the seafloor (personal communication, Karen Riedel, ODP).

The cores taken by the *Glomar Challenger* and the *JOIDES Resolution,* together with other data such as seismic records, have furnished firsthand information about the age, thickness, and character of oceanic rocks that provides spectacular confirmation of seafloor spreading. In addition, deep-sea drilling data have furnished remarkable insight into such aspects of ocean history as the origin of ocean basins (e.g., opening of the Japan Sea back-arc basin), changes in organic productivity cycles in the ancient ocean, episodes of volcanic activity, and sediment transport by mechanisms such as turbidity currents.

1.2 NEW CONCEPTS AND NEW ANALYTICAL TOOLS FOR STUDYING SEDIMENTARY ROCKS

The geologic revolution that began in the 1960s with the development of the plate tectonics concept has had an enormous impact on the evolution of sedimentology and stratigraphy as disciplines and on the methods of studying sedimentary rocks. These changes in the disciplines have been so profound that despite their long history we can look upon sedimentology and stratigraphy as virtually new sciences. Miall (1990, p. 5) categorizes these major changes under seven fundamental headings:

1. refinements in chronostratigraphy, the study of the absolute ages of rocks, particularly integration of radiometric, magnetostratigraphic, and biostratigraphic data (see Chapters 18, 16, and 17, respectively, this book)
2. evolution of sedimentology through facies studies and facies models into a mature science capable of explaining the origins of rocks
3. development of the concept of depositional systems, a complete package of environments and its sedimentary products, formulated on the basis of Walther's Law and the concept of stratigraphic sequences (Chapter 14, this book)
4. evolution of modern techniques of seismic stratigraphy (Chapter 15, this book)
5. revitalization of interest in all forms of stratigraphic cycles and cyclicity
6. emergence of powerful techniques for numerical simulation and computer modeling of basin evolution (Chapter 19, this book)
7. emergence of a suite of basin models, characterized by distinctive structural and stratigraphic geometries and paleogeographic styles

These changes in emphasis and focus of sedimentology and stratigraphy have resulted in the development of many new, advanced tools and techniques for the study of sedimentary rocks. Today's geoscientist has access to a variety of sedimentologic and stratigraphic techniques that include both long-established classic methods and sophisticated modern techniques. Students and professional geologists today can take advantage of a dazzling variety of tools and techniques for the study of sedimentary rocks. Methods of field study include the standard techniques of measuring stratigraphic sections and field mapping, as well as more sophisticated techniques such as magnetic and seismic surveying. Samples of rocks and fossils can be collected from surface outcrops and from well cores and cuttings retrieved from subsurface formations. Also, characteristics of subsurface formations—such as bed thickness, porosity, and lithology—can be evaluated by advanced well-logging techniques. In the laboratory, sedimentologists and stratigraphers have access to a variety of tools for analyzing the chemical and isotope composition of minerals and rocks; for examining minerals, fossils, and rock textures at high magnification; for measuring the grain sizes of sediment grains; and for determining the ages of rocks. Brief discussions of some of these techniques are given in appropriate sections of this book.

1.3 APPLICATIONS OF SEDIMENTOLOGIC AND STRATIGRAPHIC STUDY

Sedimentologic and stratigraphic study is aimed ultimately at developing a deeper understanding of the origin and evolution of Earth through time. What, for example, did Earth look like 100, 500, or 2000 million years ago? Where were the continents with respect to the oceans? Where were the mountains and shorelines at various times in the past? Can sedimentologic and stratigraphic study answer such questions? All earth materials hold clues to Earth history. Sedimentary rocks and the fossils they contain are particularly important because they provide significant information about Earth's past geography, climates, depositional environments, life forms, and ocean composition. As mentioned, the fossils in sedimentary rocks also provide a means for determining the relative ages of sedimentary rocks that makes possible the organization of these layered rocks into a meaningful stratigraphic succession. The following paragraphs explore a few of the ways that study of sedimentary rocks has contributed and continues to contribute to knowledge of Earth history.

Paleogeography and Environmental Analysis

Paleogeography is the study and description of the physical geography of the geologic past. It entails historical reconstruction of the patterns of Earth's surface, or of a given area, at a particular time in the geologic past, and the successive changes in geographic patterns through time. It is the science that tells us literally how the face of Earth has changed with time. Paleogeography thus involves, among other things, interpretation of the changing relationship of continents and oceans. On a global scale, great strides have been made in interpreting the changing relative positions of ancient continental masses by applying the principles of seafloor spreading and plate tectonics. On a smaller, regional scale, geologists can study the characteristics of ancient sedimentary rocks and the stratigraphic relationships of these rocks to reconstruct ancient sedimentary environments and ecological conditions. This knowledge allows geologists to fix the approximate position of shorelines at various times in the geologic past and to map advances and retreats of the ocean throughout geologic time. Interpretation of ancient depositional environments involves study of the textures, structures, fossils, and other properties of sedimentary rocks as a basis for deducing ancient environmental conditions and processes (see Chapter 9). **Paleoecology,** the science of the relationships between ancient organisms and their environment, is an integral part of environmental analysis.

Paleogeography also involves interpretation of the relative positions of major ocean basins and uplifted sediment source areas. The former presence of vanished continental highlands can be deduced from the nature of the sediments shed from these mountains and deposited in adjacent ocean basins as the mountains were lowered by erosion. For example, a chain of ancestral "Rocky Mountains" is believed to have arisen across the Colorado region of western North America in Pennsylvanian time. This ancient mountain chain is no longer present, but its former existence is postulated on the basis of minerals, rock fragments, and sedimentary structures preserved in Pennsylvanian-age sediments deposited in sedimentary basins adjacent to the mountains. The composition of the rocks allows geologists to interpret the probable composition of the ancestral mountain chain. Regional grain-size distribution patterns and paleocurrent indicators in the basins also allow geologists to reconstruct the approximate location of the mountains, because these features show the flow direction of currents that carried sediment away from the mountain. Paleocurrent indicators are sedimentary structures such as cross-bedding that indicate the direction of flow of depositing currents.

Paleoclimatology

Paleoclimatology is the study of ancient climates. Paleoclimate analysis is based on identifying paleoclimatic indicators in sedimentary rocks. These indicators include such features as poorly sorted tillites, which suggest glacial climates; distinctive fossils such as palm leaves and corals that indicate warm climates; and distinctive lithologies that point toward deposition under special climatic conditions. For example, wind-blown sands and evaporite deposits such as gypsum suggest deposition in arid to semiarid desert climates. Extensive coal beds, on the other hand, suggest moist climatic conditions under which lush swamp vegetation flourished. These very simple examples are intended only to emphasize the point that paleoclimate analysis depends almost entirely on analysis of sedimentary rocks and the minerals, textures, and fossils they contain.

Even clues to evolution of Earth's atmosphere and ocean are locked up in certain types of sedimentary rocks. For example, differences in the relative degree of oxidation of iron-bearing minerals in older and younger Precambrian rocks may indicate changes in the relative levels of oxygen in the early atmosphere. Variations in relative abundance of sulfur isotopes, particularly $^{34}S/^{32}S$ ratios, in ancient evaporite and shale deposits provide evidence of the sulfur isotope composition of the ancient ocean, and thus the sulfate and sulfide composition of the ocean. Oxygen isotope ratios ($^{18}O/^{16}O$) in marine carbonates provide information about past ocean temperature and allow geologists to identify glacial and interglacial episodes in the marine geologic record. The presence of thick, widespread deposits of salt, gypsum, or other evaporites in stratigraphic sequences of various ages not only furnishes information about past climates but also suggests temporary changes in salinity of the ocean, at least locally. Changes in salinity are inferred because deposition of large quantities of evaporites may locally deplete the ocean of salts, causing a temporary episode of lowered salinity. These examples are highly generalized, but they illustrate how we can use the chemical composition and mineralogy of sedimentary rocks to illuminate Earth history.

Applications in Industry

Finally, many of the principles of sedimentology and stratigraphy have practical application. A good example of such application is seen within the fossil fuel industry.

Almost all of the world's oil and gas and all of its coal occur in sedimentary rocks. Successful exploration for oil and gas requires the services of trained geologists who can evaluate the sedimentological characteristics and stratigraphic relationships of subsurface formations and identify favorable reservoir and trapping conditions for petroleum (Fig. 1.5). Knowledge of rock properties such as porosity and permeability, organic geochemistry, age, and stratigraphic relationships thus plays a significant role in the search for new fossil fuels. The development of the field of seismic stratigraphy as a tool for petroleum exploration is an outstanding example of the practical application of stratigraphic principles. The value of applied sedimentology and stratigraphy to industry is clearly indicated by the fact that energy companies have long been the major employers of geologists, particularly sedimentologists, stratigraphers, and paleontologists.

Sedimentology and stratigraphy also have application within the mineral industries. Certain types of ores, including uranium, vanadium, manganese, iron, lead, zinc, and copper, may become localized in sedimentary deposits from particular environments such as fluvial or reef environments. Geologists' knowledge of and ability to interpret ancient sedimentary environments on the basis of physical and biological characteristics of sedimentary units thus become of paramount importance in this type of exploration. Exploration for commercial deposits of phosphate rock, salt, gypsum, and other nonmetallic mineral deposits is equally dependent upon knowledge of environments and stratigraphy. Other examples of practical applications of sedimentological and stratigraphic principles include exploration for groundwater resources, which occur primarily but not exclusively in sedimentary rocks, and application to engineering problems. Engineering problems involving sediment transport in estuaries and other nearshore regions, shore-zone erosion, silting up of reservoirs, stream channel control, highway and dam construction, and foundation evaluation are but a few examples.

The principles of sedimentology and stratigraphy also find application in the practice of environmental geology. These principles are applied to a wide variety of environmental problems, ranging from the study of groundwater and groundwater-related problems to proper siting of sanitary landfills. Somewhat more esoteric appli-

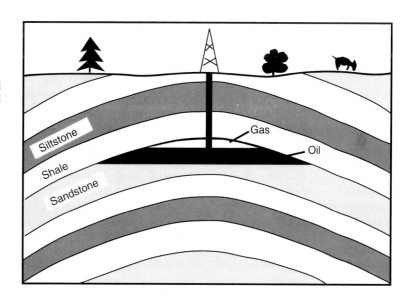

FIGURE 1.5 Schematic diagram illustrating how gas and oil are trapped in folded sedimentary rocks. The nearly impermeable shale lying above the porous and permeable sandstone reservoir bed prevents gas and oil from escaping upward from the fold trap.

cations of sedimentological principles include study of the influence of sedimentation on concrete aggregates; study of clay minerals in the field of ceramics; evaluation of the quality of foundry sands; and use of sedimentary rocks as storage sites for water, oil, natural gas, and radioactive and chemical wastes.

FURTHER READINGS

Ager, D. V., 1981, The nature of the stratigraphical record, 2nd ed.: John Wiley & Sons, New York, 122 p.

Brenchley, P. J., and B. P. J. Williams (eds.), 1985, Sedimentology: Recent developments and applied aspects: Geol. Soc. Spec. Pub. 12, Blackwell, Oxford, 320 p.

Conklin, B. A., and J. E. Conklin (eds.), 1984, Stratigraphy: Foundations and concepts: Benchmark Papers in Geology, v. 82, Van Nostrand Reinhold, New York, 365 p.

Hallam, A., 1973, A revolution in the earth sciences: From continental drift to plate tectonics: Oxford University Press, London, 127 p.

Holland, H. D., and A. F. Trendall (eds.), 1984, Patterns of change in Earth evolution: Springer-Verlag, New York, 432 p.

Tucker, M. (ed.), 1988, Techniques in sedimentology: Blackwell, Oxford, 394 p.

Warme, J. E., R. G. Douglas, and E. L. Winterer (eds.), 1981, The Deep Sea Drilling Project: A decade of progress: Soc. Econ. Paleontologists and Mineralogists Spec. Pub. 32, Tulsa, Okla., 564 p.

PART 2

WEATHERING PROCESSES AND PRODUCTS AND SEDIMENT TRANSPORT

Chapter 2 Weathering and Soils
Chapter 3 Transport and Deposition of Siliciclastic Sediment

Logan Butte, Weathered John Day Formation (Oligocene), eastern Oregon. (Photograph by Gregory Retallack.)

Sedimentary rocks are formed through a complex set of processes that include weathering of source rocks, erosion, sediment transport and deposition, and burial diagenesis. All sedimentary rocks are composed of particles or grains of some kind, which may range in size from microns to hundreds of centimeters. Many of these grains are individual mineral crystals; others are composite grains made up of aggregates of crystals bonded or cemented together. Some grains are epiclastic particles, derived by weathering of preexisting (source) rocks. Others are pyroclastic particles, formed through explosive volcanism. A third group of particles are chemically/biochemically formed particles, produced by precipitation from water in lakes or the ocean (e.g., the particles in limestones).

The process of forming sediments and sedimentary rocks begins with **weathering**—the physical disintegration and chemical decomposition of rock that produces solid (minerals, rock fragments) and dissolved chemical products. The solid products of weathering may accumulate *in situ* to form soils, or they may be removed eventually by erosion and transported to depositional basins. These transported weathering products, together with pyroclastic particles that originate through explosive volcanism, are the source materials of sandstones, conglomerates, and shales. Because they are derived from land, these clastic particles (broken fragments of preexisting rock) are referred to as terrigenous grains. Most terrigenous grains are composed in large part of silica; therefore, silicate terrigenous grains are commonly called siliciclastic grains.

Siliciclastic grains are removed from weathering sites by erosion and are transported as solids to depositional basins. Mass-transport processes such as slumps, debris flows, and mud flows are important agents in initial stages of transport of sediment from weathering sites to valley floors. Fluid-flow processes—which include moving water, ice, and wind—move sediment from valley floors to depositional basins at lower elevations. When transport processes are no longer capable of moving sediment, deposition of sands, gravels, and muds takes place—subaerially (e.g., in desert dune fields) or subaqueously in river systems, lakes, or the marginal ocean. Sediment deposited at the ocean margin may be reentrained and retransported tens to hundreds of kilometers into deeper water by turbidity currents or other transport processes. Sedimentary rocks made up of siliciclastic grains are **siliciclastic sedimentary rocks.**

Weathering processes also release from source rocks soluble constituents such as calcium, magnesium, and silica that make their way in surface water and groundwater to lakes or the ocean. When concentrations of these chemical elements become sufficiently high, they are removed by chemical and biochemical processes to form limestones, cherts, and other **chemical/biochemical sedimentary rocks.**

Organic matter is an important part of many soils and sedimentary rocks. Plant material on land is partially broken down by chemical and microbial processes to yield organic residues. Some organic residues accumulate in soils; others are deposited in swampy environments to form peats and coals; and still others may be transported along with weathering detritus to depositional basins. Both plant and animal organic residues are also generated in the ocean and may become deposited along with siliciclastic or biochemical sediment. Sedimentary rocks that contain substantial amounts of organic material are called **carbonaceous sedimentary rocks.**

Chapter 2 focuses on the processes of weathering and soil formation and on the characteristics of soils—particularly ancient soils or paleosols. Sediment transport and deposition are treated in Chapter 3 and in some subsequent chapters in Part 5.

2

Weathering and Soils

2.1 INTRODUCTION

Weathering involves chemical, physical, and biological processes, although chemical processes are by far the most important. Details of these processes may be found in several books devoted to weathering and soil formation, such as those by Birkland (1974), Drever (1985), Colman and Dethier (1986), Lerman and Meybeck (1988), and Nahon (1991). Only a brief summary of weathering processes is presented here to illustrate how weathering acts to decompose and disintegrate exposed rocks, producing particulate residues and dissolved constituents. These weathering products are the source materials of soils and sedimentary rocks.

It is important to understand how weathering attacks source rocks and what remains after weathering to form soils and be transported as sediment and dissolved constituents to depositional basins. The ultimate composition of soil and terrigenous sedimentary rock bears a relationship to the composition of the source rock, but it is clear from study of residual soil profiles that both the mineral composition and the bulk chemical composition of soils may differ greatly from those of the bedrock on which they form. Some minerals in the source rock are destroyed completely during weathering, whereas more chemically resistant or stable minerals are loosened from the fabric of the decomposing and disintegrating rock and accumulate as residues. During this process, new minerals such as ferric oxides and clay minerals may form *in situ* in the soils from chemical elements released during breakdown of the source rocks. Thus, soils are composed of survival assemblages of minerals and rock fragments derived from the parent rocks plus any new minerals formed at the weathering site. Soil composition is governed not only by the parent-rock composition but also by the nature, intensity, and duration of weathering and soil-forming processes.

In the following section we examine the principal processes of subaerial weathering and discuss the nature of the particulate residues and dissolved constituents that result from weathering. We also look briefly at the processes of submarine weathering. Submarine weathering includes both the interaction of seawater with hot oceanic rocks along mid-ocean ridges—a process that leaches important amounts of chemical constituents from hot crustal rocks—and the low-temperature alteration of volcanic rocks and sediments on the ocean floor.

2.2 SUBAERIAL WEATHERING PROCESSES

Physical Weathering

Physical (mechanical) weathering is the process by which rocks are broken into smaller fragments through a variety of causes, but without significant change in chemical or mineralogical composition. Except in extremely cold or very dry climates, physical weathering occurs together with chemical weathering, and it is difficult to separate their effects.

Frost wedging, caused by freezing and thawing of water in rock fractures, is the most important physical weathering process in climates where recurring freezing and thawing take place. Water increases in volume by about 9 percent when it changes to ice, creating enough pressure in tortuous rock fractures to crack most types of rock. To be effective, water must be trapped within the rock body, and repeated freezing and thawing are necessary to allow progressive disintegration of the rock. This process commonly produces large, angular blocks of rock but may also cause granular disintegration of coarse-grained rocks such as granites. The presence of microfractures and other microstructures exerts an important control on the sizes and shapes of shattered blocks. Mechanically weak rocks such as shales and sandstones tend to disintegrate more readily than do hard, more strongly cemented rocks such as quartzites and igneous rocks.

Early workers suggested that **alternate expansion and contraction** of rock surfaces as a result of diurnal changes in temperature caused weakening of bonds along grain boundaries and subsequent flaking off of rock fragments or dislodging of mineral grains. The quantitative importance of this postulated process is still being debated. Griggs (1936) carried out experiments in the laboratory over a temperature range of 110°C to simulate the equivalent of 244 years of heating and cooling; he showed that in the absence of water, little disintegration occurred. Ollier (1969) points out, however, that Griggs's laboratory specimens were unconfined and could expand in all directions, whereas a patch of rock on the surface of a boulder is confined by neighboring patches and can expand only outward. Such a confined area of rock is more likely to fracture under repeated stress. Ollier also suggested that small heating and cooling stresses maintained for longer periods of time than the 15-minute heating and cooling cycles in Griggs's experiments might lead to permanent strain. Thus, high temperatures and temperature changes from day to night may cause fracturing of rocks owing to thermal changes alone (Kerr et al., 1984).

High temperatures in desert environments also tend to promote weathering caused by the **crystallization of salts** in pore spaces and fractures (Sperling and Cooke, 1980). Evaporation of water acts to concentrate dissolved salts in saline solutions that have access to rock fractures and pores. Growth of salt crystals generates internal pressures that can force cracks apart or cause disintegration of weakly cemented rocks. This process is most common in semiarid regions but can occur also along seacoasts where salt spray is blown onto sea cliffs.

Release of overburden pressure owing to erosion of overlying strata causes the development of rock fractures that are nearly parallel to the topographic surface. These fractures divide the rock into a series of layers or sheets; hence, this process of crack formation is called **sheeting.** These layers increase in thickness with depth and may exist for several tens of meters below Earth's surface. Sheeting is most conspicuous in homogeneous rocks such as granite but may occur also in layered rocks.

Other processes that may contribute to mechanical weathering under certain conditions include volume increases owing to hydration of clay minerals or other minerals (which may be more important than heating and cooling); volume changes owing to alteration of minerals such as biotite and plagioclase to clay minerals; alternate wetting and drying of rocks (causing alternate expansion and contraction); growth of plant roots in the cracks of rocks; plucking of mineral grains and rock fragments from rock surfaces by lichens as they expand and contract in response to wetting and drying; and burrowing and ingestion of soils and loosened rock materials by worms or other organisms.

The grain size of the particulate rock materials that result from physical weathering is a function of the thoroughness of the weathering process, but it is determined ultimately by the grain size and degree of cementation of the parent rock and by the abundance and spacing of large fractures and microfractures. Coarse-grained parent rocks such as granites tend to yield grains of individual minerals upon disintegration, whereas physical weathering of fine-grained sedimentary, volcanic, or metamorphic rocks is more likely to produce rock fragments as disintegration products (Boggs, 1968; Carrol, 1970).

Chemical Weathering

Chemical Weathering Processes. Chemical weathering involves changes that can alter both the chemical and the mineralogical composition of rocks. Minerals in the rocks are attacked by water and dissolved atmospheric gases (oxygen, carbon dioxide), causing some components of the minerals to dissolve and be removed in solution. Other mineral constituents recombine *in situ* and crystallize to form new mineral phases. These chemical changes, along with changes caused by physical weathering processes, disrupt the fabric of the weathered rock, producing a loose residue of resistant grains and secondary minerals. Water and dissolved gases play a dominant role in every aspect of chemical weathering. Because some water is present in almost every environment, chemical weathering processes are commonly far more important than physical weathering processes, even in very arid climates. Nevertheless, owing to the low temperatures of weathering environment (<30°C), chemical weathering occurs very slowly. The principal processes of chemical weathering are listed and briefly described in Table 2.1 along with selected examples of new minerals formed *in situ* during the weathering processes.

Hydrolysis is an extremely important chemical reaction between silicate minerals and acids (solutions containing H^+ ions) that leads to breakdown of the silicate minerals and release of metal cations and silica. If aluminum is present in the minerals undergoing weathering, clay minerals such as kaolinite, illite, and smectite may form as a byproduct of hydrolysis. Thus, orthoclase feldspar can break down to yield kaolinite or illite, albite (plagioclase feldspar) can decompose to kaolinite or smectite, and so on, as illustrated by the reactions in Table 2.1. The H^+ ions shown in Table 2.1 are commonly supplied by the dissociation of CO_2 in water, increasing its acidity ($CO_2 + H_2O \leftrightarrow H_2CO_3 \leftrightarrow H^+ + HCO_3^-$). Thus, the more CO_2 that is dissolved in water, the more

TABLE 2.1 Principal processes of chemical weathering

Name of process	Nature of process	Examples	Principal types of rock materials affected
Hydrolysis	Reaction between H^+ and OH^- ions of water and the ions of silicate minerals, yielding soluble cations, silicic acid and clay minerals (if Al present)	$2KAlSi_3O_8 + 2H^+ + 9H_2O \rightarrow H_4Al_2Si_2O_9 + 4H_4SiO_4 + 2K^+$ (silicic acid) aq (orthoclase) aq (kaolinite) $2NaAlSi_3O_8 + 2H^+ + 9H_2O \rightarrow H_4Al_2Si_2O_9 + 4H_4SiO_4 + 2Na^+$ (silicic acid) (albite) aq (kaolinite)	Silicate minerals
Hydration and dehydration	Gain or loss of water molecules from a mineral, resulting in formation of a new mineral	$CaSO_4 \cdot 2H_2O \leftrightarrow CaSO_4 + 2H_2O$ (Dehydration) (gypsum) (anhydrite) $Fe_2O_3 + H_2O \leftrightarrow 2FeOOH$ (Hydration) (hematite) (goethite)	Evaporites Ferric oxides
Oxidation	Loss of an electron from an element (commonly Fe or Mn) in a mineral, resulting in the formation of oxides or, if water is present, hydroxides	$4FeSiO_3 + O_2 \rightarrow 2Fe_2O_3 + 4SiO_2$ (pyroxene) (hematite) (quartz) $MnSiO_3 + 1/2\ O_2 + 2H_2O \rightarrow MnO_2 + H_4SiO_4$ (rhodonite) $2FeS_2 + 15/2\ O_2 + 4H_2O \rightarrow Fe_2O_3 + 4SO_4^{2-} + 8H^+$ (pyrite) (hematite)	Iron and manganese-bearing silicate minerals, sulfur
Solution	Dissolution of soluble minerals, commonly in the presence of CO_2, to yield cations and anions in solution	$H_2O + CO_2 + CaCO_3 \leftrightarrow Ca^{2+} + 2HCO_3^-$ (carbonation) (calcite) (bicarbonate) $CaSO_4 \cdot 2H_2O \rightarrow Ca^{2+} + SO_4^{2-} + 2H_2O$ (direct solution) (gypsum)	Carbonate rocks Evaporites
Ion exchange	Exchange of ions, principally cations, between solutions and minerals	$Na\text{-clay} + H^+ \rightarrow H\text{-clay} + Na^+$	Clay minerals
Chelation	Bonding of metal ions to organic molecules having ring structures	Metal ions (cations) + chelating agent [excreted by lichens] $\rightarrow H^+$ ions + chelate [in solution]	Silicate minerals

aggressive the hydrolysis reaction. Most of the silica set free during hydrolysis goes into solution as silicic acid (H_4SiO_4); however, some of the silica may separate as colloidal or amorphous SiO_2 and be left behind during weathering to combine with aluminum to form clay minerals or crystallize into minute grains of quartz (Krauskopf, 1979). Hydrolysis is the primary process by which silicate minerals decompose during weathering. A more rigorous and detailed discussion of this process is given by Nahon (1991, p. 7).

Hydration is the process by which water molecules are added to a mineral to form a new mineral. Common examples of hydration are the addition of water to hematite to form goethite, or to anhydrite to form gypsum. Hydration is accompanied by volume changes that may lead to physical disruption of rocks. Under some conditions, hydrated minerals may lose their water, a process called **dehydration,** and be converted to the anhydrous forms, with accompanying decrease in mineral volume. Dehydration is relatively uncommon in the weathering environment because some water is generally present in this environment.

Oxidation of iron and manganese in silicate minerals such as biotite and pyroxenes, owing to oxygen dissolved in water, is also an important weathering process because of the abundance of iron in the common rock-forming silicate minerals. An electron is lost from iron during oxidation ($Fe^{2+} \rightarrow Fe^{3+} + e^-$, where e^- = electron transfer), which causes loss of other cations such as Si^{4+} from crystal lattices to maintain electrical neutrality. Cation loss leaves vacancies in the crystal lattice that either bring about the collapse of the lattice or make the mineral more susceptible to attack by other weathering processes (Birkland, 1974). Oxidation of manganese minerals to form oxides and silicic acid or other soluble products is a less important but common weathering process. Another element that oxidizes during weathering is sulfur. For example, pyrite (FeS_2) is oxidized to form hematite (Fe_2O_3), with release of soluble sulfate ions. Under some conditions where material undergoing weathering is water saturated, oxygen supply may be low and oxygen demand by organisms high. These conditions can bring about **reduction** of iron (gain of an electron) from Fe^{3+} to Fe^{2+}. Ferrous iron (Fe^{2+}) is more soluble, and thus more mobile, than ferric iron (Fe^{3+}) and may be lost from the weathering system in solution.

Simple solution of highly soluble minerals such as calcite, dolomite, and gypsum owing to exposure to meteoric water (rainwater) during weathering can result in decomposition of these minerals. If carbon dioxide is dissolved in the rainwater through interaction with atmospheric or soil CO_2, the usual case in the weathering environment, the solubilizing ability of water is enhanced (because of increased acidity), particularly for carbonate minerals. Simple solution of this type is an important weathering process only in moderately wet climates where carbonate rocks or evaporites are present near the surface or at the water table.

Ion exchange is a weathering process that is particularly important in alteration of one type of clay mineral to another. It is the reaction between ions in solution and those held in a mineral, for example, the exchange of sodium for calcium. Most ion exchange takes place between cations, but anion exchange also occurs.

Chelation involves the bonding of metal ions to organic substances to form organic molecules having a ring structure (see Boggs, Livermore, and Seitz, 1985, for additional discussion of chelates). During weathering, chelation, or organic complexing, performs the dual role of removing cations from mineral lattices and also keeping the cations in solution until they are removed from the weathering site. Chelated metal ions will remain in solution under pH conditions and at concentration levels at which nonchelated ions would normally be precipitated. The bonding of aluminum or iron with a complexing agent and the subsequent removal of these elements from a rock are of particular importance. A good example of natural chelation is provided by lichens that cause an increase in the rate of chemical weathering on the rock surfaces on which they grow by secreting organic chelating agents. In addition to the role of plant organic matter as chelating agents, plants also enhance chemical weathering processes by retaining soil moisture and by acidifying waters by release of CO_2 and various types of organic acids during decay.

Rates of Chemical Weathering. Chemical weathering proceeds at different rates depending upon the climate and the mineral composition and grain size of the rocks. Weathering processes are more rapid in humid, hot climates than in cold or very dry climates. Average rainfall is known to be a controlling factor in the weatering rate (Nahon, 1991, p. 4); however, the influence of temperature on weathering rate is difficult to quantify. A general qualitative rule is that chemical reaction rates are approximately doubled with a 10°C increase in temperature (Hunt, 1979, p. 127). The rate of weathering of silicate rocks of a given grain size is related to the relative stabilities of the common rock-forming silicate minerals. Chemical stability refers to the resistance of minerals to alteration or destruction by chemical processes. Table 2.2 shows the order of relative stability to weathering of the most important mafic and felsic minerals, as determined by Goldich (1938) through empirical study of sand- and silt-size particles in soil profiles. Readers will recognize this order as the same as that in which minerals crystallize in Bowen's reaction series. Minerals that crystallize at high temperatures (e.g., olivine) have the greatest degree of disequilibrium with surface weathering temperatures and thus tend to be less stable than minerals that crystallize at lower temperatures (e.g., quartz). Furthermore, the high-temperature minerals are bonded with weaker ionic or ionic-covalent bonds, whereas quartz is bonded with strong covalent bonds. Jackson (1968) suggests that the stability of very fine-size (clay-size) particles may differ somewhat from that of larger particles (Table 2.2).

Owing to the preponderance of low-stability minerals in basic igneous rocks, these rocks tend to weather faster than acid igneous rocks of the same grain size. Thus, gabbro weathers faster than granite, and basalt weathers faster than rhyolite. Fine-

TABLE 2.2 Relative stability of common sand-size minerals and various clay-size minerals under conditions of weathering

Sand- and silt-size minerals*		Clay-size minerals**
Mafic minerals	**Felsic minerals**	1. Gypsum, halite
		2. Calcite, dolomite, apatite
Olivine		3. Olivine, amphiboles, pyroxenes
	Ca plagioclase	4. Biotite
Pyroxene		5. Na plagioclase, Ca plagioclase,
	Ca-Na plagioclase	K-feldspar, volcanic glass
Amphibole	Na-Ca plagioclase	6. Quartz
	Na plagioclase	7. Muscovite
Biotite		8. Vermiculite (clay mineral)
	K-feldspar, muscovite,	9. Smectite (clay mineral)
	quartz	10. Pedogenic (soil) chlorite
		11. Allophane (clay mineral)
		12. Kaolinite, halloysite (clay minerals)
		13. Gibbsite, boehmite (clay minerals)
		14. Hematite, goethite, magnetite
(Increasing stability)		15. Anatase, titanite, rutile, ilmenite (all, titanium-bearing minerals), zircon

Source: *Goldich (1988); **Jackson (1968).

grained basic igneous rocks such as basalt may, however, weather more slowly than coarse-grained granitic rocks (Birkland, 1974). There is no rule of weathering susceptibility that can be applied generally to sedimentary rocks. Rates of weathering of these rocks are a function of the mineralogy, the amount and type of cement in the rocks, and the climate. Limestones, for example, weather rapidly by solution in wet climates and much more slowly in very arid or very cold climates. Quartz-rich sandstones cemented with silica cement weather very slowly under most climatic conditions. The general order of weathering of some common rocks, in order of increasing stability to weathering in temperate to tropical climates, is probably limestone, dolomite, siltstone, sandstone, basalt, granite, chert, and quartzite (Birkland, 1974).

Finally, it is likely that rates of weathering have varied throughout geologic time depending upon climatic conditions and vegetative cover. Prior to the development of land plants in early Paleozoic time, absence of plant cover to hold soil moisture and contribute organic acids probably slowed rates of chemical weathering while contributing to increased rates of physical erosion. The rates of chemical weathering are now being investigated extensively through theoretical, empirical, and experimental studies; however, many unresolved problems remain, such as mechanisms and rates of individual processes, the relative roles of equilibrium and nonequilibrium kinetics, and the influence of macro- and microenvironmental variables (Colman and Dethier, 1986).

Products of Subaerial Weathering

Subaerial weathering generates three types of weathering products (Table 2.3):

1. source-rock residues consisting of chemically resistant minerals and rock fragments
2. secondary minerals formed *in situ* by chemical recombination and crystallization, largely as a result of hydrolysis and oxidation
3. soluble constituents released from parent rocks by hydrolysis and solution

Until they are removed by erosion, residues and secondary minerals accumulate at the weathering site to form a soil mantle composed of particles of various compositions and of grain sizes ranging from clay to gravel. Grain size and composition depend upon the grain size and composition of the parent rock and upon the nature and intensity of the weathering process. These characteristics of the weathering environment are in turn functions of climate, topography, and duration of the weathering process.

TABLE 2.3 Principal kinds of products formed by subaerial weathering processes and the types of sedimentary rocks ultimately formed from these products

Weathering process	Type of weathering product	Example	Ultimate depositional product
Physical weathering	Particulate residues	Silicate minerals such as quartz and feldspar; all types of rock fragments	Sandstones, conglomerates, mudrocks
Chemical weathering			
Hydrolysis	Secondary minerals	Clay minerals; fine quartz	Mudrocks; mud matrix
	Soluble constituents	Silicic acid; K^+, Na^+, Mg^{2+}, Ca^{2+}, etc.	Chert, limestones, evaporites, etc.
Oxidation	Secondary minerals	Fine-grained SiO_2 minerals; ferric oxides	Mudrocks; mud matrix
Solution	Soluble constituents	Silicic acid, SO_4^{2-}, etc.	Chert, evaporites, etc.
	Soluble constituents	Bicarbonate, SO_4^{2-}, Ca^{2+}, Mg^{2+}, etc.	Limestones, evaporites, etc.

Source-Rock Residues. The residual particles in young or immature soils developed on igneous or metamorphic rocks may include, in addition to rock fragments, assemblages of minerals with low chemical stability, for example, biotite, pyroxenes, hornblende, and calcic plagioclase. Mature soils, developed after more prolonged or intensive weathering of these rocks, tend to contain only the most stable minerals: quartz, muscovite, and perhaps potassium feldspars. Because the silicate minerals that make up siliciclastic sedimentary rocks such as sandstones have already passed through a weathering cycle before the siliciclastic rocks were formed, the weathering products of these rocks tend to be depleted in easily weathered minerals. Thus, even young soils developed on siliciclastic sedimentary rocks may have assemblages of mature minerals. Weathering of limestones by solution produces thin soils composed of the fine-size insoluble silicate and iron oxide residues of these rocks.

Secondary Minerals. Secondary minerals developed at the weathering site are dominantly clay minerals, iron oxides or hydroxides, and aluminum hydroxides. The common secondary iron minerals include goethite, limonite, and hematite. The weathering products that form are a function of both the nature and the intensity of the weathering process and the composition of the parent rock. Clay minerals formed in immature soils under only moderately intense chemical weathering conditions may be illites or smectites. More prolonged and intense leaching conditions lead to formation of kaolinite. Under extremely intense chemical weathering conditions, aluminum hydroxides such as gibbsite and diaspore are formed. These latter clay minerals are aluminum ores. As an example of the influence of parent rock composition, potassium feldspars tend to weather to kaolinite or, with more intense weathering, gibbsite. Biotite weathers to chlorite, smectite, vermiculite, and kaolinite, and plagioclase feldspars weather to sericite, vermiculite, smectite, kaolinite, and gibbsite (Tardy et al., 1973).

Comparison of the chemical composition of unweathered silicate rocks with that of the weathering products of these rocks shows a net loss owing to weathering of all major cations except aluminum and iron (Krauskopf, 1979). In the oxidized state, aluminum and ferric iron (Fe^{3+}) are both relatively insoluble. Although considerable silica is lost during weathering as soluble silicic acid, loss of Mg, Ca, Na, and K is comparatively much greater. Therefore, the relative abundance of silica, aluminum, and ferric iron in the particulate weathering residues of silicate rocks is greater than that in the parent source rocks.

Soluble Materials. Soluble materials extracted from parent rocks by chemical weathering processes are removed from the weathering site in surface water or soil groundwater more or less continuously throughout the weathering process. Ultimately these soluble products make their way into rivers and are carried to the ocean. The most abundant inorganic constituents of rivers, representing the principal soluble products of weathering, are, in order of decreasing abundance, HCO_3^- (bicarbonate), Ca^{2+}, H_4SiO_4 (silicic acid), SO_4^{2-} (sulfate), Cl^-, Na^+, Mg^{2+}, and K^+ (Garrels and McKenzie, 1971). These constituents are the raw materials from which chemically and biochemically deposited rocks such as limestones and cherts are formed in the oceans.

2.3   SUBMARINE WEATHERING PROCESSES AND PRODUCTS

Geologists have long recognized that sediments and rocks on the seafloor are altered by reaction with seawater, a process called **halmyrolysis** (submarine weathering). Halmyrolysis includes alteration of clay minerals of one type to another, formation of glau-conite from feldspars and micas, and formation of phillipsite (a zeolite mineral) and palagonite (altered volcanic glass) from volcanic ash. Dissolution of the siliceous and calcareous tests of organisms may also be considered a type of submarine weathering. Prior to the 1970s, submarine weathering processes had not received a great deal of research, and that they might have a significant effect on the overall chemical composition of the oceans was not recognized. Our concept of the importance of submarine weathering has changed dramatically since the middle 1970s because studies of volcanic rocks and weathering processes on the seafloor show that submarine weathering of basalts, particularly on mid-ocean ridges, is an extremely important chemical phenomenon. This process results in both widespread hydration and leaching of basalts and changes in composition of reacting seawater owing to ion exchange during the reaction of seawater with basalt.

Alteration of oceanic rocks occurs both at low temperatures (less than 20°C) and at higher temperatures ranging to 350°C. Low-temperature alteration takes place as seawater percolates through fractures and voids in the upper part of the ocean crust, perhaps extending to depths of 2 to 5 km. Olivine and interstitial glass in the basalts are replaced by smectite clay minerals, and further alteration may lead to formation of zeolite minerals and chlorite. As a result of these changes, an exchange of elements between rock and water takes place, and large volumes of seawater become fixed in the oceanic crust in hydrous clay minerals and zeolites.

As a result of the discovery in 1977 of submarine thermal springs along the Galapagos Rift (Corliss et al., 1979), we are now aware that large-scale hydrothermal activity is taking place along the crests of mid-ocean ridges. Since that initial discovery, scientists using submersible vehicles and water-sampling techniques have located many additional hot springs along the East Pacific Rise and the Juan de Fuca Ridge in the Pacific (e.g., McConachy et al., 1986; Baker and Massoth, 1987). Hydrothermal systems have also been reported along the Mid-Atlantic Ridge (e.g., Rona et al., 1986) and even on midplate volcanoes in the Hawaiian chain (Karl et al., 1988). These springs originate from the activity of seawater in areas of active or recent volcanism along the ridge crests. Seawater enters the ocean crust along fractures or other voids and comes in contact with hot volcanic rock. The heated water then flows out into the ocean through vents on the ocean floor and mixes with the overlying water. The heated water rises as hydrothermal plumes 100 to 300 m above the vent field. Exceptional plumes rising to heights of 1000 m have also been reported (Cann and Strens, 1989).

On the top of the East Pacific Rise, for example, investigators have found spectacular vents composed of sulfide, sulfate, and oxide deposits up to 10 m tall and discharging plumes of hot solutions. These vents or chimneys are called **black smokers** if they discharge water containing suspended, fine-grained, dark-colored minerals or **white smokers** if the water contains no suspended dark minerals (McDonald, Spiess, and Ballard, 1980). The temperature of the water when it emerges from the vents may exceed 350°C. When these hot solutions mix with seawater of ambient temperature, they precipitate various minerals, particularly pyrite (FeS) and chalcopyrite ($CuFeS_2$), to build sulfide deposits around the vents (Fig. 2.1). The deposits of fossil hydrothermal systems have now been observed in ancient oceanic ophiolite complexes exposed on land (Cann and Strens, 1989). Study of these ancient complexes suggests that the structure of seafloor hydrothermal systems may be something like that shown in Figure 2.2.

Reactions between hot basalt and seawater may play a major role in regulating the chemical composition of seawater. Magnesium and sulfate ions are removed from seawater during this exchange, whereas other elements—such as calcium, manganese, silicon, potassium, lithium, rubidium, and barium—are enriched in the seawater

FIGURE 2.1 A multiple-orifice black smoker issuing from its constructional chimney of chalcopyrite-sphalerite-anhydrite on the East Pacific Rise. The temperature of the water issuing from the chimney is 350°C. The "smoke" is caused by the presence of fine-grained sulfide precipitates that form by reaction of the hot waters with cold, ambient seawater. (Photo by Fred Spiess. Courtesy of Scripps Institution of Oceanography, University of California, San Diego.)

(Edmond et al., 1982; Von Damm et al., 1985). The magnitude of hydrothermal alteration of basalts along mid-ocean ridges is still being investigated; however, Edmond (1980) suggests that a volume of seawater equivalent to the entire ocean could be circulated through crustal rocks along ridges every 8 million years or so. This flow rate is only about 0.5 percent of that of all the world's rivers, but the concentration of many elements in these hot waters is 100 to 1000 times greater than in the average river. Therefore, the fluxes of dissolved material out of and into ridge axes could be comparable to those derived from continental weathering. Clearly, both seafloor hydrothermal reactions and continental weathering are important processes that supply salts to the ocean.

2.4   SOILS

Soil-forming Processes

The subaerial physical, chemical, and biologic weathering processes discussed in the preceding section generate a mantle of soil above bedrock. The characteristics and thickness of this soil mantle are a function of the bedrock lithology, the climate (rainfall, temperature), and the slope of the bedrock surface. These factors govern the intensity of weathering and determine which minerals survive to become part of the soil profile, what new minerals are created in the soil, and the length of time soil materials remain before being eroded and transported to depositional basins. On very steep slopes, for example, the weathered mantle may be removed so rapidly by erosion that little soil accumulates.

In addition to the weathering processes discussed, several other processes operate within soils to produce their overall characteristics. Among the more important of these are humification, gleization, podzolization, lessivage, ferrallitization,

FIGURE 2.2 Hypothetical structure of a hydrothermal black smoker system underlying a spreading ridge. (From J. R. Cann and M. R. Strens, 1989, Modeling periodic megaplume emissions by black smoker systems: Jour. Geophys. Research, v. 94, Fig. 1, p. 12, 228. Reproduced by permission.)

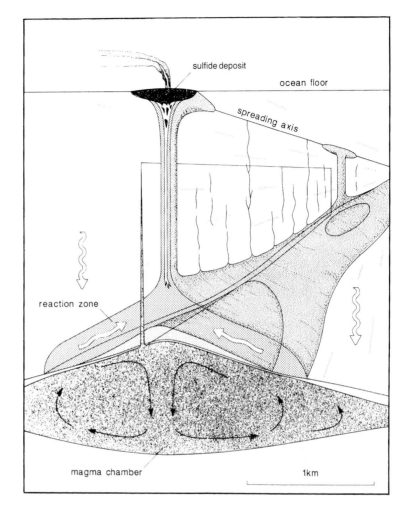

calcification, and salization and desalization. Each of these processes is explained and briefly described in Table 2.4 and illustrated graphically in Figure 2.3. Additional details of these soil-forming processes may be found in Buol, Hole, and McCracken (1989), Fanning and Fanning (1989), Fitzpatrick (1980), Nahon (1991), Retallack (1990), and Ross (1989).

Soil Profiles and Soil Classification

Soils are classified on the basis of the characteristic horizontal layers or horizons that are visible in roadcuts, pits, and so on. The thickness and nature of these **soil horizons** are determined by the various soil-forming processes discussed and may vary widely. Soil profiles can be divided crudely into three major horizons: A, B, and C. The **A horizon** is the dark-colored upper zone of organic accumulation composed of leaf litter that is decaying and mixing with mineral soil. The **B horizon** is composed dominantly of minerals with minimal organic content; most of the original rock structures have been obliterated by soil-forming processes. The **C horizon,** which lies above bedrock, can be

TABLE 2.4 Some important soil-forming processes

Process	Description
Humification	The transformation of raw organic matter into soil humus (dark, partially decomposed, more or less stable organic material) and soluble organic acids (e.g., humic and fulvic). Transformation takes place through biogenic comminution of larger organic structures to fine, amorphous organic matter and chemical oxidation of organic matter to simpler compounds such as carbon dioxide.
Gleization	The reduction of iron under anoxic or anaerobic (low-oxygen) soil conditions to produce bluish to greenish gray waterlogged soil (gley). Anaerobic microorganisms play an important role in reducing oxidized minerals. The process of gleying tends to retard mineral weathering.
Podzolization	The downward (commonly) chemical migration of aluminum, iron, and/or organic matter within a soil profile. This process results in the relative concentration of silica in the upper layer from which these elements are removed and concentration of aluminum, iron, and organic matter in a deeper layer. Podzolization causes destruction of clay minerals and leaching of exchangeable cations such as Ca^{2+}, Mg^{2+}, K^+, and Na^+.
Lessivage	The mechanical migration of clay-size mineral particles from the A (surface) to B horizons of a soil, producing in the B horizon relative enrichment in clay (argillic horizons). This process produces a near-surface horizon that is light in color and a subsurface clayey zone that is darker in color.
Ferrallitization	Intense, deep weathering resulting in thick, uniform soil profiles depleted of exchangeable cations. The soil as a whole is enriched in clay and sesquioxides (Fe_2O_3, Al_2O_3) in the form of fine-grained, weather-resistant minerals such as kaolinite, gibbsite, and hematite.
Calcification	The accumulation of calcium in subsurface horizon near the depth of average rainfall wetting in well-drained soils of semiarid to subhumid regions. Sufficient water may be present in these subsurface horizons to remove some exhchangeable cations (Na^+, K^+, Mg^{2+}) but not Ca^{2+}; thus, calcium becomes relatively enriched.
Salizination and Desalinization	The accumulation (salizination) of soluble salts such as sulfates and chlorides of calcium, magnesium sodium, and potassium in salty horizons, or the leaching and removal (desalinization) of salts from these horizons.

Source: Buol, Hole, and McCracken (1989); Retallack (1990).

deeply weathered but is relatively unaffected by soil-forming processes. Studies of soil profiles show, however, that soil layers are commonly much more complex than indicated by this simple scheme. Up to 16 different kinds of soil horizons have been described (Fanning and Fanning, 1989). Several systems for more detailed classification of soils are in existence: the British classification, the Australian handbook classification, the U.S. Soil Taxonomy, and the FAO (UNESCO) world map classification

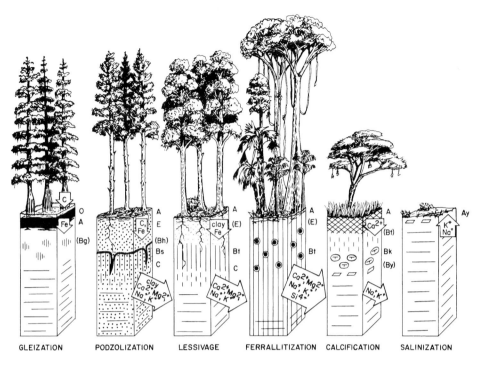

FIGURE 2.3 Schematic illustration of the major regimes in which the common soil-forming processes take place. (From G. J. Retallack, 1990, Soils of the past: Unwin Hyman (Chapman & Hall), Fig. 4.12, p. 87, reproduced by permission.)

(Courtney and Trudgill, 1984, Chapter 7; Retallack, 1990, Chapter 5). One of the more widely used soil classifications in the United States is the U.S. Soil Taxonomy (Soil Survey Staff, 1975).

Paleosols

Paleosols, sometimes referred to as *fossil soils,* are buried soils or horizons of the geologic past. Most soil horizons that developed in the past on elevated landscapes were eventually destroyed as erosion lowered the landscape. Nonetheless, some soils, presumably those formed mainly in low-lying areas, escaped erosion to become part of the stratigraphic record. Quaternary soils that formed particularly on glacial or fluvial deposits are most common (e.g., Catt, 1986). Such soils that have not been buried are called relict soils. Many buried soils of Quaternary and much older age are also known. Old paleosols occur in the stratigraphic record at major unconformities, including unconformities in Precambrian rocks, where their presence may reflect the combined processes of soil formation, erosional landscape lowering, reorganization of preexisting soil horizons, and changing flow of groundwater (Retallack, 1990, p. 14). Paleosols are also present as interbeds in sedimentary sequences that are at least as old as the Ordovician (e.g., Reinhardt and Sigleo, 1988). Geologists are becoming increasingly interested in paleosols as indicators of paleoenvironments and ancient climatic conditions.

Recognition of Paleosols

Because interbedded paleosols in sedimentary sequences superficially resemble sediments or sedimentary rocks, many paleosols have unquestionably gone unrecognized in the past. Many of us have simply identified them as gray, red, or green mudstones or shales. As awareness of paleosols has increased, however, more and more paleosols are being recognized. How can the ordinary geologist, not specifically trained in soil science, recognize paleosols in the field? Fenwick (1985) suggests a number of possible criteria for recognition, including the following:

1. the presence of surface horizons enriched in organic matter
2. red-colored horizons that become more intense in color toward the top
3. marked decline in weatherable minerals toward the top of the soil profile
4. disruption of original structures by organisms such as earthworms or by physical processes such as frost action and solifluction

Retallack (1988) discusses three main features of paleosols that are particularly useful in field recognition: root traces, soil horizons, and soil structures. **Root traces** provide diagnostic evidence that rock was exposed to the atmosphere and colonized by plants, thus forming a soil. The top of a paleosol is the surface from which root traces emanate. Root traces mostly taper and branch downward (Fig. 2.4), which helps to distinguish them from burrows. On the other hand, some root traces spread laterally over hardpans in soils, and some kinds branch upward and out of the soil. Root traces

FIGURE 2.4 An example of root traces in a paleosol. The original organic matter has been partially replaced by iron oxides. Early Miocene, Molalla Formation, western Oregon. (Photograph courtesy of G. J. Retallack.)

are most easily recognized when their original organic matter is preserved, which occurs mostly in paleosols formed in waterlogged, anoxic lowland environments. Root traces in red, oxidized paleosols consist mainly of tubular features filled with material different from the surrounding paleosol matrix.

The presence of **soil horizons** is a second general feature of paleosols. The top of the uppermost horizon of a paleosol is commonly sharply truncated by an erosional surface, but boundaries between underlying horizons are typically gradational. Differences in grain size, color, reaction with weak hydrochloric acid (to test for the presence of carbonates), and the nature of the boundaries must all be examined to detect soil horizons (Retallack, 1988).

Owing to bioturbation (disruption) by plants and animals, wetting and drying, and other soil-forming processes, paleosols develop characteristic **soil structures** at the expense of the original bedding and structures in the parent rock. One of the characteristic kinds of soil structure is a network of irregular planes (called **cutans**) surrounded by more stable aggregates of soil material called **peds**. This structure gives a hackly appearance to the soil. Peds occur in a variety of sizes and shapes (Fig. 2.5). Their recognition in the field depends upon recognition of the cutans that bound them, which commonly form clay skins around the peds. Other kinds of soil structure include concentrations of specific minerals to form hard, distinct, calcareous, ferruginous, or sideritic lumps called **glaebules** (a general term including nodules and concretions). More diffuse, irregular, or weakly mineralized concentrations are called mottles.

This short, generalized description of paleosols is intended only to pique reader interest in fossil soils. Several books, many of which are listed under Further Readings below, are available to anyone who wishes to pursue further research on paleosols.

TYPE	PLATY	PRISMATIC	COLUMNAR	ANGULAR BLOCKY	SUBANGULAR BLOCKY	GRANULAR	CRUMB
SKETCH	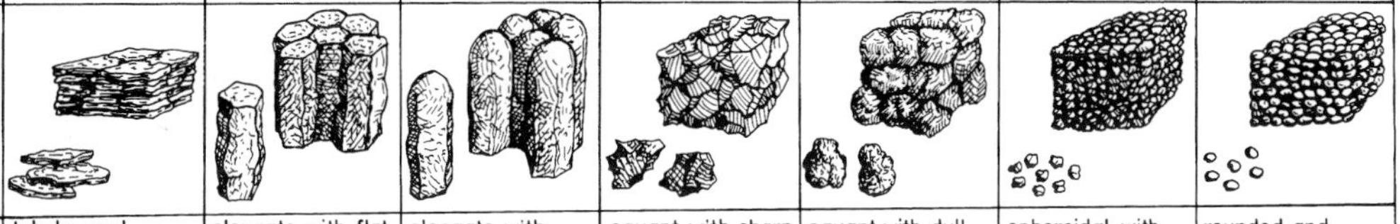						
DESCRIPTION	tabular and horizontal to land surface	elongate with flat top and vertical to land surface	elongate with domed top and vertical to surface	equant with sharp interlocking edges	equant with dull interlocking edges	spheroidal with slightly interlocking edges	rounded and spheroidal but not interlocking
USUAL HORIZON	E,Bs,K,C	Bt	Bn	Bt	Bt	A	A
MAIN LIKELY CAUSES	initial disruption of relict bedding; accretion of cementing material	swelling and shrinking on wetting and drying	as for prismatic, but with greater erosion by percolating water, and greater swelling of clay	cracking around roots and burrows; swelling and shrinking on wetting and drying	as for angular blocky, but with more erosion and deposition of material in cracks	active bioturbation and coating of soil with films of clay, sesquioxides and organic matter	as for granular; including fecal pellets and relict soil clasts
SIZE CLASS	very thin < 1 mm	very fine < 1 cm	very fine < 1 cm	very fine < 0.5 cm	very fine < 0.5 cm	very fine < 1 mm	very fine < 1 mm
	thin 1 to 2 mm	fine 1 to 2 cm	fine 1 to 2 cm	fine 0.5 to 1 cm	fine 0.5 to 1 cm	fine 1 to 2 mm	fine 1 to 2 mm
	medium 2 to 5 mm	medium 2 to 5 cm	medium 2 to 5 cm	medium 1 to 2 cm	medium 1 to 2 cm	medium 2 to 5 mm	medium 2 to 5 mm
	thick 5 to 10 mm	coarse 5 to 10 cm	coarse 5 to 10 cm	coarse 2 to 5 cm	coarse 2 to 5 cm	coarse 5 to 10 mm	not found
	very thick > 10 mm	very coarse > 10 cm	very coarse > 10 cm	very coarse > 5 cm	very coarse > 5 cm	very coarse > 10 mm	not found

FIGURE 2.5 Characteristics of various kinds of soil peds. (From G. J. Retallack, 1988, *in* J. Reinhardt and W. R. Sigleo, eds., Field recognition of paleosols: Geol. Soc. America Spec. Paper 216. Fig. 9, p. 216.)

FURTHER READINGS

Weathering

Balasubramaniam, D. S., et al. (eds.), 1989, Weathering; its products and deposits, v. 1, Processes, 462 p., v. II, Deposits, 671 p.: Theophrastus Publicatione, Athens, Greece.

Birkland, P. W., 1974, Pedology, weathering and geomorphological research. Oxford University Press, New York, 285 p.

Carroll, D., 1970, Rock weathering: Plenum Press, New York, 203 p.

Colman, S. M., and D. P. Dethier, 1986, Rates of chemical weathering of rocks and minerals: Academic Press, Orlando, 603 p.

Drever, J. I. (ed.), 1985, The chemistry of rock weathering: D. Reidel, Hingham, Mass., 336 p.

Lerman A., and M. Meybeck (eds.), 1988, Physical and chemical weathering in geochemical cycles: Kluwer Academic, Dordrecht, 375 p.

Loughnan, F. C., 1969, Chemical weathering of silicate minerals: Elsevier, New York, 154 p.

Nahon, D. B., 1991, Introduction to the petrology of soils and chemical weathering: John Wiley & Sons, New York, 313 p.

Rona, P. A. and R. P. Lowell (eds.), 1980, Seafloor spreading centers: Hydrothermal systems: Benchmark Papers in Geology, v. 56, Dowden, Hutchinson and Ross, Stroudsburg, Pa., 424 p.

Soils and Paleosols

Boardman, J. (ed.), 1985, Soils and Quaternary landscape evolution: John Wiley & Sons, Chichester, 391 p.

Bronger, A., and J. A. Catt (eds.), 1989, Paleopedology: Nature and application of paleosols: Catena Verlag, Destedt, Germany, 232 p.

Buol, W. W., F. D. Hole, and R. J. McCracken, 1989, Soil genesis and classification, 3rd ed.: Iowa State University Press, Ames, 446 p.

Catt, J. A., 1986, Soils and Quaternary geology: A handbook for field scientists: Clarendon Press, Oxford, 267 p.

Courtney, F. M., and S. T. Trudgill, 1984, The soil, 2nd ed.: Edward Arnold, London, 123 p.

Duchaufour, P., 1982, Pedology: George Allen & Unwin, London, 448 p.

Fanning, D. S., and M. C. B. Fanning, 1989, Soil: Morphology, genesis, classification: John Wiley & Sons, New York, 395 p.

Reinhardt, J., and W. R. Sigleo (eds.), 1988, Paleosols and weathering through geologic time: Principles and applications: Geol. Soc. America Spec. Paper 216, 181 p.

Retallack, G. J., 1990, Soils of the past: Unwin Hyman, Boston, 520 p.

Wright, V. P., 1986, Paleosols: Their recognition and interpretation: Princeton University Press, Princeton, N.J., 315 p.

Yaalon, D. H., 1971, Paleopedology: Origin, nature, and dating of palesols: International Society of Soil Science and Israel University Press, Jerusalem, 350 p.

3

Transport and Deposition
of Siliciclastic Sediment

3.1 INTRODUCTION

The weathering residues and pyroclastic particles discussed in Chapter 2 are ultimately eroded from highlands and transported to depositional basins where they may undergo additional transport before final deposition. Sediment is transported either by flow of air, water, or ice or by gravity-driven sediment-flow processes that commonly involve the presence of water. During gravity-flow transport, fluids may act as a support mechanism and lubricant for moving sediment, and some sediment gravity flows, such as debris flows, behave like fluids. Study of sediment transport thus requires some understanding and application of the principles of fluid flow. The fundamental laws of fluid dynamics are complex when applied to fluid flow alone. These complexities are greatly magnified when particles are entrained in the flow, as during sediment transport. The problem of understanding sediment transport by fluid flow is further magnified by the fact that geologists are interested primarily in understanding conditions that existed in the past. They must attempt to interpret ancient flow conditions on the basis of the depositional products of sediment transport, long after the fluid flows or sediment gravity flows themselves have disappeared. In short, they must infer from the preserved characteristics of the sediment the nature of flows that took place millions of years ago. Thus, geologists seek to find a relationship between physical properties of sedimentary rocks, such as sedimentary structures and texture, and parameters of fluid flow, such as velocity and water depth, that lend insight into depositional mechanisms and environments. This task is a formidable one and requires application of principles and knowledge gained from theoretical studies, experimental laboratory research, and study of sediment transport and depositional processes in modern environments.

In this chapter, we investigate sediment transport processes by first examining some of the properties of fluids and the basic concepts of fluid flow. We then consider the problems involved in entrainment and transport of particles by fluid-flow and gravity-flow processes. Fluid mechanics is a subject of particular importance to engineers, and much of the theory of fluid flow has been developed for application to engineering problems. Unfortunately, many of these principles of fluid dynamics are not readily applicable to the more pragmatic problem of understanding sediment transport. For example, engineers are usually not greatly concerned with such aspects of fluid flow as the decay of flow velocity, which is of special interest to geologists studying sediment deposition. No attempt is made here to give a comprehensive review of fluid mechanics. Only those concepts of fluid flow that are important to understanding sediment transport and deposition are discussed, and these concepts are presented in very simplified form. Introduction of some terminology peculiar to fluid dynamics is unavoidable, but new terminology is kept to a minimum. The emphasis in this chapter is on flow processes as they relate to sediment transport. I have not attempted to fully explore the relationships between fluid-flow parameters and the physical properties of sedimentary rocks, although the relationship of fluid-flow parameters to bedforms (ripples) and cross-beds is examined in moderate detail. The relationship between flow parameters and sediment properties cannot be developed completely until sedimentary textures and structures have been discussed. Therefore, this relationship will be discussed further in Chapter 5 and subsequent chapters of the book, as appropriate.

3.2 FUNDAMENTALS OF FLUID FLOW

Fluids are substances that change shape easily and continuously as external forces are applied. Thus, they have negligible resistance to shearing forces. **Shearing forces** set up conditions within a body that create a tendency for parts of the body to slide over other parts along a series of parallel shear planes. Natural fluids include crude petroleum, natural gas, air, and water; however, air, water, and water containing various amounts of suspended sediment are the fluids of primary geological interest. The basic physical properties of these fluids are density and viscosity. Differences in these properties markedly affect the ability of fluids to erode and transport sediment.

The fluid density (ρ), defined as mass per unit fluid volume, affects the magnitude of forces that act within a fluid and on the bed as well as the rate at which particles fall or settle through a fluid. Density particularly influences the movement of fluids downslope under the influence of gravity. Density increases with decreasing temperature of a fluid. The density of water (0.998 g/mL at 20°C) is more than 700 times greater than that of air. This density difference influences the relative abilities of water and air to transport sediment. Water can transport particles of much larger size than those transported by wind.

Fluid viscosity is a measure of the ability of fluids to flow. Fluids with low viscosity flow readily, and vice versa. For example, air has very low viscosity and ice has very high viscosity. To appreciate the significance of viscosity, imagine a simple experiment in which a fluid is trapped between two parallel plates. The lower plate is stationary, and the upper plate is moving over it with a constant velocity (V). The fluid can be thought of as forming parallel sheets between the plates. As the upper plate moves over the lower, the fluid in between is put in motion with a velocity that varies linearly from zero at the lower plate to velocity (V) at the upper plate (Fig. 3.1). The shearing force per unit area needed to produce a given rate of shearing, or a given velocity gradient normal to the shear planes, is determined by the viscosity. **Dynamic viscosity** (μ) is thus the measure of resistance of a substance to change in shape taking place at finite speeds during flow. It is force per unit velocity gradient and is defined as the ratio of shear stress (τ) to the rate of deformation (du/dy) sustained across the fluid:

$$\mu = \frac{\tau}{du/dy} \tag{3.1}$$

The velocity gradient du/dy is the rate of change of the local fluid velocity u in the direction y normal to the shearing surface. **Shear stress** is the shearing force per unit area (e.g., dynes/cm^2) exerted across the shearing surface at some point in a fluid. It acts on the fluid parallel to the surface of the fluid body. Shear stress is generated at the boundary of two moving fluids, and it is a function of the extent to which a slower moving mass retards a faster moving one. Thus, as a faster moving layer moves over a slower moving layer, the shear stress is the force that produces a change in velocity (du) relative to height (dy) (Fig. 3.1). The greater the viscosity, the greater the shear stress must be to produce the same rate of deformation. Shear stress at the bed is a function of the **shear velocity** (U^*), which is equal to $\sqrt{\tau_0/\rho}$. Shear velocity, expressed in velocity units such as cm/s, is a very important factor in sediment transport. It plays a critical role in the erosion or entrainment of sediment on a stream bed or ocean floor and in the continued downcurrent transport of sediment.

Viscosity decreases with temperature; thus, a given fluid flows more readily at higher temperatures. Because both density and dynamic viscosity strongly affect fluid behavior, fluid dynamicists commonly combine the two into a single parameter called **kinematic viscosity** (ν), which is the ratio of dynamic viscosity to density:

$$\nu = \frac{\mu}{\rho} \tag{3.2}$$

Kinematic viscosity is an important factor in determining the extent to which fluid flows exhibit turbulence.

FIGURE 3.1 Geometric representation of the factors that determine fluid viscosity. A fluid is enclosed between two rigid plates, A and B. Plate A moves at a velocity (V) relative to Plate B. A shear force (τ) acting parallel to the plates creates a steady-state velocity profile, shown by the inclined line, where fluid velocity (u) is proportional to the length of the arrows. The shear stress may be thought of as the force that produces a change in velocity (du) relative to height (dy) as one fluid layer slides over another. The ratio of shear stress to du/dy is the viscosity (μ).

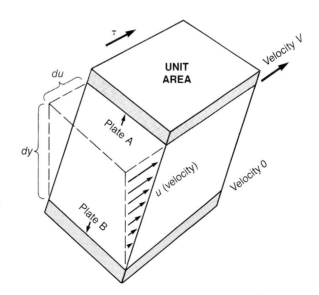

Types of Fluids

Air and water are the only fluids of importance in sediment transport, but water can display variable properties as a fluid medium if it contains substantial concentrations of sediment or is frozen into ice. Because these fluid properties affect the way fluids flow and transport sediment, it is important to understand the behavior of various types of fluids. Depending upon the extent to which dynamic viscosity (μ) changes with shear or strain (deformation) rate, three general types of fluids can be defined. **Newtonian fluids** have no strength and do not undergo a change in viscosity as the shear rate increases. Thus, ordinary water which does not change viscosity as it is stirred or agitated is a Newtonian fluid. Additional resistance to flow does arise during fluid turbulence owing to movement of eddies, which absorb energy. This resistance is called eddy viscosity, but it does not represent a change in the dynamic viscosity. **Non-Newtonian fluids** have no strength but show variable viscosity (μ) with change in shear or strain rate. Water that contains dispersions of sand in concentrations greater than about 30 percent by volume—or even lower concentrations of cohesive clay—behaves as a non-Newtonian fluid. Therefore, highly water-saturated, noncompacted muds display non-Newtonian behavior. Such muds may flow very sluggishly at low flow velocities, but they display much less viscous flow at higher velocities.

Some extremely concentrated dispersions of sediment may behave as plastic substances, which have an initial strength that must be overcome before yield occurs. If the plastic material behaves as a substance with constant viscosity after the yield strength is exceeded, it is called a **Bingham plastic.** Debris flows in which large cobbles or boulders are supported in a matrix of interstitial fluid and fine sediment are examples of natural substances that behave as Bingham plastics. Water with dispersed sediment and other plastic materials (such as ice), which behave as substances with variable viscosity after yield strength is exceeded and they start to flow, are called **pseudoplastics. Thixotropic substances,** a special type of pseudoplastic, have strength until sheared. Shearing destroys their strength; the substances behave like a fluid (commonly non-Newtonian) until allowed to rest a short while, after which strength is regained. Freshly deposited muds commonly display thixotropic behavior. Shearing resulting from earthquake tremors, for example, can cause liquefaction and failure of such muds. Such momentary liquefaction may result in downslope movement of sediment that otherwise would not undergo transport. It may also lead to formation of certain kinds of deformation structures. Differences in behavior of Newtonian fluids, non-Newtonian fluids, and plastic substances in response to shear stress are illustrated in Figure 3.2.

Laminar vs. Turbulent Flow

Fluids in motion display two modes of flow depending upon the velocity and viscosity of the fluids. Experiments with dyes show that a thin stream of dye injected into a slowly moving, unidirectional fluid will persist as a straight, coherent stream of nearly constant width. This type of movement is **laminar flow.** It can be visualized as a series of parallel sheets or filaments by which movement is occurring on a molecular scale owing to constant vibration and translation of the fluid molecules (Fig. 3.3A and B). If velocity of flow is increased or viscosity of the fluid decreased, the dye stream is no longer maintained as a coherent stream but breaks up and becomes highly distorted. It moves as a series of constantly changing and deforming masses in which there is sizable transport of fluid perpendicular to the mean direction of flow (Fig. 3.3C). This

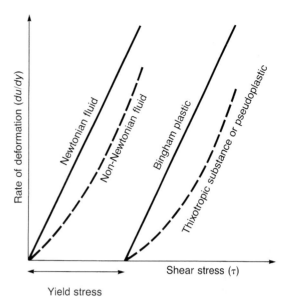

FIGURE 3.2 Rates of deformation vs. shear stress for fluids and plastics. (After Blatt, H., G. V. Middleton, and R. Murray, Origin of sedimentary rocks, 2nd ed., © 1980. Fig. 5.26, p. 187. Reprinted by permission of Prentice-Hall, Englewood Cliffs, N.J.)

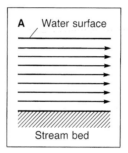
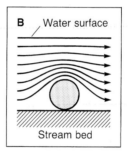
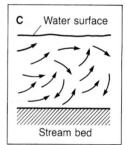

FIGURE 3.3 Schematic representation of laminar vs. turbulent fluid flow: A. Laminar flow over a smooth stream bed. B. Laminar flow over a spherical particle on a smooth bed. C. Turbulent flow over a smooth bed. The arrows indicate flow paths of the fluid.

type of flow is called **turbulent flow** because of the transverse movement of these masses of fluid. Turbulence is thus an irregular or random component of fluid motion. Highly turbulent water masses are referred to as **eddies.** Most flow of water and air under natural conditions is turbulent, although flow of ice and of mud-supported debris (non-Newtonian fluids) is essentially laminar.

The upward motion of water particles in turbulent water masses slows the fall of settling particles and thus decreases their settling velocity. Also, fluid turbulence tends to increase the effectiveness of fluid masses in eroding and entraining particles from a sediment bed. Because of the significance of turbulence in sediment transport, it is important to develop a fuller understanding of this property. Velocity measured over a period of time at a particular point in a laminar flow is constant. By contrast, velocity measured at a point in turbulent flow tends toward an average value when measured over a long period of time, but it varies from instant to instant about this average value. As we shall see, a calculated variable called the Reynolds number can be used to predict the boundary conditions separating laminar and turbulent flow. Turbulent flow resists distortion to a much greater degree than does laminar flow. Thus, a fluid undergoing turbulent flow appears to have a higher viscosity than the same fluid undergoing laminar flow. As mentioned, this apparent viscosity, which varies with the character of the turbulence, is called **eddy viscosity.** Eddy viscosity results from turbulent momentum transfer, and it is the rate of exchange of fluid mass between adjacent water bodies. It is necessary in dealing with fluids undergoing turbulence to rewrite the equation for shear stress to include a term for eddy viscosity. Thus, for laminar flow, shear stress is given by the relationship

$$\tau = \mu \frac{du}{dy} \tag{3.3}$$

but for turbulent flow

$$\tau = (\mu + \eta) \frac{du}{dy} \tag{3.4}$$

where η is eddy viscosity, which is commonly several orders of magnitude higher than dynamic viscosity.

Reynolds Number

The fundamental differences in laminar and turbulent flow arise from the ratio of inertial forces that tend to cause fluid turbulence and viscous forces that tend to suppress turbulence. Inertial forces are related to the scale and velocity of fluids in motion. Viscous forces arise from the viscosity of the fluid, and these forces resist deformation of the fluid. The relationship of inertial to viscous forces can be shown mathematically by a dimensionless value called the **Reynolds number** (R_e), which is expressed as

$$R_e = \frac{UL\rho}{\mu} = \frac{UL}{\nu} \tag{3.5}$$

where U is the mean velocity of flow, L is some length (commonly water depth) that characterizes the scale of flow, and ν is kinematic viscosity. When viscous forces dominate, as in highly concentrated mud flows, Reynolds numbers are small and flow is laminar. Very low flow velocity or shallow depth also produces low Reynolds numbers and laminar flow. When inertial forces dominate and flow velocity increases, as in the atmosphere and most flow in rivers, Reynolds numbers are large and flow is turbulent. Thus, most flow under natural conditions is turbulent. Note from equation 3.5 that an increase in viscosity can have the same effects as a decrease in flow velocity or flow depth. The transition from laminar flow to turbulent flow takes place above a critical value of Reynolds number, which commonly lies between 500 and 2000 and which depends upon the boundary conditions such as channel depth and geometry. Thus, under a given set of boundary conditions, the Reynolds number can be used to predict whether flow will be laminar or turbulent and to derive some idea of the magnitude of turbulence. Because the Reynolds number is dimensionless, it is of particular value when used to compare scaled-down models of flow systems to natural flow systems. If the length value in equation 3.5 is replaced by grain diameter, the Reynolds number becomes an important parameter in evaluating sediment erosion and entrainment from the bed, a subject discussed subsequently in this chapter.

Velocity Profile and Bed Roughness

Because of the greater shear stress required to maintain a particular velocity gradient in turbulent flow, both the vertical velocity profile above the bed and the velocity profile in a flow channel, as observed from above, have a different shape than laminar-flow velocity profiles (Fig. 3.4). Owing to variations in flow velocity during turbulent flow, the shape of the turbulent-flow vertical profile is determined by time-averaged values of velocity. Under conditions of turbulent flow, laminar or near-laminar flow occurs only very near the bed. The exact shape of the turbulent profile depends upon the nature of the bed over which the flow takes place. For smooth beds, there is a thin layer close to the bed boundary where molecular viscous forces dominate. Molecular adhesion causes the fluid immediately at the boundary to remain stationary. Successive overlying layers of fluid slide relative to those beneath at a rate dependent upon the fluid viscosity (Fig. 3.1). Flow within this thin boundary layer tends toward laminar, although it is characterized by streaks of faster and slower moving fluid and is not truly laminar. This layer is the **viscous sublayer,** or **laminar sublayer.** Over a very rough or irregular bed such as coarse sand or gravel, the viscous sublayer is destroyed by these irregularities, which extend through the layer into the turbulent flow. The flow of fluid over a boundary is thus affected by the roughness of the boundary. Obstacles on the bed generate eddies at the boundary of a flow; the larger and more abundant the obstacles, the more turbulence is generated. The presence or absence of a viscous sublayer may be an important factor in initiating grain movement.

Boundary Shear Stress

As a fluid flows across its bed, a stress that opposes the motion of the fluid exists at the bed surface. This stress, called the **boundary shear stress** (τ_0) to differentiate it from fluid shear stress (τ), is defined as force per unit area parallel to the bed, that is, tangential force per unit area of surface. It is a function of the density of the fluid, slope of the bed, and water depth. Boundary shear stress is expressed as

$$\tau_0 = \gamma R_h S \tag{3.6}$$

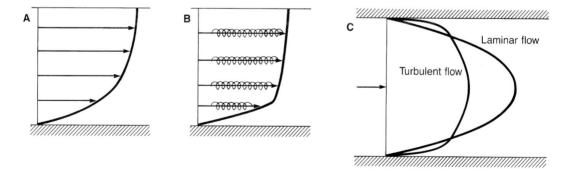

FIGURE 3.4   Comparison of vertical velocity profiles for (A) laminar and (B) turbulent flow in a wide channel. Velocities in the turbulent profile are time-averaged values. C. General form of laminar and turbulent velocity profiles as observed from above. (After Collinson, J. D., and D. B. Thompson, 1982, Sedimentary structures. Fig. 3.3, p. 22, reprinted by permission of George Allen & Unwin, London; and Leeder, M. R., 1982. Sedimentology: Process and product. Fig. 5.8, p. 53, reprinted by permission of George Allen & Unwin, London.)

where γ is density of the fluid, R_h is hydraulic radius (cross-sectional area divided by wetted perimeter), and S is the slope (gradient). The boundary shear stress is also a function of velocity of flow, a complex mathematical relationship not shown here. It tends to increase as velocity increases, although not in a direct way.

Because boundary shear stress is determined by the force that a flow is able to exert on the sediment bed and is related to flow velocity, it is an extremely important variable in determining the erosion and transport of sediment on the bed below a flow. Equation 3.6 indicates that boundary shear stress increases directly with increasing density of the moving fluid, increasing diameter and depth of the stream channel, and increasing slope of the stream bed. Other factors being equal, we would thus expect to see greater boundary shear stress developed, and greater ability to erode and transport sediment, in water flows than in air flows, in larger stream channels than in smaller channels, and in high-gradient streams than in low-gradient streams.

Froude Number

In addition to the effects of fluid viscosity and inertial forces, gravity forces also play an important role in fluid flow because gravity influences the way in which a fluid transmits surface waves. The velocity with which small gravity waves move in shallow water is given by the expression $\sqrt{gL}$, in which g is gravitational acceleration and L is water depth. The ratio between inertial and gravity forces is the **Froude number** (F_r), which is expressed as

$$F_r = \frac{U}{\sqrt{gL}} \tag{3.7}$$

where U is again the mean velocity of flow and L is water depth, in the case of water flowing in an open channel. The Froude number, like the Reynolds number, is a dimensionless value.

When the Froude number is less than 1, the velocity at which waves move is greater than flow velocity, and waves can travel upstream. That is, waves in a stream move upstream in the opposite direction to current flow. Flow under these conditions is called tranquil, streaming, or subcritical. If the Froude number is greater than 1, waves cannot be propagated upstream, and flow is said to be rapid, shooting, or supercritical. Thus, the Froude number can be used to define the critical velocity of water (but not air) at which flow at a given depth changes from tranquil to rapid, or vice versa. The Froude number also has a relationship to flow regimes, which are defined by characteristic bedforms such as ripples, that develop during fluid flow over a sediment bed. This relationship is discussed further in the following section.

3.3   PARTICLE TRANSPORT BY FLUIDS

Having established some of the fundamentals of fluid behavior during flow of fluids alone, we are now at a point where we can consider the more complicated processes of transport of sediment by fluid flow. Transport of sediment by fluid flow involves two fundamental steps: (1) erosion and entrainment of sediment from the bed and (2) subsequent, sustained downcurrent or downwind movement of sediment along or above the bed. The term **entrainment** refers to the processes involved in lifting resting grains from the bed or otherwise putting them in motion. More energy is commonly required to initiate particle movement than to keep particles in motion after entrainment. Thus, a great deal of experimental and theoretical work has been directed toward study of the conditions necessary for particle entrainment. Once particles are lifted from the sediment bed into the overlying water or air column, the rate at which they fall back to the bed—the settling velocity—is an important factor in determining how far the particles travel downcurrent before they again come to rest on the sediment bed. Like particle entrainment, the settling velocity of particles has been studied extensively. We will now examine some of these fundamental aspects of particle transport by fluids, beginning with a look at the factors involved in entrainment of sediments by a moving body of fluid.

Particle Entrainment

Entrainment by Currents. As the velocity and shear stress of a fluid moving over a sediment bed increase, a critical point is reached at which grains begin to move downcurrent. Commonly the smallest and lightest grains move first. As shear stress increases, larger grains are put into motion until finally grain motion is common everywhere on the bed. This **critical threshold** for grain movement is a direct function of several variables, including the boundary shear stress; fluid viscosity; and particle size, shape, and density. Indirectly, it is also a function of the velocity of flow, which varies as the logarithm of the distance above the bottom.

To understand the problems involved in lifting particles from the bed and initiating motion, let us consider the opposing forces that come into play as a fluid moves across its bed. As shown in Figure 3.5A, forces owing to gravity act downward to resist motion and hold particles against the bed. The gravity forces result from the weight of the particles and are aided in resisting grain movement by frictional resistance between particles. Fine, clay-size particles have added resistance to movement owing to cohesiveness that arises from electrochemical bonds between these small grains. The motive forces that must be generated by fluid flow to overcome the resistance to movement imposed by these retarding factors include a **drag force** that acts parallel to the bed and is related to the boundary shear stress and a **lift force** due to the Bernoulli effect of fluid flow over projecting grains. The drag force (F_D) depends upon the boundary shear stress (τ_0) and the drag exerted on each grain exposed to this stress. Thus,

$$F_D = \frac{\tau_0}{N} \tag{3.8}$$

where N is the number of exposed grains per unit area. The hydraulic lift force known as the Bernoulli effect is caused by the convergence of fluid streamlines over a projecting grain. The Bernoulli effect results from an increase in flow velocity in the zone where the streamlines converge over the grain. This velocity increase causes pressure to decrease above the grain. Hydrostatic pressure from below then tends to push the grain up off the bed into this low-pressure zone (Fig. 3.5B). The drag force and lift force combine to produce the total fluid force, represented by the fluid-force vector in Figure 3.5A. For grain movement to occur, the fluid force must be large enough to overcome the gravity and frictional forces.

The preceding discussion is greatly simplified and generalized, and a number of factors complicate calculation of critical thresholds of grain movement under natural conditions. These factors include variations in shape, size, and sorting of grains; bed roughness, which controls the presence or absence of a viscous sublayer; and cohesion of small particles. Because of these complicating factors, the critical conditions for par-

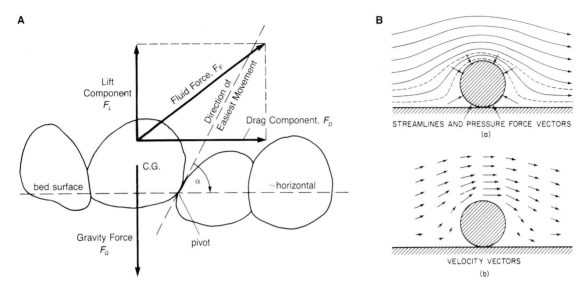

FIGURE 3.5   A. Forces acting during fluid flow on a grain resting on a bed of similar grains. B. Flow pattern of fluid moving over a grain, illustrating the lift forces generated owing to the Bernoulli effect: (a) streamlines and the relative magnitude of pressures acting on the surface of the grain, (b) direction and relative velocity of velocity vectors; higher velocities occur where streamlines are closer together. (A, after Middleton, G. V., and J. B. Southard, 1978, Mechanisms of sediment movement: Eastern Section, Soc. Econ. Paleontologists and Mineralogists Short Course No. 3. Fig. 6.1, p. 6.3, reprinted by permission of SEPM, Tulsa, Okla. B, from Blatt, H., G. V. Middleton, and R. Murray, Origin of sedimentary rocks, 2nd ed., © 1980, Fig. 4.9, p. 107. Reprinted by permission of Prentice-Hall, Englewood Cliffs, N.J.)

ticle entrainment must be determined experimentally. Two widely used plots that show experimentally derived threshold graphs for initiation of grain movement are the **Hjulström** and **Shields diagrams.**

In the Hjulström diagram (Fig. 3.6), the velocity at which grain movement begins as flow velocity increases above the bed is plotted against mean grain size (grain diameter). This diagram shows the critical velocity for movement of quartz grains on a plane bed at a water depth of 1 m. The curve separates the graph into two fields. Points above the graph indicate the conditions under which grains are in motion, and points below indicate no motion. Note from this figure that critical entrainment velocity for grains larger than about 0.5 mm increases gradually with increasing mean grain size, whereas the entrainment velocity for grains smaller than 0.05 mm increases with decreasing grain size. This seemingly anomalous behavior at smaller grain sizes is apparently due mainly to increasing cohesion of finer-size particles, making them more difficult to erode than larger, noncohesive particles. Also, extremely small grains may lie within the viscous sublayer, where little grain movement takes place.

The Shields diagram (Fig. 3.7) is widely used by sedimentologists and is well established by experimental work; however, it is more complex and difficult to understand than the Hjulström diagram because it involves two dimensionless relationships. In Figure 3.7, dimensionless shear stress (θ_t) (called β by some workers) is used instead of flow velocity as a measure of critical shear, and the mean grain-size parameter used in the Hjulström diagram is replaced by the **grain Reynolds number** (R_{eg}), another dimensionless quantity. The dimensionless bed shear stress (θ_t) is given by

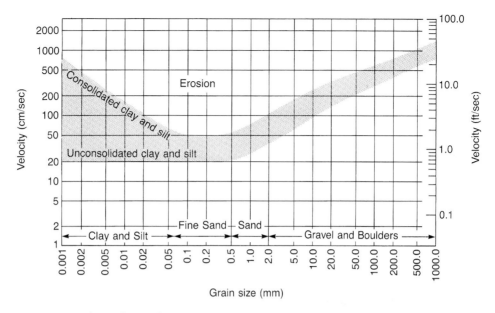

FIGURE 3.6 The Hjulström diagram, as modified by Sundborg, showing the critical current velocity required to move quartz grains on a plane bed at a water depth of 1 m. The shaded area indicates the scatter of experimental data, and the increased width of this area in the finer grain sizes shows the effect of sediment cohesion and consolidation on the critical velocity required for sediment entrainment. (After Sundborg, A., 1956, The River Klarälven, a study of fluvial processes: Geografiska Annaler, Ser. A, v. 38. Fig. 16, p. 197, reprinted by permission.)

$$\theta_t = \frac{\tau_t}{(\rho_S - \rho)gD} \tag{3.9}$$

where τ_t is boundary shear stress, ρ_S is density of the particles, ρ is density of the fluid, g is gravitational acceleration, and D is particle diameter. The value of dimensionless shear stress increases with increasing bed shear stress and increasing velocity, and it decreases with increasing density and size of the particles. Unlike the separate velocity and grain-size parameters in Hjulström's diagram, the dimensionless shear stress thus incorporates shear stress (velocity), grain size, and grain and fluid density into a single term. An increase in dimensionless shear stress indicates either an increase in flow velocity and shear stress or a decrease in grain size or density.

The grain Reynolds number (R_{eg}) differs from the ordinary Reynolds number previously discussed. The length or water depth value (L) of the ordinary Reynolds number is replaced by particle diameter (d) and the flow velocity (U) by shear velocity (U^*). The grain Reynolds number is a measure of turbulence at the grain-fluid boundary. It is thus expressed as

$$R_{eg} = \frac{U^*d}{\nu} \tag{3.10}$$

The grain Reynolds number is clearly not the same thing as mean grain size; however, it can be seen from Figure 3.7 that the grain Reynolds number increases with increas-

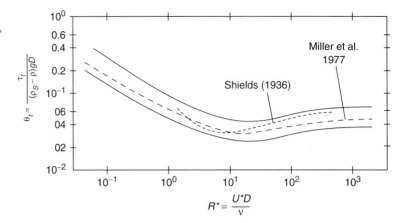

ing grain size if friction velocity and kinematic viscosity remain constant. Thus, an increase in grain Reynolds number means an increase in grain size, an increase in friction velocity and turbulence, or a decrease in kinematic viscosity.

The Shields diagram is more difficult to interpret than Hjulström's diagram, but, as in the Hjulström diagram, points above the curve indicate that noncohesive grains on the bed are fully in motion and points below indicate no motion. Beginning of motion is determined by the dimensionless shear stress, which increases with increasing bed shear stress under a given set of conditions for grain density, fluid density, and grain size. The critical dimensionless shear stress required to initiate grain movement thus depends upon the grain Reynolds number, which in turn is a function of grain size, kinematic viscosity, and turbulence. Note from the Shields diagram in Figure 3.7 that the dimensionless bed shear stress increases slightly with increasing grain Reynolds number above about 5 to 10, although it remains mainly between 0.03 and 0.05. At lower Reynolds numbers, the value increases steadily up to a value of 0.1 or higher. This greater rate of increase at lower Reynolds numbers is related to the presence of the viscous sublayer. When the bed is composed of small particles on the order of fine sand or smaller, a smooth boundary to flow results; the particles lie entirely within the viscous sublayer, where flow is essentially nonturbulent and instantaneous velocity variations are less than in the lower part of the overlying turbulent boundary layer. For coarser particles, the viscous sublayer is so thin that the grains project through the layer into the turbulent flow.

The reason that many sedimentologists use the Shields diagram in preference to the simpler Hjulström diagram is that it has more general application. For example, it can be used for wind as well as water and for a variety of conditions in water. By contrast, the Hjulström diagram is valid only in water in which fluid and grain density and dynamic viscosity are constant, as in freshwater streams in a given season during average flow.

Several complicating factors not covered in the Hjulström and Shields diagrams make prediction of the onset of grain movement difficult. Instantaneous fluctuations in boundary shear stress may arise from local eddies or from wave action superimposed on current flow, and these fluctuations may cause some particles to move before the general onset of grain movement. Fine muds and silts may not erode to yield individual grains owing to the tendency of such cohesive materials to be removed as chunks or aggregates of grains. Entrainment of grains by wind action can be strongly affected by impact of moving grains hitting the bed. At a value of wind velocity below the critical velocity needed to initiate grain movement, grain motion can be started and propagated downwind by throwing grains onto the bed, a process referred to as seeding. This lower threshold for grain movement is called the **impact threshold.**

Many of the processes involved in sediment entrainment and transport can be effectively modeled by computer simulation. An excellent recent book on this subject, *Simulating Clastic Sedimentation* (Tetzlaff and Harbaugh, 1989), takes the reader through the computations and reasoning of erosion, transportation, and deposition of clastic sediment. The book further provides detailed instructions on the techniques and computer programs used to simulate clastic sedimentation.

Role of Settling Velocity in Grain Transport

As soon as grains are lifted above the bed during the entrainment process, they begin to fall back to the bed. The distance that they travel downcurrent before again coming to rest on the bed depends upon the drag force exerted by the current and the settling velocity of the particles. A particle initially accelerates as it falls through a fluid, but acceleration gradually decreases until a steady rate of fall, called the **terminal fall velocity,** is achieved. For small particles, terminal fall velocity is reached very quickly. The rate at which particles settle after reaching fall velocity is a function of the viscosity of the fluid and the size, shape, and density of the particles. The settling rate is determined by the interaction of upwardly directed forces—owing to buoyancy of the fluid and viscous resistance (drag) to fall of the particles through the fluid—and downwardly directed forces arising from gravity. The drag force exerted by the fluid on a falling spherical grain is proportional to the density of the fluid (ρ_f), the diameter (d) of the grains, and the fall velocity (V), as given by the relationship

$$C_D \pi \frac{d^2}{4} \frac{\rho_f V^2}{2} \tag{3.11}$$

where C_D is a drag coefficient that depends upon the grain Reynolds number and the particle shape. The upward force resulting from buoyancy of the fluid is given by

$$\frac{4}{3} \pi \left(\frac{d}{2} \right)^3 \rho_f g \tag{3.12}$$

where ρ_f is fluid density and g is gravitational acceleration. [*Note:* $4/3\pi(d/2)^3$ is the volume of a sphere.] The downward force owing to gravity is given by

$$\frac{4}{3} \pi \left(\frac{d}{2} \right)^3 \rho_s g \tag{3.13}$$

where ρ_s is particle density. As the particle stops accelerating and achieves fall velocity, the drag force of the liquid on the falling particle is equal to the downward force due to gravity minus the upward force resulting from buoyancy of the liquid. Thus,

$$C_D \pi \frac{d^2}{4} \frac{\rho_f V^2}{2} = \frac{4}{3} \pi \left(\frac{d}{2} \right)^3 \rho_s g - \frac{4}{3} \pi \left(\frac{d}{2} \right)^3 \rho_f g \tag{3.14}$$

Rearranging terms, we can express this relationship in terms of fall velocity (V) as

$$V^2 = \frac{4gd}{3C_D} \frac{(\rho_s - \rho_f)}{\rho_f} \tag{3.15}$$

For slow laminar flow at low concentrations of particles and low grain Reynolds numbers (R_{eg}), C_D has been determined to equal $24/R_{eg}$ (Rouse and Howe, 1953, p. 182). Substituting this value ($24/U^*d/\mu/\rho_f$) for C_D yields

$$V = \frac{1}{18} \frac{(\rho_s - \rho_f)gd^2}{\mu} \tag{3.16}$$

which is **Stokes's Law** of settling, with particle size expressed as diameter in centimeters. This law, formulated by Stokes in 1845, is often simplified to

$$V = CD^2 \quad (cm/s) \tag{3.17}$$

where C is a constant equaling $(\rho_s - \rho_f)g/18\mu$ and D is the diameter of particles (spheres) expressed in centimeters. Values of C have been calculated for a range of common laboratory temperatures (e.g., Galehouse, 1971); thus settling velocity (V) can be determined quickly for any value of particle diameter (D). Note that the Reynolds number is a distinguishing factor in grain settling behavior just as it is in laminar and turbulent flow.

Experimental determination of particle fall velocity shows that Stokes's Law accurately predicts settling velocity of particles in water only for particles less than about 0.1 to 0.2 mm in diameter. Larger particles have fall velocities lower than those predicted by Stokes's Law, apparently owing to inertial (turbulent) effects caused by the increased rates of fall of these larger grains. Thus, the Stokes equation cannot be used for determining the settling velocity of sand, a very important component of most sediment. The fall velocity is also decreased by decrease in temperature (which increases viscosity), decrease in particle density, and decrease in the sphericity (the degree to which the shape of a particle approaches the shape of a sphere) of the particles. Most natural particles are not spheres, and departure from spherical shape decreases fall velocity. Fall velocity is also decreased by increasing concentration of suspended sediment in the fluid, which increases the apparent viscosity and density of the fluid.

Relationship of Unidirectional Current Flow to Bedforms

Anyone who has examined the sandy bed of a clear, shallow stream has certainly noticed that the bed is rarely perfectly flat and even. Instead, it is commonly marked by ripples and similar bedforms of various sizes. Such bedforms also occur in eolian and submarine environments where they range in size from small ripples a few centimeters in length and a fraction of a centimeter in height to gigantic eolian sand dunes and undersea sand waves tens to hundreds of meters in length and several meters to several tens of meters in height. If we carefully dissect a ripple exposed on the dry bed of a stream to reveal its internal structure, we almost invariably find internal fine-scale cross-laminae that dip in a downcurrent direction. Clearly, there is a close genetic relationship between fluid-flow mechanisms, ripple bedforms, and cross-lamination.

The preservation potential of ripples is relatively low; they are therefore not extremely common features on the bedding planes of ancient sedimentary rocks. On the other hand, cross-beds are exceedingly common in many ancient sandstone sequences. In an attempt to better understand the origin of bedforms and cross-stratification, many investigators have turned to the study of sediment transport in flumes. Flumes are long, slightly sloping troughs fitted with glass sides to allow observation. Sand or other sediment is placed on the floor of the flume, and water is constrained to flow over the floor at various depths and velocities.

Numerous flume experiments have established that under unidirectional flow, small ripples begin to develop in sandy sediment as soon as the critical entrainment velocity for the sediment is reached. The exact sequence of other kinds of bedforms that develop with increasing velocity depends upon the grain size of the material. If flow is over a bed of sediment ranging in size from about 0.25 to 0.7 mm (medium to coarse sand), for example, the succession of bedforms illustrated in Figure 3.8 is generated, beginning with **ripples.** The relationship between bedform type, water depth, grain size, and flow velocity is further illustrated in Figure 3.12.

The shape of ripples and the terminology used to describe ripples and larger bedforms are illustrated in Figure 3.9. Ripples are the smallest bedform, ranging in length from about 5 to 20 cm and in height from about 0.5 to 3 cm. Thus, they have a ripple

FIGURE 3.8 The succession of bedforms that develops during unidirectional flow of sandy sediment (0.25–0.7 mm) in shallow water as flow velocity increases. (After Blatt, H., G. V. Middleton, and R. Murray, Origin of sedimentary rocks, 2nd ed., 1980, Fig. 5.3, p. 137. Reprinted by permission of Prentice-Hall, Englewood Cliffs, N.J.)

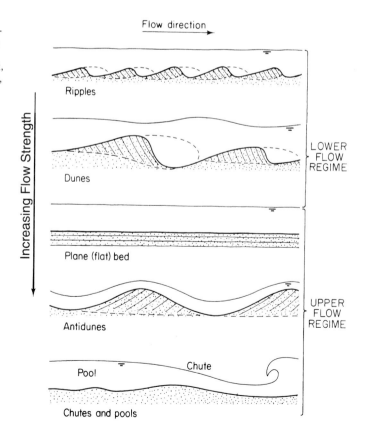

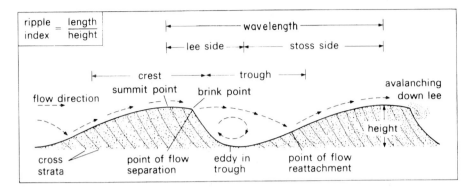

FIGURE 3.9   The terminology used to describe asymmetric ripples (From Tucker, M. E., Sedimentary petrology, an introduction. © 1981 by John Wiley & Sons, Inc. Fig. 2.18, p. 28, reproduced by permission of Open University Educational Enterprises Ltd., Stony Stratford, England.)

index (ratio of ripple length/ripple height) ranging from about 8 for coarse sand to 20 for fine sand. They form in sediment ranging in size from silt (0.06 mm) to sand as coarse as 0.7 mm. Prior to the 1980s, similar bedforms of larger size were referred to as sand waves and dunes, and earlier workers attempted to distinguish between sand waves and dunes on the basis of wave length and height. More recent work shows, however, that these large bedforms constitute a continuum with spacing, or wave length, ranging from under 1 m to over 1000 m. A panel convened to examine bedform nomenclature (Ashley, 1990) recommended that a single name, **dune,** be used for all large bedforms (larger than ripples). Dunes are similar in general appearance to ripples except for size. They form at higher flow velocities in sediment ranging in grain size from fine sand to gravel. The ripple index of dunes ranges from about 5 in finer sands to 50 in coarser sediment. In the lower part of the dune stability field, ripples may be superimposed on the backs of dunes.

   The characteristics of bedforms that develop under unidirectional flow are summarized in Table 3.1. In this table, 2D and 3D refer to two-dimensional and three-dimensional. Two-dimensional dunes are generally straight-crested dunes whose shapes can be adequately described in a two-dimensional plane oriented parallel to the flow direction (see Fig. 5.13). Three-dimensional dunes are characterized by curved faces and scour pits, and their shapes must be described in three dimensions (see Fig. 5.14).

   During the formation of ripples and dunes, either the water surface shows little disturbance, or water waves are out of phase with bedforms (Fig. 3.8). Out-of-phase waves may show slight disturbance of the water surface over large-scale ripples and large swirls or "boils" that rise to the surface. The hydraulic conditions that generate these bedforms and out-of-phase surface waves distinguish what is called the **lower-flow regime** (Simons and Richardson, 1961). Ripples and dunes generated in the lower-flow regime migrate downstream because sediment is eroded from the stoss (current-facing) side of these bedforms and carried up to the crest where it avalanches down the lee slope (Fig. 3.9). Avalanching leads to formation of cross-lamination that dips downstream at angles of up to about 30°. In addition to experimental study of bedforms, flume-generated bedforms have also been successfully modeled by computer simulation for a number of different flow and grain-size regimes (e.g., Fig. 3.10).

TABLE 3.1 Characteristics of bedforms developed under unidirectional flow

	Ripples	2D dunes	3D dunes	Lower plane bed	Upper plane bed	Antidunes
Length (spacing)	0.1–0.2 m	a few 10s of cm to 100s of m	a few 10s of cm to 10s of m (or more?)	—	—	10s of cm–m
Height	a few cm	cm to a few 10s of m	10s of cm to a few m (or more?)	—	—	cm–10s of cm
Ripple index (length/ height)	relatively low	relatively high	relatively low	—	—	relatively high
Plan geometry	strongly irregular/ short-crested	straight/sinuous and long-crested	strongly irregular/ short-crested	—	—	long-crested and short-crested
Characteristic flow velocity	low	low/moderate	moderate/ high	low	high	high
Characteristic flow depth	> a few cm	> a few dm	> a few dm	all	all	shallow flows
Characteristic sediment size	0.03–0.6 mm	>0.3 mm?	>0.2 mm	>0.6 mm	all	all

Source: Modified from Harms, J. C., J. B. Southard, and R. G. Walker, 1982, Structures and sequences in clastic rocks: Soc. Econ. Paleontologists and Mineralogists Short Course No. 9. Table 2–1, p. 2–11, reprinted by permission of SEPM, Tulsa, Okla.

With further increase in flow velocity, dunes are destroyed and give way to an **upper-flow regime** stage of flow. Sheetlike, rapid flow of water which generates symmetrical surface water waves that are **in phase** with the bedforms distinguishes the upper-flow regime. Owing to very rapid water flow, intense sediment transport takes place over an initially relatively flat bed during what is referred to as the plane-bed stage of flow. Plane-bed flow gives rise to internal planar lamination in which individual laminae range in thickness from a few millimeters to a few centimeters. The preservation potential of these plane-bed laminae appears to be low; nonetheless, upperflow-regime, plane-bed laminae have been reported in ancient sandstone deposits. The most important environments where plane beds are likely to be preserved are in stream channels, on beaches and in other nearshore areas where strong shoaling waves prevail, and under high-velocity turbidity currents (Harms et al., 1982).

At still higher velocities of flow, plane beds give way to **antidunes,** which are low, undulating bedforms up to 5 m in length with a ripple index ranging from about 7 to 100+. Antidunes form in very fast, shallow flows at Froude numbers greater than about 0.8. They migrate upstream during flow, giving rise to low-angle (<10°) crossbedding directed upstream. Antidunes have very low preservation potential, and antidune cross-bedding is probably rarely preserved; however, antidune cross-bedding has been reported at the base of some turbidite flow units.

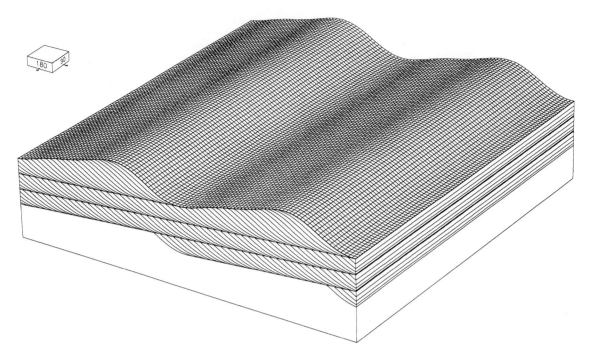

FIGURE 3.10   An example of a computer-generated bedform. (From Rubin, D. M., 1987, Cross-bedding, bedforms, and paleocurrents: Soc. Econ. Paleontologists and Mineralogists, Concepts in Sedimentology and Paleontology, v. 1. Fig. 5, p. 20, reproduced by permission of SEPM, Tulsa, Okla.)

When flow velocity increases above the antidune stage, one final type of bedform, **chute and pool structure,** develops at very high flow velocities (Fig. 3.8). These bedforms develop where shallow, rapid (supercritical) flow forms chutes that end abruptly in a deeper pool where flow is tranquil. Sediment accumulations occur in the tranquil pool region where steeply dipping backset (upstream-dipping) laminations develop (Leeder, 1982). The preservation potential of chute and pool structures is exceedingly poor, and this bedform is seldom found in nature.

Effects of Grain Size and Water Depth.   Experimental studies show that the succession of bedforms that develops at a given water depth during flow depends not only upon flow velocity but also upon grain size; therefore, the succession of bedforms shown in Figure 3.8 does not occur in sediment of all particle sizes. Figure 3.11 shows the relationship of bedforms to flow velocity and grain size at a water depth ranging from 0.25 to 0.40 m. If flow takes place over sediment coarser than about 0.9 mm, for example, the ripple phase does not develop. Instead, a lower plane-bed phase forms just prior to the formation of dunes. Note also from Figure 3.11 that below a grain size of about 0.15 mm, dunes do not form. The ripple phase is succeeded abruptly by the upper plane-bed phase. The relationships shown in Figure 3.11 are summarized in tabular form in Figure 3.12. For flow of grains of a given size in shallow water, increase in water depth has the general effect of increasing the velocity at which change from one bedform phase to another takes place. Figure 3.13 shows this relationship for fine, medium, and coarse sand. Experimental plots have also been constructed that depict bedform fields as a function of bed shear stress and sediment size. For details of these plots as well

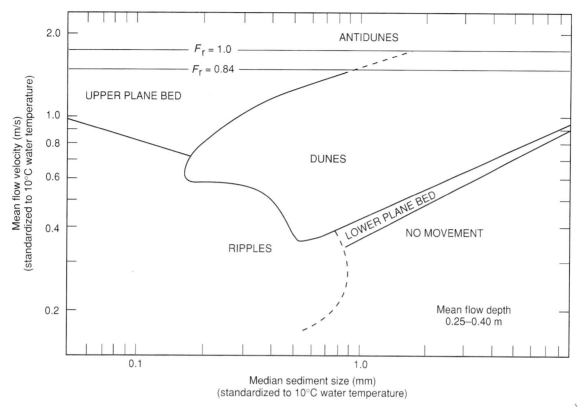

FIGURE 3.11 Plot of mean flow velocity against median sediment size showing the stability fields of bed phases. Note that the recommended terminology for bedforms is (1) lower plane bed, (2) ripples, (3) dunes (all large-scale ripples), (4) upper plane bed, and (5) antidunes. F_r = Froude number. (After Southard, J. B., and L. A. Boguchwal, 1990, Bed configurations in steady unidirectional water flows. Part 2. Synthesis of flume data: Jour. Sed. Petrology, v. 60. Fig. 3, p. 664, reprinted by permission of Society for Sedimentary Petrology, Tulsa, Okla.)

other data on bed configurations, see Southard and Boguchwal (1990) and Boguchwal and Southard (1990).

Most studies of bedforms have been carried out in laboratory flumes or under shallow-water conditions in natural environments. Therefore, most available sediment-size/velocity data pertain to the formation of bedforms under shallow-water conditions (commonly less than about 1 m). Much less is known about the development of bedforms under deeper-water conditions. Based on limited available information, Harms et al. (1982) suggest that the nature of small ripples is approximately the same in deep-water flows as in shallow-water flows; however, the larger bedforms (dunes) can grow much larger in deep-water flows. The hydraulic relationships in deep water are the same as for shallow water; that is, dunes form at higher velocities than ripples and at lower velocities than plane beds and antidunes. The exact relationship between grain size, flow velocity, and bedform phase is not well documented for deeper water, but a generalized relationship is shown in Figure 3.14. Note from Figure 3.14 that exceedingly high velocities are required to produce antidunes at a water depth greater than a few meters; therefore, it appears that antidunes are unlikely to occur under natural conditions in deep water.

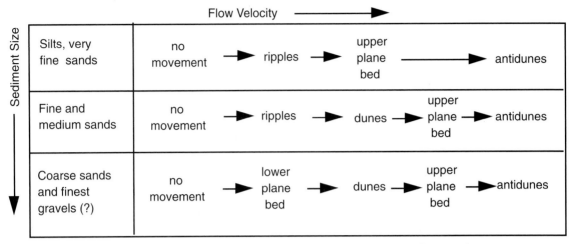

FIGURE 3.12 Sequence of bedforms that develop with increasing flow velocity, for sediments of various sizes. (Modified from Harms, J. C., J. B. Southard, and R. G. Walker, 1982, Structures and sequences in clastic rocks: Soc. Econ. Paleontologists and Mineralogists Short Course No. 9. Fig. 2–6, p. 2–15, reproduced by permission of SEPM, Tulsa, Okla.)

The mechanisms of sediment transport that are responsible for formation of the different bedforms are very complex. In general, the formation of transverse bedforms is related to a phenomenon called flow separation. Sediment is transported in suspension or by traction up the stoss side of the bedform to the brink or crest. At the brink, the flow separates from the bed to form a zone of reverse circulation or backflow, producing a separation eddy (Fig. 3.9). A zone of diffusion is present between the zone of backflow and the main flow above owing to turbulent mixing with the main flow. Downstream from the point of separation a distance several times the height of the bedform, the flow becomes reattached to the bottom. Flow separation causes separation of the transported sediment into bedload and suspended load fractions. The bedload fraction accumulates at the ripple crest until the lee slope exceeds the angle of repose and avalanching takes place. The suspended load fraction is transported downcurrent where the coarser particles in the suspended load settle through the zone of diffusion into the zone of backflow and are deposited in the lee of the ripple. It is these processes that cause development and movement of the bedforms.

The bedforms described above develop in response to unidirectional flow of water currents. They are asymmetrical in shape, with the steep or lee side facing downstream in the direction of current flow. Asymmetrical ripples formed in this fashion are called **current ripples.** Under natural conditions they form by river and stream flow, the backwash on beaches, longshore currents, tidal currents, and deep ocean bottom currents. In plan view, the crests of current ripples and dunes have a variety of shapes: **straight, sinuous, catenary, linguoid,** and **lunate** (Fig. 3.15). The plan-view shape of ripples and dunes is apparently related to water depth and velocity (Allen, 1968); however, the factors that control their shapes are not well understood. It has been observed under natural conditions that the more complex forms tend to develop in shallower water and at higher velocities than the less complex forms and that the order in which the succession of bedforms develops with decreasing water depth and velocity is straight to sinuous to symmetric linguoid to asymmetric linguoid for ripples and straight to sinuous to catenary to lunate for dunes.

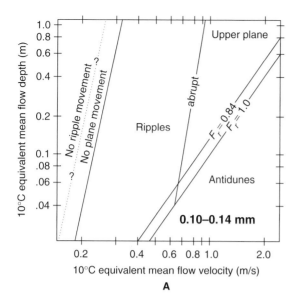

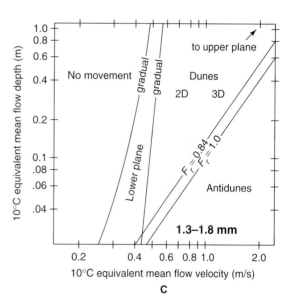

FIGURE 3.13 Dimensionless depth-velocity graphs for (A) sediment sizes 0.10–0.14 mm, (B) sediment sizes 0.40–0.50 mm, and (C) sediment sizes 1.30–1.80 mm. Data are adjusted to a reference temperature of 10°C; 2D refers to two-dimensional dunes, 3D to three-dimensional dunes. (After Southard, J. B., and L. A. Boguchwal, 1990, Bed configurations in steady unidirectional water flows. Part 2. Synthesis of flume data: Jour. Sed. Petrology, v. 60, Fig. 5, p. 667. Reproduced by permission.)

Wind Transport and Bedform Development

The fundamental principles of sediment transport by wind were set forth by Bagnold in 1941 in his classic, *The Physics of Blown Sand and Desert Dunes* (see also Pye and Tsoar, 1990, Chapter 4; and Barndorff-Nielsen and Willets, 1991). The forces involved in entrainment of grains by wind are similar to those exerted on grains by water flow (Fig. 3.5A). Transport of grains by wind is initiated when wind strength rises to the fluid threshold and also when wind blowing at greater than threshold speed over an immobile surface encounters the leading edge of a deposit of loose,

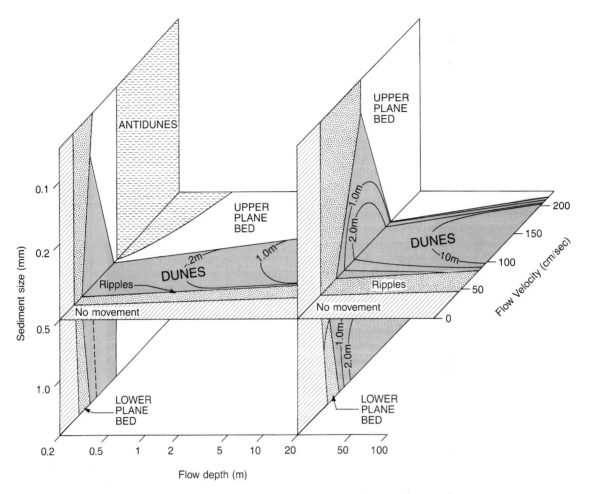

FIGURE 3.14 Generalized three-dimensional depth–velocity–grain-size diagram showing the relationship among bed phases and grain size for a wide variety of flow velocities and flow depths. Diagram based on both flume data and observations of natural flows. (After Rubin, M. D., and D. S. McCulloch, 1980, Single and superimposed bedforms: A synthesis of San Francisco Bay and fluvial observations: Sed. Geology, v. 26. Fig. 11, p. 224, reprinted by permission of Elsevier Science Publishers, Amsterdam.)

mobile material. Direct dislodgment by wind may also play a role in grain transport (Anderson, Sorenson, and Willets, 1991). Grain motion appears to cascade rapidly as those grains most susceptible to direct dislodgment collide (downwind) with and disturb less susceptible grains. The rapidity of the dislodgment depends upon the grain size, shape, sorting, and packing. At scattered locations, almost random, near-bed turbulence causes the wind flow to be seeded with low-energy ejected grains. Many of these grains translate downwind at a range of speeds, dislodging other grains as they go. A single flurry, therefore, tends to give rise to a translating and dispersing sequence of dislodgments. At a particular locality undergoing threshold wind flow, many such dislodgment sequences may be superimposed to produce overall entrainment and transport.

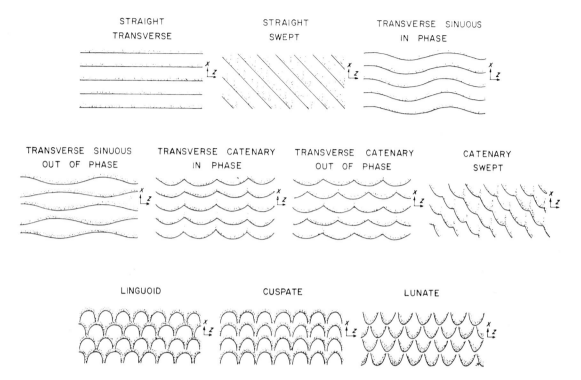

FIGURE 3.15 Idealized classification of current ripples and dunes based on plan-view shape. Flow is from the bottom to the top in each case. (After Allen, J. R. L., 1968, Current ripples: Their relation to patterns of water motion: North Holland Pub., Amsterdam. Fig. 4.6, p. 65, reprinted by permission of Elsevier Science Publishers, Amsterdam.)

The bedforms that develop during wind transport range from ripples as small as 0.01 m long and a few millimeters in height to dunes 500 to 600 m long and 100 m high. Less commonly, gigantic bedforms called **draas** that may have wave lengths measured in kilometers (up to 5.5 km) and heights up to 400 m may also form by wind transport (Wilson, 1972; McKee, 1982). The wave length of wind-transported bedforms increases with increasing wind velocity, and wave height tends to increase with increasing grain size. Wilson (1972) suggests, however, that under a given set of conditions of grain size and wind velocity, ripples, dunes, and draas can coexist. Thus, dunes exist on the backs of draas, and ripples are created on the backs of dunes. For further insight into the shapes of dunes and the processes that form them, see Pye and Tsoar (1990, Chapter 6).

The plan-view shape of eolian ripples is dominantly straight (Fig. 3.16), although sinuous forms also occur. More complex, three-dimensional forms are uncommon, although Leeder (1982) suggests that linguoid ripples analogous to those formed from water flows may occur in faster wind flows that are blowing very fine sand. The preservation potential of complete eolian bedforms is exceedingly low, and they are rarely, if ever, found in ancient sediments. The former existence of dunes and other eolian bedforms is revealed in ancient sedimentary rocks mainly by the presence of preserved cross-beds; however, considerable controversy has been generated regarding differentiation of eolian cross-bedding from subaqueous cross-bedding in ancient sedimentary rocks.

FIGURE 3.16 Straight-crested eolian ripples, northern Padre Island, Texas. Wind direction is from left to right. (Photograph by R. E. Hunter.)

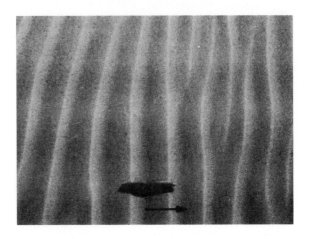

Entrainment and Transport by Waves

A special problem arises in evaluating the threshold of sediment movement under the action of orbital waves in nearshore areas of the ocean. In shallow water, the passage of waves over the ocean surface sets water in motion near the seabed; the water may have sufficient velocity to entrain and shift sediment about on the seafloor. The origin of this near-bottom water motion is related to the orbital motion generated by waves in near-surface water. If an observer watches the movement of an object floating on a lake or in the ocean during the passage of a wave, it is apparent that the object simply bobs up and down as the wave passes. It undergoes no noticeable forward motion. Numerous studies have established that water moves in orbital paths during the passage of waves (Fig. 3.17). Experiments with dyes and glass beads show that water moves forward in the crest of a wave, then downward, and finally backward under the trough and upward as the wave passes. Water particles retrace this orbit with each passing wave, returning nearly to their original position after the wave passes. Actually, a slight net forward movement of water particles occurs in shallow water during each orbit. Some forward movement takes place owing to a small "time asymmetry" in the velocity with which water moves forward under the foreshortened wave crest compared to backward movement under the longer trough (Clifton and Dingler, 1984).

Orbital motion of water dies out downward at a depth equal to about one-half the wave length. Therefore, in deep water the orbital motion is unimpeded by the bottom, and orbits are nearly circular. As waves move into shallow water, where depth is less than one-half the wave length, the bottom begins to interfere with orbital motion and thus begins to affect the shape of the orbits. By the time that waves reach very shallow water, where depth is less than about 1/20 the wave length, the motion of the particles is strongly affected by interaction with the bottom, and the orbits become much more elliptical. They become progressively flatter downward below the surface until near bottom they are essential linear, generating a to-and-fro motion as waves pass (Fig. 3.17). This motion produces bidirectional flow of water along the seafloor as each wave passes over the surface. The velocity of this bottom flow is referred to as the **orbital velocity** because it varies directly as a function of the magnitude of the orbital diameter and indirectly as a function of the wave period (the time required for passage of one

FIGURE 3.17 Orbital motion of water particles caused by passage of waves in (A) deep water and (B) shallow water. Note that in shallow water the orbits become both smaller and flatter with depth, until near the bottom they are essentially flat. (After Gross, M. G., Oceanography, 3rd ed., © 1982, Fig. 8.2, p. 214. Reprinted by permission of Prentice-Hall, Englewood Cliffs, N.J.)

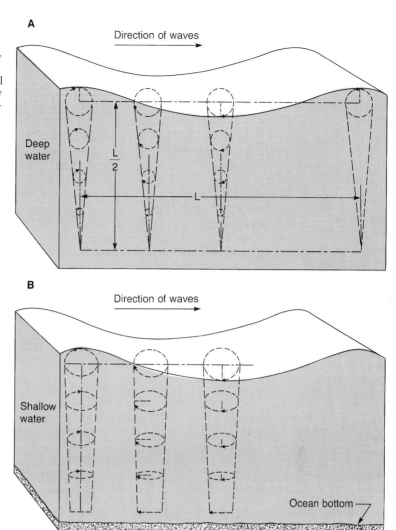

wave length). Owing to the time asymmetry referred to above, the orbital velocity is commonly greater in one direction than the other. This difference in velocity becomes particularly important when only the stronger velocity flow exceeds the threshold of movement for grains, resulting in net transport of grains in one direction. The factors that influence the threshold for grain movement are discussed in the following paragraph.

The threshold of grain movement under oscillatory waves is a function of grain diameter, wave period, and orbital velocity of the waves (Komar and Miller, 1975). Orbital velocity is in turn a function of wave height, water depth, wave period, and wave length. This relationship is shown by

$$u_t = \frac{\pi d_o}{T} = \frac{\pi H}{T \sin h \dfrac{2\pi h}{L}} \qquad\qquad (3.18)$$

in which u_t is the threshold orbital velocity, d_o is orbital diameter of the wave motion, H is wave height, L is wave length, T is wave period, and h is water depth. Note from this equation that, as mentioned, the threshold orbital velocity varies directly with the magnitude of the orbital diameter and inversely with the magnitude of the wave period. Komar and Miller (1975) used equation 3.18 to determine the threshold values for movement of grains ranging from 0.01 to 100 mm (Fig. 3.18) under waves having periods ranging from 1 to 15 s. Note the consistent, although nonlinear, increase in threshold orbital velocity with increasing grain diameter. Presumably this relationship holds only for noncohesive sediment. From the threshold values shown in Figure 3.18, Komar and Miller calculated the depths at which waves with a period of 15 s and heights up to 6 m would set sediment in motion on the seafloor. Their calculations show that particles in the size range of coarse silt (0.05 mm) to medium sand (0.3 mm) can be moved by orbital wave motion to depths of 100 m and more. Waves of longer period could entrain sediment at even greater depths. The movement of sediment under the influence of oscillatory waves has important implications to both the trans-

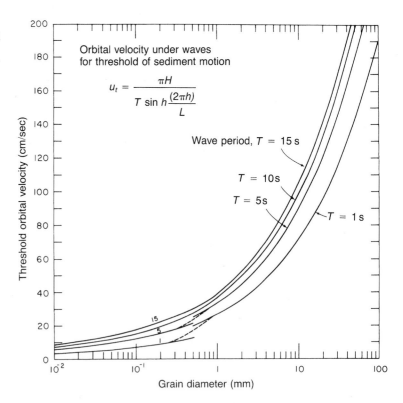

FIGURE 3.18 The near-bottom critical orbital velocity (u_t) required to initiate motion of quartz grains by waves having various wave periods. The orbital velocity in turn is related to wave height (H), water depth (h), and wave length (L). (After Komar, P. O., and M. C. Miller, 1975, Sediment threshold under oscillatory waves: Proc. 14th Conf. on Coastal Engineering. Fig. 7, p. 772, reprinted by permission of American Society of Civil Engineers.)

port of sediments in the nearshore zone and the formation of sedimentary structures such as oscillation ripples.

Bedforms Generated by Waves. Ripples and larger bedforms can form in lakes and the ocean under the influence of oscillatory wave motion. Oscillatory motion close to the bed creates oscillation ripples when near-bottom velocities become great enough to move the particles. Eddies created by these small-scale, wave-generated "orbital currents" throw sediment into suspension; the sediment alternatively moves landward as the wave crest passes and seaward as the trough passes (Fig. 3.19).

Oscillation ripples can be either symmetrical or asymmetrical (Fig. 3.19). If the forward and backward orbital wave velocities are equal, symmetrical ripples develop. If these orbital velocities are unequal, or if a unidirectional current is superimposed on the orbital motion, asymmetrical ripples develop. Clifton (1976) suggests that if the velocity asymmetry of the maximum bottom orbital velocity (the difference between forward and backward orbital velocity) is less than 1 cm/s, symmetrical bedforms develop. If the orbital velocity asymmetry exceeds 5 cm/s, asymmetrical bedforms develop, with the lee side of the bedform facing in the direction of wave movement. The issue of asymmetry of wave-formed bedforms is complex and poorly understood.

FIGURE 3.19 The relationship between orbital motion owing to wave action and sand motion on a rippled seabed. Note that wave motion with a superimposed, unidirectional current produces asymmetrical ripples, whereas wave motion alone tends to produce symmetrical ripples. (From Komar, P. D., Beach processes and sedimentation, © 1976, Fig. 11.15, p. 315. Reprinted by permission of Prentice-Hall, Englewood Cliffs, N.J. Based on D. L. Inman and A. J. Bowen, 1963, Flume experiments on sand transport by waves and currents: Proc. 8th Conf. on Coastal Engineering, p. 137–150.)

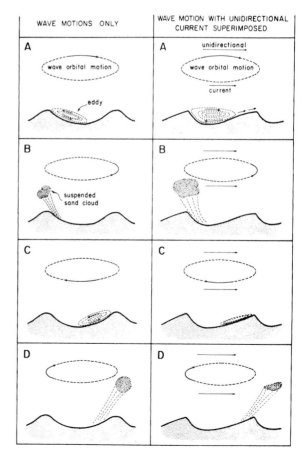

Oscillation ripples are destroyed by increasing bottom orbital velocity, and a plane-bed phase of transport occurs, called **sheet flow,** which is analogous to the plane-bed phase of upper-flow regime transport in unidirectional flow (Fig. 3.20). Under natural conditions in shallow offshore areas where waves are shoaling, a specific sequence of bedforms develops in a shoreward direction with increasing bottom orbital velocity and velocity asymmetry. This sequence (Fig. 3.21) grades from symmetrical waves in deeper water (not shown in Fig. 3.21) to asymmetrical ripples, lunate megaripples (dunes), and planar or flat beds. The abrupt increase in bedform size from ripples to lunate megaripples resembles the change from ripples to dunes in unidirectional flow (Clifton, 1976).

The crests of oscillation ripples as seen in plan view are commonly straight to sinuous and tend to bifurcate or fork (Fig. 3.22). The wave lengths of oscillation ripples generally range from about 10 cm to 2 m, and ripple heights range from about 3 to 25 cm; however, lunate megaripples can have a spacing ranging from 1 to 4 m and heights ranging between 30 to 100 cm. The internal structure of oscillation ripples varies from shoreward dipping cross-laminations in asymmetric and many symmetric ripples to both shoreward- and seaward-dipping cross-bedding in some symmetrical ripples. Shoreward dipping cross-laminations are most common, even in symmetrical ripples.

Sediment Loads and Transport Paths

Once sediment has been eroded and put into motion, the transport path that it takes during further sustained downcurrent movement is a function of the settling velocity of the particle and the magnitude of the current velocity and turbulence. Under a given set of conditions, the sediment load may consist entirely of very coarse particles, entirely of very fine particles, or of mixtures of coarse and fine particles. Coarse sediment such as sand and gravel moves on or very close to the bed during transport and is considered to constitute the **bedload.** Finer material carried higher up in the main flow above the bed makes up the **suspended load.** If the shear velocity (U^*) is greater than the settling velocity (V), material will remain in suspension.

FIGURE 3.20 Threshold velocity for initiation of grain movement in uniform quartz sand and conversion of rippled bed to flat bed (sheet flow) under orbital flow. (After Clinton, H. E., 1976, Wave-formed sedimentary structures: A conceptual model, *in* Davis, R. A., Jr., and R. L. Ethington (eds.), Beach and nearshore sedimentation: Soc. Econ. Paleontologists and Mineralogists Spec. Pub. 24. Fig. 4, p. 130, reprinted by permission of SEPM, Tulsa, Okla.)

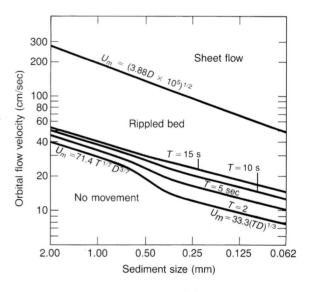

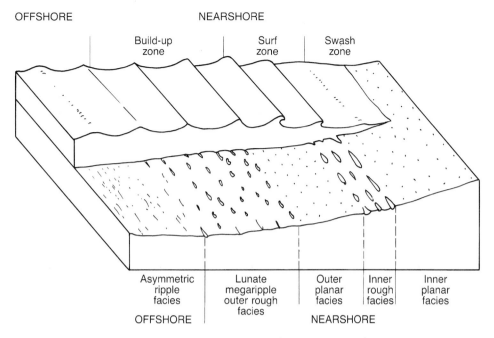

FIGURE 3.21 The sequence of bedforms that develop in the offshore and nearshore zone with increasing bottom orbital velocity and velocity asymmetry in the shoreward direction. (After Clifton, H. E., R. E. Hunter, and R. L. Phillips, 1971, Depositional structures and processes in the non-barred high-energy nearshore: Jour. Sed. Petrology, v. 41. Fig 7, p. 656, reprinted by permission of SEPM Tulsa, Okla.)

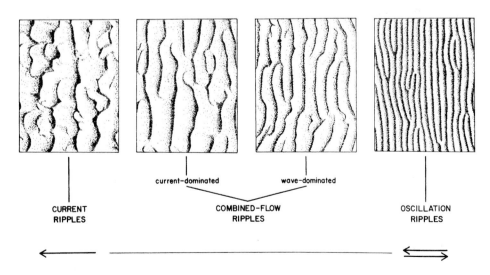

FIGURE 3.22 The crest shape of oscillation ripples, as seen in plan view, compared to the shape of current ripples and current-dominated ripples. (From Harms, J. C., J. B. Southard, and R. G. Walker, 1982. Structures and sequences in clastic rocks: Soc. Econ. Paleontologists and Mineralogists Short Course No. 9. Fig. 2–19, p. 2–48, reprinted by permission of SEPM, Tulsa, Okla.)

Bedload Transport. Particles larger than sand size are commonly transported as part of the bedload in essentially continuous contact with the bed. This type of transport, called **traction** transport, may include rolling of large or elongated grains, sliding of grains over or past each other, and creep. Creep results from grains being pushed a short distance along the bed in a downcurrent direction owing to impact of other moving grains. **Saltation** is a type of bedload transport in which grains, particularly sand-size grains, tend to move in intermittent contact with the bed. Saltating grains move by a series of jumps or hops, rising off the bed at a steep angle (~45°) to a height of a few grain diameters and then falling back along a shallow descent path of about 10°. This asymmetric saltation path may be interrupted by turbulence or by collision with another grain (Fig. 3.23). Saltation is a particularly prevalent mode of transport of fine sand by wind; the impact of saltating grains is primarily responsible for creep in eolian sands. Saltation transport may be thought of as intermediate between traction transport and suspension transport, but it is included here as part of bedload transport because most saltating grains remain relatively close to the bed during movement.

Suspended Load Transport. As the flow strength of a current increases, the intensity of turbulence increases close to the bed. Particle trajectories become longer, more irregular, and higher up from the bed than the trajectories of saltating grains. Upward components of fluid motion resulting from turbulence increase to the point that they balance downward gravitational forces on the particles, allowing the particles to stay suspended above the bed far longer than could be predicted from their settling velocities in nonturbulent water. If the lift forces arising from turbulence are erratic and do not continuously maintain this balance—a common occurrence during transport of fine-to-medium sand—the grains may drop back from time to time onto the bed. This behavior is called **intermittent suspension.** Intermittent suspension differs from saltation because the suspended particles tend to be carried higher above the bed and remain off the bed for longer periods of time. Smaller particles have settling velocities that may be so low that they remain in nearly **continuous suspension** and are carried along at almost the same velocity as the fluid flow.

Transport of sediment in suspension is common both by wind movement and by stream flow. A special type of intermittent suspension transport takes place in the

FIGURE 3.23 Schematic illustration of grain paths during bedload, suspension, and saltation transport. (From Leeder, M. R., Bedload dynamics: Grain-grain interactions in water flows: Earth surface processes, v. 4. © 1979 by John Wiley & Sons, Ltd. Fig. 5, p. 237, reprinted by permission of John Wiley & Sons, Ltd., Chichester, Sussex, England.)

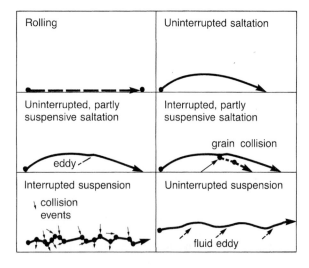

ocean within cloudy layers of near-bottom water called **nepheloid layers.** These layers were first surveyed and named by Ewing and Thorndike (1965), who discovered them using an optical nephelometer to measures light scattering at various levels in the water column. A nepheloid layer is a turbid body of suspended sediment that may reach heights of several hundred meters above the seafloor. It is denser than the surrounding ambient water but not dense enough to sink rapidly. Thus, sediment may remain suspended in such a layer for a long period of time. Most of the material of nepheloid layers consists of very fine clay particles derived initially from land. Some of this material reached the nepheloid layer by settling through the water column from the surface waters. Most of it is probably fine sediment resuspended from the ocean floor owing to erosion of the seabed by strong bottom currents, or it is fine material injected into the water column by turbidity currents or other mechanisms. Owing to its low settling velocity, fine sediment may remain in suspension in the nepheloid layer for periods ranging from days to weeks in the lowest 15 m of the water column and from weeks to months in the lowest 100 m (Kennett, 1982). Nepheloid layers in the modern ocean extend seaward for hundreds of kilometers and to water depths of 6000 m or more. Because of the relatively short residence times (weeks to months) for fine sediment in these layers, sediment transported long distances from shore in nepheloid layers must have been deposited on the ocean floor and resuspended many times.

Washload and Dustload. Much of the sediment load undergoing continuous suspension transport is composed of fine, clay-size particles with very low settling velocities. In rivers, this sediment is derived either from upstream source areas or by erosion of the bank, rather than from the stream bed, and is called the **washload.** Rivers have the capacity to transport large washloads even at very low velocities of flow. Because the washload travels in continuous suspension at about the same velocity as the water, it is transported rapidly through river systems. Similar suspended loads carried by the wind are called **dustloads.** Upward diffusion in unstable, buoyant air masses at an advancing front have been known to carry dust clouds rapidly to heights of hundreds or even thousands of meters during volcanic eruptions. Material carried to such great heights may remain in suspension for long periods of time and subsequently be spread over a very broad area, including the ocean basins (Prospero, 1981). In fact, the very fine-grained component of deep-sea pelagic sediments is believed to be largely of windblown origin.

Ice-Transport Load. Transport of sediment by ice is a kind of fluid-flow transport, although ice flows very slowly as a high-viscosity, non-Newtonian pseudoplastic. Rates of flow range from 1×10^{-6} cm/s to 2×10^{-3} cm/s (J. R. L. Allen, 1970a), which means that even a fast moving glacier moves less than 1 m per day (Sharp, 1988, p. 52). In *A Tramp Abroad,* Mark Twain describes his (fictitious) disappointment, after pitching camp on an alpine glacier in expectation of a free ride down the valley, to find that the view from his camp remained the same day after day. Glaciers advance if the rate of accumulation of snow in the upper reaches (head) of the glacier exceeds the rate of ablation (melting) of ice in the lower reaches (snout). The balance between accumulation and melting is illustrated in Figure 3.24. Ice must flow internally from the head of the glacier to replace that lost by melting at the snout. Flow of ice is laminar, and flow velocity is greatest near the top and center of the glacier. Velocity decreases toward the walls and floor, although not necessarily to zero. Glaciers retreat if the rate of melting exceeds the rate of accumulation. They reach a state of equilibrium, neither retreating nor advancing, when rates of melting and accumulation are equal, although internal movement of ice continues.

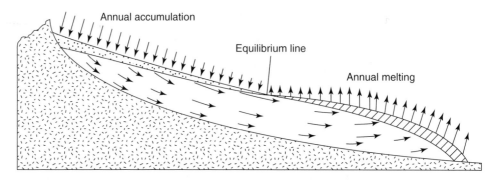

FIGURE 3.24 Diagrammatic two-dimensional illustration of the balance between glacier accumulation and melting and the movement of ice within a glacier. (After Sharp, R. P., 1988, Living ice—Understanding glaciers and glaciation: Cambridge University Press. Fig. 3.5, p. 58, and Fig. 3.6, p. 59, reproduced by permission.)

Sediment is entrained by glaciers by quarrying and abrasion by ice as the glacier erodes its bed, and by falling or sliding of material from the valley walls. Much of the sediment transported by glaciers is carried along the bottom and sides. Some of this sediment is transported in contact with the valley walls and floors and is responsible for much of the abrasion. Part of the remaining load is carried on the upper surface of the glacier, and part is carried within. The internal load is derived either from the joining of ice streams from two or more valleys or by the washing or falling of material from the surface into crevasses. Glaciers never become overloaded with debris to the point that they become immobilized. As glacial ice melts, however, the sediment load is dropped to form various kinds of glacial moraines (see Sharp, 1988, Chapters 7 and 8; Drewry, 1986, Chapters 7 and 8).

Deposits of Fluid Flows

Water and air are responsible for most sediment transport by fluid flow; however, ice may account for local transport of large volumes of sediment and particles of very large size. Sediment entrainment and transport by the various processes discussed above stop, and deposition occurs, when local hydrologic or wind conditions change sufficiently to cause decrease in bed shear stress to the point that it is no longer adequate to initiate and sustain particle movement. This decrease in bed shear stress is caused fundamentally by decrease in flow velocity. Flow velocity and shear stress may decrease below the critical level required for sediment transport owing to a variety of causes. In the case of water transport, these causes include decrease in the slope of the bed, increase in bed roughness, and loss of water volume. Decrease in wind velocity may also result from increase in bed roughness or from changes in surface topography and weather conditions. Deposition from glaciers is brought about on land when the glaciers either become stagnant or retreat owing to decrease in snow accumulation rates or increase in melting rates. Glaciers that run out to sea and calve to form icebergs eventually melt and drop their load on the seafloor.

Sediment deposition may be temporary or permanent. For example, sediment deposited in river channels and point bars, beach environments, and other very nearshore environments can be reentrained and subjected to continued transport as seasonal or longer-range changes in the hydrologic regimen take place. In fact, river sediment may be deposited and reentrained numerous times before finally reaching a depositional basin in the ocean. On the other hand, some river sediment, lake sediment, and wind-transported sediment may become deposited in continental settings and be preserved for long periods of time to become part of the geologic record. The great bulk of sediment undergoing transportation ultimately finds its way into ocean basins, where it is eventually deposited below wave base and more or less permanently immobilized until buried.

Sediments deposited by fluid flow of water or wind are commonly characterized by layers or beds of various thickness, scarcity of vertical grain-size grading, grain-size sorting ranging from poor to excellent depending upon depositional conditions, and the presence of a variety of sedimentary structures (Chapter 5). Sediments deposited from traction currents commonly preserve sedimentary structures such as cross-beds, ripple marks, and pebble imbrication that display directional features from which the direction of the ancient fluid flow can be determined. Sediments deposited from suspension lack these flow structures and are characterized instead by fine laminations. Wind is competent to transport and deposit particles in the size range of sand to dust (clay) only. By contrast, the grain size of sediment deposited by water may range from clay size to cobbles or boulders tens to hundreds of centimeters in diameter. These variations in grain size reflect the wide range of energy conditions of wind and water that prevail under natural conditions and the variations in relative competence of wind and water to initiate and sustain sediment transport. Ice does not behave as a Newtonian fluid and because of its much greater viscosity is capable of transporting particles of enormous size as well as particles of the smallest sizes. Sediments deposited by glaciers are characteristically poorly layered and extremely poorly sorted, with particles ranging from meter-size boulders to clay-size grains.

This very brief description of fluid-flow deposits is given here to illustrate the relationship between flow processes and the characteristics of the resulting sedimentary deposits. The textures and structures of sedimentary rocks are discussed in detail in Chapters 4 and 5; the characteristics of sediments deposited by fluid flows in different sedimentary environments are described in Chapters 10–12.

3.4 PARTICLE TRANSPORT BY SEDIMENT GRAVITY FLOWS

In the preceding section, we examined sediment transport resulting from the interaction of moving fluids and sediment. During fluid-flow transport, the fluids (water, wind, ice) move in various ways under the action of gravity, and the sediment is simply carried along with and by the fluid. Sediment can also be transported independently of fluid by the effect of gravity acting directly on the sediment. In this type of transport, fluids may play a role in reducing internal friction and supporting grains, but they are not primarily responsible for downslope movement of the sediment. Movement of sediment under the influence of gravity creates the flow, and flow stops when the sediment load is deposited.

Sediment transport owing to the direct action of gravity can occur in both subaerial and subaqueous environments. Gravity transport under submarine conditions has the greatest sedimentological significance. A spectrum of gravity movements exist, ranging from those in which sediment is moved en masse and fluids act mainly to reduce internal friction by lubricating the grains to those in which transport is on a grain-by-grain basis and fluids play an important role in supporting the sediment during transport. Gravity mass movements can be grouped into rock falls, slides, and sediment gravity flows, as shown in Table 3.2. **Rock fall** involves free fall of blocks or clasts from cliffs or steep slopes. **Slides** are en masse movements of rock or sediment

TABLE 3.2 Major types of mass-transport processes, their mechanical behavior, and transport and sediment support mechanisms

Mass-transport processes			Mechanical behavior	Transport mechanism and sediment support
Rock fall				Free fall and subordinate rolling of individual blocks or clasts along steep slopes
Slide	Glide		Elastic	Shear failure among discrete shear planes with little internal deformation or rotation
	Slump			Shear failure accompanied by rotation along discrete shear surfaces with little internal deformation
			— Plastic limit —	
Sediment gravity flow	Mass flow	Grain flows — Debris flow / Mud flow	Plastic	Shear distributed throughout sediment mass; strength principally from cohesion due to clay content; additional matrix support possibly from buoyancy
		Inertial / Viscous	— Liquid limit —	Cohesionless sediment supported by dispersive pressure; flow in inertial (high-concentration) or viscous (low-concentration) regime; steep slopes usually required
	Fluidal flow	Liquefied flow	Viscous fluid	Cohensionless sediment supported by upward displacement of fluid (dilatance) as loosely packed structure collapses, settling into a more tightly packed framework; slopes in excess of 3° required
		Fluidized flow		Cohesionless sediment supported by the forced upward motion of escaping pore fluid; thin (<10 cm) and short-lived
		Turbidity current		Supported by fluid turbulence

Source: Nardin, T. R., F. J. Hein, D. S. Gorsline, and B. D. Edwards, 1979. A review of mass movement processes, sediments, and acoustical characteristics, and contrasts in slope and base-of-slope systems versus canyon-fan-basin flow systems, *in* L. J. Doyle and O. R. Pilkey (eds.), Geology of continental slopes: Soc. Econ. Paleontologists and Mineralogists Spec. Pub. 27. Table 1, p. 64, reprinted by permission of SEPM, Tulsa, Okla.

owing to shear failure that take place with little accompanying internal deformation of the mass. **Sediment gravity flows** are more "fluid" types of movement in which breakdown in grain packing occurs and internal deformation of the sediment mass is intense.

Sediment gravity flows are of particular interest because they are capable of rapidly transporting large quantities of sediment, including very coarse sediment, even into very deep water in the oceans. Gravity flows that occur in subaerial environments can be considered in a broad sense to include pyroclastic flows and base surges resulting from volcanic eruptions, grain flow of dry sand down the slip face of sand dunes, and both volcanic and nonvolcanic debris flows and mud flows, in which large particles are transported in a slurrylike matrix of finer material. Subaqueous sediment gravity flows also include grain flows and debris flows, as well as turbidity currents and liquefied sediment flows.

Sediment gravity flows can occur only when grains become separated and dispersed to the point that internal friction and cohesiveness are sufficiently reduced to lower the strength of the sediment mass below the critical point required for gravity to initiate movement. Four theoretical types of dispersive and support flow mechanisms that can achieve this reduction in internal strength have been identified: turbulent flow, liquefied flow, grain flow, and plastic flow (Table 3.3). Four observed flow types can be identified that correspond to these theoretical flow types: turbidity currents, liquefied flow, grain flow, and mud and debris flows (Table 3.3 and Fig. 3.25).

TABLE 3.3 Principal kinds of sediment gravity flows and the relationship of flow type to grain-support mechanisms and fluid types (rheology)

Grain-support mechanism	Type of fluid (rheology)	Flow type (theoretical)	Flow type (observed)	
Turbulence	Turbulent fluid	Turbulent fluid flow	Turbidity current	
Upward escape of intergranular fluid	? Newtonian fluid (high viscosity)	Liquefied flow	Liquefied flow	
Dispersive pressure	Non-Newtonian fluid	Grain flow	Grain flow	▲ Debris ? flows ▼
Strength of matrix	Bingham plastic	Plastic flow	? Mud flow, muddy debris flow	

Source: Middleton (1991); Middleton and Hampton (1976).

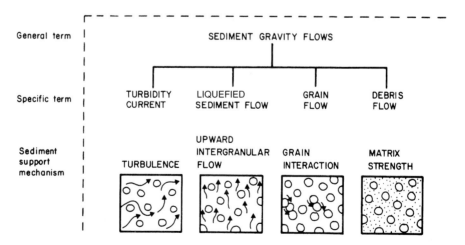

General term

Specific term

Sediment support mechanism

SEDIMENT GRAVITY FLOWS

TURBIDITY CURRENT | LIQUEFIED SEDIMENT FLOW | GRAIN FLOW | DEBRIS FLOW

TURBULENCE | UPWARD INTERGRANULAR FLOW | GRAIN INTERACTION | MATRIX STRENGTH

FIGURE 3.25 Principal types of sediment gravity flows and the types of interactions between fluids and grains that keep sediment supported during transport. (After Middleton, G. V., and M. A. Hampton, Subaqueous sediment transport and deposition by sediment gravity flows, *in* D. H. Stanley and D. J. P. Swift (eds.), Marine sediment transport and environmental management. © 1976 by John Wiley & Sons, Inc. Fig. 1, p. 198, reprinted by permission of John Wiley & Sons, Inc., New York.)

Turbidity currents are sediment gravity flows in which sediment is supported by the upward component of fluid turbulence. The presence of this suspended sediment in the flow causes its density to increase above that of the ambient water, resulting in downslope flow. Flow can occur quite rapidly even on very low slopes. **Liquefied flows** are concentrated dispersions of grains in which the sediment is supported either by the upward flow of pore water escaping from between the grains as they settle downward by gravity or by pore water that is forced upward by injection from below. Liquefaction can occur by sudden shocking of the sediment mass, greatly reducing friction between the grains. These flows can move rapidly down moderate slopes (3°–10°). **Grain flows** are dispersions of cohesionless sediment in which the sediment is supported (in air) by dispersive pressures owing to direct grain-to-grain collisions and in water by collisions and close approaches. Flow can occur rapidly under both subaerial and subaqueous conditions, especially on steep slopes that approach the angle of repose for the sediment. **Debris flows** and **mud flows** are slurrylike flows composed of highly concentrated, poorly sorted mixtures of sediment and water. Mud flows are composed mostly of mud-size (1/256–1/16 mm) material; debris flows contain a significant fraction of coarser material that may include clasts ranging to boulder size. The cohesive mud matrix in debris flows has enough strength to support large grains, but cohesiveness is not great enough to prevent flow on an adequate slope. Debris flows are generally initiated on steep slopes (>10°), but they can flow considerable distances on gentle slopes of 5° or less; they occur in both subaerial and subaqueous environments.

These four mechanisms of gravity transport are best thought of as end members of a spectrum of gravity-flow processes. One type of process may grade into another under some conditions. For example, submarine mud flows may change into turbidity currents downslope with additional mixing and dilution by water. We will now examine each of these major sediment gravity-flow processes in greater detail.

Turbidity Currents

Density currents are generated by gravity acting on differences in density between adjacent bodies of fluids. Density differences may arise from salinity or temperature variations or from sediment suspended in the fluid. A **turbidity current** is a special type of density current that flows downhill along the bottom of an ocean or lake because of density contrasts with the ambient water caused by sediment suspended in the water owing to turbulence. Turbidity currents can be generated experimentally in the laboratory by sudden release of muddy, dense water into the end of a sloping flume filled with less dense, clear water. They have been observed to occur under natural conditions in lakes where muddy river water enters the lakes, and they are believed to have occurred throughout geologic time in the marine environment on continental margins. In this setting, they appear to originate particularly in or near the heads of submarine canyons, although they may occur on the continental margin in areas where submarine canyons are absent and in other settings such as on seamounts.

Turbidity currents can be generated by a variety of mechanisms, including sediment failure, flow of sand in canyon heads triggered by storms, bedload inflow from rivers and glacial meltwater, and flows during eruption of air-fall ash (Normark and Piper, 1991). Turbidity currents may move as uniform, steady flows or as surges. When they move as steady flows, the velocity of flow is given by

$$\overline{u} \sqrt{\frac{8g}{f_0 + f_1}\left(\frac{\Delta\rho}{\rho}\right)dS} \qquad (3.19)$$

where d is the thickness of the flow, S is the slope of the bottom, f_0 is the frictional resistance at the bottom of the flow, and f_1 is the frictional resistance at the upper interface of the flow in contact with the overlying ambient water layer. As equation 3.19 shows, the velocity of steady-flow turbidity currents is sensitive to the slope over which flow takes place, although flows may occur on slopes as low as 1° (Kersey and Hsü, 1976). Uniform, steady turbidity currents have been observed to flow along the sloping bottom of lakes where sediment-laden rivers run into the lakes. They may occur also on continental shelves where muddy rivers discharge; however, they are less likely to occur in this setting because the density contrast between muddy river water and ocean water is less than that between muddy river water and freshwater.

Surges, or spasmodic turbidity currents, are initiated by some short-lived catastrophic event, such as earthquake-triggered massive sediment slumping or storm waves acting on a continental shelf. Such an event creates intense turbulence in the water overlying the seafloor, resulting in extensive erosion and entrainment of sediment, which is rapidly thrown into suspension. The sediment then remains suspended, supported in the water column by turbulence. This process generates a dense, turbid cloud that moves downslope, eroding and picking up more sediment as it increases in speed. Middleton and Hampton (1976) suggest that surge flows develop into three main parts as they move away from the source:

1. the **head,** which is about twice as thick as the rest of the flow and in which turbulence is intense
2. the **body,** of almost uniform thickness, in which nearly steady, uniform flow occurs
3. the **tail,** where the flow thins abruptly and becomes more dilute

The head is overhanging and is divided transversely into lobes and clefts (Fig. 3.26). The body of a surge-type turbidity current, which moves at a velocity similar to that of

FIGURE 3.26 Postulated structure of the head and body of a turbidity current advancing into deep water. The tail is not shown. (After Allen, J. R. L., 1985, Loose boundary hydraulics and fluid mechanics: Selected advances since 1961, *in* P. J. Brenchley and D. J. P. Swift (eds.), Sedimentology: Recent developments and applied aspects. Fig. 8, p. 20, reprinted by permission of Blackwell Scientific Publications Limited, Oxford.)

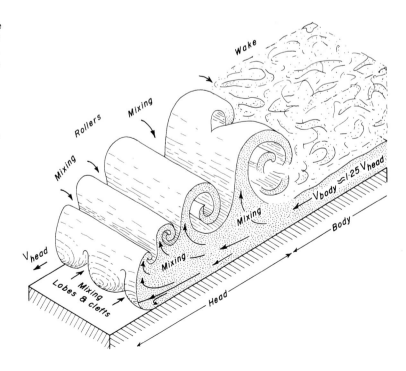

uniform, steady flows, moves faster in deep water than the head. This difference in velocity causes the forward part of the body to consume itself within the head in the process of mixing with the ambient water (Allen, 1985).

Once sediment is suspended in a turbidity current, the turbidity current continues to flow for some time under the action of gravity and inertia. Flow will stop when the sediment-water mixture that produces the density contrast with the ambient water is exhausted by settling of the suspended load. Rapid deposition of coarser particles from suspension appears to occur in regions near the source owing to early decay of the extremely intense turbulence generated by the initial event. As the flow continues to move forward, the remaining coarser material will be progressively concentrated in the head of the flow—denser fluid must be continuously supplied to the head to replace that lost to eddies that break off from the head and rejoin the body of the flow. Owing to differences in turbulence in the head and body, the head may be a region of erosion while deposition is taking place from the body.

Theoretically, sediment remaining in suspension after initial deposition of coarse material in the proximal area can, during further transport, be maintained in suspension for a very long time in a state of dynamic equilibrium called **autosuspension** (Bagnold, 1962; Pantin, 1979; Parker, 1982). A condition of autosuspension is presumably maintained because turbulence continues to be generated in the bottom of the flow owing to gravity-generated downslope flow of the turbidity current over the bed. Thus, loss of energy by friction of the flow with the bottom is compensated for by gravitational energy. The distance that turbidity currents can travel in the ocean is not known from unequivocal evidence. A presumed turbidity current triggered by the 1929 Grand Banks earthquake off Nova Scotia appears to have traveled south across the floor of the Atlantic for a distance of more than 300 km at velocities up to 67 km/h (19 m/s), as timed by breaks in submarine telegraph cables (Piper, Stow, and Normack, 1988). Transport of sediment over this distance suggests that autosuspension actually works; nonetheless, some geologists remain skeptical of the autosuspension process (e.g., review by Middleton, 1993).

The velocity of a turbidity current eventually diminishes owing to flattening of the canyon slope, overbank flow of the current along a submarine channel, or spreading of the flow over the flat ocean floor at the base of the slope. As the flow slows, turbulence generated along the sole of the flow also diminishes, and the current gradually becomes more dilute owing to mixing with ambient water around the head and along the upper interface. The remaining sediment carried in the head eventually settles out, causing the head to sink and dissipate. The exact process by which deposition takes place from various parts of a turbidity current is still not thoroughly understood, although it seems clear from experimental results that deposition does not occur in all parts of the current at the same time. As mentioned above, for example, the head may be a region of potential erosion at the same time deposition is taking place from the body behind the head. Sediment that is deposited very rapidly from some parts of the flow, such as the head, may undergo little or no subsequent traction transport before being quickly buried. On the other hand, in more distal parts of the flow or in areas where the head overflows the channel, a period of scouring by the head may be followed by slow deposition from the body and tail, during which additional traction transport of the deposited sediment takes place. Final deposition from the tail may take place after movement of the current is too weak to produce traction transport.

Depending upon position within the turbidity flow and the initial amount of sediment put into suspension by the flow, turbidity currents may contain either high concentrations of sediment or relatively low concentrations. Two principal types of turbidity currents, based on suspended particle concentration, can be considered: **low-density flows,** containing less than about 20 to 30 percent grains, and **high-density flows,** containing greater concentrations (Lowe, 1982). Low-density flows are made up largely of clay, silt, and fine- to medium-grained sand-size particles that are supported in suspension entirely by turbulence. High-density flows may include coarse-grained sands and pebble- to cobble-size clasts as well as fine sediment. Support of coarse particles during flow is provided by turbulence aided by hindered settling resulting from their own high sediment concentrations and the buoyant lift provided by the interstitial mixture of water and fine sediment. (High-density flows differ from debris flows in that debris flows are not turbulent and are less fluid.) The heads of turbidity currents may be high-density flows, whereas the tails may be dilute, low-density flows.

Turbidity current deposits, commonly called **turbidites,** are of two basic types. Turbidites deposited from high-density flows with high sediment concentrations tend to form thick-bedded turbidite successions containing coarse-grained sandstones or gravels. Individual flow units typically have relatively poor grading and few internal laminations, and basal scour marks are either poorly developed or absent. Some turbidites with thick, coarse-grained basal units may grade upward to finer-grained deposits that display traction structures such as laminations and small-scale cross-bedding. In the uppermost part of the flow units, the sediments may consist of very fine-grained, nearly homogeneous muds deposited from the tail of the flow. The deposits of more dilute, low-concentration turbidity current flows generally form thin-bedded turbidite successions. Individual flow units are fine grained at the base, with good vertical size grading, well-developed laminations, and small-scale cross-bedding (Fig. 3.27A). Scour marks may be present on the soles or bottoms of the beds.

Bouma (1962) proposed an ideal turbidite sequence, now commonly called a **Bouma sequence.** This ideal sequence consists of five structural units (Fig. 3.28–1) that include the characteristics of both types of turbidites. These structural subdivisions presumably record the decay of flow strength of a turbidity current with time and the progressive development of different sedimentary structures and bedforms in adjustment to different flow regimes (upper-flow regime to lower-flow regime) as current-flow velocity wanes. Most turbidites do not contain all of these structural units. Thick, coarse-grained turbidites tend to show well-developed A and B units, but C through E units are commonly poorly developed or absent. Thin, finer-grained turbidites may display well developed C through E units and poorly developed or absent A and B units. In fact, Hsü (1989, p. 116) claims that Bouma's D unit rarely occurs and that most turbidites can be divided into only two units: a lower, horizontally laminated unit (unit A + B, Fig. 3.28–2) and an upper, cross-laminated unit (unit C). Unit E is a problem because it may be pelagic shale and thus may not be part of a turbidite flow unit.

Turbidites laid down near the source, particularly within the main transport channel where suspended sediment concentrations are high, are generally the coarse-grained, massive, or poorly laminated type. Some very high concentration flows may also deposit coarse-grained turbidites within the main channel at considerable distances from the source. On the other hand, thin, fine-grained turbidites can also be deposited near the source where turbidity currents overflow the banks of a channel and become more dilute as they spread out over the seafloor, as well as in areas more distant from the source. The deposits of a single turbidity current flow typically display horizontal size grading in addition to vertical size grading. That is, thick, coarse-grained deposits generally grade laterally to thinner, finer-grained sediments.

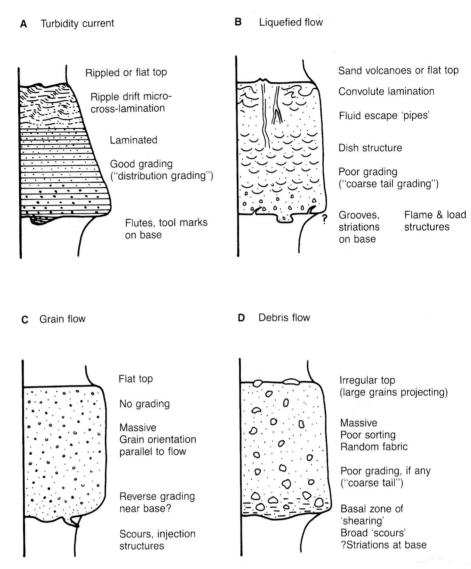

FIGURE 3.27 Comparison of sedimentary structures in different types of sediment gravity-flow deposits. (After Middleton, G. V., and M. A. Hampton, Subaqueous sediment transport and deposition by sediment gravity flows, *in* D. H. Stanley and D. J. P. Swift (eds.), Marine sediment transport and environmental management. © 1976 by John Wiley & Sons, Inc. Fig. 9, p. 213, reprinted by permission of John Wiley & Sons, Inc., New York.)

Liquefied Flows

Loosely packed, cohesionless sediment such as sand can become temporarily liquefied owing to a sudden shock, or series of shocks, that causes the grains to momentarily lose contact with each other and become suspended in their own pore fluid. Grain contact may also be lost if a fluid is introduced into the base of a mass or column of cohesionless sediment and injection is continued until the grains are pushed apart, with

FIGURE 3.28 Ideal sequence of sedimentary structures in graded-bed units as proposed by Bouma (1) and Hsü (2). Note that in Hsü's model, Bouma's units A and B are combined and unit D is omitted. (After Hsü, K. J., 1989, Physical principles of sedimentology. Fig. 7.8, p. 116, reprinted by permission of Springer-Verlag, Berlin.)

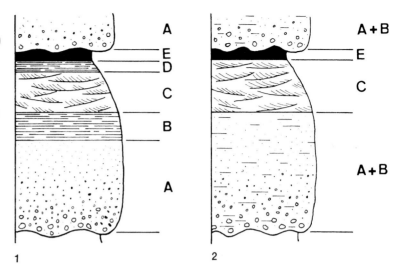

their weight being supported by the rising fluid. This process is called fluidization. Once the cohesionless sediment has become liquefied (or fluidized), it loses its strength and behaves like a high-viscosity fluid that can, nonetheless, flow quite rapidly down slopes as low as 3°.

Liquefied flow can occur only as long as grain dispersion is maintained. As soon as the grains settle out of the fluid and reestablish grain-to-grain contact, the flowing layer will "freeze up" and stop moving. "Freezing" begins at the base of the flow; a surface of settled grains rises up through the dispersion at a rate determined by the settling velocity of the particles (Fig. 3.29). The time required for settling to occur is on the order of hours for thick, fine-grained flows (Lowe, 1976); therefore, liquefied flows may travel short, though potentially important, distances before deposition occurs. The upward movement of pore waters through the settling grains as deposition occurs leads to formation of a number of fluid escape structures such as dish structures. Some liquefied flows may become turbulent as the flowing sediment mass is accelerated downslope and thus change into turbidity currents. The deposits of liquefied flows are typically thick, poorly sorted sand units that are characterized particularly by fluid

FIGURE 3.29 Schematic representation of grain settling and water expulsion during deposition of sand from a liquefied flow. (After Allen, J. R. L., and N. L. Banks, 1972, An interpretation and analysis of recumbent-folded deformed cross bedding: Sedimentology, v. 19. Fig. 3, p. 267, reprinted by permission of Elsevier Science Publishers, Amsterdam.)

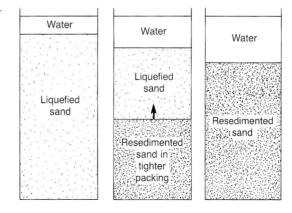

escape structures (Chapter 5), such as the dish structures, pipes, and sand volcanoes shown in Figure 3.27B.

Grain Flows

Grain flow is the movement of cohesionless sediments down steep slopes owing to sudden loss of internal shear strength of the sediment. Grain flow begins when traction processes cause cohesionless sediment, commonly sand, to be piled up beyond the critical angle of repose. This angle is a function of grain packing and grain shape and tends to be greatest in deposits with angular grains of low sphericity. When the angle of repose for a particular sediment is exceeded, avalanching occurs; flow quickly begins when the internal shear stresses owing to gravity exceed the internal shear strength of the sediment. The **dispersive pressures** needed to force the grains apart and keep them suspended during flow are provided not by fluid but by grain-to-grain collisions in air and grain collisions and close encounters in water as the failed mass of sediments moves down a slope. During the interaction of grains, dispersive pressure is the force normal to the plane of shearing which tends to expand or disperse the grains in that direction. Bagnold (1956) suggested that the relation between the shear stress (T) acting on grains and the dispersive pressure (P) is

$$\frac{T}{P} = \tan a \qquad\qquad (3.20)$$

where a is the angle of internal friction. This formula suggests that the minimum slope on which sustained grain flow in air is possible is about 30°; under water, greater slopes may be required for flow to occur. Although dispersion or dilation of sand grains is achieved and maintained during flow primarily by grain collisions, dispersion may be aided under some conditions by upward flow of pore fluids as grains settle or possibly by buoyancy of a dense mud matrix. Grain flow is similar to liquefied flow in many respects and may, in fact, grade into these flows. In contrast to liquefied flows, grain flow can occur under subaerial conditions as well as subaqueous conditions.

Grain flow is a common occurrence on the lee slopes of sand dunes. Flows of cohesionless sand have also been observed and photographed in the ocean as they moved down steep slopes in submarine canyons (Shepard, 1961; Dill, 1966; Shepard and Dill, 1966). Grain flows over the floors of Norwegian fjords are reported to be responsible for breaking submarine telephone cables. Grain flows may be of limited geological significance because of the steep slopes required to initiate flow, although it has been suggested that grain flow may accompany turbidity currents on less steep slopes, moving beneath but independently of the turbidity currents. Deposition of grain-flow sediment occurs quickly and en masse by sudden "freezing" owing primarily to reduction of slope angle.

A grain-flow origin has been suggested by some workers for very thick, almost massive sandstone beds; however, Lowe (1976) concludes that the deposits of a single grain flow in any environment cannot be thicker than a few centimeters for sand-size grains. Reverse grading—that is, grading from fine size to coarse size upward—that occurs in some sandstones has been attributed to grain-flow processes. Reverse grading is assumed to occur during grain flow as a result of smaller particles filtering down through larger particles while they are in the dispersed state, a process called **kinetic sieving**. Grain-flow deposits are massively bedded with little or no internal laminations and grading except possible reverse grading in the base (Fig. 3.27C). Deposits of a single grain flow are commonly less than about 5 cm thick.

Debris Flows and Mud Flows

Debris flows and mud flows are sediment gravity flows that behave as Bingham plastics; that is, they have a yield strength that must be overcome before flow begins. They consist of poorly sorted mixtures of particles, which may range to boulder-size, in a fine gravel, sand, or mud matrix. Those composed predominantly of mud-size grains are **mud flows,** and those with a lower but substantial mud fraction (>5 percent by volume) are **muddy debris flows** (Middleton, 1991). The grains in these mud-bearing debris flows are supported in a matrix of mud and interstitial water that has enough cohesive strength to prevent larger particles from settling but not enough strength to prevent flow. Debris flows that have a matrix composed predominantly of cohesionless sand and gravel are **mud-free debris flows** (Middleton, 1991). The support mechanism for these mud-free debris flows is poorly understood.

Subaerial debris flows occur under many climatic conditions but are particularly common in arid and semiarid regions where they are usually initiated after heavy rainfalls. **Lahars** are debris flows composed largely of volcanic particles that become water saturated during heavy rains that accompany volcanic eruptions or from melting ice and snow that accumulate on volcanic cones between eruptions.

After the yield strength of a debris flow has been overcome owing to water saturation, and movement begins, the flow may continue to move over slopes as low as $1°$ or $2°$ (Curray, 1966). Debris flows are believed to occur also in subaqueous environments, possibly as a result of mixing at the downslope ends of subaqueous slumps. As subaqueous debris flows move rapidly downslope and are diluted by mixing with more water, their strength is reduced, and they may pass into turbidity currents. Deposition of the entire mass of debris flows and mud flows occurs quickly. When the shear stress owing to gravity no longer exceeds the yield strength of the base of the flow, the mass "freezes" and stops moving.

Debris-flow deposits are thick, poorly sorted units that lack internal layering (Fig. 3.27D). They typically consist of chaotic mixtures of particles that may range in size from clay to boulders. The large particles commonly show no preferred orientation. They are generally poorly graded, but if grading is present, it may be either normal or reverse.

FURTHER READINGS

Allen, J. R. L., 1970, Physical processes of sedimentation: George Allen & Unwin, London, 248 p.

Barndorff-Nielsen, O. E., and B. B. Willetts (eds.), 1991, Aeolian grain transport 1—Mechanics: Acta Mechanica Supplementum 1, Springer-Verlag, Wien, New York, 181 p.

Garde, R. J., and K. G. Ranga Raju, 1978, Mechanics of sediment transport and alluvial stream problems: Halsted Press, New York, 483 p.

Komar, P. D., 1976, Beach processes and sedimentation: Prentice-Hall, Englewood Cliffs, N.J., 429 p.

Lowe, D. R., 1982, Sediment gravity flows: II. Depositional models with special reference to the deposits of high-density turbidity currents: Jour. Sed. Petrology, v. 52, p. 279–297.

Middleton, G. V., and M. A. Hampton, 1976, Subaqueous sediment transport and deposition by sediment gravity flows, in D. J. Stanley and D. J. P. Swift (eds.), Marine sediment transport and environmental management: John Wiley & Sons, New York, p. 197–218.

Middleton, G. V., and J. B. Southard, 1984, Mechanics of sediment movement: Soc. Econ. Paleontologists and Mineralogists Short Course Notes 3, 2nd ed., 401 p.

Pye, K., and H. Tsoar, 1990, Aeolian sand and sand dunes: Unwin Hyman, London, 396 p.

Rubin, D. M., 1987, Cross-bedding, bedforms, and pale-ocurrents: Soc. Econ. Paleontologists and Mineralogists, Concepts in Sedimentology and Paleontology, v. 1, Tulsa, Okla., 187 p.

Saxov, S. and J. K. Nieuwenhuis (eds.), 1982, Marine slides and other mass movements: NATO Conference Series IV: Marine Sciences, v. 6, Plenum Press, New York, 353 p.

Stanley, D. J. and D. J. P. Swift (eds.), 1976, Marine sediment transport and environmental management: John Wiley & Sons, New York, 602 p.

Tetzlaff, D. M., and J. W. Harbaugh, 1989, Simulating clastic sedimentation: Van Nostrand Reinhold, New York, 202 p.

Thorne, C. R., J. C. Bathurst, and R. D. Hey (eds.), 1987, Sediment transport in gravel-bed rivers: John Wiley & Sons, Chichester, N.Y., 995 p.

Yalin, M. S., 1977, Mechanics of sediment transport, 2nd ed.: Pergamon Press, New York, 298 p.

PART 3

PHYSICAL PROPERTIES
OF SEDIMENTARY ROCKS

Chapter 4   Sedimentary Textures
Chapter 5   Sedimentary Structures

Cross-bedded sandstones of the Navajo Formation, Zion National Park, Utah.

The transport and depositional processes described in Chapter 3 generate a wide variety of siliciclastic and nonsiliciclastic sedimentary rocks, each characterized by distinctive textural and structural properties. **Sedimentary texture** refers to the features of sedimentary rocks that arise from the size, shape, and orientation of individual sediment grains. Geologists have long assumed that the texture of sedimentary rocks reflects the nature of transport and depositional processes and that characterization of texture can aid in interpreting ancient environmental settings and boundary conditions. An extensive literature has thus been published dealing with various aspects of sediment texture, particularly methods of measuring and expressing grain size and shape and interpretation of grain size and shape data. The textures of siliciclastic sedimentary rocks are produced primarily by physical processes of sedimentation and are considered to encompass grain **size, shape** (form, roundness, surface texture), and **fabric** (grain orientation and grain-to-grain relations). The interrelationship of these primary textural properties controls other, derived, textural properties such as **bulk density, porosity,** and **permeability.** The texture of some nonsiliciclastic sedimentary rocks such as certain limestones and evaporites is also generated partly or wholly by physical transport processes. The texture of others is due principally to chemical or biochemical sedimentation processes. Extensive recrystallization or other diagenetic changes may destroy the original textures of nonsiliciclastic sedimentary rocks and produce crystalline textural fabrics that are largely of secondary origin. Obviously the textural features of chemically or biochemically formed sedimentary rocks, and of rocks with strong diagenetic fabrics, have quite different genetic significance from that of unaltered siliciclastic sedimentary rocks.

Whereas the term texture applies mainly to the properties of individual grains, **sedimentary structures,** such as cross-bedding and ripple marks, are features formed from aggregates of grains. These structures are generated by a variety of sedimentary processes, including fluid flow, sediment gravity flow, soft-sediment deformation, and biogenic activity. Because sedimentary structures reflect environmental conditions that prevailed at or very shortly after the time of deposition, they are of special interest to geologists as a tool for interpreting ancient depositional environments. We can use sedimentary structures to help evaluate such aspects of ancient sedimentary environments as sediment transport mechanisms, paleocurrent flow directions, relative water depths, and relative current velocities. Some sedimentary structures are also used to identify the tops and bottoms of beds and thus to determine if sedimentary sequences are in depositional stratigraphic order or have been overturned by tectonic forces. Sedimentary structures are particularly abundant in coarse siliciclastic sedimentary rocks that originate through traction transport or turbidity current transport. They occur also in nonsiliciclastic sedimentary rocks such as limestones and evaporites.

4

Sedimentary Textures

4.1 INTRODUCTION

This chapter focuses primarily on the physically produced textures of siliciclastic sedimentary rocks. Some of the special textural features that are important to understanding the classification and genesis of limestones and other nonsiliciclastic sedimentary rocks are discussed in Chapters 7 and 8. In this chapter we examine the characteristic textural properties of grain size and shape, particle surface texture, and grain fabric, and we discuss the genetic significance of these properties. Although the study of sedimentary textures may not be the most exciting aspect of sedimentology, it is nonetheless an important field of study. A thorough understanding of the nature and significance of sedimentary textures is fundamental to interpretation of ancient depositional environments and transport conditions, although much uncertainty still attends the genetic interpretation of textural data. Some long-standing ideas about the genetic significance of sediment textural data are now being challenged, while new ideas and techniques for studying and interpreting sediment texture continue to emerge. No textbook on sedimentology would be complete without some discussion of sediment texture and its genetic significance.

4.2 GRAIN SIZE

Grain size is a fundamental attribute of siliciclastic sedimentary rocks and thus one of the important descriptive properties of such rocks. Sedimentologists are particularly concerned with three aspects of particle size:

1. techniques for measuring grain size and expressing it in terms of some type of grain-size or grade scale

2. methods for summarizing large amounts of grain-size data and presenting them in graphical or statistical form so that they can be more easily analyzed
3. the genetic significance of these data

We will now examine each of these concerns.

Grain-Size Scales

Particles in sediments and sedimentary rocks range in size from a few microns to a few meters. Because of this wide range of particle sizes, logarithmic or geometric scales are more useful for expressing size than are linear scales. In a geometric scale there is a succession of numbers such that a fixed ratio exists between successive elements of the series. The grain-size scale used almost universally by sedimentologists is the **Udden-Wentworth scale.** This scale, first proposed by Udden in 1898 and modified and extended by Wentworth in 1922, is a geometric scale in which each value in the scale is either twice as large as the preceding value or one-half as large, depending upon the sense of direction (Table 4.1). The scale extends from <1/256 mm (0.0039 mm) to >256 mm and is divided into four major size categories (clay, silt, sand, and gravel), which can be further subdivided as illustrated in Table 4.1.

A useful modification of the Udden-Wentworth scale is the logarithmic **phi scale,** which allows grain-size data to be expressed in units of equal value for the purpose of graphical plotting and statistical calculations. This scale, proposed by Krumbein in 1934, is based on the following relationship:

$$\phi = -\log_2 d \qquad \textbf{(4.1)}$$

where ϕ is phi size and d is the grain diameter in millimeters. Some equivalent phi and millimeter sizes are shown in Table 4.1. Other phi sizes can be derived from equation 4.1. Note that the phi scale yields both positive and negative numbers. The real size of particles, expressed in millimeters, decreases with increasing positive phi values and increases with decreasing negative values. Because sand-size and smaller grains are the most abundant grains in sedimentary rocks, Krumbein chose the negative logarithm of the grain size in millimeters so that grains of this size will have positive phi values, avoiding the bother of constantly working with negative numbers. This usage is also consistent with the common practice of plotting coarse sizes to the left and fine sizes to the right in graphs.

Measuring Grain Size

The size of siliciclastic grains can be measured by several techniques (Table 4.2). The choice of methods is dictated by the purpose of the study, the range of grain sizes to be measured, and the degree of consolidation of sediment or sedimentary rock. Large particles (pebbles, cobbles, boulders) in either unconsolidated sediment or lithified sedimentary rock can be measured manually with a caliper. Grain size is commonly expressed in terms of either the long dimension or the intermediate dimension of the particles. Granule- to silt-size particles in unconsolidated sediments or sedimentary rocks that can be disaggregated are commonly measured by sieving through a set of nested, wire-mesh screens. The sieve numbers of U.S. Standard Sieves that correspond to various millimeter and phi sizes are shown in Table 4.1. Sieving techniques measure the intermediate dimension of particles because the intermediate particle size generally determines whether or not a particle can go through a particular mesh.

Granule- to silt-size particles can also be measured by sedimentation techniques on the basis of the settling velocity of particles. In these techniques, grains are allowed

TABLE 4.1 Grain-size scale for sediments, showing Wentworth size classes, equivalent phi (ϕ) units, and sieve numbers of U.S. Standard Sieves corresponding to various millimeter and ϕ sizes

U.S. Standard sieve mesh	Millimeters		Phi (ϕ) units	Wentworth size class
	4096		−12	
	1024		−10	Boulder
	256	256	−8	
	64	64	−6	Cobble
	16		−4	Pebble
5	4	4	−2	
6	3.36		−1.75	
7	2.83		−1.5	Granule
8	2.38		−1.25	
10	2.00	2	−1.0	
12	1.68		−0.75	
14	1.41		−0.5	Very coarse sand
16	1.19		−0.25	
18	1.00	1	0.0	
20	0.84		0.25	
25	0.71		0.5	Coarse sand
30	0.59		0.75	
35	0.50	½	1.0	
40	0.42		1.25	
45	0.35		1.5	Medium sand
50	0.30		1.75	
60	0.25	¼	2.0	
70	0.210		2.25	
80	0.177		2.5	Fine sand
100	0.149		2.75	
120	0.125	⅛	3.0	
140	0.105		3.25	
170	0.088		3.5	Very fine sand
200	0.074		3.75	
230	0.0625	1⁄16	4.0	
270	0.053		4.25	
325	0.044		4.5	Coarse silt
	0.037		4.75	
	0.031	1⁄32	5.0	
	0.0156	1⁄64	6.0	Medium silt
	0.0078	1⁄128	7.0	Fine silt
	0.0039	1⁄256	8.0	Very fine silt
	0.0020		9.0	
	0.00098		10.0	Clay
	0.00049		11.0	
	0.00024		12.0	
	0.00012		13.0	
	0.00006		14.0	

(Left-margin vertical labels: GRAVEL, SAND, MUD, SILT, CLAY)

to settle through a column of water at a specified temperature in a settling tube (Fig. 4.1), and the time required for the grains to settle is measured. The settling time of the particles is related empirically to a standard size-distribution curve (calibration curve) to obtain the equivalent millimeter or phi size. As mentioned in Chapter 3, settling velocity of particles is affected by particle shape. Spherical particles settle faster than nonspherical particles of the same mass. Therefore, determining the grain sizes of natural, nonspherical particles by sedimentation techniques may not yield exactly the same values as those determined by sieving.

TABLE 4.2 Methods of measuring sediment grain size

Type of sample	Sample grade	Method of analysis
Unconsolidated sediment and disaggregated sedimentary rock	Boulders Cobbles Pebbles	Manual measurement of individual clasts
	Granules Sand Silt	Sieving, settling-tube analysis, image analysis
	Clay	Pipette analysis, sedimentation balances, photohydrometer, Sedigraph, laser-diffractometer, electro-resistance (e.g., Coulter counter)
Lithified sedimentary rock	Boulders Cobbles Pebbles	Manual measurement of individual clasts
	Granules Sand Silt	Thin-section measurement, image analysis
	Clay	Electron microscope

The grain size of fine silt and clay particles can be determined by sedimentation methods based on Stokes's Law (equations 3.16 and 3.17). As explained in Chapter 3, Stokes's Law cannot be applied to sand-size and larger particles. If the settling velocity of small particles can be measured at a particular temperature, the diameter of the particles can be calculated by a simple mathematical rearrangement of equation 3.17 to yield

$$D = \frac{\sqrt{V}}{\sqrt{C}} \qquad (4.2)$$

where D is the diameter of the particles in centimeters, V is settling velocity of the particles, and C is a constant that depends upon the density of the particles and the density and viscosity of the fluid (usually water). The standard sedimentation method for measuring the sizes of these small particles is **pipette analysis.** To do a pipette analysis, fine sediment is stirred into suspension in a measured volume of water in a settling tube. Uniform-size aliquots of this suspension are withdrawn with a pipette at specified times and at specified depths, evaporated to dryness in an oven, and weighed. The data obtained can then be used in a modified version of equation 4.2 to calculate grain size:

$$D = \frac{\sqrt{C}}{\sqrt{x/t}} \qquad (4.3)$$

where x is the depth in centimeters to which particles have settled in a given time (the withdrawal depth), t is elapsed time in seconds, and $x/t = V$ (settling velocity). Tables of withdrawal times and depths for specific particle sizes are available in many books and articles dealing with sedimentation analysis (e.g., Galehouse, 1971). Note that Stokes's Law is based on the assumption that the settling particles are spheres. Because most natural particles are not spheres, use of Stokes's Law yields grain sizes that are commonly finer than the actual particle sizes.

FIGURE 4.1 Large automated settling tube for rapid size analysis of coarse-grained sediment; located in the Department of Geology, Portland State University, Oregon. An electronic signal from the settling tube, which is approximately 2.5 m high and 20 cm wide, is fed through the microprocessor (on table to left) to a small desk computer that calculates grain-size statistics and plots graphical representations of grain-size distributions.

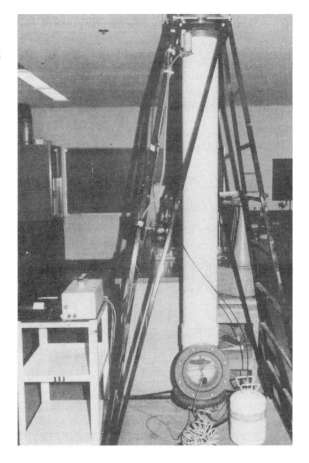

Pipette analyses as well as manual settling-tube analyses of coarser sediment are very laborious processes because of the many operations involved. To simplify these procedures, automatic-recording settling tubes have now been developed that allow the sizes of both sand-size and clay-size sediment to be more easily and rapidly determined (e.g., Fig. 4.1). Most automatic-recording settling tubes, commonly called rapid sediment analyzers, are **sedimentation balances** that function by measuring changes with time in the weight of sediment that collects on a pan suspended in a column of water in the settling tube or by measuring changes in the pressure of the water column as sediment settles out of the column. Grain size can then be determined by comparing these weight- or pressure-vs.-time curves to a calibration curve.

Several other kinds of automated particle-size analyzers are also available, each based on a slightly different principle. A **photohydrometer** is a settling tube that empirically relates changes in intensity of a beam of light passed through a column of suspended sediment to particle settling velocities and thus to particle size (Jordan, Freyer, and Hemmen, 1971). The **sedigraph** determines particle size by measuring the attenuation of a finely collimated X-ray beam as a function of time and height in a settling suspension (Stein, 1985; Jones, McCave, and Patel, 1988). A **laser-diffracter size analyzer** operates on the principle that particles of a given size diffract light through a given angle, with the angle increasing with decreasing particle size (McCave et al.,

1986). **Electro-resistance size analyzers,** such as the Coulter counter or the Electrozone particle counter, measure grains size on the basis of the principle that a particle passing through an electrical field maintained in an electrolyte will displace its own volume of the electrolyte and thus cause a change in the field. These changes are scaled and counted as voltage pulses, with the magnitude of each pulse being proportional to particle volume (Swift, Schubel, and Sheldon, 1972; Muerdter, Deuphin, and Steele, 1981). Semiautomated **image analysis** techniques involve use of TV cameras to capture and digitize grain images from which, with the aid of appropriate computer software, grain-size diameters can be calculated (Kennedy and Mazzullo, 1991). For a comparison of some of these analytical techniques, see Singer et al. (1988). Additional information is available in *Principles, Methods, and Applications of Particle Size Analysis,* a monograph edited by Syvitski (1991).

The grain size of particles in consolidated sedimentary rocks that cannot be disaggregated must be measured by techniques other than sieving or sedimentation analysis. The size and sorting of sand- and silt-size particles can be estimated by using a reflected-light binocular microscope and a standard size-comparison set, which consists of grains of specific sizes mounted on a card. More accurate size determination can be made by measuring grains in thin sections of rock by using a transmitted-light petrographic microscope fitted with an ocular micrometer or by the image analysis technique mentioned above. Both microscopic and image analysis techniques tend to yield grain sizes that are smaller than the maximum diameter of the grains because the plane of a thin section does not cut exactly through the centers of most grains. Grain sizes measured by these methods are commonly corrected mathematically in some way to make them agree more closely with sieve data (Burger and Skala, 1976; Piazzola and Cavaroc, 1991). Fine silt- and clay-size grains in consolidated rocks may be studied using an electron microscope, although the electron microscope is not commonly used for grain-size measurements.

Graphical and Mathematical Treatment of Grain-Size Data

Measurement of grain size by the techniques described generates a large number of data that must be reduced to more condensed form before they can be used. Tables of data showing the weights of grains in various size classes must be simplified to yield such average properties of grain populations as mean grain size and sorting. Both graphical and mathematical data-reduction methods are in common use. Graphical plots are simple to construct and provide a readily understandable visual representation of grain-size distributions. On the other hand, mathematical methods, some of which are based on initial graphical treatment of data, yield statistical grain-size parameters that may be useful in environmental studies.

Figure 4.2 illustrates three common graphical methods for presenting grain-size data. Figure 4.2A shows typical grain-size data obtained by sieve analysis. Raw sieve weights are first converted to individual weight percents by dividing the weight in each size class by the total weight. Cumulative weight percent may be calculated by adding the weight of each succeeding size class to the total of the preceding classes. Figure 4.2B shows how individual weight percent can be plotted as a function of grain size to yield a grain-size **histogram**—a bar diagram in which grain size is plotted along the abscissa of the graph and individual weight percent along the ordinate. Histograms provide a quick, easy pictorial method for representing grain-size distributions because the approximate average grain size and the sorting—the spread of grain-size values around the average size—can be seen at a glance. Histograms have limited application, however, because the shape of the histogram is affected by the

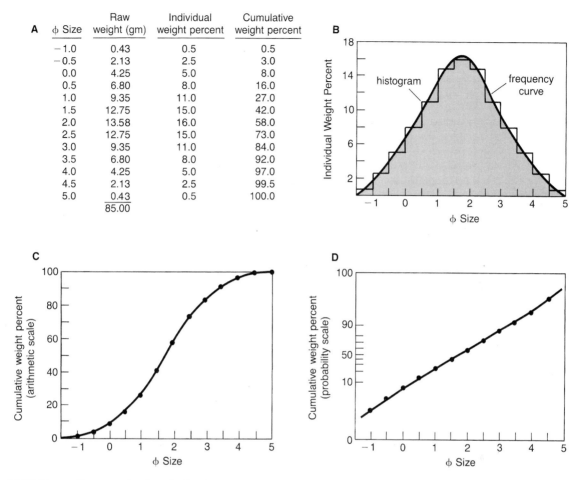

A ɸ Size	Raw weight (gm)	Individual weight percent	Cumulative weight percent
−1.0	0.43	0.5	0.5
−0.5	2.13	2.5	3.0
0.0	4.25	5.0	8.0
0.5	6.80	8.0	16.0
1.0	9.35	11.0	27.0
1.5	12.75	15.0	42.0
2.0	13.58	16.0	58.0
2.5	12.75	15.0	73.0
3.0	9.35	11.0	84.0
3.5	6.80	8.0	92.0
4.0	4.25	5.0	97.0
4.5	2.13	2.5	99.5
5.0	0.43	0.5	100.0
	85.00		

FIGURE 4.2 Common visual methods of displaying grain-size data. A. Grain-size data table. B. Histogram and frequency curve plotted from data in A. C. Cumulative curve with an arithmetic ordinate scale. D. Cumulative curve with a probability ordinate scale.

sieve interval used. Also, they cannot be used to obtain mathematical values for statistical calculations.

A **frequency curve** (Fig. 4.2B) is essentially a histogram in which a smooth curve takes the place of a discontinuous bar graph. Connecting the midpoints of each size class in a histogram with a smooth curve gives the approximate shape of the frequency curve. A frequency curve constructed in this manner does not, however, accurately fix the position of the highest point on the curve; this point is important for determining the modal size, to be described. A grain-size histogram plotted from data obtained by sieving at exceedingly small sieve intervals would yield the approximate shape of a frequency curve, but such small sieve intervals are not practical. Accurate frequency curves can by derived from cumulative curves by special graphical methods described in detail by Folk (1974).

A grain-size **cumulative curve** is generated by plotting grain size against cumulative weight percent frequency. The cumulative curve is the most useful of the grain-size plots. Although it does not give as good a pictorial representation of the grain-size distribution as a histogram or frequency curve, its shape is virtually independent of the sieve interval used. Also, data that can be derived from the cumulative curve allow calculation of several important grain-size statistical parameters. A cumulative curve can be plotted on an arithmetic ordinate scale (Fig. 4.2C) or on a log probability scale in which the arithmetic ordinate is replaced by a log probability ordinate (Fig. 4.2D). When phi-size data are plotted on an arithmetic ordinate, the cumulative curve typically has the S-shape shown in Figure 4.2C. The slope of the central part of this curve reflects the sorting of the sample. A very steep slope indicates good sorting, and a very gentle slope poor sorting. If the cumulative curve is plotted on log probability paper, the shape of the curve will tend toward a straight line if the population of grains has a normal distribution (actually log-normal, as illustrated in Fig. 4.2D). In a normal distribution, the values show an even distribution, or spread, about the average value. In conventional statistics, a normally distributed population of values yields a perfect bell-shaped curve when plotted as a frequency curve. Deviations from normality of a grain-size distribution can thus be easily detected on log probability plots by deviation of the cumulative curve from a straight line. Most natural populations of grains in siliciclastic sediments or sedimentary rocks do not have a normal (or log-normal) distribution; the nearly normal distribution shown in Figure 4.2 is not typical of natural sediments. Some investigators believe that the shape of the log probability curve reflects conditions of the sediment transport process and thus can be used as a tool in environmental interpretation. We shall return to this point subsequently.

Graphical plots permit quick, visual inspection of the grain-size characteristics of a given sample; however, comparison of graphical plots becomes cumbersome and inconvenient when large numbers of samples are involved. Also, average grain-size and grain-sorting characteristics cannot be determined very accurately by visual inspection of grain-size curves. To overcome these disadvantages, mathematical methods that permit statistical treatment of grain-size data can be used to derive parameters that describe grain-size distributions in mathematical language. These statistical measures allow both the average size and the average sorting characteristics of grain populations to be expressed mathematically. Mathematical values of size and sorting can be used to prepare a variety of graphs and charts that facilitate evaluation of grain-size data.

Average Grain Size. Three mathematical measures of average grain size are in common use. The **mode** is the most frequently occurring particle size in a population of grains. The diameter of the modal size corresponds to the diameter of grains represented by the steepest point (inflection point) on a cumulative curve or the highest point on a frequency curve. Siliciclastic sediments and sedimentary rocks tend to have a single modal size, but some sediments are bimodal, with one mode in the coarse end of the size distribution and one in the fine end. Some are even polymodal. The **median size** is the midpoint of the grain-size distribution. Half of the grains by weight are larger than the median size, and half are smaller. The median size corresponds to the 50th percentile diameter on the cumulative curve (Fig. 4.3). The **mean size** is the arithmetic average of all the particle sizes in a sample. The true arithmetic mean of most sediment samples cannot be determined because we cannot count the total number of grains in a sample or measure each small grain. An approximation of the arithmetic mean can be arrived at by picking selected percentile values from the cumulative curve and averaging these values. As shown in Figure 4.3 and Table 4.3, the 16th, 50th, and 84th percentile values are commonly used for this calculation. The mode, median, and mean may or may not be the same, as subsequently discussed under the topic of skewness (e.g., Fig. 4.6).

FIGURE 4.3 Method of calculating percentile values from the cumulative curve.

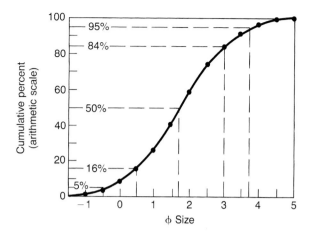

TABLE 4.3 Formulas for calculating grain-size statistical parameters by graphical methods

Graphic mean	$M_z = \dfrac{\phi_{16} + \phi_{50} + \phi_{84}}{3}$	(1)
Inclusive graphic standard deviation	$\sigma_i = \dfrac{\phi_{84} - \phi_{16}}{4} + \dfrac{\phi_{95} - \phi_5}{6.6}$	(2)
Inclusive graphic skewness	$SK_i = \dfrac{(\phi_{84} + \phi_{16} - 2\phi_{50})}{2(\phi_{84} - \phi_{16})} + \dfrac{(\phi_{95} + \phi_5 - 2\phi_{50})}{2(\phi_{95} - \phi_5)}$	(3)
Graphic kurtosis	$K_G = \dfrac{(\phi_{95} - \phi_5)}{2.44(\phi_{75} - \phi_{25})}$	(4)

Source: Folk, R. L., and W. C. Ward, 1957, Brazos River bar: A study in the significance of grain-size parameters: Jour. Sed. Petrology, v. 27, p. 3–26.

Sorting. The sorting of a grain population is a measure of the range of grain sizes present and the magnitude of the spread or scatter of these sizes around the mean size. Sorting can be estimated in the field or laboratory by use of a hand-lens or microscope and reference to a visual estimation chart (Fig. 4.4). More accurate determination of sorting requires mathematical treatment of grain-size data. The mathematical expression of sorting is **standard deviation.** In conventional statistics, one standard deviation encompasses the central 68 percent of the area under the frequency curve (Fig. 4.5). That is, 68 percent of the grain-size values lie within plus or minus one standard deviation of the mean size. A formula for calculating standard deviation by graphical-statistical methods is shown in Table 4.3. Note that the standard deviation calculated by this formula is expressed in phi (ϕ) values and is called phi standard deviation. The symbol ϕ must always be attached to the standard deviation value. Verbal terms for sorting corresponding to various values of standard deviation are as follows (after Folk, 1974):

Standard Deviation

<table>
<tr><td>$<0.35\phi$</td><td>very well sorted</td></tr>
<tr><td>0.35 to 0.50ϕ</td><td>well sorted</td></tr>
<tr><td>0.50 to 0.71ϕ</td><td>moderately well sorted</td></tr>
</table>

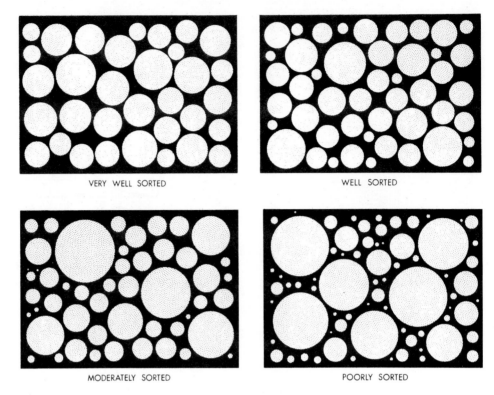

VERY WELL SORTED   WELL SORTED

MODERATELY SORTED   POORLY SORTED

FIGURE 4.4 Grain-sorting images for sediments with different degrees of sorting. (From Anstey, R. L., and T. L. Chase, 1974, Environments through time: Burgess, Minneapolis, Minn. Fig. 1.2, p. 2, reprinted by permission of Burgess Publishing Co.)

FIGURE 4.5 Frequency curve for a normal distribution of values showing the relationship of standard deviation to the mean. One standard deviation (1σ) on either side of the mean accounts for 68 percent of the area under the frequency curve. (After Friedman, G. M., and J. E. Sanders, Principles of sedimentology. © 1978 by John Wiley & Sons, Inc. Fig. 3.12, p. 70, reprinted by permission of John Wiley & Sons, Inc., New York.)

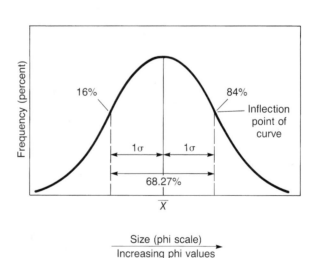

0.71 to 1.00ϕ   moderately sorted
1.00 to 2.00ϕ   poorly sorted
2.00 to 4.00ϕ   very poorly sorted
>4.00ϕ   extremely poorly sorted

As mentioned above, most natural sediment grain-size populations do not exhibit a normal or log-normal grain-size distribution. The frequency curves of such nonnormal populations are not perfect bell-shaped curves, such as the example shown in Figure 4.6A. Instead, they show some degree of asymmetry, or **skewness.** The mode, mean, and median in a skewed population of grains are all different, as illustrated in Figures 4.6B and 4.6C. Skewness reflects sorting in the "tails" of a grain-size popula-

FIGURE 4.6 Frequency curves illustrating the mode, median, and mean, and the difference between normal frequency curves and asymmetrical (skewed) curves. (After Friedman, G. M., and J. E. Sanders, Principles of sedimentology. © 1978 by John Wiley & Sons, Inc. Fig. 3.18, p. 75, reprinted by permission of John Wiley & Sons, Inc., New York.)

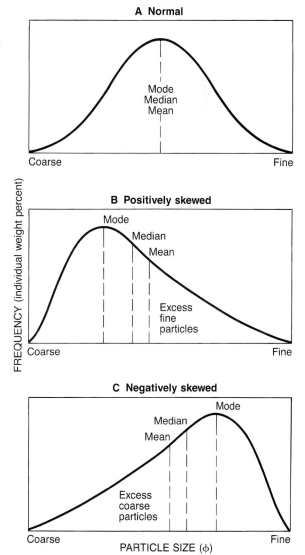

tion. Populations that have a tail of excess fine particles (Fig. 4.6B) are said to be positively skewed or fine skewed, that is, skewed toward positive phi values. Populations with a tail of excess coarse particles (Fig. 4.6C) are negatively skewed or coarse skewed. Graphic skewness can be calculated by equation 3 in Table 4.3. Verbal skewness is related to calculated values of skewness as follows (Folk, 1974):

Skewness

>+0.30	strongly fine skewed
+0.30 to +0.10	fine skewed
+0.10 to −0.10	near symmetrical
−0.10 to −0.30	coarse skewed
<−0.30	strongly coarse skewed

Grain-size frequency curves can show various degrees of sharpness or peaked-ness. The degree of peakedness is called **kurtosis.** A formula for calculating kurtosis is shown in Table 4.3. Although kurtosis is commonly calculated along with other grain-size parameters, the geological significance of kurtosis is unknown, and it appears to have little value in interpretive grain-size studies.

Grain-size statistical parameters can be calculated directly without reference to graphical plots by the mathematical **method of moments** (Krumbein and Pettijohn, 1938). The method had not been used extensively until comparatively recently because of the laborious calculations involved and because it had not been definitely proven that moment statistics are of greater value than graphical statistics in application to geologic problems. With the advent of modern computers, lengthy calculations no longer pose a problem, and moment statistics are now in common use. The computations in moment statistics involve multiplying a weight (weight frequency in percent) by a distance (from the midpoint of each size grade to the arbitrary origin of the abscissa). Formulas for computing moment statistics are given in Table 4.4, and a sample computation form using $\frac{1}{2}\phi$ size classes is given in Table 4.5. Millimeter values can also be used in computing moment statistics; that is, the size data need not be transformed to phi units. The formulas shown in Table 4.4 can easily be programmed into a computer or programmable calculator to calculate moment statistics from sieve data.

TABLE 4.4 Formulas for calculating grain-size parameters by the moment method

Mean (1st moment)	$\bar{x}_\phi = \dfrac{\Sigma fm}{n}$	(1)
Standard deviation (2nd moment)	$\sigma_\phi = \sqrt{\dfrac{\Sigma f(m - \bar{x}_\phi)^2}{100}}$	(2)
Skewness (3rd moment)	$Sk_\phi = \dfrac{\Sigma f(m - \bar{x}_\phi)^3}{100\sigma_\phi^{\,3}}$	(3)
Kurtosis (4th moment)	$K_\phi = \dfrac{\Sigma f(m - \bar{x}_\phi)^4}{100\sigma_\phi^{\,4}}$	(4)

where f = weight percent (frequency) in each grain-size grade present
m = midpoint of each grain-size grade in phi values
n = total number in sample; 100 when f is in percent

TABLE 4.5 Form for computing moment statistics using ½ϕ classes

Class interval (ϕ)	m Midpoint (ϕ)	f Weight %	fm Product	m − x̄ Deviation	(m − x̄)² Deviation squared	f(m − x̄)² Product	(m − x̄)³ Deviation cubed	f(m − x̄)³ Product	(m − x̄)⁴ Deviation quadrupled	f(m − x̄)⁴ Product
0–0.5	0.25	0.9	0.2	−2.13	4.54	4.09	−9.67	− 8.70	20.60	18.54
0.5–1.0	0.75	2.9	2.2	−1.63	2.66	7.71	−4.34	−12.59	7.07	20.50
1.0–1.5	1.25	12.2	15.3	−1.13	1.28	15.62	−1.45	−17.69	1.63	19.89
1.5–2.0	1.75	13.7	24.0	−0.63	0.40	5.48	−0.25	− 3.43	0.16	2.19
2.0–2.5	2.25	23.7	53.3	−0.13	0.02	0.47	0.00	0.00	0.00	0.00
2.5–3.0	2.75	26.8	73.7	0.37	0.13	3.48	0.05	1.34	0.02	0.54
3.0–3.5	3.25	12.2	39.7	0.87	0.76	9.27	0.66	8.05	0.57	6.95
3.5–4.0	3.75	5.6	21.0	1.37	1.88	10.53	2.57	14.39	3.52	19.71
>4.0	4.25	2.0	8.5	1.87	3.50	7.00	6.55	13.10	12.25	24.50
Total		100.0	237.9			63.65		− 5.53		112.82

Source: McBride, E. F., Mathematical treatment of size distribution data, *in* R. E. Carver (ed.), Procedures in sedimentary petrology. © 1971 by John Wiley & Sons, Inc. Table 2, p. 119, reprinted by permission of John Wiley & Sons, Inc., New York.

Application and Importance of Grain-Size Data

Grain size is a fundamental physical property of sedimentary rocks and, as such, is a useful descriptive property. Because grain size affects the related derived properties of porosity and permeability, the grain size of potential reservoir rocks is of considerable interest to petroleum geologists and hydrologists. Grain-size data are used in a variety of other ways (summarized by Syvitski, 1991):

1. to interpret coastal stratigraphy and sea-level fluctuations
2. to trace glacial sediment transport and the cycling of glacial sediments from land to sea
3. (by marine geochemists) to understand the fluxes, cycles, budgets, sources, and sinks of chemical elements in nature
4. to understand the mass physical (geotechnical) properties of seafloor sediment

Finally, because the size and sorting of sediment grains may reflect sedimentation mechanisms and depositional conditions, grain-size data are assumed to be useful for interpreting the depositional environments of ancient sedimentary rocks.

It is this assumption that grain-size characteristics reflect conditions of the depositional environment that has sparked most of the interest in grain-size analyses. Geologists have studied the grain-size properties of sediments and sedimentary rocks for more than a century; research efforts since the 1950s have focused particularly on statistical treatment of grain-size data. This prolonged period of grain-size research has generated hundreds of learned papers in geological journals and has swelled the bibliographies of numerous sedimentologists. Thus, it would be logical to assume that by now the relationship between grain-size characteristics and depositional environments has been firmly established. Alas, that is not the case!

Many techniques for utilizing grain-size data to interpret depositional environments have been tried; however, little agreement exists regarding the reliability of these methods. For example, Friedman (1967, 1979) used two-component grain-size variation diagrams, in which one statistical parameter is plotted against another (e.g., skewness vs. standard deviation, or mean size vs. standard deviation). This method putatively allows separation of the plots into major environmental fields such as beach environments and river environments (Fig. 4.7). Passega (1964, 1977) developed a

FIGURE 4.7 Grain-size bivariate plot of moment skewness vs. moment standard deviation showing the fields in which most beach and river sands plot. (Redrawn from Friedman, G. M., 1967, Dynamic processes and statistical parameters for size frequency distribution of beach and river sands: Jour. Sed. Petrology, v. 37. Fig. 5, p. 334, reproduced by permission of Society of Sedimentary Geologists, Tulsa, Okla.)

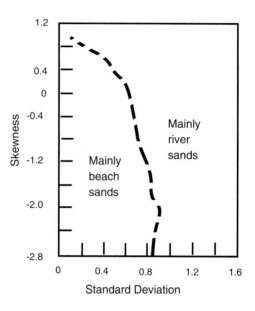

graphical approach to interpreting grain-size data that makes use of what he calls C-M and L-M diagrams, in which grain diameter of the coarsest grains in the deposit (C) is plotted against either the median grain diameter (M) or the percentile finer than 0.031 mm (L). Passega maintains that most samples from a given environment will fall within a specific environmental field in these diagrams. Visher (1969), Sagoe and Visher (1977), and Glaister and Nelson (1974) have all suggested that depositional environments can be interpreted on the basis of the shapes of grain-size cumulative curves plotted on log probability paper. Cumulative curves plotted in this way commonly display two or three straight-line segments rather than a single straight-line (Fig. 4.8). Each segment of the curve is interpreted to represent different subpopulations of grains that were transported simultaneously but by different transport modes, that is, by suspension, saltation, and bedload transport. Sediments from different environments (dune, fluvial, beach, tidal, nearshore, turbidite) can putatively be differentiated on the basis of the general shapes of the curves, the slopes of the curve segments, and the positions of the truncation points (breaks in slope) between the straight-line segments (e.g., Visher 1969). Although success using these various techniques has been reported by some investigators, all of the techniques have been criticized for yielding incorrect results in an unacceptably high percentages of cases (e.g., Tucker and Vacher, 1980; Sedimentation Seminar, 1981; Vandenberghe, 1975; Reed, LeFever, and Moir, 1975).

More sophisticated multivariate statistical techniques, such as factor analysis and discriminant function analysis (e.g., Chambers and Upchurch, 1979; Taira and Scholle, 1979; Stokes et al., 1989) and the so-called **log-hyperbolic distribution** (Barndorff-Nielsen, 1977; Bagnold and Barndorff-Nielsen (1980); Barndorff-Nielson et al., 1982), have also been used. These techniques have not yet been widely applied, however, and some of the assumptions used in these statistical methods have come under criticism (e.g., Forrest and Clark, 1989). Also, some investigators (e.g., Fieller et al., 1984; Wyroll and Smith, 1985) report no better results with log-hyperbolic distributions than with results based on the normal probability function. Thus, it appears that

FIGURE 4.8 Relation of sediment transport dynamics to populations and truncation points in a grain-size distribution as revealed by plotting grain-size data as a cumulative curve on log probability paper. (After Visher, G. S., 1969, Grain size distributions and depositional processes: Jour. Sed. Petrology, v. 39. Fig. 4, p. 1079, reprinted by permission of Society of Economic Paleontologists and Mineralogists, Tulsa, Okla.)

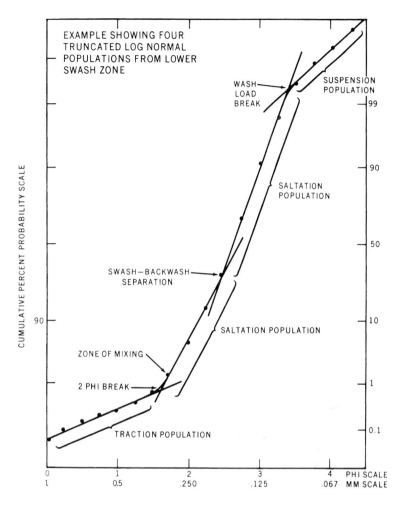

after several decades of intensive research into the techniques and significance of grain-size analysis, during which the techniques for interpreting grain-size data have come to demand increasingly more sophisticated statistical applications, there is little consensus as to their reliability. Such is science!

Reasons why grain-size techniques for identifying depositional environments of sediments fail to work consistently are probably related to variability in depositional conditions within major environmental settings. The energy conditions and sediment supply within river systems, for example, can differ considerably from one river to another and even within different parts of the same river system. Thus, in some cases, the grain-size characteristics of sediments may show as much variability within different parts of the same environmental setting as between different environments. Grain-size distributions reflect processes, not environments, and sediment transport processes are not unique to a particular environment. In any case, grain-size data should be considered as only one of the available tools for environmental interpretation and should never be used alone for this purpose.

4.3 PARTICLE SHAPE

Particle shape is defined by three related but different aspects of grains. **Form** refers to the gross, overall configuration of particles and reflects variations in their proportions (lengths of major axes). Form is often confused with **roundness,** which is a measure of the sharpness of grain corners, that is, smoothness. **Surface texture** refers to small-scale, microrelief markings such as pits, scratches, and ridges that occur on the surface of grains. Form, roundness, and surface texture are independent properties, as illustrated in Figure 4.9, and each can theoretically vary without affecting the other. Actually, form and roundness tend to be positively correlated in sedimentary deposits—particles that are highly spherical in shape also tend to be well rounded. Surface texture can change without significantly changing form or roundness, but a change in form or roundness will affect surface texture because new surfaces are exposed. The three aspects of shape can be thought of as constituting a hierarchy, where form is a first-order property, roundness a second-order property superimposed on form, and surface texture a third-order property superimposed on both the corners of a grain and the surfaces between corners (Barrett, 1980).

Particle Form

Sphericity and Form Indices. Form reflects variations in the proportions of particles. Therefore, most parameters for estimating form require that the relative lengths of the three major axes (long, intermediate, short) of a particle be known. Numerous mathematical measures for expressing form have been proposed, but **sphericity** is probably the most widely used. The concept of sphericity was introduced by Wadell (1932), who defined sphericity mathematically as the ratio of the diameter of a sphere with the same volume as a particle to the diameter of the smallest circle that would just enclose or circumscribe the outline of the particle. Wadell determined the volumes of large particles by immersing the particle in water and measuring the volume change of the water. Krumbein (1941) modified Wadell's sphericity concept slightly to express sphericity ψ by the relationship

$$\psi = \sqrt[3]{\frac{\text{volume of the particle}}{\text{volume of the circumscribing sphere}}} \tag{4.4}$$

The volume of a sphere is given by $(\pi/6)D^3$, where D is the diameter of the sphere. The approximate volume of natural particles can be calculated by assuming that the particles are triaxial ellipsoids having three diameters D_L, D_I, and D_S, where L, I, and S refer to the lengths of the long, intermediate, and short axes of the ellipsoid. By substituting appropriate values for volume into equation 4.4, we can express sphericity by

$$\psi_I = \sqrt[3]{\frac{\dfrac{\pi}{6} D_L D_I D_S}{\dfrac{\pi}{6} D_L^3}} = \sqrt[3]{\frac{D_S D_I}{D_L^2}} \tag{4.5}$$

The sphericity of a particle determined by this relationship is called intercept sphericity and can be calculated by measuring the long, intermediate, and short axes of a particle and substituting these values into formula 4.5.

Sneed and Folk (1958) suggest that Krumbein's intercept sphericity does not correctly express the behavior of particles as they settle in a fluid or are acted on by fluid flow. A rod-shaped particle, for example, settles faster than a disc, although the inter-

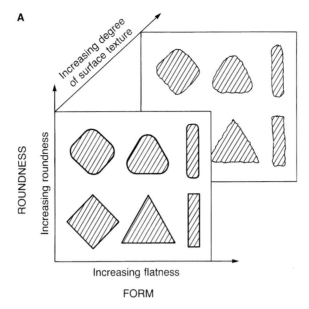

A

Increasing degree of surface texture

ROUNDNESS

Increasing roundness

Increasing flatness

FORM

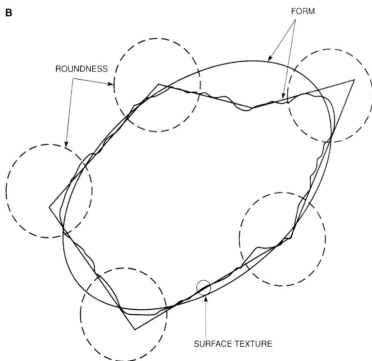

B

FORM

ROUNDNESS

SURFACE TEXTURE

FIGURE 4.9 Simplified graphical representation of particle form, roundness, and surface texture. Part A demonstrates the independence of these parameters, and B shows their hierarchical relationship; the irregular solid line is the particle outline. (After Barrett, P. J., 1980, The shape of rock particles, a critical review: Sedimentology, v. 27. Figs. 1 and 2, p. 293, reprinted by permission of Elsevier Science Publishers, Amsterdam.)

cept sphericity formula suggests the opposite. Particles falling through water tend to settle with maximum projection areas (the plane of the long and intermediate axes) perpendicular to the direction of motion, and small particles resting on the bottom orient themselves perpendicular to current flow direction. Sneed and Folk thus proposed a different sphericity measure called maximum projection sphericity (ψ_P), which they claim better expresses particle behavior. Maximum projection sphericity is defined mathematically as the ratio between the maximum projection area of a sphere with the same volume as the particle and the maximum projection area of the particle:

$$\psi_P = \sqrt[3]{\frac{D_S^2}{D_L D_I}} \tag{4.6}$$

Maximum projection sphericity has gained favor for expressing the shape of particles deposited by water. Conceptually, it is not necessarily more valid than intercept sphericity when applied to other modes of particle transport and deposition, (e.g., transport by ice and sediment gravity flows).

Regardless of the sphericity measure used, experience has shown that particles having the same mathematical sphericity can differ considerably in their overall shape. Some additional measure or index is needed to more specifically define the form of particles. Two additional form indices that permit a more graphic representation of form are in wide use. Zingg (1935) proposed the use of two shape indices, D_I/D_L and D_S/D_I, to define four shape fields on a bivariate plot: oblate (disc), equant (spheres), bladed, and prolate (rollers) (Fig. 4.10A). Lines of equal intercept sphericity can be drawn on the Zingg shape fields (Fig. 4.10B), illustrating that particles of quite different form can have the same mathematical sphericity.

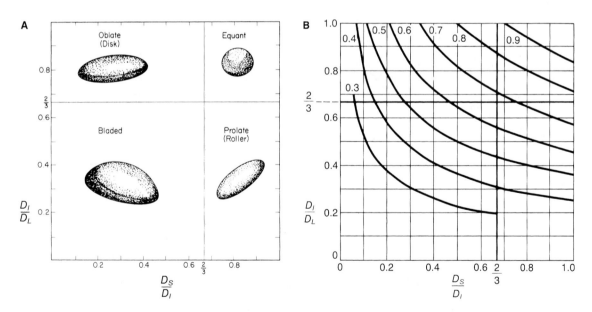

FIGURE 4.10   A. Classification of shapes of pebbles after Zingg (1935). B. Relationship between mathematical sphericity and Zingg shape fields. The curves represent lines of equal sphericity. (A, from Blatt, H., G. V. Middleton, and R. Murray, Origin of sedimentary rocks, 2nd ed. © 1980, Fig. 3.20, p. 80. Reprinted by permission of Prentice-Hall, Englewood Cliffs, N.J. B, from Sedimentary rocks, 3rd ed., by Francis J. Pettijohn. Fig. 3.19. Copyright 1949, 1957 by Harper & Row, Publishers, Inc. Copyright © 1975 by Francis J. Pettijohn. Reprinted by permission of Harper-Collins Publishers, Inc.)

FIGURE 4.11 Classification of pebble shapes, after Sneed and Folk. The symbol V refers to the adjective *very* (e.g., very platy, very bladed, very elongated). (After Sneed, E. D., and R. L. Folk, 1958, Pebbles in the Lower Colorado River, Texas, a study in particle morphogenesis: Jour. Geology, v. 66. Fig. 2, p. 119, reprinted by permission of University of Chicago Press.)

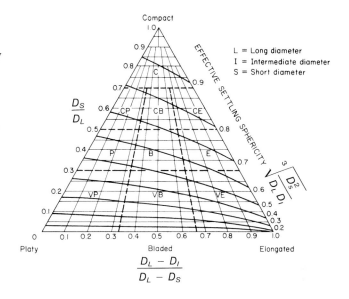

Sneed and Folk (1958) used two somewhat different shape indices to construct a triangular form diagram (Fig. 4.11), in which D_S/D_L is plotted against $D_L - D_I/D_L - D_S$ to create ten form fields (compact, C; platy, P; bladed, B; elongate, E; etc.). Lines of maximum projection sphericity drawn across the field (Fig. 4.11) again illustrate the disparity between mathematical sphericity and actual form.

Significance of Form. The form of sand-size and smaller mineral grains in sedimentary deposits is a function mainly of the original shapes of the minerals. Because of its superior hardness and durability, as well as its high average abundance in siliciclastic sedimentary rocks, quartz is commonly the only mineral examined in shape studies. Numerous studies have shown that the form of small quartz grains is not significantly modified during transport, although very slight changes may occur in the early stages of transport. The form of pebbles and larger fragments is also a function of the shape inherited from source rocks, although, owing to abrasion and breakage, pebbles are modified during transport to a greater extent than are sand-size grains.

The form of particles has a very pronounced effect on their settling velocity; in general, the more nonspherical the particle, the lower the settling velocity. Particle form thus affects relative transportability of particles traveling in suspension. That is, nonspherical particles tend to stay in suspension longer than spherical particles. Form also affects the transportability of larger particles that move by traction along the bed. In general, spheres and rollers are transported more readily than are blades and discs having the same mass. Thus, owing to preferential transport of spheres and rollers, downstream changes in pebble form may occur in rivers. Such changes can be difficult or impossible to differentiate from changes caused by pebble abrasion. A long-standing and as yet unresolved controversy has to do with the flattened, disc shapes of many pebbles found on beaches. The prevalence of disc-shaped beach pebbles has been attributed alternatively to selective transport which leaves flattened pebbles behind on the beach as a lag deposit (Kuenen, 1964) and to flattening of pebbles on the beach by abrasion in the surf (Dobkins and Folk, 1970).

It has not yet been demonstrated that the form of particles, as expressed by sphericity values or the Zingg/Folk shape measures, can be used alone as a reliable tool for interpreting depositional environments. Although empirical studies show some relative differences in sphericity or form indices of grains from different environments, these differences have not proven to be sufficiently distinctive to permit environmental discrimination. Furthermore, it is very difficult to accurately measure the three-dimensional form of sand-size grains, particularly in consolidated sediments. In a thin section, for example, we can see only two dimensions of a particle, and neither dimension may represent the true length of the particle axis. More accurate means of expressing particle form are clearly needed, as subsequently discussed under "Fourier Shape Analysis."

Roundness

Definition and Measurement. Wadell (1932) defined mathematical roundness as the arithmetic mean of the roundness of the individual corners of a grain in the plane of measurement. Roundness of individual corners is given by the ratio of the radius of curvature of the corners to the radius of the maximum-size circle that can be inscribed within the outline of the grain in the plane of measurement. The degree of Wadell roundness (R_w) is thus expressed as

$$R_w = \frac{\Sigma(r/R)}{N} = \frac{\Sigma(r)}{RN} \tag{4.7}$$

where r is the radius of curvature of individual corners, R is the radius of the maximum inscribed circle, and N is the number of corners. The relationship of r to R is illustrated in Figure 4.12.

Owing to the numerous radius measurements that must be made, it is very time-consuming to determine the Wadell roundness of large numbers of grains. Simpler roundness measures have been proposed that require only that the radius of the sharpest corner be divided by the radius of the inscribed circle (Wentworth, 1919; Dobkins and Folk, 1970); however, Wadell's roundness measure is still used by most

FIGURE 4.12   Diagram of enlarged grain image illustrating the method of measuring the radius (R) of the maximum inscribed circle and the radii of curvature (r) of the corners of the grain. (After Boggs, S., Jr., 1967, Measurement of roundness and sphericity parameters using an electronic particle size analyzer: Jour. Sed. Petrology, v. 37. Fig. 3, p. 912, reprinted by permission of Society of Economic Paleontologists and Mineralogists, Tulsa, Okla.)

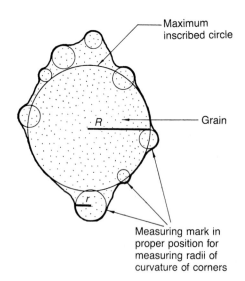

workers. Even if the simpler roundness formula is used, measuring the radii of large numbers of small grains is a very laborious process, requiring use of either a circular protractor or an electronic particle size analyzer to measure enlarged images of grains (Boggs, 1967a). Consequently, visual comparison scales or charts consisting of sets of grain images of known roundness are often used to make rapid visual estimates of grain roundness. The visual charts of Krumbein (1941) and Powers (1953) are the most widely used of these comparison scales. The Powers visual and verbal roundness scale is shown in Figure 4.13, and the mathematical limits of each Powers verbal roundness class are given in Table 4.6. The interval between roundness classes is approximately equal to $\sqrt{2}$; that is, the interval of each class is 1.41 times greater than the preceding class. Folk (1955) developed a logarithmic transformation of this scale, called the rho (ρ) scale, also shown in Table 4.6. Operator error in estimating grain roundness from visual charts is very high, and reproducibility of results even by the same operator is very low. Visual estimation methods yield only a rough approximation of the true roundness distribution in a population of grains and should not be used for serious research studies.

Significance of Roundness. The roundness of grains in a sedimentary deposit is a function of grain composition, grain size, type of transport process, and distance of transport. Hard, resistant grains such as quartz and zircon are rounded less readily

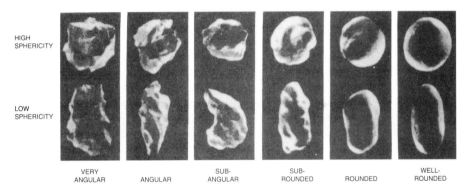

FIGURE 4.13 Powers's grain images for estimating the roundness of sedimentary particles. (After Powers, M. C., 1953, A new roundness scale for sedimentary particles: Jour. Sed. Petrology, v. 23, Fig. 1, p. 118, reprinted by permission of Society of Economic Paleontologists and Mineralogists, Tulsa, Okla.)

TABLE 4.6 Relation of Powers's verbal rounding classes to Wadell roundness and Folk's rho (ρ) scale

Powers's verbal class	Corresponding Wadell class interval	Folk's rho (ρ) scale
Very angular	0.12–0.17	0.00–1.00
Angular	0.17–0.25	1.00–2.00
Subangular	0.25–0.35	2.00–3.00
Subrounded	0.35–0.49	3.00–4.00
Rounded	0.49–0.70	4.00–5.00
Well rounded	0.70–1.0	5.00–6.00

Source: Powers, M. C., 1953, A new roundness scale for sedimentary particles: Jour. Sed. Petrology, v. 23, p. 117–119. Folk, R. L., 1955, Student operator error in determination of roundness, sphericity, and grain size: Jour. Sed. Petrology, v. 25, p. 297–301.

during transport than are weakly durable grains such as feldspars and pyroxenes. Pebble- to cobble-size grains commonly are more easily rounded than sand-size grains. Resistant mineral grains smaller than 0.05 to 0.1 mm do not appear to become rounded by any transport process. Because of these factors, it is always necessary to work with particles of the same size and composition when doing roundness studies.

Experimental studies in flumes and wind tunnels of the effects of abrasion on transport of sand-size quartz grains show that transport by wind is 100 to 1000 times more effective in rounding these grains than transport by water (Kuenen, 1959, 1960). In fact, almost no rounding occurs in as much as 100 km of transport by water. Most roundness studies of small quartz grains in rivers have corroborated these experimental results. For example, Russell and Taylor (1937) observed no increase in rounding of quartz grains over a distance of 1100 mi (1775 km) in the Mississippi River between Cairo, Illinois, and the Gulf of Mexico. The effectiveness of surf action on beaches in rounding sand-size quartz grains is not well understood. In general, surf processes appear to be less effective in rounding grains than wind transport but more effective than river transport.

Once acquired, the roundness of quartz grains is not easily lost and may be preserved through several sedimentation cycles. The presence of well-rounded quartz grains in an ancient sandstone may well indicate an episode of wind transport in its history, but it may be difficult or impossible to determine if rounding took place during the last episode of transport or during some previous cycle.

The roundness of transported pebbles is strongly related to pebble composition and size (Boggs, 1969). Soft pebbles such as shale and limestone become rounded much more readily than quartzite or chert pebbles, and large pebbles and cobbles are commonly better rounded than smaller pebbles. Although stream transport is relatively ineffective in rounding small quartz grains, pebble-size grains can become well rounded by stream transport. Depending upon composition and size, pebbles can become well rounded (roundness about 0.6) by stream transport in distances ranging from 11 km (7 mi) for limestones to 300 km (186 mi) for quartz (Pettijohn, 1975).

The presence of well-rounded pebbles in ancient sedimentary rocks is generally indicative of fluvial transport. The degree of rounding cannot, however, be depended upon to give reliable estimates of the distance of transport. The greatest amount of rounding takes place in the early stages of transport, generally within the first few kilometers. Also, the roundness of pebbles is not an unequivocal indicator of fluvial environments because pebbles can also become rounded in beach environments and possibly on lakeshores. Furthermore, rounded fluvial pebbles may eventually be transported into nearshore marine environments where they may be reentrained by turbidity currents and resedimented in deeper parts of the ocean.

Surface Texture

The surface of pebbles and mineral grains may be polished, frosted (dull, matte texture like frosted glass), or marked by a variety of small-scale, low-relief features such as pits, scratches, fractures, and ridges. These surface textures originate in diverse ways, including mechanical abrasion during sediment transport; tectonic polishing during deformation; and chemical corrosion, etching, and precipitation of authigenic growths on grain surfaces during diagenesis and weathering. Gross surface textural features such as polishing and frosting can be observed with an ordinary binocular or petrographic microscope; however, detailed study of surface texture requires high magnifications. Krinsley (1962) pioneered use of the electron microscope for studying grain surface texture at high magnifications.

Most investigators who study the surface texture of sediment grains carry out their studies on quartz grains because the physical hardness and chemical stability of quartz grains allow these particles to retain surface markings for geologically long periods of time. Through study of thousands of quartz grains, investigators have now been able to fingerprint the markings on grains from various modern depositional environments. More than 25 different surface textural features have been identified, including conchoidal fractures, straight and curved scratches and striations, upturned plates, meander ridges, chemically etched V's, mechanically formed V's, and dish-shaped concavities (Bull, 1986). Examples of these markings are shown in Figure 4.14, illustrating the kinds of markings that can be seen at high magnification. Many other excellent electron micrographs of quartz surfaces may be found in Krinsley and Doornkamp's (1973) *Atlas of Quartz Sand Surface Textures.*

Surface texture appears to be more susceptible to change during sediment transport and deposition than do sphericity and roundness. Removal of old surface textural features and generation of new features are more likely to occur than marked changes in sphericity and roundness, and surface texture is more likely to record the last cycle of sediment transport or the last depositional environment. Therefore, geologists are interested in surface textural features as possible indicators of ancient transport conditions and depositional environment. The usefulness of surface texture in environmental analysis is limited, however, because similar types of surface markings can be produced in different environments. Also, the markings produced on grains in one

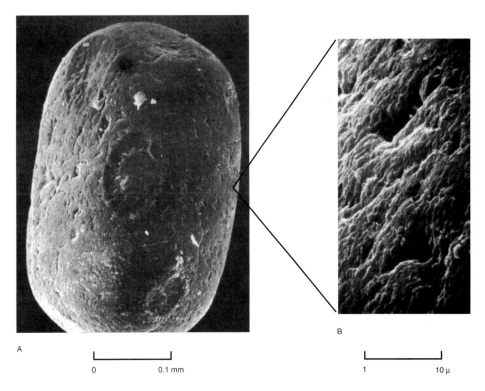

A

0   0.1 mm

B

1   10 μ

FIGURE 4.14 Electron micrograph of surface markings on a quartz grain, St. Peters Sandstone (Ordovician), U.S. midcontinent region. A. Quartz grain magnified 260×. B. Section of the grain surface magnified 2400×. (Photographs by M. B. Shaffer.)

environment may be retained on the grains when they are transported into another environment. Although less abrasion is required to remove surface markings from grains than is required to change roundness or sphericity, markings inherited from a previous environment may remain on grains for a long time before they are removed or replaced by different markings produced in the new environment. For example, grains on an arctic marine shelf may still retain surface microrelief features acquired during glacial transport of the grains to the shelf.

With care and the use of statistical methods, it has proven possible on the basis of surface textures to distinguish quartz grains from at least three major modern environmental settings: littoral (beach and nearshore), eolian (desert), and glacial. Quartz grains from littoral environments are characterized especially by V-shaped percussion marks and conchoidal breakage patterns. Grains deposited in eolian environments show surface smoothness and rounding, irregular upturned plates, and silica solution and precipitation features. Grains from glacial deposits have conchoidal fracture patterns and parallel to semiparallel striations. Techniques for studying the surface texture of quartz grains in modern environments have also been extended to study of ancient sedimentary deposits. Interpretation of paleoenvironments on the basis of surface texture is complicated by the fact that surface microtextures can be changed during diagenesis by addition of cementing overgrowths or by chemical etching and solution. Krinsley and Trusty (1986) and Marshall (1987) provide additional details and discuss applications of the study of quartz grain surface textures by using the scanning electron microscope (SEM).

Fourier Shape Analysis

Although sphericity and roundness are good general descriptors of particle shape, they do not describe the shape with sufficient accuracy for successful environmental and source-rock interpretation. The two-dimensional shape of particles can be described with a high degree of precision using a method based on Fourier analysis, which is a method of representing periodic mathematical functions as an infinite series of summed sine and cosine terms (Ehrlich and Weinberg, 1970). If the outline of a grain is "cut" and "unrolled," as illustrated in Figure 4.15, the unrolled outline is a periodic function somewhat resembling the shape of a sine wave. The shape of this unrolled grain can be represented by a series of terms called harmonics. Harmonics are periodic functions (e.g., a sine wave) that can have various shapes owing to differences in wave amplitude, frequency, and initial phase (initial position of a point x). Figure 4.16 shows some of the harmonics of a sine wave. By adding together (graphically) the shapes of various harmonics (done by computer), we can faithfully reproduce the shape of any periodic function such as that shown in Figure 4.15B. Lower-order harmonics summarize gross form (e.g., sphericity), whereas roundness and surface texture are measured by progressively higher-order harmonics. The method involves digitizing the periphery of grains by projecting grains onto a grid and recording, either manually or with an automatic digitizer, intercepts of the grain outline with the grid. Digitized data are reduced by computer to obtain the harmonics and produce computer-generated grain outlines.

Fourier analysis has been used both to study the source of sediment grains and to characterize grains from particular depositional environments. Results of such studies suggest that it has the potential to differentiate quartz grains derived from sedimentary sources from those derived from granitic or volcanic rocks and to differentiate between beach, dune, fluvial, and glaciofluvial sediments. It has also been used to map the

FIGURE 4.15 Method of "unrolling" a grain outline to produce a periodic wave. A. Grain outline showing radii measured from grain center to points on the grain perimeter. B. Unrolled grain outline as constructed from radii measurements. Note that the form of the unrolled grain is a crude sine wave. (From Boggs, S., 1992, Petrology of sedimentary rocks: Macmillan, New York. Fig. 2.9, p. 53.)

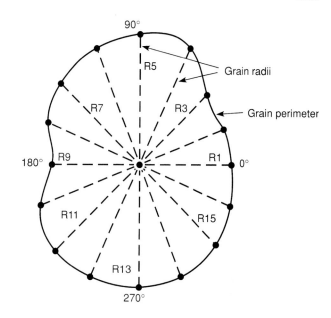

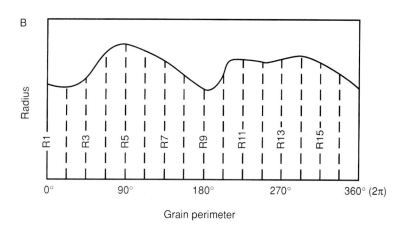

source and the transport paths of abyssal silts in the ocean. Because the technique is relatively new, having been introduced in the 1970s, additional work is needed to confirm and extend these early results.

4.4 FABRIC

The fabric of sedimentary rocks is a function of grain orientation and packing and is thus a property of grain aggregates. Orientation and grain packing in turn control such physical properties of sedimentary rocks as bulk density, porosity, and permeability.

FIGURE 4.16 Harmonics of a sine wave. A. Normal sine wave. B. Sine wave deformed by compression along the x-axis. C. Sine wave modified by a change in initial phase angle. D. Sine wave with increased amplitude. (After Tolstov, G. P., 1976, Fourier series. Fig. 4, p. 4, reprinted by permission of Dover Publications, Inc., New York.)

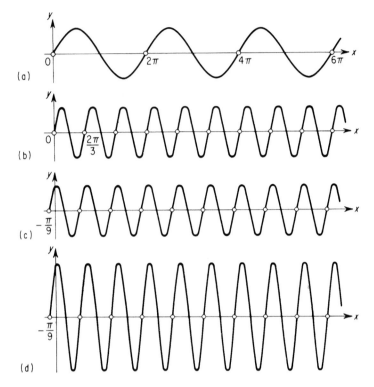

Grain Orientation

Particles in sedimentary rocks that have a platy (blade or disc) shape or an elongated (rod or roller) shape commonly show some degree of preferred orientation (Fig. 4.17). Platy particles tend to be aligned in planes that are roughly parallel to the bedding surfaces of the deposits. Elongated particles show a further tendency to be oriented with their long axes pointing roughly in the same direction. The preferred orientation of these particles is caused by transport and depositional processes and is related particularly to flow velocities and other hydraulic conditions at the depositional site. Most orientation studies have shown that sand-size particles deposited by fluid flows tend to become aligned parallel to the current direction (Fig. 4.17A; Parkash and Middleton, 1970), although a secondary mode of grains oriented normal to current flow may be present. If the grains have a streamline or teardrop shape, the blunt end of the grains commonly points upstream because this is the most stable orientation within a current. Sand grains can also show well-developed imbrication with long axes generally dipping upcurrent at angles less than about 20°. Imbrication refers to the overlapping arrangement of particles like that of shingles on a roof (Fig. 4.17C). Particles of sandy sediment deposited by turbidity currents and grain-flow or sandy debris-flow processes also tend to be aligned parallel to flow directions and display upstream imbrication at angles exceeding 20° (Hiscott and Middleton, 1980); however, in some gravity-flow deposits, orientation and imbrication directions can be variable or polymodal. Particles deposited under quiet water conditions may also show various orientations and a lack of imbrication (Fig. 4.17D) Fabric inconsistencies or bimodality

FIGURE 4.17 Schematic illustration of the orientation of elongated particles in relation to current flow. A. Particles oriented parallel to current flow. B. Particles oriented perpendicular to current flow. C. Imbricated particles. D. Random orientation of particles, characteristic of deposition in quiet water.

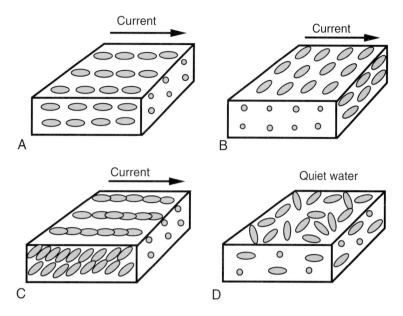

appear to be related mainly to very rapid deposition from suspension or from sandy debris flow.

Pebbles in many gravel deposits and ancient conglomerates also display preferred orientation and imbrication. River-deposited pebbles are commonly oriented with their long axes normal to flow direction (Fig. 4.17B) and may display upstream imbrication of up to 10° to 15° (Fig. 4.18). Orientation can also be parallel to flow or even bimodal. Increasing flow intensity appears to favor orientation with long axes parallel to current flow rather than normal to flow (Johansson, 1976). Pebbles deposited by turbidity currents or other gravity-flow processes also take up orientations with their long axes mainly parallel to flow direction, although orientation in some deposits can be random. Pebbles in glacial tills show preferred orientation parallel to flow, with a minor mode oriented normal to flow.

FIGURE 4.18 Well-developed imbrication in river cobbles, Kiso River, Japan. The imbrication was produced by river currents flowing from right to left.

Grain Packing, Grain-to-Grain Relations, and Porosity

Grain packing refers to the spacing or density patterns of grains in a sedimentary rock and is a function mainly of grain size, grain shape, and the degree of compaction of the sediment. Packing strongly affects the bulk density of the rocks as well as their porosity and permeability. The effects of packing on porosity can be illustrated by considering the change in porosity that takes place when even-size spheres are rearranged from loosest packing (cubic packing) to tightest packing (rhombohedral packing), as shown in Figure 4.19. Cubic packing yields porosity of 47.6 percent, whereas the porosity of rhombohedrally packed spheres is only 26.0 percent. The packing of natural particles is much more complex because of variations in size, shape, and sorting and is further complicated in lithified sedimentary rocks by the effects of compaction.

Poorly sorted sediments tend to have lower porosities and permeabilities than well-sorted sediments because grains are packed more tightly in these sediments owing to the filling of pore spaces among larger grains by finer sediment. Petroleum and groundwater geologists are especially concerned with the porosity of sedimentary rocks because porosity determines the volume of fluids (oil, gas, groundwater) that can be held within a particular reservoir rock. Compaction causes major reduction in porosity. For example, a sandstone having an original porosity of about 40 percent may have porosity reduced during burial to less than 10 percent owing to compaction resulting from the weight of overlying sediment. Compaction forces grains into closer contact and causes changes in the types of grain-to-grain contacts. Taylor (1950) identified four types of contacts between grains that can be observed in thin sections: **tangential contacts,** or point contacts; **long contacts,** appearing as a straight line in the plane of a thin section; **concavoconvex contacts,** appearing as a curved line in the plane of a thin section; and **sutured contacts,** caused by mutual stylolitic interpenetration of two or more grains (Fig. 4.20). In very loosely packed fabrics, some grains may not make contact with other grains in the plane of the thin section and are referred to

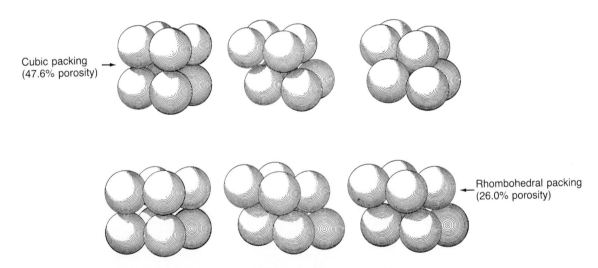

Cubic packing
(47.6% porosity)

Rhombohedral packing
(26.0% porosity)

FIGURE 4.19   Progressive decrease in porosity of spheres owing to increasingly tight packing. (After Graton, L. C., and H. J. Fraser, 1935, Systematic packing of spheres with particular relation to porosity and permeability: Jour. Geology, v. 43. Fig. 3, p. 796, reprinted by permission of University of Chicago Press.)

FIGURE 4.20 Diagrammatic illustration of principal types of grain contacts. A. Tangential. B. Long. C. Concavoconvex. D. Sutured. (Based on Taylor, J. M., 1950, Pore-space reduction in sandstones: Am. Assoc. Petroleum Geologists Bull., v. 34, p. 701–716.)

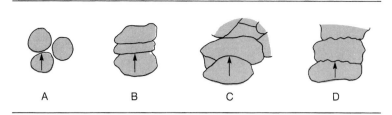

as "floating grains." Contact types are related to both particle shape and packing. Tangential contacts occur only in loosely packed sediments or sedimentary rocks, whereas concavoconvex contacts and sutured contacts occur in rocks that have undergone considerable compaction during burial. The relative abundance of these various types of contacts can be used as a rough measure of the degree of compaction and packing, and thus the depth of burial of sandstones. Several other methods for expressing the packing of sediment have been proposed; see Boggs (1992, p. 68–69) for details.

The sand-size grains in sandstones are commonly in continuous grain-to-grain contact when considered in three dimensions, and thus they form a **grain-supported** fabric. Conglomerates deposited by fluid flows also generally have a grain-supported fabric. On the other hand, conglomerates in glacial deposits, mud-flow deposits, and debris-flow deposits commonly have a **matrix-supported** fabric. In this type of fabric, the pebbles are not in grain-to-grain contact but "float" in a matrix of sand or mud. Matrix-supported conglomerates indicate deposition under conditions where fine sediment is abundant and deposition occurs by mass-transport processes or by processes that cause little reworking at the depositional site.

FURTHER READINGS

Carver, R. E. (ed.), 1971, Procedures in sedimentary petrology: John Wiley & Sons, New York, 653 p.

Folk, R. L., 1974, Petrology of sedimentary rocks: Hemphill, Austin, Tex., 182 p.

Griffith, J. C., 1967, Scientific methods in analysis of sediments: McGraw-Hill, New York, 508 p.

Krinsley, D., and J. Doornkamp, 1973, Atlas of quartz sand surface textures: Cambridge University Press, Cambridge, 91 p.

Lewis, D. W., 1984, Practical sedimentology, p. 58–108: Dowden, Hutchinson and Ross, Stroudsburg, Pa., 229 p.

Marshall, J. R. (ed.), 1987, Clastic particles: Scanning electon microscopy and shape analysis of sedimentary and volcanic clasts: Van Nostrand Reinhold, New York, 346 p.

Syvitski, J. P. M., 1991, Principles, methods, and applications of particle size analysis: Cambridge University Press, Cambridge, 368 p.

5

Sedimentary Structures

5.1 INTRODUCTION

Sedimentary structures are large-scale features of sedimentary rocks such as bedding units, ripples, and mudcracks that are best studied in the field. They are generated by a variety of sedimentary processes, including fluid flow, sediment gravity flow, soft-sediment deformation, and biogenic activity. Because they reflect environmental conditions that prevailed at, or very shortly after, the time of deposition, they are of special interest to geologists as a tool for interpreting ancient depositional environments. We know enough about the origin of sedimentary structures from experimental investigations and field studies to use them for evaluating such aspects of ancient sedimentary environments as sediment transport mechanisms, paleocurrent flow directions, relative water depth, and relative current velocity. Some sedimentary structures can also be used to identify the tops and bottoms of beds, and thus to determine if sedimentary sequences are in depositional stratigraphic order or have been overturned by tectonic forces. Sedimentary structures are particularly abundant in siliciclastic sedimentary rocks, but they occur also in nonsiliciclastic sedimentary rocks such as limestones and evaporites.

A very large body of literature on sedimentary structures has developed since the 1950s owing to their potential usefulness in environmental interpretation and paleocurrent analysis. These publications include several important monographs that contain excellent photographs and drawings illustrating a large variety of primary sedimentary structures. Books that deal with all types of sedimentary structures include those of Allen (1982), Collinson and Thompson (1982, 1989), Conybeare and Crook (1968), Pettijohn and Potter (1964), Potter and Pettijohn (1977), and Reineck and Singh (1980). Allen (1968) gives a more specialized treatment of current ripples and associated structures. McKee (1982) discusses the kinds of structures that occur in dune sands. Dzulynski and Walton (1965) discuss sole markings on the base of sandstone beds, particularly turbidite sandstones, and Picard and High (1973) cover the special sedimentary structures of ephemeral streams. Basan (1978), Crimes and Harper (1970), Curran (1985), Ekdale, Bromley, and Pemberton (1984), and Frey (1975) discuss and illustrate biogenic sedimentary structures. Bouma (1969) deals mainly with methods of studying sedimentary structures.

This chapter describes and discusses the more important sedimentary structures. The discussions are brief, but they include a summary of current ideas on mechanisms of formation and, where appropriate, an analysis of the usefulness and limitations of the structures in environmental interpretation. We begin study by examining the most commonly used names for the major primary sedimentary structures. Primary structures are those formed at, or very shortly after, the time of sediment deposition.

5.2 CLASSIFICATION OF PRIMARY SEDIMENTARY STRUCTURES

The most common primary sedimentary structures are listed in Table 5.1. The classification of sedimentary structures in this table is fundamentally descriptive; that is, it is based primarily on observable properties. Sedimentary structures are classified broadly as stratification structures and bedforms, bedding-plane markings, and other structures. Stratification structures and bedforms are further subdivided into four descriptive categories: (1) bedding and lamination, (2) bedforms, (3) cross-lamination, and (4) irregular stratification. Table 5.1 also includes a genetic classification that categorizes structures into four broad groups according to their probable origin: (1) structures formed by sedimentation processes, (2) structures formed by erosion, (3) structures formed by soft-sediment deformation (penecontemporaneous deformation), and (4) structures of biogenic origin. In the following discussion, sedimentary structures are listed and described under the descriptive headings shown in Table 5.1, although the discussion does not in all cases follow the exact order shown in the table. In some parts of the discussion, structures listed under a particular descriptive heading in Table 5.1 are further subdivided into genetic categories.

5.3 STRATIFICATION AND BEDFORMS

Bedding and Lamination

Concept of Bedding

Bedding is a fundamental characteristic of sedimentary rocks. **Beds,** or **strata,** are tabular or lenticular layers of sedimentary rock that have lithologic, textural, or structural unity that clearly distinguishes them from layers above and below. The upper and lower surfaces of beds are known as **bedding planes** or **bounding planes.** Otto (1938) regarded beds as sedimentation units, that is, the thickness of sediment deposited under essentially constant physical conditions. It is not always possible, however, to identify individual **sedimentation units.** Many beds defined by the criteria above may contain several true sedimentation units. Beds are strata thicker than 1 cm (McKee and Weir, 1953); layers less than 1 cm are **laminae.** Terms used for describing the thickness of beds and laminae are shown in Figure 5.1.

Beds can be differentiated internally into a number of informal units (Fig. 5.2). Blatt, Middleton, and Murray (1980) suggest use of the term **layers** for parts of a bed

TABLE 5.1 Classification of common primary sedimentary structures

MORPHOLOGICAL CLASSIFICATION \ GENETIC CLASSIFICATION	Depositional structures			Erosional structures		Deformation structures						Biogenic structures	
	Suspension-settling and current- and wave-formed structures	Wind-formed structures	Chemically and biochemically precipitated structures	Scour marks	Tool marks	Slump structures	Load and founder structures	Injection (fluidization) structures	Fluid-escape structures	Desiccation structures	Impact structures (rain, hail, spray)	Bioturbation structures	Biostratification structures
STRATIFICATION AND BED-FORMS													
Bedding and lamination													
Laminated bedding	X	X	X									X	
Graded bedding	X											X	
Massive (structureless) bedding	X												
Bedforms													
Ripples	X	X											
Dunes	X	X											
Antidunes	X												
Cross-lamination													
Cross-bedding	X	X											
Ripple cross-lamination	X	X											
Flaser and lenticular bedding	X												
Hummocky cross-bedding	X												
Irregular stratification													
Convolute bedding and lamination							X						
Flame structures							X						
Ball and pillow structures							X						
Synsedimentary folds and faults						X							
Dish and pillar* structures									X				
Channels				X									
Scour-and-fill structures				X									
Mottled bedding												X	
Stromatolites													X
BEDDING-PLANE MARKINGS													
Groove casts; striations; bounce, brush, prod, and roll marks					X								
Flute casts				X									
Parting lineation	X												
Load casts							X						
Tracks, trails, burrows†												X	
Mudcracks and syneresis cracks										X			
Pits and small impressions											X		
Rill and swash marks	X												
OTHER STRUCTURES													
Sedimentary sills and dikes								X					

*Not wholly stratification structures †Not wholly bedding-plane markings

FIGURE 5.1 Terms used for describing the thickness of beds and laminae. (Modified from McKee, E. D., and G. W. Weir, 1953, Terminology for stratification and cross-stratification in sedimentary rocks: Geol. Soc. America Bull., v. 64. Table 2, p. 383; and Ingram, R. L., 1954, Terminology for the thickness of stratification and parting units in sedimentary rocks: Geol. Soc. America Bull., v. 65, Fig. 1. p. 937.)

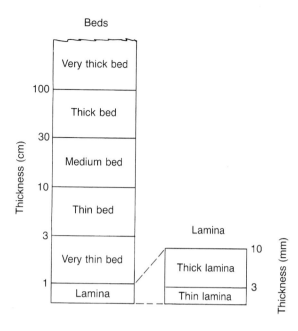

thicker than laminae that are separated by minor but distinct discontinuities in texture or composition. Note, however, that layer is also used in a much looser and more informal sense for any bed or stratum of rock. Marked discontinuities within beds are called **amalgamation surfaces. Divisions** are subunits that do not have distinct discontinuities but are characterized by a particular association of sedimentary structures. **Bands** and **lenses** are subdivisions of a bed based on color, composition, texture, or

FIGURE 5.2 Informal subdivisions of beds based on internal structures. (From Blatt, H., G. V. Middleton, and R. Murray, Origin of sedimentary rocks, 2nd ed., © 1980, Fig. 5.1, p. 130. Reprinted by permission of Prentice-Hall, Englewood Cliffs, N.J.)

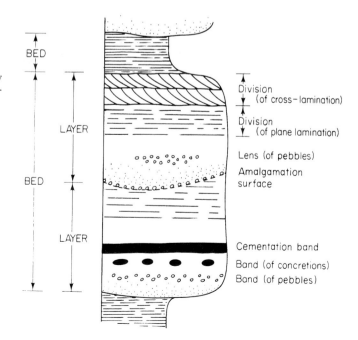

cementation. The term lens is also used less formally for any body of rock that is thick in the middle and thin at the edges.

Beds are separated by bedding planes or bedding surfaces, most of which represent planes of nondeposition, abrupt change in depositional conditions, or an erosion surface (Campbell, 1967). Some bedding surfaces may be postdepositional features formed by diagenetic processes or weathering. The gross geometry of a bed depends upon the relationship between bedding-plane surfaces, a relationship that may be either parallel or nonparallel. The bedding surfaces themselves may be even, wavy, or curved (Fig. 5.3). Depending upon the combination of these characteristics, beds can have a variety of geometric forms such as uniform-tabular, tabular-lenticular, curved-tabular, wedge-shaped, and irregular. Internal layers and laminae of a bed that are essentially parallel to the bedding planes constitute laminated bedding or **planar stratification.** Layers and laminae that make up the internal structure of some beds are deposited at an angle to the bounding surfaces of the bed and are therefore called **cross-strata** or **cross-laminae.** Beds composed of cross-laminated or cross-stratified units are called **cross-beds.**

Groups of similar beds or cross-beds are called **bedsets. A simple bedset** consists of two or more superimposed beds characterized by similar composition, texture, and internal structures. A bedset is bounded above and below by bedset (bedding) surfaces. A **composite bedset** refers to a group of beds differing in composition, texture, and internal structures but associated genetically, representing a common type of deposited sequence (Reineck and Singh, 1980). The teminology of bedsets is illustrated in Figure 5.4.

Beds are characterized by lateral continuity, and some beds can be traced for many kilometers. Others may terminate within a single outcrop. Beds terminate laterally by one of the following:

1. convergence and merging of upper and lower bounding surfaces (pinch-out)
2. lateral gradation of the composition of a bed into another bed of different composition so that the bounding bed surfaces die out
3. meeting a cross-cutting feature such as a channel, fault, or unconformity

FIGURE 5.3 Descriptive terms used for the configuration of bedding surfaces. (From Campbell, C. V., 1967, Lamina, laminaset, bed and bedset: Sedimentology, v. 8. Fig. 2, p. 18, reprinted by permission of Elsevier Science Publishers, Amsterdam.)

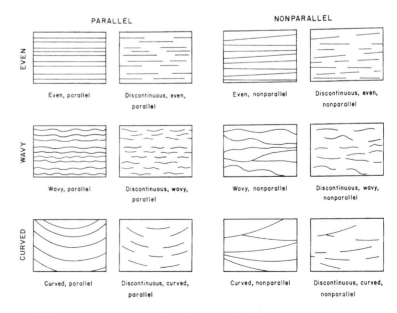

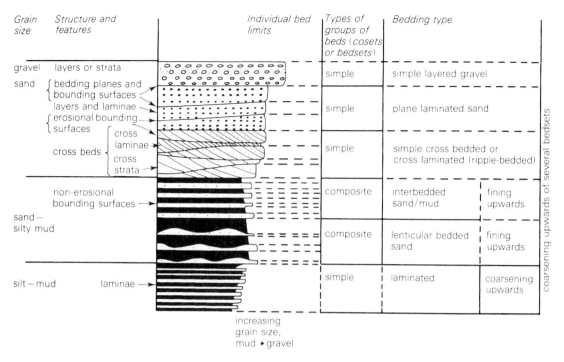

FIGURE 5.4 Diagram illustrating the terminology of bedsets. (From Collinson, J. D., and D. B. Thompson, 1982, Sedimentary structures: George Allen & Unwin, London, Fig. 2.2, p. 8.)

Origin of Bedding

Individual beds are produced under essentially constant physical, chemical, or biological conditions. Many beds must have been produced very rapidly by a single event such as a flood that lasted only a few hours or days. Even more rapid deposition lasting perhaps only seconds or minutes, such as deposition of sand laminae by grain flow down the slip face of a sand dune, occurs in some environments. On the other hand, suspension deposition of beds of very fine clay could take months or years. Thus, an individual bed may be produced rapidly by an event such as a single flood or a single sediment gravity flow, or more slowly by a single episode of sedimentation of fine sediment from suspension.

The true bedding planes or bounding surfaces between beds represent periods of nondeposition, erosion, or changes to completely different depositional conditions. Many beds are not preserved to become part of the geologic record but are destroyed by succeeding erosional episodes. The preservation potential for beds appears to be greater for those deposited by a depositional event of great magnitude, such as a very large flood, than for those formed by very small-scale events.

Origin of Lamination

Laminae are produced by less severe, or shorter-lived, fluctuations in sedimentation conditions than those that generate beds. They result from changing depositional conditions that cause variations in (1) grain size, (2) content of clay and organic material, (3) mineral composition, or (4) microfossil content of sediments. Laminae produced by alternating layers of finer- and coarser-grained sediment are probably the most common kind. The size of grains within individual laminae may be uniform or may show either normal or reverse vertical grain-size grading. Boundaries between laminae owing to grain-size differences can be either sharp or gradational. Changes in clay content of layers that otherwise have even-sized, coarser grains can also create laminae. Laminations may be produced by such differences in mineral composition as those that occur in alternating mica-enriched and mica-poor laminae; alternating heavy-mineral (black sand) laminae and light-mineral laminae, as in some beach deposits; and alternating laminae of anhydrite and dolomite in evaporite deposits. Alternations of detrital minerals and the tests or shells of pelagic organisms are also known to produce laminae. Color changes can accentuate the presence of laminae. Color changes may result from variations in the content of distinctively colored minerals, such as black, heavy minerals; the content of fine, dark-colored organic matter; and the oxidation state of iron in iron-bearing minerals. Reduced iron yields green colors; oxidized iron gives red or brown colors.

Parallel laminae (Fig. 5.5), as opposed to cross-laminae, are produced both by deposition from suspension and by traction currents. They form in a variety of sedimentary environments. Because the burrowing and feeding activities of organisms in many environments quickly destroy laminations, laminae have the greatest potential for preservation in reducing or toxic environments, where organic activity is minimal, or in environments where deposition is so rapid that the sediment is buried below the depth of active organic reworking before organisms can destroy stratification.

Deposition of Laminae by Suspension Mechanisms. Parallel laminae composed of clay or fine silt can be generated by deposition of sediment from suspension in a number of different environmental settings. The most important depositional mechanisms and settings include the following:

1. slow suspension settling in lakes, where levels of organic reworking are commonly low
2. sedimentation on some parts of deltas, where abundant fine sediment periodically supplied by distributaries leads to rapid deposition

FIGURE 5.5 Laminated fine-grained sandstone, Elkton Formation (Tertiary), southern Oregon coast. The knife is approximately 9 cm long.

3. deposition on tidal flats in response to fluctuations in energy levels and sediment supply during tidal cycles

4. deposition in subtidal shelf areas where thin sand layers that accumulate owing to storm activity may alternate with very thin mud laminae formed during periods of slower accumulation

5. slow sedimentation in deep-sea environments, where deposition takes place from nepheloid layers

6. chemical sedimentation in evaporite basins, such as deposition of laminated anhydrites

Deposition of Laminae by Traction Mechanisms. The formation of parallel laminae in sand-size sediment during traction transport has been attributed to a variety of mechanisms, most of which are based on deductive reasoning rather than on actual observation. **Swash and backwash on beaches** is one of the most common mechanisms responsible for formation of evenly laminated sands. This process leads to generation of laminae that may show reverse size grading and concentration of fine heavy minerals in the base of the laminae (Clifton, 1969). **Steady flow of currents** may also produce laminae under three different types of conditions:

1. during the plane-bed phase (Fig. 3.11) of upper-flow regime transport (Harms and Fahnestock, 1965; Allen, 1984)

2. under shallow-flow conditions in the lower-flow regime by migration of low-relief ripples in which lack of avalanche faces prevents cross-laminae from forming (McBride, Shepard, and Crawley, 1975)

3. at low velocities below the critical velocity of ripple formation, at least for coarser particles (Fig. 3.13C)

Laminae formed by process (1) are probably much more common (e.g., in fluvial sediments) than those formed by processes (2) and (3). Laminated sands may also develop owing to **wind transport** (McKee, Douglass, and Rittenhouse, 1971; Hunter, 1977). Hunter observed that parallel laminae formed by (1) traction transport and deposition at very high wind velocities; (2) grainfall deposition in zones of flow separation that occur leeward of dune crests; and (3) deposition accompanying the migration of wind ripples, an eolian analog of the subaqueous process described by McBride, Shepard, and Crawley (1975). **Phases of upper-flow-regime transport during turbidity current flow** that generate Bouma B subdivisions of turbidites (Fig. 3.28) are another mechanism by which laminated sands can form. Finally, **sheet flow** in shallow-marine environments (the equivalent of plane bed transport in the upper-flow regime; Clifton, 1976), and possibly migration of ripple forms accompanied by a very slow rate of deposition, can also produce laminations in sandy deposits (Newton, 1968).

Graded Bedding

Graded beds are sedimentation units characterized by distinct vertical gradations in grain size. They range in thickness from a few centimeters to a few meters or more. They are commonly devoid of internal laminations, although the upper part of graded turbidite sequences (Bouma B, C, D divisions, Fig. 3.28) may show parallel or wavy laminae. Beds that show gradation from coarser particles at the base to finer particles at the top have **normal grading** (Fig. 5.6A). Normally graded beds commonly occur in thin, repetitious sequences (rhythmic bedding), as illustrated in Figure 5.6B. More rarely, beds display **reverse grading**, with coarser particles at the top grading downward to finer particles. Graded beds commonly have sharp basal contacts.

FIGURE 5.6 A. Graded bedding in Miocene deep-sea sandstone (core) from Ocean Drilling Program Leg 127, Site 797, Japan Sea. Note the nearly complete Bouma sequence (units A through E) in this core. B. Rhythmically bedded, graded turbidites from the Tyee Formation (Eocene), northern Oregon Coast Range. (Photograph A courtesy of the Ocean Drilling Program, Texas A & M University.)

Normal graded bedding can form by several processes (Klein, 1965); however, the origin of most graded beds in the geologic record has been attributed to turbidity currents. Differences in the rate at which particles of different sizes settle from suspension during the waning stages of turbidity current flow appear to account for the grading, but the exact manner in which the grading process operates is not well understood. The graded materials may be mud, sand, or, more rarely, gravel. As discussed in Chapter 3, some graded turbidite units display an ideal sequence of sedimentary structures, called a Bouma sequence (Fig. 3.28), but more commonly the sequence is truncated at the top or bottom. The basal A division may be present, but some or all of the overlying divisions may be absent; or the A division itself may be missing. Graded beds occur also in shallower-water environments than those in which turbidites form. Suggested mechanisms of formation of shallow-water graded beds include sedimentation from suspension clouds generated by storm activity on the shelf, periodic silting of delta distributaries, deposition in the last phases of a heavy flood, settling of volcanic ash after an eruption, deposition by waning currents on intertidal flats, and mixing of an underlying coarser-sediment layer with an overlying mud layer owing to the bioturbation activities of burrowing and feeding organisms.

Reverse size grading is much less common than normal grading. It is known to occur in the individual laminae of beach sediments owing to segregation of fine-size heavy minerals and coarser-grained light minerals (Clifton, 1969); in some pyroclastic flows or volcanic base surge deposits; in some grain-flow deposits; and in laminae formed by the migration of wind ripples. It is also alleged to occur in some turbidite deposits deposited from high-concentration flows that decelerated rapidly. Reverse grading has been attributed to two types of mechanisms: (1) dispersive pressures and (2) kinetic sieving. Dispersive pressures (Chapter 3) are believed to be proportional to grain size: In a sediment of mixed grain size, the higher dispersive pressures acting on the larger particles tend to force them up into the zone of least shear. Alternatively, reverse grading may be explained by a kinetic sieve mechanism. In a mixture of grains undergoing agitation, the smaller grains presumably fall down through the larger grains as grain motion opens up spaces between the larger particles. Overall, reverse grading is a relatively rare phenomenon, and its origin is still poorly understood.

Massive (Structureless) Bedding

The term **massive bedding** is used to describe beds that appear to be homogeneous and lacking in internal structures (Fig. 5.7). Use of X-radiography techniques (Hamblin, 1965) or etching and staining methods often reveals that such beds are not truly massive but rather that they contain very faintly developed structures. Nonetheless, one occasionally finds beds, particularly thick sandstone beds, in which internal structures cannot be recognized even with the aid of X-ray or staining techniques. Such beds are rare, which is fortunate for us because they are very difficult to explain. Reported occurrences of massive beds include both graded bed units in turbidites, which may lack internal structures other than size grading, and certain thick, nongraded sandstones.

FIGURE 5.7 Massive-bedded sandstone (upper part of photograph) lying above thin, parallel-bedded siltstone and shale, Fluornoy Formation (Eocene), southwestern Oregon.

Some massive bedding may be a secondary feature produced by extensive bioturbation by organisms, although bioturbation commonly produces recognizable mottled structures. Liquefaction of sediment by sudden shocking or other mechanisms shortly after deposition has also been suggested as a means of destroying original stratification. Otherwise, it is assumed that lack of stratification is a primary feature that occurs in the absence of traction transport and results from very rapid deposition from suspension or deposition from very highly concentrated sediment dispersions during sediment gravity flows. Presumably, the sediment is dumped very rapidly without subsequent reworking to form a more or less homogeneous mass.

Ripples and Cross-Bedding

Ripples (Fig. 5.8) are common sedimentary structures in modern environments, where they occur in both siliciclastic and carbonate sediments. They can form by both water and wind transport. The flow conditions that generate ripples and larger bedforms (dunes) are discussed in Chapter 3 and illustrated in Figure 3.11. Note from Figure 3.11 that ripples can form in shallow water (<1 m depth) at flow velocities ranging from ~0.2 to 1.0 m/s. Note also that the flow conditions that cause ripple formation in very fine sand (in the lower-flow regime) change abruptly to those that generate plane beds in the upper-flow regime. Thus, during waning current flow, a phase of plane-bed transport of very fine sand may be succeeded by a phase of ripple formation as current velocity drops, creating a deposit in which plane-bed laminae may be overlain by ripples. In sediment coarser than about 1.0 mm, dunes form instead of ripples.

Ripples can develop in granular material under either unidirectional current flow or oscillatory flow (wave action), as discussed in Chapter 3. Figure 3.15 shows various plan-view shapes of current ripples, and Figure 3.22 illustrates some differences in crest shapes of current and oscillation ripples. Ripples are most common in shallow-water environments; however, they have been photographed on the floor of the modern ocean at depths of a few thousand meters. Ripples have relatively low preservation potential because they tend to be eroded and destroyed by current erosion before burial; therefore, ancient ripples such as those shown in Figure 5.8 are not abundant in the sedimentary record. Dunes are even less commonly preserved; nonetheless, ancient dunes do occur (Fig. 5.9).

Because the shape of ripples is related to the direction of current flow (the steep side of current ripples faces downcurrent), ripples in ancient sediments provide extremely useful information about paleocurrent directions and paleoflow conditions of the depositional environment. By determining the paleoflow direction from ancient ripples exposed on several outcrops within a region, geologists can reconstruct the flow pattern of an ancient stream or streams (Section 5.6). Paleocurrent information allows the directions of sediment transport to be determined and the location of the sediment source area to be estimated.

Ripples provide information about sedimentary processes and paleocurrent directions, but they are not unique indicators of depositional environment. Inasmuch as they can be formed under unidirectional currents (in both shallow and deep water), by wave action, and by wind transport, considerable care must be exercised in interpreting depositional environments on the basis of ripples.

Cross-bedding (Figs. 5.10–5.12) forms primarily by migration of ripples and dunes (in water or air). Ripple or dune migration leads to formation of dipping foreset laminae owing to avalanching or suspension settling in the zone of separation on the lee sides of these bedforms, as described in Chapter 3 (Fig. 3.9). If most of the sediment is too coarse to be transported in suspension, avalanching of the bedload sediment

FIGURE 5.8 A. Plan view of oscillation (wave-formed) ripples on the upper surface of a fine-grained sandstone bed, Elkton Siltstone (Eocene), southwestern Oregon. The knife is approximately 9 cm long. B. Cross-sectional view of nearly symmetrical ripples in Holocene sand, deposited in a side canyon of the Colorado River (Arizona) as a result of waves from the river washing into standing water at the mouth of the canyon. (Photograph B courtesy of John E. Bircher.)

A

B

down the lee side of the ripple will cause formation of laminae that are steep and straight. These inclined **foreset** laminae make contact with the nearly horizontal, thin **bottomset** laminae (deposited from suspension) at a distinct angle, which is approximately the same as the angle of repose. Roughly the same effect is achieved if the height of the lee slope is large compared to total flow depth, so that the suspended load falls mainly on the lee slope. If the suspended load is large, or if the height of the lee slope is small compared to flow depth, suspended sediment will pile up at the base of the lee slope rapidly enough to keep pace with growth of the avalanche deposits. This process causes the lower part of the foreset laminae to curve outward and

FIGURE 5.9   Large dunes on the surface of a sandstone bed, Tyee Formation (Tertiary), exposed along the Umpqua River, southern Oregon Coast Range. The dunes are about 15 cm high and 70 cm from crest to crest. (Photograph courtesy of Ewart Baldwin.)

approach the bottomset laminae asymptotically (Blatt, Middleton, and Murray, 1980). Thus, the cross-laminae are said to be tangential rather than angular.

The preservation potential of cross-laminae is much higher than that of the bedforms themselves (because the tops of bedforms tend to be planed off by subsequent current or wind erosion); therefore, cross-bedding is a very common type of sedimentary structure in ancient sedimentary rocks. Cross-stratification can be formed also by filling of scour pits and channels, by deposition on the point bars of meandering streams, and by deposition on the inclined surface of beaches and marine bars. Cross-bedding formed under different environmental conditions can be very similar in appearance, and it is often difficult in field studies of ancient sedimentary rocks to differentiate cross-bedding formed in fluvial, eolian, and marine environments.

Cross-beds commonly occur in sets (Fig. 5.4). Cross-bedding in sets less than about 5 cm thick is called small-scale cross-bedding; that in sets thicker than 5 cm is large-scale cross-bedding. Because of their diverse origin, many types of cross-beds occur. Allen (1963) proposed a very elaborate classification of cross-bedding based upon such properties as grouping of cross-bed sets, scale, nature of the bounding surface of the beds, angular relation of cross-strata in a set or coset to the bounding surfaces, and degree of grain-size uniformity in different laminae. The much simpler scheme of McKee and Weir (1953), as modified by Potter and Pettijohn (1977), is adopted herein. Cross-beds are divided into two principal types on the basis of overall geometry and the nature of the bounding surfaces of the cross-bedded units (Fig. 5.10).

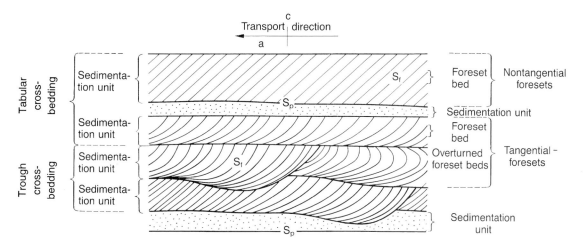

FIGURE 5.10 The terminology and defining characteristics of cross-bedding. Symbols: a, direction parallel to the average sediment transport direction; c, direction perpendicular to (a) and the transport plane (bed) in which (a) lies; S_p, the principal bedding surface or bedding plane; S_f, the foreset surface of cross-bedding. (After Potter, P. E., and F. J. Pettijohn, 1977, Paleocurrents and basin analysis, 2nd ed. Fig. 4.1, p. 91, reprinted by permission of Springer-Verlag, Heidelberg.)

FIGURE 5.11 Tabular cross-bedding in pebbly sands of the Coquille Formation (Pleistocene), southern Oregon coast. Note the opposing dip directions in the two cross-bedded units, suggesting possible deposition by reversing tidal currents (current direction from right to left in the lower unit and from left to right in the upper unit).

FIGURE 5.12 Trough cross-bedding in pebbly sands of the Coquille Formation (Pleistocene), southern Oregon coast. Note the erosion of basal, parallel-bedded sands to produce the first trough cross-bedded unit (on the right), which was truncated in turn by current scour to form the second trough cross-bedded unit.

Tabular cross-bedding consists of cross-bedded units that are broad in lateral dimensions with respect to set thickness and that have essentially planar bounding surfaces (Fig. 5.11). The laminae of tabular cross-beds are also commonly planar, but curved laminae that have a tangential relationship to the basal surface, as explained above, also occur. **Trough cross-bedding** consists of cross-bedded units whose bounding surfaces are curved (Fig. 5.12). The units are trough-shaped sets consisting of an elongate scour filled with curved laminae that commonly have a tangential relationship to the base of the set.

Tabular cross-bedding is formed mainly by the migration of large-scale ripples and dunes (Fig. 5.13); thus, it forms during lower-flow-regime conditions. Individual beds range in thickness from a few tens of centimeters to a meter or more, but bed thicknesses up to 10 m have been observed (Harms et al., 1975). Trough cross-bedding can originate both by migration of small current ripples, which produces small-scale cross-bed sets, and by migration of large-scale ripples (Fig. 5.14). Trough cross-bedding formed by migration of large-scale ripples commonly ranges in thickness up to a few tens of centimeters and in width from less than 1 m to more than 4 m.

Cross-bedding is one of the most useful sedimentary structures for determining paleocurrent direction. Because the foreset laminae in cross-beds are generated by avalanching on the downcurrent (lee) side of ripples, as mentioned, the foresets dip in the downcurrent direction. To measure paleocurrent direction from cross-beds requires that they be exposed in a three-dimensional outcrop. The strike of the foreset laminae is determined first; the dip direction is 90° to the strike. If cross-beds have been tilted by tectonic uplift after deposition, a correction must be made for this tilt (Collinson and Thompson, 1989).

Ripple Cross-Lamination

Ripple cross-lamination (climbing ripples) forms when deposition takes place very rapidly during migration of current or wave ripples (McKee, 1965; Jopling and Walker,

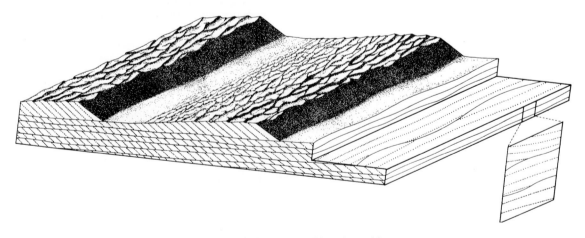

FIGURE 5.13   Diagram illustrating large-scale tabular cross-bedding formed by migrating straight-crested dunes (with rippled surfaces). Flow is from left to right. (From Harms, J. C., J. B. Southard, and R. G. Walker, 1982, Structures and sequences in clastic rocks: Soc. Econ. Paleontologists and Mineralogists Short Course No. 9. Fig. 3–11, p. 3–21, reprinted by permission of SEPM, Tulsa, Okla.)

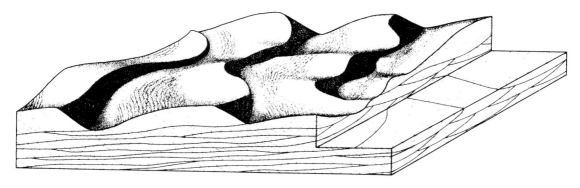

FIGURE 5.14 Diagram illustrating large-scale trough cross-bedding formed by migrating, trough-shaped dunes. Flow is from left to right. (From Harms, J. C., J. B. Southard, and R. G. Walker, 1982, Structures and sequences in clastic rocks: Soc. Econ. Paleontologists and Mineralogists Short Course No. 9. Fig. 3–10, p. 3–19, reprinted by permission of SEPM, Tulsa, Okla.)

1968). A series of cross-laminae are produced by superimposing migrating ripples (Fig. 5.15). The ripples climb one on another such that the crests of vertically succeeding laminae are out of phase and appear to be advancing upslope. This process results in cross-bedded units that have the general appearance of waves (Fig. 5.16) in outcrop sections cut normal to the wave crests. In sections with other orientations, the laminae may appear horizontal or trough-shaped, depending upon the orientation and the shape of the ripples.

The formation of ripple cross-lamination appears to require an abundance of sediment, especially sediment in suspension, which quickly buries and preserves original rippled layers. Abundant suspended sediment supply must be combined with just enough traction transport to produce rippling of the bed, but not enough to cause complete erosion of laminae from the stoss side of ripples. Some ripple laminae may be in phase (one ripple crest lies directly above the other), indicating that the ripples did not migrate. In-phase ripple laminae form under conditions where a balance is achieved

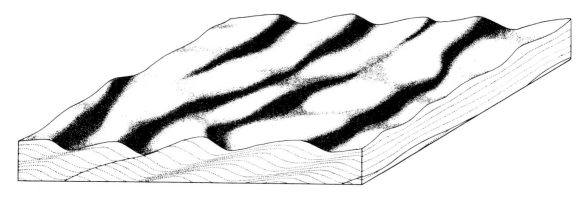

FIGURE 5.15 Diagram showing ripple cross-lamination produced by small current ripples climbing at a large angle. (From Harms, J. C., J. B. Southard, and R. G. Walker, 1982, Structures and sequences in clastic rocks: Soc. Econ. Paleontologists and Mineralogists Short Course No. 9. Fig. 3–7, p. 3–15, reprinted by permission of SEPM, Tulsa, Okla.)

FIGURE 5.16   Ripple cross-lamination (below ballpoint pen) in flood deposits of the Illinois River, southwestern Oregon. Parallel laminae at the bottom of the photograph developed during a plane-bed phase of upper-flow-regime conditions; as current velocity diminished into the lower-flow regime, ripple cross-lamination formed on top of the laminae. A later flood pulse deposited upper-flow-regime plane beds on top of the ripple cross-lamination.

between traction transport and sediment supply so that the ripples do not migrate despite a growing sediment surface. Ripple cross-lamination occurs in sediments deposited in environments characterized by rapid sedimentation from suspension—fluvial floodplains, point bars, river deltas subject to periodic flooding, and environments of turbidite sedimentation. Figure 5.16 shows the sequence of bedforms developed in a river during waning flood stage. Laminae at the bottom of Figure 5.16 developed during a plane-bed phase of upper-flow-regime transport at high flood velocity. As velocity waned into the lower-flow regime, ripple cross-lamination formed on top of the plane-bed laminae.

Flaser and Lenticular Bedding

Flaser bedding is a type of ripple bedding in which thin streaks of mud occur between sets of cross-laminated sandy or silty sediment (Fig. 5.17). Mud is concentrated mainly in the ripple troughs but may also partly cover the crests. Flaser bedding suggests deposition under fluctuating hydraulic conditions. Periods of current activity, when traction transport and deposition of rippled sand take place, alternate with periods of quiescence, when mud is deposited. Repeated episodes of current activity result in erosion of previously deposited ripple crests, allowing new rippled sand to bury and preserve rippled beds with mud flasers in the troughs (Reineck and Singh, 1980). **Lenticular bedding** is a structure formed by interbedded mud and ripple cross-laminated sand in which the ripples or sand lenses are discontinuous and isolated in both a vertical and a horizontal direction (Fig. 5.18). Reineck and Singh (1980) suggest that flaser bedding is produced in environments in which conditions for deposition and preservation of sand are more favorable than for mud, but that lenticular bedding is produced in environments in which conditions favor deposition and preservation of mud over sand. Flaser and lenticular bedding appear to form particularly on tidal flats

FIGURE 5.17 Flaser bedding in the Elkton Siltstone (Eocene), near Cape Arago, southern Oregon coast.

FIGURE 5.18 Lenticular bedding, Elkton Siltstone (Eocene), near Cape Arago, southern Oregon coast. Lenses of light-colored fine sand-stone are interbedded with dark mudstone.

and in subtidal environments where conditions of current flow or wave action that cause sand deposition alternate with slack-water conditions when mud is deposited. They also form in marine delta-front environments, where fluctuations in sediment supply and current velocity are common; in lake environments in front of small deltas; and possibly on the shallow-marine shelf owing to storm-related transport of sand into deeper water.

Hummocky Cross-Stratification

The name for **hummocky cross-stratification** was introduced by Harms et al. in 1975, although the structures had been recognized and described under different names by earlier workers. Hummocky cross-stratification is characterized by undulating sets of cross-laminae that are both concave-up (swales) and convex-up (hummocks)

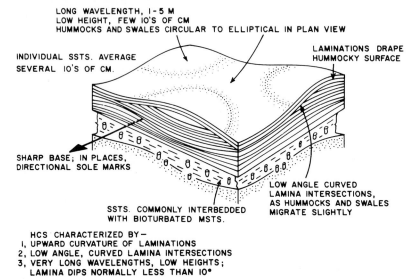

HUMMOCKY CROSS STRATIFICATION – HCS

LONG WAVELENGTH, 1 - 5 M
LOW HEIGHT, FEW 10'S OF CM
HUMMOCKS AND SWALES CIRCULAR TO ELLIPTICAL IN PLAN VIEW

LAMINATIONS DRAPE
HUMMOCKY SURFACE

INDIVIDUAL SSTS. AVERAGE
SEVERAL 10'S OF CM.

SHARP BASE; IN PLACES,
DIRECTIONAL SOLE MARKS

LOW ANGLE CURVED
LAMINA INTERSECTIONS,
AS HUMMOCKS AND SWALES
MIGRATE SLIGHTLY

SSTS. COMMONLY INTERBEDDED
WITH BIOTURBATED MSTS.

HCS CHARACTERIZED BY –
1, UPWARD CURVATURE OF LAMINATIONS
2, LOW ANGLE, CURVED LAMINA INTERSECTIONS
3, VERY LONG WAVELENGTHS, LOW HEIGHTS;
   LAMINA DIPS NORMALLY LESS THAN 10°

FIGURE 5.19 Schematic diagram of hummocky cross-stratification, which typically occurs interbedded with bioturbated mudstone. (From Walker, R. G., 1984, Shelf and shallow marine sands, *in* R. G. Walker (ed.), Facies models, 2nd ed.: Geoscience Canada Reprint Ser. 1. Fig. 11, p. 149, reprinted by permission of Geological Association of Canada. Originally after R. G. Walker, 1982, Hummocky and swaley cross stratification, *in* R. G. Walker (ed.), Clastic units of the Front Range between Field, B. C., and Drumheller, Alberta: Internat. Assoc. Sedimentologists, 11th Internat. Congress on Sedimentology (Hamilton, Canada), Guidebook to Excursion 21A, p. 22–30.)

(Fig. 5.19). The cross-beds sets cut gently into each other with curved erosion surfaces (Fig. 5.20). Hummocky cross-bedding commonly occurs in sets 15 to 50 cm thick with wavy erosional bases and rippled, bioturbated tops (Harms et al., 1975). Spacing of hummocks and swales is from 50 cm to several meters. The lower bounding surface of a hummocky unit is sharp and is commonly an erosional surface. Current-formed sole marks may be present on the base. Hummocky cross-stratification occurs most typically in fine sandstone to coarse siltstone that commonly contains abundant mica and fine carbonaceous plant debris (Dott and Bourgeois, 1982).

Hummocky cross-stratification has not yet been produced in flumes or reported from modern environments, but it has been reported in ancient strata from numerous localities. Harms et al. (1975, 1982) suggest that this structure is formed by strong surges of varied direction (oscillatory flow) that are generated by relatively large storm waves. Strong storm-wave action first erodes the seabed into low hummocks and swales that lack any significant orientation. This topography is then mantled by laminae of material swept over the hummocks and swales. More recently, Duke, Arnott, and Cheel (1991) suggest that hummocky cross-stratification originates by a combination of unidirectional and oscillatory flow related to storm activity. See also Figure 12.10 and discussion of hummocky cross-stratification in Chapter 12. Although hummocky cross-stratification is commonly confined to shallow-marine sedimentary rocks, Duke (1985) reports the occurrence of this structure in some lacustrine sedimentary rocks.

FIGURE 5.20 Hummocky cross-stratification in fine-grained sandstone of the Lower Member of the Coaledo Formation (Eocene), near Sunset Bay, southern Oregon coast. Arrows point to the hummocky erosional surface.

Irregular Stratification

Deformation Structures

Convolute Bedding and Lamination. Convolute bedding is a structure formed by complex folding or intricate crumpling of beds or laminations into irregular, generally small-scale anticlines and synclines. It is commonly, but not necessarily, confined to a single sedimentation unit or bed, and the strata above and below this bed may show little evidence of deformation (Fig. 5.21). Convolute bedding is most common in fine sands or silty sands, and the laminae can typically be traced through the folds. Faulting generally does not occur, but the convolutions may be truncated by erosional surfaces which may also be convoluted. The convolutions increase in complexity and

FIGURE 5.21 Convolute lamination in fine-grained sandstone, Coaledo Formation (Eocene), southern Oregon Coast. Note truncation of convolute laminae by parallel-laminated sandstone above.

amplitude upward from undisturbed laminae in the lower part of the unit. They may either die out in the top part of the unit or be truncated by the upper bedding surface. Beds containing convolute lamination commonly range in thickness from about 3 to 25 cm (Potter and Pettijohn, 1977), but convoluted units up to several meters thick have been reported in both eolian and subaqueous deposits.

Convolute lamination is most common in turbidite sequences. It also occurs in intertidal-flat sediments, river floodplain sediments, and point-bar deposits. The origin of convolute bedding is still not thoroughly understood, but it appears to be caused by plastic deformation of partially liquefied sediment soon after deposition. The axes of some convoluted folds have a preferred orientation which commonly coincides with the paleocurrent direction, suggesting that the process that produces convolutions occurs during deposition, at least in these cases. Liquefaction of sediment can be caused by such processes as differential overloading, earthquake shocks, and breaking waves.

Flame Structures. Flame structures are wavy or flame-shaped tongues of mud that project upward into an overlying layer, which is commonly sandstone (Fig. 5.22). The crests of some flames are bent over or overturned and tend to all point in the same direction. Flame structures are commonly associated with other structures caused by sediment loading. They are probably caused mainly by loading of water-saturated mud layers which are less dense than overlying sands and are consequently squeezed upward into the sand layers. The orientation of overturned crests suggests that loading may be accompanied by some horizontal drag or movememt between the mud and sand bed.

Ball and Pillow Structures. Ball and pillow structures are found in the lower part of sandstone beds, and less commonly in limestone beds, that overlie shales (Fig. 5.23). They consist of hemispherical or kidney-shaped masses of sandstone or limestone that show internal laminations. In some hemispheres, the laminae may be gently curved or deformed, particularly next to the outside edge of the hemispheres where they tend to conform to the shape of the edge. The balls and pillows may remain connected to the

FIGURE 5.22 Flame structures in the base of a fine-grained sandstone unit overlying laminated mudstones and siltstones, Coaledo Formation (Eocene), near Sunset Bay, southern Oregon coast. Some flames are overturned toward the left, suggesting slight downslope movement of the sand during loading. (Photograph by E. M. Baldwin.)

FIGURE 5.23 Ball and pillow structures (arrows) on the base of a thin, steeply dipping sandstone bed, Lookingglass Formation (Eocene), near Illahe, southwest Oregon.

overlying bed, or they may be completely isolated from the bed and enclosed in the underlying mud. Ball and pillow structures are believed to form as a result of foundering and breakup of seimconsolidated sand, or limey sediment, owing to partial liquefaction of underlying mud, possibly caused by shocking. Liquefaction of the mud causes the overlying sand beds or limey sediment to deform into hemispherical masses which may subsequently break apart from the bed and sink into the mud. Kuenen (1958) experimentally produced structures that closely resemble natural ball and pillow structures by applying a shock to a layer of sand deposited over a thixotropic clay.

Synsedimentary Folds and Faults. The general term **slump structures** has been applied to structures produced by penecontemporaneous deformation resulting from movement and displacement of unconsolidated or semiconsolidated sediment, mainly under the influence of gravity. Potter and Pettijohn (1977) describe slump structures as being the products of either of the following:

1. pervasive movement involving the interior of the transported mass, producing a chaotic mixture of different types of sediments, such as broken mud layers embedded in sandy sediment
2. a décollement type of movement in which the lateral displacement is concentrated along a sole, producing beds that are tightly folded and piled into nappelike structures (Fig. 5.24)

Slump structures may involve many sedimentation units, and they are commonly faulted. Thicknesses of slump units have been reported to range from less than 1 m to more than 50 m. Slump units may be bounded above and below by strata that show no evidence of deformation. It may be difficult in some stratigraphic sequences, however, to differentiate between slump units and incompetent beds such as shale that were deformed between competent sandstone or limestone beds during tectonic folding.

FIGURE 5.24 Décollement-type synsedimentary folds (arrows) in thin, fine-grained sandstone layers interbedded with mudstone, Elkton Siltstone (Eocene), near Cape Arago, southern Oregon coast.

Slump structures typically occur in mudstones and sandy shales, and less commonly in sandstones, limestones, and evaporites. They are generally found in units that were deposited rapidly, and they have been reported from a variety of environments where rapid sedimentation and oversteepened slopes lead to instability. They occur in glacial sediments, varved silts and clays of lacustrine origin, eolian dune sands, turbidites, delta and reef-front sediments, and subaqueous dune sediments, and in sediments from the heads of submarine canyons, continental shelves, and the walls of deep-sea trenches.

Dish and Pillar Structures. Dish structures are thin, dark-colored, subhorizontal, flat to concave-upward, clayey laminations (Fig. 5.25) that occur principally in sandstone and siltstone units (Lowe and LoPiccolo, 1974; Rautman and Dott, 1977). The laminations are commonly only a few millimeters thick, but individual dishes may range from 1 cm to more than 50 cm in width. They typically occur in thick beds where dish and pillar structures may be the only structures visible. They also occur in beds less than about 0.5 m thick, where they commonly cut across primary flat laminations and other laminations. **Pillar structures** generally occur in association with dish structures (Fig. 5.25). Pillars are vertical to near-vertical, cross-cutting columns and sheets of structureless or swirled sand that cut through either massive or laminated sands that also commonly contain dish structures and convolute laminations. They range in size from tubes a few millimeters in diameter to large structures greater than 1 m in diameter and several meters in length. Pillars are not actually stratification structures. They are discussed here with dish structures because of their close association with these structures and because they form by a similar mechanism, discussed below.

Dish and pillar structures were first observed in sediment gravity-flow deposits (turbidites and liquefied flows) and are most abundant in such deposits; however, they have now also been reported in sediments from deltaic, alluvial, lacustrine, and shallow-marine deposits, as well as from volcanic ash layers. They are indicative of rapid deposition and form by escape of water during consolidation of sediment. During gradual compaction and dewatering, semipermeable laminations act as partial barriers to upward-moving water carrying fine sediment. The fine particles are retarded by the laminations and are added to them, forming the dishes. Some of the water is forced horizontally beneath the laminations until it finds an easier escape route upward. This

FIGURE 5.25 Strongly curved to nearly flat dish structures (large arrow) and pillars (small arrow) formed by dewatering of siliciclastic sediment, Jackfork Group (Pennsylvanian), southeast Oklahoma. (From Lowe, D. R., 1975, Water escape structures in coarse-grained sediment: Sedimentology, v. 22. Fig. 8, p. 175, reprinted by permission of Elsevier Science Publishers, Amsterdam.)

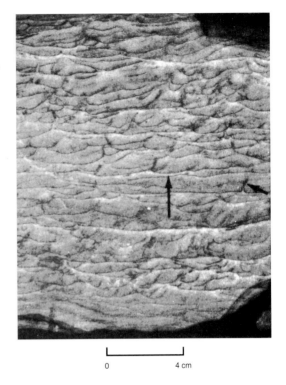

0   4 cm

forceful upward escape of water forms the pillars. Therefore, both dish structures and pillars are dewatering structures.

Erosion Structures

Channels are structures that show a U-shape or V-shape in cross section and cut across earlier formed bedding and laminations (Fig. 5.26). They are formed by erosion, principally by currents but in some cases by mass movements. Channels are commonly filled with sediment that is texturally different from the beds they truncate. Channels visible in outcrop range in width and depth from a few centimeters to many meters. Even larger channels may be definable by mapping or drilling. It is seldom possible to trace their length in outcrop, but they can presumably extend for distances many times their width. They are very common in fluvial and tidal sediments. They also occur in turbidite sediments, where the long dimensions of the channels tend to be oriented parallel to current direction as shown by other directional structures.

Scour-and-fill structures are similar to channels but are commonly smaller. They consist of small, filled, asymmetrical troughs a few centimeters to a few meters in size, with long axes that point downcurrent and that commonly have a steep upcurrent slope and a more gentle downcurrent slope. They may be filled with either coarser-grained or finer-grained material than the substrate. These structures are most common in sandy sediment and are thought to form as a result of scour by currents and subsequent backfilling as current velocity decreases. In contrast to channels, several scour-and-fill structures may occur together closely spaced in a row. They are primarily structures of fluvial origin that can occur in river, alluvial-fan, or glacial outwash-plain environments.

FIGURE 5.26 Large channel (upper left) in gravelly, cross-bedded sands of the Coquille Formation (Pleistocene) near Bandon, southern Oregon coast. The channel is filled at the base with gravels, overlain by cross-bedded sand.

5.4 BEDDING-PLANE MARKINGS

Markings Generated by Erosion and Deposition

Many bedding-plane makings occur on the underside of beds as positive-relief casts and irregular markings. Owing to their location on the bases or soles of beds, they are often referred to as **sole markings.** Sole markings are preserved particularly well on the undersides of sandstones and other coarser-grained sedimentary rocks that overlie mudstone or shale beds. Many sole markings are formed by erosional processes; consequently, they commonly show directional features that make them very useful for interpreting the flow directions of ancient currents.

These so-called erosional sole markings are actually formed by a two-stage process that involves both erosion and deposition. First, a cohesive, fine-sediment bottom is eroded by some mechanism to produce grooves or depressions. Because of the cohesiveness of the sediment, the depressions may be preserved long enough to be filled in and buried during subsequent deposition, typically by coarser-grained sediment than the bottom mud. This coarser sediment is probably deposited very shortly after erosion of the depression, possibly in some cases by the same current that formed the depression. After burial and lithification, a positive relief feature is left attached to the base of the overlying bed. If the beds subsequently undergo tectonic uplift, these structures may be exposed by weathering and subaerial erosion (Fig. 5.27). The initial erosional event that creates the depressions in a mud bottom can take the form of current scour, or the depression can result from the action of objects called **tools** that are carried by the current and intermittently or continuously make contact with the bottom. These tools can be pieces of wood, the shells of organisms, or any similar object that can be rolled or dragged along the bottom. Erosional structures may thus be classified genetically as either **current-formed structures** or **tool-formed structures.**

Erosional sole markings are most common on the soles of turbidite sandstones, but they are also present in sedimentary rocks deposited in other environments. They can form in any environments where the requisite conditions of an erosive event followed reasonably quickly by a depositional event are met. They have been reported in both fluvial and shelf deposits in addition to turbidites.

FIGURE 5.27 Postulated stages of development of sole markings owing to erosion of a mud bottom followed by deposition of coarser sediment. The diagram also illustrates how the sole markings appear as positive-relief features on the base of the infilling bed after tectonic uplift and subaerial weathering; it suggests how sole markings can be used to tell top and bottom of overturned beds. (From Collinson, J. D., and D. B. Thompson, 1982, Sedimentary structures. Fig. 4.1, p. 37, reprinted by permission of George Allen & Unwin, London. After Ricci-Lucchi, F., 1970, Sedimentografia: Zanchelli, Bologna, Italy.)

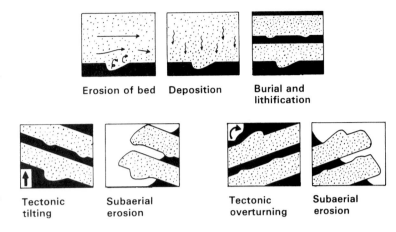

Erosion of bed   Deposition   Burial and lithification

Tectonic tilting   Subaerial erosion   Tectonic overturning   Subaerial erosion

Groove Casts. Groove casts are elongate, nearly straight ridges (Fig. 5.28) that result from infilling of erosional relief produced as a result of a pebble, shell, piece of wood, or other object being dragged or rolled across the surface of cohesive sediment. They typically range in width from a few millimeters to tens of centimeters and have a relief of a few millimeters to a centimeter or two; however, much larger groove casts also occur. Groove casts are greatly elongated in comparison to their width. They are directional features that are oriented parallel to the flow direction of the ancient currents that produced them; thus, they have paleocurrent significance. Groove casts on the same bed commonly have the same general orientation, although they may diverge at slight angles and even cross. Most groove casts do not have features that show the unique flow direction; we cannot tell from them which direction was downcurrent and

FIGURE 5.28 Large intersecting groove casts on the base of a turbidite sandstone bed, Fluornoy Formation (Eocene), southwestern Oregon.

which upcurrent. **Chevrons** are a variety of groove casts made up of continuous V-shaped crenulations with the V pointing in a downstream direction: thus, this type of groove cast can be used to determine the true direction of flow. Dzulynski and Walton (1965) suggest that chevrons are formed by tools moving just above the sediment surface, but not touching the surface, causing rucking-up of the sides of the groove. Groove casts are especially common on the soles of turbidite beds owing to shell fragments, pieces of wood, or other tools that are carried in the base of turbidity current flows being dragged across a mud bottom. They occur also on the soles of beds deposited in shallow-water environments such as tidal flats and floodplains where floating tools may touch bottom and leave grooves.

Bounce, Brush, Prod, Roll, and Skip Marks. Small gouge marks are produced by tools that make intermittent contact with the bottom, creating small gouge marks. **Brush** and **prod marks** are asymmetrical in cross-sectional shape, with the deeper, broad part of the mark oriented downcurrent. **Bounce marks** are roughly symmetrical. **Roll** and **skip marks** are formed by a tool bouncing up and down or rolling over the surface to produce a continuous track. The genesis of these structures is illustrated in Figure 5.29.

Flute Casts. Flute casts are elongated welts or ridges that have a bulbous nose at one end that flares out in the other direction and merges gradually with the surface of the bed (Fig. 5.30). They occur singly or in swarms in which all of the flutes are oriented in the same general direction. On a given sole, the flutes tend to be about the same size; however, flute casts on different beds can range in width from a centimeter or two to 20 cm or more, in height (relief) from a few centimeters to 10 cm or more, and in length from a few centimeters to a meter or more. The plan-view shape of flutes varies from nearly streamline, bilaterally symmetrical forms to more elongate and irregular forms, some of which are highly twisted.

Flute casts are formed by filling of a depression scoured in cohesive sediment by current eddies created behind some obstacle, or by chance eddy scour. This type of current scour produces asymmetrical depressions in which the steepest and deepest part of the depression is oriented upstream (Fig. 5.27). Therefore, when such depressions are filled, the filling forms a positive-relief structure with a bulbous nose oriented upstream (Fig. 5.30). Flute casts thus make excellent paleocurrent indicators because they show the unique direction of current flow. Flutes are particularly prevalent on the soles of turbidite sequences, but they are also present in sediments deposited in shallow marine and nonmarine environments. They have been reported on the soles of limestone beds as well as sandstone beds.

Current Crescents. Current crescents, also called obstacle scours, occur in modern environments as narrow, semicircular or horseshoe-shaped troughs that form around a small obstacle such as a pebble or shell owing to current scour on a mud or sand bottom (Fig. 5.31). In sandy sediment, they form on the downcurrent side of an obstacle as small ridges. In ancient sedimentary rocks, they commonly occur as casts on the underside of sandstone beds. Although they are very common in modern beach environments, in ancient sedimentary rocks they are most characteristic of fluvial sandstones with shale interbeds. They have also been reported from turbidite sequences. Somewhat similar structures are formed by sand blown around obstacles by wind, forming a ridge or tail of sand downwind from the object. Such wind-produced structures are rarely preserved in ancient sedimentary rocks.

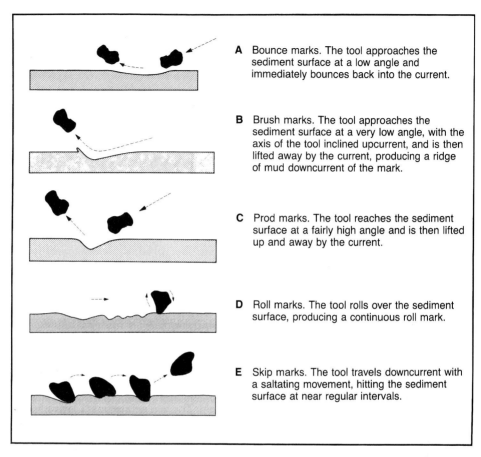

A Bounce marks. The tool approaches the sediment surface at a low angle and immediately bounces back into the current.

B Brush marks. The tool approaches the sediment surface at a very low angle, with the axis of the tool inclined upcurrent, and is then lifted away by the current, producing a ridge of mud downcurrent of the mark.

C Prod marks. The tool reaches the sediment surface at a fairly high angle and is then lifted up and away by the current.

D Roll marks. The tool rolls over the sediment surface, producing a continuous roll mark.

E Skip marks. The tool travels downcurrent with a saltating movement, hitting the sediment surface at near regular intervals.

FIGURE 5.29 Development in a cohesive mud bottom of (A) bounce marks, (B) brush marks, (C) prod marks, (D) roll marks, and (E) skip marks by the action of tools making contact with the bottom in various ways. These tool-formed depressions are subsequently filled with coarser sediment to produce positive-relief casts. (After Reineck H. E., and I. B. Singh, 1980, Depositional sedimentary environments, 2nd ed. Figs. 127, 129, 125, 132, 131, p. 82, 83, reprinted by permission of Springer-Verlag, Heidelberg.)

Markings Generated by Deformation: Load Casts

Load casts are described by Potter and Pettijohn (1977) as "swellings ranging from slight bulges, deep or shallow rounded sacks, knobby excrescences, or highly irregular protuberences." They commonly occur on the soles of sandstone beds that overlie mudstones or shales, and they tend to cover the entire bedding surface (Fig. 5.32). They range in diameter and relief from a few centimeters to a few tens of centimeters. Load casts may superficially resemble flute casts; however, they can be distinguished from flutes by their greater irregularity in shape and their lack of definite upcurrent and downcurrent ends. Also, load casts do not display a preferred orientation with respect to current direction.

Although they are called casts, load casts are not true casts because they are not fillings of a preexisting cavity or mold. They are formed by deformation of uncom-

FIGURE 5.30 Flute casts on the base of a turbidite unit, Tyee Formation (Eocene), near Valsetz, northwestern Oregon. The bulbous terminations of the flute casts indicate that paleocurrents moved from the bottom toward the top. (Photograph by P. D. Snavely, Jr.)

FIGURE 5.31 Current crescent formed around a pebble on a modern beach, southern Oregon coast. Current flow was from top to bottom.

pacted, hydroplastic mud beds owing to unequal loading by overlying sand layers. Uncompacted muds with excess fluid pore pressures, or muds liquefied by an externally generated shock, can be deformed by the weight of overlying sand, which may sink unequally into the incompetent mud. Loading owing to unequal weight of the sand forces protrusions of sand down into the mud, creating positive-relief features on

FIGURE 5.32 Large load casts on the underside of a thin-bedded sandstone, Tar Springs Sandstone (Mississippian). (From Pettijohn, F. J., and P. E. Potter, 1964, Atlas and glossary of primary sedimentary structures. Pl. 52A. Reprinted by permission of Springer-Verlag, New York. Photograph courtesy of P. E. Potter.)

the base of the sandstone beds that may resemble some erosional structures, as mentioned. Load casts are closely related genetically to ball and pillow structures and flame structures. Flute and groove casts may be modified by loading, which tends to exaggerate their relief and destroy their original shapes.

Load casts can form in any environment where water-saturated muds are quickly buried by sand before dewatering can take place. They are not indicative of any particular environment, although they tend to be most common in turbidite sequences. Their presence on the bases of some beds and not on others seems to reflect the hydroplastic state of the underlying mud. They apparently will not form on the bases of sand beds deposited on muds that have already been compacted or dewatered prior to deposition of the sand.

Biogenic Structures

Trace Fossils

The burrowing, boring, feeding, and locomotion activities of organisms can produce a variety of trails, depressions, and open burrows and borings in mud or semiconsolidated sediment bottoms. Filling of these depressions and burrows with sediment of a different type or with different packing creates structures that may be either positive-relief features, such as trails on the base of overlying beds, or features that show up as burrow or bore fillings on the tops of the underlying mud bed. Burrows and borings commonly extend down into beds; therefore, these structures are not exclusively bedding-plane structures.

Tracks, trails, burrows, borings, and other structures made by organisms on bedding surfaces or within beds are known collectively as **trace fossils** or **ichnofossils.** Although geologists have long been aware of the presence of burrows, trails, and other biogenic structures in sedimentary rocks, the recognition and naming of the many varieties of trace fossils now known, as well as a fuller understanding of the environmental significance of these structures, have come about to a large extent since the mid-1950s. Numerous research papers dealing with trace fossils have been published since that time, in addition to several full-length monographs. Only a very brief summary of the classification, occurrence, and significance of trace fossils is presented here. Additional details may be found in the books of Basan (1978); Bromley (1990); Crimes and

Harper (1970, 1977); Curran (1985); Ekdale, Bromley, and Pemberton (1984); Frey (1975); Häntzschel (1975); and Seilacher (1964).

Classification of Trace Fossils. Trace fossils are not true bodily preserved fossils but are simply biogenic structures that originated through the locomotion, feeding, burrowing, or resting activities of organisms. Interpreted broadly, biogenic structures can be considered to include the following:

1. bioturbation structures (burrows, tracks, trails, root penetration structures)
2. biostratification structures (algal stromatolites, graded bedding of biogenic origin)
3. bioerosion structures (borings, scrapings, bitings)
4. excrement (coprolites, such as fecal pellets or fecal castings)

Not all geologists regard biostratification structures as trace fossils, and these structures are not commonly included in published discussions of trace fossils.

Trace fossils can be classified in several ways on the basis of morphology (taxonomy), presumed behavior of the organism that produced the structures, and preservational process (Simpson, 1975; Frey, 1978). On the basis of morphology, they can be grouped into such categories as tracks, trails, burrows, borings, and bioturbate texture, as shown in Table 5.2. Tracks, trails, burrows, and bioturbate texture are features formed in soft sediments. Borings are formed in hard substrates. Figures 5.33, 5.34, and 5.35 illustrate some of these features. Classification of trace fossils on the basis of the behavior of the generating organism is referred to as ethological classification. Classified this way, trace fossils are divided into resting traces, crawling traces, grazing traces, feeding traces or structures, and dwelling structures (Fig. 5.36). Further description of these behavioral structures and the processes by which they are assumed to form is given in Table 5.3. Trace fossils can be classified on the basis of type of preservation, using such terms as full-relief, semirelief, concave, and convex (Fig. 5.37). Traces formed at the sediment surface are called exogenic (outside) traces, and those formed within strata are called endogenic (inside) traces.

Environmental Significance. Trace fossils are produced by a variety of organisms such as crabs, flatfish, clams, molluscs, worms, shrimp, and eel. Because different organisms engage in similar types of behavior (crawling, grazing, feeding, etc.), essentially identical traces may be produced by quite different organisms. Therefore, it is not always possible to identify the organism that produced a particular type of structure. It has been determined, however, that certain associations of biogenic structures tend to characterize particular sedimentary facies. These facies, in turn, can be related to depositional environments. The term **ichnofacies** was introduced by Seilacher (1964) for sedimentary facies characterized by a particular association of trace fossils. Salinity, water depth, and consistency of the substrate (soft or hard bottom) appear to exert the primary controls on the distribution of trace fossils. Trace fossils occur in sediments deposited in environments ranging from subaerial continental to deep marine. In subaerial environments, organisms such as insects, spiders, worms, millipeds, snails, and lizards can produce a variety of burrows and tunnels; vertebrate organisms leave tracks; and plants leave root traces. Freshwater fluvial and lacustrine environments are inhabited by organisms such as worms, crustaceans, insects, bivalves, gastropods, fish, birds, amphibians, mammals, and reptiles that can produce various kinds of traces. Trace fossils in freshwater, continental deposits are grouped into what is called the *Scoyenia* ichnofacies (Frey, Pemberton, and Fagerstrom, 1984). This ichnofacies is rather nondistinctive, consisting of a low-diversity suite of invertebrate and vertebrate tracks, trails, and burrows (Ekdale, Bromley, and Pemberton, 1984).

TABLE 5.2   Descriptive-genetic classification of trace fossils

A   Tracks and Trails

Track—impression left in underlying sediment by an individual foot or podium

Trackway—succession of tracks reflecting directed locomotion by an animal

Trail—trace produced during directed locomotion and consisting either of a surficial groove
made by an animal having part of its body in continuous contact with the substrate surface
or of a continuous subsurface structure made by a mobile endobenthic organism

B   Burrows and Borings

Boring—excavation made in consolidated or otherwise firm substrates, such as rock, shell,
bone, or wood

Burrow—excavation made in loose, unconsolidated sediments

Burrow or *boring system*—highly ramified and/or interconnected burrows or borings, typically
involving shafts and tunnels

Shaft—dominantly vertical burrow or boring or a dominantly vertical component of a burrow
or boring system having prominent vertical and horizontal parts

Tunnel (= gallery)—dominantly horizontal burrow or boring or a dominantly horizontal
component of a burrow or boring system having prominent vertical and horizontal parts

Burrow lining—thickened burrow wall constructed by organisms as a structural reinforcement;
may consist of (1) host sediments retained essentially by mucus impregnation,
(2) pelletoidal aggregates of sediment shoved into the wall, like mud-daubed chimneys,
(3) detrital particles selected and cemented like masonry, or (4) leathery or felted tubes
consisting mostly of chitinophosphatic secretions by organisms. Burrow linings of types (3)
and (4) are commonly called "dwelling tubes."

Burrow cast—sediments infilling a burrow (= burrow fill); may be either "active," if done by
animals, or "passive," if done by gravity or physical processes; active fill termed "back fill"
wherever ∪ in ∪ laminae, etc., show that the animal packed sediment behind itself as it
moved through the substrate

C   Bioturbation

Bioturbate texture—gross texture or fabric imparted to sediments by extensive bioturbation;
typically consists of dense, contorted, truncated, or interpenetrating burrows or other traces,
few of which remain distinct morphologically. Where burrows are somewhat less crowded
and thus are more distinct individually, the sediment is said to be "burrow mottled."

D   Miscellaneous

Configuration—in ichnology, the spatial relationships of traces, including the disposition of
component parts and their orientation with respect to bedding and (or) azimuth

Spreite—bladelike to sinuous, ∪-shaped, or spiraled structure consisting of sets or cosets of
closely juxtaposed, repetitious parallel or concentric feeding or dwelling burrows or grazing
traces. Individual burrows or grooves comprising the spreite commonly anastomose into a
single trunk or stem (as in *Daedalus*) or are strung between peripheral "support" stems (as
in *Rhizocorallium*). "Retrusive" spreiten extended upward, or proximal to the initial point
of entry by the animal, and "protrusive" spreiten extended downward, or distal to the point
of entry.

Source: After Frey, R. W., 1978, Behavioral and ecological implications of trace fossils, *in* P. B. Basan (ed.),
Trace fossil concepts: Soc. Econ. Paleontologists and Mineralogists Short Course No. 5. Table 2, p. 49, reprinted
by permission of SEPM, Tulsa, Okla.

Most interest in ichnofacies has focused on those facies that occur in marine sedimentary rocks. Marine trace fossils are produced by a wide variety of mainly invertebrate organisms, such as worms, shrimp, lobsters, crabs, gastropods, and pelecypods. Some traces are also produced by fish. Seven marine ichnofacies are now recognized, each named for a representative trace fossil: *Terodolites, Trypanites, Glossifungites, Skolithos, Cruziana, Zoophycos,* and *Nereites* (Fig. 5.38). The *Terodolites* ichnofacies (not shown in Fig. 5.38) occurs only in woody materials. The *Trypanites* ichnofacies is characteristic of hard, fully indurated substrates, and the *Glossifungites* ichnofacies typically occurs in firm, but uncemented, substrates. The remaining marine ichnofacies are all soft-sediment ichnofacies whose distribution appears to be controlled mainly by water depth.

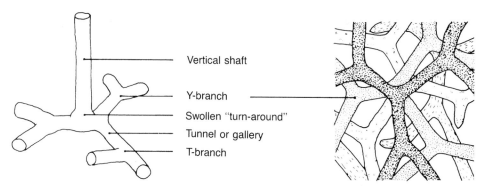

FIGURE 5.33   Diagrams illustrating standard ichnologic terminology. The diagram on the left illustrates an open burrow; the one on the right represents a composite burrow composed of successive cross-cutting mazes. (From Ekdale, A. A., R. G. Bromley, and S. B. Pemberton, 1984, Ichnology: Trace fossils in sedimentology and stratigraphy: Soc. Econ. Paleontologists and Mineralogists Short Course No. 15. Fig. 2.1, p. 14, reprinted by permission of SEPM, Tulsa, Okla.)

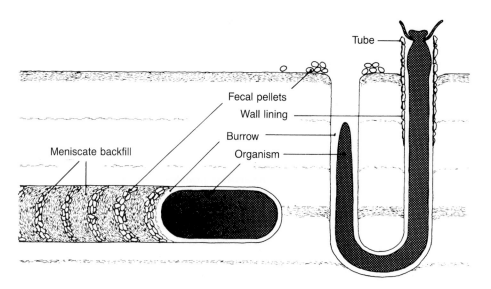

FIGURE 5.34   Two kinds of burrows: a mobile crawling trace or track of an irregular echinoid (left) and a permanent, U-shaped dwelling burrow of an unidentified worm (right). Where the wall lining consists of cemented grains, it is called a tube. (From Ekdale, A. A., R. G. Bromley, and S. B. Pemberton, 1984, Ichnology: Trace fossils in sedimentology and stratigraphy: Soc. Econ. Paleontologists and Mineralogists Short Course No. 15. Fig. 2.3, p. 15, reprinted by permission of SEPM, Tulsa, Okla.)

The supratidal and intertidal zone, subtidal zone, and deeper zones of the marine realm are each distinguished by characteristic associations of trace fossils (Fig. 5.38). In general, the biogenic structures that characterize the *Trypanites* ichnofacies of rocky coasts and pebbly shores are rock borings, most of which are dwelling structures for suspension-feeding organisms (Fig. 5.38, 1–4). Other structures in this ichnofacies include rasping and scraping traces made by feeding organisms, holes drilled by predatory gastropods, and microborings made by algae and fungi. The *Glossifungites*

FIGURE 5.35 Examples of spreite structure, which is a type of structure produced as a burrow is shifted broadside by a burrower. An animal that moves its subhorizontal burrow up or down creates a wall-like, vertical spreite (e.g., *Teichichnus*). Lateral migration produces a bladelike, horizontal spreite (e.g., *Rhizocorallium*). Migration of a U-shaped burrow toward its aperture at the sediment surface causes the spreite to develop on the outside of the U; lengthening of the burrow downward produces a spreite on the inside of the U (e.g., the *Diplocraterion* specimens above). (From Ekdale, A. A., R. G. Bromley, and S. B. Pemberton, 1984, Ichnology: Trace fossils in sedimentology and stratigraphy: Soc. Econ. Paleontologists and Mineralogists Short Course No. 15. Fig. 2.4, p. 16, reprinted by permission of SEPM, Tulsa, Okla.)

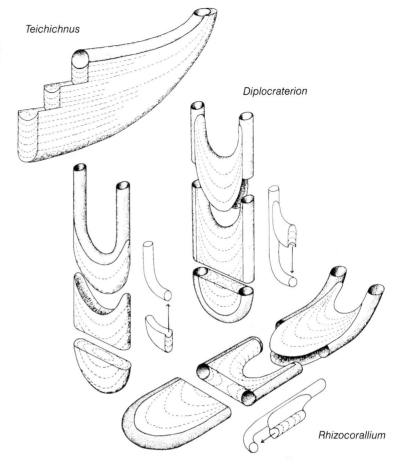

Teichichnus

Diplocraterion

Rhizocorallium

FIGURE 5.36 Classification of trace fossils on the basis of presumed behavior of the organisms producing the structures and the relationship of these traces to body fossils. Note the overlap of some categories of traces. Escape structures overlap several categories of behavioral traces and are not included here. (From Simpson, S., 1975, Classification of trace fossils, *in* R. W. Frey (ed.), The study of trace fossils. Fig. 3.2, p. 49, reprinted by permission of Springer-Verlag, Heidelberg. As translated from Seilacher, A., 1953: Neues Jahrb. Geologie u. Paläontologie, Abh. 96, p. 421–452.)

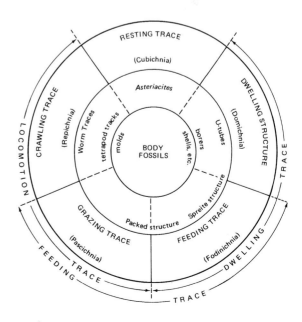

TABLE 5.3 Ethological classification of trace fossils

Categories of trace fossils	Definition	Characteristic morphology
Resting traces (Cubichnia)	Shallow depressions made by animals that temporarily settle onto, or dig into, the substrate surface; emphasis on reclusion	Troughlike relief, recording to some extent the lateroventral morphology of the animal; structures isolated, ideally, but may intergrade with crawling traces or escape structures
Crawling traces (Repichnia)	Trackways, surficial trails, and shallow horizontal structures made by organisms traveling from one place to another; emphasis on locomotion	Linear or sinuous overall structures, some branched; footprints or continuous grooves, commonly annulated; complete form may be preserved or may appear as cleavage reliefs
Grazing traces (Pascichnia)	Grooves, pits, and furrows, many of them discontinuous, made by mobile deposit feeders at or near the substrate surface; emphasis on feeding behavior analogous to "strip mining"	Unbranched, nonoverlapping, curved to tightly coiled patterns or delicately constructed spreiten dominate; patterns reflect maximum utilization of surficial feeding area; complete form may be preserved
Feeding structures (Fodinichnia)	More or less temporary burrows constructed by deposit feeders; the structures may also provide shelter for the organisms; emphasis on feeding behavior analogous to "underground mining"	Single, branched or unbranched, cylindrical to sinuous shafts or U-shaped burrows, or complex, parallel to concentric burrow repetitions (spreiten structures); walls not commonly lined, unless by mucus; oriented at various angles with respect to bedding; complete form may be preserved
Dwelling structures (Domichnia)	Burrows or dwelling tubes providing more or less permanent domiciles, mostly for hemisessile suspension feeders or, in some cases, carnivores; emphasis on habitation	Simple, bifurcated, or U-shaped structures perpendicular or inclined at various angles to bedding, or branching burrow systems having vertical and horizontal components; walls typically lined; complete form may be preserved
Escape structures	Trace fossils of various kinds modified or made anew by animals in direct response to substrate degradation or aggradation; emphasis on readjustment, or equilibrium between relative substrate position and the configuration of contained traces	Vertically repetitive resting traces; biogenic laminae either in echelon or as nested funnels or chevrons; U-in-U spreiten burrows; and other structures reflecting displacement of animals upward or downward with respect to the original substrate surface; complete form may be preserved, especially in aggraded substrates

Source: Frey, R. W., 1978, Behavioral and ecological implications of trace fossils, *in* P. B. Basan (ed.), Trace fossil concepts: Soc. Econ. Paleontologists and Mineralogists Short Course No. 5. Table 3, p. 51, reprinted by permission of SEPM, Tulsa, Okla.

Seilacher   **Martinsson**

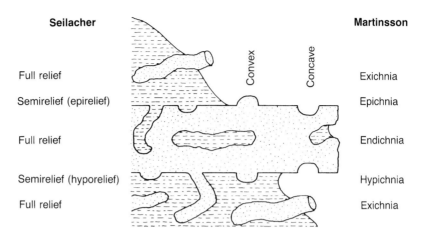

Full relief   Exichnia

Semirelief (epirelief)   Epichnia

Full relief   Endichnia

Semirelief (hyporelief)   Hypichnia

Full relief   Exichnia

FIGURE 5.37 Terminology for preservational classification of trace fossils used by Seilacher and Martinsson. (From Ekdale, A. A., R. G. Bromley, and S. B. Pemberton, 1984, Ichnology: Trace fossils in sedimentology and stratigraphy: Soc. Econ. Paleontologists and Mineralogists Short Course No. 15. Fig. 2.6, p. 22, reprinted by permission of SEPM, Tulsa, Okla. From Seilacher, A., 1964, Sedimentological classification and nomenclature of trace fossils: Sedimentology, v. 3, p. 253–256; Martinsson, A., 1970, Toponomy of trace fossils, p. 109–130, *in* T. P. Crimes and J. C. Harper (eds.), The study of trace fossils: Springer-Verlag, New York.)

ichnofacies, redefined by Frey and Seilacher (1980), is now considered to be restricted to firm, uncemented surfaces that typically consist of dewatered, cohesive muds. The trace fossils produced in this environment are mainly vertical, U-shaped, and branched dwelling burrows of suspension feeders or carnivores such as shrimp, crabs, worms, and pholadid bivalves (Fig. 5.38, 5–8). The littoral zone or intertidal zone of sandy coasts is distinguished by harsh condition resulting from high-energy waves and currents, desiccation, and large temperature and salinity fluctuations. Organisms adapt to these harsh conditions by burrowing into the sand to escape. Thus, vertical and U-shaped dwelling burrows, some with protective linings, such as the *Skolithos, Diplocraterion, Arenicolites,* and *Ophiomorpha* burrows shown in Figure 5.38, 9–13, characterize the *Skolithos* ichnofacies of this zone. The neritic zone or subtidal zone extending from the low-tide zone to the edge of the continental shelf (at about 200-m water depth) is a less demanding environment, although erosive currents may be present. Vertical dwelling burrows and protected, U-shaped burrows are less common in this zone. Burrows tend to be shorter, and surface markings made by organisms such as crustaceans (or trilobites during early Paleozoic time) are more common. In the deeper part of the neritic zone, organic matter becomes abundant enough for sediment feeders to become established and produce feeding burrows. In these deeper waters, vertical escape burrows thus tend to give way to horizontal feeding burrows. This zone of the ocean is distinguished by the *Cruziana* ichnofacies, characterized by such traces as those shown in Figure 5.38, 14–18.

The deep bathyl and abyssal zones of the ocean exist below wave base where low-energy conditions generally prevail, although erosion and deposition can occur in these zones owing to turbidity currents or deep-bottom currents. Complex feeding burrows, such as those of *Zoophycos* and *Lorenzinia* (Fig. 5.38, 19–21), are particularly common in the bathyl zone. These traces make up the *Zoophycos* ichnofacies. In the even deeper waters of the abyssal zone, where bottom sediment is almost exclusively fine-grained clay, more complex spiral, winding, and meandering forms, such as

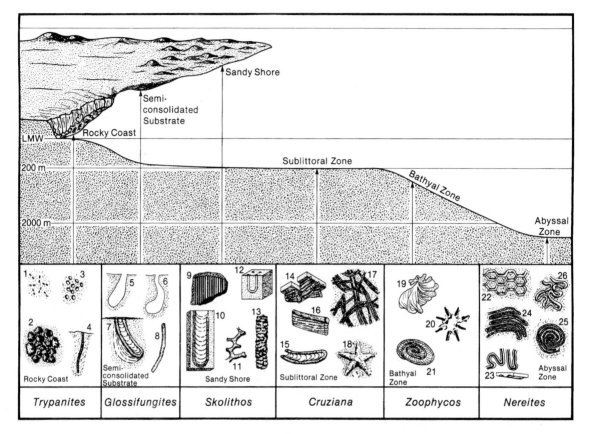

FIGURE 5.38   Schematic representation of the relationship of characteristic trace fossils to sedimentary facies and depth zones in the ocean. Borings of 1, *Polydora;* 2, *Entobia;* 3, echinoid borings; 4, *Trypanites;* 5, 6, pholadid burrows; 7, *Diplocraterion;* 8, unlined crab burrow; 9, *Skolithos;* 10, *Diplocraterion;* 11, *Thalassinoides;* 12, *Arenicolites;* 13, *Ophiomorpha;* 14, *Phycodes;* 15, *Rhizocorallium;* 16, *Teichichnus;* 17, *Crossopodia;* 18, *Asteriacites;* 19, *Zoophycos;* 20, *Lorenzinia;* 21, *Zoophycos;* 22, *Paleodictyon;* 23, *Taphrhelminthopsis;* 24, *Helminthoida;* 25, *Spirorhaphe;* 26, *Cosmorhaphe.* (From Ekdale, A. A., R. G. Bromley, and S. B. Pemberton, 1984, Ichnology: Trace fossils in sedimentology and stratigraphy: Soc. Econ. Paleontologists and Mineralogists Short Course No. 15. Fig. 15.2, p. 187, reprinted by permission of SEPM, Tulsa, Okla. Modified from Crimes, T. P., 1975, The stratigraphical significance of trace fossils, *in* T. P. Crimes and J. C. Harper (eds.), The study of trace fossils, Fig. 7.2, p. 118: Springer-Verlag, New York.)

Spirorhaphe (Fig. 5.38, 25) or patterned trace fossils such as *Paleodictyon* (Fig. 5.38, 22), occur. This association of trace fossils constitutes the *Nereites* ichnofacies, named for *Nereites,* a type of horizontal grazing trail.

Although each of these marine ichnofacies tends to be characteristic of a particular bathymetric zone of the ocean as shown in Figure 5.38, we now know that individual trace fossils can overlap depth zones. No single biogenic structure is an infallible indicator of depth and environment. The basic controls on the formation of trace fossils include nature of the substrate (sea bottom), water energy, rates of deposition, water turbidity, oxygen and salinity levels, toxic substances, and quantity of available food (Pemberton, MacEachern, and Frey, 1992). Trace fossils should be studied as assemblages of structures in conjunction with other physical, chemical, and biological characteristics of the same substrates. Trace fossils occur in rocks of all ages, including some Precambrian rocks. They have been reported in most types of sedimentary rocks except evaporites and rocks deposited in highly reducing (euxinic) environments. Highly saline environments or euxinic environments, where toxic conditions are caused by lack of oxygen and the presence of hydrogen sulfide gas, preclude or greatly reduce organic activity. Studies of bioturbation in modern open-ocean environments show that organisms may rework sediment so thoroughly that primary laminations and other physically produced structures are completely destroyed in most environments where enough oxygen is present for organisms to flourish. Exceedingly intense bioturbation can produce bedding that is so homogenized that it has a mottled or stirred appearance or is completely devoid of all structures. For bedding and other physically produced sedimentary structures to escape destruction by biogenic activity and become preserved in the geologic record, they must be formed either in an environment where sedimentation rates are so high that organisms do not have time to rework sediments and destroy original structures or in euxinic environments or highly saline environments, as mentioned, where organic activity is limited.

Other Applications of Trace Fossils. In addition to their usefulness as environmental indicators, trace fossils are also useful in several other ways. They may, for example, serve as an indicator of relative sedimentation rates based on the assumption that rapidly deposited sediments contain relatively fewer trace fossils than slowly deposited sediments. They can also help to show whether sedimentation was continuous or marked by erosional breaks, and they provide a record of the behavior patterns of extinct organisms. They may even be useful in paleocurrent analysis; study of the orientation of resting marks of organisms that may have preferred to face into the current while resting establishes the paleocurrent-flow direction. Some trace fossils such as U-shaped burrows, which opened upward when formed, can be used to tell the top and bottom orientation of beds. Trace fossils also have biostratigraphic and chronostratigraphic significance for zoning and correlation, and they may be useful for recognition of bounding discontinuities between stratigraphic sequences (Pemberton, MacEachern, and Frey, 1992; also see Frey and Pemberton, 1985, and Frey and Wheatcroft, 1989).

Stromatolites

Stromatolites are organically formed, laminated structures composed of fine silt- or clay-size sediment or, more rarely, sand-size sediment. Most ancient stromatolites occur in limestones; however, stromatolites have also been reported in siliciclastic sediments. Stromatolitic bedding ranges from nearly flat laminations that may be difficult to differentiate from sedimentary laminations of other origins to hemispherical forms in which the laminae are crinkled or deformed to varying degrees (Fig. 5.39). The hemispherical forms range in shape from biscuit- and cabbagelike forms to columns. Logan, Rezak, and Ginsburg (1964) classified these hemispherical stromatolites into three basic types: (1) laterally linked hemispheroids; (2) discrete, vertically stacked hemispheroids; and (3) discrete spheroids, or spheroidal structures (Fig. 5.40). Laterally linked hemispheroids and discrete, vertically stacked hemispheroids can combine in various ways to create several different kinds of compound stromatolites. The term **thrombolite** was proposed by Aitken (1967) for structures that resemble stromatolites in external form and size but lack distinct laminations. The laminations of stromatolites are generally less than 1 mm in thickness and are caused by concentrations of fine calcium carbonate minerals, fine organic matter, and detrital clay and silt. Stromatolites composed of quartz grains have also been reported (Davis, 1968).

FIGURE 5.39 Algal stromatolites in the Snowslip Formation (Precambrian), west of Logan Pass, Glacier National Park, Montana. (Photograph by G. J. Retallack.)

Stromatolites were considered to be true body fossils by early workers, but they are now known to be organosedimentary structures formed largely by the trapping and binding activities of blue-green algae (cyanobacteria). Although some geologists consider stromatolites to be a trace fossil, they are included here as a type of irregular stratification structure owing to their distinctive lamination. They are forming today in many localities where they occur mainly in the shallow subtidal, intertidal, and supratidal zones of the ocean. They have also been found in lacustrine environments. Because they are related to the activities of blue-green algae, which carry out photosynthesis, they are restricted to water depths and environments where enough light is available for photosynthesis. The laminated structure forms as a result of trapping of fine sediment in the very fine filaments of algal mats. Once a thin layer of sediment covers the mat, the algal filaments grow up and around sediment grains to form a new mat that traps another thin layer of sediment. This successive growth of mats produces the laminated structure. The shapes of the hemispheres are related to water energy and scouring effects in the depositional environment. Laterally linked hemispheroids tend to form in low-energy environments where scouring effects are minimal. In higher-energy environments, scouring by currents prevents linking of the stromatolite heads; thus, vertically stacked or discrete hemispheroids form.

Bedding-Plane Markings of Miscellaneous Origin

Mudcracks and Syneresis Cracks

Mudcracks in modern sediment are downward-tapering, V-shaped fractures that display a crudely polygonal pattern in plan view. The area between the cracks is commonly curved upward into a concave shape. Mudcracks form in both siliciclastic and carbonate mud owing to desiccation. Subsequent sedimentation over a cracked surface fills the cracks. In ancient sedimentary rocks, mudcracks are commonly preserved on the tops of bedding surfaces as positive-relief fillings of the original cracks (Fig. 5.41). The mudcrack polygons range in diameter from a few centimeters to a few meters. The cracks themselves commonly range in width up to a few centimeters and in depth to a

Types	Description	Vertical section of stromatolite structure
Laterally linked hemispheroids	Space-linked hemispheroids with close-linked hemispheroids as a microstructure in the constituent laminae	
Discrete, vertically stacked hemispheroids	Discrete, vertically stacked hemispheroids composed of close-linked hemispheroidal laminae on a microscale	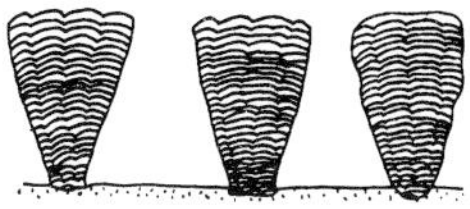
Discrete spheroids	Spheroidal structures consisting of inverted, stacked hemispheroids	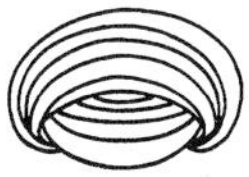
	Spheroidal structures consisting of concentrically stacked hemispheroids	
	Spheroidal structures consisting of randomly stacked hemispheroids	
Combination forms	Initial space-linked hemispheroids passing into discrete, vertically stacked hemispheroids with upward growth of structures	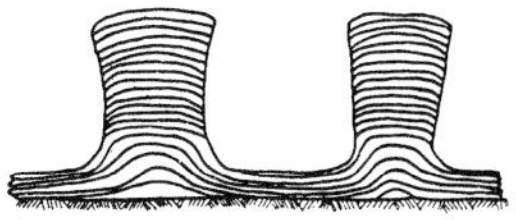
	Initial discrete, vertically stacked hemispheroids passing into close-linked hemispheroids by upward growth	
	Alternation of discrete, vertically stacked hemispheroids and space-linked hemispheroids due to periodic sediment infilling of interstructure spaces	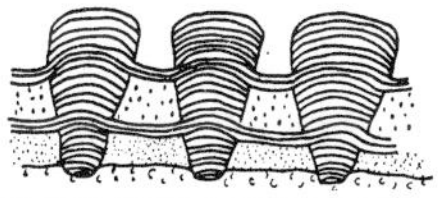
	Initial space-linked hemispheroids passing into discrete, vertically stacked hemispheroids; both with laminae of close-linked hemispheroids	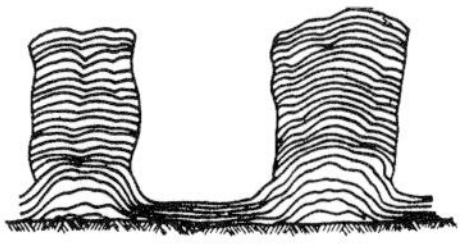
	Initial discrete, vertically stacked hemispheroids passing into close-linked hemispheroids; both with laminae of close-linked hemispheroids	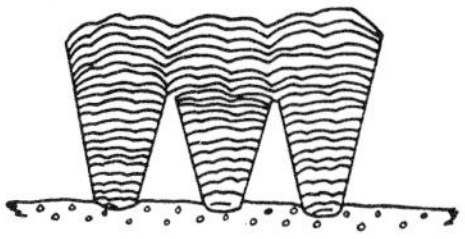

FIGURE 5.40 Structure of hemispherical stromatolites showing examples of laterally linked hemispheroids, vertically stacked hemispheroids, and discrete spheroids. (After Logarm, B. W., R. Rezak, and R. N. Ginsburg, 1964, Classification and environmental significance of algal stromatolites: Jour. Geology, v. 72. Fig. 4, p. 76, and Fig. 5, p. 78, reprinted by permission of University of Chicago Press.)

FIGURE 5.41 Mudcracks on the upper surface of a Miocene mudstone bed, Bangladesh. (Photograph by E. M. Baldwin.)

few tens of centimeters, but cracks up to a few meters in depth have been reported. The presence of undoubted mudcracks indicates intermittent subaerial exposure; however, mudcracks can be confused with syneresis cracks (below), which form under water. Mudcracks occur in estuarine, lagoonal, tidal-flat, river floodplain, playa-lake, and other environments where muddy sediment is intermittently exposed and allowed to dry. They may be associated with raindrop or hailstone imprints, bubble imprints and foam impressions, flat-topped ripple marks, and vertebrate tracks (Plummer and Gostin, 1981).

In contrast to the continuous, polygonal network of mudcracks that occurs on bedding surfaces, **syneresis cracks** tend to be discontinuous and vary in shape from polygonal to spindle-shaped or sinuous (Plummer and Gostin, 1981). They commonly occur in thin mudstones interbedded with sandstones as either positive-relief features on the base of the sandstones or negative-relief features on the top of the mudstones. Syneresis cracks are subaqueous shrinkage cracks that form in clayey sediment by loss of pore water from clays that have flocculated rapidly or that have undergone shrinkage of swelling-clay mineral lattices owing to changes in salinity of surrounding water (Burst, 1965). They are known in ancient sedimentary rocks from both marine and nonmarine environments. They may be confused with mudcracks and even some trace fossils. For example, the lenticular shape of the crack fill in plan view may resemble burrow traces. Because some syneresis cracks do closely resemble mudcracks, it is important in trying to differentiate them to look for features associated with mudcracks that indicate subaerial exposure—features such as raindrop imprints and vertebrate tracks.

Pits and Small Impressions

Small craterlike pits with slightly raised rims commonly occur together with mudcracks and are thought to be impressions made by the impact of rain **(raindrop imprints)** or hail **(hailstone imprints).** They are commonly only a few millimeters deep and less than 1 cm in diameter, and they may occur as either widely scattered pits or very closely spaced impressions. When they can be unambiguously recognized, their presence indicates subaerial exposure; however, small circular depressions created by bubbles breaking on the surface of sediment **(bubble imprints)**, escaping gas, and some types of organic markings can be confused with raindrop or hailstone imprints.

Rill and Swash Marks. Rill marks are small dendritic channels or grooves that form on beaches by the discharge of pore waters at low tide, or by small streams debouching onto a sand or mud flat. They have very low preservation potential and are seldom found in ancient sedimentary rocks. **Swash marks** are very thin, arcuate lines or small ridges on a beach formed by concentrations of fine sediment and organic debris. They are caused by wave swash and mark the farthest advance of wave uprush. They likewise have low preservation potential, but when found and recognized in ancient sedimentary rocks, they indicate either a beach or a lakeshore environment.

Parting Lineation. Parting lineation, sometimes called **current lineation,** forms on the bedding surfaces of parallel laminated sandstones. It consists of subparallel ridges and grooves a few millimeters wide and many centimeters long (Fig. 5.42). Relief on the ridges and hollows is commonly on the order of the diameter of the sandstone grains. The grains in the sandstone generally have a mean orientation of their long axes parallel to the lineation. The lineation is oriented parallel to current flow, and thus its presence in ancient sandstones is useful in paleocurrent studies—although it shows only that the current flowed parallel to the parting lineations and does not show which of the two diametrically opposed directions was the flow direction. Parting lineation occurs in newly deposited sands on beaches and in fluvial environments. It is most common in ancient deposits in thin, evenly bedded sandstones. Its origin is obviously related to current flow and grain orientation, probably owing to flow over upper-flow-

FIGURE 5.42 Parting lineation in sandstone, Haymond Formation, Texas. Current flow is parallel to the hammer. (Photograph courtesy of E. F. McBride.)

regime plane beds, but the exact mechanism by which parting lineation forms is poorly understood.

5.5   OTHER STRUCTURES

Sandstone Dikes and Sills

Sandstone dikes and sills are tabular bodies of massive sandstone that fill fractures in any type of host rock. They range in thickness from a few centimeters to more than 10 m. They lack internal structures except for oriented mica flakes and other elongated particles that are commonly aligned parallel to the dike walls. Sandstone dikes are not common structures, but they have have been reported from numerous localities in rocks ranging in age from Precambrian to Pleistocene. They occur in a wide variety of depositional environments, ranging from deep marine to subaerial.

Sandstone dikes are formed by forceful injection of liquefied sand into fractures, commonly in overlying rock; however, injection appears to have been downward in some rocks. Sandstone sills are similar features that formed by injection parallel to bedding. These sills may be difficult or impossible to distinguish from normally deposited sandstone beds unless they can be traced into sandstone dikes or be traced far enough to show a cross-cutting relationship with other beds. Suggested causes of liquefaction of sand include shocks owing to earthquakes or triggering effects related to slumps, slides, or rapid emplacement of sediment by mass flow.

Structures of Secondary Origin

Most of the structures discussed above (with the possible exception of some sandstone dikes and convolute lamination) formed during, or very shortly after, deposition of their host sediment; thus, they are **primary sedimentary structures.** A few kinds of sedimentary structures exhibit characteristics which indicate that they formed some time after deposition. These **secondary sedimentary structures** are largely of chemical origin, formed by precipitation of mineral substances in the pores of semiconsolidated or consolidated sedimentary rock or by chemical replacement processes.

Concretions are probably the most common kind of secondary structure. Most concretions are composed of calcite, but concretions composed of dolomite, hematite, siderite, chert, pyrite, and gypsum are also known. They form by precipitation of mineral matter around some kind of nucleus, such as a shell fragment, gradually building up a globular mass (Fig. 5.43). Shapes of these masses range from spherical to disc-shaped, cone-shaped, and pipe-shaped, and they may range in size from less than 1 cm to as much as 3 m. Concretions are especially common in sandstones and shales but can occur in other sedimentary rocks.

Sand crystals are very large euhedral or subhedral crystals of calcite, barite, or gypsum that are filled with detrital sand inclusions (Fig. 5.44). They appear to form during sediment burial by growth in incompletely cemented sands. **Stylolites** are suturelike seams (Fig. 5.45) that are most common in limestones. These seams are typically only a few centimenters thick, and they are generally marked by concentrations of insoluble constituents such as clay minerals, iron oxides, and organic matter. They to form as a result of pressure-solution processes. **Cone-and-cone structures** are unusual structures that consist of nested sets of small concentric cones (Fig. 5.46), composed, in most examples, of calcium carbonate minerals. They are most common in shales and marly limestones, where they apparently form by growth of fibrous crystals in the enclosing sediment while it is still in a plastic state.

FIGURE 5.43 Large concretions weathering out on the surface of a laminated sandstone bed, Coaledo Formation (Eocene), southern Oregon coast. Calcite, precipitated around some kind of nucleus, filled pore spaces in the sandstone, gradually building up the globular masses. Note the sandstone lamination preserved in the concretions. (Photograph courtesy of Robert Q. Oaks, Jr.)

FIGURE 5.44 Sand crystals, Miocene sandstone, Badlands, South Dakota. The length of the specimen is about 15 cm. (From Sedimentary rocks, 3rd ed., by Francis J. Pettijohn. Fig. 12.3. Copyright 1949, 1957 by Harper & Row, Publishers, Inc. Copyright © 1975 by Francis J. Pettijohn. Reprinted by permission of HarperCollins Publishers, Inc.)

5.6 · PALEOCURRENT ANALYSIS
FROM SEDIMENTARY STRUCTURES

As mentioned, many sedimentary structures yield directional data that show the direction ancient current flowed at the time of deposition. The dip direction of cross-bed foresets; the asymmetry and orientation of the crests of current ripples; and the orientation of flute casts, groove casts, and current lineation are all examples of directional data that can be obtained from sedimentary structures.

The orientation of directional sedimentary structures is determined in the field with a Brunton compass by taking measurements from as many different outcrops and

FIGURE 5.45 Well-developed, sutured stylolites in Cretaceous limestone, Calcare Massicio, Tuscany, Italy. (Photograph courtesy of E. F. McBride.)

FIGURE 5.46 A large specimen showing typical cone-in-cone structure. The age and locality of specimen are unknown.

individual beds as possible and practical. The orientation of directional structures determined from a particular bed or stratigraphic unit commonly shows considerable scatter. Therefore, directional data must be treated statistically in some manner to reveal primary and secondary directional trends. For example, the dip direction of cross-bed foresets in the ancient deposits of a meandering-river system may range from N 20° W to N 20° E owing to variations in flow direction of the stream in different parts of the meandering-river system. By examining the orientation data statistically, we may be able to determine that the primary flow direction of the stream was approximately due north. Because all of the cross-beds foresets in this example indicate flow in the same general direction, in spite of some scatter, we say that flow was **unidirectional.** By contrast, the cross-bed foresets in sandy deposits of marine tidal channels may display two opposing dip directions owing to formation of cross-beds during both incoming and outgoing tides. This type of opposing flow is referred to as **bidirectional.** In some environments, such as the eolian environment, depositing currents may flow in several directions **(polydirectional)** at various times during deposition of a particular sedimentary unit.

The paleocurrent data collected from stratigraphic units that have undergone little or no tectonic deformation or tilting can be compiled and summarized directly. If the rocks have undergone considerable tilting, it is necessary to correct the measured orientation by restoring directions to their original attitude before tilting. A simple procedure using a stereogram can be used to reorient directional data collected from tilted stratigraphic units (Collinson and Thompson, 1989, p. 200). After any necessary reorientation of data has been done, the data are commonly plotted as a circular histogram or "rose diagram" (Fig. 5.47). (Commercial computer software programs are available for plotting rose diagrams, e.g., *Rose,* from Rockware, Wheat Ridge, Colorado.) Such diagrams show the principal direction of paleocurrent flow and any secondary or tertiary modes of flow. If the paleocurrent flow as revealed by the rose diagram is dominantly in a single direction, the paleocurrent vector is said to be **unimodal.** If two principal directions of flow are indicated, it is **bimodal,** and if three or more directions of flow are revealed by the directional data, the paleocurrent flow is called **polymodal.**

Local paleocurrent directions may have environmental significance. For example, sediments from alluvial and deltaic environments tend to have unimodal paleo-

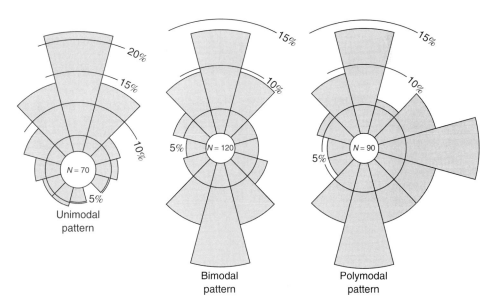

FIGURE 5.47 Hypothetical paleocurrent data plotted as rose diagrams; unimodal, bimodal, and polymodal patterns of paleoflow directions are shown. N = number of measurements (directions) taken in the field.

current vectors patterns, whereas bimodal paleocurrents patterns are more common in shoreline and shelf sediments. Paleocurrent data have their greatest usefulness when plotted on a regional scale to reveal regional paleocurrent patterns.

FURTHER READINGS

Allen, J. R. L., 1968, Current ripples: Their relation to patterns of water and sediment motion: North Holland Pub., Amsterdam, 433 p.

Allen, J. R. L., 1982, Sedimentary structures: Their character and physical basis, v. 1–2: Elsevier, Amsterdam, 664 p.

Basan, P. B. (ed.), 1978, Trace fossil concepts: Soc. Econ. Paleontologists and Mineralogists Short Course No. 5, Tulsa, Okla., 181 p.

Bouma, A. H., 1969, Methods for the study of sedimentary structures: John Wiley & Sons, New York, 457 p.

Bromley, R. G., 1990, Trace fossils, biology and taphonomy: Special topics in paleontology 3: Unwin Hyman, London, 280 p.

Collinson, J. D., and D. B., Thompson, 1989, Sedimentary structures, 2nd ed.: Chapman and Hall, London, 207 p.

Conybeare, C. E. B., and K. A. W. Crook, 1968, Manual of sedimentary structures: Department of National Development, Bureau of Mineral Resources, Geology and Geophysics, Bull. 102, 327 p.

Crimes, T. P., and J. C. Harper (eds.), 1970, Trace fossils: Steel House Press, Liverpool, 547 p.

Crimes, T. P., and J. C. Harper (eds.), 1977, Trace fossils 2: Steel House Press, Liverpool, 351 p.

Curran, H. A. (ed.), 1985, Biogenic structures: Their use in interpreting depositional environments: Soc. Econ. Paleontologists and Mineralogists Spec. Pub. 35, Tulsa, Okla., 347 p.

Dzulnyski, S., and E. K. Walton, 1965, Sedimentary features of flysch and greywackes: Developments in sedimentology, v. 7: Elsevier, Amsterdam, 274 p.

Ekdale, A. A., R. G. Bromley, and S. G. Pemberton, 1984, Ichnology, trace fossils in sedimentology and stratigraphy: Soc. Econ. Paleontologists and Mineralogists Short Course No. 15, Tulsa, Okla., 317 p.

Frey, R. W. (ed.), 1975, The study of trace fossils: Springer-Verlag, New York, 562 p.

Harms, J. C., J. B. Southard, D. R. Spearing, and R. G. Walker, 1975, Depositional environments as interpreted from primary sedimentary structures and stratification sequences: Soc. Econ. Paleontologists and Mineralogists Short Course No. 2, Tulsa, Okla., 161 p.

Harms, J. C., J. B. Southard, and R. G. Walker, 1982, Structures and sequences in clastic rocks: Soc. Econ. Paleontologists and Mineralogists Short Course No. 9, Tulsa, Okla..

Middleton, G. V. (ed.), 1965, Primary sedimentary structures and their hydrodynamic interpretation: Soc. Econ. Paleontologists and Mineralogists Spec. Pub. 12, Tulsa, Okla., 265 p.

Miller, M. F., A. A. Ekdale, and M. D. Picard (eds.), 1984, Trace fossils and paleoenvironments: Marine carbonate, marginal marine terrigenous and continental terrigenous settings: Jour. Paleontology, v. 58, p. 283–597.

Pettijohn, F. J., and P. E. Potter, 1964, Atlas and glossary of primary sedimentary structures: Springer-Verlag, New York, 370 p.

Picard, M. D., and L. R. High, Jr., 1973, Sedimentary structures of ephemeral streams: Elsevier, New York, 223 p.

Potter, P. E., and F. J. Pettijohn, 1977, Paleocurrents and basin analysis, 2nd ed.: Springer-Verlag, New York, 460 p.

Reineck, H. E., and I. B. Singh, 1980, Depositional sedimentary environments, 2nd ed.: Springer-Verlag, New York, 439 p.

Sarjeant, W. A. S. (ed.), 1983, Terrestrial trace fossils: Benchmark papers in geology, v. 76: Dowden, Hutchinson and Ross, Stroudsburg, Pa., 415 p.

Walter, M. R. (ed.), Stromatolites: Elsevier, New York, 790 p.

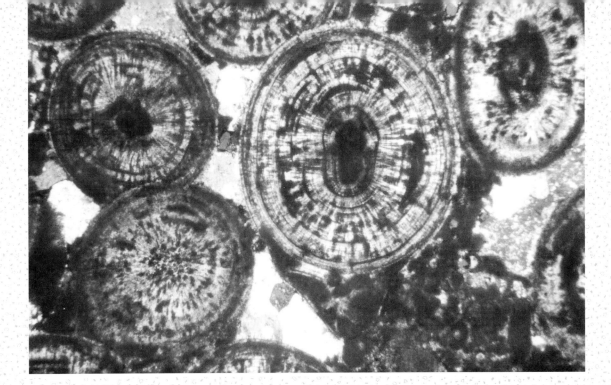

PART 4

COMPOSITION, CLASSIFICATION, ORIGIN, AND DIAGENESIS OF SEDIMENTARY ROCKS

Carbonate ooids, 0.5 to 1.0 mm in diameter, in Upper Devonian Limestones, Alberta, Canada.

In Chapters 4 and 5 we examined sedimentary textures and structures and discussed some of the ways that these properties are used to interpret the genesis of sedimentary rocks. Particle and chemical composition are also fundamental properties of sedimentary rocks, properties that allow us to distinguish one kind of sedimentary rock from another and that provide additional important information about the history of the rocks. Geologists commonly use the term **mineralogy** to refer to the identity of all the particles or grains in rocks. When we talk about the mineralogy of a sandstone, for example, we mean the quartz, feldspar, micas, and other particulate mineral grains that make up the sandstone. Bulk **chemistry** is another aspect of the overall composition of sedimentary rocks and is directly related to the mineralogy of the rocks. Traditionally, sedimentologists have been interested more in mineralogy than in bulk chemistry because they consider mineralogy to be more useful than chemical composition in characterizing and classifying sedimentary rocks and in interpreting their geologic history. That perception is changing, and we are now seeing an increasing number of published papers that are concerned in some way with sedimentary geochemistry.

The next three chapters deal with the composition of siliciclastic sedimentary rocks; carbonate rocks (limestones and dolomites); other biochemical/chemical sedimentary rocks (evaporites, cherts, iron-rich sedimentary rocks, phosphorites); and carbonaceous sedimentary rocks (oil shales and coals). Among other things, we will examine the way that composition is used in classifying these rocks. Classifications are simply arbitrary schemes for grouping rocks into "pigeonholes" on the basis of some characteristic property of the rocks. Classifications based solely upon observable, measurable rock properties such as mineral composition, without regard to the genetic significance of these properties, are called **descriptive** classifications. Classifications that provide some insight into the genesis of the rocks are **genetic** classifications. Although genetic classifications may be conceptually superior, it is often difficult or impossible to make the subjective, deductive interpretations necessary to effectively use such classifications. Therefore, descriptive classifications have gained favor in recent years.

We will also examine the various ways that mineralogy and chemical composition are used to interpret sediment provenance (source), depositional process, depositional environments, and diagenesis (burial alteration). Many of the biochemical/chemical sedimentary rocks, such as dolomite, iron-rich sedimentary rocks, and phosphorites, are enigmatic rocks whose origins are still poorly understood. Ideas concerning the origin of these rocks are discussed, and controversial and conflicting hypotheses are presented and analyzed. Detailed study of the composition of sedimentary rocks is commonly referred to as **sedimentary petrography.** I have not attempted in this volume to give a comprehensive treatment of sedimentary petrography. Additional details of the petrography of sedimentary rocks can be found in a host of papers and monographs that examine specialized aspects of sedimentary petrography. Several of these reference sources are listed under Further Readings at the ends of Chapters 6, 7, and 8.

6

Siliciclastic Sedimentary Rocks

6.1 INTRODUCTION

As discussed in Chapter 2, sedimentary rocks composed mainly of silicate particles derived by the weathering breakdown of older rocks and by pyroclastic volcanism are called siliciclastic sedimentary rocks. Sandstones, conglomerates, and shales comprise the members of this group. The siliciclastic sedimentary rocks make up roughly three-fourths of all sedimentary rocks in the geologic record, and they are present in sedimentary sequences ranging in age from Precambrian to Holocene. They are of special interest to geologists as indicators of Earth history. Many of the textures and structures discussed in Chapters 4 and 5 are particularly well developed in these rocks. These textures and structures provide important information about ancient sediment transport and depositional conditions. In addition, the minerals and rock fragments in siliciclastic sedimentary rocks furnish our most definitive clues to the nature and location of vanished ancient mountain systems such as the ancestral Rocky Mountains and Appalachian Mountains of the United States. Petroleum geologists are especially interested in sandstones because more than half of the world's reserves of oil and gas occur in these rocks. Shales are likewise of great interest because the organic matter contained in shales is believed to be the source material of oil and gas.

In this chapter, we focus on the mineralogy and chemical composition of sandstones, shales, and conglomerates and explore the ways that composition can be used to classify these rocks and interpret aspects of their origin. We will also briefly examine the changes in composition and texture that take place owing to burial diagenesis.

6.2 SANDSTONES

Sandstones make up 20 to 25 percent of all sedimentary rocks. They are common rocks in geologic systems of all ages, and they are distributed throughout the continents of

Earth. Sandstones consist mainly of particles ranging in size from 1/16 to 2 mm. These particles make up the **framework fraction** of the sandstones. Sandstones may also contain various amounts of cement and very fine-size (<~0.03 mm) material called **matrix,** which are present within interstitial pore space among the framework grains. Because of their coarse size (relative to the sizes of particles in shales), the mineralogy of sandstones can generally be determined with reasonable accuracy with a standard petrographic microscope. Bulk chemical composition can be measured by instrumental techniques such as X-ray fluorescence and ICP (inductively coupled argon plasma emission spectrometry). The chemistry of individual mineral grains can be determined by the electron probe microanalyzer and by backscattered electron microscopy.

In this section, we examine the mineralogy and chemical composition of sandstones, discuss the classification of sandstones on the basis of mineral composition, and evaluate the usefulness of particle and chemical composition in interpreting the genesis of sandstones.

Mineralogy

The particles that make up sandstones are mainly sand-size and silt-size silicate minerals and rock fragments. Only a few principal kinds of minerals make up the bulk of all sandstones. These common minerals and rock fragments are shown in Table 6.1 and are discussed in greater detail in the following paragraphs.

Quartz. Quartz (SiO_2) is the dominant mineral in most sandstones, making up on average about 50 to 60 percent of the framework fraction. The dominant mineral in most sandstones, it is a comparatively easy mineral to identify, both megascopically in hand specimens and by petrographic examination in thin sections, although it can be confused with feldspars. Because of its superior hardness and chemical stability, quartz can survive multiple recycling. The quartz grains in many sandstones display some degree of rounding acquired by abrasion during one or more episodes of transport, particularly eolian (wind) transport.

Quartz can occur as single (monocrystalline) grains or as composite (polycrystalline) grains (Fig. 6.1). Composite quartz grains that consist of exceedingly small crystals, referred to as microcrystalline quartz, are called **chert.** Chert grains are actually rock fragments, derived by weathering of bedded chert or chert nodules in limestone (Chapter 8). When examined under crossed polarizing prisms with a petrographic microscope, quartz grains commonly display sweeping patterns of extinction as the stage is rotated. This property is called **undulatory extinction.** Some authors (Folk, 1974; Basu et al., 1975) believe that the properties of polycrystallinity and undulatory extinction can be used to distinguish quartz derived from different sources. Quartz is derived from plutonic rock, particularly felsic plutonic rocks such as granites, metamorphic rocks, and older sandstones.

Feldspars. Feldspar minerals make up about 10 to 20 percent of the framework grains of average sandstones. They are the second most abundant mineral in most sandstones. Several different varieties of feldspars (Fig. 6.2) are recognized on the basis of differences in chemical composition and optical properties. They are divided into two broad groups: alkali feldspars and plagioclase feldspars.

Alkali feldspars constitute a group of minerals in which chemical composition can range through a complete solid solution series from $KAlSi_3O_8$ through $(K,Na)AlSi_3O_8$ to $NaAlSi_3O_8$. Because potassium-rich feldspars are such common members of this group, it has become widespread practice to call the alkali feldspars

TABLE 6.1   Common minerals and rock fragments in siliciclastic sedimentary rocks

---

Major minerals (abundance >~1%–2%)
  Stable minerals (greatest resistance to chemical decomposition):
    Quartz—makes up approximately 65% of average sandstone, 30% of average shale, 5% of average carbonate rock.
  Less stable minerals:
    Feldspars—include K-feldspars (orthoclase, microcline, sanidine, anorthoclase) and plagioclase feldspars (albite, oligoclase, andesine, labradorite, bytownite, anorthite); make up about 10%–15% of average sandstone, 5% of average shale, <1% of average carbonate rock.
    Clay minerals and fine micas—clay minerals include the kaolinite group, illite group, smectite group (montmorillonite a principal variety), and chlorite group; fine micas are principally muscovite (sericite) and biotite; make up approximately 25%–35% of total siliciclastic minerals but may comprise >60% of the minerals in shales.
Accessory minerals (abundances <~1%–2%)
  Coarse micas—principally muscovite and biotite.
  Heavy minerals (specific gravity >~2.9):
    Stable nonopaque minerals—zircon, tourmaline, rutile, anatase.
    Metastable nonopaque minerals—amphiboles, pyroxenes, chlorite, garnet, apatite, staurolite, epidote, olivine, sphene, zoisite, clinozoisite, topaz, monazite, plus about 100 others of minor importance volumetrically.
    Stable opaque minerals—hematite, limonite.
    Metastable opaque minerals—magnetite, ilmenite, leucoxene.
Rock fragments (make up about 10%–15% of the siliciclastic grains in average sandstone and most of the gravel-size particles in conglomerates; shales contain few rock fragments)
  Igneous rock fragments—may include clasts of any igneous rock, but fragments of fine-crystalline volcanic rock and volcanic glass are most common in sandstones.
  Metamorphic rock fragments—include metaquartzite, schist, phyllite, slate, argillite, and less commonly gneiss clasts.
  Sedimentary rock fragments—any type of sedimentary rock possible in conglomerates; clasts of fine sandstone, siltstone, shale, and chert are most common in sandstones; limestone clasts are comparatively rare in sandstones.
Chemical cements (abundance variable)
  Silicate minerals—predominantly quartz; others may include chalcedony, opal, feldspars, and zeolites.
  Carbonate minerals—principally calcite; less commonly aragonite, dolomite, siderite.
  Iron oxide minerals—hematite, limonite, goethite.
  Sulfate minerals—anhydrite, gypsum, barite.

Note: Stability refers to chemical stability.

potassium feldspars, often shortened to simply **K-spars.** Common members of the potassium-feldspar group include orthoclase, microcline, and sanidine. **Plagioclase feldspars** form a complex solid solution series ranging in composition from $NaAlSi_3O_8$ (albite) through $CaAl_2Si_2O_8$ (anorthite). A general formula for the series is $(Na,Ca)(Al,Si)Si_2O_8$. Potassium feldspars and plagioclase feldspars can usually be distinguished on the basis of optical properties such as twinning by examination with a petrographic microscope, but they may be difficult or impossible to differentiate megascopically. Potassium feldspars are generally considered to be somewhat more abundant overall in sedimentary rocks than plagioclase feldspars; however, plagioclase is more abundant in sandstones derived from volcanic rocks.

Feldspars are chemically less stable than quartz and are more susceptible to chemical destruction during weathering and diagenesis. Because they are also softer than quartz, feldspars become more readily rounded during transport. They also appear to be somewhat more prone to mechanical shattering and breakup owing to their cleavage. Feldspars are less likely than quartz to survive several episodes of recycling, although they can survive more than one cycle if weathering occurs in a moderately arid or cold climate. Owing to this possibility of recycling, the presence of a few feldspar grains in a sedimentary rock does not necessarily mean that the rock is com-

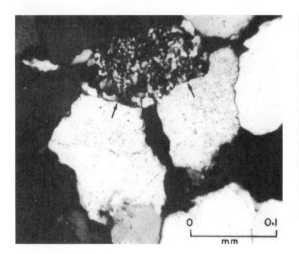

FIGURE 6.1 Photomicrograph of a
polycrystalline quartz grain (arrow)
surrounded by monocrystalline
quartz grains. Polarized light.

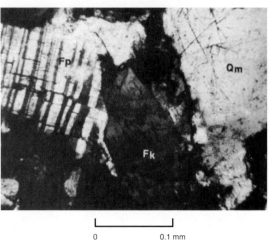

FIGURE 6.2 Plagioclase feldspar
grain (Fp), potassium feldspar grain
(Fk), and monocrystalline quartz
(Qm). Polarized light.

posed of first-cycle sediments derived directly from crystalline igneous or metamorphic rocks. On the other hand, a high content of feldspars, particularly on the order of 25 percent or more, probably indicates derivation directly from crystalline source rocks.

Clay Minerals. Clay minerals make up only a small percentage ($<{\sim}5$ percent) of most sandstones, where they are present as part of the matrix. Because of their small size, clay minerals cannot be identified by routine petrographic microscopy. They must be identified by X-ray diffraction techniques, electron microscopy (Fig. 6.3), or other nonoptical methods. Clay minerals are compositionally diverse. They belong to the phyllosilicate mineral group, which is characterized by two-dimensional layer structures arranged in indefinitely extending sheets.

The most common clay mineral groups are **illite** [$K_2(Si_6Al_2)Al_4O_{20}(OH)_4$], **smectite** (montmorillonite) [$(Al,Mg)_8(Si_4O_{10})_3(OH)_{10}{\cdot}12H_2O$], **kaolinite** [$Al_2Si_2O_5(OH)_4$], and **chlorite** [$(Mg,Fe)_5(Al,Fe^{3+})_2Si_3O_{10}(OH)_8$]. Kaolinite is a two-layer clay; the others are three-layer clays. Smectite is a clay-mineral group, with **montmorillonite** a principal variety of smectite. Clay minerals form principally as secondary minerals during subaerial weathering and hydrolysis, although they can also form by subaqueous weathering in the marine environment and during burial diagenesis.

Accessory Minerals. Minerals that have an average abundance in sedimentary rocks less than about 1 to 2 percent are called accessory minerals. These minerals include the common **micas,** muscovite (white mica) and biotite (dark mica), and a large number of so-called heavy minerals, which are denser than quartz.

The average abundance of coarse micas in siliciclastic sedimentary rocks is less than about 0.5 percent, although some sandstones may contain 2 to 3 percent. Micas are distinguished from other minerals by their platy or flaky habit (Fig. 6.4). Muscovite is chemically more stable than biotite and is commonly much more abundant in sand-

FIGURE 6.3 Electronmicrograph of kaolinite clay minerals magnified 4700×. Note the arrangement of the pseudohexagonal kaolinite plates in distinct "books." (Photograph by M. B. Shaffer.)

0   10 μ

stones than biotite. Micas are derived particularly from metamorphic source rocks as well as from some plutonic igneous rocks.

Minerals that have a specific gravity greater than about 2.9 are called **heavy minerals.** These minerals include both chemically stable and unstable or labile varieties as shown in Table 6.1. Stable heavy minerals such as zircon and rutile can survive multiple recycling and are commonly rounded, indicating that the last source was sedimentary. Less stable minerals, such as magnetite, pyroxenes, and amphiboles, are less likely to survive recycling. They are commonly first-cycle sediments that reflect the composition of proximate source rocks. Thus, heavy minerals are useful indicators of sediment source rocks because different types of source rocks yield different suites of heavy minerals. Heavy minerals are derived from a variety of igneous, metamorphic, and sedimentary rocks.

Because of their low abundance in sandstones, heavy minerals are commonly concentrated for study by separating them from the light mineral fraction by using heavy liquids such as bromoform or sodium polytungstate (e.g., Lindholm, 1987, p. 214). In this separation process, disaggregated sediment is stirred into a heavy liquid contained in a funnel placed inside a fume hood. (All heavy liquids are toxic and must be handled with extreme care.) The light minerals float on the surface of the heavy liquid, but the heavy minerals gradually sink into the stem of the funnel where they can be drawn off and separated from the light fraction. After the heavy liquid has been washed off in a suitable solvent, these heavy mineral concentrates can then be mounted on a glass microscope slide (Fig. 6.5) and studied with a petrographic microscope.

Rock Fragments. Pieces of ancient source rocks that have not yet disintegrated to yield individual mineral grains are called rock fragments or clasts. Rock fragments make up about 15 to 20 percent of the grains in the average sandstone; however, the

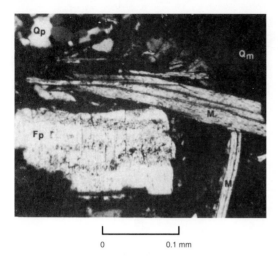

FIGURE 6.4 Photomicrograph showing a coarse muscovite mica grain (M). Polycrystalline quartz (Qp), monocrystalline quartz (Qm), and plagioclase feldspar (Fp) grains are also visible. Polarized light.

FIGURE 6.5 Heavy mineral grains from a modern shelf sand as viewed in a grain mount. The minerals shown here include orthopyroxene (Op), hornblende (Hb), and magnetite (Ma).

rock-fragment content of sandstones is highly variable, ranging from zero to more than 95 percent. Fragments of any kind of igneous, metamorphic, or sedimentary rock can occur in sandstones (Fig. 6.6); however, clasts of fine-grained source rocks are most likely to be preserved as sand-size fragments (Boggs, 1968). Coarse-grained source rocks such as granites do not yield fine sand-size clasts. The most common rock fragments in sandstones are clasts of volcanic rocks, volcanic glass (in younger rocks), and fine-grained metamorphic rocks such as slate, phyllite, and schist. Sand-size fragments of silica-cemented siltstone, very fine-grained sandstone, and shale are less common.

Rock fragments are particularly important in studies of sediment source rocks. They are moderately easily identified, and they are more reliable indicators of source rock types than are individual minerals such as quartz or feldspar, which can be derived from different types of source rocks.

Mineral Cements

The framework grains in most siliciclastic sedimentary rocks are bound together by some type of mineral cement. These cementing materials may be either silicate minerals such as quartz and opal or nonsilicate minerals such as calcite and dolomite. Quartz is the most common silicate mineral that acts as a cement. In most sandstones, the quartz cement is chemically attached to the crystal lattice of existing quartz grains, forming rims of cement called **overgrowths** (Fig. 6.7). Such overgrowths that retain crystallographic continuity of a grain are said to be **syntaxial**. Because syntaxial overgrowths are optically continuous with the original grain, they go to extinction in the same position as the original grain when rotated on the stage of a polarizing microscope. Overgrowths can be recognized by a line of impurities or bubbles that mark the surface of the original grain. Quartz overgrowths are particularly common in quartz-

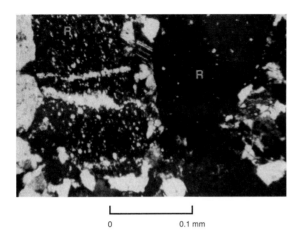

0   0.1 mm

FIGURE 6.6   Two large volcanic (?) rock fragments (R) in a fine-grained lithic sandstone. Polarized light.

0   0.1 mm

FIGURE 6.7   Quartz overgrowths (arrows) on monocrystalline quartz grains. The original surfaces of the grains are marked by a dark band caused by the presence of tiny bubbles or mineral inclusions. Polarized light.

rich sandstones. Less commonly, quartz cement is present as microcrystalline quartz, which has a fine-grained, crystalline texture similar to that of chert. When silica cement is deposited as microcrystalline quartz, it forms a mosaic of very tiny quartz crystals that fill the interstitial spaces among framework silicate grains (Fig. 6.8). Not uncommonly, the crystals next to the framework grains are slightly elongated and are oriented normal to the surfaces of the framework grains. More rarely, opal occurs as a cement in sandstones, particularly in sandstones rich in volcanogenic materials. Like quartz and microcrystalline quartz (chert), opal is also composed of SiO_2, but, unlike these minerals, opal contains some water and lacks a definite crystal structure. Thus, it is said to be amorphous. Opal is metastable and crystallizes in time to microcrystalline quartz.

Carbonate minerals are the most abundant nonsilicate mineral cements in siliciclastic sedimentary rocks. Calcite is a particularly common carbonate cement. It is precipitated in the pore spaces among framework grains, typically forming a mosaic of smaller crystals (Fig. 6.9). These crystals adhere to the larger framework grains and bind them together. Less common carbonate cements are dolomite and siderite (iron carbonate). Other minerals that act as cements include the iron oxide minerals hematite and limonite, feldspars, anhydrite, gypsum, barite, clay minerals, and zeolite minerals. Zeolites are hydrous aluminosilicate minerals that occur as cements primarily in volcaniclastic sedimentary rocks (below).

Chemical Composition

Sedimentologists have traditionally placed relatively little emphasis on study of the chemical composition of siliciclastic sedimentary rocks, in sharp contrast to geologists who study igneous and metamorphic rocks. This lack of interest is probably due mainly to the common belief that chemical composition is less useful than mineral composition for interpreting depositional history and provenance. Also, the present

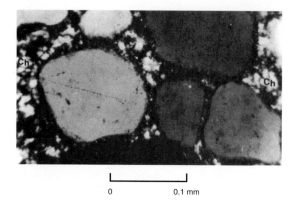

FIGURE 6.8   Rounded, monocrystalline quartz grains cemented by
microcrystalline quartz (chert) (Ch).
Polarized light.

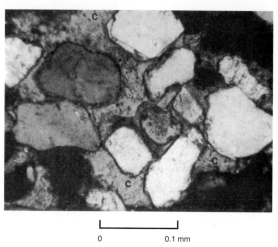

FIGURE 6.9   Rounded, monocrystalline quartz grains cemented by
calcite (C). Polarized light.

chemical composition of sedimentary rocks may not accurately reflect their composition at the time of deposition because crystallization of new minerals during sediment
burial and diagenesis can change the original chemical composition. The formerly
high cost of doing chemical analyses is an additional factor that helped to discourage
extensive chemical studies of sedimentary rocks. Several new tools, such as the electron probe microanalyzer and X-ray fluorescence equipment, now enable rapid and
comparatively inexpensive chemical analyses of rock composition. These new tools, as
well as changing attitudes regarding the significance of chemical composition, are
causing sedimentologists to develop a deeper interest in the chemistry of sedimentary
rocks. For example, the trace-element (Ti, Cr, Mn) content of individual heavy minerals such as ilmenite has been used successfully to discriminate among minerals
derived from different kinds of source rocks (e.g., Darby and Tsang, 1987).

Because most grains in siliciclastic sedimentary rocks are derived from various
types of igneous, metamorphic, and sedimentary rocks, the mineralogy and chemical
composition of siliciclastic rocks are clearly a function of parent-rock composition.
Nonetheless, sedimentary rocks display distinct chemical differences from parent
source rocks owing to chemical changes that occur during weathering and diagenesis.
For example, they tend to be enriched in silica and depleted in iron, magnesium, calcium, sodium, and potassium compared to the parent rocks. This enrichment in silica
occurs because siliceous minerals are resistant to chemical weathering; thus, silica is
concentrated in weathering residues with respect to more soluble cations. Also, the
superior chemical stability of SiO_2 minerals such as quartz and chert causes sedimentary rocks to become progressively enriched in these minerals during multiple recycling. Thus, overall silica content increases at the expense of less stable, iron- and
magnesium-rich minerals.

The average chemical composition of some sandstones reported in the literature
is shown in Table 6.2. I stress that this table shows average composition of a few sandstones. Specimens of sandstone from other formations may have chemical compositions that deviate considerably from these average values. Silicon, expressed as SiO_2,

TABLE 6.2 Average chemical composition (weight percent) of sandstones from some North American formations*

	(1) 11	(2) 23	(3) 30	(4) 16	(5) 18	(6) 12	(7) 119	(8) 12	(9) 59
$n =$									
SiO_2	86.5	67.8	65.6	56.9	56.2	68.4	70.6	37.3	50.3
TiO_2	0.53	0.95	0.91	1.42	0.89	0.69	0.64	0.34	0.64
Al_2O_3	5.71	15.4	15.1	12.3	15.3	13.5	12.6	7.91	14.0
Fe_2O_3 (t)	2.69	6.46	6.09	6.18	6.48	5.30	4.97	3.18	6.40
MnO	0.02	0.07	0.15	0.11	0.07	0.09	0.08	0.10	0.13
MgO	0.69	1.73	1.82	4.20	2.35	1.68	1.51	1.07	3.25
CaO	0.05	0.42	1.94	5.82	5.74	2.38	1.61	26.0	9.90
Na_2O	0.02	1.07	0.87	1.92	1.28	3.15	2.76	0.92	—
K_2O	1.55	2.74	3.03	1.90	2.80	2.62	2.20	0.51	2.09
P_2O_5	0.02	0.16	0.17	0.17	0.17	0.18	0.02	0.10	0.21
V (ppm)	51	123	159	100	126	71	79	103	—
Cr	55	82	88	225	71	55	44	31	—
Ni	19	231	58	130	49	30	8	5	49
Zn	29	52	104	84	114	69	—	66	91
Rb	60	123	133	72	125	93	—	10	79
Sr	29	134	113	233	168	310	110	879	267
Y	17	31	40	21	35	36	37	15	29
Zr	417	238	260	191	187	333	413	58	118

Source: Argast and Donnelly, 1987, Jour. Sed. Petrology, v. 57, p. 813–823: Society Economic Mineralogists and Paleontologists, Tulsa, Okla.

Note: Iron is reported as total Fe_2O_3; n is the number of samples in each average. A dash indicates no reported value.

*(1) Shawangunk Formation near Ellenville, New York (quartz arenite)

(2) Millport Member of the Rhinestreet Formation, Elmira, New York (lithic arenite/wacke)

(3) Oneota Formation, Unadilla, New York (lithic arenite/wacke)

(4) Cloridorme Formation, St. Yvon and Gros Morne, Quebec (lithic arenite/wacke)

(5) Austin Glen Member of Normanskill Formation, Poughkeepsie, New York (lithic arenite/wacke)

(6) Renesselaer Member of the Nassau Formation, near Grafton, New York (feldspathic arenite/wacke)

(7) Renesselaer Member, averages of analyses from Ondrick and Griffiths (1969) (feldspathic arenite/wacke)

(8) Rio Culebrinas Formation, La Tosca, Puerto Rico (fossiliferous volcaniclastics)

(9) Turbidites from DSDP site 379A (lithic arenites/wacke)

is the most abundant chemical constituent in all types of sandstones owing to the abundance of quartz in sandstones and the presence of silicon in all silicate minerals. Aluminum (Al_2O_3) is moderately abundant in sandstones containing abundant feldspars or in rock-fragment-rich sandstones that contain a matrix of clay minerals. It is much less abundant in quartz-rich sandstones, which commonly do not have a clay matrix. On average, iron, magnesium, calcium, sodium, and potassium are all less abundant in sandstones than is aluminum. Relative concentrations of these elements vary as a function of the mineralogy of the sand-size grains and the types of matrix clay minerals and diagenetic cements in the rock. For example, sandstones with abundant calcium carbonate cement or carbonate fossils may have anomalously high calcium content.

Classification of Sandstones

Descriptive classification of sandstones is based fundamentally on mineralogy, although grain size (matrix) plays a role in some classifications. Chemical composition is not used in sandstone classification. Although mineralogy is the principal basis for classifying sandstones, finding a classification that is suitable for all types of sandstones and is acceptable to most geologists has proven to be an elusive goal. In fact, more than 50 different classifications for sandstones have been proposed (Friedman and Sanders, 1978), but none has received widespread acceptance. Classifications that are all-inclusive tend to be too complicated and unwieldy for general use, and classifications that are oversimplified may convey too little useful information.

Textural Nomenclature of Mixed Sediments. Unconsolidated siliciclastic sediment is called gravel (dominance of >2 mm-size grains), sand (1/16–2 mm), or mud (<1/16 mm), depending upon grain size. The lithified rock equivalents of these sediments are conglomerate, sandstone, and shale (mudrock). Because many siliciclastic sedimentary rocks are composed of grains of mixed sizes, it is not always easy to decide whether a rock should be called a conglomerate, a sandstone, or a shale. It might be difficult to decide, for example, whether a sedimentary rock composed of nearly equal portions of sand-size and mud-size particles should be called a sandstone or a shale. Various classification schemes have been devised for naming texturally mixed sediments and sedimentary rocks, most of which make use of triangular texture diagrams such as those shown in Figure 6.10.

The textural classification illustrated in Figure 6.10A includes particles ranging in size from mud (clay and fine silt) to gravel. Note that the textural boundaries in this classification scheme are not entirely symmetrical. Ideally, we might expect the boundary between gravel and mud-sand to be set at 50 percent; however, this is not always done, as Figure 6.10A shows. Because particles of gravel size are commonly less abundant than sand and mud particles, many geologists consider a sediment with as little as 30 percent gravel-size fragments to be a gravel. If sediments contain only particles of sand size and smaller, a textural classification scheme such as in Figure 6.10B or 6.10C that uses sand, silt, and clay as end members of the classification is more appropriate. Once the textural nomenclature of siliciclastic sediment or sedimentary rocks has been established, further classification within each textural group can be made on the basis of composition.

Mineralogical Classification. Most sandstones are made up of mixtures of a very small number of dominant components. Quartz plus chert, feldspars, and rock fragments such as volcanic clasts are the only framework constituents that are commonly abun-

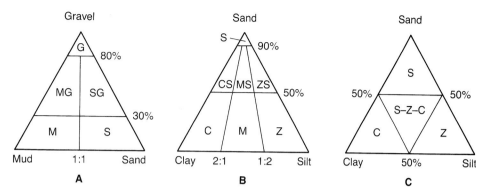

FIGURE 6.10 Nomenclature of mixed sediments. A, B. Simplified from Folk. C. After Robinson. G = gravel, S = sand, M = mud, C = clay, Z = silt, MG = muddy gravel, SG = sandy gravel, CS = clayey sand, MS = muddy sand, ZS = silty sand. (From Folk, R. L., 1954, The distinction between grain size and mineral composition in sedimentary rock nomenclature: Jour. Geology, v. 62. Fig. 1a, p. 346, and 1b, p. 349, reprinted by permission of University of Chicago Press. Robinson, G. W., 1949, Soils, their origin, constitution, and classification, 3rd ed.: Murby, London.)

dant enough to be important in sandstone classification. In addition to framework grains, matrix (consisting of clay minerals and other fine-size, <0.03 mm, minerals) may fill interstitial spaces among these grains. In spite of the very simple composition of sandstones, geologists have not been able to agree on a single acceptable sandstone classification. Published classifications range from those that have a strong genetic orientation to those based strictly on observable, descriptive properties of sandstones. Most authors of sandstone classifications use a classification scheme that involves a QFR or QFL plot. These plots are triangular diagrams on which quartz (Q), feldspars (F), and rock fragments (R or L) are plotted as end members at the poles of the classification triangle. There are numerous possible ways that such a triangle can be subdivided into classification fields, and geologists have explored the full range of these possibilities (see reviews by Klein, 1963; Boggs, 1967b; Okada, 1971; Yanov, 1978).

One of the simplest and easiest to use classifications is that of Gilbert (Williams, Turner, and Gilbert, 1982), shown in Figure 6.11. In this classification, sandstones that are effectively free of matrix (<5 percent) are classified as **quartz arenites, feldspathic arenites,** or **lithic arenites,** depending upon the relative abundance of QFL constituents. If matrix can be recognized (at least 5 percent), the terms **quartz wacke, feldspathic wacke,** and **lithic wacke** are used instead. I recommend this classification for its simplicity and its utility for field study as well as for more detailed petrographic studies in the laboratory.

Sandstone classifications that are somewhat more complex than Gilbert's and that have been rather widely used by American geologists include the classifications of McBride (1963) and of Folk, Andrews, and Lewis (1970), shown in Figure 6.12. These classifications do not include matrix as part of the classification scheme; however, a separate textural maturity classification such as that shown in Figure 6.13 can be used to supplement these classifications.

The name **arkose,** which appears in only one of the classifications shown here, is often used informally by geologists for any feldspathic arenite that is particularly rich (>~25 percent) in feldspars. Another term in general use is **graywacke.** This name is commonly applied to matrix-rich sandstones of any composition that have undergone deep burial, have a chloritic matrix, and are dark gray to dark green, very hard, and

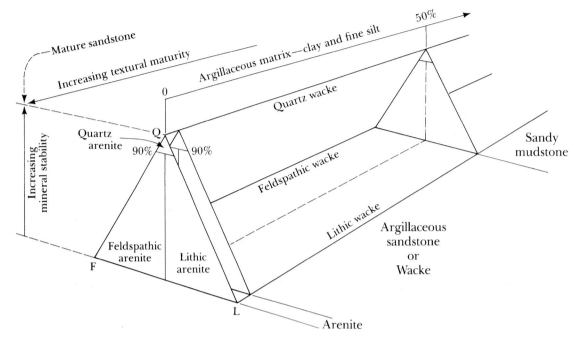

FIGURE 6.11 Classification of sandstones on the basis of three mineral components: Q = quartz, chert, quartzite fragments; F = feldspars; L = unstable, lithic grains (rock fragments). Points within the triangles represent relative proportions of Q, F, and R end members. Percentage of argillaceous matrix is represented by a vector extending toward the rear of the diagram. The term arenite is restricted to sandstones essentially free of matrix; sandstones containing matrix are wackes. (From Williams, H., F. J. Turner, and C. M. Gilbert, Petrography, an introduction to the study of rocks in thin sections, 2nd ed. W. H. Freeman and Company, San Francisco. © 1982, Fig. 13.1, p. 327. Modified from Dott, R. H., Jr., 1964, Wacke, graywacke, and matrix—what approach to immature sandstone classification?: Jour. Sed. Petrology, v. 34, Fig. 3, p. 629, reprinted by permission of Society of Economic Paleontologists and Mineralogists, Tulsa, Okla.)

dense. This term has been much misused, and its continued used is controversial. Some geologists think that the term should be abandoned entirely and that we should substitute the word wacke for graywacke. In any case, the name is best restricted to field use and should not be used as a petrographic term.

General Characteristics of Major Classes of Sandstones

The preceding discussion indicates that on the basis of framework composition, sandstones can be divided by mineralogy into three major groups: quartz arenites, feldspathic arenites, and lithic arenites. Some general characteristics of each of these major sandstone clans are discussed below.

Quartz Arenites. Quartz arenites are composed of more than 90 percent siliceous grains (quartz, chert, quartzose rock fragments). They are commonly white or light gray but may be stained red, pink, yellow, or brown by iron oxides. They are generally well lithified and well cemented with silica or carbonate cement; however, some are porous and friable. Quartz arenites typically occur in association with assemblages of rocks deposited in stable cratonic environments such as eolian, beach, and shelf environ-

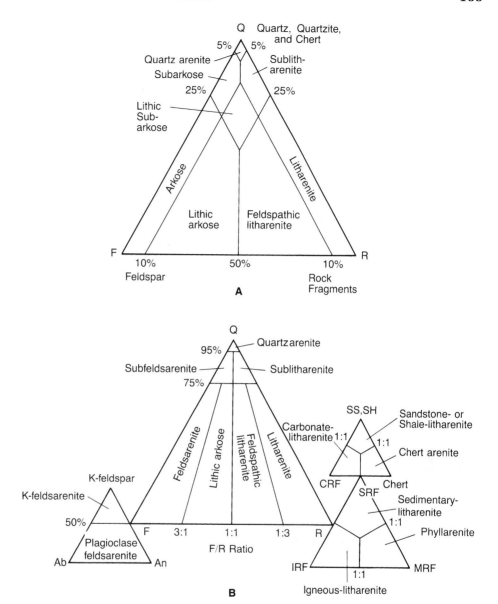

FIGURE 6.12 Classification of sandstones according to (A) McBride and (B) Folk, Andrews, and Lewis. In Folk et al.'s classification, chert is included with rock fragments at the R pole, and granite and gneiss fragments are included with feldspars at the F pole. SS = sandstone, SH = shale, CRF = carbonate rock fragments, SRF = sedimentary rock fragments, IRF = igneous rock fragments, MRF = metamorphic rock fragments (A, from McBride, E. F., 1963, A classification of common sandstones: Jour. Sed. Petrology, v. 34, Fig. 1, p. 667, reprinted by permission of Society of Economic Paleontologists and Mineralogists, Tulsa, Okla. B, from Folk, R. L., P. B. Andrews, and D. W. Lewis, 1970, Detrital sedimentary rock classification and nomenclature for use in New Zealand: New Zealand Jour. of Geology and Geophysics, v. 13, Fig. 8, p. 955, and Fig. 9, p. 959, British Crown copyright, reprinted by permission.)

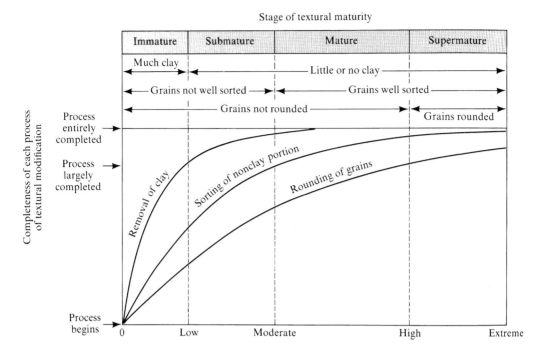

FIGURE 6.13 Textural maturity classification of Folk. Textural maturity of sands is shown as a function of input in kinetic energy. (From Folk, R. L., 1951, Stages of textural maturity in sedimentary rocks: Jour. Sed. Petrology, v. 21, Fig. 1, p. 128, reprinted by permission of Society of Economic Paleontologists and Mineralogists, Tulsa, Okla.)

ments. Thus, they tend to be interbedded with shallow-water carbonates and, in some cases, with feldspathic sandstones. Most quartz arenites are texturally mature to supermature according to Folk's (1951) textural maturity classification (Fig. 6.13). Cross-bedding is particularly characteristic of these sandstones, and ripple marks are moderately common. Fossils are rarely abundant, possibly owing to poor preservation or to the eolian origin of some quartz arenites, but fossils may be present. Also, trace fossils such as burrows of the **Skolithos** facies may be locally abundant in some shallow-marine quartz arenites. Quartz arenites are common in the geologic record. Pettijohn (1963) estimates that they make up about one-third of all sandstones.

Quartz arenites can originate as first-cycle deposits derived from primary crystalline or metamorphic rocks, but they are more likely to be the product of multiple recycling of quartz grains from sedimentary source rocks. If they are first-cycle deposits, they must have formed under weathering, transport, and depositional conditions so vigorous that most grains chemically less stable than quartz were eliminated (e.g., Johnsson, Stallard, and Mead, 1988). Conceivably, extensive chemical leaching under hot, humid, low-relief weathering conditions; prolonged transport by wind; intensive reworking in the surf zone; or a combination of these processes might be adequate to generate a first-cycle quartz arenite. Most quartz arenites are probably polycyclic, and their history may have included at least one episode of eolian transport, although not necessarily during the last depositional cycle.

Quartz arenites are very common rocks in the geologic record, particularly in Mesozoic and Paleozoic stratigraphic sequences. Some well-known examples of quartz arenites in North America include the Ordovician St. Peter Sandstone in the midcontinent United States, the Jurassic Navajo Sandstone of the Colorado Plateau, the Ordovician Eureka Quartzite in Nevada and California, parts of the Cretaceous Dakota Sandstone in the Colorado Plateau and Great Plains, and many Cambro-Ordovician sandstones in the Upper Mississippi Valley. Numerous examples of quartz arenites are also known from other continents. Pettijohn, Potter, and Siever (1987, p. 179–184) list additional examples of quartz arenites from North America, Europe, and other parts of the world.

Feldspathic Arenites. Feldspathic arenites contain less than 90 percent quartz, more feldspar than unstable rock fragments, and minor amounts of other minerals such as micas and heavy minerals. Some feldspathic arenites are colored pink or red owing to the presence of potassium feldspars or iron oxides; others are light gray to white. They are typically medium to coarse grained and may contain high percentages of subangular to angular grains. Matrix content may range from trace amounts to more than 15 percent, and sorting of framework grains can range from moderately well sorted to poorly sorted. Thus, feldspathic sandstones are commonly texturally immature or submature.

Feldspathic arenites are not characterized by any particular kinds of sedimentary structures. Bedding may range from essentially structureless to parallel-laminated or cross-laminated. Fossils may be present in marine beds. Feldspathic arenites typically occur in cratonic or stable shelf settings, where they may be associated with conglomerates, shallow-water quartz arenites or lithic arenites, carbonate rocks, or evaporites. Less typically, they occur in sedimentary sequences that were deposited in unstable basins or other deeper-water, mobile-belt settings. Feldspathic arenites of the latter types, which are commonly matrix-rich and well indurated owing to deep burial, are often called feldspathic graywackes. The abundance of feldspathic arenites in the geologic record is not well established. Pettijohn (1963) estimates that arkoses make up about 15 percent of all sandstones. If feldspathic graywackes are included, feldspathic arenites are probably more abundant than 15 percent.

Some arkoses originate essentially *in situ* by disintegration of granite and related rocks to produce a granular sediment called grus. These residual arkosic materials may be shifted a short distance downslope and deposited as fans or aprons of waste material, commonly referred to as clastic wedges. These fans may extend into basins and become intercalated or interbedded with better-stratified and better-sorted sediments. Other feldspathic arenites undergo considerable transport and reworking by rivers or the sea before they are deposited. These reworked sandstones commonly contain less feldspars than do residual arkoses, and they are better sorted and better rounded.

Most feldspathic sandstones are derived from granitic-type primary crystalline rocks, such as coarse granite or metasomatic rocks containing abundant potassium feldspar. Feldspathic arenites containing feldspars that are dominantly plagioclase, derived from igneous rocks such as quartz diorites or from volcanic rocks, are also known. The preservation of large quantities of feldspars during weathering appears to require that feldspathic arenites originate in either of the following:

1. very cold or very arid climates, where chemical weathering processes are inhibited
2. warmer, more humid climates where marked relief of local uplifts allows rapid erosion of feldspars before they can be decomposed

Although some feldspars may survive recycling from a sedimentary source, it appears unlikely that sedimentary source rocks can furnish enough feldspar to produce a feldspathic arenite or arkose.

Feldspathic arenites occur in sedimentary sequences of all ages, although they appear to be particularly abundant in Mesozoic and Paleozoic strata. Some common examples include the Old Red Sandstone (Carboniferous) in Scotland, the Triassic Newark Group in the New Jersey area, the Pennsylvanian Fountain and Lyons formations of the Colorado Front Range, and the Paleocene Swauk Formation of Washington. The Swauk Formation is particularly interesting because it is a plagioclase arkose.

Lithic Arenites. Lithic arenites are an extremely diverse group of rocks that are characterized by generally high content of unstable rock fragments. They contain less than 90 percent quartzose grains and more unstable rock fragments than feldspars. Colors may range from light gray, salt-and-pepper to uniform medium to dark gray. Many lithic arenites are poorly sorted; however, sorting ranges from well sorted to very poorly sorted. Quartz and many other framework grains are generally poorly rounded. Lithic arenites tend to contain substantial amounts of matrix, much of which may be of secondary origin. Thus, most lithic arenites are texturally immature to submature. Lithic arenites may range from irregularly bedded, laterally restricted, cross-stratified fluvial units to evenly bedded, laterally extensive, graded, marine turbidite units. They may occur in association with fluvial conglomerates and other fluvial deposits or in association with deeper-water marine conglomerates, pelagic shales, cherts, and submarine basalts. Lithic arenites include sandstones that many geologists refer to as **graywackes.** As mentioned, these graywackes differ from normal lithic arenites in that they are dark gray to dark green, are well indurated or lithified, and commonly have a matrix consisting of secondary chlorite. Pettijohn (1963) estimates that lithic arenites and graywackes together make up nearly one-half of all sandstones.

Lithic arenites are compositionally immature sandstones that originate under conditions favoring the production and deposition of large volumes of relatively unstable materials. The mechanically weak character of many of the lithic fragments in these sandstones suggests that they are probably derived from rugged, high-relief source areas. Lithic arenites may be deposited in nonmarine settings in proximal alluvial fans or other fluvial environments. Alternatively, they may be deposited in marine foreland basins (Chapter 18) adjacent to fold-thrust belts, or they may be transported by large rivers off the continent into deltaic or shallow shelf environments. Lithic sediments deposited in coastal areas may be retransported into deeper water by turbidity currents or by other sediment gravity-flow mechanisms. These deeper-water sediments are particularly likely to undergo deep burial and incipient metamorphism, leading to development of characteristics generally ascribed to graywackes.

Common examples of lithic sandstones include the Paleozoic sandstone sequences of the central Appalachians in the eastern United States (e.g., Ordovician Juniata Formation, Mississippian Pocono Formation, Pennsylvanian Pottsville Formation); many sandstones associated with the Coal Measures throughout the world; many Jurassic and Cretaceous sandstones of the U.S. and Canadian Rocky Mountains and the U.S. West Coast (e.g., Cretaceous Belly River Sandstone of Canada, Jurassic Franciscan Formation of California); and Tertiary sandstones of the Gulf Coast, the West Coast, and the Alps.

Volcaniclastic sandstones are a special kind of lithic arenite composed primarily of volcanic detritus. Volcaniclastic sandstones may be made up largely of pyroclastic materials that have been transported and reworked, or they may contain volcanic detritus derived by weathering of older volcanic rocks. They are especially characterized by the presence of euhedral feldspars, pumice fragments, glass shards, and volcanic rock fragments, and they generally have a very low quartz content (e.g., Boggs, 1992, p. 197–209).

Other Sandstones. The sandstones discussed above are composed of constituents derived primarily by weathering of preexisting rocks or by explosive volcanism. A few less abundant types of "sandstones" are known whose constituents formed largely within the depositional basin by chemical or biochemical processes. These rocks, called hybrid sandstones by some authors, include such uncommon varieties of sandstones as greensands (glauconitic sands), phosphatic sandstones, and calcarenaceous sandstones (composed of sand-size carbonate grains). These rocks are not true sandstones (siliciclastic rocks) but rather are chemical/biochemical sedimentary rocks (Chapters 7 and 8).

6.3   CONGLOMERATES

The term conglomerates is used in this book as a general class name for sedimentary rocks that contain a substantial fraction (at least 30 percent) of gravel-size (>2mm) particles (Fig. 6.14). Breccias, which are composed of very angular, gravel-size fragments, are not distinguished from conglomerates in the succeeding discussion. Conglomerates are common in stratigraphic sequences of all ages but probably make up less than 1 percent by weight of the total sedimentary rock mass (Garrels and Mackenzie, 1971, p. 40). They are closely related to sandstones in terms of origin and depositional mechanisms, and they contain some of the same kinds of sedimentary structures (e.g., tabular and trough cross-bedding, graded bedding).

Particle Composition

Conglomerates may contain gravel-size pieces of individual minerals such as quartz; however, most of the gravel-size framework grains are rock fragments (clasts). Individ-

FIGURE 6.14   Poorly sorted fluvial (river-deposited) conglomerate, Holocene, Oregon. Note hammer for scale.

ual sand- or mud-size mineral grains are commonly present as a matrix. Any kind of igneous, metamorphic, or sedimentary rock may be present in a conglomerate, depending upon source rocks and depositional conditions. Some conglomerates are composed of only the most stable and durable kinds of clasts (quartzite, chert, vein-quartz). Stable conglomerates composed mainly of a single clast type are referred to by Pettijohn (1975) as **oligomict conglomerates.** Most oligomict conglomerates were probably derived from mixed parent-rock sources that included less stable rock types. Continued recycling of mixed ultrastable and unstable clasts through several generations of conglomerates ultimately led to selective destruction of the less stable clasts and concentration of stable clasts. Conglomerates that contain an assortment of many kinds of clasts are **polymict conglomerates.** Polymict conglomerates that are made up of a mixture of largely unstable or metastable clasts such as basalt, limestone, shale, and metamorphic phyllite are commonly called **petromict conglomerates** (Pettijohn, 1975). Almost any combination of these clast types is possible in a petromict conglomerate. The matrix of conglomerates commonly consists of various kinds of clay minerals and fine micas and/or silt- or sand-size quartz, feldspars, rock fragments, and heavy minerals. The matrix may be cemented with quartz, calcite, hematite, clay, or other cements.

Classification

Conglomerates can originate by several processes, as shown in Table 6.3. We are interested most in epiclastic conglomerates, which form by breakdown of older rocks

TABLE 6.3 Fundamental genetic types of conglomerates and breccias

Major types	Subtypes	Origin of clasts
Epiclastic conglomerate and breccia	Extraformational conglomerate and breccia	Breakdown of older rocks of any kind through the processes of weathering and erosion; deposition by fluid flows (water, ice) and sediment gravity flows
	Intraformational conglomerate and breccia	Penecontemporaneous fragmentation of weakly consolidated sedimentary beds; deposition by fluid flows and sediment gravity flows
Volcanic breccia	Pyroclastic breccia	Explosive volcanic eruptions, either magmatic or phreatic (steam) eruptions; deposited by air-falls or pyroclastic flows
	Autobreccia	Breakup of viscous, partially congealed lava owing to continued movement of the lava
	Hyaloclastic breccia	Shattering of hot, coherent magma into glassy fragments owing to contact with water, snow, or water-saturated sediment (quench fragmentation)
Cataclastic breccia	Landslide and slump breccia	Breakup of rock owing to tensile stresses and impact during sliding and slumping of rock masses
	Tectonic breccia: fault, fold, crush breccia	Breakage of brittle rock as a result of crustal movements
	Collapse breccia	Breakage of brittle rock owing to collapse into an opening created by solution or other processes
Solution breccia		Insoluble fragments that remain after solution of more soluble material; e.g., chert clasts concentrated by solution of limestone
Meteorite impact breccia		Shattering of rock owing to meteorite impact

Source: Modified from Pettijohn, F. J., 1975, Sedimentary rocks, 3rd ed., Harper & Row, New York, p. 165.

TABLE 6.4 Classification of conglomerates and diamictites on the basis of clast stability and fabric support

Percentage of ultrastable clasts	Type of fabric support	
	Clast-supported	Matrix-supported
>90 <90	Quartzose conglomerate Petromict conglomerate	Quartzose diamictite Petromict diamictite

through the processes of weathering and erosion. Epiclastic conglomerates that are so rich in gravel-size framework grains that the gravel-size grains touch and form a supporting framework are called **clast-supported** conglomerates. Clast-poor conglomerates that consist of sparse gravels supported in a mud/sand matrix are called **matrix-supported** conglomerates. Boggs (1992, p. 212) suggests that clast-supported conglomerates be referred to simply as conglomerates and that matrix-supported conglomerates be called **diamictites.** Conglomerates and diamictites can be further divided on the basis of clast stability into **quartzose conglomerate/diamictite** and **petromict conglomerate/diamictite** on the basis of relative abundance of these clast types (Table 6.4). Further classification on the basis of clast type (igneous, metamorphic, sedimentary) can be made if desired (Fig. 6.15); however, such classification may not be necessary in many cases.

Origin and Occurrence of Conglomerates

Quartzose conglomerates are derived from metasedimentary rocks containing quartzite beds, igneous rocks containing quartz-filled veins, and sedimentary sequences, particularly limestone, containing chert beds. As mentioned, less stable rock types must have been destroyed by weathering, erosion, and sediment transport, perhaps through several cycles of transport, to produce a residuum of stable clasts. Because quartzose clasts represent only a small fraction of a much larger original body of rock, the total volume of quartzose conglomerates is small. They tend to occur as thin, pebbly layers or lenses of pebbles in dominantly sandstone units. They may be either clast-supported or matrix-supported. Although their overall volume is small, quartzose conglomerates are common in the geologic record ranging from the Precambrian to the Tertiary. Most quartzose conglomerates appear to be of fluvial origin (Chapter 10) and are probably deposited mainly in braided streams. Marine, wave-worked quartzose conglomerates that were deposited in the littoral (beach) environment are also known.

Most **petromict conglomerates** are polymict conglomerates that consist of a variety of metastable clasts. They can be derived from many kinds of plutonic igneous, volcanic, metamorphic, and sedimentary rocks, although the clasts in a particular conglomerate may be dominantly one or another of these rock types. Thus, a particular conglomerate may be a limestone conglomerate, a basalt conglomerate, a schist conglomerate, and so on. Conglomerates composed dominantly of plutonic igneous clasts appear to be uncommon, probably because plutonic rocks such as granites tend to disintegrate into sand-size fragments rather than forming larger blocks. The volume of ancient petromict conglomerates is far greater than that of quartzose conglomerates. They form the truly great conglomerate bodies of the geologic record and may reach thicknesses of thousands of meters. Preservation of such great thickness of conglomerate implies rapid erosion of sharply elevated highlands or areas of active volcanism.

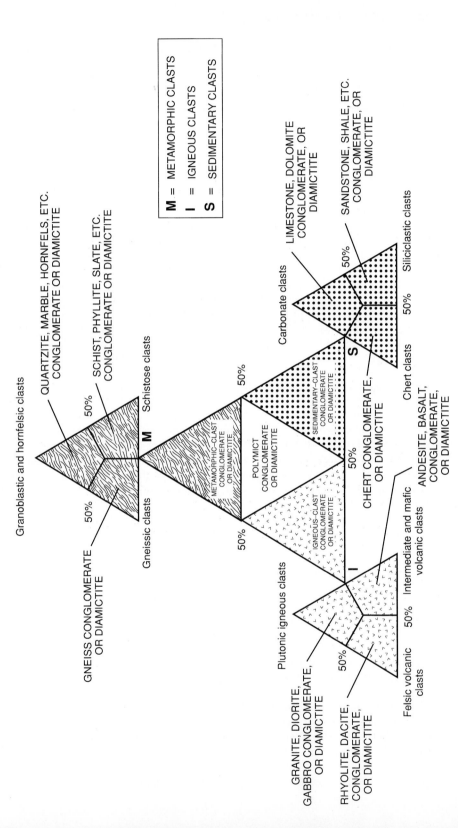

FIGURE 6.15 Classification of conglomerates on the basis of clast lithology and type of fabric support.

Petromict conglomerates may be transported by fluid-flow and sediment gravity-flow mechanisms; they are deposited in environments ranging from fluvial through shallow-marine to deep-marine. Deep-marine conglomerates are so-called **resedimented conglomerates** that were retransported from nearshore areas by turbidity currents or other sediment gravity-flow processes. The bulk of the truly thick conglomerate bodies (>20 m) were probably deposited in nonmarine (alluvial fan/braided river) settings or deep-sea fan settings.

Intraformational Conglomerates. Intraformational conglomerates are composed of clasts of sediments believed to have formed within depositional basins, in contrast to the clasts of extraformational conglomerates that are derived from outside the depositional basin. Intraformational conglomerates originate by penecontemporaneous deformation of semiconsolidated sediment and redeposition of the fragments fairly close to the site of deformation. Penecontemporaneous breakup of sediment to form clasts may take place subaerially, such as by drying out of mud on a tidal flat, or under water. Subaqueous rip-ups of semiconsolidated muds by tidal currents, storm waves, or sediment gravity flows are possible causes. In any case, sedimentation is interrupted only a short time during this process. The most common types of fragments found in intraformational conglomerates are mud clasts and lime clasts. The clasts are commonly angular or only slightly rounded, suggesting little transport. In some beds, flattened clasts are stacked virtually on edge, apparently owing to unusually strong wave or current agitation, to form what is called **edgewise conglomerates** (Pettijohn, 1975, p. 184).

Intraformational conglomerates commonly form thin beds, a few centimeters to a meter in thickness, that may be laterally extensive. Although much less abundant than extraformational conglomerates, they nonetheless occur in rocks of many ages. So-called flat-pebble conglomerates composed of carbonate or limy siltstone clasts are particularly common in Cambrian-age rocks in various parts of North America. They also occur in many other early Paleozoic limestones of the Appalachian region. Intraformational conglomerates composed of shale rip-up clasts embedded in the basal part of sandstone units are very common in sedimentary sequences deposited by sediment gravity-flow processes.

6.4 SHALES

Shales are siliciclastic sedimentary rocks composed of mud-size particles (silt and clay). Shale is an historically accepted class name for this group of rocks (Tourtelot, 1960), equivalent to the class name sandstone; however, some authors prefer to restrict the usage of shale to fine-grained rocks that show lamination (Fig. 6.16). They use the term mudrock for nonlaminated, fine-grained rocks. In this book, I use the term shale for all sedimentary rocks composed dominantly of mud-size (<0.06 mm) particles. Shales are abundant rocks in sedimentary sequences, making up roughly 50 percent of all the sedimentary rocks in the geologic record. Historically, shales have been an understudied group of rocks, mainly because their fine grain size makes them difficult to study with an ordinary petrographic microscope. This perspective is changing, however, owing to the development of instruments such as the scanning electron microscope and electron probe microanalyzer that allow study of fine-size grains at high magnification.

FIGURE 6.16 Well-laminated, dark gray shale. The "crinkled" appearance of the shale is the result of folding during incipient metamorphism. Lincoln Peak Formation (Cambrian), Nevada. Note hammer for scale.

TABLE 6.5 Average percent mineral composition of shales of different ages

Age	Number of analyses	Clay minerals	Quartz	Potassium feldspar	Plagioclase feldspar	Calcite	Dolomite	Siderite	Pyrite	Other minerals	Organic carbon
Quaternary	5	29.9	42.3	12.4	—	6.6	2.4	—	5.6	—	0.9
Pliocene	4	56.5	14.6	5.7	11.9	3.2	—	2.9	1.8	<1.0	2.6
Miocene	9	25.3	34.1	7.4	11.7	14.6	1.2	—	1.9	2.4	1.4
Oligocene	4	33.7	53.5	3.0	—	5.5	—	—	—	4.0	0.4
Eocene	11	40.2	34.6	2.0	8.1	3.8	4.6	1.7	1.6	—	3.5
Cretaceous	9	27.4	52.9	3.6	1.6	2.9	7.9	0.1	1.6	—	2.0
Jurassic	10	34.7	21.9	0.6	4.4	14.6	1.6	0.4	10.9	—	10.9
Triassic	9	29.4	45.9	10.7	0.7	3.7	4.1	5.1	—	—	0.3
Permian	1	17.0	28.0	4.0	8.0	—	1.0	—	—	42.0	0.2
Pennsylvanian	7	48.9	32.6	0.8	6.2	1.4	2.1	3.4	3.5	—	1.0
Mississippian	3	57.2	29.1	0.4	2.9	—	—	0.6	5.1	—	4.7
Devonian	22	41.8	47.1	0.6	—	2.0	1.3	0.3	3.3	—	3.7
Ordovician	2	44.9	32.2	<1.0	6.3	9.8	0.5	0.5	3.4	—	1.5
Misc. ages	29	47.8	33.1	<1.0	5.5	5.2	2.3	0.8	3.1	—	4.5

Source: O'Brien, N. R., and R. M. Slatt, 1990, Argillaceous rock atlas, Springer-Verlag, New York. Table 1, p. 124–125.

Note: Values adjusted to 100% for shales of each age.

Composition

Mineralogy. Shales are composed primarily of clay minerals and fine-size quartz and feldspars (Table 6.5). They also contain various amounts of other minerals, including carbonate minerals (calcite, dolomite, siderite), sulfides (pyrite, marcasite), iron oxides (goethite), and heavy minerals, as well as a small amount of organic carbon. The data in Table 6.5 show mineral composition as a function of age. No discernible trend of mineralogy vs. age is evident from this table, except possibly a slight trend of decreasing feldspar with increasing age. Many factors affect the composition of shales, including tectonic setting and provenance (source), depositional environments, grain size, and burial diagenesis. Some minerals, such as carbonate minerals and sulfides, form in the shales during burial as cements or replacement minerals. Quartz, feldspars, and clay minerals are mainly detrital (terrigenous) minerals, although some fraction of these minerals may also form during burial diagenesis. In particular, clay minerals appear to be strongly affected by diagenetic processes. As shown in Table 6.1, the principal clay mineral groups are kaolinite, illite, smectite, and chlorite. The relative proportions of these clay-mineral groups have been reported to change systematically with age (Fig. 6.17). With time, particularly in rocks older than the Mesozoic, the proportion of illite and chlorite increases at the expense of kaolinite and smectite. These

FIGURE 6.17 Systematic changes in relative abundance of major clay minerals as a function of geologic age. (After Singer, A., and G. Müller, 1983, Diagenesis in argillaceous sediments, *in* Larson, G. and Chilingarian, G. V., eds., Diagenesis in sediments and sedimentary rocks, v. 2. Fig. 3.28, p. 176, reproduced by permission of Elsevier Science Publishers, New York.)

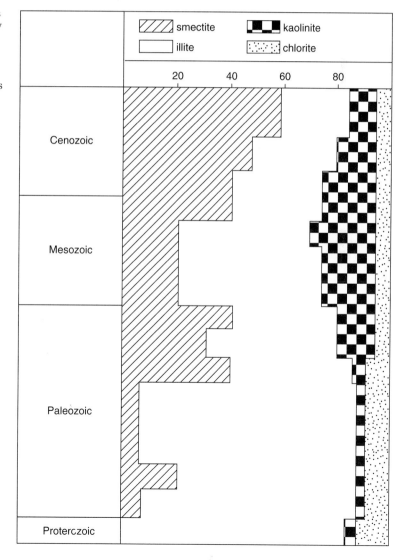

TABLE 6.6 Average chemical composition of selected shales reported in the literature

	1	2	3	4	5	6	7	8	9	10	11	12	13
SiO$_2$	60.65	64.80	59.75	56.78	67.78	64.09	66.90	63.04	62.13	65.47	64.21	64.10	63.31
Al$_2$O$_3$	17.53	16.90	17.79	16.89	16.59	16.65	16.67	18.63	18.11	16.11	17.02	17.70	17.22
Fe$_2$O$_3$	7.11	—	—	—	—	—	—	—	—	—	—	2.70	0.82?
FeO	—	5.66	5.59	6.56	4.11	6.03	5.87	7.66	7.33	5.85	6.71	4.05	5.45
MgO	2.04	2.86	4.02	4.56	3.38	2.54	2.59	2.60	3.57	2.50	2.70	2.65	3.00
CaO	0.52	3.63	6.10	8.91	3.91	5.65	0.53	1.31	2.22	4.10	3.44	1.88	3.52
Na$_2$O	1.47	1.14	0.72	0.77	0.98	1.27	1.50	1.02	2.68	2.80	1.44	1.91	1.48
K$_2$O	3.28	3.97	4.82	4.38	2.44	2.73	4.97	4.57	2.92	2.37	3.58	3.60	3.64
TiO$_2$	0.97	0.70	0.98	0.92	0.70	0.82	0.78	0.94	0.78	0.49	0.72	0.86	0.81
P$_2$O$_5$	0.13	0.13	0.12	0.13	0.10	0.12	0.14	0.10	0.17	—	—	—	0.10
MnO	0.10	0.06		0.08		0.07	0.06	0.12	1.10	0.07	0.05	—	0.06

Source:

1 Moore, 1978 (Pennsylvanian shale, Illinois Basin)
2 Gromet et al., 1984 (North American shale composite)
3 Ronov and Migdisov, 1971 (average North American Paleozoic shale)
4 Ronov and Migdisov, 1971 (average Russian Paleozoic shale)
5 Ronov and Migdisov, 1971 (average North American Mesozoic shale)
6 Ronov and Migdisov, 1971 (average Russian Mesozoic shale)
7 Cameron and Garrels, 1980 (average Canadian Proterozoic shale)
8 Ronov and Migdisov, 1971 (average Russian Proterozoic shale)
9 Cameron and Garrels, 1980 (average Canadian Archean shale)
10 Ronov and Migdisov, 1971 (average Archean shale)
11 Clarke, 1924 (average shale)
12 Shaw, 1956 (compilation of 155 analyses of shale)
13 Average of values in columns 1 through 12
13 Average of values in columns 1 through 12

trends are attributed to the diagenetic alteration of kaolinite and smectite to form illite and chlorite.

Chemical Composition. The chemical composition of shales is a direct function of their mineral composition. Compositions of some average North American and Russian shales are shown in Table 6.6. SiO_2 is the most abundant chemical constituent in these shales (57–68 percent), followed by Al_2O_3 (16–19 percent). The SiO_2 content of shales is affected by all silicate minerals present but particularly by the presence of quartz. Thus, shales tend to contain less SiO_2 than do sandstones, which commonly are enriched in quartz. Al_2O_3 is derived mainly from clay minerals and feldspars. It is more abundant in shales than in sandstones because of the greater clay mineral content of shales. Fe in shales is supplied by iron oxide minerals (hematite, goethite), biotite, and a few other minerals such as siderite, ankerite, and smectite clay minerals. K_2O and MgO abundance is related mainly to clay mineral abundance, although some Mg may be supplied by dolomite and K is present in some feldspars. Na abundance is related to the presence of clay minerals (e.g., smectites) and sodium plagioclase. Ca is supplied by calcium-rich plagioclase and carbonate minerals (calcite, dolomite).

Classification

Because special analytical techniques are required to determine the mineral composition of shales, and because such techniques are time-consuming and expensive, many geologists do not routinely determine the mineral composition of shales. Therefore, most classifications that have been proposed for shales have not been based on mineral composition, or at least not entirely on mineral composition. These classifications, none of which has been widely accepted, commonly emphasize the relative amounts of silt and clay, the hardness or degree of induration of the shales, and the presence or absence of fissile lamination. (Fissility is defined as the property of a rock to split easily along thin, closely spaced, approximately parallel layers.) Exceptions to this general practice of classification are the classification of Picard (1971), which emphasizes mineral composition of the silt-size grains in shales, and the classification of Lewan (1978), which requires semiquantitative X-ray diffraction analysis to determine mineralogy.

The classification of Potter, Maynard, and Pryor (1980), shown in Table 6.7, is based on grain size, lamination, and degree of induration. It is similar to the field classification of shales proposed by Lundgard and Samuels (1980). This classification emphasizes the importance of clay-size constituents and bedding thickness, that is, whether bedded or laminated. For example, a shale containing more than two-thirds clay-size particles is called a **claystone** if bedded (layers thicker than 10 mm) or a **clayshale** if laminated (layers thinner than 10 mm). Additional informal terms can be used with this classification to provide further information about the properties of the shales. These may include terms that express color, type of cementation (calcareous, or limy; ferruginous, or iron-rich; siliceous); degree of induration (hard, soft); mineralogy if known (quartzose, feldspathic, micaceous, etc.); fossil content (fossiliferous, foram-rich, etc.); organic matter content (carbonaceous, kerogen-rich, coaly, etc.); type of fracturing (conchoidal, hackly, blocky); or nature of bedding (wavy, lenticular, parallel, etc.).

Origin and Occurrence of Shales

Shales form under any environmental conditions in which fine sediment is abundant and water energy is sufficiently low to allow settling of suspended fine silt and clay.

TABLE 6.7   Classification of shales

Percentage clay-size constituents			0–32	33–65	66–100
Field adjective			Gritty	Loamy	Fat or slick
NONINDURATED	Beds	Greater than 10 mm	Bedded silt	Bedded mud	Bedded claymud
	Laminae	Less than 10 mm	Laminated silt	Laminated mud	Laminated claymud
INDURATED	Beds	Greater than 10 mm	Bedded siltstone	Mudstone	Claystone
	Laminae	Less than 10 mm	Laminated siltstone	Mudshale	Clayshale
METAMORPHOSED		Degree of metamorphism Low ↓ High	Quartz argillite	Argillite	
			Quartz slate	Slate	
			Phyllite and/or mica schist		

Source: Potter, P. E., J. B. Maynard, and W. A. Pryor, 1980, Sedimentology of shales. Table 1.2, p. 14, reprinted by permission of Springer-Verlag, New York.

Shales are particularly characteristic of marine environments adjacent to major continents where the seafloor lies below storm wave base, but they can form also in lakes and quiet-water parts of rivers, and in lagoonal, tidal-flat, and deltaic environments. The fine-grained siliciclastic products of weathering greatly exceed coarser particles; thus, fine sediment is abundant in many sedimentary systems. Because fine sediment is so abundant and can be deposited in a variety of quiet-water environments, shales are by far the most abundant type of sedimentary rock. They make up roughly 50 percent of the total sedimentary rock record. They commonly occur interbedded with sandstones or limestones in units ranging in thickness from a few millimeters to several meters or tens of meters. Nearly pure shale units hundreds of meters thick also occur. Shale units in marine sequences tend to be laterally extensive.

A few shales that are particularly well known owing to their thickness, widespread areal extent, stratigraphic position, or fossil content include the Cambrian Burgess Shale of western Canada, which is famous for its well-preserved imprints of soft-bodied animals; the Eocene Green River (oil) Shale of Colorado; the Cretaceous Mancos Shale of western North America, which forms a thick, eastward-thinning wedge stretching from New Mexico to Saskatchewan and Alberta; the Devonian-Mississippian Chattanooga Shale and equivalent formations that cover much of North America and whose widespread extent is still poorly explained; the Silurian Gothlandian shales of western Europe, northern Africa, and the Persian Gulf region, which contain a pelecypod and graptolite faunal association; and the Precambrian Figtree Formation of South Africa, well known for studies of its early fossils. The origin and occurrence of shales are discussed in detail by Potter et al. (1980).

6.5   PROVENANCE SIGNIFICANCE OF MINERAL COMPOSITION

The silicate mineralogy and rock-fragment composition of siliciclastic sedimentary rocks are fundamental properties of these rocks that set them apart from other sedimentary rocks. Mineralogy is a particularly important property for studying the origin of siliciclastic sedimentary rocks because it provides almost the only available clue to the nature of vanished source areas. The kinds of siliciclastic minerals and rock fragments preserved in sedimentary rocks furnish important evidence of the lithology of the source rocks. Rock fragments provide the most direct lithologic evidence, but feldspars and other minerals are also important source-rock indicators. For example, potassium feldspars suggest derivation mainly from alkaline plutonic igneous or metamorphic rocks, whereas sodic plagioclase is derived principally from alkaline volcanic rocks, and calcic plagioclase comes mainly from basic volcanic rocks. Suites of heavy minerals are also used for source-rock determination. Thus, a suite of heavy minerals consisting of apatite, biotite, hornblende, monazite, rutile, titanite, pink tourmaline, and zircon is indicative of alkaline igneous source rocks. A suite consisting of augite, chromite, diopside, hypersthene, ilmenite, magnetite, and olivine suggests derivation from basic igneous rocks. Andalusite, garnet, staurolite, topaz, kyanite, sillimanite, and staurolite constitute a mineral suite diagnostic of metamorphic rocks, whereas a suite of heavy minerals consisting of barite, iron ores, leucoxene, rounded tourmaline, and rounded zircon suggests a recycled sediment source. Even quartz may have some value as a provenance indicator. For example, Basu et al. (1975) suggest that a high percentage of quartz grains with undulose extinction greater than 5°, combined with a high percentage of polycrystalline grains containing more than three crystal units per grain, are typical of low-rank metamorphic source rocks. By contrast, nonundulose quartz and polycrystalline quartz containing less than three crystal units per grain indicate derivation from high-rank metamorphic or plutonic igneous source rocks.

In addition to providing information about source-rock lithology, the relative chemical stabilities and the degree of weathering and alteration of certain minerals can be used as tools for interpreting the climate and relief of source areas (e.g., Folk, 1974, p. 85). For example, the presence of large, fresh, angular feldspars in a sandstone suggests derivation from a high-relief source area where grains were eroded rapidly before extensive weathering occurred. Alternatively, they may have been derived from a source area having a very arid or extremely cold climate that retarded chemical weathering. Small, rounded, highly weathered feldspar grains indicate a source area of low relief and/or a warm, humid climate where chemical weathering was moderately intense. Absence of feldspars may indicate either that weathering was so intense that all feldspars were destroyed or that no feldspars were present in the source rocks. Such analyses of mineral constituents provide only tentative conclusions about climate and relief. Also, they are subject to misinterpretations owing to diagenetic alteration or destruction of source-rock minerals.

Geologists are also interested in the tectonic setting of source areas and associated depositional sites. With development of the theory of seafloor spreading and plate tectonics, this interest has focused on interpreting the tectonic setting in terms of plate tectonic provinces (Dickinson and Suczek, 1979; Dickinson, 1982; Dickinson et al., 1983). In other words, geologists want to know if a particular deposit was derived from source rocks located within a continent, in a volcanic arc associated with a subduction zone, or in other tectonic settings. Three principal types of tectonic settings, or **provenances** as they are called, have been identified: (1) continental block provenances, (2) magmatic arc provenances, and (3) recycled orogen provenances (Dickinson and Suczek, 1979).

Continental block provenances are located within continental masses, which may be bordered on one side by a passive continental margin and on the other by an orogenic belt or zone of plate convergence. Source rocks consist of plutonic igneous, metamorphic, and sedimentary rocks but include few volcanic rocks. Sediment eroded from these sources typically consists of quartzose sand, feldspars with high ratios of potassium feldspar to plagioclase feldspar, and metamorphic and sedimentary rock fragments. Sediment eroded from continental sources may be transported off the continent into adjacent marginal ocean basins, or it may be deposited in local basins within the continent.

Magmatic arc provenances are located in zones of plate convergence where sediment is eroded mainly from volcanic arc sources consisting of volcanogenic highlands (undissected arcs). Volcaniclastic debris shed from these highlands consists largely of volcanic lithic fragments and plagioclase feldspars. Quartz and potassium feldspars are commonly very sparse except where the volcanic cover is dissected by erosion to expose underlying plutonic rocks (dissected arcs). Sediment shed from volcanic highlands may be transported to an adjacent trench or deposited in fore-arc and back-arc basins.

Recycled orogen provenances are zones of plate convergence, where collision of major plates creates uplifted source areas along the collision suture belt. Where two continental masses collide, source rocks in the collision uplifts are typically sedimentary and metamorphic rocks that were present along the continental margins prior to their collision. Detritus stripped from these source rocks commonly consists of abundant sedimentary-metasedimentary rock fragments, moderate quartz, and a high ratio of quartz to feldspars. Where a continental mass collides with a magmatic arc complex, uplifted source rocks may include deformed ultramafic rocks, basalts, and other oceanic rocks, and a variety of other rock types such as greenstone (weakly metamorphosed basic igneous rock), chert, argillite (weakly metamorphosed shale), lithic sandstones, and limestones. Sediment derived from these sources may include many types of rock fragments, quartz, feldspars, and chert. Chert is a particularly abundant constituent of sediments derived from this provenance.

To differentiate sediment derived from these three major tectonic provenances, Dickinson and Suczek (1979) and Dickinson et al. (1983) suggest the use of triangular composition diagrams showing framework proportions of monocrystalline quartz, polycrystalline quartz, potassium feldspars and plagioclase feldspars, and volcanic and sedimentary-metasedimentary rock fragments. Through study of sandstone compositions from many parts of the world, they generated the provenance diagrams shown in Figure 6.18. To use these diagrams as a guide to provenance determination of other sandstones, geologists determine the compositions of the sand-size grains in a sandstone and plot them on one or both of the diagrams shown in Figure 6.18. The field in which most of the plotted points fall (e.g., craton interior, recycled orogen) is the probable tectonic setting of the source rocks. Students should keep in mind that the diagrams in Figure 6.18 are simply models. Like any model, they should be used only as a guide, and exceptions may occur. Interpretations based on modal composition may not always agree with interpretations made on the basis of stratigraphic and structural relationships.

The discussion above provides only the barest introduction to the topic of provenance interpretation. The application of provenance study to basin analysis is explored further in Chapter 18. For additional information on this important subject, including discussion of the provenance of conglomerates and shales, see Boggs (1992, Chapter 8).

6.6 DIAGENESIS

Siliciclastic sedimentary rocks form as unconsolidated deposits of gravels, sand, or mud. The mineral and chemical compositions of these deposits are functions of provenance, sediment transport, and environmental conditions. Newly deposited sediments are characterized by loosely packed, uncemented fabrics; high porosities; and high interstitial water content. As sedimentation continues in subsiding basins, older sediments are progressively buried by younger sediments to depths that may reach tens of kilometers. Sediment burial is accompanied by physical and chemical changes that take place in the sediments in response to increase in pressure from the weight of overlying sediment, downward increase in temperature, and changes in pore-water composition. These changes act in concert to bring about compaction and **lithification** of sediment, ultimately converting it into consolidated sedimentary rock. Thus, unconsolidated gravel is eventually lithified to conglomerate, sand is lithified to sandstone, and siliciclastic mud is hardened into shale.

The process of lithification is accompanied by physical, mineralogical, and chemical changes. Loose grain packing gives way with burial to more tightly packed fabrics having greatly reduced porosity. Porosity may be further reduced by precipitation of cements into pore spaces. Minerals that were chemically stable at low surface temperatures and in the presence of environmental pore waters become altered at higher burial temperatures and changed pore-water compositions. Minerals may be completely dissolved or may be partially or completely replaced by other minerals.

Thus, porosity, mineralogy, and chemical composition may all be changed to various degrees during burial diagenesis. Diagenesis is the final stage in the process of

FIGURE 6.18 Relationship between framework composition of sandstones and tectonic setting. (After Dickinson, R. W., et al., 1983, Provenance of North American Phanerozoic sandstones in relation to tectonic setting: Geol. Soc. America Bull., v. 94. Fig. 1, p. 223.)

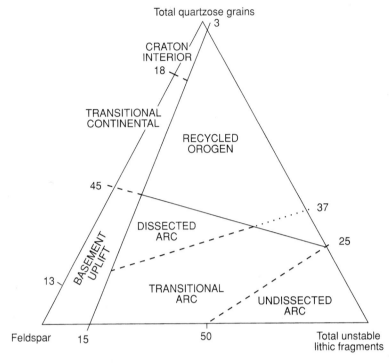

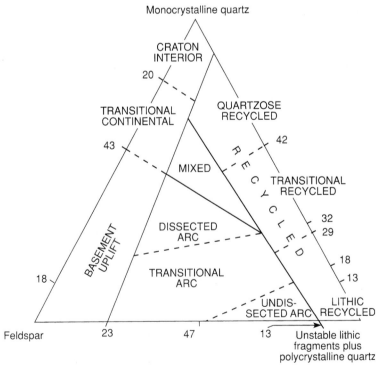

forming conglomerates, sandstones, and shales, a process that begins with weathering of source rocks and continues through sediment transport, deposition, and burial. To properly interpret the provenance, transport, and depositional history of sedimentary rocks, we must recognize and distinguish between features of sediment that were present at the time of deposition and features of sedimentary rocks that resulted from burial alteration. Only a short description of diagenetic processes can be included here. For more details, see Boggs (1992, Chapter 9).

Stages and Realms of Diagenesis

Diagenesis takes place at temperatures and pressures higher than those of the weathering environment but below those that produce metamorphism. There is no clear boundary between the realms of diagenesis and metamorphism; however, we commonly consider diagenesis to occur at temperatures below about 300°C. Diagenesis can begin almost immediately after deposition, while sediment is still on the ocean or other basin floor, and may continue through deep burial and eventual uplift. Burial subjects sediments to conditions of pressure and temperature markedly different from those that exist in the depositional environment. Increases in geostatic (rock) pressure, hydrostatic (fluid) pressure, and temperature as a function of depth are shown in Figure 6.19. Pore-fluid composition changes also. There is both a general increase in salinity of pore waters with increasing burial depth (Fig. 6.20) and a change in pore-water chemistry (Boggs, 1992, p. 373). Changes in pore-water chemistry are difficult to generalize and differ from basin to basin, but they include variations in abundance of such important mineral-forming ions as Si^{4+}, Al^{3+}, Ca^{2+}, K^+, Mg^{2+}, Na^+, and HCO_3^- (bicarbonate).

Various authors have suggested that sediments go through from three to six stages of diagenesis. Perhaps the most widely accepted stages of diagenesis are those

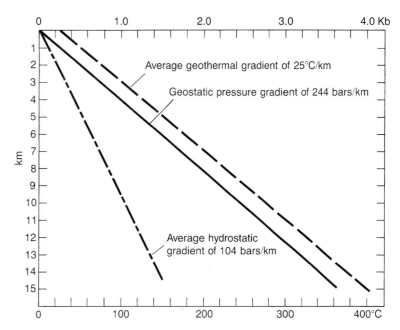

FIGURE 6.19 Average geothermal gradient, geostatic pressure gradient, and hydrostatic pressure gradient in sedimentary basins.

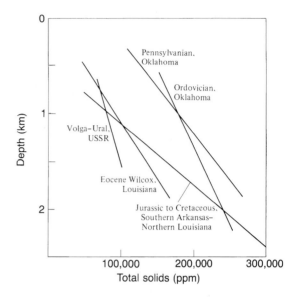

FIGURE 6.20 Changes in total salinity of subsurface waters with increasing depth in selected depositional basins. (After Hunt, J. M., Petroleum, geochemistry and geology, W. H. Freeman and Company, San Francisco. Fig. 6.3, p. 194. Copyright © 1979. Based on data from Dickey, P. A., 1969, Increasing concentrations of subsurface brines with depth: Chem. Geology, v. 4.)

proposed by Choquette and Pray (1970). **Eodiagenesis** refers to the earliest stage of diagenesis, which takes place at very shallow depths (a few meters to tens of meters) largely under the conditions of the depositional environment. **Mesodiagenesis** is diagenesis that takes place during deeper burial, under conditions of increasing temperature and pressure and changed pore-water compositions. **Telodiagenesis** refers to late-stage diagenesis that accompanies or follows uplift of previously buried sediments into the regime of meteoric waters. Sedimentary rocks that are still deeply buried in depositional basins have not, of course, undergone telodiagenesis. The most important diagenetic processes that take place in each of these diagenetic regimes, and the effects of these processes, are summarized in Table 6.8. These processes and effects are discussed in greater detail below.

Major Diagenetic Processes and Effects

Shallow Burial (Eodiagenesis)

The principal diagenetic changes that take place in the eodiagenetic regime include reworking of sediments by organisms (bioturbation), minor compaction and grain repacking, and mineralogical changes.

Organisms rework sediment at or near the depositional interface through various crawling, burrowing, and sediment-ingesting activities. Bioturbation can destroy primary sedimentary structures such as lamination and create in their place a variety of traces that may include mottled bedding, burrows, tracks, and trails. Organic reworking commonly has little effect on the mineralogical and chemical composition of sediments. Owing to very shallow burial depth, sediments undergo only very slight compaction and grain rearrangement during early diagenesis.

Early diagenesis does bring about some important mineralogical changes in siliciclastic sediments. Most of these changes involve the precipitation of new minerals. In marine environments where reducing (low-oxygen) conditions can prevail, the formation of pyrite is particularly characteristic. Pyrite may form a cement or may replace

TABLE 6.8 Principal diagenetic processes and changes that occur in siliciclastic sedimentary rocks during burial

	Diagenetic stage	Diagenetic process	Result
Burial	Eodiagenesis	Organic reworking (bioturbation)	Destruction of primary sedimentary structures; formation of mottled bedding and other traces
		Cementation and replacement	Formation of pyrite (reducing environments) or iron oxides (oxidizing environments); precipitation of quartz and feldspar overgrowths, carbonate cements, kaolinite, or chlorite
	Mesodiagenesis	Physical compaction	Tighter grain packing; porosity reduction and bed thinning
		Chemical compaction (pressure solution)	Partial dissolution of silicate grains; porosity reduction and bed thinning
		Cementation	Precipitation of carbonate (calcite) and silica (quartz) cements with accompanying porosity reduction
		Dissolution by pore fluids	Solution removal of carbonate cements and silicate framework grains; creation of new (secondary) porosity by preferential destruction of less stable minerals
		Mineral replacement	Partial to complete replacement of some silicate grains and clay matrix by new minerals (e.g., replacement of feldspars by calcite)
		Clay mineral authigenesis	Alteration of one kind of clay mineral to another (e.g., smectite to illite or chlorite, kaolinite to illite)
Uplift	Telodiagenesis	Dissolution, replacement, oxidation	Solution of carbonate cements, alteration of feldspars to clay minerals, oxidation of iron carbonate minerals to iron oxides, oxidation of pyrite to gypsum, solution of less stable minerals (e.g., pyroxenes, amphiboles)

other materials such as woody fragments. Other important reactions include formation of chlorite, glauconite (greenish iron-silicate grains), illite/smectite clays, and iron oxides in oxygenated pore waters (e.g., red clays on the deep ocean floor), and precipitation of potassium feldspar overgrowths, quartz overgrowths (Fig. 6.7), and carbonate cements (Fig. 6.9). In nonmarine environments, where oxidizing conditions commonly prevail, little pyrite forms. Instead, iron oxides (goethite, hematite) are commonly produced, creating redbeds. Formation of kaolinitic clay minerals and precipitation of quartz and calcite cements may take place also in this environment.

Deep Burial (Mesodiagenesis)

Compaction. The load pressures caused by deeper burial bring about a dramatic increase in the tightness of grain packing with concomitant loss of porosity and thinning of beds. Increased pressure at the contact point between grains also causes the solubility of the grains at the contact to increase, leading to partial dissolution of the grains. This process is referred to as **pressure solution** or **chemical compaction.** Chemical compaction further reduces porosity and increases bed thinning. Thus, under the influence of physical and chemical compaction, the primary porosity of both sands and muds is reduced dramatically during deep burial (Fig. 6.21). Compaction also causes bending of flexible grains such as micas and squeezing of soft grains such as rock fragments (Fig. 6.22).

Chemical Processes and Changes. An increase in temperature of 10°C during burial can cause chemical reaction rates to double or triple. Thus, mineral phases that were

FIGURE 6.21 Approximate best-fit curves showing changes in porosity of sediments mainly as a function of burial compaction in some California (sandstone) and Louisiana (shale) basins. (Sandstone curve based on Wilson, J. C., and E. F. McBride, 1988, Compaction and porosity evolution of Pliocene sandstones, Ventura Basin, California: Am. Assoc. Petroleum Geologists Bull., v. 72, Fig. 4, p. 669; shale curve based on Dzevanshir, R. D., et al., 1986, Sed. Geology, v. 46, Fig. 1, p. 170.)

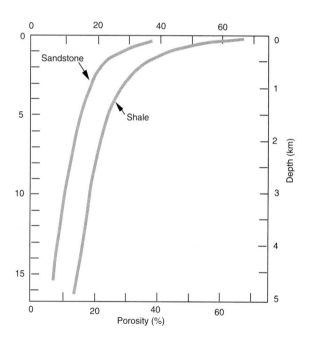

FIGURE 6.22 Schematic representation of textural criteria used to estimate volume loss in sandstones owing to compaction. The hachured areas indicate rock volume lost by grain deformation and pressure solution. (From Wilson, J. C., and E. F. McBride, 1988, Compaction and porosity evolution of Pliocene sandstones, Ventura Basin, California: Am. Assoc. Petroleum Geologists Bull., v. 72, Fig. 10. p. 679, reprinted by permission of AAPG, Tulsa, Okla.)

Plastic and Ductile
Grain Deformation

Flexible Grain
Deformation

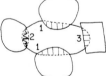

Pressure Solution
1 Concavo-Convex Contact
2 Sutured Contact
3 Long Contact

stable in the depositional environment may become unstable during deep burial. Increasing temperature favors the formation of denser, less hydrous minerals and also causes an increase in solubility of most common minerals except the carbonate minerals. Thus, silicate minerals show an increasing tendency to dissolve with greater burial depths (and temperatures), whereas carbonate minerals such as calcite are more likely to precipitate. On the other hand, decrease in pH (increase in acidity) of pore waters with depth may bring about dissolution of carbonates. For example, organic materials may decompose during deep burial diagenesis to release CO_2. Increase in the CO_2 content of pore waters results in a decrease in pH (increase in acidity) that can bring about dissolution of carbonate minerals. As discussed above, increased pressure during deep burial causes an increase in solubility of minerals at point contacts, resulting in partial dissolution of the minerals. This process, which releases silica into pore waters, is an important mechanism for furnishing silica that can later precipitate as new silicate minerals. Several kinds of chemical/mineralogical diagenetic processes take place in siliciclastic sedimentary rocks during deep burial. The most important of these processes are cementation, dissolution, replacement, and clay-mineral authigenesis.

Cementation refers to the precipitation of minerals into the pore space of a sediment, thereby reducing porosity and bringing about lithification of the sediment. Carbonate and silica cements are most common; however, feldspars, iron oxides, pyrite, anhydrite, zeolites, and many other minerals can also form as cements. Calcite is the dominant carbonate cement (Fig. 6.9); aragonite, dolomite, siderite, and ankerite are less common. Carbonate cementation is favored by increasing concentration of calcium carbonate in pore waters and increasing burial temperature. Precipitation is inhibited by increased levels of CO_2 in pore waters, which may result from decomposition of organic matter in sediments during burial. Increased CO_2 levels (partial pressure) cause pore waters to become acidic and corrosive to carbonate minerals.

Quartz precipitated as overgrowths around existing detrital quartz grains (Fig. 6.7) is the most common kind of silica cement. Quartz overgrowth cements are particularly abundant in many quartz arenites. Less commonly, silica precipitates as microcrystalline quartz (chert) cement (Fig. 6.8) or opal. Quartz cementation is favored by high concentrations of silica in pore waters and by low temperatures. Silica may be supplied by pressure solution or by dissolution of the siliceous skeletons of fossil organisms such as diatoms and radiolarians. Quartz cementation is particularly likely to occur in sedimentary basins where waters that circulated downward deeply into the basin, and dissolved silica at higher temperatures, rise upward and cool along basin edges.

Dissolution of framework silicate grains and previously formed carbonate cements may occur during deep burial under conditions that are essentially the opposite of those required for cementation. For example, carbonate minerals are dissolved in cooler pore waters with high carbon dioxide partial pressures. Rock fragments and low-stability silicate minerals, such as plagioclase feldspars, pyroxenes, and amphiboles, may dissolve as a result of increasing burial temperatures and the presence of organic acids in pore waters. The selective dissolution of less stable framework grains or parts of grain during diagenesis is called **intrastratal solution.** Dissolution of framework grains and cements leads to increase in porosity, particularly in sandstones. Petroleum geologists, who are especially interested in the porosity of sandstones, now believe that much of the porosity that exists in sandstones below a burial depth of about 3 km is secondary porosity, created by dissolution processes.

Mineral **replacement** refers to the process whereby one mineral dissolves and another is precipitated in its place—essentially simultaneously (Fig. 6.23). Replacement appears to take place without any volume change between the replaced and

FIGURE 6.23 Plagioclase feldspar (P) partially replaced by clay minerals (CM), Miocene sandstone, Ocean Drilling Program Leg 128, Site 799, Japan Sea. Crossed nicols. Scale bar = 0.1 mm.

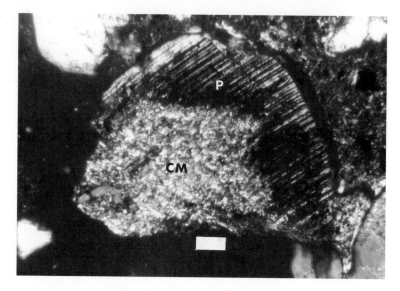

replacing mineral. Thus, delicate textures present in the original mineral may, in some cases, be faithfully preserved in the replacement mineral. Well-known examples of such preserved textures can be found in petrified wood and carbonate fossils replaced by chert.

Common replacement events include replacement of carbonate minerals by microcrystalline quartz (chert), replacement of chert by carbonate minerals, replacement of feldspars and quartz by carbonate minerals, replacement of feldspars by clay minerals, replacement of clay matrix by carbonate minerals, replacement of calcium-rich plagioclase by sodium-rich plagioclase (albitization), and replacement of feldspars and volcanic rock fragments by clay or zeolite minerals. Replacement may be partial or complete. Complete replacement destroys the identity of the original minerals or rock fragments and thereby gives a biased view of the original mineralogy of a rock. Porosity may also be affected by replacement, particularly replacement of framework grains by clay minerals, which tend to plug pore space and reduce porosity. Much of the clay matrix in sandstones may be produced diagenetically by alteration of unstable framework grains to clay minerals.

In addition to these common replacement processes, one kind of clay mineral may alter to another during diagenesis. For example, smectite clays may alter to illite at temperatures ranging from about 55°C to 200°C, with concomitant release of water. This process is particularly common in shales and is referred to as shale dewatering. Smectite may also alter to chlorite within about the same temperature range, and kaolinite typically alters to illite at temperatures between about 120°C and 150°C. It is these diagenetic processes that are believed to account for the trend of changing clay-mineral abundance with age shown in Figure 6.17.

Telodiagenesis

Sedimentary rocks that have undergone deep burial diagenesis may subsequently be uplifted by mountain-building activities and unroofed by erosion. This process brings mineral assemblages, including new minerals formed during mesodiagenesis, into an environment of lower temperature and pressure and in which mesogenetic pore waters are flushed and replaced by oxygen-rich, acidic meteoric (rain) waters of low salinity. Under these changed conditions, previously formed cements and framework grains may undergo dissolution (creating secondary porosity) or alteration of framework grains to clay minerals, e.g., potassium feldspar to kaolinite (reducing porosity). Alternatively, depending upon the nature of the pore waters, silica or carbonate cements can be precipitated. Other changes may include oxidation of iron carbonate minerals and other iron-bearing minerals to form iron oxides (goethite and hematite), oxidation of sulfides (pyrite) to form sulfate minerals (gypsum) if calcium is present in pore waters, and dissolution of less stable minerals such as pyroxenes and amphiboles. The processes of telodiagenesis grade into those of subaerial weathering as sedimentary rocks are exposed at Earth's surface.

FURTHER READINGS

Adams, A. E., W. S. Mackenzie, and C. Guilford, 1984, Atlas of sedimentary rocks under the microscope: John Wiley & Sons, New York, 104 p.

Boggs, S., Jr., 1992, Petrology of sedimentary rocks: Macmillan, New York, 707 p.

Folk, R. L., 1974, Petrology of sedimentary rocks: Hemphill, Austin, Tex., 182 p.

Koster, E. H., and R. H. Steel (eds.), 1984, Sedimentology of gravels and conglomerates: Canadian Soc. Petroleum Geologists Mem. 10, 441 p.

McDonald, D. A., and R. C. Surdam (eds.), 1984, Clastic diagenesis: Am. Assoc. Petroleum Geologists Mem. 37, 434 p.

Milner, H. B., 1962, Sedimentary petrography, v. 2: Principles and applications: Macmillan, New York, 715 p.

O'Brien, N. R., and R. M. Slatt, 1990, Argillaceous rock atlas: Springer-Verlag, New York, 141 p.

Pettijohn, F. J., P. E. Potter, and R. Siever, 1987, Sand and sandstone, 2nd ed.: Springer-Verlag, New York, 618 p.

Potter, P. E., J. B. Maynard, and W. A. Pryor, 1980, Sedimentology of shale: Springer-Verlag, New York, 553 p.

Scholle, P. A., 1979, A color illustrated guide to constituents, textures, cements, and porosities of sandstones and associated rocks: Am. Assoc. Petroleum Geologists Mem. 28, Tulsa, Okla., 201 p.

Tickell, F. G,., 1965, The techniques of sedimentary mineralogy, Elsevier, New York, 220 p.

Zuffa, G. G. (ed.), 1984, Provenance of arenites: D. Reidel, Dordrecht, 408 p.

7

Carbonate Sedimentary Rocks

7.1 INTRODUCTION

Chemical/biochemical sedimentary rocks originate by precipitation of minerals from water through various chemical or biochemical processes. They are distinguished from siliciclastic sedimentary rocks by their chemistry, mineralogy, and texture. They can be divided on the basis of mineralogy and chemistry into five fundamental types: (1) carbonates, (2) evaporites, (3) siliceous sedimentary rocks (cherts), (4) iron-rich sedimentary rocks, and (5) phosphorites. Carbonaceous sedimentary rocks, such as coals and oil shales, make up a further, special group of rocks that contain abundant nonskeletal organic matter in addition to various amounts of siliciclastic or chemical (e.g., carbonate) constituents.

The carbonate rocks, by far the most abundant kind of chemical/biochemical sedimentary rock, are described in this chapter. Other chemical/biochemical and carbonaceous sedimentary rocks are discussed in Chapter 8. Carbonate rocks can be divided on the basis of mineralogy into limestones and dolomites (dolostones). Limestones are composed mainly of the mineral calcite, and dolomites are composed mainly of the mineral dolomite. Carbonate sedimentary rocks make up 20 to 25 percent of all sedimentary rocks in the geologic record. They are present in many Precambrian assemblages and in all geologic systems from the Cambrian to the Quaternary. Precambrian and Paleozoic carbonate successions consist dominantly of dolomite, whereas Mesozoic and Cenozoic carbonates are mainly limestone. Limestones contain richly varied textures, structures, and fossils that yield important information about ancient marine environments, paleoecological conditions, and the evolution of life forms through time, particularly marine organisms. Carbonate sedimentary rocks are also an economically important group of rocks because limestones and dolomites are useful for agricultural and industrial purposes; they make good building stones; and, most importantly, they act as reservoir rocks for more than one-third of the world's petroleum reserves. Because of their environmental and economic significance, they have been extensively studied, and their mineralogy, chemistry, and textural characteristics are described in hundreds of research papers. The characteristic properties of carbonate rocks have also been summarized in several books, such as those of Bathurst (1975), Chilingarian, Bissell, and Fairbridge (1967), Lippman (1973), MacQueen (1983), Milliman (1974), Morse and Mackenzie (1990), Reeder (1983), Scholle (1978), Scoffin (1987), and Tucker and Wright (1990).

7.2 CHEMISTRY

The elemental chemistry of carbonate rocks is dominated by calcium, magnesium, carbon, and oxygen. The relative abundance of these elements (expressed as oxides) in average limestone and nearly pure dolomite is shown in Figure 7.1B. Numerous other elements are present in carbonate rocks in minor or trace amounts. Many of the elements that occur in minor concentrations are contained in noncarbonate impurities. For example, Si, Al, K, Na, and Fe occur mainly in silicate minerals such as quartz, feldspars, and clay minerals that are present in minor amounts in most carbonate rocks (Fig. 7.1A). Trace elements that are common in carbonate rocks include B, Be, Ba, Sr, Br, Cl, Co, Cr, Cu, Ga, Ge, and Li. The concentration of these trace elements is controlled not only by the mineralogy of the rocks but also by the type and relative abundance of fossil skeletal grains in the rock. Many organisms concentrate and incorporate trace elements into their skeletal structures.

7.3 MINERALOGY

The chemistry and structure of the principal carbonate minerals, only a few of which are important components of limestones and dolomites, are shown in Table 7.1. A

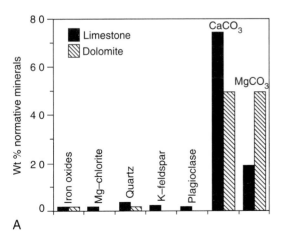
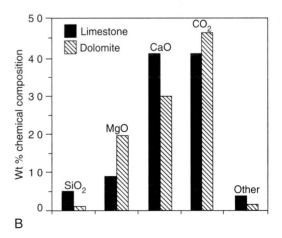

FIGURE 7.1 Normative mineral composition (A) and chemical composition (B) of an average limestone and a nearly pure (stoichiometric) dolomite (dolostone). "Other" chemical composition includes TiO_2, Al_2O_3, Fe_2O_3, FeO, MnO, Na_2O, K_2O, P_2O_5, SO_3, S, and Cl. (Data from Garrels, R. M., and F. T. Mackenzie, 1971, Evolution of sedimentary rocks: W. W. Norton, New York. Fig. 8.3, p. 211.)

TABLE 7.1 Principal carbonate minerals. The minerals marked with an asterisk (*) are important minerals in limestones and dolomites.

Mineral	Crystal system	Formula	Remarks
Calcite group			
*Calcite	Rhombohedral	$CaCO_3$	Dominant mineral of limestones, especially in rocks older than the Tertiary
Magnesite	Rhombohedral	$MgCO_3$	Uncommon in sedimentary rocks but occurs in some evaporite deposits
Rhodochrosite	Rhombohedral	$MnCO_3$	Uncommon in sedimentary rocks; may occur in Mn-rich sediments associated with siderite and Fe-silicates
Siderite	Rhombohedral	$FeCO_3$	Occurs as cements and concretions in shales and sandstones; common in ironstone deposits; also in carbonate rocks altered by Fe-bearing solutions
Smithsonite	Rhombohedral	$ZnCO_3$	Uncommon in sedimentary rocks; occur in association with Zn ores in limestones
Dolomite group			
*Dolomite	Rhombohedral	$CaMg(CO_3)_2$	Dominant mineral in dolomites; commonly associated with calcite or evaporite minerals
Ankerite	Rhombohedral	$Ca(Mg,Fe,Mn)(CO_3)_2$	Much less common than dolomite; occurs in Fe-rich sediments as disseminated grains or concretions
Aragonite group			
*Aragonite	Orthorhombic	$CaCO_3$	Common mineral in recent carbonate sediments; alters readily to calcite
Cerussite	Orthorhombic	$PbCO_3$	Occurs in supergene lead ores
Strontionite	Orthorhombic	$SrCO_3$	Occurs in veins in some limestones
Witherite	Orthorhombic	$BaCO_3$	Occurs in veins associated with galena ore

much more detailed analysis of the crystal chemistry of the carbonates is given by Reeder (1983) and by Tucker and Wright (1990, p. 284). Modern carbonate sediments are composed mainly of aragonite, but they also include calcite (especially in deep-sea calcareous ooze) and dolomite. Calcite ($CaCO_3$) can contain several percent magnesium in its formula because magnesium can readily substitute for calcium in the lattice of calcite crystals owing to the fact that magnesium ions and calcium ions are similar in size and charge. Thus, we recognize both low-magnesian calcite (called simply calcite) containing less than about 4 percent $MgCO_3$ and high-magnesian calcite containing more than 4 percent $MgCO_3$. High-magnesian calcite still retains the crystal structure of calcite in spite of the presence of Mg ions, which randomly substitute for Ca ions in the calcite crystal lattice. By contrast, true, so-called stoichiometric, dolomite is a totally different mineral in which Mg ions occupy half of the cation sites in the crystal lattice and are arranged in well-ordered planes that alternate with planes of CO_3 ions and Ca ions. Dolomite occurs in a few restricted modern environments, particularly in certain supratidal environments and freshwater lakes, but it is much less abundant in modern carbonate environments than aragonite and calcite. Other carbonate minerals such as magnesite, ankerite, and siderite are even less common in modern sediments. The mineralogy and chemistry of carbonate sediments can be strongly influenced by the composition of calcareous fossil organisms present in the sediments (e.g., Scholle, 1978, p. xi).

In contrast to the dominance of aragonite in modern shallow-water carbonate sediments, ancient carbonate rocks older than about the Cretaceous contain little aragonite. Aragonite is the metastable polymorph (having the same chemical composition but different crystal structure) of $CaCO_3$ and is converted fairly rapidly under aqueous conditions to calcite. The ratio of dolomite to calcite is much greater in ancient carbonate rocks than in modern carbonate sediments presumably because $CaCO_3$ minerals exposed to magnesium-rich interstitial waters during burial and diagenesis are converted to dolomite by replacement.

7.4 LIMESTONE TEXTURES

As discussed, ancient limestones are composed mainly of calcite. Calcite can be present in at least three distinct textural forms:

1. **carbonate grains,** such as ooids and skeletal grains, which are silt-size or larger aggregates of calcite crystals
2. **microcrystalline calcite,** or carbonate mud, which is texturally analogous to the mud in siliciclastic sedimentary rocks but which is composed of extremely fine-size, calcite crystals
3. **sparry calcite,** consisting of much coarser-grained calcite crystals that appear clear to translucent in plane (nonpolarized) light

Before classification of limestones can be considered, a fuller understanding of these contrasting carbonate textural elements must be developed.

Carbonate Grains

Early geologists tended to regard limestones as simply crystalline rocks that commonly contained fossils and that presumably formed largely by passive precipitation from seawater. We now know that many, and perhaps most, carbonate rocks are not simple crystalline precipitates. Instead, they are composed in part of aggregate particles or grains that may have undergone mechanical transport before deposition. Folk (1959) suggested use of the general term **allochems** for these carbonate grains to emphasize that they are not normal chemical precipitates. Carbonate grains typically range in size from coarse silt (0.02 mm) to sand (up to 2 mm), but larger particles such as fossil shells also occur. They can be divided into five basic types, each characterized by distinct differences in shape, internal structure, and mode of origin: carbonate clasts, skeletal particles, ooids, peloids, and aggregate grains.

Carbonate Clasts (Lithoclasts). Carbonate clasts are rock fragments that were derived either by erosion of ancient limestones exposed on land or by erosion of partially or completely lithified carbonate sediments within a depositional basin. If carbonate clasts are derived from older limestones in land sources located outside the depositional basin, they are called **extraclasts.** If they are derived from within the basin by erosion of semiconsolidated carbonate sediments from the seafloor, adjacent tidal flats, or a carbonate beach (beach rock), they are called **intraclasts.** The distinction between extraclasts and intraclasts has important implications for interpreting the transport and depositional history of limestones. Extraclasts may have iron-stained rims resulting from weathering, may contain recrystallized veins inherited from the parent rock, or may display other properties that distinguish them from intraclasts (Boggs, 1992, p. 425). Nonetheless, the distinction between fragments of ancient, weathered limestones and penecontemporaneously produced intraclasts is often difficult to make. **Lithoclast** (or limeclast) is a nonspecific term that can be used for carbonate clasts when this distinction cannot be made.

Lithoclasts range in size from very fine sand to gravel, although sand-size fragments are most common. They generally show some degree of rounding, indicative of transport, but subangular or even angular clasts are not unusual. Some clasts display internal textures or structures such as lamination, older clasts, siliciclastic grains, fossils, ooids, or pellets, but others are internally homogeneous. A limestone composed of gravel-size limeclasts is a kind of intraformational conglomerate. Clasts are not the most abundant type of carbonate grain in ancient limestones, but they occur with sufficient frequency in the geologic record to show that the clast-forming mechanism was a common process. Examples of clasts are shown in Figure 7.2.

Skeletal Particles. Skeletal fragments occur in limestones as whole microfossils, whole larger fossils, or broken fragments of larger fossils. They are by far the most common kind of grain in carbonate rocks, and they are so abundant in some limestones that they make up most of the rock. Fossils representing all of the major phyla of calcareous marine invertebrates are present in limestones. The specific kinds of skeletal particles that occur depend upon both the age of the rocks and the paleoenvironmental conditions under which they were deposited. Owing to evolutionary changes in fossil assemblages through time, different kinds of fossil remains dominate rocks of different ages. For example, trilobite skeletal remains characterize early Paleozoic age rocks, but they do not occur in Cenozoic rocks, which instead commonly contain abundant foraminifers. Likewise, certain kinds of skeletal particles characterize

FIGURE 7.2 Angular limeclasts in a matrix of dark, organic-rich micrite, Calville Limestone (Permian), Nevada. The large clast in the middle is partially replaced by microcrystalline quartz (chert). Crossed nicols. Figure 7.9 shows another example of limeclasts.

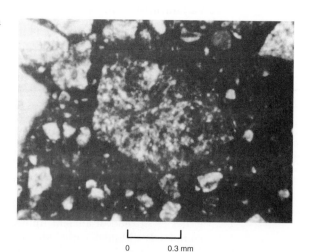

0   0.3 mm

limestones formed in different environments. To illustrate, the remains of colonial corals, which build rigid, wave-resistant skeletal structures, are commonly restricted to limestones deposited in shallow-water, high-energy environments where the water was well agitated and oxygen levels were high. By contrast, branching types of bryozoa are fragile organisms that cannot withstand the rigors of high-wave-energy environments. Thus, their remains are found mainly in limestones deposited under quiet-water conditions (Chapter 9).

Depending upon paleoenvironmental conditions, skeletal remains in a given specimen of limestone may consist entirely or almost entirely of one species of organism; however, they commonly include several species. An example of a mixed assemblage of skeletal particles is shown in Figure 7.3. The serious student of carbonate rocks must learn to identify the many kinds of fossils and fossil fragments that occur in limestones because fossils have special significance for paleoenvironmental and paleoecological interpretation (Chapter 9). Several atlases illustrating whole fossils and fossil fragments as they appear in microscope thin sections are available (e.g., Horowitz and Potter, 1971; Scholle, 1978). By use of these atlases, students should be able to identify many of the kinds of fossil remains commonly found in limestones.

Ooids. The term **ooid** is applied as a general name to coated carbonate grains that contain a nucleus of some kind—a shell fragment, pellet, or quartz grain—surrounded by one or more thin layers or coatings (the cortex) consisting of fine calcite or aragonite crystals. (In some ooids, the nucleus may be too small to be easily seen.) These coated grains are sometimes referred to as **ooliths;** however, the term ooid is preferred. Carbonate rocks formed mainly of ooids are called **oolites.** Spherical to subspherical ooids that exhibit several internal concentric layers with a total thickness greater than that of the nucleus are called normal or mature ooids (Fig. 7.4). Ooids form where strong bottom currents and agitated-water conditions exist and where saturation levels of calcium bicarbonate are high (Section 7.7). The coatings on modern ooids are composed mainly of aragonite, whereas ancient ooids are composed principally of calcite. Many of these ancient ooids were composed originally of aragonite which later transformed to calcite; however, petrographic evidence suggests that other ancient ooids originated as calcite. Precipitation of ancient calcitic ooids appears to have been particularly important during middle Paleozoic and middle Mesozoic time (Morse and Mackenzie, 1990, p. 538). Variations in ooid mineralogy appear to be related to sea levels. High

FIGURE 7.3 Skeletal grains cemented with sparry calcite cement, Salem Formation (Mississippian), Missouri. The skeletal material includes foraminifers (F), echinoderms (E), and molluscs (M). Crossed nicols.

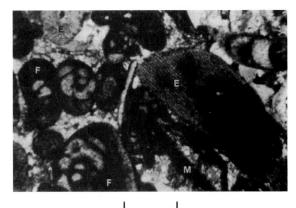

0   0.3 mm

FIGURE 7.4 Large normal ooids cemented with sparry calcite cement, Miami Oolite (Pleistocene), Florida. Note that each ooid has a distinct nucleus, probably a shell fragment, surrounded by well-developed concentric layers of calcite that formed by precipitation around the nucleus.

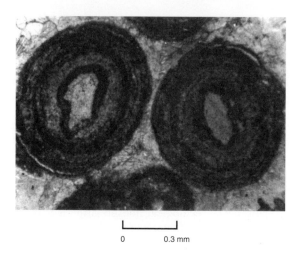

stands of the sea apparently favor formation of calcite ooids because CO_2 levels tend to be higher and Mg/Ca ratios lower during such times; low stands favor aragonite ooids because of lowered CO_2 levels and elevated Mg/Ca ratios (Wilkinson, Owen, and Carroll, 1985). As discussed in Section 7.7, high Mg/Ca ratios favor precipitation of aragonite over calcite, and vice versa.

Although most ooids display an internal structure consisting of concentric layers, some ooids show a radial internal structure (Fig. 7.5). Radial ooids that display relict concentric layers, such as those shown in Figure 7.5, probably form by recrystallization of normal ooids. Radial ooids may form also by primary sedimentation processes. The coating on some ooids consists only of one or two very thin layers which have a total thickness less than that of the nucleus. Such ooids have been called **superficial ooids** or **pseudo-ooids.** Coated grains that have an internal structure similar to that of ooids but that are much larger—that is, greater than 2 mm—are called pisoids

FIGURE 7.5 Radial ooids of marine origin, Devonian limestone, Canada. Note that the ooids preserve some of the concentric layers, suggesting that the radial fabric is secondary. Crossed nicols. Scale bar = 0.5 mm.

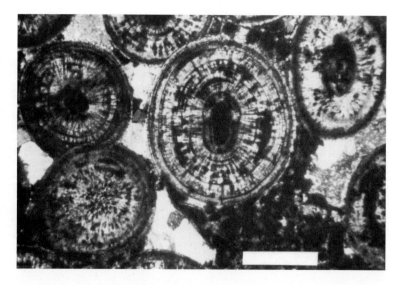

(a rock composed of pisoids is a pisolite). Pisoids are generally less spherical than ooids and are commonly crenulated. Some pisoids are of algal origin, formed by the trapping and binding activities of blue-green algae (cyanobacteria) in the same way stromatolites are formed (Chapter 5). Spheroidal stromatolites that reach a size exceeding 1 to 2 cm are called **oncoids.**

Peloids. Peloid is a nongenetic term for carbonate grains that are composed of microcrystalline or cryptocrystalline calcite or aragonite and that do not display distinctive internal structures (Fig. 7.6). Peloids are smaller than ooids and are generally of silt to fine sand size (0.03–0.1 mm), although some may be larger. The most common kind of peloids are fecal pellets, produced by organisms that ingest calcium carbonate muds. Fecal pellets tend to be small, oval to rounded in shape, and uniform in size. They commonly contain enough fine organic matter to make them appear opaque or dark colored. Pellets can be differentiated from ooids by their lack of concentric or radial internal structure and from rounded intraclasts by their uniformity of shape, good sorting, and small size. Because they are produced by organisms, their sizes and shapes are not related to current transport, although pellets may be transported by currents and redeposited after initial deposition by organisms.

Peloids may also be produced by other processes, such as micritization of small ooids or rounded skeletal fragments owing to the boring activities of certain organisms, particularly endolithic (boring) algae. These boring activities convert the original grains into a nearly uniform, homogeneous mass of microcrystalline calcite. Some peloids may simply be very small, well-rounded intraclasts formed by reworking of semiconsolidated mud or mud aggregates.

Aggregate Grains. Aggregate grains are irregularly shaped carbonate grains that consist of two or more carbonate fragments (pellets, ooids, fossil fragments) joined together by a carbonate-mud matrix that is generally dark colored and rich in organic matter. The shapes of the aggregate grains in some modern carbonate-forming environments, such as the Bahama Banks, resemble a bunch of grapes and are commonly called **grapestones** (Illing, 1954). Other aggregate grains with a somewhat smoother appearance have been referred to by the rather inelegant name of **lumps.** Tucker and Wright (1990, p. 12) suggest that lumps evolve from grapestones by continued cemen-

FIGURE 7.6 Small, even-sized peloids (probably fecal pellets) cemented with sparry calcite cement, Quaternary–Pleistocene limestone, Bahama Banks. A few ooids (larger grains) are also visible. Note the lack of internal structure in the peloids compared to that of the ooids shown in Figure 7.4. Plane (nonpolarized) light.

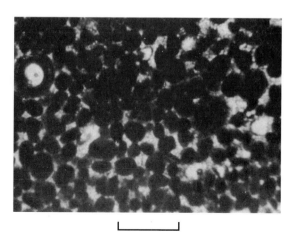

0   0.3 mm

tation and micritization of the grains (Fig. 7.7). Aggregate grains in modern carbonate environments are composed mainly of aragonite, but such grains in ancient limestones are dominantly calcite. Aggregate grains in modern environments can commonly be recognized by their botryoidal shapes and lack of internal structures; however, they can be confused with intraclasts. In fact, they are considered a type of intraclast by

FIGURE 7.7 Stages in the formation of grapestones and lumps. Stage 1: carbonate grains are bound together by foraminifers, microbial filaments, and mucilage. Chasmolithic microorganisms occur between the grains, whereas endolithic forms bore into the carbonate substrates. Stage 2: calcification of the microbial braces occurs, typically by high-magnesian calcite, to form a cemented grapestone. Progressive micritization of carbonate grains takes place. Stage 3: increased cementation at grain contacts, by microbially induced precipitation, fills depressions to create smoother relief. Stage 4: filling of any central cavity to form a dense, heavily micritized and matrix-rich aggregate. Some replacement of the high-magnesian calcite components by aragonite may also occur. (Based on Geblein, 1974; Winland and Matthews, 1974; and Fabricius, 1977.)

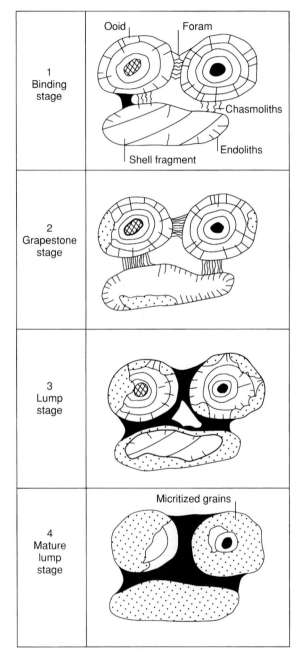

some geologists (Scholle, 1978). Aggregate grains are only rarely reported in ancient limestones, possibly because their shapes become distorted beyond recognition owing to compaction during diagenesis.

Microcrystalline Calcite

Carbonate mud composed of very fine-size calcite crystals is present in many ancient limestones in addition to sand-size carbonate grains. Carbonate mud or lime mud occurs also in modern environments where it consists dominantly of needle-shaped crystals of aragonite about 1 to 5 μm (0.001–0.005 mm) in length. The carbonate mud in ancient limestones is composed of similar-size crystals of calcite. Lime muds may also contain small amounts of fine-grained detrital minerals such as clay minerals, quartz, feldspar, and fine-size organic matter. They have a grayish to brownish, sub-translucent appearance under the microscope, and they are easily distinguished from carbonate grains and sparry calcite crystals (discussed below) by their extremely small crystal size (Fig. 7.8). Folk (1959) proposed the contraction **micrite** for microcrystalline calcite, a term that has been universally adopted to signify very fine-grained carbonate sediments.

Micrite may be present as matrix among carbonate grains, or it may make up most or all of a limestone. A limestone composed mostly of micrite is analogous texturally to a siliciclastic mudrock or shale. The presence of micrite in an ancient limestone is commonly interpreted to indicate deposition under quiet-water conditions where little winnowing of fine mud took place. By contrast, carbonate sediments deposited in environments where bottom currents or wave energy are strong are commonly mud-free because carbonate mud is selectively removed in these environments. On the basis of purely chemical considerations, carbonate mud or micrite can theoretically form by inorganic precipitation of aragonite, later converted to calcite, from surface waters supersaturated with calcium bicarbonate. Geologists are uncertain, however, about how much aragonite is actually being generated by inorganic processes in the modern ocean. Much modern carbonate mud appears to originate through organic processes (Section 7.7). These processes include breakdown of calcareous algae in shallow water to yield aragonite mud, and deposition of carbonate nannofossils (<35 μm in size) such as coccoliths in deeper water to yield calcite muds (chalks).

FIGURE 7.8 Dark, organic-rich micrite with a few small fossil fragments, Louisiana Limestone (Devonian), Illinois. Plane light.

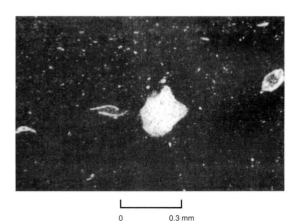

0   0.3 mm

FIGURE 7.9 Sparry calcite cementing rounded, dark-colored limeclasts, Devonian limestone, Canada. Note that the cement displays drusy texture: the small calcite crystals around the margins of the clasts are oriented with their long dimensions perpendicular to the clast surfaces; these small, oriented crystals grade toward the center of the pores to larger, randomly oriented calcite crystals. Crossed nicols.

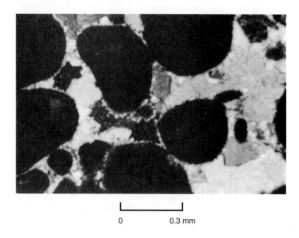

```
L_____J
0          0.3 mm
```

Sparry Calcite

Many limestones contain large crystals of calcite, commonly on the order of 0.02 to 0.1 mm, that appear clear or white when viewed with a hand lens or in plane light under a polarizing microscope. Such crystals are called sparry calcite. They are distinguished from micrite by their larger size and clarity and from carbonate grains by their crystal shapes and lack of internal texture. Some sparry calcite can be seen under the microscope to fill interstitial pore spaces among grains or to fill solution cavities as a cement (Fig. 7.9). The presence of sparry calcite cement in intergranular pore spaces indicates that grain framework voids were empty of lime mud at the time of deposition, suggesting deposition under agitated-water conditions that removed fine mud, as mentioned.

Sparry calcite can also form in ancient limestones by recrystallization of primary depositional grains and micrite during diagenesis (Section 7.8). Sparry calcite formed by recrystallization may be very difficult in some cases to differentiate from sparry calcite cement (Boggs, 1992, p. 549). It is important to make a distinction between the two types of sparry calcite because incorrectly identifying recrystallized spar as sparry calcite cement can cause errors in both environmental interpretation and limestone classification.

7.5   DOLOMITE TEXTURES

Dolomite (dolostone) is composed mainly of the mineral dolomite [$CaMg(CO_3)_2$]. Unlike limestone, which is characterized by the presence of grains, micrite, and/or sparry cement, dolomite has a largely crystalline (granular) texture. On the basis of crystal shape, two kinds of dolomite are recognized. **Planar** (or idiotopic) dolomite consists of rhombic, euhedral (well-formed) to anhedral (poorly formed) crystals. **Nonplanar** (or xenotopic) dolomite is made of nonplanar, commonly anhedral crystals (Sibley and Gregg, 1987). Each of these major kinds of dolomite can be divided into subtypes as shown in Figure 7.10. Many dolomites form by replacement of a precursor limestone. Original limestone textures may be preserved in such dolomites to various degrees, ranging from virtually unreplaced to totally replaced (Fig. 7.10). That is, the replacing dolomite may preserve the original texture as a "ghost" (mimicking replacement), or the original texture may be completely destroyed (nonmimicking replacement).

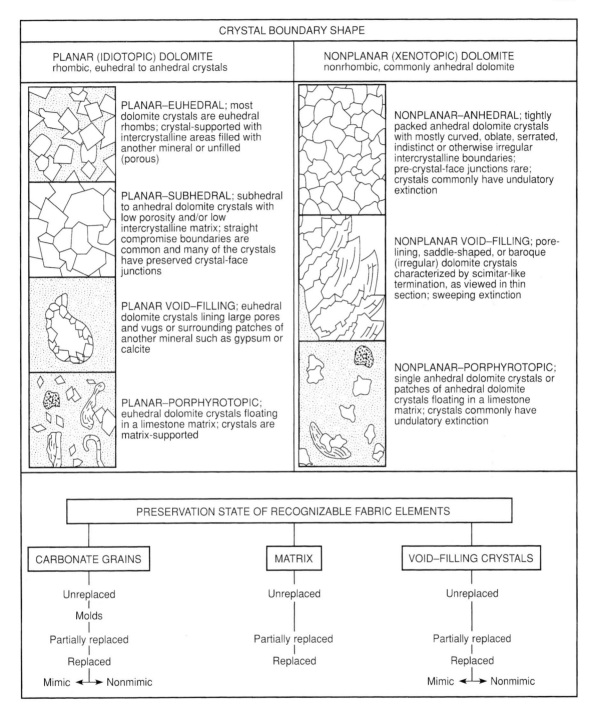

FIGURE 7.10 Classification of dolomite textures. (After Gregg, J. M., and D. F. Sibley, 1984, Epigenetic dolomitization and the origin of xenotopic dolomite: Jour. Sed. Petrology, v. 54, Fig. 6, p. 913, and Sibley, D. F., and Gregg, J. M., 1987, Classification of dolomite rock textures: Jour. Sed. Petrology, v. 57, Fig. 1, p. 968; figures reprinted by permission of Society of Economic Paleontologists and Mineralogists, Tulsa, Okla.)

7.6 CLASSIFICATION OF CARBONATE ROCKS

Attempts to classify carbonate rocks date back to at least 1904 with the publication of Grabeau's classic textbook on the classification of sedimentary rocks. Additional classifications were proposed by other authors in the 1930s, 1940s, and 1950s. Most of these early classifications were basically genetic schemes in which names such as "fore-reef talus limestone" or "low-energy limestone" were used to identify limestones according to their presumed environment of deposition (Ham and Pray, 1962). These classifications failed to recognize the clear distinction between carbonate grains and carbonate mud or to exploit difference in identity of the various kinds of carbonate grains. Publication in 1959 of Folk's largely descriptive *Practical Petrographic Classification of Limestones* marked the beginning of the modern period of limestone classification. In 1962, several additional classifications appeared (Ham, 1962) which, with one exception, are mainly descriptive classifications. Unlike the confusion attending the proliferation of sandstone classifications, the appearance of several descriptive limestone classifications seems to have had a largely positive effect because it has forced geologists to become more keenly aware of the varied constituents that make up limestones, as well as the environmental significance of these constituents.

Mineralogy plays only a small role in classification of carbonate rocks because most carbonate rocks are essentially monomineralic. Mineralogy is used primarily to differentiate dolomite from limestone or carbonate rocks from noncarbonate rocks, as shown in Figure 7.11. The principal constituents or parameters used in carbonate classification are (l) the types of carbonate grains or allochems and (2) the grain/micrite ratio. The nature of the grain packing or fabric is also used in some classifications, in which the fabric is referred to as either grain-supported or mud-supported. A grain-supported fabric is one in which grains are in contact, creating an intact grain framework in which voids may or may not be filled with mud (matrix). In a mud-supported fabric, most grains do not touch, and they appear to float in the carbonate mud.

Folk's (1959, 1962) classification has probably been the most widely accepted limestone classification because of its applicability to a wide range of carbonate rock types and the ease with which its terms can be utilized and understood. The classification is based on the relative abundance of three major types of constituents: (1) carbonate grains or allochems, (2) microcrystalline carbonate mud (micrite), and (3) sparry calcite cement. As illustrated in Table 7.2, classification is made by first determining the relative abundance of total allochems vs. micrite plus sparry calcite cement. Further subdivision is then made on the basis of the relative abundance of the various types of carbonate grains (Fig. 7.12) and the relative abundance of micrite compared to sparry calcite cement. This classification approach yields a bipartite name that reflects both the major type of carbonate grain in the limestone and the relative abundance of micrite and sparry calcite cement. Thus, an **oosparite** is an ooid-rich rock cemented with sparry calcite that contains little micrite, whereas an **oomicrite** is an ooid-rich limestone in which micrite is abundant and sparry calcite is subordinate. Additional textural information can be added by use of the textural maturity terms shown in Figure 7.13. Thus, a **packed oomicrite** indicates a grain-supported oolitic limestone, and a **sparse oomicrite** is an oolitic rock with a mud-supported fabric. Note that Folk's classification can also be used to classify dolomite rock, if "ghosts" of the original allochems are still identifiable in the dolomite.

The terms used by Folk (1959, 1962) to differentiate depositional textures are purely descriptive; however, they also have environmental significance. The term **biomicrite,** for example, conveys an interpretation of deposition under quiet-water conditions where micrite is abundant and winnowing of the lime mud is minimal. Thus,

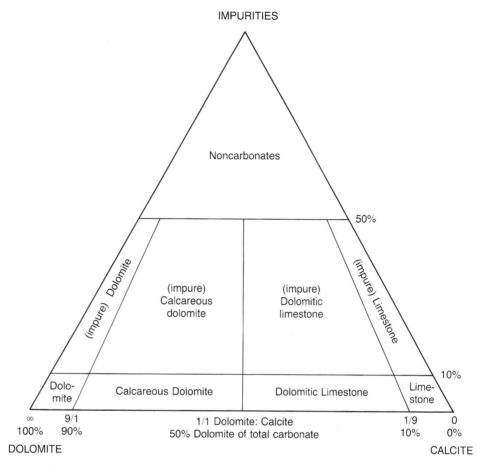

FIGURE 7.11 Terminology of carbonate rocks based on relative percentages of calcite, dolomite, and noncarbonate impurities. (After Leighton, M. W., and C. Pendexter, 1962, Carbonate rock types, *in* W. E. Ham (ed.), Classification of carbonate rocks: Am. Assoc. Petroleum Geologists Mem. 1. Fig. 2, p. 51, reprinted by permission of AAPG, Tulsa, Okla.)

micrite accumulates along with skeletal particles. On the other hand, the term **biosparite** suggests deposition in a wave-agitated environment where micrite is removed by winnowing currents, allowing mud-free carbonate grains to accumulate. These grains are subsequently cemented with sparry calcite during diagenesis.

Dunham (1962) has a somewhat different type of classification (Table 7.3A) that stresses the relative abundance of allochems and micrite but does not consider the identity of different kinds of carbonate grains. Dunham's classification is based solely upon depositional texture and considers two aspects of texture: (1) grain packing and the relative abundance of grains to micrite and (2) depositional binding of grains. By depositional binding, I mean whether or not carbonate grains show evidence of having been bound together at the time of deposition, as in a colonial reef complex, a stromatolite (cyanobacteria) bed, or a calcareous algae mat. Dunham's classification separates components that were not bound together at the time of deposition into those that lack lime mud and those that contain lime mud. Rocks that contain no mud are obviously

TABLE 7.2   Classification of carbonate rocks

Volumetric allochem composition		Limestones, partly dolomitized limestones				Replacement dolomites	
		>10% Allochems — Allochemical rocks		<10% Allochems — Microcrystalline rocks		Allochem ghosts	No allochem ghosts
		Sparry calcite cement > microcrystalline ooze matrix (Sparry allochemical rocks)	Microcrystalline ooze matrix > sparry calcite cement (Microcrystalline allochemical rocks)	1%–10% Allochems (Most abundant allochem)	<1% Allochems		
>25% Intraclasts		Intrasparrudite / Intrasparite	Intramicrudite* / Intramicrite*	Intraclasts: intraclast-bearing micrite*	*(Micrite, if disturbed, dismicrite; if primary, dolomite, dolomicrite)*	Finely crystalline intraclastic dolomite, etc.	Medium crystalline dolomite
<25% Intraclasts	<25% Ooids, >25% Ooids	Oosparrudite / Oosparite	Oomicrudite* / Oomicrite*	Oolites: ooid-bearing micrite*		Coarsely crystalline oolitic dolomite, etc.	Finely crystalline dolomite
	<25% Ooids — Volume ratio of fossils to pellets >3:1	Biosparrudite / Biosparite	Biomicrudite / Biomicrite	Fossils: fossiliferous micrite		Aphanocrystalline biogenic dolomite, etc.	
	3:1–1:3	Biopelsparite	Biopelmicrite				
	<1:3	Pelsparite	Pelmicrite	Pellets: pelletiferous micrite		Very finely crystalline pellet dolomite, etc.	etc.
(Biolithite — Undisturbed bioherm rocks)			Biolithite			Evident allochem	

Source: Folk, R. L., 1962. Spectral subdivision of limestone types, *in* W. E. Ham (ed.), Classification of carbonate rocks: Am. Assoc. Petroleum Geologists Mem. 1. Table 1, p. 70, reprinted by permission of AAPG, Tulsa, Okla.

Note: Names and symbols in the body of the table refer to limestones. If the rock contains more than 10 percent replacement dolomite, prefix the term "dolomitized" to the rock name. The upper name in each box refers to calcirudites (median allochem size larger than 1.0 mm); the lower name refers to all rocks with median allochem size smaller than 1.0 mm. Grain size and quantity of ooze matrix, cements, or terrigenous grains are ignored.

*Designates rare rock types.

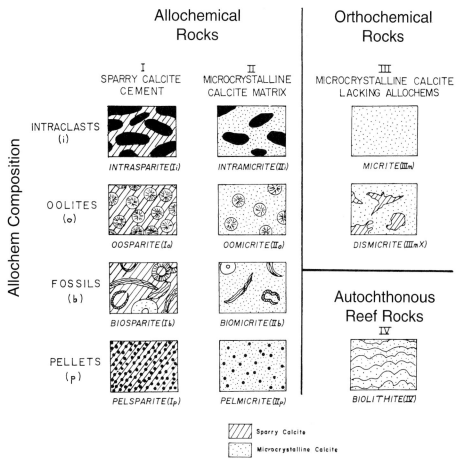

FIGURE 7.12 Schematic representation of the constituents that form the basis for Folk's classification of carbonate rocks (Table 7.2). (After Folk, R. L., 1962, Spectral subdivision of limestone types, *in* W. E. Ham (ed.), Classification of carbonate rocks: Am. Assoc. Petroleum Geologists Mem. 1. Fig. 3, p. 71, reprinted by permission of AAPG, Tulsa, Okla.)

grain-supported. Rocks that contain mud may be either grain-supported or mud-supported. Note, however, that grain support does not depend upon the absolute grain-to-mud ratio, because grain support is also a function of shapes of the carbonate grains. Platy or elongate grains such as bivalve shells may form a grain-supported fabric at much lower grain abundances than more spherical particles such as ooids. Therefore, Dunham's boundary between grain-supported and mud-supported limestones is not based on a fixed grain/micrite ratio. Dunham's classification was modified, with addition of two new names (floatstone, rudstone), by Embry and Klovan (1972) to better reflect the presence of gravel-size (>2 mm) carbonate grains (Table 7.3B). These authors also divided Dunham's boundstone into three types (framestone, bindstone, bafflestone) on the basis of the presumed kinds of organisms that bound the sediment together.

Because Dunham's classification does not consider the identity of the carbonate grains, it may be desirable to use it in conjunction with another classification such as

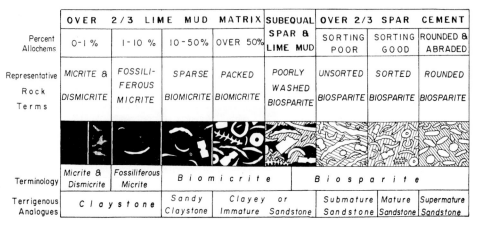

Percent Allochems	OVER 2/3 LIME MUD MATRIX				SUBEQUAL SPAR & LIME MUD	OVER 2/3 SPAR CEMENT		
	0-1 %	1-10 %	10-50%	OVER 50%		SORTING POOR	SORTING GOOD	ROUNDED & ABRADED
Representative Rock Terms	MICRITE & DISMICRITE	FOSSILI-FEROUS MICRITE	SPARSE BIOMICRITE	PACKED BIOMICRITE	POORLY WASHED BIOSPARITE	UNSORTED BIOSPARITE	SORTED BIOSPARITE	ROUNDED BIOSPARITE
Terminology	Micrite & Dismicrite	Fossiliferous Micrite	B i o m i c r i t e			B i o s p a r i t e		
Terrigenous Analogues	C l a y s t o n e		Sandy Claystone	Clayey or Immature Sandstone		Submature Sandstone	Mature Sandstone	Supermature Sandstone

■ LIME MUD MATRIX   ▨ SPARRY CALCITE CEMENT

FIGURE 7.13   Textural classification of carbonate sediments based on relative abundance of lime mud matrix and sparry calcite cement and on the abundance and sorting of carbonate grains (allochems). (After Folk, R. L., 1962, Spectral subdivision of limestone types, *in* W. E. Ham (ed.), Classification of carbonate rocks: Am. Assoc. Petroleum Geologists Mem. 1. Fig. 4, p. 76, reprinted by permission of AAPG, Tulsa, Okla.)

Folk's. Thus, a limestone identified as a packed oomicrite using Folk's classification could alternatively be called an **oomicrite packstone** using a combination of Folk's and Dunham's classifications. Additional limestone classifications are discussed in the symposium volume edited by Ham (1962), and classifications for mixed carbonate and siliciclastic sediment are discussed by Mount (1985) and Zuffa (1980).

The terms coquina, chalk, and marl are commonly used, informal names for carbonate rocks. A **coquina** is a mechanically sorted and abraded, poorly consolidated carbonate sediment consisting predominantly of fossil debris; **coquinite** is the consolidated equivalent. **Chalk** is soft, earthy, fine-textured limestone composed mainly of the calcite tests of floating microorganisms such as foraminifers. **Marl** is an old, rather imprecise term for an earthy, loosely consolidated mixture of siliciclastic clay and calcium carbonate.

7.7   ORIGIN OF CARBONATE ROCKS

The chemical weathering processes discussed in Chapter 2 release chemical ions from source rocks that eventually make their way, dissolved in groundwater and surface water, to lakes and the ocean. Bicarbonate ions (HCO_3^-) may be added also by interaction of water with atmospheric and soil CO_2. Most dissolved ions end up in the ocean, where they remain dissolved in seawater for periods ranging from hundreds to millions of years. The average time that a particular chemical element remains in solution in the ocean before precipitating is called its residence time. The residence times of some common elements in seawater are given in Table 7.4.

Table 7.5 shows the abundance of the principal ion species dissolved in both mean river water and ocean water. Note that the actual concentration (in ppm) of most constituents is greater in ocean water than in river water; however, the relative abun-

TABLE 7.3 Classification of limestones according to depositional textures

A

DEPOSITIONAL TEXTURE RECOGNIZABLE					DEPOSITIONAL TEXTURE NOT RECOGNIZABLE
Original components not bound together during deposition				Original components were bound together during deposition . . . as shown by intergrown skeletal matter, lamination contrary to gravity, or sediment-floored cavities are roofed over by organic or questionably organic matter and are too large to be interstices.	CRYSTALLINE CARBONATE
Contains mud (particles of clay and fine silt size)			Lacks mud and is grain-supported		
Mud-supported		Grain-supported			(Subdivide according to classifications designed to bear on physical texture or diagenesis.)
Less than 10% grains	More than 10% grains				
MUDSTONE	WACKESTONE	PACKSTONE	GRAINSTONE	BOUNDSTONE	

B

ALLOCHTHONOUS LIMESTONE Original components not organically bound during deposition						AUTOCHTHONOUS LIMESTONE Original components organically bound during deposition		
Less than 10% >2 mm components				Greater than 10% >2mm components		By organisms that build a rigid framework	By organisms that encrust and bind	By organisms that act as baffles
Contains lime mud (<0.03 mm)			No lime mud	Matrix-supported	>2 mm component-supported			
Mud-supported		Grain-supported	Grain-supported				B O U N D S T O N E	
Less than 10% grains (>0.03 mm <2 mm)	Greater than 10% grains							
MUD-STONE	WACKE-STONE	PACK-STONE	GRAIN-STONE	FLOAT-STONE	RUD-STONE	FRAME-STONE	BIND-STONE	BAFFLE-STONE

Source: A. After Dunham, R. J.: 1962. Classification of carbonate rocks according to depositional textures, *in* Ham, W. E., ed., Classification of carbonate rocks: Am. Assoc. Petroleum Geologists Mem. 1. Table 1, p. 117, reprinted by permission of AAPG, Tulsa, Okla. B, after Dunham, R. J.: 1962, as modified by Embry, E. F., III and J. E. Klovan, 1972, Absolute water depth limits of late Devonian paleoecological zones: Geol. Rundschau, v. 61. Fig. 5, p. 676, reprinted by permission.

TABLE 7.4   Residence times of selected elements in seawater

Element	Residence time (yr)
Cations	
Sodium (Na$^+$)	260,000,000
Magnesium (Mg^{2+})	12,000,000
Potassium (K$^+$)	11,000,000
Calcium (Ca^{2+})	1,000,000
Silicon (Si^{4+})	8,000
Manganese (Mn^{2+})	7,000
Iron (Fe^{2+}, Fe^{3+})	140
Aluminum (Al^{3+})	100
Anions	
Chlorine (Cl$^-$)	∞
Sulfate (SO$_4{}^{2-}$)	11,000,000
Carbonate (CO$_3{}^{2-}$)	110,000

Source: Ross, D. A., 1982, Introduction to oceanography: Prentice-Hall, Englewood Cliffs, N.J., and Stowe, K. S., 1979, Ocean science: John Wiley & Sons, New York.

TABLE 7.5   Dissolved ion species in mean world river water and ocean water

Ionic species	A Mean river water		B Ocean water	
	ppm	% of total dissolved solids	ppm	% of total dissolved solids
HCO$_3{}^-$, CO$_3{}^{2-}$	58.7	48.6	140	0.4
Ca^{2+}	15.0	12.4	400	1.2
H$_4$SiO$_4$	13.1	10.8	1	<0.01
SO$_4{}^{2-}$	11.2	9.3	2,649	7.7
Cl$^-$	7.8	6.5	18,980	55.0
Na$^+$	6.3	5.2	10,556	30.6
Mg^{2+}	4.1	3.4	1,272	3.7
K$^+$	2.3	1.9	380	1.1
NO$_3{}^-$	1.0	0.8	0.5	<0.01
Fe^{2+}, Fe^{3+}	0.67	0.6	0.01	<0.01
Al(OH)$_4{}^-$	0.24	0.2	0.01	<0.01
F$^-$	0.09	0.07	1.3	<0.01
Sr^{2+}	0.09	0.07	8	0.02
B(OH)$_4{}^-$	0.01–0.01	0.08–<0.01	26	0.07
Mn^{2+}	0.02	0.02	—	—
Br$^-$	—	—	65	0.02
Total	120.8		34,479	

Source: A, Livingston, D. A., 1963, Data of geochemistry. Chap. G, Chemical composition of rivers and lakes: U.S. Geol. Survey Prof. Paper 440–G. B, Mason, B., 1966, Principles of geochemistry: John Wiley & Sons, New York.

dance (percent of total dissolved solids) of certain major constituents is much greater in river water than in ocean water. For example, bicarbonate ions (HCO$_3{}^-$) and carbonate ions (CO$_3{}^-$) together make up almost 49 percent of the total dissolved solids in average river water but less than 1 percent of the dissolved solids in ocean water. The relative abundances of silica, expressed as H$_4$SiO$_4$, and calcium are also several orders of magnitude higher in river water than in the ocean. On the other hand, chlorine and sodium are relatively much more abundant in the ocean.

The explanation for these differences in relative abundance of ions in river water and ocean water is provided by examination of the relative abundance of various kinds of nonsiliciclastic sedimentary rocks preserved in the geologic record. Measurements of the thickness and volume of different types of sedimentary rocks show that the chemically and biochemically deposited sedimentary rocks make up nearly one-fourth of all sedimentary rocks. The majority of these rocks are carbonate rocks, composed of calcium carbonate and calcium-magnesium carbonate minerals. Carbonate ions, bicarbonate ions, calcium ions, and possibly magnesium ions have been preferentially removed from the oceans throughout geologic time to form carbonate rocks, thus accounting for the low relative abundance of these ions in the modern ocean. In this section, we explore the chemical/biochemical processes that control the removal of dissolved ions from the ocean to form carbonate rocks.

Limestones

Chemistry of Calcium Carbonate Deposition. The dissolution and precipitation of the calcium carbonate ($CaCO_3$) minerals calcite and aragonite are controlled chiefly by pH. Figure 7.14 shows the relationship between pH and the solubility of aragonite and calcite. Solution pH is linked in turn to the partial pressure of dissolved carbon dioxide in the water, as illustrated by the following reactions:

$$CO_2 + H_2O \leftrightarrow H_2CO_3 \quad \text{(carbonic acid)} \tag{7.1}$$

$$H_2CO_3 \leftrightarrow H^+ + HCO_3^- \quad \text{(bicarbonate ion)} \tag{7.2}$$

$$HCO_3^- \leftrightarrow H^+ + CO_3^{2-} \quad \text{(carbonate ion)} \tag{7.3}$$

These reactions show that the dissociation of carbonic acid to hydrogen ions and bicarbonate ions (equation 7.2) and the further dissociation of bicarbonate ions to hydrogen ions and carbonate ions (equation 7.3) release free hydrogen ions, thus lowering the pH of the solution.

FIGURE 7.14 Effect of pH at approximately 25°C on the solubility of calcium carbonate. Solubility decreases with increasing temperature. (After Friedman, G. M., and J. E. Sanders, Principles of sedimentology: © 1978 by John Wiley & Sons, Inc. Fig. 5.23, p. 136, reprinted by permission of John Wiley & Sons, Inc., New York.)

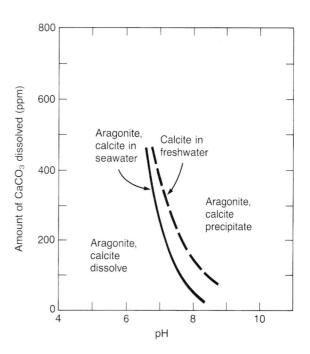

If calcite or aragonite crystals are allowed to react with a carbonic acid solution, these minerals are readily dissolved. This reaction can be summarized as

$$H_2O + CO_2 + CaCO_3 \quad \leftrightarrow \quad Ca^{2+} + 2HCO_3^- \qquad (7.4)$$
$$\text{(calcite or}$$
$$\text{aragonite)}$$

Note by the presence of double arrows that this reaction is reversible. If equilibrium conditions are disturbed by loss of carbon dioxide, the concentration of hydrogen ions decreases and the pH increases. The reaction shifts toward the left, resulting in precipitation of solid $CaCO_3$. The partial pressure of carbon dioxide thus clearly exerts a major control on calcium carbonate precipitation. Anything that causes loss of carbon dioxide should theoretically trigger the onset of precipitation, although we shall see subsequently that inorganic precipitation of calcium carbonate owing to loss of CO_2 may not be as important under natural conditions in the open ocean as suggested by equation 7.4.

Mechanisms that can cause loss of carbon dioxide from water include increase in temperature and salinity and decrease in water pressure. An increase in temperature or salinity causes a decrease in the solubility of carbon dioxide (and other gases) in water; that is, an increase in temperature reduces the capacity of water to dissolve and retain carbon dioxide, resulting in the escape of carbon dioxide. Because dissolved carbon dioxide increases the acidity of water by releasing H^+ ions, loss of carbon dioxide owing to increase in temperature or salinity causes an increase in pH (decreased acidity). Decrease in water pressure can also allow carbon dioxide to escape. Under natural conditions, pressure may be lowered by wave agitation caused by storm activity or breaking of waves in the surf zone or over shallow banks. Circulation of deep, pressurized waters to the surface can likewise release carbon dioxide, and even lowering of atmospheric pressure may cause slight loss of carbon dioxide from ocean water.

In addition to its effect on CO_2 solubility, increase in temperature causes a decrease in the solubility of calcium carbonate minerals; that is, the calcium carbonate solubility product deceases with increasing temperature. Decrease in solubility means that a mineral will be more likely to precipitate under a given set of conditions. Thus, calcium carbonate deposition is favored in the more tropical areas of the ocean, where surface water temperatures may reach almost 30°C, compared to about 0°C in the polar regions.

The solubility of chemical constituents is affected also by salinity and the ionic strength of seawater. Ionic strength is a function of the concentration of ions in solution and the charges on these ions; thus, ionic strength increases as salinity increases. The solubility of calcium carbonate minerals is markedly enhanced at higher values of salinity because increase in ionic strength causes an increase in the concentration of foreign ions (e.g., Mg^{2+}) other than Ca^{2+} and CO_3^-. These foreign ions interfere with the formation of the calcium carbonate crystal structure, making it more difficult for calcite or aragonite minerals to grow and precipitate. In general, the amount of solubility increase will depend upon the concentration of added salts (Krauskopf, 1979). Therefore, the solubility of calcium carbonate is several orders of magnitude higher in seawater than in freshwater (Degens, 1965). On the other hand, the influence of salinity on the solubility of calcium carbonate, as well as on the solubility of carbon dioxide, in the surface waters of the open ocean may be slight because these waters range in salinities only from about 32 to 36 parts per thousand (o/oo).

This discussion of carbonate solubility relationships is quite elementary and is intended only to provide a very basic understanding of carbonate solubility and the factors that govern the precipitation of carbonate minerals. For a much more rigorous discussion of carbonate geochemistry, see Morse and Mackenzie (1990).

Importance of Inorganic Precipitation of Calcium Carbonate. According to the theoretical considerations illustrated in equation 7.4, significant loss of carbon dioxide by any of the mechanisms noted should lead to precipitation of calcium carbonate minerals. On the other hand, near-surface water in the modern ocean is oversaturated by over six times with respect to calcite and over four times with respect to aragonite (Morse and Mackenzie, 1990, p. 217). Such gross oversaturation indicates a reluctance of calcium carbonate minerals to precipitate. Considerable debate has thus been generated among geologists about the actual significance of inorganic precipitation of calcite or aragonite in the modern ocean. Does precipitation of calcium carbonate owing to loss of carbon dioxide occur on an important scale in the open ocean environment today? One process considered by some investigators to constitute evidence of large-scale inorganic precipitation of $CaCO_3$ is the formation of whitings in such warm-water areas as the Bahamas, the Persian Gulf, and the Dead Sea. The sudden appearance of these **whitings,** which are milky patches of surface and near-surface water caused by dense concentrations of suspended aragonite crystals, has been suggested to result from the spontaneous nucleation of aragonite crystals in waters supersaturated with calcium bicarbonate. This view has been challenged by other workers who propose that mechanisms such as resuspension of aragonite mud from the shallow seafloor by wave action or by stirring up of mud by bottom-feeding fish are responsible for whitings rather than spontaneous precipitation of aragonite. Recent studies by Shinn et al. (1989) in the Bahama Banks area show that whitings do contain some (perhaps as much as 25 percent) newly precipitated aragonite; presumably the remainder is older sediment physically resuspended by storms. This study, while providing evidence that some aragonite does precipitate in the modern ocean during the formation of whitings, failed to provide conclusive evidence of the exact method of precipitation and the overall importance of inorganic precipitation of aragonite in modern carbonate environments.

Robbin and Blackwelder (1992) have thrown new light on the whiting problem through biochemical study of whiting sediment. They propose that whitings are in part the result of biological precipitation of calcium carbonate induced by exceedingly small algal phytoplankton (picoplankton) and cellular components. The cells act as nucleation sites for crystallization from seawater supersaturated with $CaCO_3$. Mineralization occurs on the surface of phytoplankton cells and degrading organic cellular components. These authors suggest that algal cells induce precipitation of $CaCO_3$ owing to photosynthetic removal of CO_2 from seawater. They further report that the observed carbonate crystals are distinct from skeletal debris. Macintyre and Reid (1992) also report that the crystals in whitings differ from aragonite crystals secreted by calcareous algae. On the other hand, Morse and He (1993) suggest, on the basis of experimental study of seawater, that whitings on the Bahama Banks are unlikely to be due to nucleation of calcium carbonate—although whitings in the Persian Gulf could result from direct nucleation of calcium carbonate in water with lowered CO_2 content, associated with phytoplankton blooms. So the debate goes on!

Although the extent and the overall importance of inorganic precipitation of calcium carbonate minerals under natural conditions in the ocean are still being debated, experimental work has demonstrated that neither calcite nor aragonite precipitates readily from seawater. There appear to be at least three reasons why inorganic precipitation of calcium carbonate may not be a widespread, quantitatively significant phenomenon in the modern ocean.

First, the magnitude of the pH changes that occur in the open ocean owing to loss of carbon dioxide is relatively small because seawater is a well-buffered solution. Buffering occurs because a considerable portion of the carbon dioxide dissolved in seawater forms undissociated H_2CO_3 rather than dissociating to H^+ ions, HCO_3^- ions, and CO_3^{2-} ions as predicted by equations 7.2 and 7.3. This buffering reaction is caused by the high alkalinity of ocean water; that is, the high concentrations of bicarbonate and carbonate ions already present in surface waters of the ocean inhibit breakdown of H_2CO_3 to form still more of these ions. Therefore, the actual change in pH in seawater owing to either gain or loss of carbon dioxide is comparatively small, and the pH values of seawater in the open ocean rarely fall outside the range of 7.8 to 8.3 (Bathurst, 1975).

Second, the presence of Mg^{2+} ions at the concentration levels found in seawater has been shown experimentally to have a strong inhibiting effect on the precipitation of calcite ($CaCO_3$). Experiments by Berner (1975) show that Mg^{2+} is readily adsorbed onto the surface of calcite crystals and incorporated into their crystal structure. This nonequilibrium incorporation of Mg^{2+} into growing calcite crystals was interpreted by Berner to decreases their stability, resulting in an increase in calcite solubility. Thus, calcite crystals do not readily nucleate and grow in the presence of Mg^{2+} in seawater concentrations. If they form at all, the crystallization process takes place very slowly. Berner found no retardation of precipitation in similar experiments at low Mg^{2+} concentrations. Subsequent researchers have interpreted the inhibiting influence of Mg^{2+} in different ways. They have suggested difficulties in rapidly dehydrating the Mg^{2+} ion, which surrounds itself with water molecules (Mucci and Morse, 1983), and crystal poisoning by adsorption of Mg^{2+} at reactive sites (Reddy and Wang, 1980).

Aragonite is also composed of $CaCO_3$ but has a crystal structure different from that of calcite (Table 7.1). Mg^{2+} ions appear to be less prone to sorb to aragonite nuclei and disrupt crystal growth. Therefore, aragonite is less affected by Mg^{2+} and has a tendency in the presence of Mg^{2+} to precipitate in preference to calcite. Nonetheless, aragonite does not precipitate completely freely in ocean water, even when surface waters are supersaturated with respect to calcium carbonate. Experiments by Berner et al. (1978) show that the reluctance of aragonite to precipitate in ocean water may be due to the influence of organic compounds found in natural humic and fulvic acids or in phosphates. These compounds can apparently form thin organophosphatic coatings on aragonite seed nuclei, inhibiting their growth and preventing or significantly delaying aragonite precipitation.

One process that does involve inorganic precipitation of calcium carbonate, at least in part, is the formation of ooids. As mentioned, ooids in modern environments consist mainly of aragonite, whereas many ancient ooids may have precipitated as calcite. Ooids form mainly under high-energy, agitated-water conditions in warm waters that are supersaturated with calcium carbonate. Warming and evaporation of cold ocean water driven onto shallow banks by tidal currents result in supersaturation of the water. Currents and waves keep the grains moving and intermittently suspended, allowing more or less even precipitation of calcium carbonate on all sides of the grains. Both supersaturation of the water and intermittent burial and resuspension of the ooids owing to agitation appear to be necessary for most ooids to form, although some ooids are known to form in quiet water. Cyanobacteria or other microorganisms may influence the formation of ooids—possibly by trapping carbonate grains on organic films or by mediating carbonate precipitation through removal of CO_2. The quantitative importance of organic influences on the formation of ooids is not well understood (Tucker and Wright, 1990, p. 6).

The Role of Organisms in Precipitation of Calcium Carbonate. Although precipitation of minerals from water is fundamentally a chemical process, chemical processes are apparently aided in a variety of ways by organisms. Although purely inorganic precipitation of carbonate minerals from normal-salinity seawater or freshwater apparently can occur, it may be less common today than precipitation aided in some way by organic processes. Furthermore, geologic evidence suggests that organisms may have played a significant role in carbonate sedimentation throughout at least most of Phanerozoic (post-Precambrian) time.

The most important role of organisms in chemical sedimentation is probably the direct removal of dissolved constituents to build skeletal structures. The exact mechanisms by which organisms remove dissolved substances to build their shells is not well understood, but the process is very common. Numerous marine invertebrates build protective shells or other skeletal structures from calcium carbonate. These organisms range from freely drifting, planktonic species such as foraminifers and pteropods (winged snails) to bottom-dwelling benthonic organisms such as calcareous algae, corals, and molluscs. They can remove $CaCO_3$ not only from calcium carbonate–saturated surface waters in tropical regions but also from less saturated waters in temperate and colder regions. For example, shell sands and gravels are important deposits of the modern seafloor in shallow, cool water at high latitudes (e.g., Farrow, Allen, and Akpan, 1984). It is not definitely known that organisms can precipitate $CaCO_3$ from waters that are highly undersaturated in calcium carbonate (Krauskopf, 1979). The importance of biologic removal of calcium carbonate from the oceans is demonstrated by the fact that most Phanerozoic limestones contain some recognizable calcium carbonate fossils, and many are composed dominantly of such remains. Also, large areas of the modern ocean floor are covered by calcareous oozes composed dominantly of the shells of foraminifers, one-celled algal coccolithophores, and pteropods.

Another type of organic activity that is very important to the formation of carbonate rocks is removal of carbon dioxide from water by photosynthesizing plants. As mentioned, any process that removes carbon dioxide from the water facilitates carbonate precipitation by increasing the pH. Aquatic plants, particularly blue-green algae (cyanobacteria) and planktonic algae such as diatoms, remove carbon dioxide from water during the process of photosynthesis as shown by the following relationship:

$$6H_2O + 6CO_2 \longrightarrow C_6H_{12}O_6 + 6O_2 \qquad (7.5)$$

(water + carbon dioxide $\longrightarrow$ carbohydrates + oxygen)

Blue-green algae and small phytoplankton such as diatoms, dinoflagellates, and coccoliths are the most important users of carbon dioxide in the marine realm. The activities of photosynthesizing plants are at a peak in sunlight and at a minimum in the dark; therefore, the carbon dioxide content of water in which active photosynthesis is taking place can vary measurably from day to night. Removal of CO_2 by organisms thus decreases the acidity of the water (increases pH). Decay of dead organisms can also affect the pH. As mentioned, however, buffering owing to the high alkalinity of seawater may mute the magnitude of pH change. Decay processes commonly cause acidity to increase (decrease pH) owing to release of various organic acids and carbon dioxide to the water; however, some decay products can be alkaline.

Because organisms can affect carbonate precipitation in a variety of ways, organisms likely play an important role in the precipitation of calcium carbonate minerals in the modern ocean. The ability of many marine organisms to extract calcium carbonate from seawater to form skeletal hard parts is particularly significant, as mentioned. Because these organisms exist in large numbers in some parts of the ocean, disintegration of their skeletal remains to form fine particles has the potential to supply large quantities of fine calcium carbonate crystals to the ocean floor. It is known, for example, that some green and red algae (Fig. 7.15) have calcareous skeletal elements com-

FIGURE 7.15   Fresh specimens of erect calcareous green algae from the seafloor of the Great Bahama Banks. The upper part of the plant secretes an internal skeleton of aragonite, and the basal rhizoids anchor the algae within the bottom sediment. From left to right, the specimens illustrated are *Halimeda incrassata, Penicillus capitatus, Rhipocephalus phoenix,* and *Udotea flabellum.* (From Neuman, A. C., and L. S. Land, 1975, Lime mud deposition and calcareous algae in the Bight of Abaco, Bahamas: A budget: Jour. Sed. Petrology, v. 37. Fig. 6, p. 773, reprinted by permission of Society of Economic Paleontologists and Mineralogists, Tulsa, Okla.)

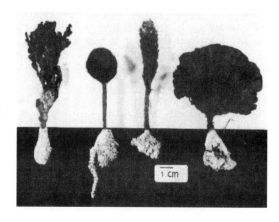

posed of tiny, needlelike aragonite crystals that act as stiffeners for soft tissue. When these organisms die, bacterial and chemical decomposition of the binding tissue releases the skeletal particles. This decay process yields a fine lime mud composed of elongated aragonite crystals, 3 to 10 μm long (Fig. 7.16), and very small (<1 μm) equant crystals (Macintyre and Reid, 1992). Many carbonate workers have suggested that the aragonite needles released from red and green algae are indistinguishable from inorganically precipitated aragonite needles. Macintyre and Reid (1992) point out, however, that algally produced crystals are blunt-ended with well-developed crystal faces

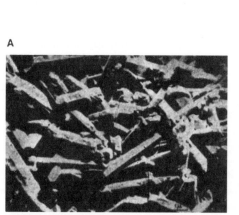

FIGURE 7.16   Electron photomicrographs of (A) aragonite crystals from the stem of *Penicillus* sp. and (B) aragonite crystals in fine lime mud taken from the seafloor west of Andros Island, Bahamas. (From Stockman, K. W., R. N. Ginsburg, and E. A. Shinn, 1967, The production of lime mud by algae in south Florida: Jour. Sed. Petrology, v. 37. Figs. 1 and 2b, p. 634–35, reprinted by permission of Society of Economic Paleontologists and Mineralogists, Tulsa, Okla.)

and tend to be larger than the majority of grains in bottom sediment and whitings west of Andros Island (Bahama Banks), which are typically pointed with poorly developed crystal faces.

Quantitative studies of the rate of production of aragonite by disintegration of calcareous algae in the Florida Reef Tract (Stockman, Ginsburg, and Shinn, 1967) and in the Bahamas (Neumann and Land, 1975) led to the suggestion that much or all of the aragonite mud deposited in these areas in the recent geologic past could have been supplied by skeletal disintegration of calcareous algae. On the other hand, these findings are somewhat at odds with more recent observations by Shinn et al. (1989), who suggest that only 10 to 20 percent of the carbonate mud in the Bahamas is algal carbonate. This suggestion by Shinn et al. appears to be supported by the observations on crystal shape by Macintyre and Reid (1992) and the biochemical studies of Robbin and Blackwelder (1992). Thus, it now appears that the contribution of lime mud brought about by disintegration of calcareous algae may be quantitatively less important than once thought.

Precambrian Limestone Deposition. Judging from the abundance of calcareous skeletal fragments and whole fossils in Phanerozoic limestones, the removal of calcium carbonate from seawater owing to organic activity may have been an important mechanism for precipitating calcium carbonates since at least early Paleozoic time—although the relative importance of biotic and abiotic precipitation of carbonates may have varied throughout this time. We have a more difficult time explaining the formation of Precambrian limestones. The Precambrian record contains impressive thicknesses of carbonate rocks (e.g., as much as 1000 m in Glacier National Park, Montana and Alberta) that, as far as we know, were deposited before the widespread appearance of calcium carbonate–secreting organisms. Thus, it does not seem likely, on the basis of available evidence, that shelled organisms were directly responsible for deposition of large volumes of Precambrian limestone. Few, if any, Precambrian organisms could extract $CaCO_3$ to build shells. Blue-green algae (cyanobacteria) may have played an indirect role in the precipitation of calcium carbonate through photosynthetic removal of carbon dioxide and by trapping and binding of fine carbonate sediment. Cyanobacteria appear to have been particularly abundant in Precambrian time, probably owing to fewer numbers of grazing organisms that fed on the algal mats. The recent suggestion by Robbin and Blackwelder (1992) that extremely tiny algal cells (picoplankton) may initiate precipitation of aragonite by acting as nucleation sites for crystallization could prove to be the answer to the problem of Precambrian limestone deposition—if subsequent work confirms this model. If algae are not involved in some way with Precambrian carbonate deposition, we are left with little choice but to conclude that purely inorganic processes account for the formation of Precambrian carbonate rocks.

Physical Processes in Carbonate Deposition. Calcium carbonate fossils, skeletal fragments, ooids, and other carbonate grains are subject to the same physical transport processes in the ocean as terrigenous grains. Thus, ultimate deposition of most limestones occurs through fluid-flow and sediment gravity-flow processes. Limestones may, therefore, display many of the same bedding characteristics and sedimentary structures as terrigenous sedimentary rocks.

Depth Control of Calcium Carbonate Production. When precipitation of a mineral phase just equals dissolution, the solution is in equilibrium with the solid and is said to be **saturated** with this mineral phase. A solution that precipitates a mineral is **supersaturated,** and a solution that dissolves the mineral is **undersaturated.** Much of the warm surface water of the modern ocean is supersaturated with calcium carbonate. Little inorganic precipitation of $CaCO_3$ may actually be occurring in these waters, however, owing to Mg inhibition or other factors discussed. This condition of supersaturation changes rapidly with depth. The degree of calcium carbonate saturation drops off abruptly in waters below the surface layer; at depths greater than a few hundred meters, seawater is undersaturated. The saturation of surface waters likewise decreases in colder waters of high latitudes.

The undersaturated state of deeper waters is a function of several factors, although increase in carbon dioxide partial pressure is one of the most important variables. In shallower water, CO_2 production is increased closer to the ocean floor by the respiration of benthonic organisms. Oxidation of organic matter on the seafloor in both shallow and deeper water also increases CO_2 production. Furthermore, colder water found at depth can contain more dissolved CO_2 than warmer surface waters. Both decrease in temperature and increase in hydrostatic pressure with depth cause an increase in the solubility of calcium carbonate and thus the corrosiveness of seawater.

Because of decreasing calcium carbonate saturation of seawater with depth, calcium carbonate production is confined mainly to the very shallow-water areas of the ocean and to the supersaturated surface waters of the deeper ocean. These are the waters in which most calcium carbonate–secreting organisms live. Calcium carbonate dissolution prevails in the deeper, undersaturated waters. The rate of dissolution does not, however, increase in a linear fashion with depth. Experiments in which calcite spheres were suspended on moorings at different depths in the ocean have demonstrated that only slight corrosion of the spheres occurred above a depth of about 3500 m, but solution of the spheres increased abruptly at that depth (Fig. 7.17). Effective solution of calcium carbonate thus occurs only at relatively great depths in the ocean

FIGURE 7.17 Weight loss of polished calcite spheres owing to solution at various depths in the ocean. Note the sharp increase in weight loss, indicating increased rate of solution, at about 3500 m. (After Peterson, M. N. A., 1966, Calcite: Rates of dissolution in a vertical profile in the Central Pacific: Science, v. 154. Fig. 2, p. 1543, reprinted by permission of American Association for the Advancement of Science, Washington, D.C. Copyright 1966 by the AAAS.)

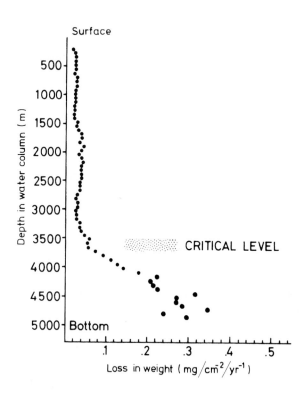

(and to some extent in very cold, high-latitude surface waters). The particular depth at any locality at which the rate of dissolution of calcium carbonate equals the rate of supply of calcium carbonate to the seafloor, so that no net accumulation of carbonate takes place, is called the **calcium carbonate compensation depth (CCD).** The position of the calcium carbonate compensation depth has been compared to the snowline of mountain ranges. Where biogenic oozes are accumulating in the modern ocean, white carbonate oozes cover elevated areas of the seafloor above the CCD but give way to brown or gray pelagic clays or siliceous oozes below. The CCD ranges in depth in different parts of the modern ocean from about 3500 m to 5500 m, owing to differences in rates of production of $CaCO_3$ in surface waters and variations in the factors that control carbonate saturation. The average depth of the CCD in today's ocean is about 4500 m. (The average depth of the modern ocean is about 3800 m; depth ranges to slightly more than 11,000 m.)

Dolomites

General Statement. Dolomites are calcium carbonate rocks composed of more than 50 percent of the mineral dolomite [$CaMg(CO_3)_2$]. To differentiate the rock from the mineral, dolomites are sometimes referred to as dolostone or dolomite rock. They are abundant and widely distributed in the geologic record, ranging in age from Precambrian to Holocene, although most dolomites are Paleozoic and older. Dolomites occur in close association with limestones and in many stratigraphic units as interbeds in the limestones; they are also commonly associated with evaporites.

Because dolomites recur so frequently in the stratigraphic record, they must have formed under environmental conditions that were relatively common and that were repeated over and over in various localities. Dolomites have been studied very extensively; therefore, in theory, we ought to understand their origin quite well. On the contrary, the origin of dolomites remains one of the most thoroughly researched but poorly understood problems in sedimentary geology. Although it is clear from the presence of relict limestone textures and structures that many coarsely crystalline dolomites are secondary rocks, formed by diagenetic replacement of older limestones, many fine-crystalline dolomites lack such textural evidence of replacement and cannot be proven to have originated by diagenetic alteration of limestones. It is these fine-crystalline dolomites that have created the so-called **dolomite problem,** which geologists have not been able to satisfactorily solve since dolomites were first recognized by the French naturalist Déodat Dolomieu in 1791.

The dolomite problem arises from the fact that scientists have not yet been successful in the laboratory in precipitating perfectly ordered dolomite at the normal temperatures (~25°C) and pressure (~1 atm) that occur at Earth's surface. Perfectly ordered dolomite has 50 percent of the cation sites filled by Mg and 50 percent filled by Ca (stoichiometric dolomite). Elevated temperatures, exceeding 100°C, are required to produce a perfectly ordered dolomite in the laboratory. In laboratory experiments carried out at the normal temperatures found in natural environments, only a dolomite-like material called **protodolomite** forms. Protodolomite contains excess $CaCO_3$ in its structure and is not a true (stoichiometric) dolomite. Thus, geochemists have been unable to determine directly from low-temperature experimental work what geochemical conditions, if any, favor the precipitation of dolomite in natural environments.

Prior to the middle 1940s, few occurrences of dolomite had been reported from modern depositional environments; consequently, geologists had little direct support for the hypothesis of primary dolomite. Since about 1946, however, recent dolomite sediments have been reported from numerous localities, including some in Russia,

South Australia, the Persian Gulf, the Bahamas, Bonaire Island off the Venezuela main-land, the Florida Keys, the Canary Islands, and the Netherlands Antilles (Table 7.6).

The ages of these dolomites have been estimated by radiocarbon methods to range from a few years to about 3000 years (Table 7.6). Most are not perfectly ordered dolomites. The mole percent $MgCO_3$ in these dolomites ranges from about 30 to 50 percent but falls mainly between 40 to 46 percent. Discovery of dolomite in modern environments was initially hailed as evidence that dolomite can be precipitated naturally as a primary deposit. Efforts were made to confirm primary precipitation through a variety of research efforts such as oxygen isotope studies of coexisting dolomite and calcite in a sample from a given environment. For example, experimental data extrapolated from high temperature to low temperature (e.g., Degens and Epstein, 1964) suggested that dolomite precipitated at 25°C should be enriched, with respect to coexisting calcite, by detectable amounts in heavy oxygen (^{18}O). Unfortunately, studies of dolomite-calcite pairs from natural environments failed to consistently produce the results predicted by these experimental data. Owing to these equivocal results, Degens and Epstein suggested that most, if not all, modern dolomites probably form by rapid alteration of an initial precipitate of $CaCO_3$. That is, the $CaCO_3$ is replaced by dolomite, a process called **dolomitization.** Whether or not dolomite that formed by such rapid dolomitization should be called primary dolomite becomes a matter of semantics. One solution to the semantics problem is to label all early-formed dolomite as **penecontemporaneous dolomite,** regardless of its exact mode of formation. Early-formed dolomites thus include all those formed at or near the surface in the unconsolidated state, as opposed to diagenetic dolomites that formed during burial and uplift by replacement of older, consolidated limestones. From a pragmatic point of view, it is probably not possible to distinguish between ancient dolomite precipitated directly from solution and dolomite precipitated initially as $CaCO_3$ and then quickly dolomitized. The important factor is to understand the conditions that favor the early formation of dolomite in

TABLE 7.6   Distribution and chemical properties of some modern marine dolomites

Location	Environment	Maximum (molar) Mg/Ca in brines	Dolomite crystal size (μm)	Mole % $MgCO_3$ in dolomite	Age (yr)
Pekelmeer, Bonaire	Salt pan		2	44–46	1480 ± 140
Quatar, Persian Gulf	Supratidal flat		1–5	45–47	2450 ± 130
Andros Island, Bahamas	Supratidal flat	>40/1	1–2	44	0–160
Great Inagua Is., Bahamas	Salt pan	>600/1	<140 <5	40	<8 (?) 2930—3420
Sugarloaf Key, Florida	Supratidal flat	>40/1	2–3	30–44	250–600
Coorong, South Australia	Supersaline lagoon and lakes	4–16/1	<20	45–50	300 ± 250
Abu Dhabi, Persian Gulf	Supratidal flats	>35/1	1–2		
Jarvis Atoll, Pacific	Brine pool	>35/1	?	calcium-rich	2650 ± 200

Source: Milliman, J. D., 1974, Marine carbonates: Springer-Verlag, New York.

modern environments and thus, by extension, the formation of fine-grained, penecontemporaneous dolomite in ancient deposits.

Requirements for Dolomite Formation. The chemical reactions of interest with respect to formation of dolomite are as follows:

$$Ca^{2+} (aq) + Mg^{2+} (aq) + 2CO_3^{2-} (aq) = CaMg(CO_3)_2 (solid) \qquad \textbf{(7.6)}$$

$$2CaCO_3 (solid) + Mg^{2+} (aq) = CaMg(CO_3)_2 (solid) + Ca^{2+} (aq) \qquad \textbf{(7.7)}$$

As suggested at the beginning of this discussion, the problem with the reaction shown in equation 7.6, which illustrates the direct precipitation of dolomite from aqueous solution, is that this reaction requires temperatures in excess of 100°C. The reasons why such high temperatures are necessary are far from well understood, but the problem is certainly related to kinetics (reaction rates). For example, it has been pointed out by a number of workers, such as Gains (1980), that the Mg^{2+} ion is strongly bound by water (hydrated) in solution and must be separated from the attached water before it can be incorporated into the solid dolomite crystal lattice. At low temperatures, Ca^{2+} ions, which are much less strongly bound by water, are more likely to enter the lattice and form $CaCO_3$ minerals. At elevated temperatures, Mg^{2+} ions are less strongly hydrated and thus more easily desolvated, allowing the naked Mg^{2+} ion to enter into the crystal lattice to form dolomite. The highly ordered state of dolomite also creates a kinetics problem at low temperatures. The nucleation and growth of the highly ordered dolomite lattice in a solution saturated in calcium bicarbonate are so slow that in competition for calcium ions and carbonate ions, well-ordered dolomite is prevented from forming, and minerals such as aragonite or cation-disordered, high-magnesium calcites form instead. For additional discussion of the kinetics of dolomite formation, see Machel and Mountjoy (1986).

Penecontemporaneous Dolomite Models. Many geologists now believe that the majority of ancient dolomites were formed by replacement processes (dolomitization). For example, Zenger and Dunham (1980) suggest that only a relatively insignificant amount of ancient dolomite is truly the product of primary precipitation at or above the sediment-water interface. Whether or not this is true, much of the recent and current research on dolomites has focused on attempts to understand the mechanisms of dolomitization. Theoretical considerations suggest that dolomite formation is favored kinetically by high Mg^{2+}/Ca^{2+} ratios, low Ca^{2+}/CO_3^{2-} ratios, and low salinity (Machel and Mountjoy, 1986). It is favored also by higher temperatures, as mentioned. In fact, at temperatures exceeding about 100°C, most kinetic inhibitors, such as Mg^{2+} hydration, become ineffective. Four models for dolomite formation that meet, in one way or another, these favorable conditions for dolomite formation have been proposed:

1. the hypersaline (sabkha) model
2. the mixing-zone model
3. the low-sulfate model
4. the shallow-subtidal (seawater) model

The Hypersaline or Sabkha Model. Many known occurrences of modern or Holocene dolomite are in hypersaline environments such as the sabkhas of the Persian Gulf and the supratidal zones of arid climates (Fig. 7.18). (Sabkhas are coastal plains characterized by the presence of evaporites.) Under strongly evaporative conditions, where rates of evaporation exceed rates of precipitation, seawater beneath the sediment surface

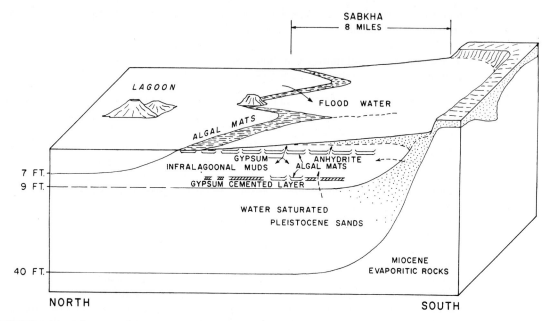

FIGURE 7.18 Schematic diagram of the sabkha environment at Abu Dhabi in the Persian Gulf. Penecontemporaneous dolomite (or protodolomite) is commonly associated with gypsum and anhydrite deposits in such environments. (From Butler, G. P., 1969, Modern evaporite deposition and geochemistry of coexisting brines, the sabkha, Trucial Coast, Arabian Gulf: Jour. Sed. Petrology, v. 39. Fig. 2, p. 72, reprinted by permission of Society of Economic Paleontologists and Mineralogists, Tulsa, Okla.)

becomes concentrated by evaporation. This concentation process leads to precipitation of aragonite and gypsum, which preferentially removes Ca^{2+} from the water and increases the Mg/Ca ratio. The Mg/Ca ratio in normal seawater is about 5:1. Dolomite is believed to form when this ratio rises to sufficiently high levels, possibly in excess of 10:1. One mechanism by which brines are concentrated involves evaporation of capillary water in the sediments of the sabkhas. Upward flow of water from the saturated groundwater zone replaces the water lost by capillary evaporation, a process called **evaporative pumping.** Brines may also be concentrated in surface ponds or bays by surface evaporation of water. These concentrated brines have higher density than that of normal seawater, causing them to sink downward. Flushing of large volumes of Mg-rich brine downward through calcium carbonate sediment can putatively bring about dolomitization, a process referred to as **seepage refluxion.**

The overall volume of dolomite that forms in sabkha environments is believed to be relatively small, and there is still considerable controversy with regard to the exact mechanism by which the dolomite forms. It is not definitely known if it forms by replacement of aragonite or high-magnesian calcite (dolomitization) or if it forms as a primary precipitate of disordered protodolomite, which presumably later develops better ordering to become true dolomite. See Hardie (1987) for additional discussion of this topic.

The Mixing-Zone Model. Some modern/Holocene dolomites and most ancient dolomites are not associated directly with evaporites. Several studies published since the early 1970s (e.g., Hanshaw, Back, and Deike, 1971; Badiozamani, 1973; Folk and Land,

1975) have suggested that brackish groundwaters produced by mixing of seawater with meteoric water could be saturated with respect to dolomite at Mg^{2+}/Ca^{2+} ratios much lower than those required under hypersaline conditions. Mixing of freshwater and saline water in environments such as the subsurface zones of coastal areas where meteoric waters come in contact with seawater is suggested to lower salinities sufficiently so that dolomites can form at Mg^{2+}/Ca^{2+} ratios ranging from normal seawater values to as low as 1:1 (Fig. 7.19). Presumably, dolomite can form at lower Mg^{2+}/Ca^{2+} ratios in these mixed waters compared to seawater owing to less competition by other ions in the less saline water. The mixing-zone model, or variations thereof, has been referred to also as the **Dorag model** (Badiozamani, 1973) and the **schizohaline model** (Folk and Land, 1975).

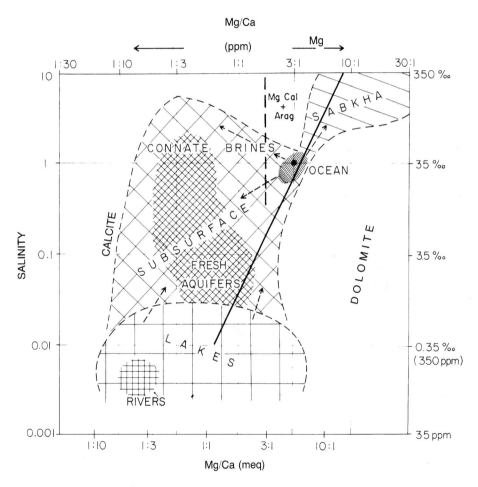

FIGURE 7.19 Graph of salinities vs. Mg/Ca ratios of common natural waters. The preferred fields of occurrence of dolomite, calcite, magnesian calcite, and aragonite are also shown. Note that the dolomite field lies to the right of the heavy, inclined line. Owing to slow crystallization rates and the scarcity of competing foreign ions at low salinities, dolomites can form at Mg/Ca ratios near 1:1. (From Folk, R. L., and L. S. Land, 1975, Mg/Ca ratio and salinity: Two controls over crystallization of dolomite: Am. Assoc. Petroleum Geologists Bull., v. 59. Fig. 1, p. 61, reprinted by permission of AAPG, Tulsa, Okla.)

Although the mixing-zone dolomite model has attracted many proponents, it has also come under some fairly devastating attacks. For example, Hardie (1987) points out that in Badiozamani's original (1973) calculations for the model, he used the solubility values of less soluble, ordered dolomite when he should have used the values of more soluble, less ordered, Ca-rich dolomite (which is the kind of dolomite that actually forms under surface temperatures). Hardie also maintains that there is no actual documentation that dolomite can form at Mg/Ca ratios of 1:1; nor is there hard evidence that demonstrates the special power of low-salinity conditions to produce cation-ordered dolomite. Machel and Mountjoy (1986) further point out that dolomite does not form in most modern freshwater/seawater mixing zones, and where it does form, the volume of dolomite is small.

The Low-Sulfate Model. Experimental work on the formation of dolomite at 200°C by Baker and Kastner (1981) demonstrated that the presence of dissolved SO_4^{2-} strongly inhibits the formation of dolomite. Extrapolating their experimental results to lower temperatures, they suggest that the reason for the scarcity of dolomite in open-marine environments is the presence of dissolved SO_4^{2-} in seawater. Dissolved SO_4^{2-} ions can allegedly inhibit the dolomitization of calcite at SO_4^{2-} values as low as 5 percent of their seawater value. Dolomitization of aragonite can occur at somewhat higher concentrations of SO_4^{2-}. Baker and Kastner (1981) propose that dolomite can form rapidly in nature only where the SO_4^{2-} concentration is low. They indicate that the most effective process for SO_4^{2-} removal in marine pore waters is its microbial reduction in organic-rich sediments. The bacterial reduction of SO_4^{2-} promotes dolomitization by (1) removal of the SO_4^{2-} inhibitor, (2) production of alkalinity, and (3) production of ammonia (NH_4^+). The NH_4^+ produced during reduction may exchange with magnesium held in exchange sites in marine silicate-rich sediments, freeing the magnesium for dolomitization. The concentration of SO_4^{2-} can also be reduced by precipitation of calcium sulfate in evaporative environments to form gypsum. Mixing of freshwater with seawater in groundwater environments likewise lowers the SO_4^{2-} concentration of seawater. Thus, it is the reduction in SO_4^{2-} in these environments, rather than the influence of the Mg/Ca ratio, that allows the formation of dolomite.

The low-sulfate model has also been critized by some authors (e.g., Hardie, 1987) on the grounds that several examples of Holocene dolomites are known that formed from brines with very high sulfate concentrations. Also, it is possible to precipitate disordered protodolomite in the laboratory at 25°C from sulfate-rich solutions. On the other hand, the model has received some support from additional experimental work. For example, Morrow and Ricketts (1988) confirmed that concentrations of sulfate as low as 0.004M prevented the dolomitization of calcite. Their experiments also demonstrated that the presence of sulfate in solution slowed the rate of calcite dissolution. Therefore, they suggest that a major reason why sulfate in solution may inhibit dolomitization is that it retards the rate of calcite dissolution.

The Shallow Subtidal (Seawater) Model. In all of the above models, some kind of "special" water (i.e., seawater modified in some way) is required for dolomitization. On the other hand, a few workers have proposed that penecontemporaneous dolomitization can take place in normal, unmodified seawater if a sufficient volume of seawater can be passed through the sediment. According to the concept embodied in this model, dolomitization can take place in normal seawater if enough water is forced through the sediment so that each pore volume of water in the sediment is constantly being renewed with new seawater. Thus, new Mg^{2+} is constantly being supplied while replaced Ca^{2+} ions and other ions that might "poison" the dolomite crystal structure are removed. As an example, Carballo, Land, and Miser (1987) report an area of Sugarload Key, Florida, where seawater is forced upward and downward through Holocene carbonate mud during rise and fall of seawater accompanying spring tides, a process they call **tidal pumping.** Owing to the large volume of seawater driven through the sediment by this mechanism, large quantities of Mg^{2+} are imported into the sediment, and pore fluids are constantly being replaced by new fluids. Under these conditions, dolomite is forming in the sediment even though little or no evaporation of the seawater has occurred. Carballo, Land, and Miser suggest that dolomite forms both by precipitation as a cement and by later replacement of preexisting crystallites.

Subsurface (Late-Stage) Dolomite. As mentioned, much dolomite in the geologic record has relict textures that indicate that the dolomite was formed by replacement (dolomitization) of a precursor limestone. Such dolomitization appears to have taken place in the subsurface much later (perhaps millions to hundreds of millions of years later) than the time of formation of penecontemporaneous dolomite. Reasoning from the concepts presented in the seawater model above, the problem of understanding late-stage, large-scale, subsurface dolomitization reduces mainly to finding a mechanism for circulating large volumes of seawater deep into the subsurface. At the higher temperatures present in the subsurface, dolomitization can apparently take place readily in any buried limestone that has sufficient porosity and permeability to allow circulation of large volumes of seawater. Several mechanisms can drive circulation of seawater down through buried limestones in a basin, including the following:

1. gravity-driven flow owing to the presence of a hydraulic head, the magnitude of which is determined by the elevation of the meteoric recharge area for the subsurface formations (Garven and Freeze, 1984).
2. thermal convection resulting from the presence of a geothermal heat source below the basin (Kohout, Henry, and Banks, 1977).
3. buoyant circulation caused by circulation within freshwater lenses along the mixing zone with saline waters. The resulting discharge of brackish water at the coast causes a compensating inflow of saline waters at depth (Whitaker and Smart (1990).

7.8 DIAGENESIS

Most carbonate sediments are deposited under marine conditions, although carbonate rocks can also form under nonmarine conditions (Chapter 12). After deposition, carbonate sediments are subjected to a variety of diagenetic processes that bring about changes in porosity, mineralogy, and chemistry. Carbonate sediments are generally more susceptible to dissolution, recrystallization, and replacement than are most silicate minerals. Thus, they may experience pervasive alteration of mineralogy. For example, an original aragonitic mud may alter entirely to calcite during early diagenesis or burial. In turn, the calcite may be replaced completely or nearly completely by dolomite at a later time. Such changes may also destroy or modify original depositional textures such as carbonate grains and micrite. Porosity of carbonate sediments may be either reduced by compaction and cementation or enhanced by dissolution.

Regimes of Carbonate Diagenesis

Carbonate sediments may go through the same general stages of diagenesis as siliciclastic sediments, that is, shallow burial (eodiagenesis), deep burial (mesodiagenesis), and uplift and unroofing (telodiagenesis). Diagenesis takes place in three major regimes or realms (Fig. 7.20): the marine, the meteoric, and the subsurface.

The **marine realm** includes the seafloor and very shallow-marine subsurface. The diagenetic environment here is characterized by seawater temperatures and marine waters of normal salinity. The principal diagenetic processes in this environment involve bioturbation of sediments, modification of carbonate shells and other grains by boring organisms, and cementation of grains in warm-water areas, particularly in reefs, platform-margin sand shoals, and carbonate beach deposits (beachrock).

Marine carbonate sediments may be brought from the seafloor realm into the **meteoric realm** in two ways: (1) by falling sea level and (2) by progressive sediment filling of a shallow carbonate basin. Older carbonate rock can also be brought into the meteoric realm by late-stage uplift and unroofing of a deeply buried carbonate complex (telodiagenesis). The meteoric realm is characterized by the presence of freshwater; it includes the unsaturated (sediment pores not filled with water) vadose zone above the water table and the phreatic zone, or water-saturated zone, below the water table. Meteoric waters are typically highly charged with CO_2; thus, they are chemically very aggressive. Because aragonite and high-magnesian calcite are more soluble than calcite, they dissolve readily in these corrosive waters. On the other hand, dissolution of aragonite and high-magnesian calcite may saturate the waters in calcium carbonate with respect to calcite, causing calcite to precipitate. This dissolution-reprecipitation process causes less stable aragonite and high-magnesian calcite to be replaced by more stable calcite. Calcite may also precipitate into open spaces as a cement. Thus, dissolu-

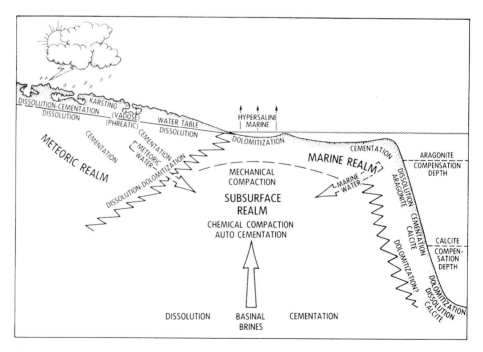

FIGURE 7.20   The principal environments in which postdepositional modification of carbonate sediments occurs. The dominant diagenetic processes that occur in each of the major diagenetic realm are also indicated. See text for details. (From Moore, C. H., 1989, Carbonate diagenesis and porosity. Fig. 3.1, p. 44, reprinted by permission of Elsevier Science Publishers, Amsterdam.)

tion, neomorphism, and calcite cementation are the principal diagenetic processes in the meteoric realm.

After an initial period of diagenesis on the seafloor, and possibly in the meteoric realm, carbonate sediments are gradually buried and subjected to increased pressures, higher temperatures, and changed pore fluids in the **subsurface realm.** Under these changed conditions, carbonate sediments may undergo physical compaction, chemical compaction, and additional chemical/mineralogical changes that may include dissolution, cementation, neomorphism, and replacement. The exact nature of the changes that take place during deep subsurface diagenesis depends upon the specific conditions (temperature, pore-fluid composition, pH) of the burial environment.

Major Diagenetic Processes and Changes

Biogenic Alteration. Organisms in carbonate depositional environments rework sediment by boring, burrowing, and sediment-ingesting activities, just as they do in siliciclastic environments. These activities may destroy primary sedimentary structures in carbonate sediment and leave behind mottled bedding and various kinds of organic traces. In addition, many kinds of small organisms, such as fungi, bacteria, and algae, create microborings in skeletal fragments and other carbonate grains. Fine-grained (micritic) aragonite or high-magnesian calcite may then precipitate into these holes. This boring and micrite-precipitation process may be so intensive in some warm-water environments that carbonate grains are reduced almost completely to micrite, a process called **micritization.** If boring is less intensive, only a thin **micrite rim,** or micrite envelope, may be produced around the grain (Fig. 7.21). Larger organisms such as sponges and molluscs create macroborings in skeletal grains and carbonate substrate, and other organisms such as fish, sea cucumbers, and gastropods may break down carbonate grains in various ways to smaller pieces.

Cementation. Cementation is an important process in all diagenetic realms. On the ocean floor, cementation takes place mainly in warm-water areas within the pore

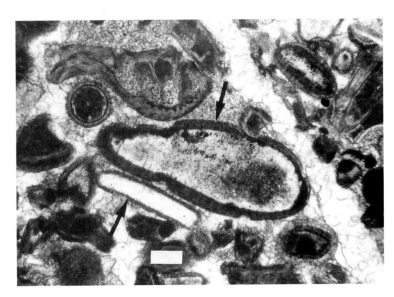

FIGURE 7.21 Dark, micrite envelopes (arrows) around bioclasts (fossils). Renault Formation (Mississippian), Missouri. Crossed nicols. Scale bar = 0.2 mm.

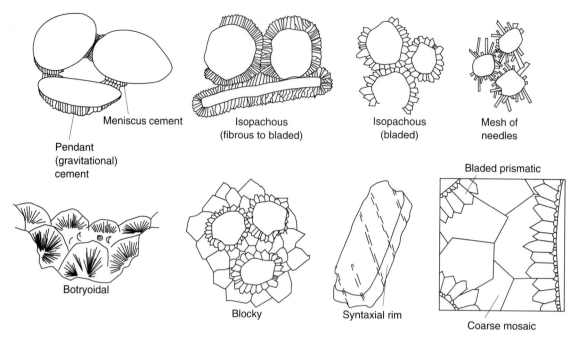

FIGURE 7.22 Principal kinds of cements that form in carbonate rocks during diagenesis. Seafloor diagenetic environments are characterized particularly by aragonitic meniscus and pendant cements (in beachrock), isopachous cement, needle cement, and botryoidal cement. Meteoric-realm cements are composed dominantly of calcite and include meniscus and pendant cements in the vadose zone and isopachous, blocky, and syntaxial rim cements in the phreatic zone. Cements of the subsurface burial realm are also mainly calcite and include syntaxial rim, bladed prismatic, and coarse mosaic types. (Modified from James, N. P., and P. W. Choquette, 1983, Geoscience Canada, v. 10, Fig. 3, p. 165; 1984, Geoscience Canada, v. 11, Fig. 24, p. 177; 1987, Geoscience Canada, v. 14, Fig. 21, p. 16.)

spaces of grain-rich sediments or cavities. Reefs, carbonate sand shoals on the margins of platforms, and carbonate beach sands are favored areas for early cementation. Areas of the seafloor along platform margin where sediments become well cemented are referred to as **hardgrounds.** Cemented carbonate beach sand is called **beachrock.** Seafloor cement is commonly aragonite, less commonly high-magnesian calcite. Seafloor cement can take several different textural forms, as shown in Figure 7.22. Beachrock may contain meniscus cements that form where water is held by capillary forces as interstitial water drains from beaches during low tide. Because beach sediments are not constantly bathed in water, pendant cements may also form in beachrock along the bottoms of grains where drops of water are held. Isopachous rinds, which completely surround grains, form under subaqueous conditions where grains are constantly surrounded by water. Aragonite cements may also occur as a mesh of needles or as fibrous radial crystals that have a botryoidal form.

In the meteoric realm, dissolution is a more important process than cementation; however, cementation does occur. The cement is almost exclusively calcite. As mentioned, the calcium carbonate that forms this cement is derived by dissolution of less stable aragonite and high-magnesian calcite. In the (water) unsaturated vadose zone, calcite cements are commonly meniscus and pendant cements. In the water-saturated phreatic zone, they are isopachous, blocky, or syntaxial rim cements. Syntaxial rims form by precipitation of optically continuous calcite around single-crystal fossil echinoderm fragments, in much the same way that cement overgrowths form around quartz grains.

Calcite cementation may also form during deep burial, although the conditions that control cementation at depth are poorly understood. Factors that have been cited to favor carbonate cementation during deep burial include unstable mineralogy (presence of aragonite and high-magnesian calcite); pore waters highly oversaturated in calcium carbonate; high porosity and permeability (which enable high rates of fluid flow); increase in temperature; and decrease in carbon dioxide partial pressure. The calcium carbonate needed for cementation at depth may be supplied, at least in part, by pressure solution of carbonate sediment in much the same way that pressure solution of quartz grains supplies silica to pore waters in siliciclastic sediment. Coarse mosaic calcite and bladed prismatic calcite (Fig. 7.22) are common kinds of deep-burial cements. The combination of bladed prismatic and coarse mosaic cement shown in Figure 7.22 is called **drusy** cement. These calcite cements are commonly coarse- grained and clear or white in appearance. They are usually referred to as sparry calcite cement.

Dissolution. Cementation is a very common diagenetic process in carbonate rocks, yet, somewhat paradoxically, so is dissolution. Dissolution of carbonate minerals requires conditions essentially opposite to those that lead to cementation. Dissolution is favored by unstable mineralogy (presence of aragonite or high-magnesian calcite), cool temperatures, and low-pH (acidic) pore waters that are undersaturated with calcium carbonate. Dissolution takes place particularly in chemically aggressive pore waters highly charged with CO_2 and/or organic acids. Dissolution is relatively unimportant on the seafloor but is particularly prevalent in the meteoric realm where chemically aggressive meteoric waters percolate or flow down through the vadose zone into the phreatic zone. Extensive dissolution of aragonite and high-magnesian calcite takes place in this environment, and even calcite may be dissolved if pore waters are sufficiently aggressive. Dissolution tends to be concentrated particularly along the water table (the boundary between the vadose and phreatic zones), which accounts for the common presence of caves in carbonate rocks at the level of the water table. Dissolution is less intensive in the deep-burial (subsurface) realm than in the meteoric realm for two reasons. First, most aragonite and high-magnesian calcite may already have been converted to more stable calcite in the meteoric realm (see "Neomorphism" below). Second, increasing temperature at depth (Fig. 6.19) decreases the solubility of all carbonate minerals. Dissolution may occur at depth if enough CO_2 is added to pore waters as a result of burial decay of organic matter (decarboxylation) to overcome the decrease in solubility resulting from increased temperature. Buried carbonate sediments that are brought back into the meteoric zone after uplift may undergo extensive dissolution of both previously formed cements and other carbonate minerals under the influence of chemically aggressive, CO_2-charged meteoric waters.

Neomorphism. Neomorphism is a term used by Folk (1965) to cover the combined processes of inversion and recrystallization. Inversion refers to the change of one mineral to its polymorph, such as aragonite to calcite. Strictly speaking, inversion takes place only in the solid (dry) state. When the transformation of aragonite to calcite takes place in the presence of water, it occurs by means of dissolution of the less stable aragonite and nearly simultaneous precipitation replacement by more stable calcite. Many geologists refer to this process as **calcitization**. During diagenesis, most aragonite is eventually calcitized. **Recrystallization** indicates a change in size or shape of a crystal, with little or no change in chemical composition or mineralogy. Calcitization and recrystallization commonly go hand in hand.

Neomorphism may occur in all three diagenetic realms but is particularly important in the meteoric and subsurface diagenetic environments. Neomorphism may affect both carbonate grains and micrite and commonly results in increase in crystal size. This process destroys original textures and fabrics and, when pervasive, may cause the entire rock to becomes recrystallized. Thus, a fine-grained (micritic) limestone can be converted into a coarse-grained sparry rock. On a smaller scale, recrystallization results in the formation of large, clear crystals of calcite that closely resemble sparry calcite cement. In fact, one of the most difficult problems in the microscopic study of carbonate rocks is to differentiate between sparry calcite cement and neomorphic spar. Figure 7.23 shows an example of neomorphic spar.

Replacement. As described under "Siliciclastic Diagenesis," replacement involves the dissolution of one mineral and the nearly simultaneous precipitation of another mineral of different composition in its place. Replacement of calcium carbonate minerals by other minerals is a common diagenetic process. Dolomitization of of $CaCO_3$ sediment is one kind of replacement process. In addition, many other kinds of noncarbonate minerals may replace carbonate minerals during diagenesis, including microcrystalline quartz (chert), pyrite (iron sulfide), hematite (iron oxide), apatite (calcium phosphate), and anhydrite (calcium sulfate). Replacement can occur in all diagenetic environments. We have already discussed the replacement of $CaCO_3$ by dolomite in seafloor and burial environments. In carbonate-evaporite sequences, replacement of carbonate minerals by anhydrite at depth is a common process. Figure 7.24 shows an example of anhydrite replacing calcium carbonate in a fossil brachiopod shell. Replacement of carbonates by microcrystalline quartz (chert) is also common in the meteoric and deep-burial environments. For example, Maliva and Siever (1989) report replacement of Paleozoic carbonates by chert at burial depths ranging from 30 to 1000 m. Replacement of carbonate minerals by silica may be very selective, with silica replacing fossils and other carbonate grains such as ooids in preference to micrite (Fig. 7.25).

Physical and Chemical Compaction. Newly deposited, watery carbonate sediments have initial porosities ranging from 40 to 80 percent. As burial into the subsurface pro-

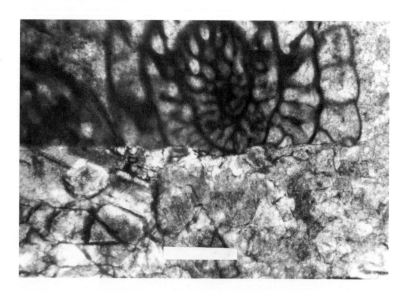

FIGURE 7.23 A fossil fusulinid foraminifer (top) cut in half owing to recrystallization to neomorphic spar (bottom), Pennsylvanian limestone, Paradox Basin, Utah. Plane light. Scale bar = 0.15 mm.

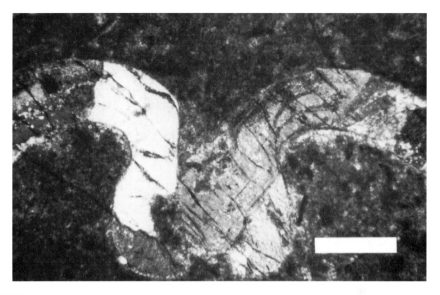

FIGURE 7.24 Section of a brachiopod shell almost completely replaced by coarse anhydrite, Pennsylvanian limestone, Paradox Basin, Utah. Crossed nicols. Scale bar = 0.5 mm.

FIGURE 7.25 A fusulinid foraminifer replaced along one edge by microcrystalline quartz (chert, Ch), Morgan Formation (Pennsylvanian), Colorado. Note the ragged boundary between the foraminifer caused by the replacement process. Crossed nicols.

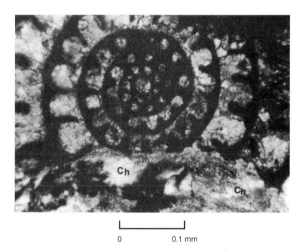

ceeds, the pressure of overlying sediments brings about grain reorientation and tighter packing. As with siliciclastic sediments, compaction results in loss of porosity and thinning of beds at fairly shallow burial depth. With deeper burial to depths of about 1000 ft (305 m) and with progressively higher overburden pressures, grains may also deform by brittle fracturing and breaking and by plastic or ductile squeezing. Even at burial depths as shallow as 100 m, compaction can reduce the depositional thickness of carbonate sediments by as much as one-half, with accompanying porosity losses of 50 to 60 percent of original pore volumes (Shinn and Robbin, 1983).

At burial depths ranging from about 200 to 1500 m, chemical compaction of carbonate sediments is also initiated. As discussed under "Siliciclastic Diagenesis," pressure solution at grain-to-grain contacts can result in interpenetrating or sutured contacts between grains. On a larger scale, pressure solution seams called **stylolites** develop. Stylolites are particularly common in carbonate rocks. The stylolite seams are marked by the presence of clay minerals and other fine-size noncarbonate minerals (commonly referred to as an insoluble residue) that accumulate as carbonate minerals dissolve. Stylolites range in size from microstylolites between grains, in which the amplitude of the interpenetrating contacts between grains is less than 0.25 mm, to stylolites with amplitudes exceeding 1 cm (e.g., Fig. 5.45). Pressure solution, with accompanying stylolite formation, causes significant loss of porosity (perhaps as much as 30 percent of original pore volume) and thinning of beds.

Summary Results of Carbonate Diagenesis

The combined effects of organic, physical, and chemical diagenesis produce significant changes in the depositional characteristics of carbonate sediments. Organisms destroy

FIGURE 7.26 Hypothetical curves illustrating (1) a "normal" porosity-depth relationship for fine-grained sediments with marine pore waters; (2) cementation in the meteoric zone (horizontal segments) alternating with burial in marine pore waters; (3) reversal of normal porosity-depth trend owing to dissolution in the deep subsurface, followed by resumption of normal burial; and (4) arrested porosity reduction owing to abnormally high pore pressure. (From Choquette, P. W., and N. P. James, 1987, Diagenesis 12. Diagenesis in limestones—3. The deep burial environment: Geoscience Canada, v. 14. Fig. 33, p. 23, reprinted by permission of Geological Association of Canada.)

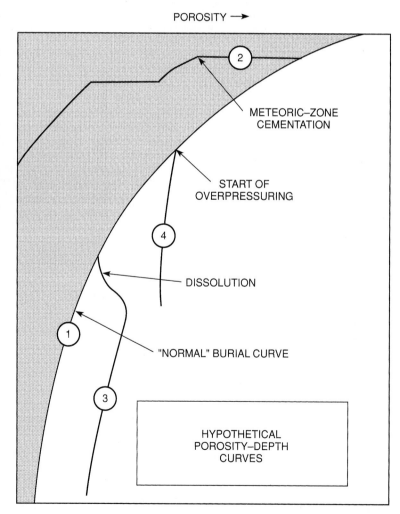

POROSITY ⟶

METEORIC–ZONE CEMENTATION

START OF OVERPRESSURING

DISSOLUTION

"NORMAL" BURIAL CURVE

HYPOTHETICAL POROSITY–DEPTH CURVES

primary sedimentary structures such as lamination and attack carbonate grains by boring and sediment ingestion. Physical and chemical compaction resulting from overburden pressure causes extensive porosity reduction and bed thinning. Cementation further reduces porosity. On the other hand, dissolution processes in the meteoric realm, and to a lesser extent the subsurface realm, create new porosity. Even previously deposited cements may be dissolved during telodiagenesis. The various ways by which diagenetic processes can combine to affect the final porosity of carbonate sediments are illustrated in Figure 7.26. Finally, calcitization (neomorphism) converts most aragonite and high-magnesian calcite to calcite, and subsurface dolomitization can bring about pervasive replacement of carbonate minerals by dolomite.

FURTHER READINGS

Bathurst, R. G. C., 1975, Carbonate sediments and their diagenesis, 2nd ed.: Elsevier, Amsterdam, 658 p.

Flügel, E., 1977, Microfacies analysis of limestones: Springer-Verlag, Berlin, 633 p.

Ham, W. E. (ed.), 1962, Classification of carbonate rocks: Am. Assoc. Petroleum Geologists Mem. 1, 279 p.

Milliman, J. D., 1974, Marine carbonates: Springer-Verlag, New York, 375 p.

Moore, C. H., 1989, Carbonate diagenesis and porosity: Elsevier, Amsterdam, 338 p.

Morse, J. W., and F. T. Mackenzie, 1990, Geochemistry of sedimentary carbonates: Elsevier, Amsterdam, 707 p.

Scoffin, T. P., 1987, Carbonate sediments and rocks: Blackie, Glasgow, 274 p.

Tucker, M. E., and R. G. C. Bathurst (eds.), 1990, Carbonate diagenesis: Internat. Assoc. Sedimentologists Reprint Series, v. 1, Blackwell, Oxford, 320 p.

Tucker, M. E., and V. P. Wright, 1990, Carbonate sedimentology: Blackwell, Oxford, 482 p.

8

Other Chemical/Biochemical and Carbonaceous Sedimentary Rocks

8.1 INTRODUCTION

In addition to the carbonate rocks discussed in Chapter 7, the chemical/biochemical sedimentary rocks include evaporites, siliceous sedimentary rocks (cherts), iron-rich sedimentary rocks, and phosphorites. Although volumetrically less significant than carbonate rocks, these sedimentary rocks are extremely important both as economic resources and as indicators of specialized paleoenvironments. Evaporite deposits such as gypsum, halite (rock salt), and trona (sodium carbonate) are mined for industrial and agricultural purposes. The iron-rich sedimentary rocks are iron ores that have enormous worldwide economic significance as a source for most of our iron. Sedimentary phosphorites are the major source of commercial phosphates used for fertilizers and chemical purposes. Even the siliceous sedimentary rocks may have some economic value in the semiconductor industry.

Aside from their economic value, these chemical/biochemical rocks are extremely interesting because they indicate past environmental conditions that appear to be uncommon on Earth today—or perhaps we simply do not recognize the modern counterparts of these ancient environments. For example, we know of no place on Earth where massive sedimentary iron deposits, of the type common in late Precambrian rocks, are forming today. Nor do we fully understand the mechanism of iron deposition or the source of the enormous amount of iron present in iron-rich sedimentary rocks. Likewise, the mechanism by which low levels of phosphorus in ocean water become concentrated a millionfold to form phosphorite deposits is only partially understood. In this chapter, we look at the characteristics of these enigmatic sedimentary rocks and discuss some of the more interesting aspects of their origin.

We also briefly examine the characteristics and origins of the carbonaceous sedimentary rocks—rocks that contain significant amounts (>~10 percent) of organic carbon. Together with petroleum and natural gas, these organic-rich rocks, which include coals and oil shales, are the source of our fossil fuels. Fossil fuels currently supply most of the world's energy needs; therefore, commercial interest in exploiting carbonaceous sedimentary rocks is understandably high. Geologists are further interested in the origin of these rocks, particularly the processes and conditions that allow preservation of such high levels of organic matter. The average organic content of other sedimentary rocks is only about 1.5 percent.

8.2 EVAPORITES

The term **evaporites** is used for all deposits, such as salt deposits, that are composed of minerals that precipitated from saline solutions concentrated by evaporation. Evaporites occur in rocks of all ages, including the Precambrian, but they are particularly common in Cambrian, Permian, Jurassic, and Miocene sequences (Ronov et al., 1980). Although the total volume of evaporites in the geologic record is much less than that of carbonate rocks, some individual evaporite deposits, such as the Miocene Messinian of the Mediterranean region, reach thicknesses exceeding 1 km. Evaporites form under both marine and nonmarine conditions; however, marine evaporites tend to be thicker and more laterally extensive than nonmarine evaporites and are of greater geologic interest.

Composition

Evaporite deposits are composed dominantly of varying proportions of halite (rock salt), anhydrite, and gypsum. Although approximately 80 minerals have been reported from evaporite deposits (Stewart, 1963), only about a dozen of these minerals are common enough to be considered important evaporite rock formers. Evaporite minerals are commonly classified into those of marine origin and those of nonmarine origin, although Hardie (1991) suggests that identifying the marine or nonmarine origin of evaporites on the basis of their mineralogy and chemistry may not be entirely justified. If carbonate minerals, most of which are not evaporites, are excluded, the most common minerals generally considered to characterize marine evaporites are the calcium sulfate minerals gypsum and anhydrite. Halite is next in abundance, followed by the potash salts, sylvite, carnellite, langbeinite, polyhalite, and kainite, and the magnesium sulfate, kieserite (Table 8.1). The marine evaporite minerals can be grouped by chemical composition into chlorides, sulfates, and carbonates. Marine evaporites commonly contain mixtures of minerals, although gypsum (or anhydrite) and halite predominate in most horizons. Deposits may range from those that are composed almost entirely of anhydrite or gypsum to those that are mainly halite. Gypsum is more abundant than anhydrite in modern evaporite deposits, but anhydrite is more abundant in ancient deposits. Marine evaporite deposits may also contain various amounts of impurities such as clay minerals, quartz, feldspar, and sulfur.

Nonmarine evaporites are characterized by evaporite minerals that are not common in marine evaporites because the water from which nonmarine evaporites precipitate generally has proportions of chemical elements different from those of marine water (e.g., more bicarbonate and magnesium and little or no chlorine). These minerals may include trona [$Na_3H(CO_3)2 \cdot 2H_2O$], mirabilite ($Na_2SO_4 \cdot 10H_2O$), glauberite [$Na_2Ca(SO_4)$], borax [$Na_2B_4O_5(OH)_4 \cdot 8H_2O$], epsomite ($MgSO_4 \cdot 7H_2O$), thenardite ($NaSO_4$), gaylussite ($Na_2CO_3 \cdot CaCO_3 \cdot 5H_2O$), and bloedite ($Na_2SO_4 \cdot MgSO_4 \cdot 4H_2O$). Nonmarine deposits may also contain anhydrite, gyspum, and halite and may even be dominated by these minerals.

TABLE 8.1   Classification of marine evaporites on the basis of mineral composition

Mineral class	Mineral name	Chemical composition	Rock name
Chlorides	Halite	NaCl	Halite; rock salt
	Sylvite	KCl	Potash salts
	Carnallite	$KMgCl_3 \cdot 6H_2O$	
Sulfates	Langbeinite	$K_2Mg_2(SO_4)_3$	
	Polyhalite	$K_2Ca_2Mg(SO_4)_6 \cdot H_2O$	
	Kainite	$KMg(SO_4)Cl \cdot 3H_2O$	
	Anhydrite	$CaSO_4$	Anhydrite
	Gypsum	$CaSO_4 \cdot 2H_2O$	Gypsum
	Kieserite	$MgSO_4 \cdot H_2O$	—
Carbonates	Calcite	$CaCO_3$	Limestone
	Magnesite	$MgCO_3$	—
	Dolomite	$CaMg(CO_3)_2$	Dolomite; dolostone

Source: Data from Stewart, F. H., 1963, Marine evaporites, *in* M. Fleischer (ed.), Data of geochemistry: U.S. Geol. Survey Prof. Paper 440–Y; Borchert, H., and R. O. Muir, 1964, Salt deposits: The origin, metamorphism, and deformation of evaporites: Van Nostrand, London.

Classification

General. Evaporites may be classified as chlorides, sulfates, or carbonates on the basis of their chemical composition (Table 8.1); however, few rock names have been applied to evaporite deposits. Rocks composed predominantly of the mineral halite are called halite or rock salt. Rocks made up dominantly of gypsum or anhydrite are simply called gypsum or anhydrite, although some geologists use the names **rock gypsum** or **rock anhydrite**. Few evaporite beds are composed dominantly of minerals other than the calcium sulfates and halite. No formal names have been proposed for rocks enriched in other evaporite minerals, although the term potash salts is used informally for potassium-rich evaporites.

Anhydrite. No other classification scheme for evaporites is in general use; however, Maiklem, Bebolt, and Glaister (1969) divided anhydrites into several subtypes on the basis of structural characteristics. Calcium sulfates are deposited dominantly as gypsum. Gypsum can be altered into, and pseudomorphosed by, anhydrite while the sediments are still in their general depositional environment (Shearman, 1978). Gypsum is also dehydrated to anhydrite with burial to a few hundred meters, and this loss of water is accompanied by a 38 percent decrease in solid volume of the gypsum. Because of this rapid dehydration with burial, most ancient calcium sulfate deposits are composed of anhydrite. Anhydrite can be hydrated back to gypsum after uplift and exposure to low-salinity surface waters, with an accompanying increase in volume. These volume changes resulting from dehydration and hydration can cause distortion of original depositional structures and textures, and many calcium sulfate deposits are characterized by distorted fabrics. On the basis of fabric, bedding, and the presence or absence of distortion, Maiklem, Bebolt, and Glaister (1969) divided anhydrites into about two dozen structural types, which can be placed into three fundamental structural groups: nodular anhydrites, laminated anhydrites, and massive anhydrites.

Nodular anhydrites are irregularly shaped lumps of anhydrite that are partly or completely separated from each other by a salt or carbonate matrix (Fig. 8.1). Maiklem, Bebolt, and Glaister (1969) make a distinction between nodular anhydrite and **mosaic anhydrite** in which the anhydrite masses or lumps are approximately equidimensional and are separated by very thin stringers of dark carbonate mud or clay. Many other authors do not make this distinction and use the term nodular anhydrite in a general sense to include mosaic anhydrite. The term **chickenwire structure** is used for a particular type of mosaic or nodular anhydrite that consists of slightly elongated, irregular polygonal masses of anhydrite separated by thin, dark stringers of other minerals such as carbonate or clay minerals (Fig. 8.2).

The formation of nodular anhydrite is initiated by displacive growth of gypsum in carbonate or clayey sediments. Gypsum crystals subsequently alter to anhydrite pseudomorphs, which continue to enlarge by addition of Ca^{2+} and SO_4^{2-} from an external source, ultimately growing displacively into anhydrite nodules. Chickenwire anhydrite forms when, with increasing size, the nodules ultimately coalesce and interfere. Most of the enclosing sediment is pushed aside, and what remains forms thin stringers between the nodules (Shearman, 1978).

Nodular anhydrites have been observed in many modern coastal **sabkha** environments (salt-encrusted supratidal zones in areas such as the Trucial Coast of the Persian Gulf where climates are arid and evaporation rates high). Nodular anhydrites can also

FIGURE 8.1 Nodular anhydrite in a core sample of the Buckner Anhydrite (Jurassic), Texas. Dark-colored carbonate separates and surrounds the lighter-colored anhydrite nodules.

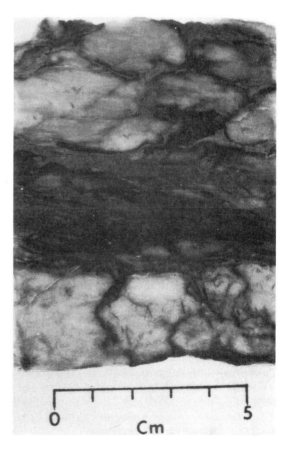

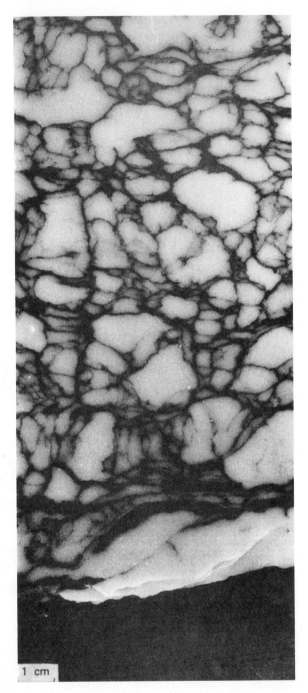

FIGURE 8.2 Chickenwire structure in anhydrite. Evaporite series of the Lower Lias (Jurassic), Aquitaine Basin, southwest France. (From Bouroullec, J., 1981, Sequential study of the top of the evaporitic series of the Lower Lias in a well in the Aquitaine Basin (Auch 1), southwestern France, *in* Chambre Syndical de la Recherche et de la Production due Pétrole et du Gaz Naturel (eds.), Evaporite deposits: Illustration and interpretation of some environmental sequences. Pl. 36, p. 157, reprinted by permission of Editions Technip, Paris, and Gulf Publishing Co., Houston, Tx. Photograph courtesy of J. Bouroullec.)

1 cm

form in deeper-water environments. In fact, all that is needed for formation of nodular anhydrite is growth of crystals in mud in contact with highly saline brines, which can occur in deep or shallow standing water as well as in a sabkha environment.

Laminated anhydrites consist of thin, nearly white anhydrite or gypsum laminations that alternate with dark gray or black laminae rich in dolomite or organic matter (Fig. 8.3). These laminae are commonly only a few millimeters thick and rarely reach 1 cm. Many of the thin laminae are remarkably uniform, with sharp planar contacts. Many laminae can be traced laterally for long distances—some nearly 300 km (Anderson and Kirkland, 1966). Also, they may comprise vertical sequences hundreds of meters thick in which hundreds of thousands of laminae may be present (e.g., the Permian Castile Formation, southeast New Mexico and west Texas). Alternating light and dark pairs of bands have been suggested to be annual varves resulting from seasonal changes in water chemistry and temperature; however, they might equally well represent cyclic changes or disturbances of longer duration. Laminae of anhydrite can also alternate with thicker layers of halite, producing laminated halite.

Because of the lateral persistence of laminated evaporites, which indicates uniform depositional conditions over a wide area, laminites are commonly interpreted to form by precipitation of evaporites in quiet water below wave base. They could presumably form either in a shallow-water area protected in some manner from strong bottom currents and wave agitation or in a deeper-water environment. Some laminated anhydrite may form by coalescing of anhydrite nodules, which by continued growth in

FIGURE 8.3 Laminated anhydrite from the Prairie Evaporite (Devonian), Canada.

a lateral direction merge into one another to produce a layer. Layers formed by this mechanism are thought to be thicker and less distinct and continuous than laminae formed by precipitation. A special type of contorted layering that has resulted from coalescing nodules has been observed in some modern sabkha deposits where continued growth of nodules creates a demand for space. The lateral pressures that result from this demand cause the layers to become contorted, forming ropy bedding or **enterolithic structures** (Fig. 8.4).

Massive anhydrite is anhydrite that lacks perceptible internal structures. True massive anhydrite appears to be less common than nodular and laminated anhydrite, and little information is available regarding its formation. Presumably, it represents sustained, uniform conditions of deposition. Haney and Briggs (1964) suggest that massive anhydrite forms by evaporation at brine salinities of approximately 200 to 275 o/oo (parts per thousand), just below the salinities at which halite precipitation begins. (Seawater has an average salinity of 35 o/oo.)

Origin of Evaporite Deposits

Evaporation Sequence. When ocean water is evaporated in the laboratory, evaporite minerals are precipitated in a definite sequence that was first demonstrated by Usiglio in 1848 (reported in Clarke, 1924). Minor quantities of carbonate minerals begin to form when the original volume of seawater is reduced by evaporation to about one-half. Gypsum appears when the original volume has been reduced to about 20 percent, and halite forms when the water volume reaches approximately 10 percent of the original volume. As discussed in Chapter 7, the precipitation of gypsum causes an increase in the Mg/Ca ratio in remaining water, which favors the process of dolomitization. Dolomite occurs in association with evaporites in many ancient sedimentary successions, as well as in some modern environments. Magnesium and potassium salts are deposited when less than about 5 percent of the original volume of seawater remains. The same general sequence of evaporite minerals occurs in natural evaporite deposits, although many discrepancies exist between the theoretical sequences predicted on the basis of laboratory experiments and the sequences actually observed in the rock

FIGURE 8.4 Enterolithic structure in nodular gypsum of the Grenada Basin, southern Spain. (Photograph courtesy of J. M. Rouchy.)

record. In general, the proportion of CaSO$_4$ (gypsum and anhydrite) is greater and the proportion of Na-Mg sulfates is less in natural deposits than predicted from theoretical considerations (Borchert and Muir, 1964).

Depositional Models for Evaporites. Modern evaporite deposits accumulate in a variety of subaerial and shallow subaqueous environments, as illustrated in Figure 8.5. Subaerial environments include both coastal and continental sabkhas, or salt flats, and interdune environments. Shallow subaqueous environments are mainly in saline coastal lakes called **salinas.** With the possible exception of the Dead Sea in the Middle East, no modern examples of a deep-water evaporite basin exist; however, geologists believe that many of the thick, laterally extensive ancient evaporite deposits did accumulate in deep-water basins.

Some ancient evaporite deposits such as the Permian Zechstein of the North Sea area exceed 2 km in thickness, yet evaporation of a column of seawater 1000 m thick will produce only about 15 m of evaporites. Evaporation of all the water of the Mediterranean Sea, for example, would yield a mean thickness of evaporites of only about 60 m. Obviously, special geologic conditions operating over a long period of time are required to deposit thick sequences of natural evaporites. The basic requirements for deposition of marine evaporites are a relatively arid climate, where rates of evaporation exceed rates of precipitation, and partial isolation of the depositional

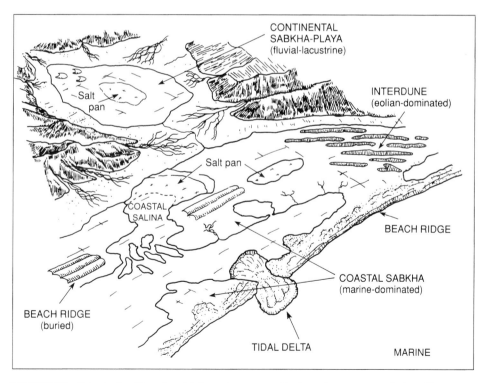

FIGURE 8.5 Principal settings in which modern evaporite deposits are accumulating. (From Kendall, A. C., 1984, Evaporites, *in* R. G. Walker (ed.), Facies models: Geoscience Canada Reprint Ser. 1. Fig. 1, p. 260, as modified slightly by Warren, 1989, reprinted by permission of Geological Association of Canada.)

basin from the open ocean. Isolation is achieved by means of some type of barrier that restricts free circulation of ocean water into and out of the basin. Under these restricted conditions, the brines formed by evaporation are prevented from returning to the open ocean, causing them to become concentrated to the point where evaporite minerals are precipitated.

Although geologists agree on these general requirements for formation of evaporites, considerable controversy still exists regarding deep-water vs. shallow-water depositional mechanisms for many ancient evaporite deposits. Figure 8.6 shows three possible models for deposition of thick sequences of marine evaporites. The deep-water, deep-basin model assumes existence of a deep basin separated from the open ocean by some type of topographic sill. The sill acts as a barrier to prevent free interchange of water in the basin with water in the open ocean, but it allows enough water into the basin to replenish that lost by evaporation. Seaward escape of some brine allows a particular concentration of brine to be maintained for a long time, leading to thick deposits of certain evaporite minerals such as gypsum. The shallow-water, shallow-basin model assumes concentration of brines in a shallow, silled basin, but it allows for accumulation of great thicknesses of evaporites owing to continued subsidence of

FIGURE 8.6 Schematic diagram illustrating three models for deposition of marine evaporites in basins where water circulation is restricted by the presence of a topographic sill. (From Kendall, A. C., 1979, Subaqueous evaporites, *in* R. G. Walker (ed.), Facies models: Geoscience Canada Reprint Ser. 1. Fig. 17, p. 170, reprinted by permission of Geological Survey of Canada, Ottawa.)

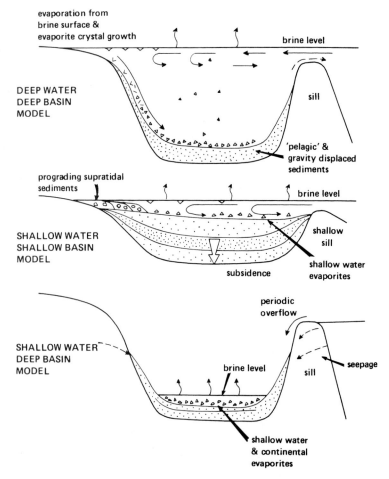

the floor of the basin. The shallow-water, deep-basin model requires that the brine level in the basin be reduced below the level of the sill, a process called **evaporative drawdown,** with recharge of water from the open ocean taking place only by seepage through the sill or by periodic overflow of the sill. Total desiccation of the floors of such basins could presumably occur periodically, allowing the evaporative process to go to completion and thereby deposit a complete evaporite sequence, including magnesium and potassium salts. Thick evaporite deposits could accumulate under these conditions owing to continued subsidence.

Application of these models to ancient evaporite deposits is a challenging task, and geologists have not always agreed upon the environmental interpretation of ancient evaporite deposits. Over time, the concept of evaporite deposition has swung from deep-water to shallow-water deposition; then it swung away from the tidal sabkha regime to very moderate water depths (Sonnenfeld and Kendall, 1989). For additional insight into the deposition of evaporites, see Melvin (1991); Schreiber, Tucker, and Till (1986); Warren and Kendall (1985); and Warren (1989).

Physical Processes in Deposition of Evaporites. Although we tend to regard evaporite deposits as simply the products of chemical precipitation owing to evaporation, many evaporite deposits are not just passive chemical precipitates. The evaporite minerals have, in fact, been transported and reworked in the same way as the constituents of siliciclastic and carbonate deposits. Transport can occur by normal fluid-flow processes or by mass-transport processes such as slumps and turbidity currents. Turbidity current transport mechanisms may have been particularly important in deposition of ancient deep-water evaporite deposits (Schreiber, Tucker, and Till, 1986). Therefore, evaporite deposits may display clastic textures, including both normal- and reverse-size grading and various types of sedimentary structures such as cross-bedding and ripple marks.

Diagenesis of Evaporites

As mentioned, gypsum is converted to anhydrite with burial, particularly above 60°C, with a volume loss in water of about 38 percent. If exhumation subsequently occurs, anhydrite will hydrate to gypsum, with accompanying increase in volume. These volume changes are in part responsible for the formation of nodules and enterolithic structure, as discussed. Also, evaporite deposits respond to burial and tectonic pressures by plastic deformation—resulting in folding and diapirism (formation of salt diapirs or salt domes). During burial, evaporites may in addition undergo dissolution, cementation, replacement, and calcitization of sulfates (Schreiber, 1988; Warren, 1989).

8.3   SILICEOUS SEDIMENTARY ROCKS (CHERTS)

Siliceous sedimentary rocks are fine-grained, dense, very hard rocks composed dominantly of the SiO_2 minerals quartz, chalcedony, and opal, with minor impurities such as siliciclastic grains and diagenetic minerals. **Chert** is the general term used for siliceous rocks as a group. Cherts are common rocks in geologic sequences ranging in age from Precambrian to Tertiary; however, they probably make up less than 1 percent of all sedimentary rocks. They are particularly abundant in Jurassic to Neogene (Tertiary) rocks, moderately abundant in Devonian and Carboniferous rocks, and least abundant in Silurian and Cambrian deposits (Hein and Parrish, 1987). Geologists are particularly interested in cherts because of the information they provide about such aspects of Earth history as paleogeography, paleooceanographic circulation patterns, and plate tectonics. Cherts may also have minor economic significance. Silicon is used in the semiconductor and computer industries and for making glass and related products such as fire bricks, although much of this silica may come from quartz sand. Furthermore, siliceous deposits occur in association with important economic deposits of other minerals, including Precambrian iron ores; uranium, manganese, and phosphorite deposits; and petroleum accumulations.

Mineralogy and Texture

Chert is composed mainly of microcrystalline quartz, with minor chalcedony and perhaps opal, depending upon age. Opal is metastable and alters to quartz with age. Cherts can be divided into three main textural types (Folk, 1974):

1. **microquartz,** consisting of nearly equidimensional grains of quartz 1 to 5 μm in size
2. **chalcedonic quartz,** forming sheaflike bundles of radiating, extremely thin crystals about 0.1 mm long
3. **megaquartz,** composed of equant to elongated grains greater than 20 μm

Figure 8.7 illustrates the texture of microquartz and megaquartz. The silica that makes up the tests of siliceous organisms is amorphous silica or opal, commonly called **opal-A.** Because the remains of siliceous organisms contribute to the formation of chert, opal-A is present in some cherts, particularly those of Tertiary and younger age. Opal-A is metastable and crystallizes in time to opal-CT (below) and finally to quartz (chert). All gradations may be present in siliceous deposits, from nearly pure opal to nearly pure quartz chert, depending upon the age of the deposits and the conditions of burial.

Chemical Composition

Chert is composed dominantly of SiO_2 but may also include minor amounts of Al, Fe, Mn, Ca, Na, K, Mg, and a few other elements such as the rare earth elements cerium

FIGURE 8.7 Fine-textured, nearly equigranular microquartz (chert) (CH) cut by a vein of much coarser megaquartz (Q). Source of specimen unknown. Crossed nicols. Scale bar = 0.1 mm.

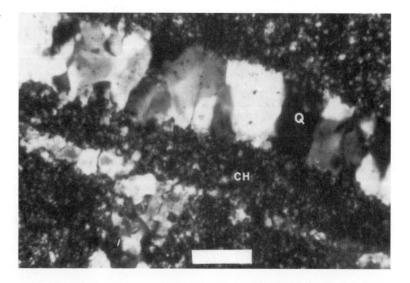

(Ce) and europium (Eu). Many of these additional elements are contained in impurities such as authigenic hematite and pyrite, detrital siliciclastic minerals, and pyroclastic particles. A few elements, including Fe, Mn, Ni, and Cu, may have precipitated from seawater as chert was formed. The amount of SiO_2 varies markedly in different types of cherts, ranging from more than 99 percent in very pure cherts such as the Arkansas Novaculite to less than 65 percent in some nodular cherts (Cressman, 1962). Aluminum is commonly the second most abundant element in cherts, followed by Fe, Mg or K, Ca, and Na. See Jones and Murchey (1986) for additional details.

Classification and Occurrence

Several informal names are applied to chert, depending upon color, inclusions, and texture. **Flint** is a term used both as a synonym for chert and for a variety of chert, particularly chert nodules, that occurs in Cretaceous chalks. **Jasper** is a variety of chert colored red by impurities of disseminated hematite. Jasper interbedded with hematite in Precambrian iron-formations is called **jaspilite. Novaculite** is a very dense, fine-grained, even-textured chert that occurs mainly in mid-Paleozoic rocks of the Arkansas, Oklahoma, and Texas region of south central United States. **Porcelanite** is a term used for fine-grained siliceous rocks with a texture and a fracture resembling those of unglazed porcelain. **Siliceous sinter** is porous, low-density, light-colored siliceous rock deposited by waters of hot springs and geysers. Although most siliceous rocks consist dominantly of chert, some contain abundant detrital clays or micrite. These impure cherts grade into siliceous shales or siliceous limestones.

Cherts can be divided on the basis of gross morphology into two principal types: (1) bedded cherts and (2) nodular cherts. Bedded cherts are further distinguished by their content of siliceous organisms of various kinds. Mineralogy cannot be used as a basis for classifying siliceous sedimentary rocks because these rocks are all composed mainly of microcrystalline quartz. The principal distinguishing characteristics of bedded and nodular cherts are described below.

Bedded Chert

Bedded chert consists of layers of nearly pure chert ranging to several centimeters in thickness that are commonly interbedded with millimeter-thick partings or laminae of siliceous shale (Fig. 8.8). Bedding may be even and uniform or may show pinching and swelling. Most chert beds lack internal sedimentary structures; however, graded bedding, cross-bedding, ripple marks, and sole markings have been reported in some cherts. The presence of these kinds of structures indicates that some mechanical transport was involved in the deposition of these rocks. Bedded cherts are commonly associated with submarine volcanic rocks, pelagic limestones, and siliciclastic or carbonate turbidites.

Many bedded cherts are composed dominantly of the remains of siliceous organisms, which are commonly altered to some degree by solution and recrystallization. Bedded cherts can be subdivided on the basis of type and abundance of siliceous organic constituents into four principal kinds: (1) diatomaceous deposits, (2) radiolarian deposits, (3) siliceous spicule deposits, and (4) bedded cherts containing few or no siliceous skeletal remains.

Diatomaceous Deposits. Diatomaceous deposits include both diatomites and diatomaceous cherts. **Diatomites** are light-colored, soft, friable siliceous rocks composed chiefly of the opaline (opal-A) frustules of diatoms, a unicellular aquatic alga. Thus,

FIGURE 8.8   Thin, well-bedded cherts in the Mino Belt Group (Triassic), near Inuyama, Honshu, Japan.

they are fossil diatomaceous oozes. Diatomites of both marine and lacustrine origin are recognized. Marine diatomites are commonly associated with sandstones, volcanic tuffs, mudstones or clay shales, clayey limestones (marls), and, less commonly, gypsum. Lacustrine diatomites are almost invariably associated with volcanic rocks. **Diatomaceous chert** consists of beds and lenses of diatomite that have well-developed silica cement or groundmass that has converted the diatomite into dense, hard chert. Beds of marine diatomaceous chert comprising strata several hundred meters in thickness have been reported from sedimentary sequences such as the Miocene Monterey Formation of California (Garrison et al., 1981) and occur in rocks as old as the Cretaceous. (Diatoms evolved in the Cretaceous.) Nonmarine diatomaceous deposits have been reported in rocks as old as the Eocene (Barron, 1987). When diatomaceous deposits are converted to quartz chert during diagenesis, the diatom tests are generally destroyed by dissolution and recrystallization.

Radiolarian Deposits. Radiolarian deposits consist dominantly of the remains of radiolarians, which are marine planktonic protozoans with a latticelike skeletal framework of opal. **Radiolarian** deposits can be divided into radiolarite and radiolarian chert. Radiolarite is the comparatively hard, fine-grained, chertlike equivalent of radiolarian ooze, that is, indurated radiolarian ooze. **Radiolarian chert** is well-bedded, microcrystalline radiolarite that has a well-developed siliceous cement or groundmass. Radiolarian cherts are commonly associated with tuffs, mafic volcanic rocks such as pillow basalts, pelagic limestones, and turbidite sandstones that indicate a deep-water origin. On the other hand, some radiolarian cherts are associated with micritic limestones and other rocks that suggest deposition in water perhaps as shallow as 200 m (Iijima, Inagaki, and Kakuwa, 1979). Radiolarians tend to survive silica diagenesis more effectively than do diatoms; therefore, they are common components of many quartz cherts (Fig. 8.9; Hein, Yeh, and Barron, 1990).

FIGURE 8.9   Radiolarian chert from the Otter Point Formation (Jurassic), southwestern Oregon. Most of the small, rounded bodies in this sample are radiolarians. The large fracture at the lower right is filled with silica (quartz) cement. Ordinary light. Scale bar = 0.2 mm. (Photograph courtesy of Shelia A. Monroe.)

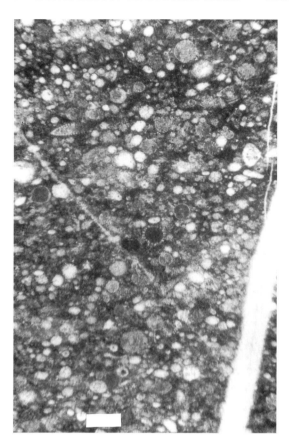

Siliceous Spicule Deposits. Spicularite (spiculite) is a siliceous rock composed principally of the siliceous spicules of invertebrate organisms, particularly sponges. Spicularite is loosely cemented, in contrast to **spicular chert** which is hard and dense. Spicular cherts are mainly marine in origin and occur associated with glauconitic sandstones, black shales, dolomite, argillaceous (clayey) limestones, and phosphorites. They are not generally associated with volcanic rocks and are probably deposited mainly in relatively shallow water a few hundred meters deep.

Skeletal-poor Cherts. Many bedded chert deposits have been described that contain few or no recognizable remains of siliceous organisms. Some of these reported occurrences of fossil-barren cherts may simply be the result of inadequate microscopic examination of the cherts, which might be found upon closer examination to contain siliceous organisms. Others have been examined closely and clearly contain few siliceous organisms. Cherts in this latter group include most cherts associated with Precambrian iron-formations, as well as many Phanerozoic-age cherts such as the Arkansas Novaculite (Mississippian–Devonian) of Arkansas and Oklahoma and the Caballos Novaculite of Texas. Except for the absence of skeletal remains, these cherts resemble radiolarian cherts both megascopically and in their lithologic associations (Cressman, 1962).

The origin of cherts that do not contain siliceous organic remains is poorly understood. Direct, inorganic precipitation of amorphous silica has been reported in some ephemeral Australian lakes (Peterson and von der Borch, 1965); however, no similar occurrences have been reported in the open-marine environment that could help explain the presence of widespread nonfossiliferous chert deposits such as the Arkansas Novaculite. The scarcity of radiolarians and sponge spicules in the Arkansas Novaculite and similar Phanerozoic-age cherts does not, however, preclude the possibility that these cherts were formed by organisms. They could have been derived from siliceous oozes that were subsequently almost completely dissolved and recrystallized, leaving few recognizable siliceous organic remains (Weaver and Wise, 1974). Murray, Jones, and ten Brink (1992) suggest that some bedded chert-shale couplets may form by diagenetic processes. According to these authors, biogenic SiO_2 in shales dissolves, migrates out of the shales, and then reprecipitates adjacent to the shale to form bedded chert.

Nodular Cherts

Nodular cherts are subspheroidal masses, lenses, or irregular layers or bodies that range in size from a few centimeters to several tens of centimeters (Fig. 8.10). They commonly lack internal structures, but some nodular cherts contain silicified fossils or relict structures such as bedding. Colors of these cherts vary from green to tan and black. Nodular cherts typically occur in shelf-type carbonate rocks, where they tend to be concentrated along certain horizons parallel to bedding. They occur also in some sandstones, shales, deep-sea clays, lacustrine sediments, and evaporites. Nodular cherts originate by diagenetic replacement (see "Diagenesis of Chert"). Diagenetic origin is clearly demonstrated in many nodules by the presence of partly or wholly silicified remains of calcareous fossils or ooids.

Origin of Cherts

Source of Silica. In average river water, the concentration of silica in transport (from continental weathering sites) as H_4SiO_4 is about 13 ppm (Table 8.2). In addition to silica transported to the oceans by rivers, silica is added to the oceans through reaction of seawater with hot volcanic rocks along mid-ocean ridges and by low-temperature alteration of oceanic basalts and detrital silicate particles on the seafloor, as described in Chapter 2. Some silica may also escape from silica-enriched pore waters of pelagic sed-

FIGURE 8.10 Nodular chert (arrows) in limestones of the Onondaga Formation (Devonian), New York. (Photograph by E. M. Baldwin.)

TABLE 8.2 Dissolved ion species in mean world river water and ocean water

Ionic species	A Mean river water		B Ocean water	
	ppm	% of total dissolved solids	ppm	% of total dissolved solids
HCO_3^-, CO_3^{2-}	58.7	48.6	140	0.4
Ca^{2+}	15.0	12.4	400	1.2
H_4SiO_4	13.1	10.8	1	<0.01
SO_4^{2-}	11.2	9.3	2,649	7.7
Cl^-	7.8	6.5	18,980	55.0
Na^+	6.3	5.2	10,556	30.6
Mg^{2+}	4.1	3.4	1,272	3.7
K^+	2.3	1.9	380	1.1
NO_3^-	1.0	0.8	0.5	<0.01
Fe^{2+}, Fe^{3+}	0.67	0.6	0.01	<0.01
$Al(OH)_4^-$	0.24	0.2	0.01	<0.01
F^-	0.09	0.07	1.3	<0.01
Sr^{2+}	0.09	0.07	8	0.02
$B(OH)_4^-$	0.1–0.01	0.08–<0.01	26	0.07
Mn^{2+}	0.02	0.02	—	—
Br^-	—	—	65	0.02
Total	120.8		34,479	

Source: A, Livingston, D. A., 1963, Data of geochemistry. Chap. G, Chemical composition of rivers and lakes: U.S. Geol. Survey Prof. Paper 440–G. B, Mason, B., 1966, Principles of geochemistry: John Wiley & Sons, New York.

iments on the seafloor. These silica sources are summarized in Figure 8.11. Despite contributions of silica from these various sources, the silica concentrations in different parts of the ocean range from less than 0.01 ppm to a maximum of about 11 ppm; the average dissolved silica content of the ocean is only 1 ppm (Heath, 1974). Clearly, silica is constantly being removed by some process and has a relatively short residence time in the ocean.

Silica Solubility. Solubility studies show that the solubility of silica in seawater differs for different silicate minerals. The solubility of SiO_2 at 25°C and normal ocean pH (~7.8–8.3) ranges from ~6 to 10 ppm for quartz to ~60 to 130 ppm for amorphous or noncrystalline varieties of silica such as opal (Krauskopf, 1959; Morey, Fournier, and Rowe, 1962, 1964; Iller, 1979). The solubility of amorphous silica appears to control the precipitation of silica. That is, silica concentrations must reach values on the order of 115 ppm or higher at 25°C for seawater to be saturated with silica and precipitation to occur. Therefore, the ocean, with an average dissolved silica content of only 1 ppm, is grossly undersaturated with respect to silica, in sharp contrast to the saturated state of the surface ocean with respect to calcium carbonate. This fact raises a very intriguing question: What mechanism (or mechanisms) is (are) capable of removing silica from highly undersaturated ocean water to form chert beds and maintain the low concentrations of dissolved silica in the ocean?

The solubility of silica is affected by both pH and temperature. Change in solubility of silica with pH is illustrated in Figure 8.12A. Solubility changes only slightly with increase in pH up to about 9, but it rises sharply at pH values above 9. Solubility increases with increasing temperature in essentially a linear fashion, and solubility at 100°C is nearly 3 to 4 times that at 25°C (Fig. 8.12B). Once silica is in solution under a

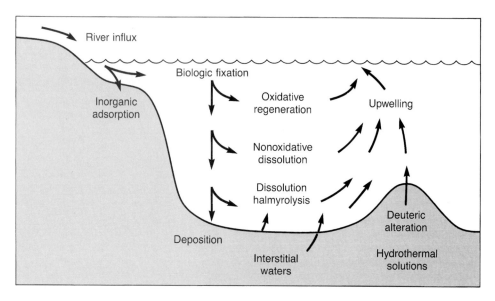

FIGURE 8.11   Sources of dissolved silica in ocean water. (After Heath, G. R., 1974, Dissolved silica and deep-sea sediments, *in* W. W. Hay (ed.), Studies in paleooceanography: Soc. Econ. Paleontologists and Mineralogists Spec. Pub. 20. Fig. 7, p. 81, reprinted by permission of SEPM, Tulsa, Okla.)

given set of temperature and pH conditions, it does not readily crystallize to form quartz, even from solutions that have silica concentrations greatly exceeding the solubility of quartz (6–10 ppm). Therefore, it is unlikely that chert, which consists of microcrystalline quartz, can be precipitated by inorganic processes from highly undersaturated ocean water. Chert might be precipitated in some local basins where waters are saturated with silica, owing perhaps to dissolution of volcanic ash. Also, some silica may be removed from seawater in the open ocean by adsorption onto clay minerals or other silicate particles; however, such processes cannot account for the many bedded successions of nearly pure chert present in the geologic record.

Silica Extraction from Seawater.   Removal of silica from ocean water by silica-secreting organisms to build opaline skeletal structures appears to be the only mechanism capable of large-scale silica extraction from undersaturated seawater. This biologic process has operated since at least early Paleozoic time to regulate the balance of silica in the ocean. **Radiolarians** (Cambrian/Ordovician–Holocene), **diatoms** (Cretaceous–Holocene), and **silicoflagellates** (Cretaceous–Holocene) are microplankton that build skeletons of opaline silica (opal-A). These siliceous microplankton (particularly diatoms and radiolarians) have apparently been abundant enough in the ocean during Phanerozoic time to extract most of the silica delivered to the oceans by rock weathering and other processes. Diatoms are probably responsible for the bulk of silica extraction from ocean waters in the modern ocean (Calvert, 1983); however, radiolarians were the important users of silica in the Phanerozoic oceans of Jurassic and older ages. Heath (1974) calculates that the residence time for dissolved silica in the ocean ranges from 200 to 300 years for biologic utilization to 11,000 to 16,000 years for incorporation into the geologic record—a very short time from a geologic point of view.

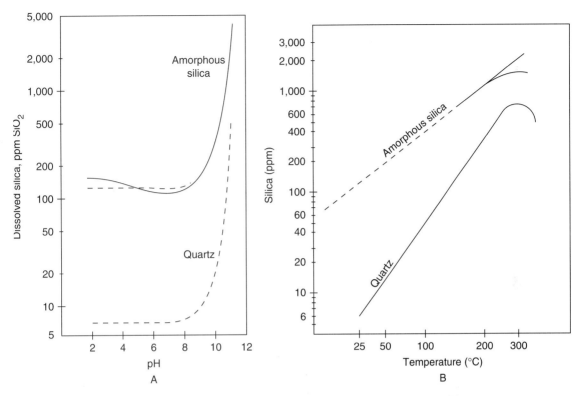

FIGURE 8.12 The solubility of silica as a function of (A) pH and (B) temperature. The solid line in A shows the variation in solubility of amorphous silica as determined experimentally. The upper dashed curve shows the calculated solubility of amorphous silica, based on an assumed constant solubility of 120 ppm SiO_2 at a pH below 8. The lower dashed line is the calculated solubility of quartz based on the approximate known solubility of 6 ppm SiO_2 in neutral and acid solutions. (A, after Krauskopf, K. B., 1979, Introduction to geochemistry, 2nd ed. Fig. 6.3, p. 133, McGraw-Hill, New York. B, after Fournier, R. O., 1970, Silica in thermal waters: Laboratory and field investigations: Proc. International Symposium on Hydrochemistry and Biochemistry, Tokyo, p. 122–139.)

While silica-secreting organisms are alive, their siliceous (opal) skeletons are protected by an organic coating that prevents them from dissolving in highly undersaturated and corrosive seawater. After death, this coating is destroyed by biochemical decomposition, and the opaline skeletons begin to undergo dissolution; however, in areas of the ocean where silicious organisms flourish, the rate of production of siliceous skeletons may be so high that they cannot all be dissolved as rapidly as they are produced. Under such conditions, a sufficient number of the siliceous skeletons may survive total dissolution to accumulate on the seafloor as siliceous oozes (sediments containing at least 30 percent siliceous skeletal material, commonly more than 60 percent). It is postulated that after burial by additional siliceous ooze or clayey sediment, these opal skeletal materials continue to undergo solution; however, after burial, the dissolving silica is trapped in the pore spaces of the sediment and cannot escape back to the open ocean. The pore waters thus become increasingly enriched in silica, perhaps to concentrations of 120 ppm or more. Opal-CT (defined below) chert then slowly precipitates from these concentrated interstitial solutions. Thus, the formation of chert is in part a sedimentation process involving the depositional concentration of biogenic opaline tests and in part a diagenetic process with crystallization and recrystallization of the chert taking place after sediment burial.

Precambrian Cherts

Biogenic removal of silica from seawater is believed to account for deposition of most, if not all, Phanerozoic-age bedded cherts, although siliceous fossil fragments are not preserved in all of these cherts. Bedded cherts are also very common in stratigraphic sequences of Precambrian age. The existence of large populations of silica-secreting organisms during Precambrian time has not yet been proven; however, a few workers (e.g., LaBerge, Robbins, and Han, 1987) have reported the existence in some Precambrian cherts of possible siliceous organic remains. Until the presence of silica-secreting Precambrian organisms is definitely proven, we assume that deposition took place by inorganic processes. How these processes operated given the geochemical constraints discussed above is not fully understood; nor is the immediate source of the silica known. Perhaps the silica content of the Precambrian ocean was higher than that of the Phanerozoic ocean, possibly related to volcanism; or perhaps it was higher simply because silica concentrations could build in the absence of silica-secreting organsims. Thus, the origin of Precambrian cherts, like the origin of Precambrian limestones, remains something of an enigma.

Diagenesis of Chert

Transformation of Opal to Quartz Chert. As mentioned, biogenic silica deposited in marine environments is principally in the form of opal, commonly called **opal-A** to indicate amorphous silica. During the process of transformation of biogenic opal to chert, opal-A may not convert directly to quartz chert, which is microcrystalline quartz, but commonly goes through an intermediate, metastable phase: opal-CT. Although called **opal-CT,** it is composed mainly of low-temperature cristobalite disordered by interlayered tridymite lattices. Cristobalite and tridymite are metastable varieties of quartz that alter with time to quartz. Opal-CT may occur in open spaces in sediments as **lepispheres,** which are microcrystalline aggregates of blade-shaped crystals. It can also form as nonspherulitic blades, rim cements, and overgrowths, and as a massive cement (Maliva and Siever, 1988a). The rates of diagenetic evolution of silica from biogenic opal-A to opal-CT and finally to quartz chert are controlled by several physicochemical factors. Temperature is commonly considered to be a particularly important control, with increasing temperature promoting an increased rate of transformation (Siever, 1983). The time required for transformation of opal-A to quartz chert is believed to be a function of sedimentation and burial rates and the rates of heat flow (a function of the geothermal gradient) in regions of burial. Transformation can take place at rates ranging from a few million years to hundreds of millions of years. The transformation is fastest where rates of sedimentation and burial are high (sediment is rapidly buried to depths where high temperatures prevail) or where a high geothermal gradient exists in a region (Fig. 8.13).

Kastner and Gieske (1983) have demonstrated that the rate of transformation of opal-A to quartz chert depends also upon the nature of the opal starting material and the presence of magnesium hydroxide compounds, which serve as a nucleus for the crystallization of opal-CT. Williams, Parks, and Crerar (1985) conclude that increasing surface-to-volume ratio of siliceous particles—a ratio that increases with decreasing

Because temperature has a particularly significant effect on the diagenesis of chert, as well as on many other diagenetic processes discussed in preceding chapters, geologists are greatly interested in estimating the temperatures at which particular diagenetic reactions take place. Considerable research has been carried out to develop reliable techniques for **paleotemperature** analysis. Tools used for determining paleotemperatures are called **geothermometers.** The principal techniques now in use for determining diagenetic paleotemperatures include methods based on (1) conodont color alteration, (2) vitrinite reflectance, (3) graphitization levels in kerogen, (4) clay mineral assemblages, (5) zeolite mineral assemblages, (6) fluid inclusions, and (7) oxygen isotope ratios. These methods are described briefly in Appendix A, which also gives the useful temperature range of each method and some explanatory remarks about the materials that constitute the geothermometers. The methods have various degrees of reliability, and none can be considered an infallible estimator of the paleotemperatures of diagenesis; condont color alteration and vitrinite reflectance are generally regarded to be the most useful methods. Methods based on analyses of mineral assemblages tend to be less sensitive and more equivocal. The formation of zeolite minerals, for example, depends upon pressure and the salinity and chemical composition of sediment pore waters, as well as temperature. Two or three different methods, which generally include examination of conodont color alteration and vitrinite reflectance, are commonly used together as a cross-check on reliability.

particle size—results in greater solubility of opal and an increase in the rate of transformation. These authors also suggest that under some conditions, opal-A can transform directly to chert without going through the intermediate opal-CT stage.

Replacement Chert. In addition to occurrence as bedded chert, chert can occur in the form of small nodules, lenses, or thin, discontinuous beds, as mentioned (Fig. 8.10). Nodular cherts are especially common in limestones but may be present in evaporites and siliciclastic sedimentary rocks, particularly shales. Relict textures in nodular cherts suggest that most are formed by diagenetic replacement. Some replacement cherts form in the open ocean where they replace carbonates and clays, as shown by preservation of burrows and other sedimentary structures (Hein and Karl, 1983). The silica that forms these so-called deep-sea cherts is furnished by dissolution of local siliceous organisms, particularly diatoms (in Cenozoic occurrences). Replacement cherts are particularly common in shallow platform carbonate rocks. Silica is supplied in this environment by the dissolution of sponge spicules or other forms of biogenic opal-A within the sediment pile. This dissolution process causes the pore waters to become supersaturated in silica with respect to opal-CT and quartz. Chertification then occurs by force-of-crystallization–controlled replacement of the host carbonate owing to nonhydrostatic stresses resulting from opal-CT and quartz crystal growth that simultaneously causes calcite dissolution (Maliva and Siever, 1989). For additional insight into some of the geochemical conditions that favor the formation of replacement chert, and the mechanisms of replacement, see Maliva and Siever (1988b).

FIGURE 8.13 A. Time-tempera-ture diagram for the diagenetic alteration of siliceous sediments under conditions of variable, but moderately rapid, sedimentation in a rift zone with a high geothermal gradient. The diagram shows a time-temperature curve that is based on estimates of sedimenta-tion rates and the geothermal gradi-ent. Time-temperature regions for the conversion of opal-A to opal-CT and quartz (chert) are also shown. B. Time-temperature diagram for the diagenetic alteration of siliceous sediments under condi-tions of slow sedimentation on a spreading seafloor where the geothermal gradient is low. (From Siever, R., 1983, Evolution of chert at active and passive continental margins, *in* A. Iijimo, J. R. Hein, and R. Siever (eds.), Siliceous deposits in the Pacific regions. Fig. 6, p. 17, and Fig. 8, p. 20, reprinted by permission of Elsevier Science Publishers, Amsterdam.)

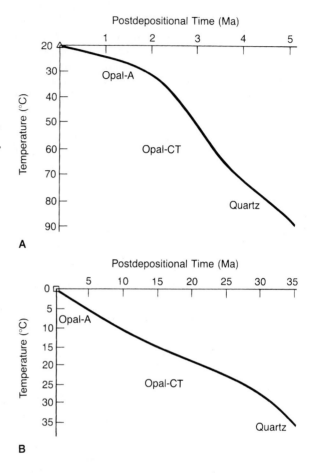

8.4   IRON-BEARING SEDIMENTARY ROCKS

Some iron is present in almost all sedimentary rocks. The average iron content of sili-ciclastic shales, for example, is 4.8 percent. Sandstones contain 2.4 percent iron on average, and limestones contain about 0.4 percent (Blatt, 1982). Iron-rich sedimentary rocks have much higher concentrations of iron than these average sedimentary rocks. The term **iron-rich** is usually reserved for sedimentary rocks that contain at least 15 per-cent iron, corresponding to 21.3 percent Fe_2O_3 or 19.4 percent FeO (Blatt, 1982, p. 400). Most iron-rich sedimentary rocks were deposited during three time periods: the Pre-cambrian, the early Paleozoic, and the middle to late Mesozoic (Jurassic–Cretaceous). They make up only a small fraction (<~1 percent) of the total sedimentary record; however, they have great economic significance as iron ores. Important iron deposits are located on all the major continents except Antarctica, and at least one deposit of sedimentary iron characterized as very large occurs in each of these continents.

Kinds of Iron-rich Sedimentary Rocks

Dimroth (1979) suggests that iron-rich sedimentary rocks can be classified into three broad groups: (1) detrital chemical iron-rich sedimentary rocks; (2) iron-rich shales;

and (3) miscellaneous iron-rich deposits (Table 8.3). Of these, the detrital chemical rocks are the only volumetrically important iron-rich sedimentary rocks. James (1966) proposed that these rocks be divided into two principal kinds: (1) iron-formation (for dominantly Precambrian iron-rich sediments) and (2) ironstone (for dominantly Phanerozoic iron-rich sediments). Some subsequent workers (e.g., Trendall, 1983; Young, 1989) have expressed dissatisfaction with this scheme, particularly labeling ironstones as solely Phanerozoic deposits. In this book, I use the term **ironstone** for the mainly nonbanded, noncherty, commonly oolitic, iron-rich sedimentary rocks. The term **iron-formation** is used for the mainly well-banded, cherty, iron-rich sedimentary rocks. Iron-formations are also referred to as banded iron-formations. Iron-formations and ironstones are differentiated on the basis of physical characteristics and mineralogy as shown in Table 8.4. Volumetrically, ironstones are much less important than iron-formations.

Iron-Formations

Iron-formations are iron-rich deposits that range in age from early Precambrian to Devonian, although they are primarily of Precambrian age (James and Trendall, 1982). They consist of distinctively banded sequences (Fig. 8.14), 50 to 600 m thick, composed of layers enriched in iron alternating with layers rich in chert. Banding occurs on a scale ranging from millimeters to tens of meters. Cherty iron-formations can grade into slightly cherty iron-rich sandstones, siltstones, and shales. The textures of iron-formations resemble those of limestones. Textural types in iron-formations equivalent to micritic, pelleted, intraclastic, peloidal, oolitic, pisolitic, and stromatolitic lime-

TABLE 8.3   Principal classes of iron-rich sedimentary rocks

I. Detrital chemical iron-rich sediments
   A. Cherty iron-formation
      Texture: analogous to limestone texture
      Composition: iron-rich chert containing hematite, magnetite, siderite, ankerite, or (predominantly alumina-poor) silicates as predominating iron minerals; relatively poor in Al and P
   B. Ironstone
      Texture: analogous to limestone texture
      Composition: aluminous iron silicates (chamosite, chlorite, stilpnomelane), iron oxides, and carbonates; relatively rich in Al and P

II. Iron-rich shales
   C. Pyritic shales
      Bituminous shales containing nodules or laminae of pyrite; grade into massive pyrite bodies by coalescence of pyrite laminae and nodules
   D. Siderite-rich shales
      Bituminous shales with siderite concretions; grade into massive siderite bodies by coalescence of concretions

III. Miscellaneous iron-rich deposits
   E. Iron-rich laterites
   F. Bog iron ores
   G. Manganese nodules and oceanic iron crusts
   H. Iron-rich muds precipitated from hydrothermal brines, Lahn-Dill type iron oxide ores, and stratiform, volcanogenic sulfide deposits
   I. Placers of magnetite, hematite, or ilmenite sand

Source: Modified slightly from Dimroth, E., 1979, Models of physical sedimentation of iron formations, *in* R. G. Walker (ed.), Facies models: Geoscience Canada Reprint Ser. 1. Table I, p. 175, reprinted by permission of Geological Association of Canada.

TABLE 8.4 Differences between ironstone and iron-formation

	Ironstone	Iron-formation
Age	Pliocene to Middle Precambrian; principal beds from Lower Paleozoic and Jurassic–Cretaceous	Devonian to Early Precambrian; principal formations approximately 2000 million years old
Thickness	Major units a few meters to a few tens of meters	Major units 50–600 m
Original areal extent	Individual depositional basins rarely more than 150 km in maximum dimension	Difficult to determine; some deposits with continuity over many hundreds of kilometers
Physical character	Massive to poorly banded; silicate and oxide facies oolitic	Thinly bedded; layers of dominantly hematite, magnetite, siderite, or silicate alternating with chert, which makes up approximately half the rock; ooids rare
Mineralogy	Dominant oxide goethite; hematite very common; magnetite relatively rare; chamosite primary silicate; calcite and dolomite common constituents	No goethite; magnetite and hematite about equally abundant; primary silicate greenalite; chert a major constituent; dolomite present in some units, but calcite rare or absent
Chemistry	Except for high iron content, no distinctive aspects	Remarkably low content of Na, K, and Al; low P
Associated rocks	Both typically interbedded with shale, sandstone, or graywacke; yet the iron-formation has few or no clastics compared to the ironstone.	
Relative abundance of facies	No gross differences apparent; probable order of abundance for ironstone: oxide, silicate, siderite, sulfide; for iron-formation the order is similar, but siderite facies may be more abundant than silicate facies	

Source: James, H. L., 1966, Chemistry of the iron-rich sedimentary rocks, *in* M. Fleischer (ed.), Data of geochemistry: U.S. Geol. Survey Prof. Paper 440–W.

stone textures are recognized (Table 8.5). Sedimentary structures reported from banded iron-formations include cross-bedding, graded bedding, load casts, ripple marks, erosion channels, shrinkage cracks, and slump structures. These structures show that many of the particles that make up iron-formations have undergone mechanical transport and deposition.

Mineralogy and Chemistry. On the basis of relative abundance of major kinds of iron-bearing minerals, James (1966) defines four different mineral facies in iron-rich sedimentary rocks: oxides, silicates, carbonates, and sulfides (Table 8.6). **Oxides** include hematite, goethite, and magnetite. Hematite is present in both iron-formations and ironstones; goethite occurs in ironstones but is absent in Precambrian iron deposits. Magnetite is most abundant in Precambrian deposits but occurs also in Phanerozoic deposits. **Silicates** refers here to the iron silicate minerals chamosite and greenalite. Chamosite (iron-rich chlorite) is the primary silicate mineral in ironstones; greenalite $[Fe_3Si_2O_5(OH)_4]$ is dominant in Precambrian iron-formations. **Carbonate** minerals in iron-rich facies include siderite, dolomite, calcite, and ankerite. Siderite is an important constituent in both Precambrian and Phanerozoic iron-rich sediments, where it commonly consists of flattened nodules or more or less continuous beds. Dolomite is also common in both iron-formations and ironstones. Calcite is common in ironstones but is rare in iron-formations, and ankerite is most common in iron-formations. Pyrite is the dominant **sulfide** mineral in the sulfide facies of iron-rich rocks, but marcasite is

FIGURE 8.14 Banded iron-formation from the Negaunee Iron Formation (Precambrian), Michigan. The light-colored bands are chert; the dark layers are the iron-rich units. Length 8 cm. (Specimen furnished by M. H. Reed.)

also common. Sulfide minerals rarely form the major iron mineral in these rocks, but locally they can predominate in some thin beds.

Iron-formations consist mainly of SiO_2 and Fe, but the chemical composition of these rocks varies over a wide range depending upon the type of deposit. It is difficult to establish a truly representative average composition; however, Gole and Klein (1981) suggest that iron-formations commonly contain 40 to 50 percent SiO_2; 29 to 32 percent total Fe; 3 to 6 percent MgO; 2 to 7 percent CaO; 1 to 2 percent Al_2O_3; and less than 1 percent each of TiO_2, MnO, Na_2O, K_2O, P_2O_5, S, and C. Although iron, expressed as Fe_2O_3, FeO, or FeS, is the dominant chemical constituent in some iron-rich sediments, the iron content of many iron-rich sedimentary rocks is commonly exceeded by that of silica. Note from Table 8.7 the quite considerable variation in chemical composition of the different iron-rich facies. Manganese concentration, in particular, may reach considerable percentages in some iron-formations. Additional references include Appel and LaBerge (1987), Melnik (1982), and Trendall and Morris (1983).

Ironstones

Ironstones are dominantly Phanerozoic-age sedimentary deposits that occur on all continents. They are mainly Early Paleozoic and Jurassic–Cretaceous in age but range in age from Pliocene to Middle Precambrian; few deposits of Cambrian and Carboniferous–Triassic age are known. They form thin, massive, or poorly banded sequences a

TABLE 8.5 Textural types of cherty iron-formation equivalent to limestone textural types

Type	Description
Micrite type	Deposited as a mud whose particles are too fine-grained to survive diagenesis; only lamination and stratification visible as depositional structures; small-scale cross-beds here and there prove depositional as particulate, noncohesive matter.
Pelleted	Fine pellet texture of silt or very fine sand.
Intraclastic	Containing gravel-size fragments (intraclasts) whose internal textures prove derivation from penecontemporaneous sediment; fragments embedded in a micrite-type matrix or bound by a cement introduced during diagenesis.
Peloidal	Containing sand-size fragments (peloids) without internal textures; peloids embedded in a micrite-type matrix or bound by a clear chert cement introduced during diagenesis.
Oolitic	Containing concentrically laminated ooids, either set in a micrite-type matrix or, more commonly, bound by a clear chert cement introduced during diagenesis.
Pisolitic	Containing pisolites either set in a micrite-type matrix or cemented by clear chert.
Stromatolitic	Wavy, columnar, or digitating stromatolites.

Source: After Dimroth, E., 1979, Models of physical sedimentation of iron formations, *in* R. G. Walker (ed.), Facies models: Geoscience Canada Reprint Ser. 1. Table II, p. 176, reprinted by permission of Geological Association of Canada.

TABLE 8.6 Principal iron-bearing minerals in iron-rich sedimentary rocks

Mineral class	Mineral	Chemical formula
Oxides	Goethite* Hematite Magnetite	$FeOOH$ Fe_2O_3 Fe_3O_4
Silicates	Chamosite Greenalite Glauconite Stilpnomelane Minnesotaite (iron talc)	$3(Fe,Mg)O \cdot (Al,Fe)_2O_3 \cdot 2SiO_2 \cdot nH_2O$ $FeSiO_3 \cdot nH_2O$ $KMg(Fe,Al)(SiO_3)_6 \cdot 3H_2O$ $2(Fe,Mg)O \cdot (Fe,Al)_2O_3 \cdot 5SiO_2 \cdot 3H_2O$ $(OH)_2(Fe,Mg)_3Si_4O_{10}$
Sulfides	Pyrite Marcasite	FeS_2 FeS_2
Carbonates	Siderite Ankerite Dolomite Calcite	$FeCO_3$ $Ca(Mg,Fe)(CO_3)_2$ $CaMg(CO_3)_2$ $CaCO_3$

*Not found in Precambrian iron-formations

TABLE 8.7   Chemical composition of sedimentary facies of iron-formations

	Oxide facies	Silicate facies	Carbonate facies	Sulfide facies
Fe	37.80	26.5	21.23	20.0
FeO	2.10	28.9	22.22	2.35
Fe_2O_3	51.69	5.6	5.74	—
FeS_2	—	—	—	38.70
SiO_2	42.89	50.7	48.72	36.67
Al_2O_3	0.42	0.4	0.15	6.90
Mn	0.3	0.4	0.50	0.001
P	0.03	—	0.07	0.09
CaO	0.1	0.1	4.60	0.13
MgO	+	4.2	0.84	0.65
K_2O	+	—	+	1.81
Na_2O	+	—	0.01	0.26
TiO_2	+	+	+	0.39
CO_2	—	5.1	14.10	—
S	—	+	2.76	—
SO_3	—	—	—	2.60
C	—	+	++	7.60
H_2O^+	0.43	5.2	2.67	1.25

Source: Eichler, J., 1976, Origin of Precambrian banded iron-formations, *in* K. H. Wolf (ed.), Handbook of strata-bound and stratiform ore deposits, v. 7. Table VI, p. 187, reprinted by permission of Elsevier Science Publishers, Amsterdam.

Note: + = trace; ++ = larger trace; — = not present

few meters to a few tens of meters thick, in sharp contrast to the much thicker, well-banded iron-formations. They commonly have an oolitic texture (Fig. 8.15), and they may contain fossils that have been partly or completely replaced by iron minerals. Sedimentary structures that include cross-bedding, ripple marks, scour-and-fill structures, clasts, and burrows are present in many ironstones, indicating that mechanical transport of grains was involved in the origin of these rocks. Ironstones are commonly interbedded with carbonates, shales, and fine-grained sandstones of shelf to shallow-marine origin. For a much more comprehensive discussion of ironstones, see Van Houten and Bhattacharyya (1982) and Young and Taylor (1989).

Iron-rich Shales

Pyritic black shales occur in association with both Precambrian iron-formations and Phanerozoic ironstones. They commonly form thin beds in which sulfide content may range as high as 75 percent. Pyrite occurs disseminated in these black carbonaceous shales and in some limestones. It may be present also as nodules, as laminae, and as a replacement of fossil fragments and other iron minerals. Pyrite-rich layers have likewise been reported in some limestones. **Siderite-rich shales** (clay ironstones) occur primarily in association with other iron-rich deposits. They are present also in the coal measures of both Great Britain and the United States. Siderite (iron carbonate) occurs disseminated in the mudrocks or as flattened nodules and more or less continuous beds.

Miscellaneous Iron-rich Sediments

Bog iron ores are minor accumulations of iron-rich sediments that occur particularly in small freshwater lakes of high altitude. They range from hard, oolitic, pisolitic, and

FIGURE 8.15 Ironstone ooids with quartz nuclei, cemented with sparry calcite cement, Clinton Formation (Silurian), New York. Ordinary light. Scale bar = 0.5 mm.

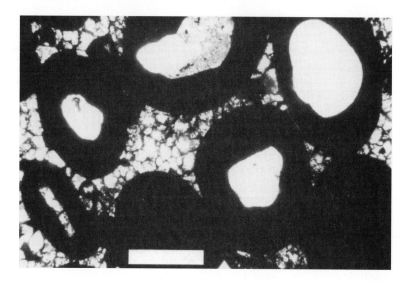

concretionary forms to soft, earthy types. **Iron-rich laterites** are residual iron-rich deposits that form as a product of intense chemical weathering. They are basically highly weathered soils in which iron is enriched. **Manganese crusts and nodules** are widely distributed on the modern seafloor in deeper parts of the Pacific, Atlantic, and Indian oceans in areas where sedimentation rates are low. They have been reported also from ancient sedimentary deposits in association with such oceanic sediments as red shales, cherts, and pelagic limestones. Both iron-rich (15–20 percent Fe) and iron-poor (< ~6 percent Fe) varieties of manganese nodules are known. These nodules contain various amounts of Cu, Co, Ni, Cr, and V in addition to manganese and iron oxide minerals. Owing to the presence of these valuable metals in manganese nodules, considerable interest has developed in the possibility of mining them from the seafloor. Recovery vessels are already being planned and designed, and political negotiations have been underway for some time among the major nations of the world regarding undersea mining rights to these potentially valuable deposits. They are not likely to be mined, however, until cheaper land sources of iron are exhausted.

  Iron-rich metalliferous sediments have been discovered in several oceanic settings, particularly near active mid-ocean spreading ridges. They form by precipitation from metal-rich hydrothermal fluids that have become enriched through contact and interaction with hot basaltic rocks. These sediments are enriched in Fe, Mn, Cu, Pb, Zn, Co, Ni, Cr, and V. Metal-enriched sediments have been reported also from some ancient sedimentary deposits in association with submarine pillow basalts and ophiolite sequences of ocean crustal rocks.

  Heavy mineral placers are sedimentary deposits that form by mechanical concentration of mineral particles of high specific gravity, commonly in beach or alluvial environments. Magnetite, ilmenite, and hematite sands are common constituents of placers, particularly beach and marine placers. Placers are local accumulations, generally less than 1 to 2 m in thickness, that occur mainly in Pleistocene–Holocene sediments. Marine placer deposits containing about 5 percent iron ore have been mined off the southern tip of Kyushu, Japan, for many years (Mero, 1965). Offshore placers containing up to 10 percent magnetite and ilmenite have been reported off the southeastern coast of Taiwan (Boggs, 1975). Beach placers containing ilmenite have been exploited commercially in Australia since about 1965 (Hails, 1976). "Fossil" placer deposits are comparatively rare, although thin, heavy-mineral laminae are common in some ancient beach deposits. Hails (1976) reports that outcrops of ilmenite- and magnetite-bearing placers of Cretaceous age are exposed discontinuously through New Mexico, Colorado, Wyoming, and Montana subparallel to the Rocky Mountains.

Origin of Iron-rich Sedimentary Rocks

Iron Deposition in Modern Environments. There are no modern counterparts to the ancient environments that presumably favored widespread deposition of iron-rich sediments to produce iron-formations and ironstones, but iron-bearing minerals are being deposited on a relatively small scale in a variety of modern environments. Iron sulfides, particularly pyrite (FeS_2), are forming in black muds that accumulate under reducing conditions in stagnant ocean basins, tidal flats, and organic-rich lakes. Iron sulfides are also accumulating around the vents of hot springs located on the crests of mid-ocean ridges, as discussed in Chapter 2. Chamosite, a complex Fe-Mg-Al silicate, has been reported in modern sediments at water depths as great as 150 m in the Orinoco and Niger deltas and on the ocean floor off Guinea, Gabon, and Sarawak and in the Malacca Straits. Glauconite is a K-Mg-Fe-Al silicate mineral that has been reported from Monterey Bay, California, and from various other parts of the ocean at water depths ranging down to about 2000 m. Iron oxides such as goethite ($FeOOH$) are accumulating in some modern lakes and bogs, as ooids on the floor of the North Sea (Pettijohn, 1975, p. 420), and in manganese nodules in both seawater and freshwater. Manganese nodules, which contain manganese, copper, cobalt, nickel, and other metals in addition to iron, are particularly widespread in the Pacific Ocean at water depths of about 4 to 5 km. Presumably, deposition of these various iron-bearing minerals in modern oceanic settings accounts for the lower iron content of the ocean compared to that of rivers entering the ocean (Table 8.2).

Iron Deposition in Ancient Environments. The transport and deposition of iron are governed by both the Eh and the pH of the environment. Eh-pH diagrams such as Figure 8.16A can be used to predict the stability of iron-bearing minerals and serve to illustrate that Eh is commonly more important than pH in determining which iron-bearing mineral will be deposited. For example, hematite (Fe_2O_3) is precipitated under oxidizing conditions at the pHs commonly found in the ocean and most surface waters; siderite ($FeCO_3$) forms under moderately reducing conditions; and pyrite (FeS_2) forms under moderate to strong reducing conditions. Figure 8.16B shows the ranges of Eh and Ph in some natural environments.

Because the iron geochemistry of natural systems is far more complex than the simplified conditions assumed in constructing Eh-pH diagrams, such diagrams are of only limited use in interpreting the actual environment of iron deposition. Many problems are associated with the formation of sedimentary iron deposits, and the mechanisms by which transport and deposition of iron occurred in the past to generate iron-formations and ironstones are still poorly understood and controversial. One of the principal problems stems from the fact that iron in the oxidized or ferric (Fe^{3+}) state is much less soluble than iron in the reduced or ferrous (Fe^{2+}) state (Fig. 8.16). Ferric iron is soluble only at pHs less than about 4; such values rarely occur under natural conditions. Thus, under oxidizing conditions iron tends to precipitate rather than undergo solution. How, then, can large quantities of iron be taken into solution and transported from subaerial weathering sites under the oxidizing conditions that commonly prevail in streams and rivers?

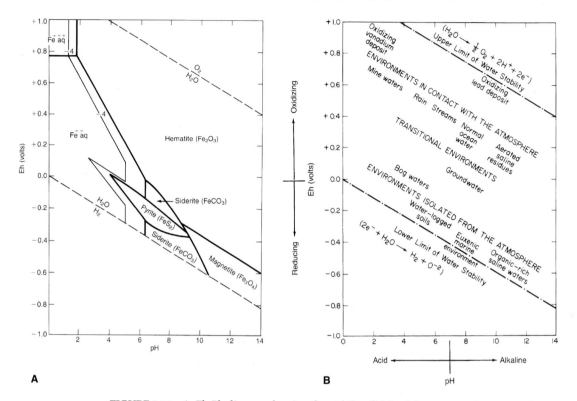

FIGURE 8.16   A. Eh-Ph diagram showing the stability fields of the common iron minerals, sulfides, and carbonates in water at 25°C and 1 atm total pressure. Total dissolved sulfur = 10^{-6}; total dissolved carbonate = 10^0. B. Graph showing Eh and pH of waters in some natural environments. (A, from Garrels, R. M., and C. L. Crist, 1965, Solutions, minerals, and equilibria. Fig. 7.21, p. 224, reprinted by permission of Harper & Row, New York. B, after Blatt, H., G. V. Middleton, and R. Murray, 1980, Origin of sedimentary rocks, 2nd ed. Fig. 6–12, p. 241, reprinted by permission of Prentice-Hall, Englewood Cliffs, N.J., based on data from Baas Becking, L. G. M. et al., 1960, Limits of the natural environment in terms of pH and oxidation-reduction potentials: Jour. Geology, v. 68, p. 243–284.)

This problem of solution and transport of iron under oxidizing conditions prompted some workers (Lepp and Goldich, 1964; Cloud, 1973; Lepp, 1987) to postulate that low oxygen levels existed during the Precambrian, allowing great quantities of iron to be transported in the soluble, reduced (Fe^{2+}) state to marine basins. Presumably, owing to photosynthetic generation of oxygen by cyanobacteria, local oxidizing conditions existed within some parts of broad, shallow-marine basins where the iron was oxidized to the insoluble ferric (Fe^{3+}) state and precipitated. Lepp (1987) suggests that the ferrous iron was stored in basinal bottom waters for long periods of time before finally precipitating. The concept of low oxygen levels during the late Precambrian has been questioned by other investigators (e.g., Dimroth and Kimberley, 1976). In any case, this argument cannot be used to explain the solution and transport of iron during Phanerozoic time when an oxidizing atmosphere clearly existed. Some workers have suggested that iron may have been transported as colloids by physical processes rather than in true solution or that it was sorbed to clay particles or organic materials and transported along with these substances. It appears unlikely, however, that mechanisms such as colloidal transport can account for transport of the large quantities of iron that occur in iron-formations or ironstones (Ewers, 1983). Reducing conditions seem to be required for transport of large amounts of iron in solution; hence the dilemma.

Three problems have to be addressed to account for the formation of iron-rich rocks:

1. the source of the iron
2. transport of the iron to the depositional basin, presumably under reducing conditions
3. precipitation of the iron within the basin, presumably under oxidizing conditions

Most workers originally assumed that the iron was derived by subaerial weathering of iron silicate minerals; however, the problem of transport of ferric iron to depositional basins remains an obstacle if iron is derived from the land. To get around this difficulty, some workers have suggested that the source of the iron may have been within the depositional basin close to the depositional site.

Three principal kinds of depositional models have been proposed to explain the origin of iron-rich deposits, particularly banded iron-formations: the **subaerial weathering model,** the **upwelling model,** and the **exhalation model** (Table 8.8). Each of these

TABLE 8.8 Postulated models for deposition of cherty, iron-rich sediments

	Depositional process	Problems	References
Subaerial weathering model	Fe derived by on-land weathering in anoxic Precambrian atmosphere; transported to ocean as ferrous (Fe^{2+}) iron and stored in ocean bottom water for long periods of time until finally precipitated as ferric (Fe^{3+}) iron under oxidizing conditions.	Not possible in the oxic environments that existed after Precambrian time; requires weathering of huge quantities of rock to furnish iron; does not explain chert deposition.	Lepp and Goldich, 1964; Cloud, 1973; Lepp, 1987
Upwelling model	Fe derived from within depositional basins by upwelling of deep bottom waters charged with ferrous (Fe^{2+}) iron resulting from reduction of ferric oxide in terrigenous siliciclastic sediments and other submarine rocks; precipitation of Fe takes place on shelves or in continent-margin basins where oxidation to Fe^{3+} and subsequent precipitation occur.	Does not adequately explain how silica concentrated in ocean to levels that would allow inorganic precipitation of chert.	Drever, 1974; Button et al., 1982
Exhalation (hydrothermal) model	Hydrothermal activity from hot springs located along mid-ocean spreading ridges furnishes Fe; hot solutes saturated in Fe and silica are dispersed from deeper water upward and outward into shallower water where precipitation takes place.	Distance over which hydrothermal solutes have to be moved is great; Fe source should be a well-mixed (ocean) reservoir; volcanic exhalation does not fit this requirement.	Gross, 1980; Simonson, 1985

models attempts to provide a rational explanation for the source of the huge amounts of iron that are present in many banded iron-formations, the origin of the interbedded chert (and thus the banding), and the mechanisms of iron transport and deposition. All of the models have weaknesses—none provides a completely satisfactory explanation for all of these concerns. Other ideas to explain some of these problems have also been advanced. For example, Garrels (1987) suggests that the small-scale bands in Australian banded iron-formations formed by evaporation of water in restricted basins. LaBerge, Robbins, and Han (1987) believe that the iron in iron-formations is precipitated as ferric hydrate and that precipitation was effected mainly by bacterial stripping and algal photosynthesis. Furthermore, these authors believe that the chert associated with iron-formations may be the result of organic precipitation of the silica in the form of siliceous frustules—in other words, that organisms capable of secreting siliceous tests existed in Precambrian time. Dimroth (1979) suggested that iron-rich sediments may form by iron and silica replacement of initial $CaCO_3$ deposits; however, the replacement theory is not widely accepted. With respect to iron transport, Schieber (1987) suggests that large quantities of iron might be moved in groundwater, where the iron could be dissolved in the reduced state. When groundwater is discharged to river systems, the iron oxidizes; however, instead of precipitating immediately, the iron may form a stable sol and be transported on the last leg of its journey as colloids.

Many puzzling aspects of the formation of sedimentary iron deposits remain. Why, for example, were chert and iron not deposited together as banded iron-formations after the Precambrian? Why was deposition of iron-rich sediments particularly prevalent during the Precambrian, early Paleozoic, and Jurassic–Cretaceous? What role, if any, did organisms play in the deposition of iron-formations and ironstones? Were silica-secreting organisms actually present in the Precambrian ocean? Was the local production of oxygen by photosynthesizing organisms such as algae important? Did low forms of life such as bacteria and algae catalyze or initiate precipitation of iron in some manner? If so, how did they cause precipitation, and how important was such biologic activity? How did the iron-rich ooids that are common in ironstones form (e.g., Young, 1989)? Obviously, much additional research will be required before the mystery surrounding the origin of iron-rich sediments is solved.

8.5 SEDIMENTARY PHOSPHORITES

The phosphorus content of rocks is expressed as percentage P_2O_5. Sedimentary **phosphorites** are rocks that are significantly enriched in phosphorus over other types of rocks. Significantly, in this context, it is commonly taken to mean that they contain more than 15 to 20 percent P_2O_5. These phosphorus-rich sedimentary rocks are called by a variety of other names—phosphate rock, rock phosphate, phosphates—in addition to phosphorites. Sedimentary rocks that contain less than 15 to 20 percent P_2O_5 but considerably more than that in average sediments (0.11–0.17 shales; 0.08–0.16 sandstones; 0.03–0.7 limestones; McKelvey, 1973) are referred to as phosphatic, e.g., phosphatic shale. The total volume of sedimentary phosphates in the geologic record is small; however, phosphorites are of special economic interest. They contribute more than 80 percent of the world's production of phosphate and make up about 96 percent of the world's total resources of phosphate rock. Total world resources of sedimentary phosphate rock are estimated to be about 158,000 million tons of all grades and types (Notholt, Sheldon, and Davidson, 1989).

Sedimentary phosphates occur in rocks of all ages from Precambrian to Holocene, but phosphorite deposition appears to have been particularly prevalent during the Precambrian and Cambrian in central and southeast Asia (China, USSR/MPR, Australia); the Permian in North America; the Jurassic and Early Cretaceous in eastern Europe; the Late Cretaceous to Eocene in the Tethyan Province of the Middle East and North Africa; and the Miocene of southeastern North America (Fig. 8.17; Cook, 1976).

Phosphorite nodules and phosphatic sediments occur also on the present ocean floor at shallow depths in the vicinity of coastlines. They are particularly common off the coasts of Peru and Chile, southwest Africa, eastern United States, and Southern and Baja California. They occur also on some seamounts and atolls (Fig. 8.18; Baturin, 1982; Burnette and Riggs, 1990). Many of these ocean-floor phosphate occurrences are older than the Holocene; however, modern phosphate nodules are present on the ocean floor in a few places, such as the Peru-Chile continental margin (Burnette and Froelich, 1988) and the Namibian shelf off southwest Africa (Bremner and Rogers, 1990).

Composition

Sedimentary phosphorites are composed of calcium phosphate minerals, all of which are varieties of apatite. The principal varieties include fluorapatite [$Ca_5(PO_4)_3F$], chlorapatite [$Ca_5(PO_4)_3Cl$], and hydroxyapatite [$Ca_5(PO_4)_3OH$]. Most are carbonate hydroxyl fluorapatites in which up to 10 percent carbonate ions can be substituted for phos-

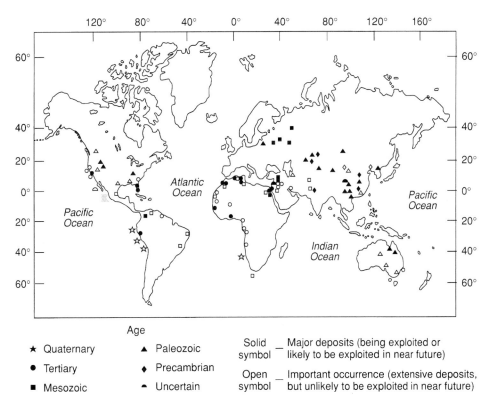

FIGURE 8.17 Worldwide distribution of major sedimentary phosphorite deposits. (After Cook, P. J., 1976, Sedimentary phosphate deposits, *in* K. H. Wolf (ed.), Handbook of strata-bound and stratiform ore deposits. Fig. 1, p. 505, reprinted by permission of Elsevier Science Publishers, Amsterdam.)

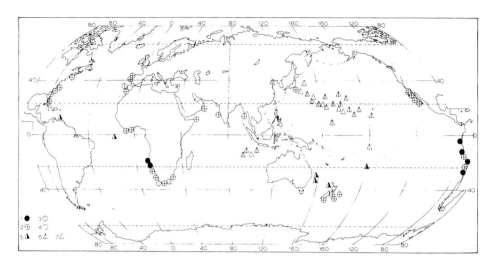

FIGURE 8.18   Distribution of phosphorite nodules and phosphatic sediments on the ocean floor: 1–4 show the locations of phosphorites on continental shelves: 5–7, phosphorites on seamounts. Ages of the phosphorites are 1, Holocene; 2 and 5, Neogene; 3 and 6, Paleogene; 4 and 7, Cretaceous. (From Baturin, G. N., 1982, Phosphorites on the sea floor: Developments in Sedimentology 33. Fig. 2.1, p. 56, reprinted by permission of Elsevier Science Publishers, Amsterdam.)

phate ions to yield the general formula $Ca_{10}(PO_4,CO_3)_6F_{2-3}$. These carbonate hydroxyl fluorapatites are commonly called **francolite.** The wastebasket term **collophane** is often used for sedimentary apatites for which the exact chemical composition has not been determined. Detrital quartz, authigenic chert (microcrystalline quartz), opal-CT, calcite, and dolomite are also common constituents of many phosphorites. Glauconite, illite, montmorillonite, and zeolites may also be present in some deposits; moderately abundant organic matter is a characteristic constituent of many phosphorites (Nathan, 1984).

The chemistry of phosphorites is dominated by phosphorus, silicon (present in minerals other than apatite), and calcium. Slansky (1986, p. 70) shows that the abundance of these elements in 20 phosphorites ranging in age from Precambrian to Holocene is P_2O_5 = 22–39 percent; SiO_2 = <1–25 percent; and CaO = 43–53 percent. Other common constituents include Al_2O_3 (<1–5 percent); Fe_2O_3 (<1–4 percent); MgO (<1–6 percent); Na_2O (<1 percent); K_2O (<1 percent); F (1–4 percent), Cl (<1 percent); SO_3 (0–11 percent); and organic carbon (0–2 percent). Many trace elements, such as Ag, Cd, Mo, Se, Sr, U, Yu, Zn, as well as the rare earth elements, may also be present in phosphorites in amounts exceeding their average compositions in seawater, the crust, and the averge shale (Nathan, 1984). For additional information on the mineralogy and chemistry of individual phosphorites of various ages throughout the world, see Notholt, Sheldon, and Davidson (1989).

Principal Kinds of Phosphate Deposits

Phosphate-rich sedimentary rocks may occur in layers ranging from thin laminae a few millimeters thick to beds a few meters thick. Some phosphate successions such as the Phosphoria Formation of the Idaho-Wyoming area may reach several hundred meters in thickness, although such successions are not composed entirely of phosphate-rich rocks. Phosphorites are generally interbedded with shales, cherts, limestones, dolomites, and, more rarely, sandstones. Phosphatic rocks commonly grade regionally into nonphosphatic sedimentary rocks of the same age.

Phosphorites have textures that resemble those in limestones. Thus, they may be made up of peloids, ooids, fossils (bioclasts), and clasts that are now composed of apatite. Some phosphorites lack distinctive granular textures and are composed instead of fine, micritelike, textureless collophane. The phosphatic grains may contain inclusions of organic matter, clay minerals, silt-size detrital grains, and pyrite. Peloidal or pelletal phosphorites are particularly common; oolitic phosphorites are somewhat less so. Phosphatized fossils or fragments of original phosphatic shells are important constituents of some deposits. Most phosphorite grains are sand size, although particles greater than 2 mm may be present. These larger grains, referred to as nodules, can range in size to several tens of centimeters.

Because the textures of phosphorites have such close resemblance to those of limestones, some geologists suggest using modified limestone classifications to distinguish different kinds of phosphorites. For example, Slansky (1986) advocates use of a classification system based to some extent on Folk's (1962) limestone classification, and Cook and Shergold (1986b) suggest modification of Dunham's (1962) carbonate classification for use in describing phosphorites. Use of this modified classification thus yields the names mudstone phosphorite, wackestone phosphorite, packstone phosphorite, grainstone phosphorite, and boundstone phosphorite.

Most major phosphorite deposits are bedded marine deposits. **Bedded phosphorites** form distinct beds of variable thickness, commonly interbedded and interfingering with carbonaceous mudrocks, cherts, and carbonate rocks. The phosphorite in bedded deposits occurs as peloids, ooids, pisoids, phosphatized brachiopods and other skeletal fragments, micritelike apatite mud, and cements. Perhaps the best-studied example of a bedded phosphate deposit is the Permian Phosphoria Formation (Fig. 8.19). This formation has a total thickness of 420 m and extends over an area of about 350,000 km^2 in the Idaho-Wyoming area (McKelvey et al., 1959; Sheldon, 1989). Bedded marine phosphorites are also common in the Precambrian and Cambrian rocks of Australia, the Cretaceous–Tertiary rocks of North Africa, and many other parts of the world (Cook and Shergold, 1986a; Notholt, Sheldon, and Davidson, 1989; Burnett and Riggs, 1990; Soudry, 1992). **Bioclastic phosphorites** are a special type of bedded phosphate deposit composed largely of vertebrate skeletal fragments such as fish bones, sharks teeth, fish scales, coprolites, and so on. The Rhaetic Bone Bed (Upper Triassic) of western England (Greensmith, 1989, p. 213) provides an example. Deposits composed mainly of invertebrate fossil remains such as phosphatized brachiopod shells are also known. These phosphate-bearing organic materials commonly become further enriched in P_2O_5 during diagenesis and may be cemented by phosphate minerals.

Nodular phosphorites are brownish to black, spherical to irregularly shaped nodules ranging in size from few centimeters to a meter or more. Internal structure of phosphate nodules ranges from homogeneous (structureless) to layered or concentrically banded. Phosphatic grains, pellets, sharks teeth, and other fossils may occur within the nodules. Nodular phosphorites are particularly common in many Neogene to Holocene phosphatic deposits of the world (Burnett and Riggs, 1990). Phosphate nodules are also forming today in zones of upwelling in the ocean, such as on the Peru continental margin (Burnett and Froelich, 1988). Many ancient nodular phosphorites may have had a similar origin under conditions of marine upwelling; however, some ancient phosphorite nodules may be of diagenetic origin.

Pebble-bed phosphorites are composed of phosphatic nodules, phosphatized limestone fragments, or phosphatic fossils that have been mechanically concentrated

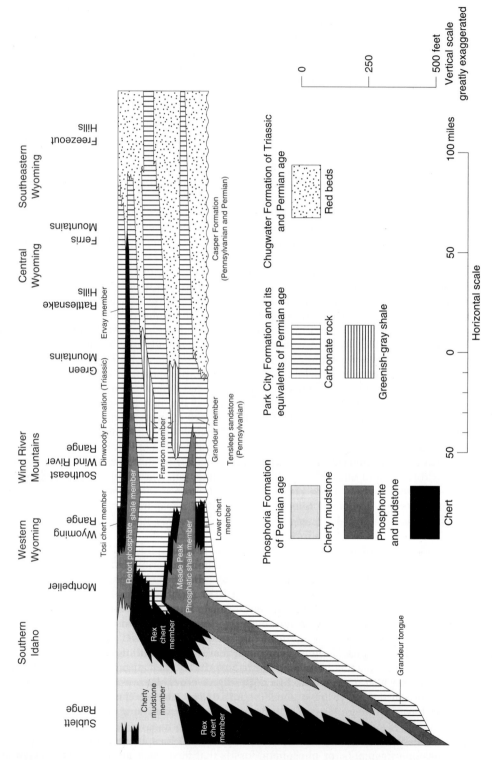

FIGURE 8.19 Stratigraphic relations of the phosphatic Phosphoria Formation (Permian age), Park City Formation (Permian), and Chugwater Formation (Triassic) of Idaho and Wyoming. Note in particular the Retort phosphate shale member and the Meade Peak phosphatic shale member. The section runs from west to east. (From Sheldon, R. P., 1986. Phosphorite deposits of the Phosphoria Formation, Western United States, *in* A. J. G. Notholt, R. P. Sheldon, and D. F. Davidson (eds.), Phosphate deposits of the world, v. 2. Fig. 8.1, p. 54, Cambridge University Press, Cambridge.)

by reworking of earlier-formed phosphate deposits. The Miocene and Quaternary river-pebble and land-pebble deposits of Florida (Cathcart, 1989) provide a good example of this type of deposit.

Guano deposits are composed of bird and bat excrement that has been leached to form an insoluble residue of calcium phosphate. Guano occurs today on small oceanic islands in the Eastern Pacific and the West Indies. Guano deposits are not important in the geologic record.

Origin of Phosphorites

Chemical/Biochemical Processes. As mentioned, the principal phosphate minerals in sedimentary rocks are various varieties of apatites, of which carbonate apatite $[Ca_{10}CO_3(PO_4)]_6$ is most important. The conditions that favor precipitation of calcium carbonate also favor formation of carbonate apatite, although carbonate apatite can precipitate at values of pH possibly as low as 7.0, whereas calcium carbonates generally do not precipitate below a pH of about 7.5 (Bentor, 1980b). Carbonate apatite precipitation also appears to be favored by conditions that are slightly reducing. The factors affecting phosphate solubility are not thoroughly understood, however, and the solubility of carbonate apatite has not been definitely established. It is not definitely known if the ocean is saturated with carbonate apatite, although it may be very near saturation (Kolodny, 1981). The average concentration of phosphorus in the ocean is 70 parts per billion (ppb) (Gulbrandsen and Roberson, 1973) compared to 20 ppb in average river water. The concentration of phosphorus in ocean water ranges from only a few ppb in surface waters, which are strongly depleted by biologic uptake, to values of 50 to 100 ppb at depths greater than about 200 to 400 m.

Phosphorus is removed from seawater in several ways. Some phosphorus is precipitated along with calcium carbonate minerals during deposition of limestones; however, the average limestone contains only 0.03 to 0.7 percent P_2O_5. Phosphorus is removed also by concentration in the tissue and bones of organisms, but this phosphorus is returned to the ocean when organisms die, unless they are quickly buried by sediment before decay of the organic tissue is complete. Some phosphorus may be removed from seawater by incorporation into metalliferous sediments as a result of adsorption onto metallic minerals such as iron hydroxides. We are particularly interested in identifying a mechanism that can explain how trace amounts of phosphorous in seawater can be concentrated to form phosphorite deposits. Assuming approximately the same average content of phosphorus in the ancient and modern ocean, how was the 70- to 100-ppb phosphorus content of ancient oceans upgraded to form widespread deposits of carbonate apatite containing as much as 40 percent P_2O_5, an enrichment of up to 2-millionfold?

A secondary, or replacement, origin has been suggested for some phosphorite deposits to account for this enrichment. On the other hand, the preservation in many other phosphorite deposits of clastic textures and primary sedimentary structures such as cross-bedding and lamination indicates that these sediments are primary deposits. Thus, the phosphorus must have been extracted in some way from ocean water. Available evidence suggests an association between phosphorite deposition and areas of upwelling in the oceans. For example, studies of the distribution of ancient phosphorites show that most occur in lower latitudes in the trade wind belts along one side of a basin where deeper water could have upwelled adjacent to a continent. Most phosphate nodule deposits on the modern ocean floor also occur in areas of upwelling.

Early ideas on upwelling and phosphorite deposition assumed that inorganic precipitation of apatite occurred as cold, deep, phosphate-rich waters upwelled onto a shallow shelf. Under these postulated conditions, carbon dioxide would be lost from the upwelling waters owing to pressure decrease, warming, or photosynthesis, causing pH to increase and, presumably, carbonate apatite precipitation to occur. Martens and Harris (1970) demonstrated, however, that Mg^{2+} ions in seawater have an inhibiting effect on the growth of carbonate apatite crystals in much the same way that they inhibit the precipitation of calcite. Also, most or all of the phosphorus brought to surface waters by upwelling currents is quickly used up by organisms that utilize phosphate as one of the essential nutrients needed for organic growth. Rapid biologic utilization prevents phosphate levels in the ocean from rising to the point of saturation. The low phosphorus content of surface seawater (owing to biologic utilization) together with Mg^{2+} inhibition appears to rule out purely inorganic precipitation of apatite from the open ocean.

Biologic utilization of phosphate to build soft body tissue appears to provide the answer to the problem of phosphate concentration in sediments. Modern phosphate nodules are forming in areas of oceanic upwelling where a steady supply of phosphate brought from the large deep-ocean reservoir allows continuous growth of organisms in large numbers. After death, organisms and organic debris not consumed by scavengers pile up on the ocean floor under reducing conditions where decay is inhibited. These organic materials include the remains of phytoplankton and zooplankton, coprolites (feces), and the bones and scales of fish. All contain phosphorus; for example, phytoplankton contain about 0.4 percent phosphorus by dry weight (Gross, 1982, p. 326). Under the reducing conditions of the seafloor, some of the soft body tissue is thus preserved long enough to be buried and incorporated into accumulating sediment. Perhaps about 1 to 2 percent of the total phosphorus involved in primary productivity in upwelling zones is ultimately incorporated into the sediments in this way (Baturin, 1982).

Slow decay of body tissue after burial releases phosphorus to the interstitial waters of the sediment. Studies of the chemistry of interstitial waters in sediments where modern phosphate nodules are forming and in other areas of the seafloor where organic-rich sediments are accumulating under reducing conditions have turned up phosphorus concentrations ranging from 1400 ppb to as much as 7500 ppb (Bentor, 1980a; Froelich et al., 1988). At such high phosphorus concentrations, the interstitial waters are supersaturated with respect to calcium phosphate. The phosphate thus begins to precipitate on the surfaces of siliceous organisms, carbonate grains, particles of organic matter, fish scales and bones, siliciclastic mineral grains, or older phosphate particles (Baturin, 1982). Phosphate may also replace skeletal grains and carbonate grains, a process called **phosphatization.** Phosphorite nodules thus form within the sediments by diagenetic reactions between organic-rich sediments and their phosphate-enriched interstitial waters. To allow phosphate precipitation to take place, Mg^{2+} ions may be removed from pore waters owing to magnesium iron replacements in clay minerals in the anoxic marine sediments (Drever, 1971). On the other hand, phosphate minerals are reported to precipitate from some sediments in which Mg concentrations are about that of seawater (e.g., Froelich et al., 1988). How phosphate precipitation under these conditions is possible is poorly understood. It may be related in some way to the presence of filamentous bacteria and certain organic compounds within the pore waters (Glenn and Arthur, 1988), which are not present in open ocean water. Thus, phosphate minerals may be able to precipitate in some Mg-bearing pore waters, in the presence of these organisms, whereas they will not precipitate in seawater.

Physical Processes. The presence of clastic textures and primary depositional sedimentary structures in many ancient phosphorite deposits seems inconsistent with a diagenetic concentration mechanism. Therefore, Kolodny (1980) suggested a two-stage process for the origin of ancient phosphorite deposits. In the first stage, apatite forms diagenetically in reducing basins by phosphorus mobilization in interstitial waters in the manner postulated for formation of modern phosphorites. The final stage involves reworking and enrichment of these diagenetically formed phosphorite grains by mechanical concentration processes under oxidizing conditions. Concentration presumably takes place in a high-energy environment, probably during lower stands of sea level. During this stage, the phosphate grains may be transported into a different depositional setting from that in which they formed. The grainstone/packstone phosphorite deposits of the Lady Annie region of Queensland, Australia, for example, are suggested to have formed in deeper water before subsequent transport into the shallow Lady Annie embayment (Fig. 8.20; Cook and Elgueta, 1986). This final stage of phosphorite formation, during which the original diagenetically formed phosphorite sediments are mechanically reworked under shallow-water conditions, accounts for the clastic textures and primary sedimentary structures in many ancient phosphorites.

Summary of Phosphorite Deposition

In summary, upwelling of phosphate-rich waters from deeper parts of the ocean and biologic utilization of the phosphate in soft body tissue appear to be important factors in the origin of phosphorite deposits. Phosphorus is deposited on the seafloor in organic detritus and is buried with accumulating sediment. Phosphate becomes concentrated in the pore waters of sediment during slow decay of the phosphate-bearing, soft-bodied organisms and other organic detritus. Carbonate apatite precipitates diage-

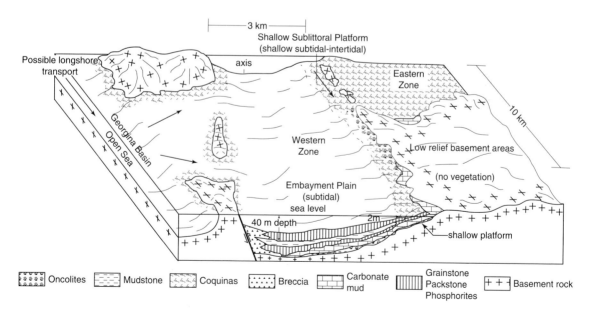

FIGURE 8.20 Reconstruction of paleogeography, depositional conditions, and facies relationships that may have existed in the Lady Annie region of Australia during late Precambrian and Cambrian time and that led to deposition of phosphorites in this region. (From Cook, P. J., and S. A. Elgueta, 1986, Proterozoic and Cambrian phosphorites-deposits: Lady Annie, Queensland, Australia, *in* P. J. Cook and J. M. Shergold (eds.), Phosphate deposits of the world, v. 1. Fig. 11.9, p. 146, Cambridge University Press, Cambridge.)

netically from these phosphate-enriched pore waters by some process not yet fully understood to form phosphate grains and cements. Apatite may also replace skeletal grains or other carbonate grains. Subsequently, these diagenetic deposits are reworked mechanically, possibly owing to lowered sea levels, allowing final concentration and deposition of phosphatic sediments by waves and currents. These processes are summarized diagrammatically in Figure 8.21.

This postulated multistage process for formation of phosphorite deposits has some limitations. It does not, for example, explain why phosphorites accumulated on a much vaster scale at some times in the geologic past than at present. A possible explanation for this phenomenon is that major episodes of phosphorite deposition were tied to climate and sea-level changes. For example, a period of glaciation may produce a large volume of cold, nutrient-rich water that, after a long residence time, will eventually be circulated into shallower areas during rising sea level (transgression). Such an event would produce a major burst of organic activity in the shallow zone (Cook and Shergold, 1986b), leading to increased phosphorite deposition. For additional discus-

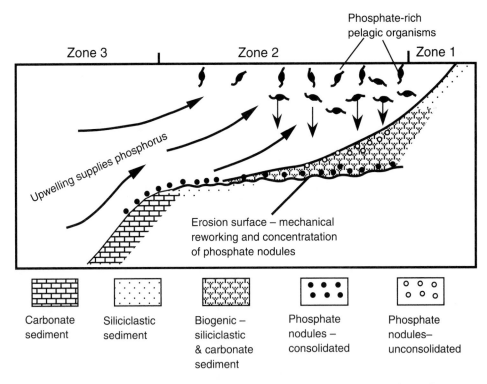

FIGURE 8.21   Schematic illustration of the formation of phosphorites in areas of upwelling on open ocean shelves. Near-shore, shallow-water siliciclastic deposits form in zone 1. Zone 2 is the zone where high contents of phosphate-rich biogenic detritus accumulate in the sediment from rain-out of pelagic organisms; phosphate nodules form in this zone by diagenetic processes, followed by reworking of phosphate-rich sediments during lowered sea level. Zone 3 is a deeper-water zone where carbonate sediments with local phosphate nodules occur. (After Baturin, G. N., 1982, Phosphorites on the sea floor: Origin, composition, and distribution. Fig. 5.4, p. 227, reprinted by permission of Elsevier Science Publishers.)

sion of the possible influence of climate on paleocirculation patterns (e.g., upwelling) and phosphorite deposition, see Parrish (1990).

8.6 CARBONACEOUS SEDIMENTARY ROCKS

Introduction

Most sedimentary rocks, even those of Precambrian age, contain at least a small amount of organic matter that consists of the preserved residue of plant or animal tissue. When the tissue of organisms decays in an oxygen-deficient environment, organic degradation is incomplete; more decay-resistant fractions of organic substances, such as cellulose, fats, resins, and waxes, are not immediately decomposed. If a depositional basin happens to be in a highly reducing environment—such as a restricted basin, stagnant swamp, or bog—decay-resistant organic matter may be preserved long enough to become incorporated into accumulating sediment where it may persist for hundreds of millions of years after burial.

The average content of organic matter in sedimentary rocks is 2.1 weight percent in shales,, 0.29 percent in limestones, and 0.05 percent in sandstones (Degens, 1965). The average in all sedimentary rocks is about 1.5 percent. Organic matter contains about 50 to 60 percent carbon; therefore, the average sedimentary rock contains about 1 percent organic carbon. A few special types of sedimentary rocks contain significantly more organic material than these average rocks. Black shales typically contain 3 to 10 percent organic matter. Oil shale or kerogen shale contains even higher percentages, ranging to 25 percent or more, and coals may be composed of more than 70 percent organic matter. Certain solid hydrocarbon accumulations—such as asphalt, formed from petroleum by oxidation and loss of volatiles—constitute another example of a sedimentary deposit greatly enriched in organic carbon.

Kinds of Organic Matter in Sedimentary Rocks

Three basic kinds of organic matter are accumulating in subaerial and subaqueous environments under present conditions: humus, peat, and sapropel. Soil **humus** is plant organic matter that accumulates in soils to form a number of decay products such as humic and fulvic acids (complex, high-molecular-weight, organic acids). Most soil humus is eventually oxidized and destroyed, and little is preserved in sedimentary rocks. **Peat** also consists of humic organic matter, but peat accumulates in freshwater or brackish-water swamps and bogs where stagnant, anaerobic conditions prevent total oxidation and bacterial decay. Therefore, some of the humus that accumulates under reducing conditions can be preserved in sediments. **Sapropel** refers to fine organic matter that accumulates in lakes, lagoons, or marine basins where oxygen levels are low. It consists of the remains of phytoplankton, zooplankton, and spores and fragments of higher plants. Phytoplankton are tiny plants such as algae that drift about in the upper water column owing to currents; zooplankton are small, drifting animals such as foraminifers.

It is often difficult to differentiate accurately between the types of organic matter found in ancient sediments; however, both humic and sapropelic types are recognized. Humic organic matter is the chief constituent of most coals, although a few are formed of sapropel. The organic matter in oil shales and other carbonaceous mudrocks and limestones originated from sapropel, but it is so finely disseminated and altered that it is difficult to identify. This type of organic matter is called **kerogen** [see "Oil Shale (Kerogen Shale)" below].

Classification of Carbonaceous Sedimentary Rocks

The predominant organic constituents of carbonaceous sediments are thus humic and sapropelic organic matter. The nonorganic constituents are mainly either siliciclastic grains or carbonate materials. Carbonaceous sediments can be classified on the basis of relative abundance of nonorganic constituents and the kind of organic matter that composes the organic constituents (humic vs. sapropelic) into three basic types of organic-rich rocks: coal, oil shale, and asphaltic substances (Fig. 8.22). Each of these types of rocks contains at least 10 to 20 percent organic constituents.

Coals

Coals are the most abundant type of carbonaceous sediment. They are composed dominantly of combustible organic matter but contain various amounts of impurities (ash), which are largely siliciclastic materials. The amount of ash that coals can contain and still retain the name of coal is not precisely fixed. Some very impure coals (bone coals) may contain 70 to 80 percent ash, but most coals have less than 50 percent ash by weight. Most coals are humic, although a few are sapropelic coals made up mostly of spores, algae, and fine plant debris. Cannel coals and boghead coals (see below) are sapropelic coals. Coals are defined in various ways, but a commonly accepted definition is that of Schopf (1956, p. 527):

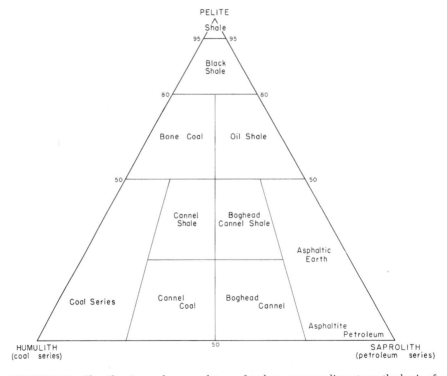

FIGURE 8.22 Classification and nomenclature of carbonaceous sediments on the basis of relative abundance of humic organic constituents (humulith), sapropelic organic constituents (saprolith), and fine-grained terrigenous constituents (pelite). (From Sedimentary rocks, 3rd ed., by Francis J. Pettijohn. Fig. 11.37. Copyright 1949, 1957 by Harper & Row, Publishers, Inc. Copyright © 1975 by Francis J. Pettijohn. Reprinted by permission of HarperCollins Publishers, Inc.)

Coal is a readily combustible rock containing more than 50 percent by weight and more than 70 percent by volume of carbonaceous material, formed from compaction or induration of variously altered plant remains similar to those of peaty deposits. Differences in the kinds of plant materials (type), in degree of metamorphism (rank), and range of impurities (grade), are characteristic of the varieties of coal.

Characteristics and Classification. A common method of classifying coals is by **rank,** which is based on the degree of coalification or carbonification (increase in organic carbon content) attained by a given coal owing to burial and metamorphism (Table 8.9). Peat is included in Table 8.9 but is actually not a true coal. **Peat** consists of unconsolidated, semicarbonized plant remains with high moisture content. **Lignite** or brown coal is the lowest-rank coal. Lignites are brown to brownish black coals that have high moisture content and commonly retain many of the structures of the original woody plant fragments. They are dominantly Cretaceous or Tertiary in age. **Bituminous coals** are hard, black coals that contain fewer volatiles and less moisture than lignite and have a higher carbon content. They commonly display thin layers consisting of alternating bright and dull bands (Fig. 8.23). **Subbituminous coal** has properties intermediate between those of lignite and bituminous coal. **Anthracite** is a hard, black, dense coal commonly containing more than 90 percent carbon. It is a bright, shiny rock that breaks with conchoidal fracture, such as the fractures in broken glass. Bituminous coals and anthracite are largely of Mississippian and Pennsylvanian (Carboniferous) age. **Cannel coal** and **boghead coal** are nonbanded, dull, black coals that also break with conchoidal fracture; however, they have bituminous rank and much higher volatile content than anthracite. Cannel coal is composed mainly of spores. Boghead coals are composed dominantly of nonspore algal remains. **Bone coal** is very impure coal containing high ash content.

Coals are also classified on the basis of megascopic textural appearance and recognizable petrographic or microscopic constituents. Stopes (1919) recognized four types of coal, now called lithotypes, on the basis of megascopic appearance. These lithotypes, which comprise millimeter-thick bands or layers of humic coal, are described in Table 8.10. Examples of these coal lithotypes are illustrated in Figure 8.24.

Under the microscope, coal can be seen to consist of several kinds of organic units that are single fragments of plant debris or, in some cases, are fragments consisting of more than one type of plant tissue. Stopes (1935) suggested the name maceral for these organic units as a parallel word for the term mineral used for the constituents of inorganic rocks. The starting materials for macerals are woody tissues, bark, fungi,

TABLE 8.9   Classification of coal on the basis of rank

Class (rank steps)	Fixed carbon limits (wt. percent), dry, mineral- and matter-free basis	Volatile matter (wt. percent), dry, mineral- and matter-free basis	Calorific value limits (Btu/lb), moist, mineral- and matter-free basis
Anthracite	86–98	2–14	—
Bituminous	69–86	22–>31	10,500–14,000
Subbituminous	<69	>31	8,300–10,500
Lignite	<69	>31	6,300–8,300
Peat	low	high	low

Source: Data from American Society for Testing Materials (ASTM), 1981, Annual book of ASTM standards, Part 26, American Society for Testing Materials.

Note: Volatile matter is that part of coal that burns as a gas, mainly hydrogen.

FIGURE 8.23   Layered and banded bituminous coal, Cedar Grove Seam (Pennsylvanian), Logan County, West Virginia. The thickness of the coal seam is about 2.3 m. (Photograph courtesy of Island Creek Coal Company.)

spores, and so on; however, these materials are not always recognizable in coals. Macerals are divided into three major groups: vitrinite, inertinite, and liptinite (Table 8.10).

Coal macerals are identified on the basis of several characteristics:

1. reflectivity—the extent to which they reflect light
2. degree of anisotopy (differences in reflectivity in different directions within a maceral) or isotopy as viewed under a petrographic microscope
3. presence or absence of fluorescence when irradiated with blue (ultraviolet) light
4. morphology
5. relief
6. size

Study of macerals is referred to as coal petrology (e.g., Bustin et al., 1985; Ward, 1984).

Origin. Coals occur in rocks ranging in age from Precambrian (algal coal) to Tertiary, and peat analogs of coal are present in Quaternary sediments. Coals originate in climates that promote plant growth under depositional conditions that favor preservation of organic matter. Although ancient coals accumulated at all latitudes, from the equator to polar regions, most were deposited in midlatitudes (McCabe, 1984). For coal to be preserved, the rate of accumulation of organic matter must exceed the rate of decomposition owing to microbial and chemical processes. Accumulating organic matter is most likely to be preserved in depositional environments where oxidation of organic matter is inhibited owing to rapid burial in swampy areas in which the water table is close to the peat surface. For thick coal deposits to form, these conditions must

TABLE 8.10 Principal coal lithotypes and macerals

Lithotypes

  Vitrain—brilliant, glossy, vitreous, black coal, bands 3–5 mm thick; breaks with a conchoidal fracture; clean to touch.

  Clarin—smooth fracture with pronounced gloss; dull intercalations or striations; small-scale sublaminations within layers give surface a silky luster; the most common macroscopic constituent of humic coals.

  Durain—occurs in bands a few cm thick; firm, somewhat granular texture; broken surfaces have a fine lumpy or matte texture; characterized by lack of luster, gray to brownish black color, and earthy appearance.

  Fusain—soft, black; resembles common charcoal; occurs chiefly as irregular wedges; friable and porous if not mineralized.

Macerals

  Vitrinites—originated as wood or bark; a major humic constituent of bright coals. Subtypes:

    Collinite—structureless or nearly structureless; commonly occurs as a matrix or impregnating material for fragments of other macerals.

    Tellinite—derived from cell-wall material of bark and wood and preserves some of the celluar texture.

  Inertinites—composed of woody tissues, fungal remains, or fine organic debris of uncertain origin; relatively high carbon content. Subtypes:

    Fusunite—cell structures composed of carbonized or oxidized cell walls and hollow lumens (the space bounded by the wall of an organ) that are commonly mineral filled; characteristic of fusain.

    Semifusinite—a transitional state between fusinite and vitrinite.

    Schlerotinite—composed of the remains of fungal schlerotia (a hardened mass of tubular filaments or threads) or altered resins; characterized by oval shape and varying size.

    Micronite (<10 μm) and **macronite** (10–100 μm)—structureless, opaque, granular macerals derived from fine-grained organic detritus.

    Inertodetrinite—finely divided, structureless, clastic form of inertinite in which fragments of various kinds of inertinite macerals occur as dispersed particles.

  Liptonites (exinites)—originate from spores, cuticles, resins, and algae; can be recognized from shapes and structures unless original constituents are compacted and squashed. Subtypes:

    Sporinite—composed of the remains of yellow, translucent bodies (spore exines) that are commonly flattened parallel to bedding.

    Cutinite—formed from macerated fragments of cuticles (layers covering the outer wall of a plant's epidermal cells).

    Resinite—the remains of plant resins and waxes; occur as isolated rounded to oval or spindle-shaped, reddish, translucent bodies, or as diffuse impregnations, or as fillings in cell cavities.

    Alginite—composed of the remains of algal bodies; serrated, oval shape; characteristic of boghead coal.

---

last for a geologically long period of time. Although land plants were moderately well established by Devonian time, swampy environments large enough to form major coal deposits have existed only since the carboniferous (Mississippian and Pennsylvanian). Since that time, only the Triassic Period appears to have been a time when coal-forming processes were at a minimum.

Compaction and volatile loss accompanying deep burial result in thinning of coal beds by as much as 30 to 1 (Ryer and Langer, 1980); that is, 30 m of original peat may produce only 1 m of coal. The rank of coal tends to increase with depth owing to increase in temperature with depth. The formation of anthracite, for example, requires temperatures in excess of about 200°C (Daniels et al., 1990). Coals occur predominantly in siliciclastic depositional sequences, although thin limestones may be associated with some coals.

Oil Shale (Kerogen Shale)

The term **oil shale** is applied to fine-grained sedimentary rocks from which substantial quantities of oil can be derived by heating. The term is actually a misnomer because relatively little free oil occurs in these rocks, although small blebs, pockets, or veins of

FIGURE 8.24 Bituminous coal showing examples of three different lithotypes: V, vitrain; C, clarain; and D, durain. The small divisions on the scale equal 1 cm. (From Bustin, R. M., et al., 1985, Coal petrology, its principles, methods, and applications: Geol. Assoc. Canada Short Course Notes, v. 3. Pl. 6A, p. 51, reprinted by permission.)

asphaltic bitumins may be present. Oil shales are of particular interest because of their potential to generate oil when refined into fuel at sufficiently high temperatures. At least 50 countries of the world have reserves of oil shale that have the potential to be exploited as a fuel resource in the future. More than 80 percent of the organic matter in oil shales is present in the form of kerogen, which yields oil when heated to a temperature of about 350°C. The principal constituents of oil shales are shown in Figure 8.25. Organic constituents commonly do not exceed about 25 percent of the rock. Kerogen is

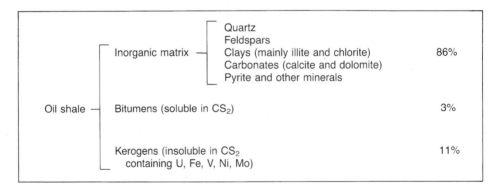

FIGURE 8.25 Principal constituents in oil shales. (After Yen, T. F., and G. V. Chilingarian, 1976, Introduction to oil shales, *in* T. F. Yen and G. V. Chilingarian (eds.), Oil shales: Developments in petroleum science 5. Fig. 1.2, p. 3, reprinted by permission of Elsevier Science Publishers, Amsterdam.)

disseminated organic matter that is insoluble in nonoxidizing acids, bases, or organic solvents. It consists of masses of almost completely macerated (disintegrated by biochemical/chemical processes) organic debris, which consists chiefly of plant remains such as algae, spores, spore coats, pollen, resins, and waxes. On the basis of the type of organic remains from which it was derived, kerogen is classified into five principal types (Hunt, 1979):

1. **algal**—composed dominantly of the remains of algae
2. **amorphous**—composed largely of sapropelic organic matter from plankton and other low forms of life
3. **herbaceous**—composed of pollen, spores, cuticles, and so on
4. **woody**
5. **coaly** (inertinite)

Not all so-called oil shales are actually shales. Some are organic-rich siltstones, limestones, and impure coals. Three basic types are recognized. **Carbonate-rich oil shales** are those in which the principal nonkerogen constituents are calcite, dolomite, ankerite, and various amounts of siliciclastic silt. They are generally hard, tough, and resistant to weathering (Duncan, 1976). **Silica-rich oil shales** are shales in which the main constituents apart from kerogen are fine-grained quartz, feldspar, and clay minerals. They also contain chert, opal, and phosphatic nodules. Siliceous oil shales are generally dark brown or black and are less resistant to weathering than the carbonate-rich shales. **Cannel shale** is an oil shale that consists predominantly of organic matter that completely encloses other mineral grains. Cannel shales are sometimes classified as impure cannel coals and are referred to as **torbanites.** Many oil shales are characterized by distinct lamination caused by alternations of millimeter-thick organic laminae, which are either siliciclastic or carbonate laminae. The amount of oil that can be extracted from oil shales through heating and retorting ranges between 10 to 150 gallons of oil per ton of rock (Duncan, 1976). The potential world supply of oil from oil shale is estimated to be 2000 trillion tons (Russell, 1990, p. 4). On the other hand, many technological problems exist with respect to mining, extracting and refining oil shale. Oil shale is not now being processed commercially because it cannot compete economically with petroleum.

Oil shales form in environments where organic matter is abundant and anaerobic or reducing conditions prevent oxidation and total bacterial decomposition. They are deposited in both lacustrine (lake) and marine environments where the above conditions are met. The principal environments are as follows:

1. large lakes
2. shallow seas or continental platforms and continental shelves in areas where water circulation was restricted and reducing or weakly oxidizing conditions existed
3. small lakes, bogs, and lagoons associated with coal-producing swamps

Oil shales formed in lakes or swamps may be associated with impure cannel or boghead-type coal, tuffs and other volcanic rocks, or even evaporites. Many oil shales deposited in large lakes are carbonate-rich types and tend to have high oil yields, apparently owing to enhanced preservation potential of organic material in lake environments. Oil shales deposited in marine environments are characteristically the silica-rich type and have lower oil yields, although some Tertiary and Mesozoic age siliceous oil shales have rich oil yields. Oil shales extend over wide geographic areas and are commonly associated with limestones, cherts, sandstones, and phosphatic deposits (Yen and Chilingarian, 1976b).

Petroleum and Solid Bitumins

Petroleum. Petroleum is not sedimentary rock but a carbon-rich organic substance that occurs as liquid and gas accumulations predominantly in sandstones and carbonate rocks. For this reason, it is included here with the carbonaceous sedimentary rocks. Petroleum forms from plant and animal organic matter by a complex maturation process during burial that involves initial microbial alteration and subsequent thermal alteration and cracking (Tissot and Welte, 1984, p. 69). The source materials for petroleum are contained primarily in organic-rich shales and carbonate rocks. After petroleum has formed from organic source materials at substantial burial depths, it migrates out of the fine-grained source rocks into coarser-grained, porous, and permeable sandstone or carbonate reservoir rocks, where it eventually accumulates in traps such as anticlines (Fig. 8.26).

Petroleum is composed dominantly of carbon (about 84 weight percent) and hydrogen (about 13 percent). It also contains an average of about 1.5 percent sulfur, 0.5 percent nitrogen, and 0.5 percent oxygen (Hunt, 1979). Despite its simple elemental chemical composition, the molecular structure of petroleum can be exceedingly complex. The molecules in petroleum range from the simple methane gas molecule (CH_4)

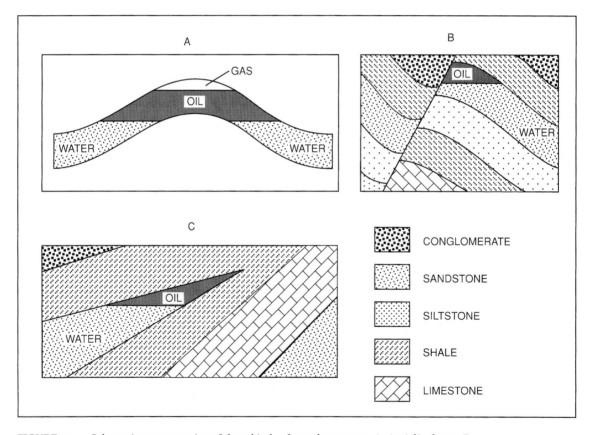

FIGURE 8.26 Schematic representation of three kinds of petroleum traps: A. Anticlinal trap. B. Fault trap, with an impermeable fault gouge or mineral seal along the fault. C. Stratigraphic (pinch-out) trap.

with a molecular weight of 16 to molecules with molecular weights in the thousands. Several hundred different hydrocarbons have been recorded in natural crude oils; however, all hydrocarbons can be grouped into a few basic classes or series having common molecular structural form. These structural forms are complex and are not explained in detail here, but the main hydrocarbon series are as follows:

1. **paraffins (alkanes)**—open chain molecules with single covalent bonds between carbon atoms (Fig. 8.27)
2. **napthenes (cycloparaffins)**—closed ring molecules with single covalent bonds between carbon atoms (Fig. 8.28)
3. **aromatics (arenes)**—one or more benzene ring structures with double covalent bonds between some carbon atoms (Fig. 8.29)

Most natural gases as well as many liquid petroleums belong to the paraffin series of hydrocarbons. Most napthene hydrocarbons are liquid petroleums, although two occur as gases at normal temperatures. The aromatics, which are named for their strong aromatic odor, are liquid petroleums. They commonly make up only a small percentage of the petroleums in natural crude oils.

Solid Bitumins. These substances are hydrocarbons such as natural asphalts and mineral waxes that occur in a semisolid or solid state. Most solid hydrocarbons probably formed from liquid petroleums that were subjected to loss of volatiles, oxidation, and biologic degradation after seepage to the surface. Others may never have existed as light oils. Solid bitumins occur as seepages, surface accumulations, and impregnations occupying the pore spaces of sandstones or other sedimentary rock (e.g., the Cretaceous Athabasca tar sands of Canada) and in veins and dikes. They are black or dark brown and have a characteristic odor of pitch or paraffin.

Solid hydrocarbons have roughly the same elemental chemical composition as liquid petroleum, but the percentage of carbon and hydrogen tends to be somewhat lower and the content of sulfur, nitrogen, and oxygen somewhat higher. They are divided into four main varieties or series on the basis of fusibility (melting temperature) and solubility in carbon disulfide (CS_2), an organic solvent (Fig. 8.30): (1) asphalts, (2) asphaltites, (3) pyrobitumins, and (4) native mineral waxes.

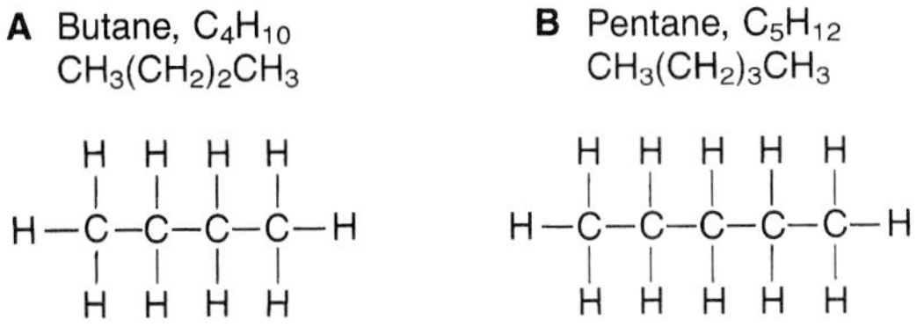

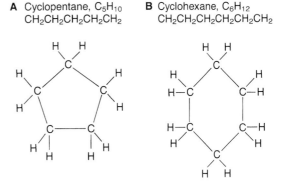

A Butane, C_4H_{10}
$CH_3(CH_2)_2CH_3$

B Pentane, C_5H_{12}
$CH_3(CH_2)_3CH_3$

A Cyclopentane, C_5H_{10}
$CH_2CH_2CH_2CH_2CH_2$

B Cyclohexane, C_6H_{12}
$CH_2CH_2CH_2CH_2CH_2CH_2$

FIGURE 8.27   Schematic structure of paraffin hydrocarbons having the general formula C_nH_{2n+2} where n refers to the number of carbon or hydrogen atoms. A. Butane. B. Pentane.

FIGURE 8.28   Schematic structure of naphthene (cycloparaffin) hydrocarbons having the general formula C_nH_{2n}. A. Cyclopentane. B. Cyclohexane.

FIGURE 8.29 Schematic structure of aromatic hydrocarbons having the general formula C_nH_{2n-6}. A. Benzene. B. Toluene.

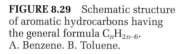

A Benzene, C_6H_6

B Toluene, $C_6H_5CH_3$

Asphalts are soft, semisolid bitumins that occur as seeps, surface pools, or viscous impregnations in sediments (tar sands). They are dark colored, plastic to fairly hard, easily fusible, and soluble in carbon disulfide. Varietal names for asphalts from different areas are shown in Figure 8.30. Asphalts are commonly associated with active oil seeps.

Asphaltites occur primarily in dikes and veins that cut sediment beds. They are harder and denser than asphalts and melt at higher temperatures. They are largely soluble in carbon disulfide. Names applied to varieties of asphaltites that differ slightly in density, fusibility, and solubility are **gilsonite, glance pitch,** and **grahamite.**

FIGURE 8.30 Terminology of principal kinds of naturally occurring solid hydrocarbons. (From Petroleum geochemistry and geology, by J. M. Hunt. W. H. Freeman and Company, © 1979, Fig. 8.28, p. 400.)

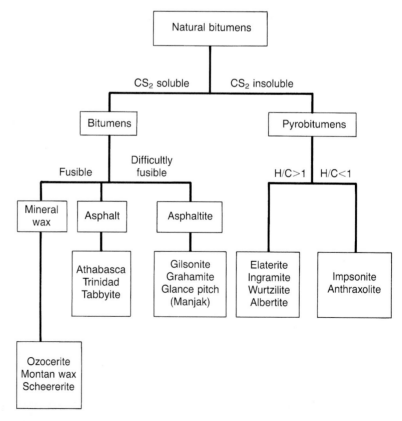

Pyrobitumins occur in dikes and veins like asphaltites but are infusible and largely insoluble in carbon disulfide. Several varieties of pyrobitumins are recognized. Softer forms include **elaterite,** a soft, elastic substance rather like India rubber, and **wurtzlite.** More indurated forms are **albertite,** a black, solid bitumin with a brilliant, jetlike luster and conchoidal fracture; **ingramite;** and the metamorphosed pyrobitumins **impsonite** and **anthraxolite.**

Native mineral waxes are solid, waxy, light-colored substances that consist largely of paraffinic hydrocarbons of high molecular weight. They represent the residuum of high-wax oils exposed at the surface. The most important native mineral wax is **ozocerite,** which consists of veinlike deposits of greenish or brown wax. **Montan wax** is an extract obtained from some kinds of brown coals or lignites.

The solid hydrocarbons are of interest to geologists because their presence at the surface is an indication of petroleum at depth in a region and because study of their occurrence may help to solve the problems related to the origin and alteration of petroleum. Also, many of the solid hydrocarbons are of commercial value themselves. Details of the geochemistry, origin, distribution, and exploitation of bitumin and other solid hydrocarbons are discussed in Chilingarian and Yen (1978) and Hunt (1979, p. 398–404).

FURTHER READINGS

Evaporites

Borchert, H., and R. O. Muir, 1964, Salt deposits: The origin, metamorphism and deformation of evaporites: Van Nostrand, New York, 338 p.

Braitsch, O., 1971, Salt deposits: Their origin and composition: Springer-Verlag, New York, 279 p.

Chambre Syndical de la Recherche et de la Production due Pétrole et du Gaz Naturel (eds.), 1981, Evaporite deposits: Illustrations and interpretation of some environmental sequences: Editions Technip, Paris, and Gulf Publishing, Houston, 284 p.

Dean, W. E., and B. C. Schreiber, 1978, Marine evaporites: Soc. Econ. Paleontologists and Mineralogists Short Course Notes No. 4, Tulsa, Okla., 193 p.

Kirkland, D. W., and R. Evans (eds.), 1973, Marine evaporites: Origin, diagenesis, and geochemistry: Dowden, Hutchinson and Ross, Stroudsburg, Pa., 444 p.

Melvin, J. L. (ed.), 1991, Evaporites, petroleum and mineral resources: Elsevier, Amsterdam, 555 p.

Schreiber, B. C. (ed.), 1988, Evaporites and hydrocarbons: Columbia University Press, New York, 475 p.

Sonnenfeld, P., 1984, Brines and evaporites: Academic Press, London, 624 p.

Warren, J. K., 1989, Evaporite sedimentology: Prentice-Hall, Englewood Cliffs, N.J., 285 p.

Siliceous Sedimentary Rocks

Aston, S. R. (ed.), 1983, Silicon geochemistry and biochemistry: Academic Press, London, 248 p.

Calvert, S. E., 1974, Deposition and diagenesis of silica in marine sediments; *in* K. J. Hsü and H. C. Jenkyns (eds.), Pelagic sediments: On land and under the sea: Internat. Assoc. Sedimentologists Spec. Pub. 1, p. 273–300.

Cressman, E. R., 1962, Nondetrital siliceous sediments: U.S. Geol. Survey Prof. Paper 440–T, 22 p.

Garrison, R. E., R. B. Douglas, K. E. Pisciotta, C. M. Isaacs, and J. C. Ingle (ed.), 1981, The Monterey Formation and related siliceous rocks of California: Pacific Section, Soc. Econ. Paleontologists and Mineralogists Spec. Pub. 1, Tulsa, Okla., 327 p.

Heath, G. R., 1974, Dissolved silica and deep-sea sediments, *in* W. W. Hay (ed.), Studies in paleooceanography: Soc. Econ. Paleontologists and Mineralogists Spec. Paper No. 20, p. 77–94.

Hein, J. R. (ed.), 1987, Siliceous sedimentary rock–hosted ores and petroleum: Van Nostrand Reinhold, New York, 304 p.

Hein, J. R., and J. Obradović (eds.), 1989, Siliceous deposits of the Tethys and Pacific regions: Springer-Verlag, New York, 244 p.

Iijima, A., J. R. Hein, and R. Siever (eds.), 1983, Siliceous deposits in the Pacific region: Elsevier, Amsterdam, 472 p.

Iller, R. K., 1979, Chemistry of silica: John Wiley & Sons, New York, 866 p.

Ireland, H. A., 1959, Silica in sediments: Soc. Econ. Paleontologists and Mineralogists, Spec. Pub. 7, Tulsa, Okla., 185 p.

McBride, E. F. (ed.), 1979, Silica in sediments: Nodular and bedded chert: Soc. Econ. Paleontologists and Mineralogists Reprint Ser. No. 8, Tulsa, Okla., 184 p.

Van der Linder, G. J. (ed.), 1977, Diagenesis of deep-sea biogenic sediments: Benchmark Papers in Geology,

v. 40, Dowden, Hutchinson and Ross, Stroudsburg, Pa., 385 p.

Iron-rich Sedimentary Rocks

Appel, P. W. U., and G. L. LaBerge, 1987, Precambrian iron-formations: Theophrastus, S.A., Athens, Greece, 674 p.

Dimroth, E., 1976, Aspects of the sedimentary petrology of cherty iron-formation, in K. H. Wolf (ed), Handbook of strata-bound ore deposits, v. 7: Elsevier, New York, p. 203–254.

Eichler, J., 1976, Origin of the Precambrian iron-formation, in K. H. Wolf (ed.), Handbook of strata-bound and stratiform ore deposits, v. 7: Elsevier, New York, p. 157–202.

James, H. L., 1966, Chemistry of the iron-rich sedimentary rocks: Data of geochemistry, 6th ed., U.S. Geol. Survey Prof. Paper 440–W, 61 p.

James, H. L., and P. K. Sims (eds.), 1973, Precambrian iron formations of the world: Econ. Geology, v. 68, p. 913–1179.

Lepp, H. (ed.), 1975, Geochemistry of iron: Benchmark Papers in Geology, v. 18: Dowden, Hutchinson and Ross, Stroudsburg, Pa., 464 p.

Melnik, Y. P., 1982, Precambrian banded iron-formations: Developments in Precambrian Geology 5: Elsevier, Amsterdam, 310 p. (Trans. from Russian by Dorothy B. Vitaliano)

Trendall, A. F., and R. C. Morris (eds.), 1983, Iron-formation facts and problems: Developments in Precambrian Geology 6: Elsevier, Amsterdam, 558 p.

Van Houten, F. B., and D. P. Bhattacharyya, 1982, Phanerozoic oolitic ironstone: Geologic record and facies models: Ann. Rev. Earth and Planetary Sci., v. 10, p. 441–457.

Young, T. P., and W. E. G. Taylor (eds.), 1989, Phanerozoic ironstones: Geol. Soc. Spec. Pub. 46, The Geological Society, London, 251 p.

Phosphorites

Baturin, G. N., 1982, Phosphorites on the sea floor: Origin, composition, and distribution: Developments in Sedimentology 33, Elsevier, Amsterdam, 343 p. (Trans. from Russian by Dorothy B. Vitaliano)

Bentor, Y. K. (ed.), 1980, Marine phosphorites—Geochemistry, occurrence, genesis: Soc. Econ. Paleontologists and Mineralogists Spec. Pub. No. 29, Tulsa, Okla., 249 p.

Burnett, W. C., and P. N. Froelich (eds.), 1988, The origin of marine phosphorites: The results of the R.V. *Robert D. Conrad* Cruise 23–06 to the Peru shelf: Special issue of Marine Geology, v. 80, p. 181–346.

Burnett, W. C., and S. R. Riggs (eds.), 1990, Phosphate deposits of the world, v. 3: Neogene to Modern phosphorites: Cambridge University Press, 464 p.

Cook, P. J., and J. H. Shergold (eds.), 1986, Phosphate deposits of the world, v. 1: Proterozoic and Cambrian phosphorites: Cambridge University Press, 386 p.

Journal Geological Society (London), 1980, v. 136, pt. 6 (an issue devoted to phosphatic and glauconitic sediments), p. 657–805.

Notholt, A. J. G., R. P. Sheldon, and D. F. Davidson (eds.), 1989, Phosphate deposits of the world, v. 2: Phosphate rock resources: Cambridge University Press, 566 p.

Nriagu, J. O., and P. B. Moore (eds.), 1984, Phosphate minerals: Springer-Verlag, New York, 434 p.

Slansky, M., 1986, Geology of sedimentary phosphates: North Oxford, Essex, Great Britain, 210 p.

Carbonaceous Sedimentary Rocks

Bustin, R. M., A. R. Cameron, D. A. Grieve, and W. D. Kalkreuth, 1985, Coal petrology, its principles, methods, and applications: Geol. Assoc. Canada Short Course Notes, v. 3, 230 p.

Chilingarian, G. V., and T. F. Yen, 1978, Bitumins, asphalts and tar sands: Elsevier, New York, 331 p.

Crelling, J. C., and R. Dutcher, 1980, Principles and applications of coal petrology: Soc. Econ. Paleontologists and Mineralogists Short Course Notes No. 8, Tulsa, Okla., 127 p.

Hunt, J. M., 1979, Petroleum, geochemistry and geology: W. H. Freeman, San Francisco, 617 p.

International Committee for Coal Petrology, 1963, International handbook of coal petrology, 2nd ed.: Centre National de la Recherche Scientifique, Paris. Supplements published in 1971, 1975.

Meyers, R. A. (ed.), 1982, Coal structure: Academic Press, New York, 340 p.

Petrakis L., and D. W. Grandy, 1980, Coal analysis, characterization and petrography: Jour. Chem. Ed., v. 57, p. 689–694.

Rahmani, R. A., and R. M. Flores (eds.), 1984, Sedimentology of coal and coal-bearing sequences: Internat. Assoc. of Sedimentologists Spec. Pub. 7, Blackwell, Oxford, 412 p.

Russell, P. L., 1990, Oil shales of the world: Their origin, occurrence and exploitation: Pergamon, Oxford, 736 p.

Stach, E., M.-Th. Mackowsky, M. Teichmüller, G. H. Taylor, D. Chandra, and R. Teichmüller, 1982, Handbook of coal petrology, 3rd ed.: Gebrüder Borntraeger, Berlin-Stuttgart, 535 p.

Tissot, B. P., and D. H. Welte, 1984, Petroleum formation and occurrence, 2nd ed.: Springer-Verlag, Berlin, 699 p.

Ward, C. R. (ed.) 1984, Coal geology and coal technology: Blackwell, Melbourne, 345 p.

Yen, T. F., and G. V. Chilingarian (eds.), 1976, Oil shales: Elsevier, New York, 292 p.

PART 5

SEDIMENTARY ENVIRONMENTS

Rippled sand dunes in central Utah. The Henry Mountains are in the background.

The characteristic properties of sedimentary rocks are generated through the combined action of the various physical, chemical, and biological processes that make up the sedimentary cycle. Weathering, erosion, sediment transport, deposition, and diagenesis all leave their impress in some way on the final sedimentary rock product. The sedimentary processes and conditions that collectively constitute the depositional environment play the primary role in determining the textures, structures, bedding features, and stratigraphic characteristics of sedimentary rocks. This close genetic relationship between depositional process and rock properties provides a potentially powerful tool for interpreting ancient depositional environments. If geologists can find ways to relate specific rock properties to particular depositional processes and conditions, they can work backwards to infer the ancient depositional processes and environmental conditions that created these particular rock properties.

Unfortunately, we can never know the exact nature of depositional processes and conditions that operated in the past. We are forced to look to the rocks themselves for clues to these conditions long after the processes that produced the rocks ceased to operate. Therefore, relating depositional process to depositional product is not a simple procedure. We must turn to study of sediments and sedimentary processes in modern environments for help in understanding the link between sedimentary processes and sedimentary rock properties. Knowledge acquired through such study can in turn be applied to environmental interpretation of ancient sedimentary rocks.

This procedure of applying knowledge gained through study of modern sedimentary environments to interpretation of ancient depositional processes and settings is the essence of environmental analysis. It does, however, have limitations. For example, the distribution of lands and seas that we see today is not typical of much of the geologic past. Furthermore, the intensity of geologic processes has very likely varied at different times in the past as well as differing from the intensity at present. Also, some geologic events of the past were probably unique. Many environments may have existed in the past that are not available for study today (Reineck and Singh, 1980). Therefore, geologists must be careful in the interpretation of ancient depositional environments not to be guided too rigidly by environmental models based on modern conditions.

Study of ancient depositional environments is important because the insight gained through environmental analysis allows us to reconstruct the paleogeography of previous geologic periods, that is, the relationship of ancient lands and seas. It also helps us to develop a proper understanding of sedimentary processes in the history of Earth and an improved ability to interpret complex stratigraphic relationships such as lateral and vertical variations in lithology and texture. Furthermore, thorough understanding of depositional environments is an essential factor in evaluating the economic significance of sedimentary rocks—their potential as reservoir rocks and source rocks for petroleum, for example.

Chapter 9 gives a brief discussion of the fundamental tools of environmental analysis and the methods geologists use to recognize and identify ancient depositional environments. Chapters 10 through 13 deal respectively with continental, marginal-marine, and marine depositional settings and the sedimentary rocks deposited in these settings.

9

Principles of Environmental
Interpretation and Classification

9.1 CONCEPT OF ENVIRONMENT

Depositional environments have been defined in various ways (e.g., Gould, 1972, p. 1; Krumbein and Sloss, 1963, p. 234; Potter, 1967); however, all definitions of environment have in common an emphasis on the physical, chemical, and biological conditions of the environment. A depositional environment is thus characterized by a particular geomorphic setting in which a particular set of physical, chemical, and biological processes operates to generate a certain kind of sedimentary deposit. For example, a beach can be considered a geomorphic unit of specified size and shape on which specific physical and biologic processes take place to produce a body of beach sand characterized by a particular geometry, sedimentary textures, structures, mineralogy, and biogenic remains (shells).

The physical environment is characterized by both **static** and **dynamic** elements. Static physical elements include basin geometry; depositional materials such as siliciclastic gravel, sand, and mud; water depth; and temperature. Dynamic physical elements include factors such as the energy and flow direction of wind, water, and ice; rainfall; and snowfall, all of which influence currents and waves. Chemical characteristics of the environment (salinity, pH, Eh, and carbon dioxide and oxygen content of waters) control chemical processes such as mineral precipitation and solution. The biological aspects of the environment can be considered to encompass both the activities of organisms (plant growth, burrowing, boring, sediment ingestion, and extraction of silica and calcium carbonate from water to form skeletal materials) and the presence of organic remains as depositional materials.

9.2   SEDIMENTARY PROCESSES AND PRODUCTS

It is important in the study of depositional environments to make a clear distinction between sedimentary environments and sedimentary facies. Each sedimentary environment is characterized by a particular suite of physical, chemical, and biological parameters that operate to produce a body of sediment characterized by specific textural, structural, and compositional properties. We refer to such distinctive bodies of sediment or sedimentary rock as **facies.** The term facies refers to stratigraphic units distinguished by lithologic, structural, and organic characteristics detectable in the field. A sedimentary facies is thus a unit of rock that, owing to deposition in a particular environment, has a characteristic set of properties. **Lithofacies** are distinguished by physical characteristics such as color, lithology, texture, and sedimentary structures. **Biofacies** are defined on the basis of paleontologic characteristics. The point emphasized here is that depositional environments generate sedimentary facies. The characteristic properties of the sedimentary facies are in turn a reflection of the conditions of the depositional environment. We return to the concept of facies in Chapter 14.

Sedimentary Process and Response

We can take a very simplistic and optimistic point of view and assume that a particular set of environmental conditions operating at a particular intensity will produce a sedimentary deposit with a unique set of properties that identify it as the product of that particular environment. Although it is probably not true that each environment produces a unique sedimentary product, the basis of environmental interpretation rests on the assumption that particular environments generate deposits that bear the impress of environmental processes and conditions to a degree sufficient to allow discrimination of the environment. This linked set of reactions between environments and facies is commonly referred to as **process** and **response** (Fig. 9.1).

As Figure 9.1 illustrates, the term process is used rather loosely to include both the dynamic and the static elements of the environment. Together, these process elements are responsible for generating a particular response, in the form of specific facies. When dealing with the depositional environments of ancient sedimentary rocks, geologists cannot, of course, observe the process elements of the environment. They have only the response element with which to work. Thus, the first step in environmental interpretation is always to characterize the facies in terms of specific physical, chemical, and biological properties. Geologists then attempt to work the process-response model backwards and infer the conditions of the ancient depositional environment. In other words, as Middleton (1978) puts it, "It is understood that (facies) will ultimately be given an environmental interpretation." Thus, environmental analysis must always begin with study of sedimentary facies. Only after the facies have been carefully and painstakingly analyzed and characterized can we make a reasonable interpretation of the depositional environment or environments in which these facies were formed.

Facies Associations

Environmental interpretation is commonly hampered by the fact that very similar facies can be produced in different environmental settings. It is often impossible to make a unique environmental interpretation on the basis of a single depositional facies. For example, cross-bedded sandstones can be formed by either wind or water

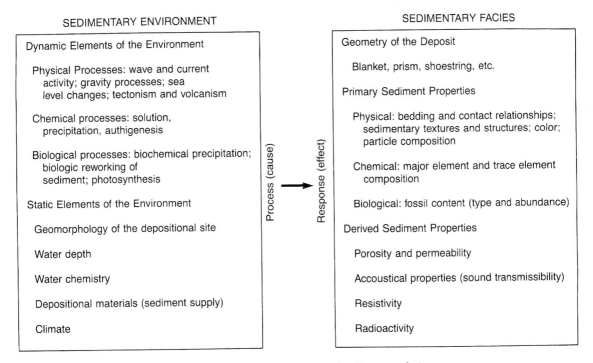

FIGURE 9.1   Relationship between sedimentary environments and sedimentary facies

transport. If deposited in water, they can originate on a beach, in a river or tidal channel, on a shallow-marine shelf, or in any other environment where traction transport occurs. Environmental interpretation is improved if we study facies associations and successions rather than individual facies. Facies associations can be thought of as groups of facies that occur together and are genetically or environmentally related. For example, if cross-bedded sandstones are closely associated with overlying or underlying peat, coal, or silty shale containing fossil roots, leaves, and stems, we can make an interpretation of deposition in a river system with some confidence. Such an interpretation might be very difficult to make on the basis of the cross-bedded sandstones alone. The key to environmental interpretation is to analyze all of the facies together, that is, to study the entire stratigraphic successions in which facies occur. The vertical succession and lateral variation of facies can contribute as much environmental information as the characteristics of the facies themselves.

In the study of facies associations and successions, careful attention must be given both to the nature of the contacts between facies and to the degree of randomness or nonrandomness of the successions themselves. By application of stratigraphic principles discussed in Chapter 14, we can infer that two facies separated by a gradational contact or boundary represent environments that were once laterally adjacent. On the other hand, facies separated by sharp or erosive boundaries may or may not represent environments that were laterally adjacent. In fact, facies overlying erosive contacts commonly indicate a significant change in depositional conditions and the beginning of a new cycle of sedimentation. The facies within a particular association of facies may be distributed vertically in an apparently random manner, or they may show a definite or preferred pattern of vertical change. Two common types of vertical facies changes are **coarsening-upward successions** and **fining-upward successions.** Coarsening-upward successions display an increase in grain size upward from a sharp or erosive base; fining-upward successions are those in which the succession becomes finer-grained upward to a sharp or erosive top. In general, fining-upward successions indicate a decrease in transporting power of currents during deposition, and coarsening-upward successions indicate an increase. Fining- and coarsening-upward successions should not be confused with graded bedding. Although such successions can occur on the scale of a single graded bed, they commonly involve many different beds that individually may not be graded. Each bed in the vertical succession is simply finer, or coarser, than the underlying bed. Figure 9.2 illustrates some typical vertical successions of facies in fluvial, deltaic, and shelf deposits. The significance of fining- and coarsening-upward successions is further discussed in appropriate parts of Chapters 10 through 13.

    In the study of facies associations, it may be possible to determine by visual inspection if facies are randomly or nonrandomly distributed; however, it is often necessary to resort to statistical techniques to detect whether or not one facies passes into another more often than would be predicted purely on a random basis. Complex statis-

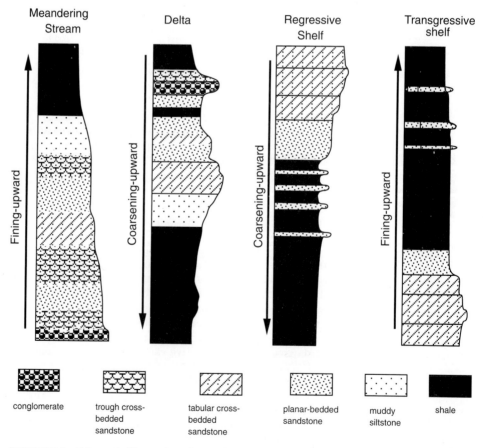

FIGURE 9.2   Schematic illustration of some fining-upward and coarsening-upward successions

tical methods involving Markov chain analysis are required to evaluate facies transitions and handle the large quantities of data involving different types of facies contacts and preferred relationships (Carr, 1982; Powers and Easterling, 1982; Harper, 1984).

9.3 BASIC TOOLS OF ENVIRONMENTAL ANALYSIS

Interpretation and reconstruction of ancient depositional environments depend upon identifying physical, chemical, and biological characteristics of sedimentary rocks that can be related to environmental parameters. The most important criteria for environmental recognition are listed in Table 9.1. No single criterion can generally be relied upon to provide unequivocal environmental interpretation; geologists commonly must make use of all available properties of the sedimentary rocks. Only when several inde-

TABLE 9.1 Criteria for recognition of ancient sedimentary environments

Criteria based on primary depositional properties
Mainly physical properties
  Geometry of facies units—useful only if very distinctive, e.g., ribbon shape of channels; lobate shape of deltaic deposits.
  Gross lithology and mineralogy of strata—a very general environmental indicator; e.g., fossiliferous limestone suggests shallow-marine shelf settings; coal indicates swampy environments; the mineral glauconite suggests marine conditions.
  Facies associations (stratigraphic successions)—e.g., fining-upward successions are characteristic of meandering-stream deposits; regressive shelf environments (shoreline advancing seaward with time) produce coarsening-upward successions.
  Sedimentary structures
    Nondirectional structures—not unique environmental indicators but suggest depositional process; e.g., ripples indicate current flow, graded bedding indicates settling of grains from suspension, mudcracks indicate subaerial exposure.
    Directional structures (paleocurrent indicators)—paleocurrent patterns may have environmental significance; e.g., bimodal patterns suggest tidal influence; unimodal patterns of high variability suggest meandering-stream environments.
  Sedimentary textures—grain-size data of limited usefulness; grain shape measured by Fourier analysis may be significant; grain orientation (e.g., imbrication) a useful paleocurrent indicator.
Mainly chemical properties
  Major-element composition—very limited usefulness.
  Trace-element composition—some application in paleosalinity interpretation; e.g., boron more abundant in marine shales than in freshwater shales.
  Isotope ratios—carbon and oxygen isotopes may be used to interpret marine vs. nonmarine conditions; oxygen isotopes a possible ocean paleotemperature indicator.
Many biologic properties
  Kinds of fossils and their ecologic characteristics—very useful indicators of salinity, temperature, depths, energy, and turbidity of ancient oceans; also an indicator of substrate type (rock, sand, mud).
  Types of trace fossils—water depth indicators.
Criteria based on derived sediment properties
Properties measured or interpreted from instrumental well logs
  Properties such as rock resistivity, velocity of sound transmission, and natural radioactivity can be measured in well bores and used, for example, to interpret coarsening- and fining-upward successions in subsurface strata.
Characteristics interpreted from seismic reflection records
  Seismic reflection characteristics identified from seismic records indicate features such as inclined bedding, truncations, and pinch-outs that have environmental significance.

pendent criteria yield the same interpretation can the environment be confidently assigned. Furthermore, because environmental interpretation can be severely hampered by diagenetic changes in sediments, geologists must be particularly careful to separate primary depositional features from postdepositional features caused by diagenesis.

The most important physical criteria for environmental interpretation include gross lithology (i.e., sandstone, shale, limestone, evaporites), facies associations, and sedimentary structures. In general, the chemical properties of sedimentary rocks are less important in environmental interpretation than are physical and biological properties; however, isotope and trace-element analyses have some application in paleosalinity and paleotemperature determinations. The fossils contained in sedimentary rocks are among the most useful of all environmental criteria because they provide information about water depth, energy, salinity, temperature, and turbidity. Instrumental well logs (discussed in Chapter 14) provide indirect information about such properties of subsurface sediments as porosity and lithology, which may have environmental significance. For example, well-log information can be used to interpret fining- and coarsening-upward successions in buried sedimentary rocks that cannot be studied by normal field techniques. Seismic reflection techniques (Chapter 15) provide information about subsurface rocks such as inclined bedding (e.g., in deltaic deposits) and truncation and pinch-out of beds that can be related to environmental conditions. Specific applications and further discussion of these various criteria of environmental analysis are given in appropriate parts of Chapters 10 through 13.

9.4   CLASSIFICATION OF DEPOSITIONAL ENVIRONMENTS

Most textbooks that discuss sedimentary environments include some kind of environmental classification, either formally stated in tabular form or otherwise provided by the organization of chapters and subheadings. We recognize three fundamental depositional environmental settings: (1) continental, (2) marginal-marine, and (3) marine. Each of these primary environmental realms has been further divided by different workers into three to five or more major environments plus numerous subenvironments. The most comprehensive listing of depositional environments is probably that of Crosby (1972), who compiled a list of 18 major environments and more than 50 subenvironments. Although such a detailed listing of modern depositional environments is useful in illustrating the wide variety of depositional conditions under which sedimentary rocks can accumulate, it is impractical as a workable guide to ancient depositional environments; we cannot discriminate ancient sedimentary environments on such a fine scale. We are fortunate indeed in some cases just to be able to recognize that an ancient sedimentary rock was deposited in a marine environment vs. a nonmarine environment. The most practical and useful kind of classification of ancient sedimentary environments is one that includes only a relatively small number of major environments and subenvironments, each of which is capable of being recognized and distinguished from other environments by the tools of environmental interpretation available to us.

Table 9.2 is a simplified classification that meets these general requirements; however, it is probably impossible to create a classification of depositional environments that is totally acceptable to all geologists. The choice of major environments and subenvironments shown in Table 9.2 may not be acceptable to all workers, some of whom may prefer to include more, fewer, or different environments. Also, the classification contains some inconsistencies and omissions. To illustrate, glacial environ-

TABLE 9.2 Simplified classification of ancient depositional environments

Primary depositional setting	Major environment	Subenvironment
Continental	*Fluvial	*Alluvial fan *Braided stream *Meandering stream
	*Desert Lacustrine *Glacial	
Marginal-marine	*Deltaic	*Delta plain *Delta front *Prodelta
	*Beach/barrier bar *Estuarine/lagoonal Tidal flat	
Marine	Neritic	Shelf **Organic reef
	Oceanic	Slope Deep-ocean floor

*Dominantly siliciclastic deposition

**Dominantly carbonate deposition

Note: Environments not marked by an asterisk(s) may be sites of siliciclastic carbonate, evaporite, or mixed sediment deposition depending upon depositional conditions.

ments may be the deposition sites of fluvial, eolian, and lake sediments as well as sediments transported and deposited directly by glaciers. Eolian deposits can form in back-beach, marginal-marine environments and on the tops of barrier islands, as well as in continental environments. Organic reefs may form in marginal-marine and possibly even freshwater environments, as well as in truly marine environments. Deltas can form in lakes in addition to shoreline or transitional environments, and turbidites can form in environments other than deep-marine. In any case, it is probably best not to get too involved in the nomenclature used in classifications because no simple classification can adequately account for all possible environments. Classifications are not particularly important in themselves, but they provide a convenient framework to which we can relate facies models, which are necessary for environmental interpretation.

9.5 FACIES MODELS

Few, if any, of the tools for environmental analysis can be used alone for interpreting depositional environments. It is not possible in most cases to examine a single property of sedimentary rocks—geometry or sedimentary structures, for example—and, on the basis of that property, to confidently deduce the depositional environment of the rock. To interpret the depositional environment of an ancient sedimentary rock, we must examine many different properties of the rock and then compare these properties to some mental picture we have of the properties of rocks deposited in known depositional environments. This mental picture constitutes an environmental model. Few of us have had enough personal field experience, have read enough books and papers, or have good enough memories to carry around a mental picture of every important depositional environment. Fortunately, we can draw on the experience of many geologists through their published data and ideas to construct facies models that will provide the reference framework we need for interpreting ancient depositional environments.

Definition of Models

A facies model is a general summary of a given depositional system, written in terms that make the summary usable (Walker, 1992). Facies models can be expressed as idealized successions of facies, as block diagrams, or as graphs and equations. Such summary models act as a norm for purposes of comparison and as a framework and guide for future observations. They also serve as a predictor of new geological situations, and they must act as an integrated basis for interpretation of the systems they represent. Facies models thus provide a method for simplifying, ordering, categorizing, and interpreting data that may otherwise seem random and confusing. They provide a means of distilling local details until only the pure essence of the environment remains. This process of distilling away local variations to arrive at a model that can serve as a norm and a predictor is illustrated in cartoon form in Figure 9.3. Facies models can be developed for each depositional system or environment. Once such models have been developed, we can use them as a frame of reference to which we can compare ancient sedimentary rocks. Each facies model can be described in terms of the properties that

FIGURE 9.3   Schematic illustration of the construction of facies models, using submarine fans as an example. (From Walker, R. G., 1992, Facies, facies models, and modern stratigraphic concepts, *in* R. G. Walker and N. P. James (eds.), Facies models, 2nd ed.: Geoscience Canada Reprint Ser. 1. Fig. 6, p. 7, reprinted by permission of Geological Association of Canada.)

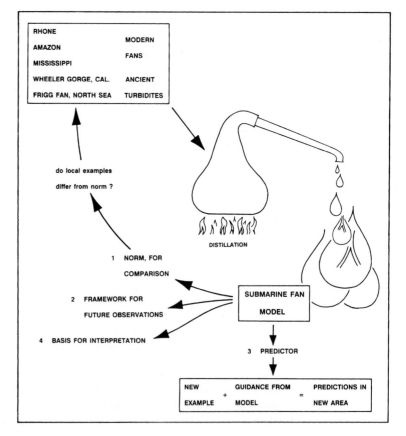

characterize the facies: geometry, sedimentary structures, sedimentary textures, particle and chemical composition, fossil content, vertical grain-size trends, and associated lithologic types. We can then use the model to infer the depositional environment.

Types of Facies Models

Models take many different forms, including descriptive, geometric, and mathematical or statistical types. Descriptive models are written summaries of the distinguishing characteristics of particular environments. Geometric models may consist of topographic maps, cross sections, three-dimensional block diagrams, and other forms that graphically illustrate the basic depositional framework. Four-dimensional geometric models that portray changes in erosion and deposition with time have also been utilized (e.g., Davis and Fox, 1972). Statistical models employ techniques such as multiple linear regression, trend-surface analysis, and factor analysis. Often, the objective of statistical models is to examine several environmental parameters simultaneously in order to predict the response of one element to another in a process-response model. The use of statistical techniques and computer-simulation models in environmental studies has the potential to yield results that in some cases may not be obtainable through less mathematically rigorous methods. Nonetheless, Hallam (1981, p. 12) warns against the danger of being "seduced away from significant geological problems by the beauty of the techniques" and of using "a sledgehammer to crack a nut" when much simpler approaches to modeling may produce equally good or better results and are generally better understood by the majority of geologists.

Construction and Use of Models

Facies models are commonly based on modern sedimentary environments. We assume that natural laws are constant in time and space and that we need not invoke hypothetical unknown processes to explain features of the rock record if presently observable processes and conditions can explain these features. Still, it is not possible in all cases to create facies models based on observation of sedimentation processes and products in modern environments. For example, no one has has ever actually observed a modern turbidity current in the ocean, and few examples of historical turbidites are known. In developing a model for turbidite environments, we have to use general knowledge of sedimentation processes in modern environments, supplemented with knowledge based on experimental laboratory investigations and studies of ancient examples of turbidites—examples identified as turbidites through the processes of reasoning rather than by comparison with modern deposits. Likewise, we cannot confidently use analogy with modern environments to develop models for carbonate sedimentation in shallow epicontinental seas that extended into the interior of continents—the Paleozoic epicontinental seas of the U.S. mid-continent region, for example—because we have no modern analogs of carbonate epicontinental seas on which to draw. Thus, it becomes necessary in some cases to turn to the ancient record for guidance in constructing facies models.

I interject a word of caution at this point about the use of facies models. Although such models are invaluable aids to the study of ancient depositional environments, too rigid adherence to a particular model can lead to problems in interpretation. As mentioned, many environments are complex and cannot be adequately represented by a simple model. Reading (1978a) points out that whereas many environments can be described by one general model, only by concentrating on differences rather than similarities can we evaluate the importance of the various processes that combine to make the facies in each environment. Common sense and good judgment must be exercised in the use of facies models, and we must be careful not to ignore or subordinate facts that do not fit in order to force interpretation of depositional environments on the basis of a particular model. The best sedimentologist is the one who has studied the most sediments, is familiar with the most facies models, and keeps an open mind when faced with data that do not fit preconceived concepts: "The conviction that one's own hypothesis is right is frequently the mark of one who is poorly informed about alternatives" (Reading, 1978a, p. 10).

The following three chapters discuss depositional environments in the continental, marginal-marine, and marine settings. Although not labeled specifically as models, facies models of various kinds are used throughout these chapters to summarize distinguishing characteristics of the principal sedimentary facies generated in these major sedimentary settings.

FURTHER READINGS

Anderton, R., 1985, Clastic facies models and facies analysis, *in* P. J. Brenchley and B. P. J. Williams (eds.), Sedimentology, recent developments and applied aspects: The Geological Society, Blackwell, Oxford, p. 31–48.

Cant, D. J., and F. J. Hein, 1987, Approaches to interpreting sedimentary environments: Soc. Econ. Paleontologists and Mineralogists Reprint Ser. 11, 259 p.

Gall, J. C., 1983, Ancient sedimentary environments and the habitats of living organisms: Springer-Verlag, Berlin, 219 p.

Hallam, A., 1981, Facies interpretation and the stratigraphic record: W. H. Freeman, San Francisco, 291 p.

Reading, H. G. (ed.), 1986, Sedimentary environments and facies, 2nd ed.: Elsevier, New York, 614 p.

Walker, R. G., and N. P. James (eds.), 1992, Facies models—Response to sea level change: Geol. Assoc. Canada, 409 p.

10

Continental Environments

10.1 INTRODUCTION

We turn now, following discussion of the principles of environmental interpretation and facies analysis in Chapter 9, to application of these principles in the study of **continental depositional systems.** Geologists recognize four major kinds of continental environments: fluvial (alluvial fans and rivers), desert, lacustrine (lake), and glacial (Table 9.2). Although treated in this book as separate depositional systems, similar kinds of sediments can be generated in more than one of these environments. For example, eolian (windblown) sediments can accumulate both in desert environments and in some parts of glacial environments. Lacustrine sediments form in lakes in any environment, including desert and glacial lakes. Fluvial sediments are deposited mainly in river systems of humid regions, but they form also in rivers within desert areas and glacial environments.

Facies deposited in continental environments are dominantly siliciclastic sediments characterized by general scarcity of fossils and complete absence of marine fossils. Nonsiliciclastic sediments such as freshwater limestones and evaporites occur also in continental environments, but they are distinctly subordinate to siliciclastic deposits. Continental sedimentary rocks are less abundant overall than are marine and marginal-marine sediments, but they nonetheless form an important part of the geologic record in some areas. Tertiary fluvial sediments of the Rocky Mountain–Great Plains region of the United States, Jurassic eolian sandstones of the Colorado Plateau, Tertiary lacustrine sediments (Green River Formation) of Wyoming and Colorado, and the late Paleozoic glacial deposits of South Africa and other parts of ancient Gondwanaland are all examples of continental deposits. Some continental sediments have economic significance. They may contain important quantities of natural gas and petroleum, coal, oil shale, and uranium. We now examine in turn each of the major continental environments.

10.2   FLUVIAL SYSTEMS

Fluvial deposits encompass a wide spectrum of sediments generated by the activities of rivers, streams, and associated sediment gravity-flow processes. Such deposits occur at the present time under a variety of climatic conditions and in various continental settings ranging from desert areas to humid and glacial regions. Although many subenvironments of the fluvial system can be recognized, most ancient fluvial deposits can be assigned to one of three broad environmental settings: alluvial fan, braided river, or meandering river. These fluvial environments may be interrelated and overlapping. To illustrate, braided rivers commonly occur on alluvial fans in humid regions and contribute to deposition of the fans; some humid alluvial fans merge downslope into braided-river systems; and some braided rivers are transformed downslope into meandering rivers. In the analysis of ancient fluvial systems, it may be extremely difficult to differentiate sediments from such overlapping environments.

Alluvial Fans

Depositional Setting

Alluvial fans are deposits with gross shape approximating a segment of a cone (Fig. 10.1). Modern alluvial fans form in areas of high relief, commonly at the base of a mountain range, where an abundant supply of sediment is available. They are particularly common in sparsely vegetated arid or semiarid regions where sediment transport occurs infrequently but with great violence during sudden cloudbursts. The alluvial fans in Death Valley, California, are good examples of modern arid fans. Individual arid or semiarid fans are typically small, commonly less than about 30 km across. In arid or semiarid settings, alluvial fans may pass downslope into desert-floor environments with internal drainage, including playa-lake environments. Alluvial fans occur also in humid regions, including proglacial settings where they develop as outwash fans in front of melting glaciers. In these humid regions, they may merge downslope with alluvial or deltaic plains and beaches or tidal flats, or they may even build into lakes or the ocean. Alluvial fans that build into standing bodies of water are called **fan deltas** (Chapter 11). Examples of modern humid fans include the giant Kosi Fan of Nepal and India, which is about 150 km from apex to toe, and the Scott Fan of Alaska, which is formed as an outwash fan from Scott Glacier. Along mountain fronts, alluvial fans developed in adjacent drainage systems may merge laterally to form an extensive piedmont, or bajada.

Sedimentary Processes and Deposits

Streamflow, debris flow, mudflow, and landsliding are all important processes in transport and deposition of alluvial-fan deposits. Streamflow processes are particularly important on humid fans and tend to dominate over other transport processes. Streamflow leads to deposition of three types of fan deposits. **Stream-channel sediments** accumulate within the channels of streams that debouch onto and flow upon fans. These deposits form long, narrow bodies consisting of the coarsest and most poorly sorted of the streamflow deposits (Nilsen, 1982). **Sheetflood deposits** are formed by surges of sediment-laden water that spread out from the end of a stream channel onto a fan. Deposition occurs because of widening of the flow into shallow bands or sheets, with concurrent decrease in water depth and velocity of flow (Bull, 1972). These processes result in sheetlike deposits of gravel, sand, or silt that tend to be well sorted

FIGURE 10.1 Aerial view of an alluvial fan at the mouth of a canyon in the steep east wall of Death Valley, California. The highway gives the scale. (Photograph by John S. Shelton.)

and that may be cross-bedded, laminated, or nearly structureless. **Sieve deposits** consist of coarse gravel lobes. They occur on fans where the source supplies relatively little sand, silt, and clay to the fan. Under these conditions, highly permeable gravel deposits are generated; they allow water to pass through rather than over the deposits, holding back only the coarser material. Streamflow in humid regions may be perennial; however, most sediment transport probably occurs as a result of periodic major flood events. As streams debouch onto areas of lower relief, they spread out, become more shallow, and lose their competence and capacity to carry sediment. Thus, sediment loads tend to be dumped quickly, causing stream channels to choke and the stream to shift sideways across the fan.

Debris flows and mudflows are more common on fans in arid or semiarid regions where rainfall is infrequent but violent, slopes are steep, and vegetation is sparse. Such flows occur also where abundant unconsolidated volcaniclastic or glacial sediment is available. The fans of arid regions may also include streamflow deposits, which may even be dominant in some arid fans. Streamflow in arid regions is markedly periodic, with intense sediment transport taking place during occasional floods and little transport during intervening periods. **Debris-flow deposits** (Chapter 3) are characteristically poorly sorted and lacking in sedimentary structures except possible reverse-graded bedding in their basal parts. They may contain blocks of various sizes, including large boulders, and they are typically impermeable and nonporous owing to their high content of muddy matrix. Debris flows commonly "freeze up" and stop flowing after relatively short distances of transport over lower slopes on the fan; however, some flows have been reported to travel distances of up to 24 km (15 mi) (Sharp and Nobles, 1953). **Mudflows** are similar to debris flows but consist mainly of sand-size and finer sediments. **Landslides** are commonly associated with debris flows, and in many cases landslide deposits form a source of sediment for the debris flows. Landslides may include rock falls, slumps, slides, and snow avalanches. Material transported by landslide processes can include mud, sand, boulders, blocks of bedrock, soil, vegetation, and an occasional automobile or house. Owing to the similarity of much of the material transported by landslides and debris flows, landslide deposits may be difficult to differentiate from debris-flow deposits in ancient alluvial fans.

Characteristics of Alluvial-Fan Sediments

Geometry. Alluvial fans are cone shaped to arcuate in plan view, with a well-developed system of sinuous to anastomosing distributary channels that cross the fan (Fig. 10.2). The radial surface profile, from fanhead to fantoe, is commonly concave upward, but the cross-fan profile is generally convex upward. In cross section, alluvial fans are typically either wedge shaped or lens shaped. If wedge shaped, the wedge may be thick near the mountain front and thin out away from the mountain. Alternatively, the wedge may be thin near the mountain front and get thicker away from it. In longitudinal (radial) cross profile, alluvial fans can be divided into three parts. The **upper fan,** also called the proximal fan or fanhead, has the steepest slope and coarsest sediment. Streamflow in the upper fan tends to be confined to a single channel, which may be entrenched as much as 20 to 30 m below the fan surface. Shifting of this channel can occur owing to clogging of the channel with streamflow or debris-flow deposits. The **midfan** is characterized by a gentler slope and sediment of intermediate size. A branching network of shallower channels typically feeds different parts of the midfan. The **distal fan,** or fanbase, makes up the toe of the fan and is distinguished by the gentlest slopes, finest sediment, and lack of well-defined channels (Fig. 10.2).

Textures, Structures, and Vertical Successions. The sediments of alluvial fans commonly show strong proximal-distal differences in grain size. The steep upper fan is characterized by coarse, extremely poorly sorted deposits with poorly developed sedimentary structures. These deposits consist largely of coarse-grained, matrix-rich conglomerates, mainly of debris-flow origin. Some clast-supported conglomerates may occur in stream channels. These proximal deposits grade downfan into somewhat finer-grained, thinner midfan deposits. Midfan deposits include both extensive streamflow sediments and debris-flow units. Sheetlike deposits of sands and gravels (formed by sheetflood and other processes) predominate and may display both planar and trough cross-bedding. Coarse-grained conglomerates are commonly present, including both debris-flow and channeled, streamflow conglomerates. Gravels in the streamflow conglomerates commonly show well-developed imbrication with the clasts dipping upfan. Distal fan sediments are largely sand and silt deposits of sheetflood origin, although thin conglomerate layers may be present. Channeled deposits are rare. Distal fan deposits tend to be better sorted than midfan and upper-fan deposits, and they may show low-angle cross-stratification and, in the more distal part of the fan, trough stratification (Figs. 10.3 and 10.4).

Many individual beds in alluvial fans display no detectable vertical grain-size trends; however, others may become either finer or coarser upward. Overall, alluvial-fan deposits tend to be characterized by strongly developed, thickening- and coarsening-upward successions, caused by active fan progradation or outbuilding. Nonethe-

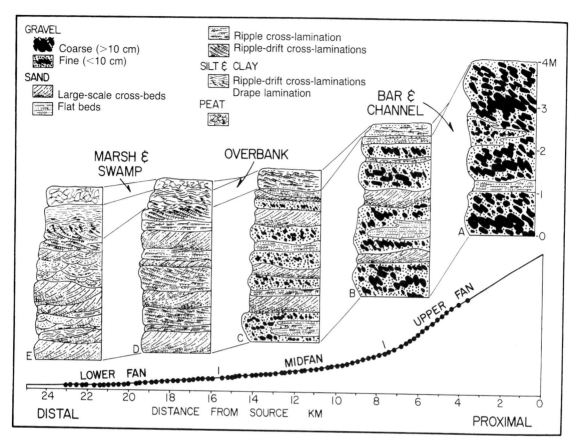

FIGURE 10.4 Downstream variations in facies and sedimentary structures in a glacial outwash fan. Bar and channel sequences do not fine upward but are capped by a finer overbank facies that becomes more important in a downfan direction. (From Boothroyd, J. C., and G. M. Ashley, 1975, Processes, bar morphology and sedimentary structures on braided outwash fans, northwest Gulf of Alaska, *in* A. V. Joplin and B. C. McDonald (eds.), Glaciofluvial and glaciolacustrine sedimentation: Soc. Econ. Paleontologists and Mineralogists Spec. Pub. 23. Fig. 25, p. 218, reprinted by permission of SEPM, Tulsa, Okla.)

ern North America, and the Neogene Ogallala Formation of Texas. They are commonly associated with lake or windblown deposits.

The Van Horn Sandstone of west Texas is a particularly well-studied example of a Precambrian wet-fan system. McGowen and Groat (1971) identify three distinct facies in this fan system (Fig. 10.3). The proximal fan or upper fan consists of massive conglomerate deposits, with boulders reaching meter size, representing channel streamflow deposition and debris flows. These coarse deposits grade downfan into alternating beds of conglomerates and thin, cross-stratified, pebbly sandstones, indicating a greater dominance of sheetflow deposition. The distal fan facies is characterized by thinner sedimentary units, an increasing proportion of trough to planar cross-beds, and increasing mudstone content.

Relatively few examples of arid or semiarid fans have been reported in the geologic record. The New Red Sandstone of Scotland and the Permian Peranera Formation

TABLE 10.1   Principal distinguishing characteristics of alluvial-fan deposits

Characteristic	Description
Texture	Sorting characteristically very poor; great range of grain sizes; clasts poorly rounded, reflecting short distance of transport; rapid downfan decrease in both average and maximum clast size.
Composition	Compositionally immature; commonly composed of a wide variety of clast types; mineralogy and clast types dependent upon source rocks; thus composition may change laterally in coalescing fans from different drainage systems.
Vertical successions	Individual beds possibly showing no vertical change in grain size or coarsening or thinning upward; overall, alluvial-fan successions displaying strong thickening- and coarsening-upward trend or thinning- and fining-upward trend, reflecting either progradation or retrogradation of the fan.
Sedimentary structures	Limited suites of sedimentary structures; mainly medium- to large-scale trough or planar cross-bedding and planar stratification; overall, poorly stratified; may display many laterally discontinuous sediment-filled channels and cut-and-fill structures, particularly in upper-fan deposits; paleocurrent patterns radiating outward downfan; complex radiating paleocurrent patterns in coalescing fans.
Geometry	Lobate in plan view; lenticular or wedge shaped in cross section; typically forming clastic wedges.
Other	Deposits commonly oxidized (red, brown, yellow); contain very little fine organic matter and few fossils except rare vertebrate bones and plant remains; generally composed of a mixture of streamflow and debris-flow deposits—may also contain sieve deposits (gravel lobes) that are unique to alluvial-fan environments.

Source: After Bull, W. B., 1972, Recognition of alluvial fan deposits in the stratigraphic record, *in* J. K. Rigby and W. K. Hamblin (eds.), Recognition of ancient sedimentary environments: Soc. Econ. Paleontologists and Mineralogists Spec. Pub. 16, p. 63–83; Nilsen, T. H., 1982, Alluvial fan deposits, *in* P. A. Scholle and D. Spearing (eds.), Sandstone depositional environments: Am. Assoc. Petroleum Geologists Mem. 31, p. 49–86.

of Spain, which abuts the Hercynian core of the Pyrenees Mountains, are redbed formations (colored red by iron oxides) interpreted to be ancient arid fans.

River Systems

Owing to extensive study by geologists, geomorphologists, and engineers, the processes that form river deposits and the characteristics of these deposits are reasonably well understood. On the basis of stream morphology, four types of rivers are recognized: braided, anastomosing, straight or almost straight, and meandering (Fig. 10.5). Of these types, straight rivers are uncommon, and anastomosing rivers can be regarded as a special type of meandering river that has a relatively permanent and stable system of high-sinuosity channels with cohesive banks and separated by large, stable, vegetation-covered islands (Miall, 1977). Therefore, braided rivers and meandering rivers are commonly regarded as the two principal types of rivers and are the only types discussed here. Actually, a nearly continuous spectrum of river systems may exist between these distinct end-member types. Both predominantly gravelly and predominantly sandy rivers occur. Sandy systems are most common and best known in the geologic record.

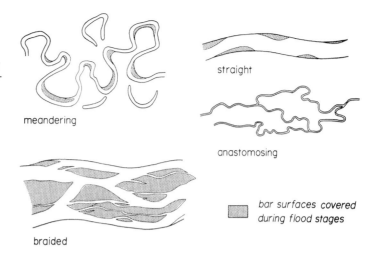

FIGURE 10.5 Principal types of rivers. (From Miall, A. D., 1977, A review of the braided-river depositional environment: Earth Science Rev., v. 13. Fig. 1, p. 5, reprinted by permission of Elsevier Science Publishers, Amsterdam.)

Braided-River Systems

Depositional Setting and Depositional Processes. Braided rivers are distinguished from meandering rivers by their lower sinuosity, which is defined as the ratio of channel length to length of the valley containing the river. That is, the longer the channel length compared to the length of the valley, the greater the sinuosity. They are also characterized by many channels separated by bars or small islands (Fig. 10.6). Gravelly braided rivers occur in areas of high relief and typically have very limited areal extent. They commonly grade downstream, by abrupt grain-size decrease, to sandy braided rivers. Sandy braided rivers are more common than gravelly braided rivers in both modern environments and the ancient record.

Braided rivers show their best development in the distal parts of alluvial fans, on glacial outwash plains, and in the mountainous reaches of river systems. In these areas, sediment is abundant, water discharge is high and commonly sporadic, and little vegetation may be present to hinder runoff. Under these conditions, the rivers are generally overloaded with sediment, leading to rapid deposition. Braiding apparently takes place because of rapid, large fluctuations in river discharge; an abundance of coarse sediment; a high rate of supply of sediment; and easily erodable, noncohesive banks (Cant, 1982). Braiding is developed by sorting action as a stream leaves behind those sizes of particles that it is incompetent to handle. Deposition of the coarser bedload causes mid-channel bars to form. Thus, during periods of high discharge, the stream channel is rapidly choked with coarse bedload detritus, creating bars around which the discharge is diverted. Repeated bar formation and channel branching generate a braided network of bars and channels over the entire stream bed. Although particles of all sizes can be transported during flood stage of the river, only very fine sediment or no sediment at all moves during periods of lower discharge. Braided rivers tend to have high slopes, or gradients, and are most characteristic of upstream reaches of river systems. In fact, many braided rivers grade or merge downslope into meandering rivers as both the river gradient and the grain size of the sediment load decrease. The processes of deposition in braided rivers are much the same as those that take place on humid fans, particularly in the lower, or distal, part of the fans. Consequently, it is often difficult to distinguish between the facies that form in these two environments, particularly in the transition zone from fans to braided rivers.

FIGURE 10.6   Braided streamflow on the gravelly bed of the Muddy River, Alaska. (Photograph by Bradford Washburn/Boston Museum of Science.)

Bedforms and Structures. Braided rivers are characterized particularly by large bedforms called bars that can be grouped into three basic types: (1) longitudinal bars; (2) linguoid and transverse bars; and (3) lateral bars, including point bars and side bars (Fig. 10.7). **Longitudinal bars** are mid-channel bars (Fig. 10.7) that form when the coarsest part of the stream load is deposited as streamflow wanes or other factors cause loss of stream competency. They are oriented with their long axis roughly parallel (longitudinal) to current flow. Bars are small when first formed but continue to grow in length and height as fine particles are trapped in the interstices of the original deposit and as more bedload sediment is deposited downstream in the lee of the bar. Coarsest material is concentrated along the central axis and bottom of the bar, and grain size tends to decrease upwards and downstream. The internal structure of longitudinal bars is characterized by massive or crude horizontal bedding that may indicate transport and deposition under upper-flow-regime conditions. **Linguoid** and **transverse bars** are oriented transverse (at an angle) to the direction of streamflow and are particularly characteristic of sandy braided streams. Linguoid bars are lobate or rhombic in shape,

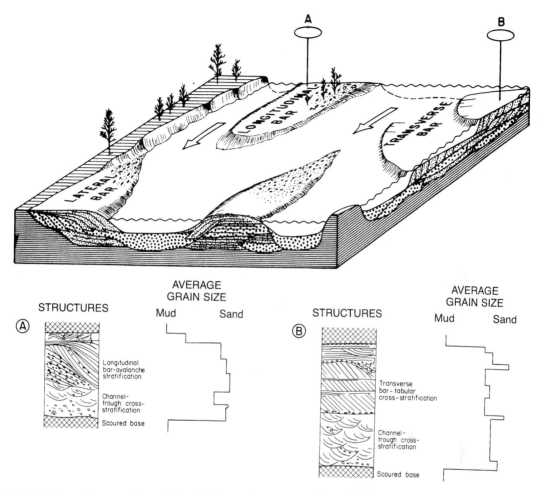

FIGURE 10.7 Structure of bars in braided rivers. Sequence A is dominated by migration of a gravelly longitudinal bar. Sequence B records deposition of successive transverse bar cross-bed sets upon a braid-channel fill. (After Galloway, W. E., and D. K. Hobday, 1983, Terrigenous clastic depositional systems. Fig. 4.4, p. 56, reprinted by permission of Springer-Verlag, Heidelberg.)

with steep downstream avalanche faces. Transverse bars are similar except that they tend to have straight crests. Linguoid and transverse bars appear to be large ripple forms (dunes) that develop under high flood conditions. **Lateral bars** are typically very large bars that develop in areas of relatively lower energy along the sides of the stream channel. They are attached to the bank, as illustrated in Figure 10.7.

Owing to the generally high-flow conditions that prevail in braided rivers, most of the sediment deposited is gravel and sand. Mud is a distinctly subordinate constituent of braided-stream deposits. Longitudinal bars tend to be composed largely of gravels or mixtures of sand and gravels. Linguoid, transverse, and lateral bars are generally more sandy. Gravelly longitudinal bars show crude planar stratification to poorly developed cross-bedding. Cross-bedding is commonly much better developed in sandy units. Linguoid and transverse bars are extensively cross-bedded, and both planar and trough cross-bed sets are common. The dip direction of cross-bed foreset is variable, but overall it is unidirectional downstream. Ripple marks are common on the surface of sandy bar deposits. The principal lithofacies and sedimentary structures that characterize braided-river deposits are summarized in Table 10.2. Both gravelly and sandy braided rivers migrate laterally, leaving sheetlike or wedge-shaped deposits of channel and bar complexes (Cant, 1982). Lateral migration combined with aggradation leads to deposition of sheet sandstones or conglomerates with thin, nonpersistent shales enclosed within coarser sediments. Paleocurrent patterns in braided-river sediments are unimodal and commonly show a fan-shaped distribution, reflecting shifting of these rivers within alluvial-fan systems.

Vertical Successions of Facies. Braided-river deposits are highly variable depending upon the sizes of bedload sediment transported, the depth of the stream channel, and the amount and variability of stream discharge. Some deposits are generated by lateral accretion, as side or point bars develop. Others form by vertical accretion on channel floors and bar tops. Channels may fill by aggradation during waning current flow, and flooding can cause beds formed under decreasing current velocity to be superimposed.

TABLE 10.2   Facies and characteristic structures of braided rivers

	Facies identifier	Lithofacies	Sedimentary structures	Interpretation
Gravel	Gm	Gravel, massive or crudely bedded; minor sand, silt, or clay lenses	Ripple marks, cross-strata in sand units, gravel imbrication	Longitudinal bars, channel-lag deposits
	Gt	Gravel, stratified	Broad, shallow trough cross-strata, imbrication	Minor channel fills
	Gp	Gravel, stratified	Planar cross-strata	Linguoid bars or deltaic growths from older bar remnants
Sand	St	Sand, medium to very coarse; may be pebbly	Solitary or grouped cross-strata	Dunes (lower-flow regime)
	Sp	Sand, medium to very coarse; may be pebbly	Solitary or grouped planar cross-strata	Linguoid bars, sand waves (upper- and lower-flow regimes)
	Sr	Sand, very fine to coarse	Ripple marks of all types, including climbing ripples	Ripples (lower-flow regime)
	Sh	Sand, very fine to very coarse; may be pebbly	Horizontal lamination, parting or streaming lineation	Planar bed flows (lower- and upper-flow regimes)
	Ss	Sand, fine to coarse; may be pebbly	Broad, shallow scours (including cross-stratification)	Minor channels or scour hollows
	Fl	Sand (very fine), silt, mud, interbedded	Ripple marks, undulatory bedding, bioturbation, plant rootlets, caliche	Deposits of waning floods, overbank deposits
Mud	Fm	Mud, silt	Rootlets, desiccation cracks	Drape deposits formed in pools of standing water

Source: Miall, A. D., 1977, A review of the braided-river depositional environment: Earth Science Rev., v. 13. Table III, p. 20, reprinted by permission of Elsevier Science Publishers, Amsterdam.

These depositional processes generate vertical successions of facies that either show no distinctive pattern or vertical grain-size change or display a fining-upward trend (e.g., Darby et al., 1990). Miall (1977) proposed four vertical-profile models (named after distinctive modern braided-river deposits) that can develop under different conditions of bedload and discharge (Fig. 10.8). The **Scott-type** model consists mainly of roughly horizontally bedded gravels and minor sand wedges; this model shows poorly developed cycles and reflects deposition in gravelly proximal streams during high river discharges. The **Donjek-type** model consists of fining-upward cycles of variable scale that reflect deposition in braided rivers with mixed bedloads of sand and gravel, where sedimentation may occur at different levels within channels or where channel aggradation is followed by channel shifting. Sandy braided rivers of more steady discharge are dominated by linguoid and transverse bars that generate largely cross-bedded, sandy, **Platte-type** deposits that do not have very distinct cycles, although some fining-upward successions may be identified. **Bijou Creek–type** deposits are characterized by superimposed flood sediments that accumulate during waning current flow. Each flood event is represented by a fining-upward deposit. These deposits form in braided streams with markedly variable discharge, owing to periodic flooding and relatively little topographic differentiation between channels and bars (Cant, 1982). See also Rust and James (1987), Fedo and Cooper (1990), and Godwin (1991).

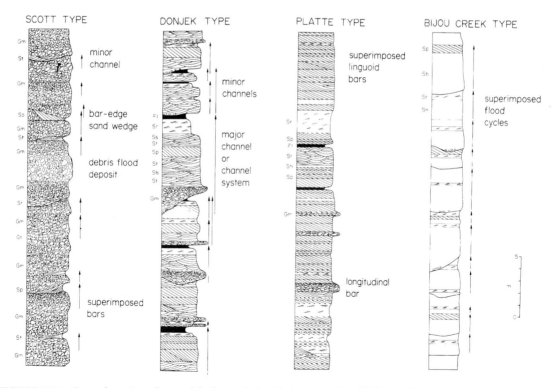

FIGURE 10.8 General stratigraphic models for sandy braided streams. See Table 10.2 for description of individual facies. (From Miall, A. D., 1977, A review of the braided-river environment: Earth Sciences Rev., v. 13. Fig. 12, p. 46, and Fig. 14, p. 47, reprinted by permission of Elsevier Science Publishers, Amsterdam.)

Braided-River Deposits in the Geologic Record. Braided-river deposits are recognized in the geologic record mainly on the basis of grain-size characteristics and sedimentary structures (Table 10.2). Although modern braided rivers are known from many areas, braided-stream deposits appear to be rather poorly represented in the geologic record. Reported examples include the Westwater Canyon Member of the Jurassic Morrison Formation of New Mexico, which is known for its uranium deposits; the Triassic Hawkesbury Formation near Sydney, Australia; and the Triassic Ivishak Formation on the Alaskan North Slope. The Ivishak Formation forms reservoir rocks for oil in the Prudhoe Bay Oil Field. A particularly well-studied example of an ancient braided-stream deposit is the Devonian Battery Point Sandstone of Quebec (Cant and Walker, 1976; Walker and Cant, 1984). A summary sequence for the Battery Point Sandstone is shown in Figure 10.9. This sequence begins with a channel-floor lag deposit lying on a scoured surface (bed SS) overlain by poorly defined trough cross-bedding (bed A). These deposits are succeeded upward by in-channel sands that display well-defined trough cross-bedding (B beds) and large sets of planar-tabular cross-bedding (C beds). Above this are bar-top deposits consisting of small sets of planar-tabular cross-bedding (bed D) with isolated scour fills (bed E), overlain by vertical accretion deposits (V.A.) of cross-laminated siltstones interbedded with mudstones (bed F), and finally low-angle cross-stratified sandstones (bed G). The sequence displays a generally fining-upward trend. The Battery Point Sandstone most nearly resembles Miall's (1977) Donjek-type model. It cannot be considered a general model for all braided-river deposits, but it illustrates many of the features of the Donjek and Platte types.

In common with alluvial-fan deposition, braided-river deposition appears to have been particularly common in Pre-Devonian time, apparently owing to scarcity of vegetation and low stability of banks (Pettijohn, Potter, and Siever, 1987, p. 362). With

FIGURE 10.9   Summary vertical succession of braided-stream facies for the Devonian Battery Point Sandstone, Quebec. The succession was developed using statistical Markov analysis, and preferred facies relationships were drawn as a stratigraphic column using average thicknesses. Arrows show paleoflow directions. The letters indicate facies, which are explained in the text. (After Walker, R. G., and D. J. Cant, 1984, Sandy fluvial systems, *in* R. G. Walker (ed.), Facies models, 2nd ed.: Geoscience Canada Reprint Ser. 1. Fig. 20, p. 83, reprinted by permission of Geological Association of Canada.)

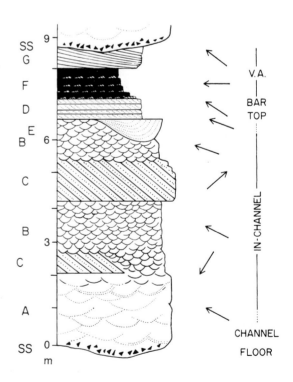

the spread of land vegetation in Devonian and later periods, the proportion of meandering streams to braided streams increased.

Meandering-River Systems

In contrast to the network of channels that characterize braided rivers, meandering rivers tend to be confined within a single major channel, characterized by cohesive banks that are difficult to erode. Meandering streams also differ from braided rivers by their much greater sinuosity, lower gradients, and finer sediment load. Many meandering rivers are simply downstream continuations of braided rivers, formed as stream slope and coarseness of bedload decrease and large-scale fluctuations in discharge become less marked. Others occur independently of braided rivers. Large meandering rivers commonly discharge into delta systems. Meandering rivers have been studied extensively, and the depositional model for these rivers is better established and understood than that for braided rivers (e.g., Allen, 1965a; Cant, 1982; Collinson, 1986a; Diemer, 1991; Reineck and Singh, 1980; Walker and Cant, 1984).

Depositional Setting and Sedimentation Processes. The morphological elements of the meandering-river system are shown in Figure 10.10. These elements consist principally of the main meander channel, point bars that build outward on the inside bend of meander loops, natural levees, floodbasins alongside the levees, and oxbow lakes and abandoned cut-off meanders. Sediments accumulate in different parts of this system owing to channel flow and periodic overbank flooding.

Channel flow within the main meander channel in response to episodes of increased streamflow is responsible for much of the sediment erosion and deposition

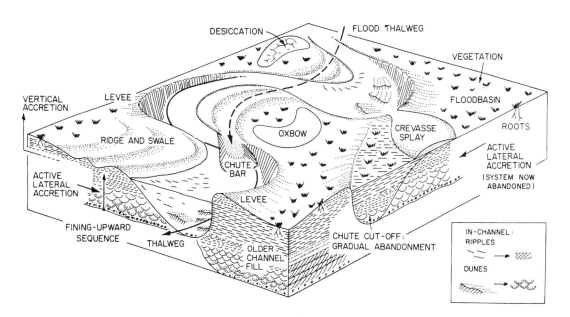

FIGURE 10.10 The morphological elements of a meandering-river system. A thalweg is a line connecting the deepest points along a stream channel. It is commonly the line of maximum current velocity. (From Walker, R. G., and D. J. Cant, 1984, Sandy fluvial systems, *in* R. G. Walker (ed.), Facies models: Geoscience Canada Reprint Ser. 1. Fig. 1, p. 72, reprinted by permission of Geological Association of Canada.)

that takes place within the meandering-river system. Periods of increased streamflow occur periodically, often seasonally. During low-flow conditions, the maximum velocity of channel flow swings back and forth across the channel, hugging the bank along the concave (outer) part of the meanders (Fig. 10.11A). During high-flow conditions, the current takes a straighter path. The lateral shifting of the current sets up a strong transverse spiral or helical flow that tends to deflect the water along the streambed from the outer concave bank toward the inner convex bank of the meander (Fig. 10.11B). Because the deflecting force is larger near the surface, where stream velocity is higher than near the bed, the transverse spiral flow moves the surface water toward the outer bank and the bottom water toward the inner bank. This helical-flow component carries sediment across the stream channel and up the sloping bank of adjacent point bars (and perhaps to the next bar downstream), where it is deposited under lower velocity conditions. Only the coarsest sediment accumulates as a lag deposit in the deeper part of the channel. The remaining sediment eroded from the concave bank of the meander bend is transported laterally across the stream, as the zone of maximum current velocity shifts back and forth from one outside bend to the next, and coarser bedload sediment is deposited by lateral accretion on the next downstream point bar (Fig. 10.11B). This process causes both lateral and downstream migration of the meanders.

During flood stage, the river becomes bank full, and overbank flooding takes place. Overbank flooding leads to deposition of fine silt and mud on the bank near the stream edge, causing build-up of natural levees if the rate of deposition exceeds the rate of bank cutting. Sediment is deposited also in adjacent floodbasins (floodplains) and oxbow lakes, largely by vertical accretion. Natural levees may occasionally become breached, allowing sediment-laden water containing both suspended-load and bedload materials to wash suddenly into floodbasins. There the sediment load is

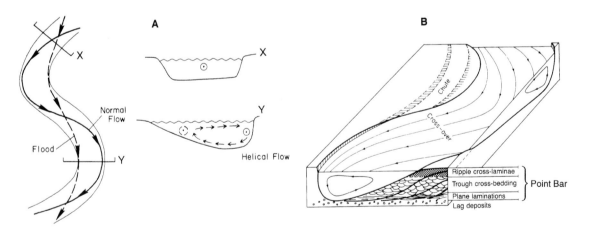

FIGURE 10.11 Flow patterns in a meandering stream. A. Pattern of the thread of maximum velocity as it shifts back and forth from one meander to the next. Cross profiles X and Y indicate differences in transverse current flow in different reaches of the stream. B. Details of helical flow and the nature of deposits formed as a result of such flow. (A, after Galloway, W. E., and D. K. Hobday, 1983, Terrigenous clastic depositional systems. Fig. 4.2, p. 53, reprinted by permission of Springer-Verlag, Heidelberg, Germany. B, from Blatt, H., G. V. Middleton, and R. Murray, Origin of sedimentary rocks, 2nd ed. © 1980, Fig. 19.5, p. 637. Reprinted by permission of Prentice-Hall, Englewood Cliffs, N.J.)

rapidly deposited. This process is called **crevasse splay.** Further details of these sedimentation processes are given below in the discussion of meandering-river deposits.

Deposits. Fluvial depositional processes lead to accumulation of sediment in five different settings within the meandering-stream system: (1) the main channel, (2) point bars, (3) natural levees, (4) the floodbasin, and (5) oxbow lakes and meander (chute) cutoffs. Each of these subenvironments of the system generates deposits with characteristic grain sizes and sedimentary structures.

Channel sediments are primarily lag deposits composed of coarse material that the river can move only at maximum stream velocity during flood stage. These deposits include coarse bedload gravels, together with waterlogged plant material and chunks of partly consolidated mud eroded from the channel wall (Walker and Cant, 1979). Bedding is indistinct in these coarse materials, but imbrication of pebbles and cobbles is common. Channel-lag deposits are typically thin and discontinuous and may be absent altogether in sandy streams that transport little gravel.

At discharge rates lower than those required to move coarse lag gravels, sand is transported over the channel-lag deposits and accreted to the sloping surface of downstream **point bars.** When the stream is bank full of water, helical flow develops; as described, this flow creates a vertical circulation cell normal to the stream bank and carries bottom water and sediment load up the sloping face of the point bar. Bed shear stress falls when velocities decrease with shallower depth as the upslope component of flow moves upward over the point-bar surface. Thus, coarsest grains tend to be deposited on the lower part of the point bar and finer grains higher up on the bar, leading to a fining-upward point-bar sequence (Fig. 10.10). Large dune bedforms are generated on the lower part of the bar, whereas ripples form on the higher parts of the bar. The preserved sedimentary structures of the point bar thus pass from large-scale trough cross-bedded coarse sands in the lower part of the bar to small-scale trough cross-beds higher on the bar. Cross-bed dip directions may be highly variable, but overall they show unidirectional dip in the downstream direction. Upper-flow-regime, plane-bed flow can be achieved at different heights on the bar depending upon flow velocity; thus, plane-bed parallel laminations may be preserved, interbedded with small-scale or large-scale trough cross-beds.

When the stream floods and overtops its banks, deposition of fine sediment occurs on natural levees, in adjacent floodbasins, and in oxbow lakes, as mentioned. Deposition from overbank waters results in upbuilding of the sediment surface and is thus called vertical accretion, in contrast to the lateral accretion that takes place on point bars. **Natural-levee deposits** form primarily on the concave or steep-bank side of meander loops immediately adjacent to the channel (Fig. 10.10) as a result of sudden loss of competence of streams as they overtop their banks. These deposits are thickest and coarsest near the channel bank and become thinner and finer grained toward the floodbasin. Sedimentary structures consist of rippled and horizontally stratified fine sands overlain by laminated mud (Davis, 1983). **Floodbasin (floodplain) deposits** are fine-grained sediments that settle out of suspension from floodwaters carried into the floodbasin, which may be a broad, low-relief plain, a swamp, or even a shallow lake. These thin, fine-grained deposits commonly contain considerable plant debris and may be bioturbated by land-dwelling organisms or plant roots. **Crevasse-splay deposits** may also occur on floodplains where rising floodwaters breach natural levees (Fig. 10.10). Sedimentation from traction and suspension occurs rapidly as water containing both coarse bedload sediment and suspended sediment debouches suddenly onto the plain, resulting in graded deposits that may resemble a Bouma turbidite sequence

(Walker and Cant, 1979). **Oxbow-lake deposits** consist of fine silt and mud introduced into the lakes from the main stream during overbank flooding. They are commonly well laminated and may contain plant remains and the shells of ostracods and fresh-water molluscs.

Vertical Succession of Facies. At any particular time in a meandering-stream system, sedimentation may take place essentially simultaneously in the lag channel, on point bars, and in the various overbank environments. As lateral shifting of these different environments takes place owing to stream meandering, sediments from laterally contiguous environments will become superimposed or vertically stacked (Fig. 10.12). As a result of meander migration, coarse lag deposits thus become overlain by sandy, fining-upward point-bar deposits, which themselves are overlain by silty and muddy overbank deposits, producing an overall fining-upward succession. Thus, the classic model for meandering-stream deposits, as proposed by J. R. L. Allen (1970b), consists of a fining-upward succession that begins with a basal lag-gravel conglomerate lying above an erosion surface. This unit is replaced upward by cross-bedded point-bar sands, which in turn are overlain by fine overbank mud and silt containing root traces, desiccation cracks, and possibly bioturbation structures (Fig. 10.13). The sands that form the basal unit decrease in size upward and display trough cross-bedding with upward reduction in cross-bed set size. Parallel laminae may be interbedded with the cross-bedded units at various levels. Allen's fining-upward model can be utilized as a very useful norm to which fluvial deposits may be compared; however, Collinson (1986a) points out that variations from this norm can occur in some meandering-stream deposits. Complications caused by deposition when discharge is less than bank full, flow patterns that diverge from the helicoidal model, or deposition on coarse, gravelly point bars may result in point-bar deposits in which vertical grain-size changes may not be clear or in which grain size may even coarsen upward. Also, the distribution of bedforms and sedimentary structures may be much less ordered than in

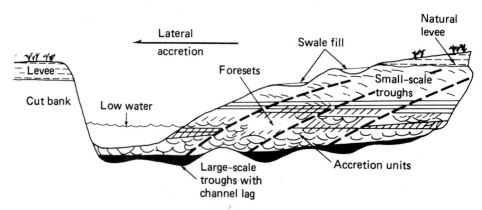

FIGURE 10.12 Diagrammatic cross section of a point bar showing contiguous environments and the structure of the bar. (From Bernard, H. A., C. F. Major, Jr., B. S. Parrott, and J. J. LeBlanc, Sr., 1970, Recent sediments of southeast Texas—A field guide to the Brazos alluvial and deltaic plains and the Galveston barrier island complex: Texas Bur. Econ. Geology Guidebook 11; described by W. F. Fisher and L. F. Brown, 1972, Clastic depositional systems—A genetic approach to facies analysis, annotated outline and bibliography: Texas Bur. Econ. Geology. Reprinted by permission of The University of Texas, Austin.)

FIGURE 10.13 Classic fining-upward sequence in a meandering-river deposit. (After Walker, R. G., and D. J. Cant, 1984, Sandy fluvial systems, *in* R. G. Walker (ed.), Facies models: Geoscience Canada Reprint Ser. 1. Fig. 2, p. 73, reprinted by permission of Geological Association of Canada. Modified from J. R. L. Allen, 1970, Studies in fluviatile sedimentation: A comparison of fining-upward cyclothems, with special reference to coarse-member composition and interpretation: Jour. Sed. Petrology, v. 40, Fig. 12, p. 331.)

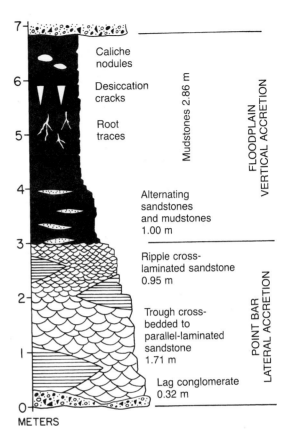

the ideal model, as seen, for example, in the development of ripples rather than dunes in the deeper part of the channel.

Ancient Examples. Meandering-stream deposits are moderately common in the geologic record, particularly in rocks of Devonian and younger age. Many examples from the United States are known, including Paleozoic successions of the Appalachian region [the Devonian Catskill Formation and the Port Hood Formation (Carboniferous) of Nova Scotia, for example]; parts of the Newark Group (Upper Triassic–Lower Jurassic) of northeastern North America and parts of the Jurassic Morrison Formation; the Cretaceous Mesaverde Group; and the Tertiary Wasatch and Fort Union formations of the Colorado Plateau. The lower part of the Devonian Old Red Sandstone of Wales and England is a particularly famous and well-studied European example of a meandering-stream deposit. The characteristics of this sandstone sequence largely formed the basis for Allen's (1970b) model of fining-upward facies shown in Figure 10.13.

Geometry of River Deposits

Owing to extensive lateral channel migration and channel aggradation, braided rivers produce sheet sandstones or conglomerates containing thin beds or lenses of shales enclosed within the thicker sediments (Fig. 10.14). By contrast, meandering streams, which are confined within the rather narrow, sandy meander belt of stream flood-

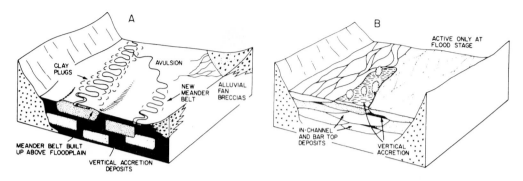

FIGURE 10.14   Contrasting geometry of meandering (A) and braided (B) rivers. (From Walker, R. G., and D. J. Cant, 1984, Sandy fluvial systems, *in* R. G. Walker (ed.), Facies models: Geoscience Canada Reprint Ser. 1. Fig. 9, p. 77, reprinted by permission of Geological Association of Canada.)

plains, generate linear "shoestring" sand bodies oriented parallel to the river course (Cant, 1982). These shoestring sands are surrounded by finer-grained, overbank flood-plain sediments. As the river aggrades its channel and builds up above the floodplain, a natural levee may eventually break at some point during flood stage, causing abandonment of the old channel and creation of a new channel, a process called **avulsion.** Successive episodes of avulsion can lead to formation of several linear sand bodies within a major stream valley (Fig. 10.14).

Recognizing Ancient Fluvial Deposits

There are no unequivocal criteria for recognition of ancient fluvial deposits. In general, they commonly contain few fossils (and no marine fossils), although fine-grained, overbank deposits may be characterized by root traces. They have poor to moderate sorting, red color (in some cases), and unidirectional paleocurrent patterns that are more variable in meandering-stream deposits than those of braided rivers; they also generally display a strong downstream decrease in particle size. Meandering-river deposits are particularly characterized by distinct fining-upward grain-size trends. By contrast, braided-stream deposits may display either no detectable vertical size trends or a fining-upward trend. Alluvial-fan deposits may also show no vertical grain-size trend, or they may display either a fining-upward or a coarsening-upward trend. Because of similarities in many of the characteristics of alluvial-fan, braided-river, and meandering-river deposits, it may be difficult to differentiate these deposits in ancient sedimentary successions. Also, fluvial deposits may be difficult to differentiate from glacial outwash deposits and delta-plain deposits. For further insight into the problem of recognizing ancient fluvial deposits, see Collinson (1986a), Collinson and Lewin (1983), Darby et al. (1990), Ethridge and Flores (1981), and Miall (1992).

10.3   EOLIAN DESERT SYSTEMS

Introduction

Deserts cover broad areas of the world today, particularly within the latitudinal belts of about 10° to 30° north and south of the equator, where dry, descending air masses create prevailing wind systems that sweep toward the equator. Deserts occur also in the interiors of continents and in the rain shadows of large mountain ranges where they are cut off from moisture from the oceans. Deserts are areas in which potential rates of evaporation greatly exceed rates of precipitation. They cover about 20 to 25 percent of the present land surface.

Owing to their generally low rainfall, commonly less than about 25 cm/yr, we tend to think of deserts as extremely dry areas dominated by wind activity and covered by sand. In reality, a variety of subenvironments exist within deserts, including alluvial fans; ephemeral streams that run intermittently in response to occasional rains; ephemeral saline lakes, also called playas or inland sabkhas; sand-dune fields; interdune areas covered by sediments, bare rocks, or deflation pavement; and areas around the fringe of deserts where windblown dust (loess) accumulates. Large areas of the desert environment may indeed be carpeted by windblown, or **eolian,** sand. Such areas that cover more than about 125 km^2 are called **sand seas** or **ergs;** smaller areas are called **dune fields.** Ergs and dune fields cover about 20 percent of modern deserts or about 6 percent of the global land surface. The remaining areas of deserts are covered by eroding mountains, rocky areas, and desert flats. The largest desert in the world, the Sahara (7 million km^2), contains several ergs arranged in belts. The larger belts cover areas as extensive as 500,000 km^2 (Walker and Middleton, 1979). The deposits of deserts have been studied extensively. See, for example, Ahlbrandt and Fryberger (1982), Bigarella (1972), Brookfield (1984, 1992), Brookfield and Ahlbrandt (1983), Collinson, (1986b), Frostick and Reid (1987), Glennie (1970, 1987), Hunter (1977), McKee (1979a), Pye and Tsoar (1990), Reineck and Singh (1980), Walker and Middleton (1979), and Wilson (1972).

Depositional Processes

Most deserts are characterized by extreme fluctuations in temperature and wind, on both a daily and a seasonal basis. Rainfall rates are low, as mentioned, and the rains are very sporadic. Vegetation is generally extremely sparse. When rains do come, they tend, owing to the lack of vegetative cover, to create flash floods. Rainwater typically drains toward the centers of desert basins, where playas or inland sabkhas may develop and become sites of deposition of carbonate and evaporite minerals. Because periodic rains create flash floods and ephemeral streams and mobilize debris flows and mudflows, they are extremely important agents of sediment transport in deserts. Nonetheless, much of the time, water plays a relatively small role in sediment transport in deserts. Most of time, wind is the dominant agent of sediment transport and deposition. Wind is much less effective than water as an agent of erosion, but it is an extremely effective medium of transport for loose sand and finer sediment. Not only does it account for the transport of vast quantities of siliciclastic sand in deserts, but it is also responsible for sediment transport in glacial environments, on river floodplains, and along many coastal areas, where both carbonate and siliciclastic sands may be transported inland. The windblown deposits of these latter environments are quite small compared to the sand seas of desert areas. Wind storms, or dust storms, may also carry silt and clay far from their sources and are responsible for transport of much of the pelagic sediment to deep ocean basins.

Wind transports sediment in much the same way as water, separating the sediment into three transport populations: traction, saltation, and suspension. Wind effectively separates sediment finer than about 0.05 mm from coarser sediment and transports this fine sediment long distances in suspension. Except at unusually high wind velocities, coarser sediment travels by traction and saltation close to the ground. Saltation is a particularly important mode of wind transport, aided by downslope creep of grains owing to the impact of saltating grains as they strike the bed. Wind appears to be especially effective in transport of medium to fine sand and finer sediment, but coarse particles (up to 2 mm or somewhat larger) may also undergo transport by rolling and surface creep under high-velocity winds. Bagnold's (1954) study dealing with the physics of blown sand remains the classic piece of research in the field of eolian sediment transport; however, several more recent workers have also investigated this subject (e.g., Barndorff-Nielsen and Willits, 1991; Pye and Tsoar, 1990). The transporting and sorting action of wind tends to produce three kinds of deposits:

1. dust (silt) deposits, sometimes referred to as **loess,** which commonly accumulate far from the source
2. sand deposits, which are commonly well sorted
3. lag deposits, consisting of gravel-size particles that are too large to be transported by wind and that form a **deflation pavement**

Wind transport and deposition generate many of the same kinds of bedforms and sedimentary structures—including ripples, dunes, and cross-beds—as those produced by water transport.

Deposits of Modern Deserts

The various environments of deserts can be grouped into three main subenvironments: dune, interdune, and sand sheet (Ahlbrandt and Fryberger, 1982). The dune environment is primarily the site of wind transport and deposition of sand, which accumulates in a variety of dune forms, many of which have steeply dipping slip faces or avalanche faces. Interdune areas can receive both windblown sediment and sediment transported and deposited by ephemeral streams in stream floodplains or playa lakes. The sheet-sand environment exists around the margins of dune fields. The deposits of this environment form a transitional facies between dune and interdune deposits and deposits of other environments.

Dunes

Many types of dunes occur in the sand seas or dune fields of modern deserts, ranging from those with no slip faces to those with three or more slip faces (Fig. 10.15). Eolian bedforms range in scale from small ripples to transverse and longitudinal dunes 0.1 to 100 m high to complex dunes, called **draas,** with heights of 20 to 450 m (Wilson, 1972). **Barchans, barchanoid ridges,** and **transverse dunes** form under the influence of unidirectional winds. They have single slip faces and appear to represent a gradational series corresponding to an increase in sand supply (McKee, 1979b). **Parabolic** and **blowout dunes** have one or more slip faces. They are related to the preceding group of dunes, but their development is controlled by vegetation cover. **Dome dunes** are circular in plan view and have no definite slip faces. They may originate by modification of barchanoid dunes by strong winds. **Linear dunes,** also called **seif** or **longitudinal dunes,** have nearly symmetrical ridges, and **reversing dunes** have asymmetrical ridges; both types have two slip faces (Fig. 10.15). Linear dunes form in areas of uniform sand accumulation under generally high-velocity winds of variable directions. Reversing dunes, in which slip faces form on opposite sides at different times, result from a close balance between opposed winds (McKee, 1979c). **Star dunes,** also called **draas,** are huge stellate dunes up to 450 m high, with a high central peak and radiating arms and commonly with superposed complex or compound dunes (Fig. 10.15). They have three

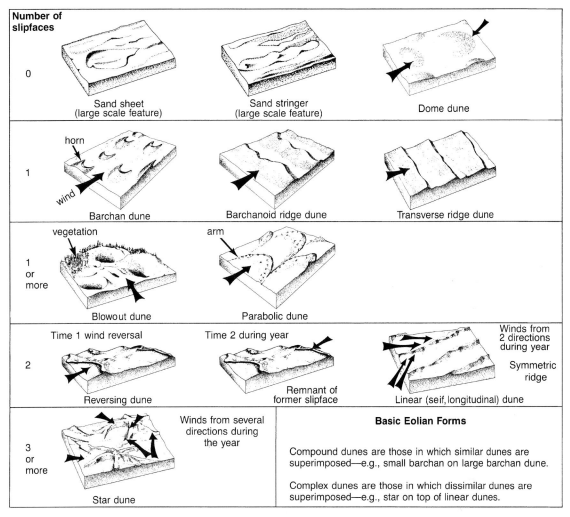

FIGURE 10.15 Basic eolian dune forms grouped by number of slip faces. (After Ahlbrandt, T. S., and S. G. Fryberger, 1982, Introduction to eolian deposits, *in* P. A. Scholle and D. Spearing (eds.), Sandstone depositional environments: Am. Assoc. Petroleum Geologists Mem. 31. Fig. 3, p. 14, reprinted by permission of AAPG, Tulsa, Okla.)

or more slip faces and apparently form under intense, multidirectional wind systems in areas of high sand-drift potential (Fryberger and Dean, 1979).

Dune deposits commonly consist of texturally mature sands that are well sorted and well rounded; however, considerable textural variation can occur. They are also typically quartz rich, although many coastal dune deposits contain high concentrations of heavy minerals and unstable rock fragments. Coastal dunes in some tropical areas may consist largely of ooids, skeletal fragments, or other carbonate grains, and dunes composed of gypsum occur in some desert areas, such as White Sands, New

Mexico. Eolian dunes are characterized particularly by the presence of cross-bedding (Fig. 10.16). The following sedimentary structures are common to most types of eolian dunes (McKee, 1979b):

1. sets of medium- to large-scale cross-strata that typically consist of foresets dipping in the direction of wind transport (leeward) at angles of repose ranging as high as 30° to 34°
2. sets of tabular-planar cross-strata that in vertical section tend to be progressively thinned from the base upward
3. bounding planes between individual sets of cross-strata (Fig. 10.18) that are mainly horizontal or dip leeward at low angles

In addition to these large-scale, gross structural features of dune deposits, several kinds of small-scale internal structures may be present: plane-bed laminae, rippleform laminae, ripple-foreset cross-laminae, climbing translatent strata (climbing ripples), grainfall laminae, and sandflow cross-strata (Table 10.3).

Owing to the variety of dune types that can form under different wind conditions, local paleocurrent vectors derived from eolian cross-bed data can range from unimodal to polymodal. Paleocurrent data may thus show a high degree of scatter that complicates calculation of ancient prevailing sediment transport directions. On a regional scale, eolian paleocurrent patterns are reported to swing around over hundreds of miles around high-pressure wind systems (Selley, 1978).

Interdunes

Interdune areas occur between dunes and are bounded by dunes or other eolian deposits such as sand sheets. Interdunes may be either deflationary (erosional) or depositional. Very little sediment accumulates in most deflationary interdunes except coarse, granule-size lag sediments that may show rippled surfaces and inverse grading. Deflationary interdunes are preserved in the rock record as a disconformity overlain by thin, discontinuous, winnowed lag deposits. Sediments deposited in depositional interdunes can include both subaqueous and subaerial deposits depending upon whether they are deposited in wet, dry, or evaporite interdunes (Ahlbrandt and Fryberger, 1981). All interdune deposits are characterized by low-angle stratification (<~10°), because they are formed by processes other than dune migration, although many deposits may be almost structureless owing to secondary processes, largely bioturbation, that destroy stratification.

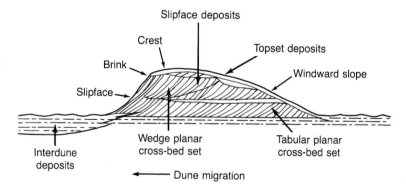

FIGURE 10.16 Typical geometry and internal structure of a barchanoid dune ridge. (From Ahlbrandt, T. S., and S. G. Fryberger, 1980, Eolian deposits in the Nebraska Sand Hills: U.S. Geol. Survey Prof. Paper 1120A, Fig. 7, p. 9.)

Slipface deposits

Crest

Topset deposits

Brink

Windward slope

Slipface

Interdune deposits

Wedge planar cross-bed set

Tabular planar cross-bed set

◄——— Dune migration

TABLE 10.3 Basic types of stratification in eolian deposits

Depositional process	Character of depositional surface	Type of stratification	Dip angle	Thickness of strata Sharpness of contacts	Segregation of grain types Size grading	Packing	Form of strata
Tractional deposition	Rippled	Subcritically climbing translatent stratification	Stratification: low (typically 0°–20°, maximum ~30°) Depositional surfaces: similarly low	Thin (typically 1–10 mm, maximum ~5 cm) Sharp, erosional	Distinct Inverse	Close	Tabular, planar
		Supercritically climbing translatent stratification	Stratification: variable (0°–90°) Depositional surface: intermed. (10°–25°)	Intermediate (typically 5–15 mm) Gradational	Distinct Inverse except in contact zones	Close	Tabular, commonly curved
		Ripple-foreset cross-lamination	Relative to translatent stratification: intermed. (5°–20°)	Individual laminae: thin (typically 1–3 mm) Sharp or gradational, nonerosional	Individual laminae and sets of laminae: indistinct Normal and inverse, neither greatly predominating	Close	Tabular, concave-up or sigmoidal
		Rippleform lamination	Generalized: intermed. (typically 10°–25°)			Close	Very tabular, wavy
	Smooth	Plane-bed lamination	Low (typically 0°–15° max.?)	Sets of laminae: intermediate (typically 1–10 cm) Sharp or gradational, nonerosional		Close	Very tabular, planar
Largely grainfall deposition	Smooth	Grainfall lamination	Intermed. (typically 20°–30°, min. 0°, max. ~40°)			Intermediate	Very tabular, follows preexistent topography
Grainflow deposition	Marked by avalanches	Sandflow cross-stratification	High (angle of repose) (typically 28°–34°)	Thick (typically 2–5 cm) Sharp, erosional or nonerosional	Distinct to indistinct Inverse except near toe	Open	Cone-shaped, tongue-shaped, or roughly tabular

Source: Hunter, R. E., 1977, Basic types of stratification in small eolian dunes: Sedimentology, v. 24. Table 1, p. 364, reprinted by permission of Elsevier Science Publishers, Amsterdam.

323

Dry interdunes or interdunes that are wetted only occasionally are most common. Deposits in dry interdunes are generated by ripple-related wind-transport processes, grainfall in the wind shadow in the lee of dunes, or sandflow (avalanching) from adjacent dunes. The deposits tend to be relatively coarse, bimodal, and poorly sorted, with gently dipping, poorly laminated layers. They are also commonly extensively bioturbated by both animals and plants.

Wet interdune areas are the sites of lakes or ponds where silts and clays are trapped by semipermanent standing bodies of water rather than being deflated and removed. These sediments may contain freshwater species of organisms such as gastropods, pelecypods, diatoms, and ostracods. They are also commonly bioturbated and may contain vertebrate footprints. Some wet interdune sediments become contorted owing to loading by dune sediments.

Evaporite interdunes, or inland sabkhas, occur where drying up of shallow ephemeral lakes or evaporation of damp surfaces causes precipitation of carbonate minerals, gypsum, or anhydrite. Growth of carbonate minerals or gypsum in sandy sediment tends to disrupt and modify primary depositional features. Desiccation cracks, raindrop imprints, evaporite layers, and pseudomorphs may characterize these sediments (Galloway and Hobday, 1983; Lancaster and Teller, 1988).

Sheet Sands

Sheet sands are flat to gently undulating bodies of sand that commonly surround dune fields (Fig. 10.17). They are typically characterized by low to moderately dipping (0°–20°) cross-stratification and may be interbedded in some parts with ephemeral-stream deposits. Sheet-sand deposits may also contain gently dipping, curved, or irregular surfaces of erosion several meters in length; abundant bioturbation traces formed by insects and plants; small-scale cut-and-fill structures; gently dipping, poorly laminated layers resulting from adjacent grainfall deposition; discontinuous, thin layers of coarse sand intercalated with fine sand; and occasional intercalations of high-angle eolian deposits (Ahlbrandt and Fryberger, 1982; Kocurek and Nielson, 1986; Schwan, 1988).

Ancient Desert Deposits

Stratification

The most striking features of modern sandy deserts are various bedforms ranging from ripples to dunes to draas; however, these bedforms are only rarely preserved in ancient eolian deposits. Instead, we see only cross-bedding and other internal features that remain as a record of bedform migration across ancient deserts. Furthermore, the preserved remains commonly represent only the lowest parts of the original eolian bedform. Brookfield (1984) suggests that migration of eolian bedforms leads to generation of three types of bounding surfaces within ancient dune deposits (Fig. 10.18). **First-order surfaces** are flat-lying bedding planes that cut across all other eolian structures. These surfaces are believed to be caused by the migration of very large bedforms (draas) across the desert. The development of first-order surfaces may be related to the position of the water table in sands. The crests of dunes tend to be blown off down to the water table, but the cohesiveness of the wet sand prevents its deflation. For example, Kocurek et al. (1992) describe cycles of accumulation and erosion in modern dune fields of Padre Island, Texas, that are controlled by rise and fall of the water table. **Second-order surfaces** lie at an angle between first-order surfaces and dip downwind.

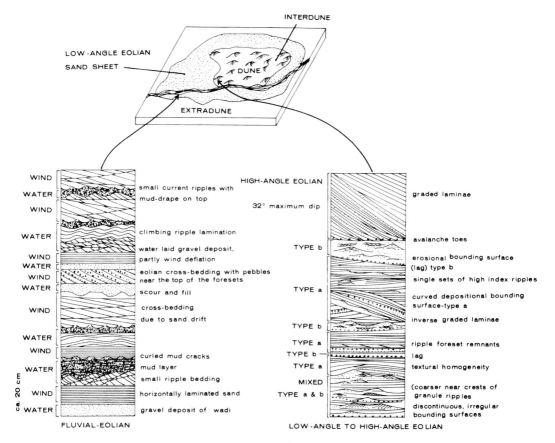

FIGURE 10.17 Areal distribution and stratigraphic relationships of sheet sands and eolian dune sands. Type a and type b beds are both low-angle, cross-laminated units. Type a beds are generally finer grained and better sorted than type b beds, which commonly contain isolated horizons of coarse-grained sediment. (From Fryberger, S. G., T. S. Ahlbrandt, and S. Andrews, 1979, Origin, sedimentary features, and significance of low-angle eolian "sand sheet" deposits, Great Sand Dunes National Monument and vicinity, Colorado: Jour. Sed. Petrology, v. 49. Fig. 12, p. 745, reprinted by permission of Society of Economic Paleontologists and Mineralogists, Tulsa, Okla.)

They are apparently caused by migration of dunes down the lee slopes or slip faces of draas or by lateral migration of longitudinal dunes across the lee slope. **Third-order surfaces** form bounding surfaces for bundles of laminae within cosets of cross-laminae. They are attributed to erosion followed by renewed deposition owing to fluctuations in wind direction and velocity.

The nature of the internal cross-stratification and bounding surfaces preserved in ancient eolian deposits thus depends upon the types of dune forms that migrated across the ancient desert floors and the migration patterns of these dunes. Some dunes may develop a single mode of cross-stratification; others may show bimodal or even more complex patterns depending upon local conditions and migration patterns. There is not a direct (one-for-one) correlation between external eolian dune morphology and the type of stratification developed, as there is in subaqueous dunes produced under unidirectional flow. Furthermore, single sets of cross-stratification thicker than 10 m

FIGURE 10.18 Relationship and origin of bounding surfaces in dune deposits. First-order surfaces form by migration of large dunes, second-order surfaces are generated by migration of dunes down the lee slopes of draas, and third-order surfaces mark discontinuities between bundles of foreset beds. (After Brookfield, M. E., 1984, Eolian sands, *in* R. G. Walker (ed.), Facies models, 2nd ed.: Geoscience Canada Reprint Ser. 1. Fig. 8, p. 97, reprinted by permission of Geological Association of Canada.)

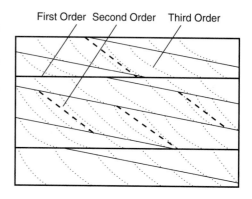

are common in eolian deposits but rare in subaqueous deposits; sets as thick as 20 m are almost certainly eolian (R. G. Walker, personal communication).

Vertical Successions

Owing to the unpredictable distribution of dune types in deserts, it has commonly been assumed that eolian deposits do not display a preferred vertical sequence of sedimentary structures or grain-size patterns or any consistent lateral change (Walker and Middleton, 1979). Brookfield (1984) suggests, however, that large-scale vertical successions of desert deposits can develop that may reflect the growth and decay of the desert environment during slow environmental change. Figure 10.19 shows a vertical sequence of structures and facies in a Permian deposit that can be interpreted to have been generated during a desert cycle that began with the onset of arid conditions followed by gradual change to a semiarid climate. Thus, the sequence begins with small-scale cross-bed units, representing onset of desert conditions. Then it passes upward through large-scale cross-bed units, generated during the main eolian phase, to small-scale eolian cross-beds and alluvial-fan deposits formed during waning desert conditions. It culminates in alluvial-fan and desert-floor fluvial deposits that signal the end of desert conditions. This model may not, of course, be applicable to all ancient eolian

FIGURE 10.19 An example of a vertical succession of eolian deposits. This section is through the Lower Permian deposits of the Thornhill and Dumfries intermontane basins, southwestern Scotland. No scale shown. (From Brookfield, M. E., 1984, Eolian sands, *in* R. G. Walker (ed.), Facies models, 2nd ed. Fig. 15, p. 101, reprinted by permission of Geological Association of Canada.)

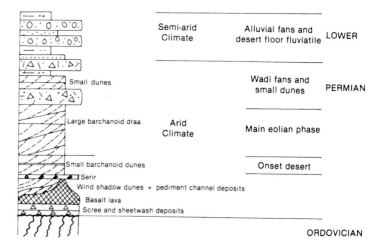

deposits, but it illustrates the point that vertical facies models for ancient eolian deposits are possible.

Summary Characteristics

We tend to think of ancient desert deposits as sedimentary units composed dominantly of eolian sands characterized by fine-size, well-sorted, well-rounded, commonly frosted (pitted by grain impact) quartz sand grains. Eolian strata are also believed to typically display large-scale, high-angle, planar-tabular or planar-wedge cross-bedding and may preserve tracks and trails of desert fauna. Although these characteristics probably are typical of many dune deposits, the discussion above shows that desert deposits as a whole can be much more complex. They can include interdune and sheet sand deposits, with structural and textural properties different from those of dune deposits, and they may also include noneolian sediments such as ephemeral-stream deposits. Therefore, the simple facies model that depicts desert deposits as high-angle cross-bedded, lithologically homogeneous eolian units must be expanded to include these more complex, lithologically heterogeneous sediments.

Criteria for recognition of eolian deposits discussed in preceding paragraphs thus include excellent sorting and rounding of grains; thick cross-bed sets with generally high-angle foresets; uni- to polymodal paleocurrents patterns; individual sandstone units that may display little vertical change in grain size but may be interbedded with other facies (e.g., stream deposits); and the presence (in Late Paleozoic and younger rocks) of vertebrate footprints and wood fragments. In spite of all these criteria, troublesome problems still exist in differentiating eolian deposits from some water-deposited sediments. Identification of eolian deposits can be difficult and equivocal, and the eolian origin of some presumed ancient sandstones is controversial. As a case in point, the Jurassic Navajo Sandstone of the Colorado Plateau has long been cited as a classic example of an ancient eolian sandstone. Yet the Navajo has been interpreted by some workers (e.g., Freeman and Visher, 1975) to be a water-laid deposit, an interpretation hotly disputed by still other workers (e.g., Picard, 1977). Future recognition of eolian deposits may have to rely more on detailed study of styles of stratification and cross-bedding, particularly study of smaller-scale structures such as those proposed by Hunter (1977).

Examples of Ancient Desert Deposits

On the basis of the kinds of criteria discussed above, ancient sandstones interpreted to be windblown deposits have been described from sedimentary successions as old as the Precambrian from many parts of the world. Particularly noteworthy are the Jurassic Navajo Sandstone, as mentioned; the Jurassic Entrada Sandstone and Aztec Formation; the Triassic/Jurassic Moenave and Wingate Sandstones; and the Permian White Rim Sandstone, Coconino Sandstone, and Lyons Sandstone of the Colorado Plateau.

Examples from other continents include the Permian Rotliegendes of northwestern Europe, the Jurassic–Cretaceous Botucatu Formation of the Parana Basin of Brazil, the Permian Lower Bunter Sandstone of Great Britain, the Permo-Triassic Hopeman Sandstone of Scotland, and the Permian Corrie Sandstone of Scotland. The characteristics of the upper Rotliegendes, summarized by Glennie (1972), illustrate well the various kinds of eolian and noneolian sediments that can occur in desert environments (Fig. 10.20). Note in the lower part of the section shown in Figure 10.20 the interbedding of cross-bedded sandstones, interpreted as eolian deposits, with pebbly conglomerates and conglomeratic sandstones, interpreted as wadi deposits. A wadi is an inter-

FIGURE 10.20 Generalized stratigraphic section from the Rotliegendes of northwestern Europe, south-central part of Rotliegendes basin, showing interbedded eolian and noneolian desert deposits. (After Glennie, K. W., 1972, Permian Rotliegendes of northwest Europe interpreted in light of modern desert sedimentation studies: Am. Assoc. Petroleum Geologists Bull., v. 56. Fig. 9, p. 1055, reprinted by permission of AAPG, Tulsa, Okla.)

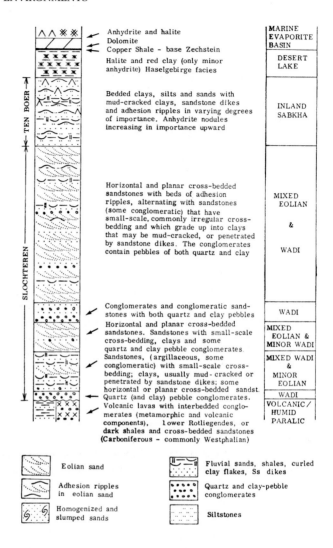

mittent or ephemeral (short-lived) stream. In the upper part of the section, cross-bedded sandstones give way to mud-cracked clays, silts, and sands containing anhydrite nodules, indicating deposition in an inland sabkha. These deposits are overlain by evaporite and red clay deposits that formed in a desert lake. Marine evaporites overlie the lake-bed deposits, indicating the end of desert conditions and encroachment of a marine basin. For additional discussion of desert environments and the characteristics of ancient desert deposits, see Chan (1989), Clemmensen and Hegner (1991), Frostick and Reid (1987), Glennie (1987), and Hesp and Fryberger (1988).

10.4   LACUSTRINE SYSTEMS

Lakes form about 1 percent of the present continental surface of Earth. Because the world's continents are presently in a higher state of emergence than was typical of much of Phanerozoic time, lake sedimentation is more prevalent today than it was during much of the geologic past. In fact, ancient lake sediments appear to be of only minor importance volumetrically in the overall stratigraphic record, although they have been reported in stratigraphic successions ranging in age from Precambrian to Holocene. Although not abundant in the geologic record, lake sediments are nonetheless important. Lake chemistry is sensitive to climatic conditions, making lake sediments useful indicators of past climates. Also, some lake deposits contain economically significant quantities of oil shales, evaporite minerals, coal, uranium, or iron. Many lake sediments also contain abundant fine organic matter that may act after burial as a source material for petroleum (Katz, 1990). Lacustrine sedimentation has not been studied overall as thoroughly as fluvial and eolian sedimentation, but the better-known lacustrine deposits have been examined in detail, and their characteristics are reasonably well known (Anadón, Cabrera, and Kelts, 1991; Collinson, 1986b; Fouch and Dean, 1982; Lerman, 1978; Matter and Tucker, 1978; Picard and High, 1972, 1981).

Origin and Size of Lakes

The basins or depressions in which lakes form can be created by a variety of mechanisms, including tectonic movements such as faulting and rifting; glacial processes such as ice scouring, ice damming, and moraine damming; landslides or other mass movements; volcanic activity such as lava damming or crater explosion and collapse; deflation by wind scour or damming by windblown sand; and fluvial activity such as the formation of oxbow lakes and levee lakes. Many existing lakes appear to have originated directly or indirectly by glacial processes (Picard and High, 1981) and thus may not be typical of ancient lakes, which formed predominantly by tectonic processes. On the other hand, we know that some large modern lakes also formed by tectonic processes (e.g., Lake Tanganyika in the East African rift system, Lake Baikal in the Baikel rift system in Siberia) and volcanic processes (e.g., Crater Lake, Oregon).

Modern lakes range in areal dimensions from a few tens of square meters to tens of thousands of square kilometers. The largest modern lake is the saline, inland Caspian Sea with a surface area of 436,000 km^2 (Van der Leeden, 1975). Other large lakes with surface areas ranging between 50,000 and 100,000 km^2 include Lake Superior, Lake Huron, and Lake Michigan in North America; Lake Victoria located between Uganda and Kenya in east-central Africa; and Lake Aral east of the Caspian Sea. Water depths of modern lakes range from a few meters in small ponds to more than 1700 m in the world's deepest lake, Lake Baikal, Siberia. Water depth and surface area are not necessarily related; thus, some of the largest lakes have very shallow depths, and vice versa. For example, Lake Victoria has a surface area of 68,000 km^2 but a maximum depth of only 79 m, whereas Crater Lake, Oregon, with a surface area of about 52 km^2, has a maximum depth exceeding 580 m.

Preserved lacustrine sediments show that ancient lakes also ranged in size from small ponds to large bodies of water exceeding 100,000 km^2. Three of the largest ancient lakes recognized are the Late Triassic Popo Agie Lake of Wyoming and Utah, which had a minimum areal extent, based on the preserved sediment record, of 130,000 km^2 (Picard and High, 1981); the Jurassic T'oo'dichi Lake of the eastern Colorado Plateau, with an area of 150,000 km^2 (Turner and Fishman, 1991); and the Eocene Green River Basin, with an area of about 100,000 km^2 (Eugster and Hardie, 1978). Reported thickness of preserved ancient lake sediments ranges from less than 20 m to as much as 9000 m (e.g., Pliocene Ridge Basin Group, California; Link and Osborne, 1978).

Sedimentation Processes in Lakes

Modern lakes occur in a variety of environmental settings, including glaciated inland plains and mountain valleys, nonglaciated inland plains and mountain regions, deserts, and coastal plains. They exist under a spectrum of climatic conditions ranging from very hot to very cold and from highly arid to very humid. Most lakes are filled with freshwater, but others, such as the Caspian Sea and many lakes in arid regions (e.g., Great Salt Lake, Utah) are highly saline. Many lakes are associated with other types of depositional systems, notably glacial, fluvial, eolian, and deltaic systems. The depositional processes that occur in lakes are influenced both by climatic conditions and by a variety of physical, chemical, and biological factors that include the chemistries of their waters and fluctuations in their shorelines and siliciclastic sediment supply.

Open lakes are those that have an outflow of water and a relatively stable (fixed) shoreline and in which inflow and precipitation are approximately balanced by outflow and evaporation. Siliciclastic sedimentation commonly predominates in open lakes; however, chemical sedimentation can occur in open lakes that have a low supply of clastic sediment. **Closed lakes** do not have a major outflow and have fluctuating shorelines, and inflow is commonly exceeded by evaporation and infiltration. These conditions lead to concentration of ions in lake water and a predominance of chemical sedimentation, although siliciclastic sediments may accumulate also.

Influence of Climate. Climatic factors affect lake sedimentation in numerous ways. First, water level in lakes is maintained by the balance between evaporation and precipitation. Also, chemical sedimentation in lakes strongly reflects climatic conditions. For example, chemical sedimentation in lakes of arid regions is dominated by precipitation of gypsum, halite, and various other salts, but in humid climates chemical sedimentation is dominated by carbonate deposition. Sediment input to lakes is influenced by the vegetation cover in the drainage area of the lakes and is greatest in arid regions with low vegetation cover. In cold climates, seasonal drops in temperature lead to freezing of lakes, causing decrease in sediment input and cessation of wave activity, allowing deposition of fine-grained suspended sediment during these quiet-water conditions. Climate and the physiography of lake settings also determine the local weather conditions over lakes. Severe, localized storms with high winds can cause considerable shore erosion, coupled with sediment transport and deposition, during short periods of time.

Physical Processes. The physical processes that interact in lakes to bring about sediment transport and deposition include wind, river inflow, atmospheric heating, surface barometric pressure, and gravity (Fig. 10.21). Surface barometric pressure and gravity effects are of least importance (Sly, 1978). Except in very large lakes, tides commonly play only a relatively minor role in lake processes. Wind processes are of major importance because winds create waves and currents and may help generate seiches (periodic rocking back and forth of water in lake basins). River inflow may generate plumes of fine sediment that extend in surface waters far out into a lake (Fig. 10.22), or it may generate density underflows, or turbidity currents, that carry sediment along the bottom toward the basin center. River inflow can also create currents that flow along the margins of lakes. Other currents may be generated by flow-through of water along the lake bottom toward a point of lake discharge. A more extended discussion of interaction of river inflow into standing bodies of water is given in Chapter 11 under the

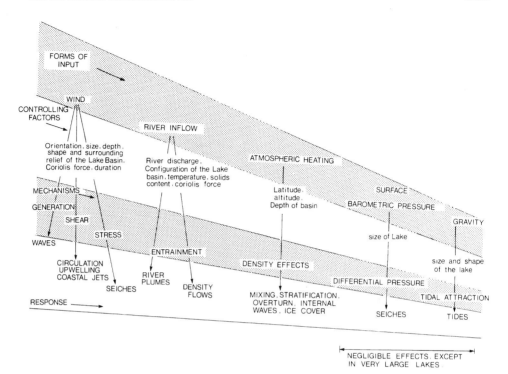

FIGURE 10.21 Lake response to various forms of physical input. (From Sly, P. G., 1978, Sedimentation processes in lakes, *in* A. Lerman (ed.), Lakes—Chemistry, geology, physics. Fig. 1, p. 68, reprinted by permission of Springer-Verlag, Heidelberg.)

heading "Deltaic Sedimentation." Atmospheric heating, which is a function of climate, is responsible for density differences in lake water. These differences can cause stratification of water on the one hand (heating of surface water) or, under some conditions, generation of density currents (by cooling of surface water) that produce mixing and lake overturn. Also, temperature variations may cause alternate freezing and melting of lake surface waters, thereby affecting sediment transport within the lake.

Thus, a variety of sediment transport and depositional mechanisms operate in lakes. Deposition of siliciclastic sediment in the calmer, deeper portion of lakes can take place by settling of fine particles that were suspended in the water column owing to wave and current activity, or deposition may occur from turbidity currents generated where sediment-laden streams discharge into lakes. Sedimentation can occur also along the shallow shoreline of lakes from wind-generated traction currents or river-inflow currents deflected along the lake margin.

The physical depositional processes that take place in lakes are analogous in many ways to those that occur in marine environments and may include turbidite, deltaic, and beach sedimentation. One type of sedimentation process that appears to be particularly characteristic of cold-climate lakes is the formation of **varves,** which are very thin, alternating light- and dark-colored sediment layers. Varves are presumably generated in so-called glacial lakes owing to seasonal freezing and melting. These seasonal changes cause periodic fluctuations in sediment input to the lakes by rivers. Dur-

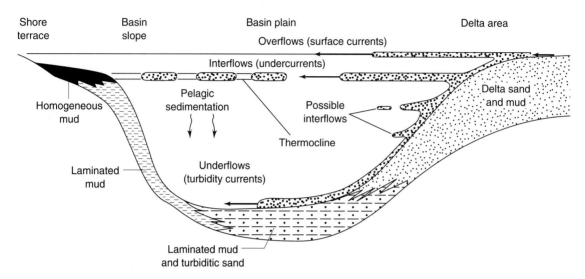

FIGURE 10.22 Distribution mechanisms and kinds of siliciclastic sediments in lakes with annual thermal stratification. Thickness of sediments and basin width are diagrammatic. (After Sturm, M., and A. Matter, 1978, Turbidites and varves in Lake Brienz (Switzerland): Deposition of clastic detritus by density currents, *in* A. Matter and M. E. Tucker (eds.), Modern and ancient lake sediments: Blackwell Scientific Publications, Oxford. Fig. 10, p. 162, reproduced by permission.)

ing summer months when lakes are open, thick, light-colored, coarse-grained laminae accumulate. Sediment input is limited during the winter, when the lakes are frozen over, resulting in deposition of thinner, finer-grained laminae that are colored dark by the presence of fine organic matter that settles slowly from suspension.

Varves may form also in nonglacial lakes (Picard and High, 1981) owing, at least in some lakes, to seasonal variations in carbonate production. Greater carbonate production in warm water during summer tends to mask fine organic matter, which accumulates very slowly throughout the year, producing a thick, light-colored lamina of carbonate sediment. A very thin, clayey, organic layer forms during the winter months when cold water inhibits carbonate production. Fine-grained clastic sediments produced during spring floods may mix with the upper part of an organic layer, also creating varves.

The physical sedimentary processes in lakes differ from those in marine environments in three other major aspects (Sly, 1978):

1. The small size of most lakes greatly limits the generation of long-period wind waves; thus, the energy level of all but the very largest lakes is well below that of marine systems. As a result, deposition of coarse sands and gravels is confined mainly to the shallow-water areas of lakes.
2. Lakes are nearly closed systems with respect to sediment transport. The ratio between the area of land drainage and lake area is commonly high, and sedimentation rates in lakes are much higher than in marine environments, generally at least 10 times that of marine environments (which range from <3 mm/1000 yr to >1 m/1000 yr).
3. Lakes are almost tideless. Tidal currents are of negligible importance in sediment transport, and littoral zones are absent or much reduced compared to marine littoral zones.

Physicochemical Processes. The chemistry of lake waters varies from lake to lake but is dominated by the presence of calcium, magnesium, sodium, potassium, carbonate, sulfate, and chloride ions. Thus, the most common chemical sediments in the lakes of humid regions are carbonates, although phosphates, sulfides, cherts, and iron and manganese oxides are present in some lakes. In arid regions, where rates of evaporation are high, chemical lake sediments are dominated by carbonates, sulfates, and chlorides. The evaporite deposits of lakes include many common marine evaporite minerals such as gypsum, anhydrite, halite, and sylvite, but they also include several minerals such as trona, borax, epsomite, and bloedite that are not common in marine evaporites (Chapter 8). The pH of lake waters commonly falls between 6 and 9; however, it can range from less than 2 in some volcanic lakes to as much as 12 in some closed desert lakes. Chemical sedimentation processes are most important in closed lakes but may predominate also in some open lakes where the clastic sediment supply is low. For additional discussion of lake chemistry, see Eugster and Hardie (1978), Eugster and Kelts (1983), Jones and Bowser (1978), Kelts and Hsü (1978), and Stumm (1985).

Organic Processes. Organisms play an important role in lake sedimentation through the following processes:

1. extraction of chemical elements from lake water to build shells, or tests, and subsequent deposition of these shells
2. CO_2 assimilation during photosynthesis
3. bioturbation
4. contribution of plant remains to form plant deposits

Many kinds of organisms live in lakes and contribute their skeletal and nonskeletal remains to lake sediments. Siliceous diatoms are particularly widespread and noteworthy. Diatoms carry out photosynthesis and are the only important type of lake organism that produces siliceous tests. Their remains form important diatomite deposits in many Pleistocene lakes. Pelecypods, gastropods, calcareous algae, and ostracods also abound in many lakes and are important contributors of calcium carbonate sediments. Blue-green algae (cyanobacteria) carry on photosynthesis and also trap fine sediment to form stromatolites. Many different types of higher plants live in lakes. Under the reducing conditions and high sedimentation rates that exist in some lakes, the remains of higher plants may be partially preserved to eventually form peat and coal. Considering the small size of many lakes and their lower alkalinity and buffering capacity, compared to those of the open ocean, the assimilation of CO_2 by plants during photosynthesis is a much more important factor in controlling the pH of lakes than that of the ocean. Thus, increase in pH owing to photosynthetic removal of CO_2 likely exerts a dominant control on carbonate sedimentation in lakes. Finally, organisms such as pelecypods, freshwater shrimp, and worms may burrow and rework lake sediments, destroying laminations and other primary sedimentary structures.

Characteristics of Lacustrine Deposits

The deposits of most hydrologically open lakes are dominated by siliciclastic deposits, derived mainly from rivers—but possibly including windblown, ice-rafted, and volcanic detritus. Much of this sediment is deposited along the shores of lakes, particularly near river mouths. Gravelly sediment may be present in the toes of alluvial fans or fan deltas that extend to the lake edge or into the lake. Sand likewise accumulates mainly along the lake shore in deltas, beaches, spits, or barriers. Sand may also be carried by turbidity currents into the middle of the lake (Fig. 10.22); however, deeper parts of the lake are characterized particularly by the presence of fine silt and clay. Some muddy sediment is transported into deeper water by surface overflows. In density-stratified lakes, muddy sediment may also be carried as a turbidity interflow above cold, denser lake water. Coarser particles in such interflows settle fairly quickly and accumulate as silt layers. Finer particles settle more slowly to form clay layers. Thus, as shown in Figure 10.22, the siliciclastic deposits of open lakes may consist of deltaic sands and muds (and possibly alluvial-fan gravels), turbidite sands and silts, and homogeneous to laminated muds.

In open lakes where the clastic sediment supply is low, chemical and biochemical processes predominate, resulting in deposition of largely chemical sediments. Primary inorganic carbonate precipitation (caused by loss of CO_2 through plant photosynthesis and/or increase in water temperature or mixing of water masses) and production of shells by calcium carbonate– or silica-secreting organisms account for most of the sedimentation. Various types of invertebrate organisms live in lakes, mostly at water depths less than about 10 m (Reineck and Singh, 1980). The principal types of invertebrates remains in lacustrine sediments include bivalves, ostracods, gastropods, diatoms, and charophytes and other algae. Chemical lake deposits consist mainly of carbonate sands and muds (less commonly siliceous diatom deposits). Stromatolites produced by blue-green algae (cyanobacteria) are common also in some lake deposits. Various amounts of noncarbonate organic matter and some siliciclastic sediment may be present. Plant life is commonly abundant in shallow water around lake margins, and plant deposits may become important during the late stages of lake filling. Carbonate sediments may interfinger along the lake margin with siliciclastic deltaic or alluvial deposits. Typical facies in an open lake with low siliciclastic sediment input are illustrated in Figure 10.23.

Hydrologically closed lakes occur in regions of interior drainage where lake levels may experience considerable fluctuation owing to seasonal flooding. Alluvial fans are commonly present around the borders of such lakes, and the sandy aprons (sandflats) of such fans may extend into the lake. During high water, the edges of these sandflats can be reworked by wave action, resulting in redeposition of wave-rippled sandy

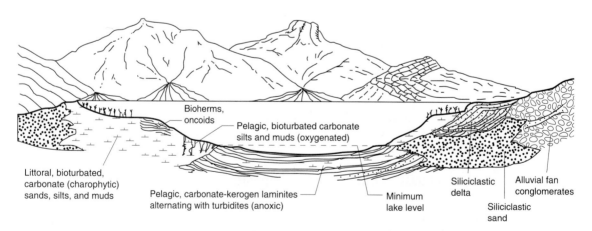

Bioherms,
oncoids
Pelagic, bioturbated carbonate
silts and muds (oxygenated)

Littoral, bioturbated,
carbonate (charophytic)
sands, silts, and muds

Pelagic, carbonate-kerogen laminites
alternating with turbidites (anoxic)

Minimum
lake level

Siliciclastic
delta

Alluvial fan
conglomerates

Siliciclastic
sand

FIGURE 10.23 Principal features and sediment types in an open lake characterized by low siliciclastic sediment input. (After Eugster, H. P., and K. Kelts, 1983, Lacustrine chemical sediments, *in* J. J. Goudie, and K. Pye (eds.), Chemical sediments and geomorphology: Academic Press, New York. Fig. 12.2, p. 333, reproduced by permission.)

sediment along the lake edge. Most sedimentation in closed lakes takes place by chemical/biochemical processes in waters made saline by high rates of evaporation. Two kinds of closed lakes are recognized. **Perennial** basins receive inflow from at least one perennial stream. They commonly do not dry up completely from year to year, although some may dry up occasionally. Most perennial lakes are saline, but some are dilute. The deposits of perennial lakes include carbonate muds, silts, and sands, commonly with intergrowths of evaporite minerals, and may include stromatolites (Fig. 10.24A). Bedded evaporites may be present in the central part of the lake. **Ephemeral** salt-pan basins are fed by ephemeral runoff, springs, and groundwater and are generally dry through part of each year. Ephemeral salt-pan deposits may also contain carbonate sediments, including spring travertine or tufa, but bedded salt deposits are much more important (Fig. 10.24B). Saline deposits interfinger with siliciclastic sand-flat deposits around the margin of the salt pan.

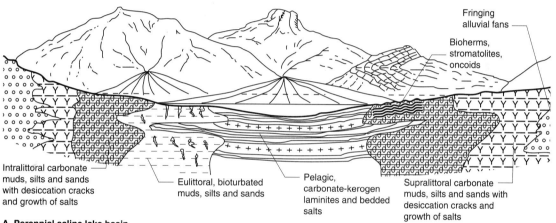

A Perennial saline lake basin

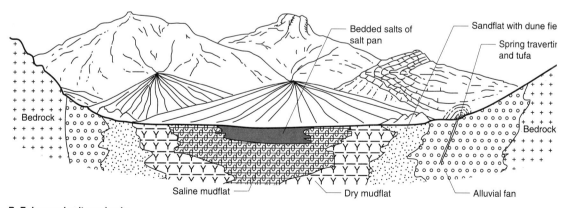

B Ephemeral salt pan basin

FIGURE 10.24 Depositional subenvironments and sediment types in hydrologically closed lake basins. A. Perennial saline lake basin. B. Ephemeral salt-pan basin. (After Eugster, H. P., and K. Kelts, 1983, Lacustrine chemical sediments, *in* J. J. Goudie and K. Pye (eds.), Chemical sediments and geomorphology: Academic Press, New York. Figs. 12.8 and 12.9, p. 351, reproduced by permission.)

Numerous kinds of sedimentary structures occur in lake sediments, including laminated bedding, varves, stromatolites, cross-bedding, ripple marks, parting lineations, graded bedding, groove casts, load casts, soft-sediment deformation structures, burrows and worm trails, raindrop and possible ice-crystal impressions, mudcracks, and vertebrate footprints. Many of these structures are illustrated by Fouch and Dean (1982) in a sequence of excellent color photographs. Varves are one of the more diagnostic characteristics of lake sediments, but light and dark laminae resembling varves have also been reported in nonlacustrine sediments (e.g., some laminated marine deposits). Another distinguishing characteristic of lake sediments is that individual lake beds tend to be thin and laterally continuous compared to associated fluvial deposits (although total lake sediments can be very thick). Otherwise, no uniquely diagnostic structures occur in lake sediments. Many sedimentary structures of lacustrine deposits are similar to those of shallow-marine sediments.

Owing to high sedimentation rates in lakes and the fact that they are closed systems with respect to sediment transport, all lakes are ephemeral features. Lake basins eventually fill with sediment, and most are converted into fluvial plains as they are overrun by fluvial systems. Therefore, lake filling is commonly regarded as a regressive process. That is, coarser, nearshore sediments are believed to gradually encroach on finer lake basin sediments and to be covered in turn with fluvial sediments. This postulated process of filling theoretically generates shallowing and coarsening-upward successions of lake facies. Figure 10.25 shows idealized facies models for lacustrine carbonates that illustrate some possible shallowing-upward successions for carbonate lake sediments deposited under low- and high-energy conditions in bench-margin and ramp-margin type lakes as basinal sedimentation gave way to marginal sedimentation during shallowing. (Bench-type margins are characterized by a nearly flat, nearshore platform that breaks abruptly offshore to a steep slope; ramp-type margins lack the flat platform and may have a lagoon separated by a shoal from deeper water.) Most ancient lake sediments are composed of successions of lake sediments that are far more complex than these ideal models. They commonly display evidence of multiple transgressive-regressive cycles, indicating more than one episode of lake development and filling, which may be related to climate or tectonism. Although the ultimate filling of lakes and their encroachment by prograding fluvial or other coarser-grained deposits may generate a gross coarsening-upward pattern of facies, ideal coarsening-upward successions of lake sediments probably rarely occur, except perhaps in some very small lakes (Picard and High, 1981).

Recognition of Ancient Lake Deposits

Recognition of ancient lake deposits requires considerable care to differentiate them from sediments of other origins, particularly shallow-marine deposits. Freshwater fossils, vertebrate tracks, and desiccation structures in lake-margin sediments can help establish them as nonmarine deposits, although some invertebrate faunas from saline lakes are similar to marine faunas. The presence of varves is moderately diagnostic, although varvelike deposits can form in some other environments. Lake sediments are commonly better sorted than fluvial sediments, and they may display a general tendency toward fining upward and inward toward the basin center. They occur in association with other continental sediments, particularly fluvial and periglacial sediments (described in Section 10.5). Lake deposits are predominantly fine-grained sediments, either siliciclastic muds or carbonate sediments and evaporites. Sandstones and conglomerates may be present also, but they are distinctly subordinate to fine-grained sed-

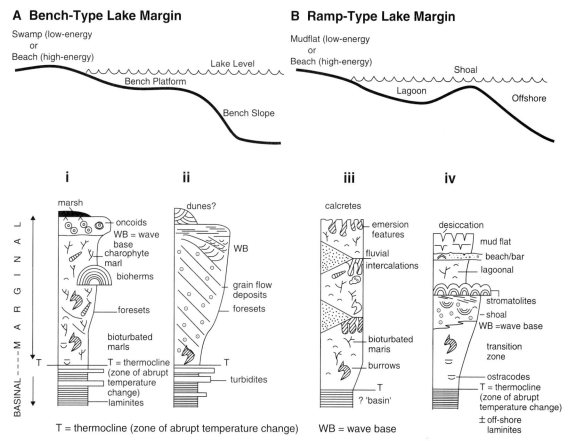

FIGURE 10.25 Shallowing-upward, progradational (regressive) facies models for lacustrine carbonates generated in (i) low-energy bench-margin; (ii) high-energy bench-margin; (iii) low-energy ramp-margin; (iv) high-energy ramp-margin type lakes. T = thermocline (zone of abrupt temperature change); WB = wave base. (After Platt, N. H., and V. P. Wright, 1991, *in* P. Anadón, L. Cabrera, and K. Kelts (eds.), Lacustrine facies analysis: Blackwell Scientific Publications, Oxford. Fig. 2, p. 59, reprinted by permission.)

iments. Finally, some evaporite minerals in closed-lake deposits may be different from evaporite minerals in marine sediments (Picard and High, 1972, 1981).

Examples of Ancient Lake Deposits

Some ancient lakes may have existed in either open (freshwater) lake basins or closed (mostly saline) basins throughout their history; however, many ancient lakes were modified with time, changing from open to closed condition. Therefore, vertical profiles of lake sediments may include the deposits of both open and closed lakes. Lake sediments are preserved in a variety of tectonic settings, including extensional rift systems, strike-slip basins, foreland basins, and cratonic basins. Ancient lake sediments are known from many parts of the world in sedimentary successions ranging from

Holocene to Precambrian; however, the number of known lacustrine deposits appears to decrease exponentially with increasing time, probably owing to chance of preservation or lack of recognition (Eugster and Kelts, 1983).

Some of the better-known lacustrine deposits in North America include the Pliocene Glenn Ferry Formation of the Snake River Plain; the Eocene Green River Formation of Utah, Colorado, and Wyoming, known for its oil-shale deposits; much of the Jurassic Morrison Formation of the Colorado Plateau; parts of the Triassic Chugwater Group of Wyoming; the Triassic Lockatong Formation of New Jersey; the Devonian Escuminac Formation of southern Quebec; and the Carboniferous Strathlorne Formation of Nova Scotia. Some well-known lake deposits from other parts of the world include the Cenozoic rift-basin deposits of East Africa; the Cretaceous rift-basin deposits of Brazil (which are source rocks for much of Brazil's oil; Abrahão and Warme, 1990); the clastics, evaporites, and carbonates of the Triassic Keuper Marl of South Wales; the Permo-Triassic Beaufort strata of the Eastern Karoo Basin, Natal, South Africa; parts of the Lower Permian Rotliegend deposits of southwest and eastern Germany; and the middle Devonian sediments from the Old Red Sandstone of the Orcadian Basin of northeast Scotland. Many of these lake deposits throughout the world include thick successions of organic-rich shales that are important source rocks for petroleum (Katz, 1990).

As mentioned, many ancient lakes evolved from open to closed systems. The deposits of the Lockatong Formation of the Triassic Newark Group of New Jersey provide an example of this kind of development. The entire formation is characterized by small-scale cycles or successions, a few meters thick, that can be traced for distances of a kilometer or so. Van Houten (1964) grouped these cycles into two major types: **detrital cycles** about 5 m thick occurring mainly in the lower part of the sequence and **chemical cycles** about 3 m thick (Fig. 10.26). The detrital cycles are generally coarsening-upward successions, 4 to 6 m thick, that begin with black pyritic mudstone and pass upward through interlaminated dolomitic mudstones to massive dolomitic mudstones, commonly mud-cracked and bioturbated. Fine-grained, cross-stratified sandstones with convolutions may cap the cycles. The cycles are interpreted as small-scale regressive units that formed in a hydrologically open lake during lake shrinkage following deepening of the lake. The chemical cycles grade upward from dark, laminated mudstone with dolomite and calcite laminae and lenses into more massive dolomite or analcime-rich (zeolite-rich) mudstone, characterized by mudcracks and small-scale syneresis cracks. They formed during periods of closed drainage resulting from progressive reduction of the lake area and subsequent shallowing as siliciclastic input was reduced. The entire Lockatong Formation is interpreted as the deposits of a perennial lake that was thermally stratified most of the time. The climate was warm, with cyclic rainfall that brought about changing water levels, leading to cyclic sedimentation.

Another example of ancient lake-basin evolution is provided by the Ridge Basin Group (Pliocene) of California. The Ridge Basin sediments range to 9000 m thick but accumulated in a basin only 15 km by 40 km (Link and Osborne, 1978). Sedimentation began under open conditions in deep water by deposition of deltaic sediments and turbidites (Fig. 10.27A). Owing to continued strike-slip displacement along the San Gabriel fault, external drainage from the Ridge Basin lake was blocked to the south, and it became a closed system. Infilling of the lake led to shallowing and deposition during this shallow phase of alluvial-fan, fluvial, deltaic, barrier-bar sediments along the margin of the lake and siliciclastic mud, zeolite mud, dolomite, and stromatolites in the central part of the basin (Fig. 10.27B).

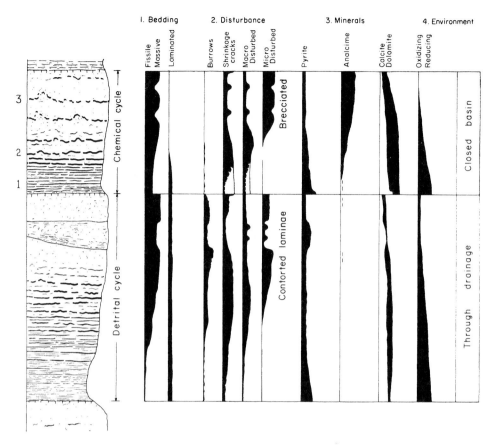

FIGURE 10.26 Depositional model for terrigenous (detrital) and chemical cycles in sediments of the Lockatong Formation, New Jersey, interpreted as lake deposits. (After Van Houten, F. B., 1964, Cyclic lacustrine sedimentation, Upper Triassic Lockatong Formation, central New Jersey and adjacent Pennsylvania, *in* D. F. Merriam (ed.), Symposium on cyclic sedimentation: Kansas Geol. Survey Bull. 169, v. 11. Fig. 6, p. 505, reprinted by permission of Kansas Geological Survey, Lawrence.)

One of the most intensely studied ancient closed-lake deposits, probably owing to the fact that it contains large reserves of oil shale and trona (sodium carbonate), is the Eocene Green River Formation of the Colorado Plateau. This formation consists of several members, which appear to record climatic change from pluvial (wet) to arid and back to pluvial through time. Eight lithofacies are recognized in various parts of the formation: (1) flat pebble conglomerate, (2) lime sandstone, (3) mudstone, (4) oil shale, (5) trona-halite, (6) siliciclastic sandstone, (7) volcanic tuffs, and (8) stromatolitic limestone. The Wilkins Peak Member, which was deposited under arid conditions, is particularly interesting because it records many changes from perennial to ephemeral conditions, as illustrated in Figure 10.28.

Additional examples of ancient lake deposits are described by Anadón, Cabrera, and Kelts (1991), Collinson (1986a), Eugster and Kelts (1983), Matter and Tucker (1978), and Picard and High (1972, 1981).

FIGURE 10.27 Paleoenvironmental reconstruction of the Pliocene Ridge Basin, California, during (A) the open, deep-water lacustrine and/or marine phase and (B) the closed, shallow-water lacustrine phase. (After Link, M. H., and R. H. Osborne, 1978, Lacustrine facies in the Pliocene Ridge Basin Group, Ridge Basin, California, *in* A. Matter, and M. E. Tucker (eds.), Modern and ancient lake sediments: Blackwell Scientific Publications, Oxford. Fig. 14, p. 185, and Fig. 15, p. 186, reproduced by permission.)

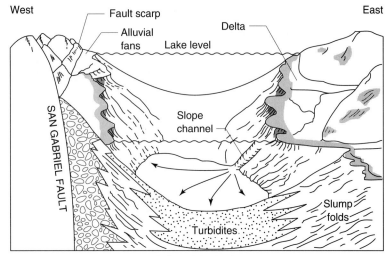

A Open lake basin

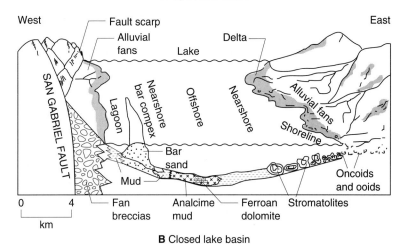

B Closed lake basin

10.5 GLACIAL SYSTEMS

Introduction

I have placed glacial systems last in this discussion of continental environments because the glacial environment, in a broad sense, is a composite environment that includes fluvial, eolian, and lacustrine environments. It may also include parts of the shallow-marine environment. Glacial deposits make up only a relatively minor part of the rock record as a whole, although glaciation was locally important at several times in the geologic past, particularly during the Late Precambrian, Late Ordovician, Carboniferous/Permian, and Pleistocene (Eyles and Eyles, 1992). Glaciers presently cover

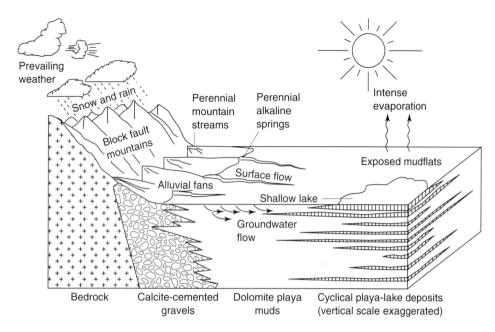

FIGURE 10.28 Schematic diagram showing the general depositional framework for the Wilkins Peak Member of the Green River Formation. (After Eugster, H. P., and L. A. Hardie, 1975, Sedimentation in an ancient playa-lake complex: The Wilkins Peak Member of the Green River Formation of Wyoming: Geol. Soc. America Bull., v. 86. Fig. 19, p. 331.)

about 10 percent of Earth's surface, mainly at high latitudes. They exist primarily as large ice masses on Greenland and Antarctica and as smaller masses on Iceland, Baffin Island, and Spitsbergen. Small mountain glaciers occur at high elevations in all latitudes. By contrast to their present distribution, ice sheets covered about 30 percent of Earth during maximum expansion of glaciers in the Pleistocene and extended into much lower latitudes and elevations than those currently affected by continental glaciation.

The glacial environment is confined specifically to those areas where more or less permanent accumulations of snow and ice exist. Such environments are present in high latitudes at all elevations (continental glaciers) and at low latitudes (mountain or valley glaciers) above the snow line—the elevation above which snow does not melt in summer. Mountain glaciers form above the snow line by accumulation of snow. They move downslope below the snow line only if rates of accumulation of snow above the snow line exceed rates of melting of ice below. The factors affecting glacier movement and the mechanisms of ice flow are not of primary interest here. Our concerns are the sediment transport and depositional processes associated with glacial movement and melting and the sediments deposited by glaciers. Useful references to these processes and deposits include Anderson and Ashley (1991); Ashley, Smith, and Shaw (1985); Brodzikowski and van Loon (1991); Dowdeswell and Scourse (1990); Drewry (1986); Easterbrook (1982); Edwards (1986); Eyles and Eyles (1992); Jopling and McDonald (1975); Molnia (1983); and Sharp (1988).

Environmental Setting

The glacial environment proper is defined as all those areas in direct contact with glacial ice. It is divided into the following zones:

1. the **basal** or **subglacial zone,** influenced by contact with the bed
2. the **supraglacial zone,** which is the upper surface of the glacier
3. the **ice-contact zone** around the margin of the glacier
4. the **englacial zone** within the glacier interior

Depositional environments around the margins of the glacier are influenced by melting ice but are not in direct contact with the ice. These environments make up the **proglacial environment,** which includes glaciofluvial, glaciolacustrine, and glaciomarine (where glaciers extend into the ocean) settings (Fig. 10.29). The area extending beyond and overlapping the proglacial environment is the **periglacial environment.**

The basal zone of a glacier is characterized by erosion and plucking of the underlying bed. Debris removed by erosion is incorporated into the bed of the glacier. This debris causes increased friction with the bed as the glacier moves and thus aids in abrasion and erosion of the bed. The supraglacial and ice-contact zones are zones of melting or ablation where englacial debris carried by the glacier accumulates as the glacier melts. The glaciofluvial environment is situated downslope from the glacier front and is characterized by fluctuating meltwater flow and abundant coarse englacial debris that is available for fluvial transport. The glaciofluvial environment is one of the characteristic environments in which braided streams develop. Extensive outwash plains or aprons may also be present along the margins of outwash glaciers. Lakes are very common proglacial features, created by ice damming or damming by glacially deposited sediments. Meltwater streams draining into these lakes may create large, coarse-grained deltas along the lake edge, while finer sediment is carried outward in the lake by suspension or as a density underflow. Glaciers that extend out to sea create an important environment of glaciomarine sedimentation where sediments are deposited close to shore by melting of the glacier in contact with the ocean or farther out on the shelf or slope by melting of ice blocks, or icebergs.

The glacial environment may range in size from very small to very large. **Valley glaciers** are relatively small ice masses confined within valley walls of a mountain. **Piedmont glaciers** are larger masses or sheets of ice formed at the base of a mountain front where mountain glaciers have debouched from several valleys and coalesced. **Ice sheets,** or continental glaciers, are huge sheets of ice that spread over large continental areas or plateaus.

Transport and Deposition in Glacial Environments

Glaciers acquire their sediment load by abrasion and plucking of material from the bedrock and adjacent valley walls and by free fall of rock debris from the valley walls above. Much of this material is carried in the base of the glacier, but other rock debris is concentrated along the glacier margin or dispersed within the body of the glacier (Fig. 10.30). Large and small blocks of rock can be quarried by moving ice and incorporated into the base or sides of the glacier. Extremely fine sediment, called **rock flour,** is also produced by grinding of the rock-studded glacier base over bedrock. Thus, the glacier sediment load typically consists of an extremely heterogeneous assortment of particles ranging from clay-size grains to meter-size boulders.

As glaciers move downslope below the snow line, they eventually reach an elevation where the rate of melting at the front of the glacier equals or exceeds the rate of

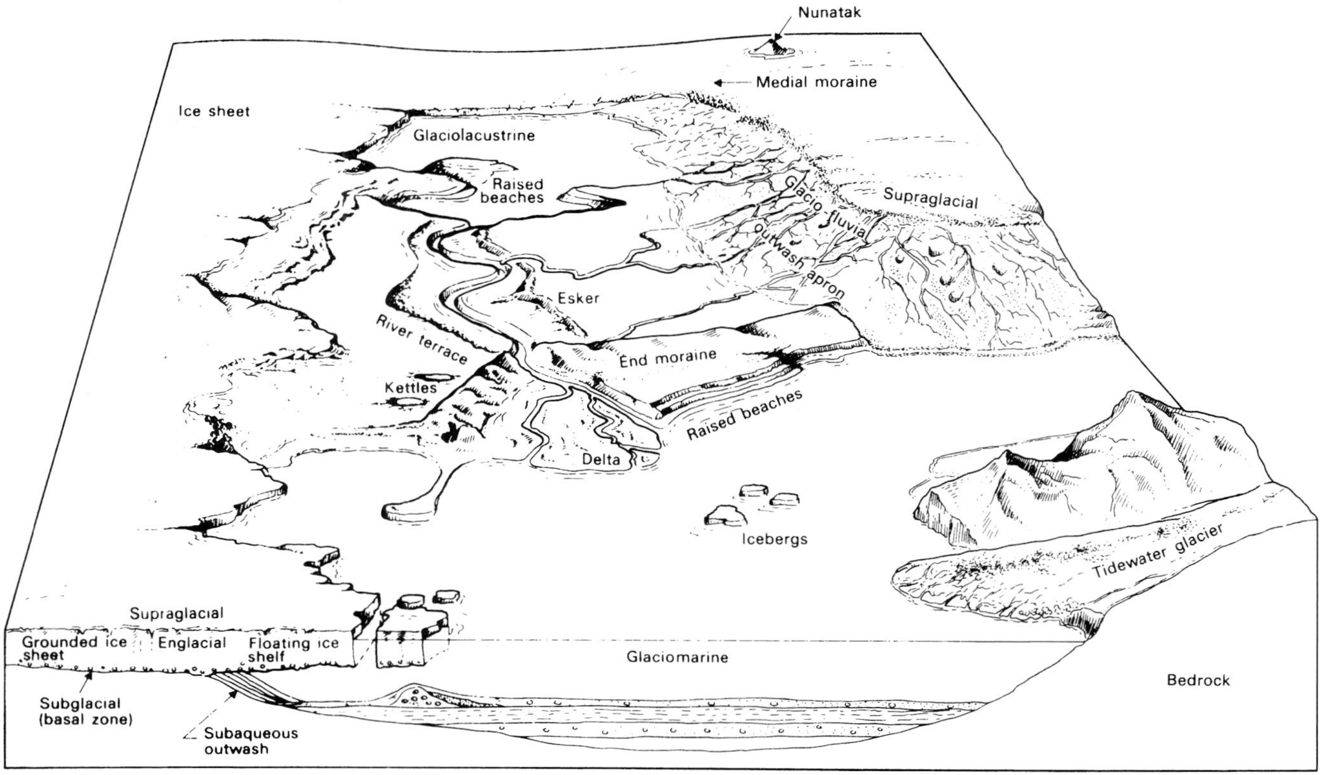
Nunatak
Medial moraine
Ice sheet
Glaciolacustrine
Raised beaches
Supraglacial
Glacio-fluvial outwash apron
Esker
River terrace
End moraine
Kettles
Raised beaches
Delta
Icebergs
Tidewater glacier
Supraglacial
Grounded ice sheet
Englacial
Floating ice shelf
Glaciomarine
Bedrock
Subglacial (basal zone)
Subaqueous outwash

FIGURE 10.29 Glacial and associated proglacial environments. (From Edwards, M. B., 1986, Glacial environments, *in* H. G. Reading (ed.), Sedimentary environments and facies, 2nd ed. Fig. 13.2, p. 448, reprinted by permission of Elsevier Science Publishers, Amsterdam.)

FIGURE 10.30 Schematic diagram illustrating sediment transport paths within glaciers and the various kinds of glacial moraines. (After Sharp, R. P., 1988, Living ice: Cambridge University Press, Cambridge. Fig. 2.5, p. 30, reproduced by permission.)

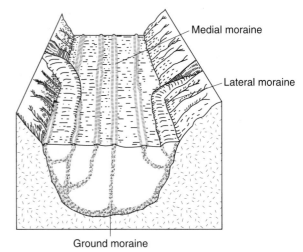

Medial moraine

Lateral moraine

Ground moraine

new snow accumulation above the snow line. If the rate of melting approximately equals the rate of accumulation, the glacier achieves a state of equilibrium in which it neither advances nor retreats. Within such an equilibrium glacier, internal movement of ice continues to carry the rock load along and supply rock debris to the melting snout of the glacier. This process causes a ridge of unsorted sediment, called an **end moraine** or terminal moraine, to accumulate in front of the glacier. **Lateral moraines,** or marginal moraines, can accumulate from concentrations of debris carried along the edges of the glacier where ice is in contact with the valley wall. **Medial moraines** may form where the lateral moraines of two glaciers join (Fig. 10.30). When the rate of melting at the snout of a glacier exceeds the rate of new snow accumulation above the snow line, the glacier retreats back up the valley. If a glacier retreats steadily, it drops its load of rock debris as lateral moraines, medial moraines, and a more or less evenly distributed sheet of **ground moraine.** If the glacier retreats in pulses, it leaves a succession of end moraines, called **recessional moraines.**

As glaciers melt on land, large quantities of water run along the margins, beneath, and out from the front of the glacier to create a meltwater stream. Such streams flow with high but variable discharge in response to seasonal and daily temperature variations. Near the glacier front, the meltwater quickly becomes choked with suspended sediment and loose bedload sand and gravel, leading to formation of branching and anastomosing braided-stream channels. (The characteristics of these braided streams are discussed in Section 10.2.) Streams that discharge into glacial lakes tend to build prograding delta systems into the lakes with steeply inclined foresets that grade downward to gently inclined bottomset beds (Edwards, 1986). Very fine sediment discharged into the lake from streams may be dispersed basinward in suspension by wind-driven waves or currents. If a large enough concentration of sediment is present in suspension to create a density difference in the water, a density underflow or turbidity current will result that can carry sediment along the lake bottom into the middle of the basin. Strong winds blowing over a glacier or an ice sheet pick up fine sand from exposed, dry outwash plains and deposit the sand downwind in nearby areas as sand dunes. Fine dust picked up by wind can be kept in suspension and transported long distances before being deposited as loess in the periglacial environment.

Where glaciers extend beyond the mouths of river valleys to enter the sea, their sediment load is dumped into the ocean to form **glacial-marine** sediments. Sedimentation under these circumstances may take place in four different ways:

1. Melting beneath the terminus of the glacier allows large quantities of glacial debris to be released onto the seafloor with little reworking (Fig. 10.31.).
2. Large blocks of ice calve off from the front of the glacier and float away as icebergs. These icebergs gradually melt, allowing their sediment load to drop onto the seafloor, either on the shelf or in deeper water.
3. Fresh glacial meltwater charged with fine sediment can rise to the surface to form a low-density overflow above denser saline water. Silt and flocculated clays then gradually settle out of suspension from this freshwater plume.
4. Mixing of fresh meltwater and seawater may produce a high-density underflow that can carry sand-size sediment seaward.

Glacial Facies

Because the broad glacial environment encompasses the proglacial and periglacial environments as well as the glacial environment proper, it is necessary, to avoid confusion, to distinguish between glacial facies deposited directly from the glacier and facies transported and reworked by processes operating beyond the margins of glaciers. Furthermore, it is desirable to distinguish between glacial facies deposited on land and

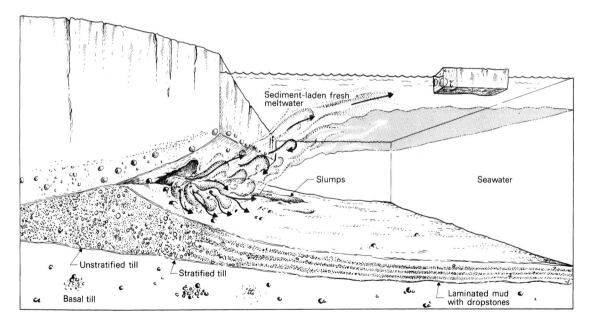

FIGURE 10.31 Hypothetical model for glaciomarine sedimentation. Most of the fresh glacial meltwater rises to the surface of the sea as a low-density overflow layer. This layer gradually mixes with seawater. Silt and flocculated clay are gradually deposited from suspension. Rapid mixing of freshwater and seawater adjacent to tunnel mouths may produce a high-density underflow capable of transporting sand-grade sediment and possibly coarser material. (From Edwards, M. B., 1978, Glacial environments, *in* H. G. Reading (ed.), Sedimentary environments and facies. Fig. 13.5, p. 423, reprinted by permission of Elsevier Science Publishers, Amsterdam.)

TABLE 10.4 Facies of glacial environments

Facies of continental glacial environments
Grounded ice facies
Glaciofluvial facies
Glaciolacustrine facies
    Facies of proglacial lakes
    Facies of periglacial lakes
Cold-climate periglacial facies
 	
Facies of marine glacial environments
Proximal facies
Continental shelf facies
Deep-water facies

Source: Eyles, N., and A. D. Miall, 1984, Glacial facies, *in* R. G. Walker (ed.), Facies models, 2nd ed.: Geoscience Canada Reprint Ser. 1, p. 15–38.

those deposited on the seafloor. Table 10.4 illustrates the range of sedimentary facies that are affected in some way by glacial processes. Many of these facies are subtypes of facies deposited in fluvial, lacustrine, eolian, shallow-marine, and deep-marine environments, which are treated elsewhere in this book. Therefore, we focus our discussion here primarily on grounded-ice facies and proximal marine-glacial facies.

Poorly sorted glacial deposits are called glacial **diamicts.** Their consolidated equivalents are **diamictites,** sometimes called diamictons. Glacial diamicts are deposited more or less directly from ice without additional winnowing or reworking by water. Easterbrook (1982) suggests that glacial diamicts be divided into two types:

1. **till,** which is deposited on land
2. **glaciomarine drift,** which is glacial debris melted out of floating ice in marine water

On the other hand, owing to some disagreement among glacial geologists about the exact meaning of till, Eyles and Miall (1984) prefer to use the nongenetic terms diamict or diamictite, as appropriate, for all poorly sorted gravel-sand-mud deposits.

Continental Ice Facies

Grounded Ice Facies

Unstratified Diamicts. Diamicts deposited directly from ice on land in various kinds of moraines consist of unstratified, unsorted pebbles, cobbles, and boulders with an interstitial matrix of sand, silt, and clay (Fig. 10.32). They are thus characterized by a bimodal particle-size distribution in which pebbles predominate in the coarser fraction, with cobbles and boulders scattered throughout (Easterbrook, 1982). Some pebbles are rounded, indicating that they are probably stream pebbles entrained by the ice. Others may be faceted, striated, or polished owing to glacial abrasion. Elongated pebbles and cobbles tend to show some preferred orientation, commonly with their long dimensions parallel to the direction of glacial advance. They may also be crudely imbricated, with long axes dipping upstream. Pebble composition can be highly diverse and may include rock types derived from bedrock located hundreds of kilometers distant. Sands and silts are commonly angular or subangular. Much of the silt in glacial deposits is produced by glacial abrasion and grinding.

Stratified Diamicts. In addition to deposition directly from melting ice, deposition of glacial debris can occur also from meltwaters flowing upon (supraglacial), within

FIGURE 10.32 Thick, poorly stratified, poorly sorted glacial diamict (Quaternary), Alaska. (Photograph courtesy of Gail M. Ashley.)

(englacial), underneath (subglacial), or marginal to the glacier. The deposits of these meltwaters form on, against, or beneath the ice and thus are commonly known as **ice-contact** sediments. They are reworked to some degree by meltwater and thus exhibit some stratification. They are also better sorted than sediments deposited directly from ice, commonly lack the characteristic bimodal size distribution of direct deposits, and may contain pebbles rounded by meltwater transport. These stratified deposits can accumulate in channels or as mounds or ridges known as kames, kame terraces, or eskers. **Kames** are small, mound-shaped accumulations of sand or gravel that form in pockets or crevasses in the ice. **Kame terraces** are similar accumulations deposited as terraces along the margins of valley glaciers. **Eskers** are narrow, sinuous ridges of sediment oriented parallel to the direction of glacial advance. They are the deposits of meltwater streams that probably flowed through tunnels within the glacier. The deposits were then let down onto the subglacial surface after the ice melted. Stratified diamicts are commonly characterized by slump or ice collapse features, including contorted bedding and small gravity faults. Stratified glacial facies can include gravels, sands, and silts, some of which may be extremely well stratified, as illustrated in Figure 10.33.

Facies of Proglacial and Periglacial Environments

As discussed, meltwaters issuing from glaciers transport large quantities of glacial debris downslope and deposit it as **glaciofluvial** sediment in braided streams or as

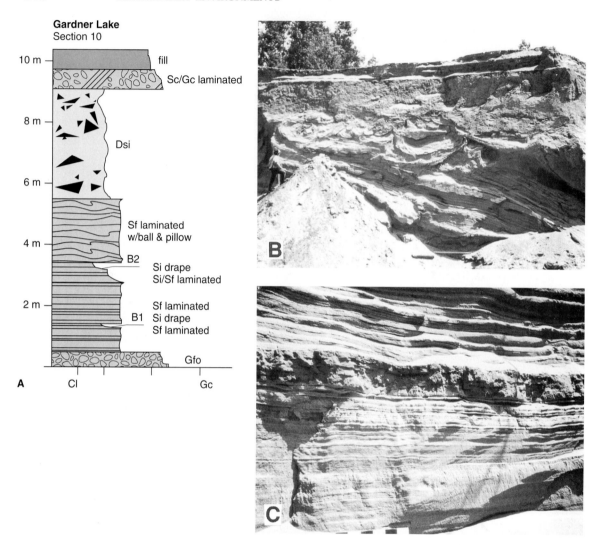

FIGURE 10.33 A. Vertical stratigraphic profile of glacial sediments deposited in an esker system. The environment is interpreted as ice-tunnel gravel overlain by laminated, marine fan sand and diamictite. Lithofacies code: G, gravel; S, sand; Si, silt; D, diamictite; Dsi, silty diamictite; Cl, clay. Grain size: c, coarse; me, medium; f, fine; vf, very fine. Other: m, massive; b, bedding; d, deformed; B1 and B2, bedding surfaces upon which dip of surface fan measured. B. Photograph showing laminated fine sand overlain by fine sand deformed into ball and pillow structures, in turn overlain by a massive diamictite (5.5 to 9 m interval in A), interpreted as a debris flow on the fan surface. Person on left gives scale. C. Close-up view of laminated sand within a thin diamictite interbed. Scale increments are 5 cm. (After Ashley, G. M., et al., 1991, Sedimentology of late Pleistocene (Laurentide) deglacial-phase deposits, eastern Maine: An example of a temperate marine grounded ice-sheet margin: Geol. Soc. America Spec. Paper 261. Fig. 5, p. 113.)

glaciolacustrine sediment in glacial lakes, formed by ice damming or moraine damming. These transported and reworked deposits take on the typical characteristics of the environment in which they are deposited; however, they may retain some characteristics that identify them as glacially derived materials. For example, the large daily to seasonal fluctuations in meltwater discharge may be reflected in abrupt changes in particle size of sediments deposited in meltwater streams or lacustrine deltas. Sediments deposited in streams or lakes very close to the glacier front may also display various slump deformation structures caused by melting of supporting ice. As mentioned in the discussion of lakes, one of the most characteristic properties of glacial lakes is the presence of varves, which form in response to seasonal variations in meltwater flow. Additional details on the characteristics of glaciofluvial and glaciolacustrine sediments are given by Brodzikowski and van Loon (1991) and Jopling and McDonald (1975).

Although sand dunes accumulate in periglacial areas adjacent to some glaciers owing to wind transport of sand from outwash plains, the primary wind deposits in these environments are silts. Deflation of rock flour and other fine sediment from outwash plains and alluvial plains provides enormous quantities of silt-size sediment that is transported by wind and deposited as widespread sheets of fine, well-sorted loess. Owing to the even size of its grains, loess typically lacks well-defined stratification. It is composed dominantly of angular grains of quartz but may also contain some clays.

Marine Glacial Facies

Proximal Facies

In environments where marine water is in direct contact with the glacier margin, substantial quantities of sediment are deposited directly from meltwater conduits or tunnels into subaqueous fans, with additional sediment supplied by melting of rafted ice (Fig. 10.34; Eyles and Miall, 1984). Coarse cobbles and gravels accumulate at the tops of fans, and sands and gravels accumulate within channels owing to sediment gravity underflows. Mud and sand are contributed by ice melting and "rain-out" of suspended sediment. Some reworking of sediment occurs by downslope sediment gravity flows and episodic traction-current activity. Proximal glacial-marine sediments may thus range from poorly sorted, poorly stratified diamicts that resemble those deposited on land to coarse-grained stratified diamicts, with a muddy sandy matrix, that may display current-produced structures. Melting of buried ice masses can cause surface subsidence and associated deformation and faulting of sediment.

Distal Facies

Away from the proximal environment in which glaciers are in direct contact with marine water, glacial sediment is supplied by floating ice masses, and deposition is dominated by marine processes. Melting of icebergs supplies both fine sediment and coarse debris to the ocean floor by "rain-out," or fallout. Ice-rafted debris deposited on the continental shelves may be reworked to some degree by marine waves and currents (and possibly turbidity currents) and may be affected by iceberg grounding (where icebergs touch bottom). In deeper water on the continental shelf, the debris fallout from floating ice may or may not be retransported by turbidity currents to deeper water. Ice-rafted debris that settles to the deep ocean floor is probably little modified by further depositional processes, except for deposition of a mantle of hemipelagic or pelagic sediment. In general, glacial-marine sediments are distinguished from grounded glacial

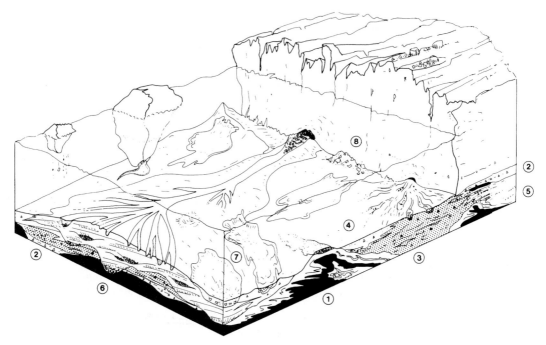

FIGURE 10.34 Proximal subaqueous sedimentation from glaciers: (1) glacially tectonized marine sediment; (2) lensate lodged diamict units; (3) coarse-grained stratified diamicts; (4) pelagic muds and diamicts; (5) coarse-grained proximal outwash; (6) interchannel cross-stratified sands with channel gravels; (7) resedimented facies (debris flows, slides, turbidites); and (8) supraglacial debris. Sediment deformation results from ice advances, melt of buried ice, and iceberg turbation. (From Eyles, N., and A. D. Miall, 1984, Glacial facies, *in* R. G. Walker (ed.), Facies models: Geoscience Canada Reprint Ser. 1. Fig. 8, p. 20, reprinted by permission of Geological Association of Canada.)

diamicts by the presence of some stratification and from all on-land glacial diamicts by the presence of marine fossils and dropstones. Dropstones are scattered cobbles or boulders that drop to the seafloor from melting ice blocks or icebergs. Fossil evidence that particularly suggests a glacial-marine origin includes fossils preserved in growth position as whole shells (i.e., fossils entombed by fallout sediment); marine molluscs or barnacles attached to glacially faceted pebbles; preservation of delicate ornamentation on shells; and the presence of forams and diatoms in the matrix material (Easterbrook, 1982).

Vertical Facies Successions

Successive advances and retreats of valley glaciers and ice sheets produce complex vertical successions of facies as ice progressively overrides proglacial environments during glacial advance, and conversely as direct ice deposits and ice-contact deposits are reworked in the proglacial environment as a glacier retreats. These facies are much too varied and complex to attempt description here; however, Figure 10.35 illustrates some typical vertical profiles that might develop in different parts of a glacial environment during a single phase of glacier advance and retreat. Additional examples of ver-

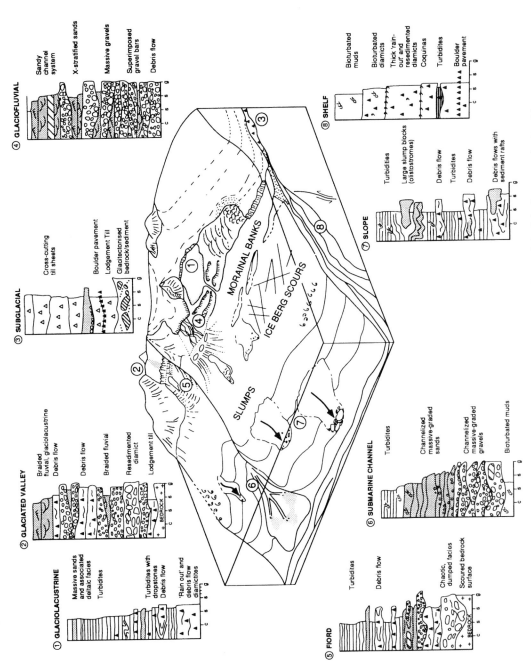

FIGURE 10.35 Typical vertical profiles of facies deposited during a single phase of glacial advance and retreat in various parts of the glacioterrestrial and glaciomarine setting; c, s, s, g = clay, silt, sand, gravel. (From Eyles, N., and C. H. Eyles, 1992. Glacial depositional systems, *in* R. G. Walker and N. P. James (eds.), Facies models: Geological Association of Canada. Fig. 3, p. 74, reproduced by permission.)

351

tical profiles in glacial sediments are given in Brodzikowski and van Loon (1991, Chapter 9) and Edwards (1986, p. 465–466).

Ancient Glacial Deposits

Glacial deposits range in size from small bodies deposited by valley glaciers to diamictite sheets, deposited by continental glaciers, that cover many thousands of square kilometers. The most characteristic feature of continental diamictites, or grounded ice facies, is their extremely poor sorting and lack of stratification. Ancient facies of glaciofluvial, glaciolacustrine, and glaciomarine environments tend to be better stratified and better sorted. Because the characteristics of these proglacial sediments reflect the environment in which they are deposited, it may be very difficult in ancient stratigraphic successions to distinguish proglacial sediments from other types of continental sediments. For example, glaciofluvial sediments may appear much the same as other fluvial sediments. Nevertheless, a few characteristics of these deposits may reveal their relationship to glacial environments. As mentioned, the presence of varves may be diagnostic of glacial lakes, and abrupt changes in sediment size related to variable meltwater discharge may be suggestive of proglacial deposition in general. Ancient glaciomarine sediments are distinguished from other types of glacial deposits by the presence of marine fossils and possibly from other marine deposits by their generally poorer sorting and stratification. They tend to display extreme variation in clast type, reflecting multiple sources. Also, the presence of dropstones, which may deform sedimentary structures such as laminae when they drop into soft sediment, suggests deposition from rafted ice. Anderson and Ashley (1991) further describe several ancient glacial-marine deposits and the paleoclimatic significance of these sediments.

Ancient glacial deposits are best known from sedimentary units of Pleistocene age, which are widespread in many parts of the world. At least four major pulses of continental glaciation occurred during the Pleistocene, plus numerous smaller pulses. Extensive continental glaciation appears to have been important also during the late Paleozoic, late Ordovician, late Precambrian, and early Proterozoic. Carboniferous- to Permian-age glacial deposits are known from South America, southern Africa, Antarctica, India, and Australia. Late Ordovician diamictites have been reported from South America, several parts of Africa, and possibly Ethiopia. Late Precambrian deposits are known on all continents except Antarctica, and early Proterozoic glacial sediments have been reported in North America in a belt extending from Wyoming to Quebec.

The late Paleozoic sediments of the Karoo Basin of southern Africa provide an example of glacial sediments deposited in a transitional valley (land) to shelf (marine) setting (Visser, 1991). During an early grounded ice sheet phase, basal lodgment diamictites (deposited by pressure melting and other mechanical processes) and diamictites generated by normal meltout were deposited on land (Fig. 10.36). On the shelf, deposition took place mainly by rain-out of basal ice debris and to a lesser extent by sediment input from subaqueous meltwater streams, suspension settling, and resedimentation by sediment gravity flows. With rising sea level and regional decoupling of the marine ice sheet from its substrate, the final phase of sedimentation (which took place in a temperate setting) was characterized by accumulation of marine muds on the shelf and formation of glacial valley fills and subaqueous outwash fans in front of tidewater glaciers along the highlands.

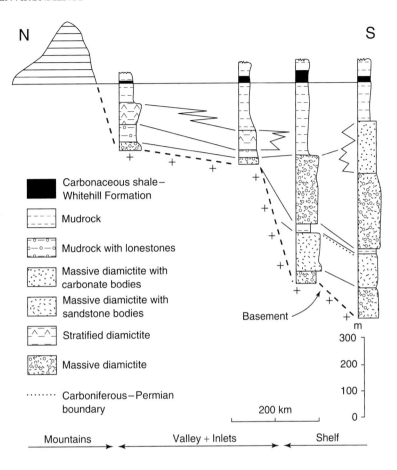

FIGURE 10.36 Stratigraphic sections illustrating the relation between valley and shelf glacial deposits in the Karoo Basin, southern Africa. (From Visser, J. N. J., 1991, The paleoclimatic setting of the late Paleozoic marine ice sheet in the Karoo Basin of southern Africa: Geol. Soc. America Spec. Paper 261. Fig. 5, p. 185.)

FURTHER READINGS

Fluvial Systems

Ethridge, F. G., and R. M. Flores (eds.), 1981, Recent and ancient nonmarine depositional environments: Part II: Alluvial fan and fluvial deposits: Soc. Econ. Paleontologists and Mineralogists Spec. Pub. 31, Tulsa, Okla., p. 49–212.

Flores, R. M., F. G. Ethridge, A. D. Miall, W. E. Galloway, and T. D. Fouch, 1985, Recognition of fluvial depositional systems and their resource potential: Soc. Econ. Paleontologists and Mineralogists Short Course Notes 19, Tulsa, Okla., 290 p.

Fraser, G. S., and L. Suttner, 1986, Alluvial fans and fan deltas: International Human Resources Development Corporation, Boston, 199 p.

Miall, A. D. (ed.), 1978, Fluvial sedimentology: Canadian Soc. Petroleum Geologists Mem. No. 5, Calgary, Alberta, 589 p.

Miall, A. D., 1982, Analysis of fluvial depositional systems: Am. Assoc. Petroleum Geologists Education Course Note Series No. 20, Tulsa, Okla., 75 p.

Nilsen, T. H. (ed.) 1984, Fluvial sedimentation and related tectonic framework, western North America: Sed. Geology, v. 36, 523 p.

Rachocki, A., 1981, Alluvial fans: John Wiley & Sons, New York, 161 p.

Rachocki, A. H., and M. Church (eds.), 1990, Alluvial fans: John Wiley & Sons, Chichester and New York, 391 p.

Schumm, S. A., 1977, The fluvial system: John Wiley & Sons, New York, 338 p.

Van Houten, F. B. (ed.), 1977, Ancient continental deposits: Benchmark Papers in Geology, Dowden, Hutchinson and Ross, Stroudsburg, Pa., 367 p.

Eolian Systems

Barndorff-Nielsen, O. E., and B. B. Willets (eds.), 1991, Aeolian grain transport 1—Mechanics: Springer-Verlag, Wien, New York, 181 p.

Brookfield, M. E., and T. S. Ahlbrandt (eds.), 1983, Eolian sediments and processes: Elsevier, Amsterdam, 660 p.

Ethridge, F. G., and R. M. Flores (eds.), 1981, Recent and ancient nonmarine depositional environments: Part IV: Eolian deposits: Soc. Econ. Paleontologists and Mineralogists Spec. Pub. 31, Tulsa, Okla., p. 279–349.

Frostick, L. E., and I. Reid, 1987, Desert sediments: Ancient and modern: Geol. Soc. Spec. Pub. 35, Blackwell, Oxford, 401 p.

Glennie, K. W., 1970, Desert sedimentary environments: Developments in sedimentology 14: Elsevier, Amsterdam, 222 p.

Hesp, P., and S. G. Fryberger (eds.), 1988, Eolian sediments: Sedimentary geology, v. 55, p. 1–163 (special issue devoted to eolian sediments).

Kocurek, G. (ed.), 1988, Late Paleozoic and Mesozoic eolian deposits of the western interior of the United States: Sed. Geology, v. 56, p. 1–413.

Kocurek, G., 1991, Interpretation of ancient eolian sand dunes: Ann. Rev. Earth and Planetary Sciences, v. 19, p. 43–75.

Mckee, E. D. (ed.), 1979a, A study of global sand seas: U.S. Geological Survey Prof. Paper 1052.

Pye, K., and H. Tsoar, 1990, Aeolian sand and sand dunes: Unwin Hyman, London, 396 p.

Lacustrine Systems

Anadón, P., Ll. Cabrera, and K. Kelts (eds.), 1991, Lacustrine facies analysis: Internat. Assoc. Sedimentologists Spec. Pub. 13, Blackwell, Oxford, 318 p.

Ethridge, F. G., and R. M. Flores (eds.), 1981, Recent and ancient nonmarine depositional environments: Part III: Lacustrine deposits: Soc. Econ. Paleontologists and Mineralogists Spec. Pub. 31, Tulsa, Okla., p. 213–278.

Hakanson, L., and M. Jansson, 1983, Lake sedimentation: Springer-Verlag, Berlin, 320 p.

Lerman, A. (ed.), 1978, Lakes: Chemistry, geology, physics: Springer-Verlag, New York, 363 p.

Matter, A., and M. E. Tucker (eds.), 1978, Modern and ancient lake sediments: Internat. Assoc. Sedimentologists Spec. Pub. 2, 290 p.

Picard, M. D., and L. R. High, Jr., 1972, Criteria for recognizing lacustrine rocks, *in* J. K. Rigby and W. K. Hamblin (eds.), Recognition of ancient sedimentary environments: Soc. Econ. Paleontologists and Mineralogists Spec. Pub. 16, Tulsa, Okla., p. 108–145.

Stumm, W. (ed.), 1985, Chemical processes in lakes: John Wiley & Sons, New York, 435 p.

Glacial Systems

Anderson, J. B., and G. M. Ashley (eds.), 1991, Glacial marine sedimentation; Paleoclimatic significance: Geol. Soc. America Spec. Paper 261, 232 p.

Ashley, G. M., N. D. Smith, and J. D. Shaw, 1985, Glacial sedimentary environments: Soc. Econ. Paleontologists and Mineralogists Short Course Notes 16, Tulsa, Okla., 246 p.

Brodzikowski, K., and A. J. van Loon, 1991, Glacigenic sediments: Elsevier, Amsterdam, 674 p.

Dowdeswell, J. D., and J. D. Scourse (eds.), 1990, Glaciomarine environments: Processes and sediments: Geol. Soc. London Spec. Pub. 53, 423 p.

Drewry, D., 1986, Glacial geologic processes: Edward Arnold, London, 276 p.

Eyles, N. (ed.), 1983, Glacial geology: An introduction for engineers and earth scientists: Pergamon, Oxford, 409 p.

Evenson, E. B., Ch. Schlüchter, and J. Rabassa, 1983, Till and related deposits: A. A. Balkema, 454 p.

Flint, R. F., 1971, Glacial and Quaternary geology: John Wiley & Sons, New York, 892 p.

Goldthwait, R. P. (ed.), 1975, Glacial deposits: Dowden, Hutchinson and Ross, Stroudsburg, Pa., 464 p.

Jopling, A. V., and B. C. McDonald, 1975, Glaciofluvial and glaciolacustrine sedimentation: Soc. Econ. Paleontologists and Mineralogists Spec. Pub. 23, 320 p.

Leggett, R. F. (ed.), 1976, Glacial till: An inter-disciplinary study: Royal Soc. Canada Spec. Pub. 12, 412 p.

Molnia, B. F., 1983, Glacial-marine sedimentation: Plenum, New York, 844 p.

Sharp, R. P., 1988, Living ice: Cambridge University Press, Cambridge, 225 p.

Wright, A. E., and F. Moseley, 1975, Ice ages: Ancient and modern: Geol. Jour. Spec. Issue 6, Seal House Press, Liverpool, 320 p.

11

Marginal-Marine Environments

11.1 INTRODUCTION

The marginal-marine setting lies along the boundary between the continental and the marine depositional realms. It is a narrow zone dominated by river, wave, and tidal processes. Salinities may range in different parts of the system from freshwater through brackish water to supersaline, depending upon river discharge and climatic conditions. Intermittent to nearly constant subaerial exposure characterizes some environments of the marginal-marine setting. Others are continuously covered by shallow water. Many marginal-marine environments are further characterized by high-energy waves and currents, although some lagoonal and estuarine environments are dominated by quiet-water conditions. Owing to intermittent exposure, high-energy conditions, or marked variations in salinity or temperature, much of the marginal-marine realm is a high-stress environment for organisms. Therefore, organisms that live in marginal-marine environments tend to be species with high tolerance for salinity or temperature changes. They adapt to high-energy conditions by burrowing into sandy bottoms or boring into rocky substrates.

A wide variety of sediment types—including conglomerates, sandstones, shales, carbonates, and evaporites—can accumulate in marginal-marine environments. Owing to the large quantities of siliciclastic sediment delivered by rivers to the coastal zone throughout geologic time, the volume of marginal-marine deposits preserved in the geologic record is significant. The principal depositional settings for marginal-marine sediments are deltas; beaches, strand plains, and barrier bars; estuaries; lagoons; and tidal flats (Fig. 11.1). Estuaries and lagoons are particularly characteristic of transgressive coasts; deltas are features of prograding coasts.

FIGURE 11.1 The principal coastal environments of the marginal-marine depositional setting. The figure is organized to show the relative influence of tidal power (increasing to the left) and wave power (increasing to the right) on each environment. Note that deltas are features of prograding (regressive) coasts, whereas estuaries and lagoons are particularly characteristic of transgressive coasts. (From Boyd et al., 1992, Classification of clastic coastal depositional environments: Sed. Geology, v. 80. Fig. 2, p. 141, reproduced by permission.)

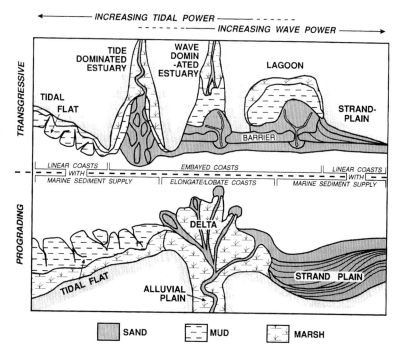

11.2   DELTAIC SYSTEMS

Introduction

The word **delta** was used by the Greek philosopher Herodotus about 490 B.C. to describe the triangular-shaped alluvial plain formed at the mouth of the Nile River by deposits of the Nile distributaries. Most other modern deltas are less triangular and more irregular in shape than the Nile delta. Nevertheless, the term (alluvial) delta is still applied to any deposit, subaerial or subaqueous, formed by fluvial sediments that build into a standing body of water. Deltas are "discrete shoreline protuberances formed where rivers enter oceans, semi-enclosed seas, lakes or lagoons and supply sediment more rapidly than it can be redistributed by basinal processes" (Elliott, 1986a). Thus, deltas can occur in lakes and inland seas as well as in the ocean, but they are most important in the open ocean. Much of the siliciclastic sediment transported to coastal zones throughout geologic time has been deposited in deltas.

Ancient deltaic deposits have been identified in stratigraphic successions of many ages, and deltaic sediments are known to be important hosts for petroleum and natural gas, coal, and some minerals such as uranium. Although ancient deltaic sediments are common in the rock record, much of what we know about delta systems comes from study of modern deltas. Deltas are particularly common in the modern ocean owing to post-Pleistocene sea-level rise coupled with high sediment loads carried by many rivers. High sea level causes increased sedimentation rates on deltas because sediment is trapped by the rising water, inhibiting sediment removal by currents. The locations, dimensions, and discharge characteristics of some modern deltas are given in Table 11.1.

TABLE 11.1 Characteristics of some modern deltas

Delta	Location	Subaerial area (km²·10³)	Average water discharge (m³/s·10³)	Annual sediment discharge (tons·10⁶)
Chao Phraya	Thailand—Gulf of Bangkok	25	1	5
Danube	Romania—Black Sea	4	6	91
Ganges-Brahmaputra	India and Pakistan—Bay of Bengal	91	39	635
Hwang Ho and N. Yellow Plain	China—Yellow Sea	127	4	2
Irrawaddy	Burma—Gulf of Martaban	31	28	272
Lena	Russia—Laptev Sea	28	9	—
Mekong	Vietnam—South China Sea	52	11	—
Mississippi	United States—Gulf of Mexico	29	17	469
Niger	Nigeria—Gulf of Guinea	19	6	23
Nile	Egypt—Mediterranean Sea	16	3	54
Orinoco	Venezuela—Atlantic Ocean	57	17	—
Po	Italy—Gulf of Venice	14	1	61
Red	Vietnam—Gulf of Tonkin	8	2	118
Rhine	Netherlands—North Sea	22	2	—
Rhône	France—Gulf of Lyons	3	2	41
Rio Grande	United States—Gulf of Mexico	8	0.1	17
Ural	Russia—Caspian Sea	9	—	—
Volga	Russia—Caspian Sea	11	—	—
Yangtze and S. Yellow Plain	China—East China Sea	124	22	544

Source: Smith, A. E., Jr., 1966, *in* M. L. Shirley and J. A. Ragsdale (eds.), Deltas: Houston Geological Society, p. 233–251; Reineck, H. E., and I. B. Singh, 1980, Depositional environments, 2nd ed.: Springer-Verlag. Table 25, p. 321.

Modern deltas occur on all continents, with the possible exception of Antarctica. (Trough-mouth, glacially influenced, submarine fans are present on the Weddell Sea continental margin of Antarctica; however, these may not be true deltas.) Deltas form where large, active drainage systems with heavy sediment loads exist. These conditions appear to be met particularly well on trailing-edge or passive coasts such as the east coasts of Asia and the Americas where tectonic activity is low. Fewer than 10 percent of major modern deltas occur on collision coasts, where tectonic activity is high and drainage divides are close to the sea (Inman and Nordstrom, 1971; Wright, 1978). Under such conditions, the large drainage systems neccessary to supply heavy sediment loads are not developed. Owing to their potential importance as oil and gas reservoirs, considerable interest has been generated in deltaic deposits since the 1950s. Consequently, the literature on deltas and deltaic deposits is extensive, for example, Bhattacharya and Walker (1992), Broussard (1975), Colella and Pryor (1990), Coleman (1981), Coleman and Prior (1982, 1983), Elliott (1986a), Galloway (1975), Miall (1984b), Morgan (1970), Nemec and Steel (1988a and 1988b), Shirley and Ragsdale (1966), Whateley and Pickering (1990), and Wright (1977, 1978). Many of the important papers on deltas published prior to 1975 are summarized by LeBlanc (1975).

Delta Classification and Sedimentation Processes

Two basic kinds of deltas exist: **alluvial deltas** and **nonalluvial deltas** (Fig. 11.2). Alluvial deltas can be further divided into the following (Nemec, 1990):

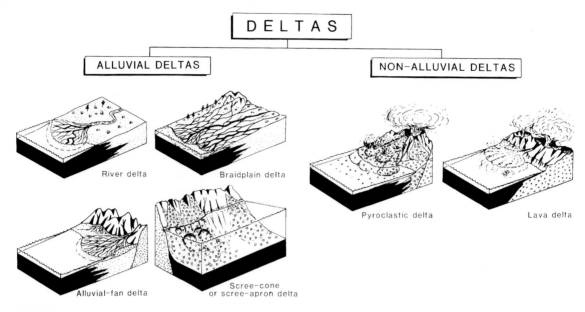

FIGURE 11.2   Subdivision of deltas into alluvial and nonalluvial varieties, with specific examples of both delta types. Only alluvial deltas are discussed in this book. (From Nemec, W., 1990, Deltas—Remarks on terminology and classification, *in* A. Colella and D. B. Prior (eds.), Coarse-grained deltas: Internat. Assoc. Sedimentologists Special Pub. 10, Blackwell Scientific Publications, Oxford. Fig. 1, p. 5, reproduced by permission.)

1. river deltas—formed from the deposits of single rivers
2. braidplain deltas—formed from the deposits of a braided-stream system
3. alluvial-fan deltas—formed where alluvial fans prograde into standing water
4. scree-apron deltas—formed where scree deposits extend into water

Nonalluvial "deltas" (fans) form by pyroclastic flows and lava flows that extend into water. We are concerned here only with alluvial deltas.

River and Braidplain Deltas

The distribution and characteristics of alluvial deltas are controlled by a complex set of interrelated fluvial and marine/lacustrine processes and environmental conditions. These factors include climate, water and sediment discharge, river-mouth processes, nearshore wave power, tides, nearshore currents, and winds (Coleman, 1981). Other factors that influence the formation of deltas are slope of the shelf, rates of subsidence and other tectonic activity at the depositional site, and geometry of the depositional basin. Among these variables, river (sediment) input, wave-energy flux, and tidal flux are the most important processes that control the geometry, trend, and internal features of the progradational framework sand bodies of deltas (Galloway and Hobday, 1983).

Alluvial deltas can be classified in several ways (Nemec, 1990); however, classification on the basis of delta-front regime (Galloway, 1975) appears to be favored by many geologists. Deltas are thus classified as (1) fluvial-dominated, (2) tide-dominated, or (3) wave-dominated (Fig. 11.3). Each of these kinds of deltas can be further distinguished on the basis of dominant grain size of sediments (Orton, 1988) as mud/silt-dominated, sand-dominated, gravel/sand-dominated, or gravel-dominated (Fig. 11.3).

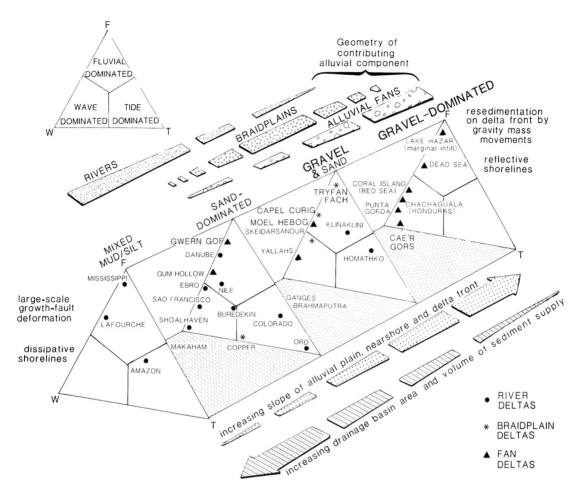

FIGURE 11.3 Classification of deltas on the basis of dominant process (fluvial, F; wave, W; tidal, T) of sediment dispersal at the delta front and the prevailing grain size of sediment delivered to the front. Names within the diagram refer to examples of modern and ancient deltas. (From Orton, G. J., 1988, A spectrum of Middle Ordovician fan deltas and braidplain deltas, North Wales: A consequence of varying fluvial clastic input, *in* W. Nemec and R. J. Steel (eds.), Fan deltas: Sedimentology and tectonic setting: Blackie, Glasgow. Fig. 18, p. 46, as modified from Galloway, 1975, reproduced by permission.)

An alternative, less subjective approach is classification on the basis of the slope of the delta face (steep or gentle) and dominant grain size (Fig. 11.4). Owing to the wide popularity of Galloway's classification, the ensuing discussion follows Galloway's approach.

Fluvial-dominated Deltas. In an early but important paper dealing with deltaic processes, C. C. Bates (1953) contrasted the behavior of sediment-laden river water as it enters equally dense, more dense, and less dense basin water. River water entering basin water of almost equal density, referred to by Bates as **homopycnal flow,** leads to rapid, thorough mixing and abrupt deposition of much of the sediment load (Fig. 11.5A). This type of jet outflow presumably causes the formation of **Gilbert-type** deltas

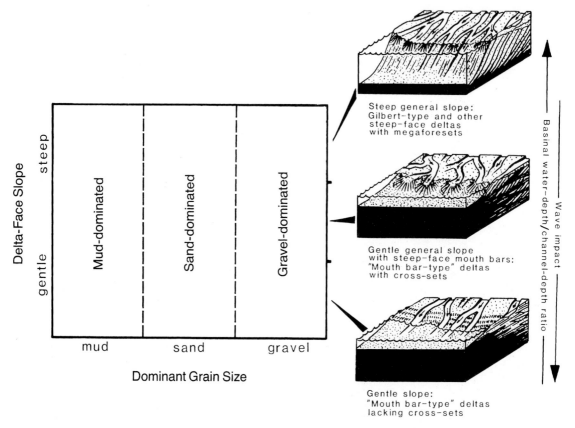

FIGURE 11.4 Classification of deltas on the basis of delta-face slope and dominant grain size. (After Orton, G. J., 1990, *in* W. Nemic (ed.), Deltas—Remarks on terminology and classification, *in* A. Colella and D. B. Prior (eds.), Coarse-grained deltas: Internat. Assoc. Sedimentologists Spec. Pub. 10, Blackwell Scientific Publications, Oxford. Fig. 2, p. 9, reproduced by permission.)

that display a topset, foreset, and bottomset arrangement of beds (Fig. 11.6), created as sediment deposition progrades basinward. River water that has higher density than basin water flows beneath the basin water, generating a vertically oriented, plane-jet flow called **hyperpycnal flow** (Fig. 11.5B). This type of jet flow moves along the bottom as a density current that may be erosive in its initial stages but eventually deposits its load along the more gentle slopes of the delta front to form turbidites. Such turbidity currents have been observed in lakes and may occur also under marine conditions. If river outflow is less dense than basin water, as when rivers flow into denser seawater, it flows outward on top of the basin water as a horizontally oriented plane jet called **hypopycnal flow** (Fig. 11.5C). Fine sediment may thus be carried in suspension some distance outward from the river mouth before it flocculates and settles from suspension. Flocculation involves aggregation of fine sediment into small lumps owing to the presence of positively charged ions in seawater that neutralize negative charges on clay particles. Hypopycnal flow tends to generate a large, active delta-front area, typically dipping at 1° or less, as contrasted with the 10° to 20° dip of most Gilbert-type deltas (Miall, 1984b). Hypopycnal flow is probably the most important type of river outflow in marine basins.

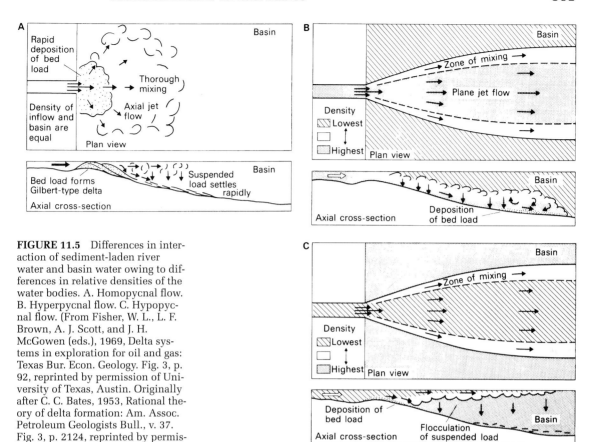

FIGURE 11.5 Differences in interaction of sediment-laden river water and basin water owing to differences in relative densities of the water bodies. A. Homopycnal flow. B. Hyperpycnal flow. C. Hypopycnal flow. (From Fisher, W. L., L. F. Brown, A. J. Scott, and J. H. McGowen (eds.), 1969, Delta systems in exploration for oil and gas: Texas Bur. Econ. Geology. Fig. 3, p. 92, reprinted by permission of University of Texas, Austin. Originally after C. C. Bates, 1953, Rational theory of delta formation: Am. Assoc. Petroleum Geologists Bull., v. 37. Fig. 3, p. 2124, reprinted by permission of AAPG, Tulsa, Okla.)

The kinds of sediments deposited by river-mouth processes in coastal areas with low tidal range and low wave energy depend upon the relative dominance of (1) outflow inertia, (2) turbulent bed friction seaward of the river mouth, and (3) outflow buoyancy (Wright, 1977). Outflows dominated by inertial forces—that is, characterized by large Reynolds numbers and high flow velocities—generate fully turbulent, homopycnal jet flows with negligible interference from the bottom. They exhibit low lateral spreading angles and progressive lateral and longitudinal deceleration, and they produce narrow river-mouth bars of the Gilbert type (Fig. 11.7). These types of outflows are considered to be rare. More commonly, water depths seaward of river mouths are shallow, and turbulent bed friction becomes dominant owing to high outflow velocities and bed shear stresses. This friction causes rapid deceleration and lateral expansion of the outflow and leads to formation of subaqueous levees, formation of triangular-shaped "middle-ground" bars that become finer grained basinward, and channel bifurcation (Fig. 11.8). Where river mouths are relatively deep and tidal range is low, fine-grained sediment loads prevail. Mixing is minimal, and strong density stratification can develop with freshwater flowing out over an underlying saltwater wedge. The outflow then spreads as a buoyant plume above the underlying saltwater, the hypopycnal inflow of Bates (1953). Buoyant outflows lead to formation of elongate distributaries with parallel banks, called subaqueous levees; few channel bifurcations;

FIGURE 11.6 Gilbert-type delta. A. Section through a Pleistocene delta in Lake Bonneville. B. Vertical facies sequence produced by delta progradation. (From Elliot, T., 1978, Deltas, *in* H. G. Reading (ed.), Sedimentary environments and facies. Fig. 6.1, p. 97, reprinted by permission of Elsevier Science Publishers, Amsterdam. Originally after G. K. Gilbert, 1885, The topographic features of lake shores: U.S. Geol. Survey Ann. Rept.; J. Barrell, 1912, Criteria for recognition of ancient delta deposits: Geol. Soc. America Bull., v. 23.)

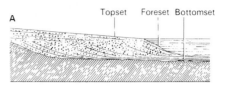

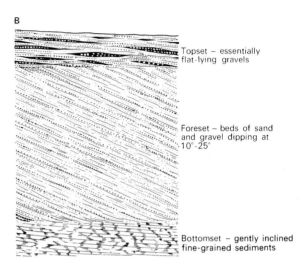

Topset – essentially flat-lying gravels

Foreset – beds of sand and gravel dipping at 10°-25°

Bottomset – gently inclined fine-grained sediments

and narrow distributary-mouth bars that grade seaward to fine-grained distal-bar deposits and prodelta clays (Fig. 11.9). Bar sands or **bar-finger** sands are typical components of such deltaic assemblages.

The modern Mississippi River delta is a classic example of a birdsfoot-type, river-dominated delta. It is among the largest deltas in the world outside of Asia (Table 11.1). The Mississippi delta consists of seven distinct sedimentary lobes that have been active during the past 5000 to 6000 years, indicating that periodic channel or distributary abandonment, to be discussed, is a common process. The generalized characteristics of the Mississippi delta system are shown in Figure 11.10. This figure illustrates the well-developed birdsfoot distributary system, typical of the delta, with bar-finger sands developed at the mouths of the distributaries. Common sediment facies on the Mississippi delta include marsh and natural-levee deposits, delta-front silts and sands, and prodelta clays. These types of deltaic deposits are described in greater detail in the discussion of sediment characteristics. Other modern fluvial-dominated deltas include the Danube (Black Sea) and the Po (Adriatic Sea).

Tide-dominated Deltas. The processes and deposits described above may be significantly modified under conditions of high tidal range or high wave energy. If tidal currents are stronger than river outflow, these bidirectional currents can redistribute river-mouth sediments, producing sand-filled, funnel-shaped distributaries. The distributary mouth bar may be reworked into a series of linear tidal ridges that replace the bar and extend from within the channel mouth out onto the subaqueous delta-front platform.

The modern Ganges-Brahmaputra delta is a well-known example (Fisher et al., 1969) of a tide-dominated delta (Fig. 11.11). The areal size of this delta is more than three times that of the Mississippi delta. It has a mean river discharge about twice that

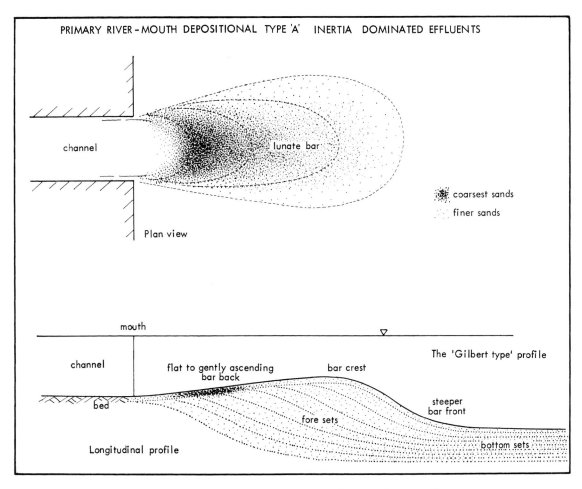

FIGURE 11.7 Idealized depositional pattern resulting from inertia-dominated river outflow. (From Wright, L. D., 1977, Sediment transport and deposition at river mouths: A synthesis: Geol. Soc. America Bull., v. 88. Fig. 4, p. 860.)

of the Mississippi, with exceedingly high discharge taking place during the monsoon season when extreme flooding is common. Mean tidal range is large, about 4 m, and wave energy is relatively low. Sand transport is intense during the monsoon season, leading to deposition of sandy deposits similar to those in braided streams. The delta is characterized by tidal-flat environments, natural levees, and floodbasins in which fine sediment is deposited from suspension. The strong tidal influence is manifested by the presence of a network of tidal sand bars and channels oriented roughly parallel to the direction of tidal current flow (Fig. 11.11). A variety of sediment types thus accumulate on the Ganges-Brahmaputra delta, including tidal-bar or tidal-ridge sands; braided, channel-fill sands; and natural-levee, tidal-flat, and floodbasin muds. Examples of other modern tide-dominated deltas include the Colorado (Gulf of California), the Ord (Timor Sea, Australia), the Fly (Gulf of Papua, New Guinea), and the Yalu (Korea Bay).

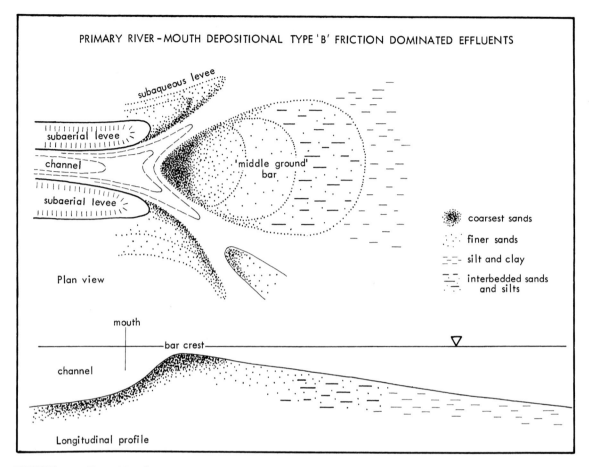

FIGURE 11.8   Depositional patterns associated with friction-dominated river-mouth outflow.
(From Wright, L. D., 1977, Sediment transport and deposition at river mouths: A synthesis: Geol.
Soc. America Bull., v. 88. Fig. 5, p. 861.)

Wave-dominated Deltas. Strong waves cause rapid diffusion and deceleration of river
outflow and produce constricted or deflected river mouths. Distributary-mouth
deposits are reworked by waves and are redistributed along the delta front by longshore currents to form wave-built shoreline features such as beaches, barrier bars, and
spits. A smooth delta front, consisting of well-developed, coalescent beach ridges, may
eventually be generated. Delta-plain geometries thus formed may range from arcuate to
cuspate (Galloway and Hobday, 1983).

The São Francisco delta of Brazil (Fig. 11.12) has long been cited as the classic
example of a wave-dominated delta. It is smaller in areal dimensions than the Mississippi and has lower discharge. Tidal range is about 2 m; however, wave power is
reported to be about 100 times that of the Mississippi delta (Coleman, 1976). Owing to
this extreme wave energy, the São Francisco delta is dominated by high-energy environments in which sand deposition takes place. Muds accumulate locally in marshes
and floodplains, but the interdistributary bay mud deposits characteristic of the Mississippi delta are absent. São Francisco delta deposits are dominated by beach-ridge

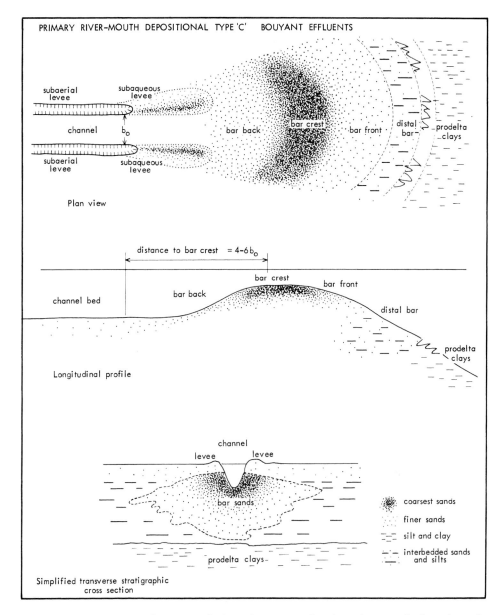

FIGURE 11.9 Depositional patterns relating to buoyant outflow from river mouths; b_0 = channel width. (From Wright, L. D., 1977, Sediment transport and deposition at river mouths: A synthesis: Geol. Soc. America Bull, v. 88. Fig. 6, p. 863.)

barrier sands that cover much of the delta surface. A belt of eolian dune sands is distributed along the outer margin of the delta.

Recent work by Dominguez et al. (1987) shows that sediments of the São Franciscan delta are not supplied dominantly by the river. A relative fall in sea level of about 5 m during the last 5000 years provided a new source of sediments from the inner shelf. These shelf sands have been transported by longshore (current) drift and

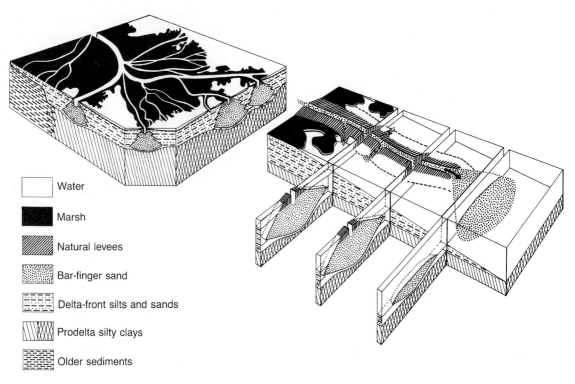

Water

Marsh

Natural levees

Bar-finger sand

Delta-front silts and sands

Prodelta silty clays

Older sediments

FIGURE 11.10 The Mississippi delta system—a fluvial-dominated delta. (From Reineck, H. E., 1970, Marine sandkörper, rezent und fossil: Geol. Rundschau, v. 60. Fig. 2, p. 305, reprinted by permission. Originally modified from H. N. Fisk, E. McFarland, C. R. Kolb, and L. J. Wilbert, 1954, Sedimentary framework of the modern Mississippi delta: Jour. Sed. Petrology, v. 24. Fig. 1, p. 77.)

FIGURE 11.11 The modern Ganges-Brahmaputra delta—a tide-dominated delta. (After Fisher, W. L., L. F. Brown, Jr., A. J. Scott, and J. H. McGowen, 1969, Delta systems in the exploration for oil and gas—A research colloquium: Texas Bur. Econ. Geology. Fig. 47, reprinted by permission of University of Texas, Austin.)

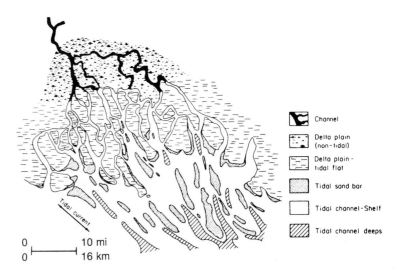

Channel

Delta plain (non-tidal)

Delta plain-tidal flat

Tidal sand bar

Tidal channel-Shelf

Tidal channel deeps

Tidal current

| 0 | | 10 mi |
| 0 | | 16 km |

FIGURE 11.12 Map of the São Francisco delta, Brazil, which is characterized by the presence of sandy beach ridges oriented roughly parallel to the shoreline. The sand in these ridges was derived mainly from the shelf rather than from the river. (After Dominguez, J. M. L., L. Martin, and A. C. S. P. Bittencourt, 1987, Sea-level history and Quaternary evolution of river mouth–associated beach-ridge plains along the east-southeast Brazilian coast: A summary, *in* D. Nummedal, O. H. Pilkey, and J. D. Howard (eds.), Sea-level fluctuation and coastal evolution: Soc. Econ. Paleontologists and Mineralogists Spec. Pub. 41. Fig. 2, p. 117, reprinted by permission of SEPM, Tulsa, Okla.)

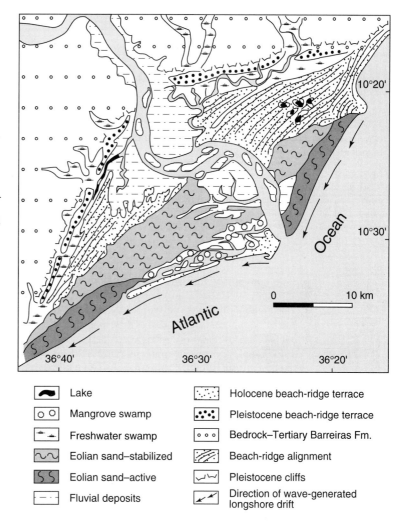

Lake		Holocene beach-ridge terrace
Mangrove swamp		Pleistocene beach-ridge terrace
Freshwater swamp		Bedrock–Tertiary Barreiras Fm.
Eolian sand–stabilized		Beach-ridge alignment
Eolian sand–active		Pleistocene cliffs
Fluvial deposits		Direction of wave-generated longshore drift

deposited around the river mouth, where the drift was interrupted by river outflow. Thus, the São Francisco sediment is not true delta sediment because most of it was not delivered by the river. The São Francisco is nonetheless a delta because it meets the definition of being a protuberance of the shoreline that forms where a river enters the ocean. Other examples of modern wave-dominated deltas include the Brazos (Gulf of Mexico), the Rhône (Mediterranean Sea), the Shoalhaven (Pacific, Australia), and the Tiber (Italy).

The examples discussed above illustrate some differences in characteristics of modern fluvial-dominated, tide-dominated, and wave-dominated deltas, which are summarized further in Table 11.2. Many deltas have characteristics that are transitional between these "end-member" types. The Nile delta, for example, is suggested in Fig. 11.3 to be a fluvial-dominated delta; however, the Nile also has many characteristics of a wave-dominated delta (Sestini, 1989).

TABLE 11.2   Characteristics of fluvial-, tide-, and wave-dominated deltas

	Fluvial-dominated	Wave-dominated	Tide-dominated
Lobe geometry	Elongate to lobate	Arcuate	Estuarine to irregular
Bulk composition	Muddy to mixed	Sandy	Muddy to sandy
Framework facies	Distributary-mouth bar and delta-front sheet sand; distributary channel-fill sand	Coastal-barrier sand; distributary-channel sand	Tidal sand ridge sand; estuarine distributary channel-fill sand
Framework orientation	Highly variable; average parallels depositional slope	Dominantly parallels depositional strike; subsidiary dip trends	Parallels depositional slope unless skewed by local basin geometry
Common channel type	Suspended load to fine mixed load	Mixed load to bedload	Variable, tidally modified geometry

Source: Galloway, W. E., and D. K. Hobday, 1983, Terrigenous clastic depositional systems. Table 5.1, p. 109, reprinted by permission of Springer-Verlag, Heidelberg.

Fan Deltas

The concept of fan deltas was introduced by Holmes (1965, p. 554). A fan delta, as defined by Holmes and modified slightly by Nemec and Steel (1988a), is a coastal prism of sediments delivered by an alluvial-fan system and deposited, mainly or entirely subaqueously, at the interface between the active fan and a standing body of water. Fan deltas were recognized first in modern settings, but fan-delta deposits have now been reported in many ancient sedimentary successions (e.g., Nemec and Steel, 1988b). Both fluvial-dominated and wave-dominated fan deltas are known; presumably, tide-dominated, or tide-influenced, fan deltas exist as well.

Fan deltas can form in both arid and humid alluvial-fan systems. Figure 11.13 illustrates some of these settings. Figure 11.14 is a simplified model that shows details of the subaqueous portion of a fan delta. Sediments are deposited downslope in the subaqueous part of fan deltas by processes such as slumping and turbidity current flow. A brief description of an ancient fan delta is given subsequently in the discussion of ancient deltaic deposits. Readers are referred to the volumes on fan deltas and deltas edited by Nemec and Steel (1988b) and Collella and Prior (1990) for additional details.

Physiographic and Sediment Characteristics of Deltaic Systems

Owing to variations in sediment input, outflow velocity, wave and current energy, and other factors, the depositional features of deltas exhibit a high degree of variability from one delta to another. Nevertheless, all deltas can be divided into subaerial and subaqueous components, each of which can be further subdivided (Fig. 11.15; Table 11.3). The subaerial component of deltas is generally larger than the subaqueous component and is divided into an **upper delta plain,** which lies largely above high-tide level, and a **lower delta plain,** lying between low-tide mark and the upper limit of tidal influence. The upper delta plain is commonly the oldest part of the delta and is dominated by fluvial processes. The lower delta plain is exposed during low tide but is covered by water during high tide. Thus, it is subjected to both fluvial and marine processes. The **subaqueous delta plain** lies seaward of the lower deltaic plain below low-tide water level and is characterized by relatively open marine faunas. The uppermost part of the subaqueous delta, lying at water depths down to 10 m or so, is commonly called the **delta front.** The remaining seaward part of the subaqueous delta is called the **prodelta,** or prodelta slope.

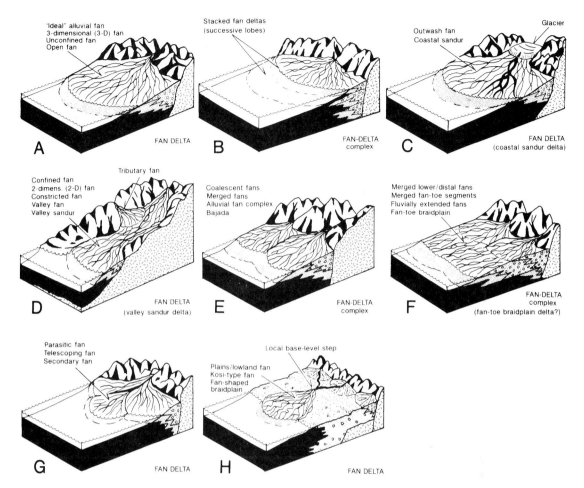

FIGURE 11.13 A–H. Various kinds of fan deltas with emphasis on their feeder components. (After Nemec, W., and R. J. Steel, 1988, What is a delta and how do we recognize it? *in* W. Nemic, and R. J. Steel (eds.), Fan deltas: Sedimentology and tectonic setting: Blackie, Glasgow. Fig. 1, p. 6, reproduced by permission.)

Upper Delta Plain Sediments. The upper delta plain lies mainly above tidal influence and is little affected by marine processes. Sedimentation on the upper delta is dominated by distributary-channel migration and associated fluvial sedimentation processes such as channel and point-bar deposition, overbank flooding, and crevassing into lake basins (Chapter 10). The principal depositional environments include braided channels, meandering channels, lacustrine delta fill, backswamps, and flood-plain environments such as swamps, marshes, and freshwater lakes (Coleman and Prior, 1982). Therefore, upper delta-plain sediments are predominantly fluvial sands, gravels, and muds that may be closely associated with lacustrine, swamp, and marsh deposits.

Lower Delta Plain Sediments. The width of the lower delta plain is greatest on deltas where tidal range is large. This plain includes the active distributary system of the delta, as well as abandoned distributary-fill deposits, and may be flanked by marginal-basin or bay-fill deposits. Distributary channels are numerous, but environments between channels make up the largest percentage of the lower delta plain. These environments include actively migrating tidal channels, natural levees, interdistributary bays, bay fills (crevasse splays), marshes, and swamps (Coleman and Prior, 1982). The

FIGURE 11.14 Simplified model of depositional environments in the Miocene Doumsan Fan-delta/deep-water system of Korea; AF = alluvial fan, GT = Gilbert-type, GF = Gilbert-type foreset, GTE = Gilbert-type bottomset, PD = prodelta, SA = slope apron, and BP = basin plain. (From Chough, S. K., et al., 1990, The Miocene Doumsan fan delta, southeast Korea: A composite fan-delta system in back-arc margin: Jour. Sed. Petrology, v. 60. Fig. 8, p. 453, reproduced by permission of Society of Economic Paleontologists and Mineralogists, Tulsa, Okla.)

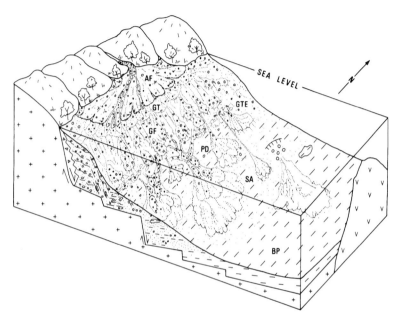

FIGURE 11.15 Principal components of a delta system. (From Coleman, J. M., and D. B. Prior, 1982, Deltaic environment of deposition, *in* P. A. Scholle and D. Spearing (eds.), Sandstone depositional environments: Am. Assoc. Petroleum Geologists Mem. 31. Fig. 1, p. 139, reprinted by permission of AAPG, Tulsa, Okla.)

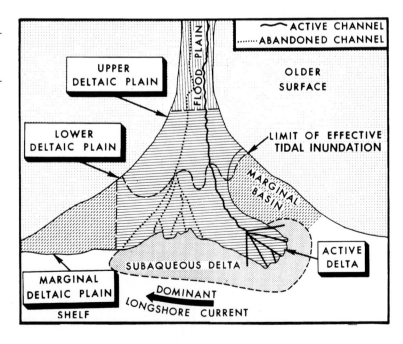

TABLE 11.3 Principal categories
of delta facies

Upper delta plain
Migratory channel deposits—braided-channel and
meandering-channel deposits
Lacustrine delta-fill and floodplain deposits
Lower delta plain
Bay-fill deposits (interdistributary bay, crevasse
splay, natural levee, marsh)
Abandoned distributary-fill deposits
Subaqueous delta plain
Distributary-mouth bar deposits (prodelta distal bar,
distributary-mouth bar)
River-mouth tidal-range deposits
Subaqueous slump deposits
Prodelta—seafloor seaward of the subaqueous delta

Source: Coleman, J. M., and D. B. Prior, 1982, Deltaic environment
of deposition, in P. A. Scholle and D. Spearing (eds.), Sandstone
depositional environments: Am. Assoc. Petroleum Geologists
Mem. 31, p. 140, reprinted by permission of AAPG, Tulsa, Okla.

major sand bodies generated in this environment are bay-fill deposits, which may form thin sand wedges stacked one on top of the other and separated by finer-grained, interdistributary-bay and marsh deposits. In very arid climates, evaporites also may be deposited in some parts of the lower delta plain. Deposits of the lower delta plain also commonly include abandoned distributary deposits. These consist of locally derived sands, muds, and organic debris that gradually fill distributary channels after they have been abandoned by the main stream owing to blocking or other processes that cause channel shifting.

Subaqueous Delta Plain. The subaqueous delta plain constitutes that area of a delta that lies seaward of low tide level and actively receives fluvial sediments. It may extend outward for distances of a few kilometers to tens of kilometers to water depths as much as 300 m. Deposits of the subaqueous delta thus form the base over which subaerial delta deposits prograde as the delta builds seaward. The deposits typically consist in part of sands, and possibly gravels, deposited near the river mouths, forming distributary–mouth-bar deposits. These deposits grade seaward to finer sands and coarse silts that settle from suspension to form the distal bar (Fig. 11.9). The uppermost part of the subaqueous delta, the delta front, may be dominated by high-energy marine processes, including waves, longshore currents, and tides in some cases. Sediment is reworked and winnowed by these processes, creating well-sorted delta-front sheet sands that are cross-bedded on a variety of scales. The finest silts and clays are transported still farther seaward and settle on the prodelta on the outermost part of the subaqueous delta. Previously deposited sediments may be reentrained, transported, and redeposited farther downslope on the subaqueous delta by gravity-driven mass-movement processes that include landslides, slumps, and mudflows. Mud diapirs are also a common feature of many subaqueous deltas (Fig. 11.16). These structures are piercement-type bodies of soft sediment that is squeezed upward and intruded into overlying sand bodies owing to sediment loading.

Constructional vs. Destructional Phases of Delta Development

During the active phases of delta outbuilding, most sedimentation processes on deltas are constructional in the sense that delta formation is dominated by sediment deposi-

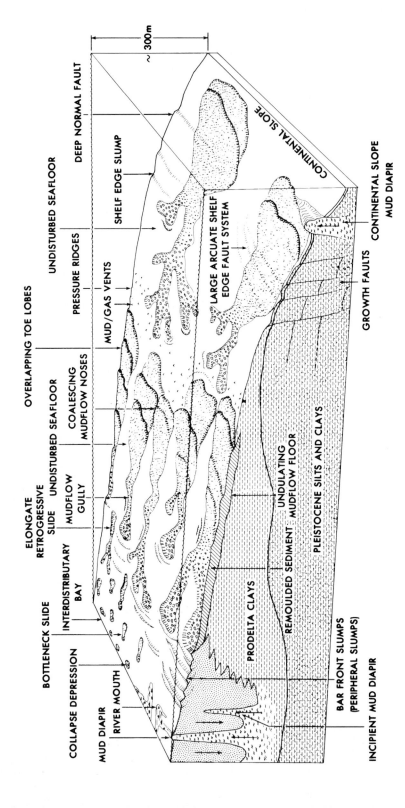

FIGURE 11.16 An example of submarine mass movements on the subaqueous delta. This example shows the major types of submarine landslides, diapirs, and contemporary faults in the Mississippi River delta. Horizontal dimensions not to scale. (From Coleman, J. M., and D. B. Prior, 1982, Deltaic environment of deposition, *in* P. A. Scholle and D. Spearing (eds.), Sandstone depositional environments: Am. Assoc. Petroleum Geologists Mem. 31. Fig. 23, p. 168, reprinted by permission of AAPG, Tulsa, Okla.)

Labels in figure:

BOTTLENECK SLIDE
COLLAPSE DEPRESSION
MUD DIAPIR — RIVER MOUTH
INTERDISTRIBUTARY BAY
ELONGATE RETROGRESSIVE SLIDE
UNDISTURBED SEAFLOOR
MUDFLOW GULLY
OVERLAPPING TOE LOBES
COALESCING MUDFLOW NOSES
UNDISTURBED SEAFLOOR
PRESSURE RIDGES
MUD/GAS VENTS
UNDISTURBED SEAFLOOR
DEEP NORMAL FAULT
SHELF EDGE SLUMP
CONTINENTAL SLOPE
LARGE ARCUATE SHELF EDGE FAULT SYSTEM
CONTINENTAL SLOPE MUD DIAPIR
GROWTH FAULTS
PLEISTOCENE SILTS AND CLAYS
UNDULATING MUDFLOW FLOOR
REMOULDED SEDIMENT
PRODELTA CLAYS
BAR FRONT SLUMPS (PERIPHERAL SLUMPS)
INCIPIENT MUD DIAPIR
~300 m

tion. On the other hand, tidal currents and waves represent destructional processes to the extent that they cause erosion and redistribution of some sediment. Destructional processes become particularly important when deltas, or portions of deltas, enter an inactive phase. Channel or distributary abandonment, foundering owing to subsidence, or marine transgression may interrupt active construction of a delta. Such an interruption leads to a phase when erosion by waves and tidal currents becomes dominant as sediment influx from the river ceases. Destructional processes cause redistribution of eroded sediment, commonly producing thin, laterally persistent beds of sand or mud that contain marine faunas and are abundantly bioturbated. Such beds form important marker units for stratigraphic studies in otherwise heterogeneous deltaic successions.

Growth of deltas thus tends to be cyclic. During active, prograding phases, prodelta fine silts and clays are progressively overlain by delta-front silts and sands, distributary-mouth sands, and finally marsh, fluvial, and possibly eolian deposits as the delta builds seaward, producing a coarsening-upward regressive succession. Interruption of progradation by delta-lobe abandonment or marine transgression brings on a destructive phase in which erosion and redistribution of river-mouth deposits predominate. Subsequent distributary shifting or regression may bring on another phase of active progradation. A complete delta cycle may range in thickness from 50 to 150 m. Smaller-scale cycles representing progradation of individual distributaries range from only about 2 to 15 m (Miall, 1984b).

Recognition of Ancient Deltaic Deposits

The subenvironments of delta systems range from normal marine (beach, barrier, lagoonal) to nonmarine (fluvial, marsh, eolian), and a variety of different sediment types can be deposited in these subenvironments. Deltaic sedimentary successions are characterized by assemblages of lithofacies, each of which can occur in other environments, such as fluvial, lacustrine, and shallow-marine environments. Identifying ancient delta deposits is best accomplished in a series of steps, eliminating other possibilities and using distinguishing characteristics of facies types, bed geometry, and types of cyclic successions to focus gradually on the correct delta model. Some general characteristics of delta deposits that can be useful in their recognition include (1) geometry, (2) lateral facies relationships, (3) vertical successions of facies, and (4) sedimentary structures and fossils.

Geometry. Ideally, deltas are triangular in areal shape; however, much variation from this ideal shape can occur, particularly with tide- and wave-dominated deltas. In cross section, deltas are typically wedge- or lens-shaped bodies extending laterally to several hundred kilometers. Modern deltas vary in their areal dimensions from small bodies a few thousand square kilometers in size to huge deltas exceeding 125,000 km^2 (Table 11.1).

Lateral Facies Relationships. A wedge- or lobe-shaped deposit of nonmarine to shallow-marine sand, silt, and mud that grades landward into totally nonmarine, largely fluvial sediments and basinward into finer-grained, deeper-water marine sediments is indicative of deltaic origin. On a smaller scale, lateral facies relationships are likely to be complex. Delta-plain deposits may range from coarse distributary-channel deposits to finer-grained marsh or interdistributary-bay or lacustrine deposits. The lateral associations of delta-front sediments can also be highly variable depending upon whether deposition was dominated by fluvial, wave, or tidal processes; it is difficult to generalize about these facies relationships. Coarser, delta-front sediments may grade seaward into prodelta silts and clays which, in turn, grade to open-shelf muds. Prodelta muds may be difficult to differentiate from open-shelf muds except perhaps by greater thickness and higher sedimentation rates. They also tend to contain more mud-turbidite units than shelf muds.

Vertical Facies Successions. Progradation of deltaic deposits during active delta growth produces a generally coarsening-upward sedimentary succession. Migration of delta-front sands over prodelta silts and clays generates fairly well-defined, large-scale (50-150 m), coarsening-upward successions. Progradation of subaerial delta-plain sediments over subaqueous delta-plain sediments tends to produce smaller-scale (2–15 m), coarsening-upward units. Locally, filling of abandoned channels may even cause fining-upward successions to develop. Major progradational coarsening-upward facies may be interrupted by thin, widespread facies generated during inactive stages of delta growth, producing cycles of progradational and channel-abandonment facies. Although the generalized coarsening-upward progradational model for deltas serves as a useful norm for comparison, readers are cautioned that variations in delta behavior can produce vertical successions that differ from this idealized model. The lithologic types, sedimentary structures, textures, and other features preserved in the progradational facies depend upon the type of delta. An example of a typical succession of progradational facies produced in a fluvial-dominated delta—the Mississippi delta—is illustrated in Figure 11.17.

Sedimentary Structures and Fossils. Numerous types of sedimentary structures—such as cross-bedding, ripple marks, bioturbation structures, root traces (marsh deposits), slump structures, and mud diapirs—occur in deltaic deposits. With the possible exception of mud diapirs, all of these structures are found also in many other environments; none is diagnostic of delta environments. Suites of structures may help to define a particular type of deltaic deposit such as fluvial deposits. Paleocurrent directions in deltaic deposits can be highly variable, ranging from unidirectional patterns in fluvial-dominated portions of deltas to bidirectional patterns in tide-dominated portions. There are no specific fossil taxa that are characteristic of deltas alone. A transition from freshwater to brackish water to saltwater fauna in facies having the characteristics described under "Geometry" and "Lateral Facies Relationships" is indicative of deltaic origin; however, such transitions can occur also in estuarine sediments.

Ancient Deltaic Systems

Both river- and fan-delta deposits have been recognized in the ancient geologic record. Ancient deltaic sediments have been reported in stratigraphic successions of most ages (Broussard, 1975; Colella and Prior, 1990; Elliott, 1986a; Morgan, 1970; Nemec and Steele, 1988b; Shirley and Ragsdale, 1966; Whateley and Pickering, 1989), but they appear to be particularly common in rocks of Carboniferous and Tertiary age. Examples of fluvial-, wave-, and tide-dominated deltas are known.

Horn et al. (1978) describe fluvial deltaic sediments from the Carboniferous of Kentucky (Fig. 11.18A). The deposits shown in Figure 11.18A are distributary-mouth bar sandstones that grade laterally into bay-fill muds. The sand bodies are 1.5 to 5 km wide and 15 to 25 m thick. They are widest at the base and have gradational lower and upper contacts. Grain size increases upward in the succession and toward the center of the bars. Fining-upward, graded beds are common on the flanks of the bars, as are oscillation- and current-rippled surfaces. Pebble-lag conglomerates are present at the bases of the channel deposits. Note by comparison with Fig. 11.10 that these ancient

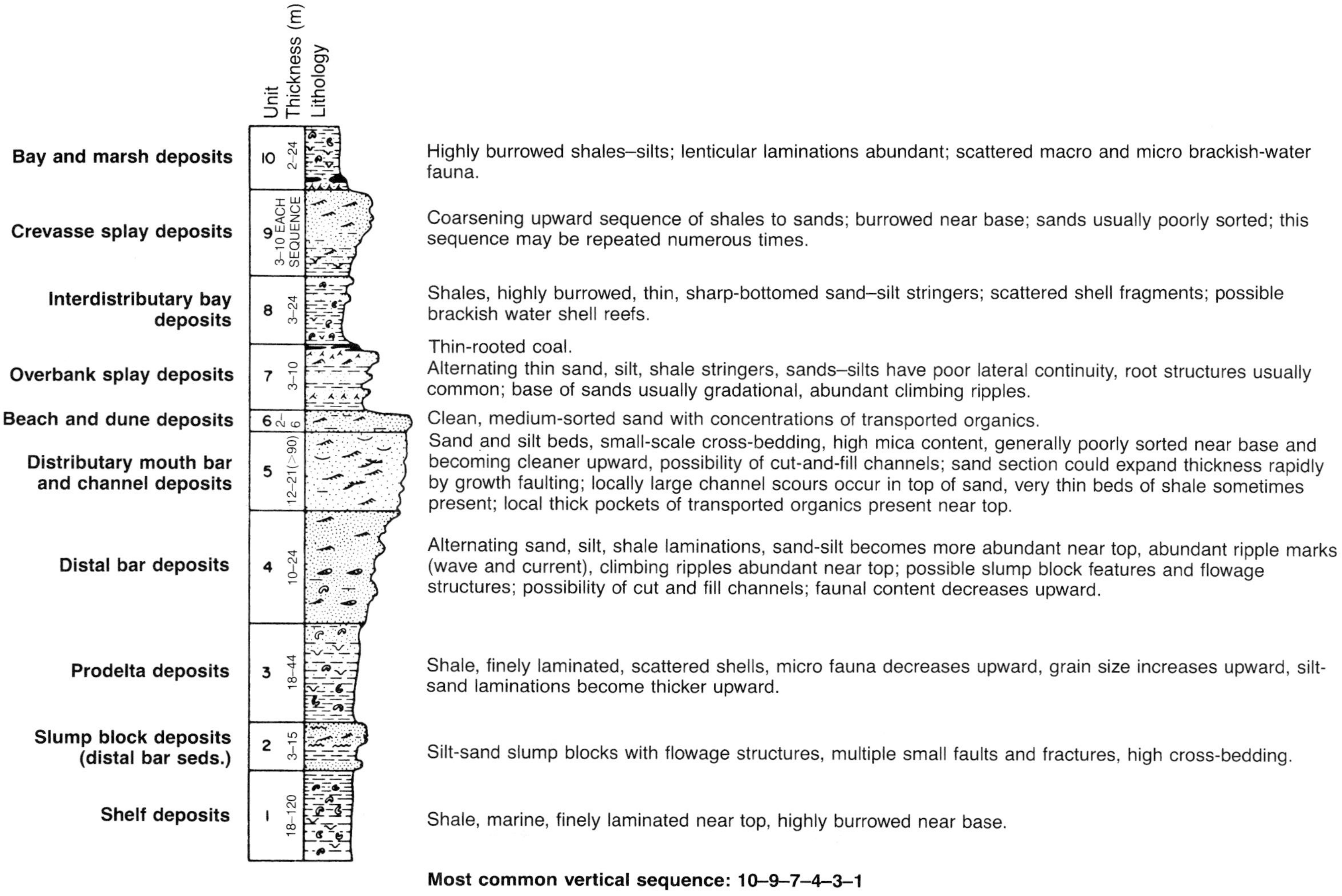

FIGURE 11.17 Idealized vertical succession of facies in a fluvial-dominated (Mississippi) delta. Note the thickness of individual units shown in the column. (From Coleman, J. M., 1981, Deltas: Processes of deposition and models for exploration, 2nd ed. Fig. 4.3, p. 91, reprinted by permission of IHRDC Publications, Boston.)

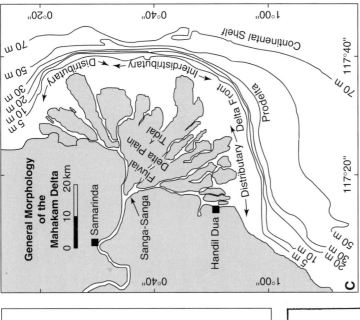

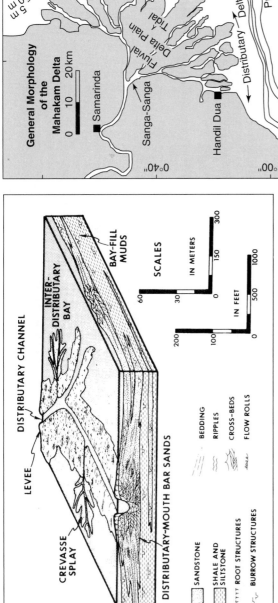

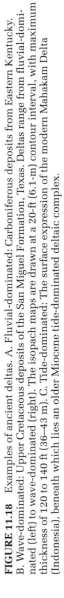

FIGURE 11.18 Examples of ancient deltas. A. Fluvial-dominated: Carboniferous deposits from Eastern Kentucky. B. Wave-dominated: Upper Cretaceous deposits of the San Miguel Formation, Texas. Deltas range from fluvial-dominated (left) to wave-dominated (right). The isopach maps are drawn at a 20-ft (6.1-m) contour interval, with maximum thickness of 120 to 140 ft (36–43 m). C. Tide-dominated: The surface expression of the modern Mahakam Delta (Indonesia), beneath which lies an older Miocene tide-dominated deltaic complex.

(A, from Horne, J. C. et al., 1978, Depositional models in coal exploration and mine planning in Appalachian Region: Am. Assoc. Petroleum Geologists Bull., v. 62. Fig. 6, p. 2387. B. After Weise, B. R. 1980, Wave-dominated deltaic systems of the Upper Cretaceous Miguel Formation, Maverick Basin, South Texas. Fig. 28, p. 23: Texas Bur. Econ. Geology, Report of Investigations 107, 39 p. C, from Verdier, A. C., T. Oki, and A. Suardy, 1980, Geology of the Handil Field (East Kalimantan—Indonesia), in M. T. Halbouty (ed.), Giant oil and gas fields of the decade 1968–1978: Am. Assoc. Petroleum Geologists, Tulsa, Okla. Fig. 12, p. 411.)

376

river-dominated deltaic sediments have geometry and sediment characteristics very similar to those of the modern Mississippi delta. Other ancient river-dominated deltas are described by Bhattacharya and Walker (1991), Fisher and McGowan (1967), Galloway (1975), and Pulham (1989).

Weise (1980) describes Upper Cretaceous sediments from the San Miguel Formation in the subsurface of Texas that are interpreted from core and well-log information to be wave-dominated deltaic sediments. He constructed thickness (isopach) maps of 10 sandy delta lobes having a variety of sand-body shapes ranging from those with the characteristics of river-dominated deposits (reflecting least influence of marine processes) to those with the characteristics of marine wave-dominated deltas (Fig. 11.18B). During periods of high sediment input and a low rate of sea-level rise, redistribution of sediment by waves was minimal, and lobate deltas resulted. During periods of low sediment input and high rates of sea-level rise, sediment was extensively reworked by waves to form elongate, strike-aligned sandstone bodies. Other ancient wave-dominated deltaic sediments are described by Elliott (1986a) and Leckie and Walker (1982).

Relatively few examples of ancient tide-dominated deltas have been reported. Verdier, Oki, and Suardy (1981) describe, on the basis of drilling and seismic data, a Miocene-age, tide-dominated delta that lies underneath the modern Mahakham delta in Indonesia (Fig. 11.18C). These authors identified delta-channel and bar sands, which form reservoir rocks for petroleum, interbedded with silty or organic shales and coal beds. Eriksson (1979) describes tide-dominated delta sediments from the Precambrian of South Africa.

The Late Carboniferous–Permian Reinodden Formation of western Spitsbergen provides an example of an ancient fan-delta depositional system (Kleinspehn et al., 1984). Several fan-delta successions are present in this area, one of which is shown in Figure 11.19. The stratigraphic succession in Figure. 11.19 begins with carbonate and siliciclastic muds, deposited in a prodelta setting. These fine-grained prodelta deposits are succeeded upward by barrier/spit and distal mouth-bar sands, cross-bedded in part, deposited by turbidity currents and other processes on the delta front. Sandy deposits are overlain by planar to cross-bedded gravels that formed in fluvial channels on the fan-delta plain. A thin unit of wave-reworked gravels caps the succession, representing a minor phase of marine transgression. The small maps in Figure 11.19 show the postulated evolution of the fan with time as it prograded over the delta-front and prodelta environments to generate the coarsening-upward succession of facies shown.

These examples illustrate the characteristics of a few specific ancient delta systems and are not intended to serve as general models for all delta deposits. Readers should consult the references given above for additional examples of the wide variety of facies and stratigraphic successions that characterize ancient delta systems.

11.3 BEACH AND BARRIER-ISLAND SYSTEMS

Introduction

Mainland beaches are long, narrow accumulations of sand aligned parallel to the shoreline and attached to land. Bodies of beach sand are typically cut across here and there by headlands and sea cliffs, estuaries, river deltas, tidal inlets, bays, and lagoons. Barrier-island beaches are similar to mainland beaches but are separated from land by a shallow lagoon, estuary, or marsh. They are also commonly dissected by tidal channels or inlets. Beaches may occur within delta systems, along depositional strike from

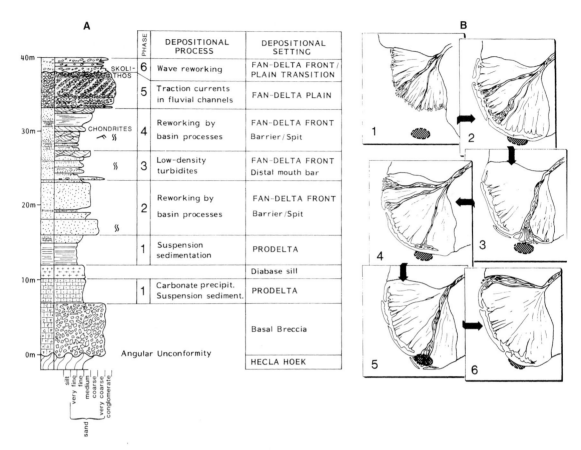

FIGURE 11.19 Vertical facies sequence (A) developed in Carboniferous–Permian fan-delta deposits of western Spitsbergen in response to various stages of fan progradation (B). Phases 1–6 in the stratigraphic column correspond to phases 1–6 in the plan-view sketches. The depositional site, represented by the striped oval area, remained fixed in space while the fan-delta geometry changed relative to that area. (After Kleinspehn, K. L., et al., 1984, Conglomeratic fan-delta sequences, Late Carboniferous–Early Permian, western Spitsbergen, *in* E. H. Koster and R. J. Steel (eds.), Sedimentology of gravels and conglomerates: Canadian Soc. Petroleum Geologists Mem. 10. Fig. 8, p. 289, reproduced by permission.)

deltas, or in other marine or even lacustrine settings that have no connection with deltas. They are the most dynamic of all depositional environments and are subject to both seasonal and longer-range changes that keep them in a state of virtually constant flux. In contrast to deltas, which are influenced by both fluvial and marine processes, beach and barrier-island systems are generated predominantly by marine processes, aided to a minor degree by eolian sand transport.

Modern and Holocene beaches have perhaps been studied more extensively than any other depositional environment owing to their recreational use; their accessibility; their economic potential as a source of placer gold, platinum, and various minerals; and their importance as an erosion buffer between the sea and the land. Much of the study of modern beaches has been carried out by coastal engineers, geographers, and geomorphologists. Geologists also have a strong scientific interest in beaches owing to the insight they provide into ancient depositional processes and environments. Moreover, beaches are a great place to do research and relax a little on the side, which may account for some of their research appeal. Ancient beach deposits also have been extensively studied. In addition to their significance as indicators of ancient nearshore processes and conditions, ancient beach and barrier-island sediments have considerable economic importance as reservoirs for petroleum and natural gas and as host rocks for uranium. Hundreds of research papers have been devoted to study of modern and ancient beaches, and several books have been published on this subject (e.g., Davis and Ethington, 1976; Fisher and Dolan, 1977; Hails and Carr, 1975; Hardisty, 1990; Hayes and Kana, 1976; Komar, 1976; Oertel and Leatherman, 1985; Pilkey, 1983; Schwartz, 1973). Useful, shorter summaries of beach and barrier-island systems are provided by Davis (1985), Elliott (1986b), Heward (1981), McCubbin (1982), Reineck and Singh (1980), and Reinson (1992), among others. Although most beaches are composed of siliciclastic sediments, some modern beaches on carbonate shelves are made up predominantly of carbonate grains consisting of skeletal fragments, ooids, pellets, and other particles. Carbonate beach deposits are known also from the geologic record (Inden and Moore, 1983).

Depositional Setting

Beach and barrier-island complexes are best developed on wave-dominated coasts where tidal range is small to moderate. Coasts are classified on the basis of tidal range into three groups: (1) **microtidal** (0–2 m tidal range), (2) **mesotidal** (2–4 m tidal range), and (3) **macrotidal** (>4 m tidal range). Hayes (1975) has shown that barrier-island and associated environments occur preferentially along microtidal coasts, where they are well developed and nearly continuous. They are less characteristic of mesotidal coasts and, when present, are typically short or stunted, with tidal inlets common. Barriers are generally absent on macrotidal coasts; extreme tidal range causes wave energy to be dispersed and dissipated over too great a width of shore zone to effectively form barriers.

Considerable difference of opinion exists about the origin of barrier-island complexes. As summarized by Reineck and Singh (1980), mechanisms of origin may include the following:

1. shoal and longshore-bar aggradation, that is, upward building and eventual emergence of offshore bars
2. spit segmentation by breaching and detachment of spits oriented parallel to the coast
3. mainland ridge engulfment owing to submergence and drowning of shoreline-attached beaches
4. welding or veneering of Holocene dune, beach, and foreshore sand into and over pre-Holocene topographic highs
5. lateral shifting of coastal sands during transgression to form the barrier islands

Mechanisms (2), (3), and (5) appear most feasible; however, composite modes of origin seem possible. The origin controversy remains unresolved because most of the evidence pertaining to origin has been destroyed by subsequent modification (Reinson, 1992).

Subenvironments. Deposits of the beach and barrier-island environment can occur as one of the following:

1. a **single beach** attached to the mainland
2. a broader beach-ridge system that constitutes a **strand plain** that consists of multiple parallel beach ridges and parallel swales but that generally lacks well-developed lagoons or marshes
3. a **barrier island** separated wholly or partly from the mainland by a lagoon or marsh (Fig. 11.20)

A type of strand plain consisting of sandy ridges elongated along the coast and separated by coastal mudflat deposits is called a **chenier plain**. As illustrated in Figure 11.21, the barrier-island setting is not a single environment but a composite of three separate environments:

1. the sandy barrier-island chain itself (the subtidal to subaerial barrier-beach complex)
2. the enclosed lagoon, estuary, or marsh behind it (the back-barrier, subtidal-intertidal region)
3. the channels that cut through the barrier and connect the back-barrier lagoon to the open sea (the subtidal-intertidal delta and inlet-channel complex)

FIGURE 11.20 Morphological relationship between beaches, strand plains, and barrier islands. (From Reinson, G. E., 1984, Barrier-island and associated strand-plain systems, *in* R. G. Walker (ed.), Facies models: Geoscience Canada Reprint Ser. 1. Fig. 1, p. 119, reprinted by permission of Geological Association of Canada.)

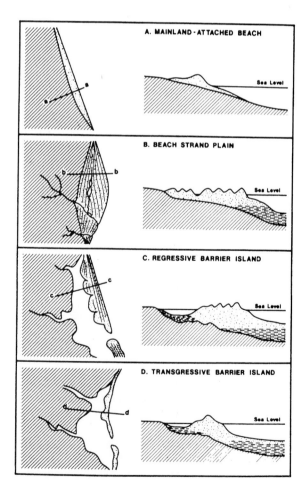

A. MAINLAND-ATTACHED BEACH

B. BEACH STRAND PLAIN

C. REGRESSIVE BARRIER ISLAND

D. TRANSGRESSIVE BARRIER ISLAND

FIGURE 11.21 Generalized model illustrating the various subenvironments in a transgressing barrier-island system. (From Reinson, G. E., 1992, Transgressive barrier island and estuarine systems, *in* R. G. Walker and N. P. James (eds.), Facies of models. Fig. 3, p. 180, reproduced by permission of Geological Association of Canada.)

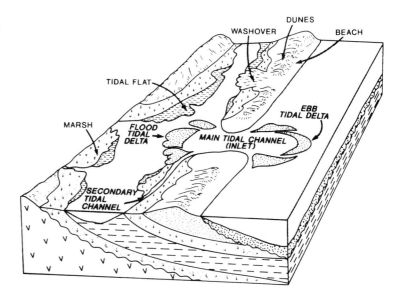

Identification and interpretation of ancient barrier-island complexes require that this intimate association of lagoonal, estuarine, and tidal-flat facies be recognized. Barrier-island systems are not simply barrier-beach complexes.

Morphology of the Beach-Nearshore Zone. The morphological features of the beach profile are similar on mainland coasts and the seaward coast of barriers. The beach is divided into the **backshore,** which extends landward from the beach berm above high-tide level and commonly includes back-beach dune deposits; the **foreshore,** which mainly encompasses the intertidal (littoral) zone between low-tide and high-tide levels; and the **shoreface,** also called the nearshore, which extends from about low-tide level to the transition zone between beach and shelf sediments (Fig. 11.22A), that is, to fair-weather wave base at a depth of about 10 to 15 m. Figure 11.22A illustrates also the approximate zones of shoaling and breaking waves and the position of the surf zone and swash zone, discussed in a succeeding section. The relationships of the beach profile to other elements of a barrier-island complex (washover fan, lagoon, mainland) are illustrated in Figure 11.22B.

Depositional Processes on Beaches

Erosion, sediment transport, and depositional processes on beaches have been studied extensively by engineers interested in coastal processes, as well as by geologists. The published results of engineering studies tend to be expressed in mathematical terms that may not be of much interest to the average geologist. Perhaps the most detailed and mathematically rigorous descriptions of beach processes written for geologists are those by Komar (1976) and Hardisty (1990). Beach processes have been summarized in less rigorous form by several other workers, including Davis (1985), Elliott (1986b), and Reineck and Singh (1980). Only a very brief description of these processes is given here. As mentioned, beaches are best developed on wave-dominated coasts where tidal ranges are small. Beaches and barrier islands are constructed primarily by wave-related processes, which include wave swash, storm waves, and nearshore currents (longshore and rip currents). Wind also plays a role in sediment transport on beaches.

FIGURE 11.22 Generalized profile of (A) the beach and nearshore zone and (B) a barrier-island complex showing major environments. (A, from Reinson, G. E., 1984, Barrier-island and associated strand-plain systems, *in* R. G. Walker (ed.), Facies models: Geoscience Canada Reprint Ser. 1, 2nd ed. Fig. 5, p. 122, reprinted by permission of Geological Association of Canada. B, from Richard A. Davis, Jr., Depositional systems: A genetic approach to sedimentary geology, © 1983, Fig. 12.2, p. 405. Reprinted by permission of Prentice-Hall, Englewood Cliffs, N.J.)

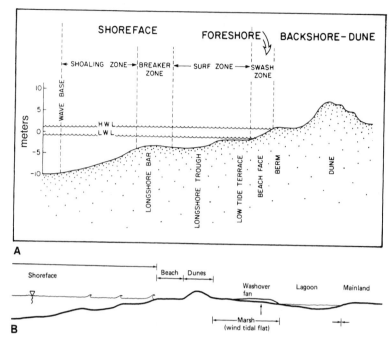

Wave Processes. The influence of orbital wave motion in generating bedforms such as ripples and dunes is discussed in Chapter 3. As deep-water orbital waves approach shallow water where depth is about one-half the wave length, the orbital motion of the water is impeded by interaction with the bottom. Orbits become progressively more elliptical and eventually develop near the bottom a nearly horizontal to-and-fro motion that can move sediment back and forth. This to-and-fro movement is important in generating ripple bedforms as well as in producing some net sediment transport. As waves progress farther shoreward into the shallow **shoaling zone** (Fig. 11.22), forward velocity of the wave slows, wave length decreases, and wave height increases. The waves eventually steepen to the point where orbital velocity exceeds wave velocity and the wave breaks, creating the **breaker zone.** Breaking waves generate turbulence that throws sediment into suspension and also brings about a transformation of wave motion to create the **surf zone.** In this zone, a high-velocity translation wave (a wave translated by breaking into a current), or bore, is projected up the upper shoreface, causing landward transport of bedload sediment and generation of a short-duration "suspension cloud" of sediment. At the shoreline, the surf zone gives way to the **swash zone,** in which a rapid, very shallow swash flow moves up the beach, carrying sediment in partial suspension, followed almost immediately by a backwash flow down the beach. The backwash begins at very low velocity but accelerates quickly. (If heavy minerals are present in the suspended sediment, they settle rapidly to generate a thin heavy-mineral lamina.) The width of the surf and swash zones is governed by the steepness of the shoreface and foreshore. Very steep shorefaces may develop no surf zone at all, and waves break very close to shore, whereas gentle shorefaces commonly have very wide surf zones.

Sediment transport on beaches is particularly important landward of the shoaling zone. In the high-energy breaker zone, coarse sediments move by saltation in a

FIGURE 11.23 Sediment transport associated with a breaking wave. Coarser sediment moves as bedload in a series of loops at position (B). Finer sediment moves in suspension in position (A). Shoreward sediment (C) and seaward sediment (D) move toward the breaking wave as shown by the arrows. (After Ingle, J. C., 1966, The movement of beach sand: Developments in sedimentology, v. 5. Fig. 46, p. 53, reprinted by permission of Elsevier Science Publishers, Amsterdam.)

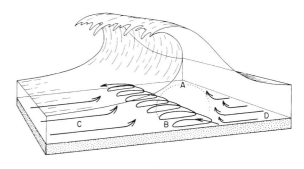

series of elliptical paths that move sediment parallel to the coast, while finer sediment is thrown into suspension (Fig. 11.23). So-called translation waves, which are actually currents, transport sediment through the surf and swash zone up the beach face. If waves approach the shoreline obliquely (a very common occurrence), sediment is transported alongshore in a zigzag manner owing to the fact that the upswash is directed across the beach at an angle, whereas the backswash flow is perpendicular to the beach face. Thus, normal waves of moderate to low energy tend to produce a net landward and alongshore transport of sediments in a largely constructive sedimentation regime in which the beach builds owing to deposition. Repeated deposition and reentrainment of sediment in the beach regime tends to winnow and remove the finest sediment, producing generally well-sorted, positively skewed deposits. Owing to high-energy conditions created by storms, steep, long-period storm waves cause considerable erosion of the beach area and a net displacement of sediment in a seaward direction (Davis, 1985). Great quantities of sediment are thrown into suspension during storms for transport by surf-zone currents, and sand bars on the inner beach may be planed off and displaced seaward considerable distances. Thus, it is quite common to observe marked seasonal changes on modern beaches, which often build in a landward direction during low-energy summer conditions but are eroded and reduced in size during winter storm conditions.

Wave-induced Currents. As breakers and winds pile water against the beach, not only do they create bidirectional translation waves that move up and down the swash zone, but they also create two different types of unidirectional currents: longshore currents and rip currents. **Longshore currents** are generated when waves that approach the shore at an angle break, and a portion of the translation wave is deflected laterally parallel to the shore. These currents move parallel to shore following longshore troughs, which are shallow troughs in the lower part of the surf zone oriented parallel to the strandline (shoreline) (Fig. 11.22). This system of parallel longshore troughs between shallow beach ridges is referred to as a ridge and runnel system. The velocity of longshore currents is related to wave height and the angle at which the waves approach shore. As water piles up between shallow sand bars and the shoreline with continued shoreward movement of waves, it cannot go back against incoming waves the way it came. It must find a different way to return seaward. Thus, it moves parallel to shore as a longshore current until it finds a topographic low between sand bars, where it converges with flow moving in the opposite direction (Fig. 11.24) and moves seaward as a narrow, near-surface current. These converging, seaward-moving currents are called **rip currents.** Longshore currents play a very important role in sediment transport and

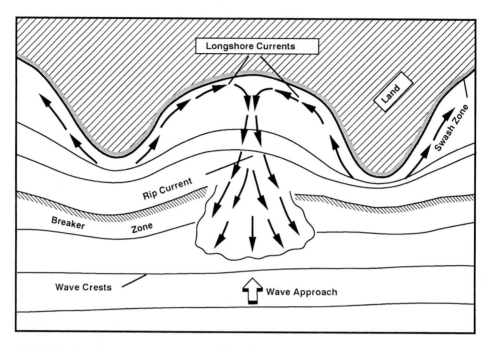

FIGURE 11.24   Schematic representation of longshore currents that move locally in opposite directions, generated owing to bending (refraction) of wave crests as they move over an irregular seafloor, leading to the formation of rip currents that flow seaward through the breaker zone.

deposition on beaches because they achieve velocities great enough to transport sand. Together with the processes producing transport in the swash zone, they are primary agents of alongshore sand movement. The transport paths of sand under longshore currents of different relative velocity are illustrated in Figure 11.25. Rip currents are primarily surface phenomena and thus are less important in near-bed sediment transport than longshore currents. Nonetheless, they can entrain considerable quantities of sediment (and perhaps an occasional unwary swimmer) and move it through the breaker zone out into shoal water.

Wind. In addition to the indirect role that wind plays in generating normal waves, storm waves, and longshore currents, wind also plays a direct role in sediment transport on beaches. The subaerial parts of beaches, above high-tide level, are more or less continuously under the influence of wind. Large quantities of sand may be transported, in a largely onshore to alongshore direction, by wind action. (Sea breezes tend to move from cool ocean water onto warmer land surfaces.) Wind may also move sand about on the lower shoreface as sands dry out during low-tide phases.

Characteristics of Beach and Barrier-Island Deposits in Modern Environments

Overall Geometry and Lithofacies. The mainland-beach and barrier-island system as a whole generates a narrow body of sediments elongated parallel to the depositional strike, or strike of the shoreline. This body of sediment is composed predominantly of sand that originates on the beach shoreface, foreshore, and backshore and is commonly

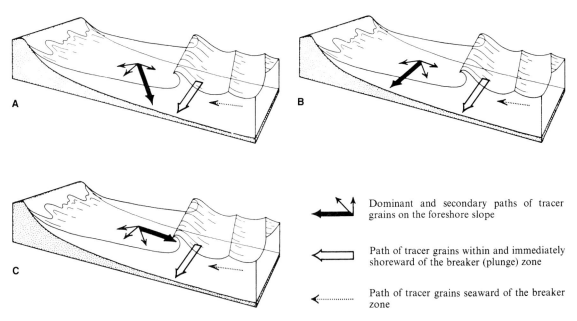

FIGURE 11.25 Transport of sand on beaches by longshore currents under different surf conditions. A. Sediment movement under surf conditions where the longshore current and the wave motion exert an equal influence. B. Sediment grain motion under conditions of a high-velocity longshore current (velocity > 60 cm/s). C. Sediment grain motion where the longshore current velocity is less than 30 cm/s, and the onshore-offshore motion of waves controls the sediment grain transport. (After Ingle, J. C., 1966, The movement of beach sand: Developments in Sedimentology, v. 5. Fig. 46, p. 53, reprinted by permission of Elsevier Science Publishers, Amsterdam.)

tens to hundreds of meters broad, up to hundreds of kilometers long, and 10 to 20 m thick (Reineck and Singh, 1980). It may be interrupted in many places along its length by deltaic, estuarine, bay, and other deposits where these features cut across the beach. Where barrier islands occur, sands of the barrier beach grade landward into back-barrier sediments that may include washover sands; tidal-delta sands and muds; lagoonal silts and muds; and sandy, muddy tidal-flat and marsh deposits (Fig. 11.21).

Beach (Foreshore and Backshore) Deposits. The beach face, or foreshore, is the intertidal zone extending from mean low-tide level to mean high-tide level, corresponding to the zone of wave swash. Sediments of the foreshore consist predominantly of fine to medium sand but may also include scattered pebbles and gravel lenses or layers. Sedimentary structures are mainly parallel laminae, formed during swash-backwash flow, that dip gently (2°–3°) seaward. Thin, heavy-mineral laminae are commonly present, alternating with layers of quartzose sand. Thin, lenticular sets of low-angle, landward-dipping laminae, possibly formed by antidune migration during backswash, may be present also. Some foreshore sands display high-angle, landward-dipping cross-beds caused by migration of foreshore ridges. The foreshore is separated from the backshore by a break in slope at the berm crest (Fig. 11.22), which is formed by sand thrown up by storm waves. The **backshore** is inundated only during storm conditions and is thus a zone dominated by intermittent storm-wave deposition and eolian sand transport and deposition. Faint, landward-dipping, horizontal laminae, interrupted locally by crustacean burrows, record deposition by storm waves. These beds may be overlain by small- to medium-scale eolian trough cross-bed sets, which are commonly disturbed by root growths and burrows of land-dwelling organisms.

Shoreface Deposits. The shoreface environment extends from mean low-tide level on the beach down to the lower limit of fair-weather wave base. Wave base is the depth below which normal waves do not react with the bottom. It is a function of wave length and wave period and thus varies with the wave conditions on the shoreface. The depth of wave base on the shoreface is commonly on the order of 10 to 15 m (Fig. 11.22A), but this depth can be lowered significantly during storms. The shoreface can be divided into the lower, middle, and upper shorefaces, which correspond roughly to the surf, breaker, and shoaling zones. Each is distinguished by characteristic facies.

Upper-shoreface (surf-zone) deposits form in an environment dominated by strong bidirectional translation waves and longshore currents. Depending upon local sediment supply and energy conditions, sediment grain size ranges from fine sand to gravel. Sedimentary structures consist predominantly of multidirectional trough cross-bed sets (which form owing to migration of ripples and dunes), but they may also include low-angle, bidirectional cross-beds and subhorizontal plane beds. Bidirectional cross-beds oriented parallel to depositional strike (shoreline) are common also. These structures may indicate deposition under strong longshore current conditions (Reinson, 1984). Trace fossils such as *Skolithos* (Fig. 5.38) are common but not abundant.

Middle-shoreface (breaker-zone) deposits also form under high-energy conditions owing to breaking waves and associated longshore and rip currents. This is the zone of longshore-bar development. Sediments consist mainly of fine- to medium-grained sand with minor amounts of silt and shell material, although gravels may accumulate also in this environment. Sedimentary structures can be highly complex, depending upon the presence or absence of longshore bars, and can include landward-dipping ripple cross-lamination; seaward-dipping low-angle planar bedding; subhorizontal plane laminations; and seaward- and landward-dipping trough cross-beds. Trace fossils consisting of vertical burrows, such as *Skolithos* and *Ophiomorpha* (Fig. 5.38), are common in this zone.

Lower-shoreface (shoaling-zone) deposits form under relatively low-energy conditions and grade seaward into open-shelf deposits. They are composed dominantly of fine to very fine sand but may also contain thin, intercalated layers of silt and mud. Small-scale cross-stratification formed by predominantly landward-migrating ripples and planar, nearly horizontal laminated bedding (probably resulting from upper-flow-regime sediment transport) are the predominant sedimentary structures present. Hummocky cross-stratification may occur also in lower shoreface sands. Hummocky cross-stratified deposits may, in fact, constitute much of the ancient record of shoreface deposits. In some deposits, hummocky stratified beds have a basal lag deposit consisting of shells or mud clasts and may be capped by a thin layer with wave-ripple stratification, suggesting that each hummocky stratified unit was deposited by a single storm event (McCubbin, 1982). Such deposits may contain abundant plant materials, mica flakes, and other hydraulically light particles. Laminae in lower shoreface deposits tend to be obliterated by bioturbation, and suspension-feeder and deposit-feeder traces such as *Thalassinoides* may be common.

Storms may strongly modify shoreface deposits formed under normal wave-energy conditions owing to lowering of effective wave base by storm-generated waves. The effects of storm waves are particularly important on the middle and lower shoreface, causing severe erosion and redeposition of sediment. Waves scour the bottom, causing suspension of sediment, which is then redeposited farther seaward as the storm wanes (Reinson, 1984). Storm deposits tend to be thicker and more lenticular than normal shoreface deposits and, as mentioned, are commonly characterized by hummocky cross-stratification.

Summary Characteristics of Modern Beach Deposits. Beach deposits are composed predominantly of fine- to medium-grained, well-sorted sand that displays subhorizontal parallel laminations and low-angle, seaward-, landward-, and alongshore-dipping cross-beds. Characteristic facies developed along different parts of the beach profile are shown in Figure 11.26. Bioturbation structures are common in middle and lower shoreface deposits and in sediments of the transition zone between the beach and open shelf. A typical vertical succession of facies developed on an idealized, low-energy, prograding (regressive) beach is illustrated in Figure 11.27. In a preserved transgressive beach-shelf deposit, the succession of facies shown in Figure 11.27 would be essentially reversed; however, transgressive beach-shelf successions appear to be less commonly preserved than are regressive successions.

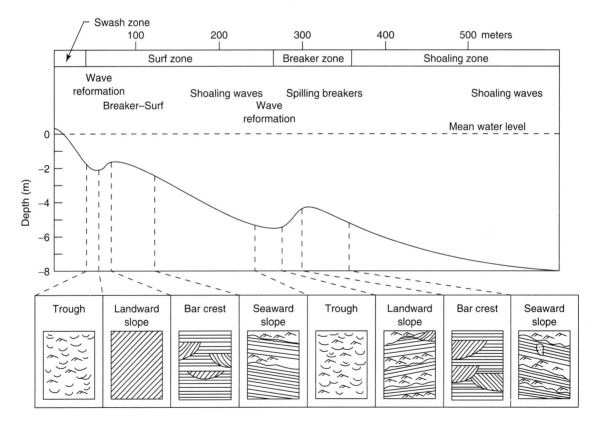

FIGURE 11.26 Facies model of near-surface sediments developed on nearshore barred topography showing principal kinds of sedimentary structures in different parts of the beach. (From Davidson-Arnott, R. G. D., and B. Greenwood, 1976, Facies relationships on a barred coast, Kouchibouguac Bay, New Brunswick, Canada, *in* R. A. Davis, Jr., and R. L. Ethington (eds.), Beach and nearshore sedimentation: Soc. Econ. Paleontologists and Mineralogists Spec. Pub. 24. Fig. 4, p. 154, reproduced by permission of SEPM, Tulsa, Okla.)

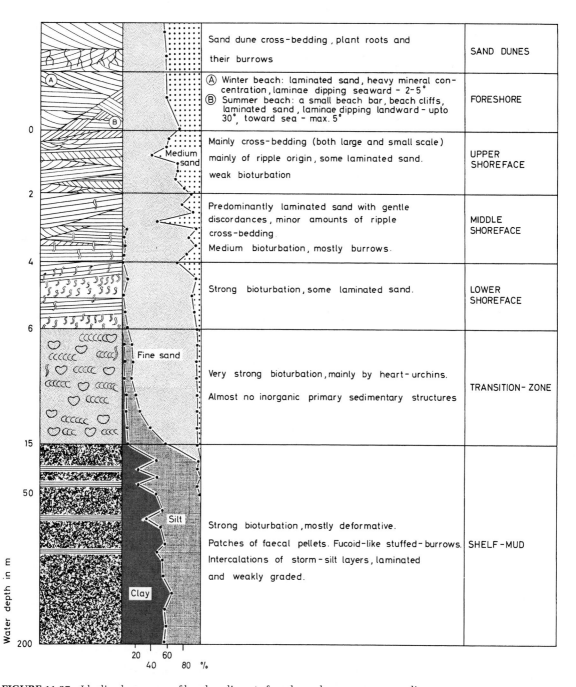

FIGURE 11.27 Idealized sequence of beach sediments found on a low-energy, prograding, Holocene beach. (From Reineck, H. E., and I. B. Singh, 1980, Depositional sedimentary environments, 2nd ed. Fig. 534, p. 387, reprinted by permission of Springer-Verlag, Heidelberg.)

Back-Barrier Deposits. Sediments are deposited in several subenvironments in the back-barrier lagoon landward of barrier beaches. **Washover deposits** occur where storm-driven waves cut through and overtop barriers, washing lobes of sandy beach sediment into the back-barrier lagoon (Fig. 11.21). Washover sediment consist dominantly of fine- to medium-grained sand that displays subhorizontal planar laminations and small- to medium-scale, landward-dipping foreset bedding. Where tidal channels cut through barriers into the inner lagoon, sediments are deposited in a number of tide-related environments, including tidal channels, tidal deltas, and tidal flats (Fig. 11.21). **Tidal-channel deposits** consist dominantly of sand, and the deposits commonly have an erosional base marked by coarse lag sands and gravels. Sedimentary structures may include bidirectional large-to small-scale planar and trough cross-beds that may display a general fining-upward textural trend. **Tidal-delta deposits** form on both the lagoonal side of the barrier (flood-tidal delta) and the seaward side of the barrier (ebb-tidal delta). They are predominantly sandy deposits to tens of meters thick with a gross parabolic shape or geometry. They are characterized by a highly varied succession of planar and trough cross-bed sets that may dip in either a landward or a seaward direction. **Tidal-flat deposits** form along the margins of the mainland coast and the back of the barrier. They grade from fine- to medium-grained ripple-laminated sands in lower areas of the tidal flats through flaser- and lenticular-bedded fine sand and mud in mid-tidal flats to layered muds in higher parts of the flats. **Lagoonal** and **marsh deposits** accumulate in the low-energy back-barrier lagoon and grade laterally into higher-energy, sandy deposits of tidal channels, deltas, and washover lobes. They consist largely of interbedded and interfingering fine sands, silts, muds, and peat deposits that may be characterized by disseminated plant remains, brackish-water invertebrate fossils such as oysters, and horizontal to subhorizontal layering. A generalized succession of facies deposited in an ancient back-barrier environment is illustrated in Figure 11.28. The cyclic nature of this deposit suggests recurring episodes of transgression and regression. Such a succession might be confused with some deltaic successions, although the absence of delta-front muddy sands and prodelta clays and mud turbidites should help to distinguish it from deltaic deposits.

Transgressive and Regressive Beach and Barrier-Island Deposits in the Geologic Record

Recognition of ancient beach and barrier-island complexes requires general knowledge of idealized models plus an understanding of how real stratigraphic successions can deviate from these norms. Facies developed during regression produce vertical successions different from those formed during transgression: Successions developed on mainland beaches without barriers lack the back-barrier lagoonal and associated deposits that characterize those deposited on coasts with barrier islands, and local variations in depositional patterns within a given beach and barrier-island environment can generate different vertical patterns of facies that may include tidal-delta deposits, tidal-channel deposits, beach deposits, and back-barrier lagoonal and marsh deposits.

The vertical succession of facies generated in beach and barrier-island systems depends upon whether changes in depositional environments with time are the result of transgression or regression. The specific characteristics of transgressive or regressive successions depend upon rates of sea-level change, rates of basin subsidence, and the sediment supply. Transgression, or movement of the shoreline in a landward direction, can take place as a result of rising sea level, either eustatic or relative rise, provided that concurrent influx of terrigenous clastic sediment is not too rapid to prevent the

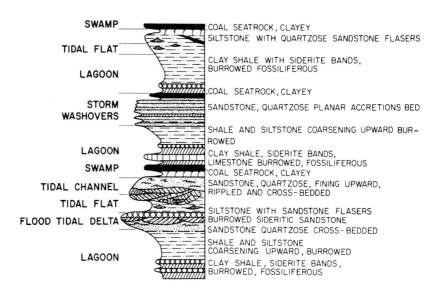

FIGURE 11.28 Generalized sequence of facies deposited in a back-barrier environment, Carboniferous of eastern Kentucky and southern West Virginia. Such successions range from 7.5 to 24 m thick. (From Horne, J. C., J. C. Ferm, F. T. Caruccio, and B. P. Baganz, 1978, Depositional models in coal exploration and mine planning in Appalachian region: Am. Assoc. Petroleum Geologists Bull., v. 62. Fig. 4, p. 2385, reprinted by permission of AAPG, Tulsa, Okla.)

shoreline from shifting landward. Regression, or seaward shift of the shoreline, occurs particularly as a result of falling sea level, but it can also occur during static sea level or even rising sea level if influx of clastic sediment is exceptionally great. These factors are discussed further in Chapter 14.

Transgression leads to the formation of barrier-island complexes in which back-barrier lagoonal and marsh deposits are overlapped by sandy deposits of the barrier-beach complex. The generation of transgressive beach and barrier-island deposits has been suggested to occur by two different mechanisms (Reinson, 1992):

1. Landward advance of the shoreline occurs owing to shoreface erosion, as might take place during slow rise of sea level.
2. Relatively sudden upward "jumps" of the shoreline occur during rapidly rising sea level.

These alternative mechanisms are illustrated in Figure 11.29. During shoreline retreat, beach and upper shoreface deposits are presumably eroded and transported to the lower shoreface or offshore, as storm beds, or to the lagoon as washover deposits (Fig. 11.29A). In-place drowning owing to "sudden" inundation during a rapid rise in sea level would cause the barrier to be covered by water, resulting in the wave zone moving landward until a new sand barrier forms on the inner side of the lagoon (Fig. 11.29B). Ancient beach and barrier deposits interpreted to be transgressive deposits have been reported (e.g., McCubbin, 1982).

Barrier islands can prograde, under conditions of high sediment supply relative to sea-level change, to produce regressive barrier-island facies. Under these conditions, barriers tend to be transformed into strand plains, producing dominantly sandy facies in which beach (backshore and foreshore) deposits overlie foreshore deposits. Galveston Island, Texas (Fig. 11.29C), provides an example of such a progradational deposit;

FIGURE 11.29 Barrier-island facies generated by transgression and regression. Part A illustrates transgression owing to shoreface retreat during gradual sea-level rise, and B shows the effects of rapid sea-level rise, producing in-place drowning. SL = sea level. Part C illustrates the facies formed as a result of progradation under conditions of high sediment supply relative to sea-level change. (A and B, after Rampino, M. R., and J. E. Sanders, 1980, Holocene transgression in south-central Long Island, New York: Jour. Sed. Petrology, v. 50. Fig. 8, p. 1075, reproduced by permission of Society of Economic Paleontologists and Mineralogists, Tulsa, Okla. Elliott, T., 1986, Siliciclastic shorelines, *in* H. G. Reading (ed.), Sedimentary environments and facies: Blackwell Scientific Publications. Fig. 7.33, p. 180. Based on Fischer, 1961; Swift, 1975; and Sanders and Kumar, 1975. C, after Galloway, W. E., and D. K. Hobday, 1983, Terrigenous clastic depositional systems: Springer-Verlag, New York. Fig. 6.10, p. 126.)

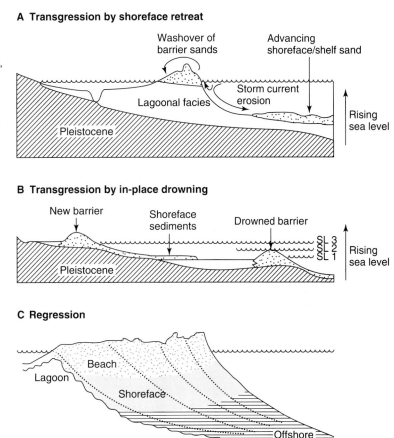

however, Galveston Island is not necessarily typical of all regressive barrier-island complexes.

Differences in vertical successions of lithofacies generated by transgression and regression are illustrated in the generalized facies succession shown in Figure 11.30. The general succession of facies deposited on a regressive, prograding beach is shown in Figure 11.30A (also see Fig. 11.27). In general, regression produces a coarsening-upward succession from fine-grained, lower-shoreface deposits to coarser foreshore and backshore deposits. Back-barrier lagoonal and marsh deposits commonly are not preserved during regression and thus do not appear in this facies model.

The model for a transgressive barrier-island complex shown in Figure 11.30B is characterized by interbedded back-barrier deposits and does not display a definite fining- or coarsening-upward trend. This model is incomplete in the sense that it does not show the deposits of the foreshore and shoreface that would lie on top of backshore-dune deposits with continued transgression. Figure 11.30C illustrates an idealized model of the vertical succession of facies generated in the barrier-island environment

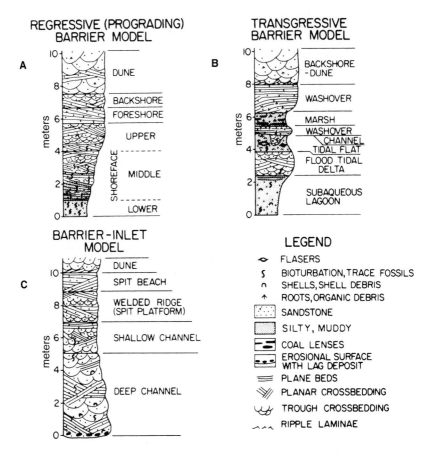

FIGURE 11.30   End-member facies models for transgressive barrier, regressive barrier, and barrier inlet stratigraphic sequences. A standard 10-m unit is shown, but thickness could range up to a few tens of meters. A spit platform (C) is the subaqueous part of a spit, formed by longshore currents. (From Reinson, G. E., 1984, Barrier island and associated strand-plain systems, *in* R. G. Walker (ed.), Facies models, 2nd ed: Geoscience Canada Reprint Ser. 1. Fig. 26, p. 133, reprinted by permission of Geological Association of Canada.)

by migration of spit and beach sands over tidal-channel deposits. (A spit is a fingerlike extension of a beach into deeper water.)

Two specific examples of ancient beach and barrier-island deposits are included here to illustrate differences in ancient regressive deposits and transgressive barrier deposits. Figure 11.31 shows a regressive, or progradational, succession in the Cretaceous Gallup Sandstone of Northwestern New Mexico (McCubbin, 1982). In this succession, sandy beach deposits overlie burrowed silty, offshore shales containing normal marine fossils. The basal unit of the succession consists of fine- to very fine-grained sandstone with mostly subhorizontal to planar stratification and hummocky cross-stratification. Burrows are abundant. This unit is overlain above a scoured surface by fine- to medium-grained sandstones with high-angle cross-stratification in trough-shaped sets. Some thin interbeds with planar stratification are also present. Burrowing is less common in this unit than in the basal section. The uppermost part of the succession consists of fine-grained, well-sorted sandstones characterized by nearly

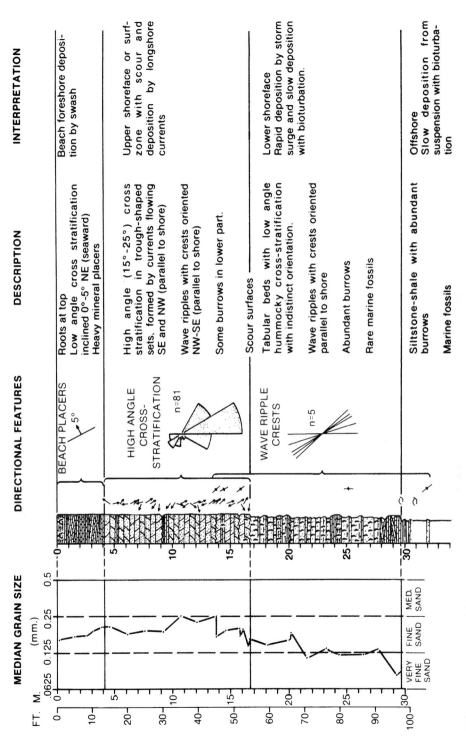

FIGURE 11.31 Vertical facies sequence in part of the Cretaceous Gallup Sandstone, northwestern New Mexico. The sequence is interpreted as a progradational beach deposit, probably formed on a nonbarred coast. (After McCubbin, D. G., 1982. Barrier island and strand-plain facies, *in* P. A. Scholle and D. Spearing (eds.). Sandstone depositional environments: Am. Assoc. Petroleum Geologists Mem. 31. Fig. 25, p. 260, reprinted by permission of AAPG, Tulsa, Okla.)

horizontal, planar stratification and very low-angle cross-stratification; root traces are present at the top. These three divisions of the Gallup Sandstone are interpreted to represent, in ascending order, lower-shoreface, upper-shoreface, and beach-foreshore deposits. The grain-size and stratification characteristics of the Gallup Sandstone suggest that it was developed on a coast with moderate to high wave energy. Because back-barrier lagoonal sediments are absent in this succession, we can infer that it is probably a mainland beach deposit, formed on a coast that lacked barriers.

The Cretaceous Cliff House Sandstone in the San Juan Basin of northwestern New Mexico has been interpreted as a transgressive barrier complex (Donselaar, 1989; McCubbin, 1982). The interpreted paleogeography of the Cliff House depositional site in Late Cretaceous time is shown in Figure 11.32. The basal part of the stratigraphic section shown in this figure consists of dark, laminated shales with abundant plant fragments and widely spaced layers with carbonized plant roots. Coal beds up to 1.8 m thick are common, and brackish-water fossils such as oysters are present. This shale and coal unit is interpreted as back-barrier lagoonal sediments. It is overlain by and interfingers with fine- to medium-grained barrier-bar sandstones characterized by high-angle trough cross-stratification, hummocky cross-stratification, and planar stratification. *Ophiomorpha* burrows are common. Although the overall succession shown in Figure 11.32 is interpreted to be a transgressive succession, accumulation of the barrier sands took place during regressive intervals (Donselaar, 1989). During subsequent transgressive intervals, part of the barrier sand was eroded; only topographically low-lying parts of the barrier (e.g., shoreface, tidal-inlet, and lagoonal sediment) were preserved. The Cliff House barrier sands interfinger in a seaward direction with the marine Lewis Shale.

Other siliciclastic sedimentary successions identified as beach and barrier-island complexes are present in rocks of widely differing ages in North America. Such successions have been reported from several Pennsylvanian formations in the Appalachian Basin of Kentucky, Virginia, West Virginia, and Tennessee; the Lower Cretaceous Muddy Sandstone of Wyoming and Montana; the Eocene Wilcox Group of east Texas; and the Quaternary of California (e.g., Davis, 1992). Several ancient carbonate deposits have also been interpreted as beach complexes. The Lower Cretaceous Edwards Formation of west Texas, the Lower Cretaceous Cow Creek Formation of central Texas, the Mississippian Newman Formation in eastern Kentucky, and the Missis-

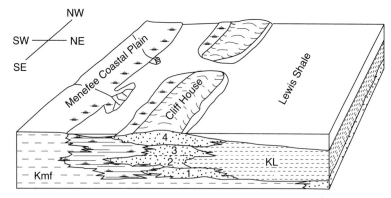

FIGURE 11.32 Diagrammatic sketch illustrating the paleogeography of the coastal area of the San Juan Basin, northwestern New Mexico, during deposition of the Late Cretaceous Cliff House Formation. Numerals 1–4 refer to different stacked, barrier-beach sandstones bodies. Kmf = Menefee Formation; KL = Lewis Shale. No scale shown. (After Donselaar, M. E., 1989, The Cliff House sandstone, San Juan Basin, New Mexico: Model for the stacking of "transgressive" barrier complexes: Jour. Sed. Petrology, v. 59. Fig. 4, p. 15, reproduced by permission of Society of Economic Paleontologists and Mineralogists, Tulsa, Okla.)

sippian Mission Canyon Formation in the Williston Basin in the Montana area are examples of ancient stratigraphic units that contain putative carbonate beach deposits. Carbonate beach deposits are further discussed in Chapter 13.

11.4 ESTUARINE AND LAGOONAL SYSTEMS

Introduction

Relatively small, semienclosed coastal embayments are loosely called coastal bays. Two broad types of coastal bays are recognized: estuaries and lagoons. **Estuaries** (term derived from the Latin word *aestus,* meaning tide, and from the adjective *aestuarium,* meaning tidal) are considered in a general sense to be the lower courses of rivers open to the sea; however, they have been defined somewhat differently by geologists, geographers, and chemists. Fairbridge (1980, p. 7) defines an estuary as "an inlet of the sea reaching into a river valley as far as the upper limit of tidal rise, usually divisible into three sectors: (a) a marine or lower estuary, in free connection with the open sea; (b) a middle estuary, subject to strong salt and freshwater mixing; and (c) an upper or fluvial estuary, characterized by fresh water but subject to daily tidal action." Dalrymple, Zaitlin, and Boyd (1992) argue that the concept of net landward movement of sediment derived from outside the estuary mouth is necessary to distinguish estuaries from deltas. Therefore, they define an estuary as "the seaward portion of a drowned valley system which receives sediment from both fluvial and marine sources and which contains facies influenced by tide, wave and fluvial processes. The estuary is considered to extend from the landward limit of tidal facies at its head to the seaward limit of coastal facies at its mouth." According to Dalrymple, Zaitlin, and Boyd, estuaries can form only in the presence of a relative sea-level rise (i.e., a transgression). Progradation tends to fill and destroy estuaries, causing them to change into deltas.

A coastal **lagoon** is defined as a shallow stretch of seawater—such as a sound, channel, bay, or saltwater lake—near or communicating with the sea and partly or completely separated from it by a low, narrow, elongate strip of land, such as a reef, barrier island, sandbank, or spit (Bates and Jackson, 1980). Lagoons commonly extend parallel to the coast, in contrast to estuaries, which are oriented approximately perpendicular to the coast. Many lagoons have no significant freshwater runoff; however, some coastal embayments that otherwise satisfy the general definition of lagoons do receive river discharge. Estuaries and lagoons may occur in close association with river deltas, barrier islands, and tidal flats.

Owing to the generally small size of coastal bays, estuarine and lagoonal sediments are volumetrically less significant in the geologic record than are deltaic sediments. When present, they provide important information about shoreline conditions and environments; further, they may have economic significance. Estuaries are among the most biologically productive environments known (Lauff, 1967); therefore, estuarine sediments may be important source rocks for petroleum. Furthermore, the association of well-sorted sandy facies and muddy facies in many estuaries and lagoons provides a favorable setting for stratigraphic traps for petroleum. The hydrologic characteristics and sediment transport conditions of estuaries and lagoons are quite variable depending upon the climate, tidal range, and wave energy. These characteristics, as well other aspects of the geologic, biologic, and chemical properties of estuaries and lagoons, are described in several monographs, including Ashley (1988), Barnes (1980), Castanares and Phleger (1969), Cronin (1975), Kjerfve (1978), Lauff (1967), Nelson (1972), Officer (1977), Olaussen and Cato (1980), van de Kreeke (1986), Ward and

Ashley (1989), and Wiley (1976). The distinguishing characteristics of estuaries and lagoons have been summarized in shorter contributions by Boothroyd (1978), Clifton (1982), Colombo (1977), Fairbridge, (1980), Hayes (1975), Nichols and Biggs (1985), Phleger (1969), and Reinson (1992).

Physiography of Estuaries and Lagoons

Based on the physiographic characteristics of relative relief and degree of channel-mouth blocking, seven basic types of modern estuaries are recognized (Fig. 11.33; Fairbridge, 1980). **Fjords** are high-relief estuaries with a U-shaped valley profile formed by drowning of glacially eroded valleys during Holocene sea-level rise. **Fjards** or firths are related to fjords but have lower relief. Estuaries developed in winding valleys with moderate relief are **rias. Coastal-plain estuaries** are low-relief estuaries, funnel-shaped in plan view, that are open to the sea. Low-relief estuaries that are L-shaped in plan view and that have lower courses parallel to the coast are **bar-built estuaries.** Similar

FIGURE 11.33 Principal types of estuaries based on physiographic characteristics. (From Fairbridge, R. W., The estuary: Its definition and geodynamic cycle, *in* E. Olausson and I. Cato (eds.), Chemistry and biochemistry of estuaries, Fig. 2, p. 9, © 1980 John Wiley & Sons, Ltd. Reprinted by permission of John Wiley & Sons, Ltd., Chichester, England.)

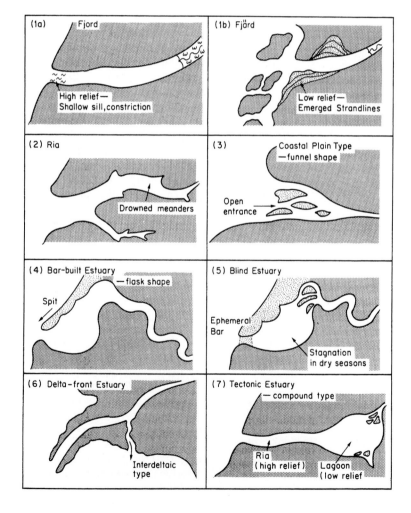

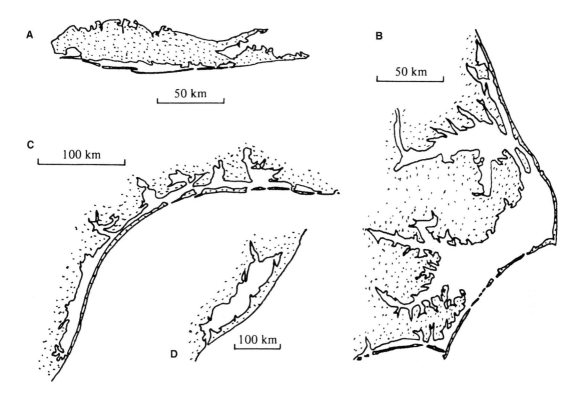

FIGURE 11.34 Lagoonal systems enclosed by barrier-island chains in the United States. A. Long Island, New York. B. Pamlico and Albermarle Sounds, North Carolina. C. Laguna Madre and associated lagoons, Texas. D. A lagoon enclosed within the land: The Lagoa dos Patos between Porto Alegre and Rio Grande, Brazil. (From Barnes, R. S. K., 1980, Coastal lagoons. Fig. 1.3, p. 5, reprinted by permission of Cambridge University Press, Cambridge.)

estuaries that are seasonally blocked by longshore drift or dune migration are called blind estuaries. Bar-built estuaries and **blind estuaries** are very similar to lagoons and might be considered lagoons by some workers. **Deltaic estuaries** occur on delta fronts as ephemeral distributaries. Flask-shaped, high-relief rias backed by a low-relief plain created by tectonic activity are called compound estuaries or **tectonic estuaries.** Most modern lagoons are formed behind spits or offshore barriers of some type and thus are elongated bodies lying parallel to the coast with a narrow connection to the open ocean (Fig. 11.34). Lagoons also form behind barrier reefs and atolls.

Hydrologic and Sediment Characteristics

Estuaries

Water circulation patterns in estuaries are affected by freshwater (river) runoff, tidal forces, and waves. Therefore, salinity distributions, water circulation patterns, and sediment transport paths within estuaries are complex and differ from estuary to estuary. Estuaries are commonly characterized on the basis of water circulation patterns as (1) salt-wedge (stratified), (2) partially mixed, or (3) fully mixed (Postma, 1980; Nichols

and Biggs, 1985). **Salt-wedge** estuaries are present where estuarine circulation is dominated by river flow, and tides play a negligible role (microtidal coasts; 0–2 m tidal
range). A wedge of saltwater extends along the bottom of the estuary and thins in a
landward direction. Freshwater spreads out on top of this saltwater and thins seaward
(e.g., the seaward zone of the Mississippi River). Incoming tidal currents are too weak
to transport most sediment; thus, sediment moves mainly out of the estuary. On
mesotidal coasts (2–4 m tidal range), tidal currents are increased so that river flow does
not dominate the circulation. Tidal mixing affects the sharp boundary between salt-
and freshwater layers, causing **partial mixing** along this interface (e.g., Chesapeake
Bay). Landward current flow is strong enough to transport sediment in a landward
direction. Where river discharge varies seasonally, however, the dominant direction of
sediment transport may reverse in response to discharge fluctuations. In some Oregon
estuaries, for example, flood-tide velocity is dominant during low river discharge in
summer, and a flood-tide delta builds upstream into the estuary. During high river discharge in winter, velocity asymmetry reverses, the flood-tide delta is eroded, and all
the sand transported into the estuary during summer is transported back to the coast
(Boggs and Jones, 1976). Where tidal range and current velocities are great enough
[macrotidal (>4 m tidal range) and some mesotidal coasts], vertical salinity stratification breaks down completely to generate **well-mixed** estuaries (e.g., Delaware Bay).
There is little landward flow but mainly lateral movement and mixing. Suspended sediment is dispersed away from lateral or weak river sources and, also, above the bed by
tidal resuspension (Nichols and Biggs, 1985).

   Dalrymple, Zaitlin, and Boyd (1992) present a model for estuaries, based on
dominant hydrologic characteristics and the kinds of sediment and sediment bodies
formed in the estuary, that is perhaps more geologically relevant than the above mixing
model. They suggest that estuaries are wave dominated, tide dominated, or mixed
wave and tide dominated.

Wave-dominated Estuaries. In a wave-dominated estuary, tidal influence is small, and
the mouth of the estuary experiences high wave energy. Hydrologic conditions may
range from partially mixed to well-stratified. Sediments tend to move alongshore and
onshore into the mouth of the estuary, where a subaerial barrier/spit or submerged bar
develops. This barrier prevents most of the wave energy from entering the estuary;
thus, only internally generated waves are present behind the barrier. Depending upon
tidal range and current velocity, a small number of inlets may be kept open in this barrier. Alternatively, the barrier may close the estuary entirely at times to produce a
blind estuary or coastal lake. The relative influence of marine and river processes and
the kinds of sediment bodies formed in different parts of a wave-dominated estuary are
illustrated in Figure 11.35.

   Note that marine-derived sands are carried landward only a short distance
beyond the barrier as washover deposits or flood-tidal delta sands. Muddy sediments,
supplied mainly by the river, accumulate in the central part of the estuary where total
energy is lowest. Deposition of mud is enhanced by mixing of river and marine water,
which causes flocculation of clay particles owing to the presence of positively charged
ions in seawater that neutralize the negative charges on clay particles. Muddy sediment may include biogenic debris such as molluscan shells, wood fragments, and
organically produced pellets. Horizontal and subhorizontal stratification and bioturbation traces are common structures. In the head of the estuary, coarse, river-derived sediments are deposited in channels; marsh deposits may form adjacent to the channels.
Examples of modern wave-dominated estuaries include San Antonio Bay, the United
States; the Miramichi River, Canada; and Hawksbury Estuary, Australia.

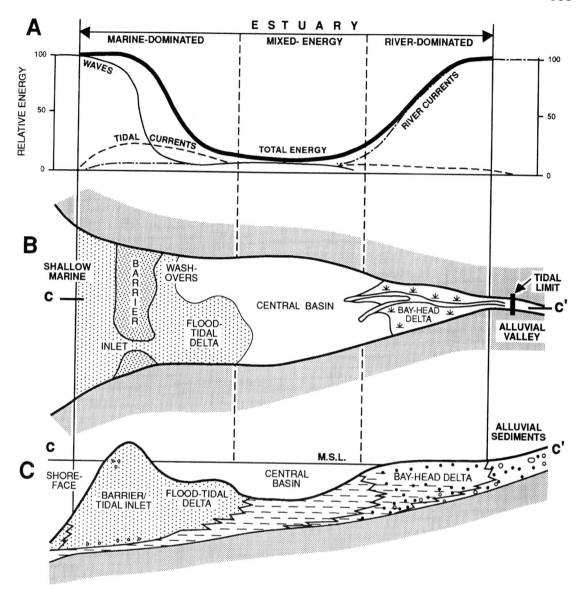

FIGURE 11.35 Distribution of (A) energy types, (B) morphological components in plan view, and (C) sedimentary facies in longitudinal section within an idealized wave-dominated estuary. The shape of the estuary is schematic. The barrier/sand plug is shown as headland-attached; however, on low-gradient coasts, it may be separated from the mainland by a lagoon. The section in Part C represents the onset of estuary filling following a period of transgression. (From Dalrymple, R. W., B. A. Zaitlin, and R. Boyd, 1992, Estuarine facies models: Conceptual basin and stratigraphic implications: Jour. Sed. Petrology, v. 62. Fig. 4, p. 1134, reproduced by permission of Society of Economic Paleontologists and Mineralogists, Tulsa, Okla.)

Tide-dominated Estuaries. Tide-dominated estuaries occur mainly on macrotidal coasts where tidal-current energy exceeds wave energy at the mouth of the estuary, creating energy conditions in the estuary higher than those typical of wave-dominated estuaries. Water in the estuary is commonly well mixed. Elongate sand bars develop parallel to the length of the estuary from sand carried into the estuary from marine sources. The presence of these bars tends to dissipate tidal energy. On the other hand, constriction of incoming tidal currents between the tidal bars causes an increase in their velocity for some distance up the estuary. Figure 11.36 shows the hydrologic conditions and sediment bodies developed in tide-dominated estuaries.

In addition to forming tidal bars in the mouth of the estuary, sand may be transported landward through the estuary in the tidal-fluvial channel. Bedforms ranging in size from ripples to large dunes develop on sandy sediment in the bars and tidal channels, and cross-bedding generated by migration of these bedforms can dip in either a landward or a seaward direction. Flaser bedding may form during slack water owing to deposition of suspended mud over sand ripples. Muddy sediment is deposited also in lower-energy parts of the estuary floor and in salt marshes adjacent to the channel along the edges of the estuary. Muddy sediments are characterized by nearly planar alternations of silt, clay, very fine sand, and carbonaceous (plant) debris. Bioturbation by burrowing and feeding organisms may locally mix and homogenize these layers. Estuarine sediments typically contain a brackish-water fauna that may include oysters, mussels, other pelecypods, and gastropods. Examples of tide-dominated estuaries include Cook Inlet, Alaska; Ord River, Australia; Gironde Estuary, France; and the Severn River, the United Kingdom.

Mixed Wave- and Tide-dominated Estuaries. Many estuaries have characteristics that are intermediate between wave-dominated and tide-dominated types. For example, as tidal energy increases relative to wave energy, the barrier system of wave-dominated estuaries becomes progressively more dissected by tidal inlets, and elongate sand bars develop in locations previously occupied by barrier segments and the channel-margin linear bars of ebb-tidal deltas. Marine-derived sand is transported greater distances up estuary, and the generally muddy central basin is replaced by sandy tidal channels flanked by marshes. Examples of mixed-energy estuaries include the St. Lawrence River, Canada; Willipa Bay, the United States; and Oosterschelde Estuary, the Netherlands.

Vertical Successions of Estuarine Facies. The vertical succession of facies developed in estuaries depends upon the kind of estuary (wave- or tide-dominated) and the location within the estuary. Facies dominated by cross-bedded, bioturbated sand are present near the mouths of estuaries and in fluvial-tidal channels, whereas laminated to well-bioturbated muds occupy the nonchannel middle and upper parts of the estuary. Many estuaries are subjected in time to transgression. Transgression brings about a landward shifting of environments, resulting in vertical stacking of estuary-mouth sands on top of middle-estuary muds and/or fluvial-tidal channel sands (Fig. 11.37). By contrast, a regression causes filling and destruction of the estuary and seaward progradation, which may lead to deposition of fluvial sediments on top of estuarine deposits.

Lagoons

Many factors affect water flow, water mixing, and sediment transport in lagoons, including tides, wind waves, freshwater runoff, episodic storms, density gradients, sea-level changes, and changes in climate and temperature. Even so, water circulation

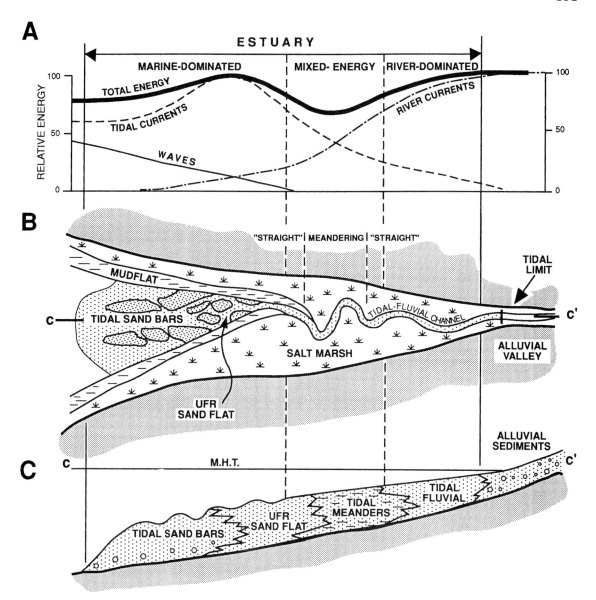

FIGURE 11.36 Distribution of (A) energy types, (B) morphological components in plan view, and (C) sedimentary facies in longitudinal section within an idealized tide-dominated estuary. UFR = upper-flow regime; M.H.T. = mean high tide. The section in Part C is taken along the axis of the channel and does not show the marginal mudflat and salt-marsh facies; it illustrates the onset of progradation following transgression. (From Dalrymple, R. W., B. A. Zaitlin, and R. Boyd, 1992, Estuarine facies models: Conceptual basin and stratigraphic implications: Jour. Sed. Petrology, v. 62. Fig. 7, p. 1136, reproduced by permission of Society of Economic Paleontologists and Mineralogists, Tulsa, Okla.)

FIGURE 11.37 Schematic facies model of a composite transgressive vertical sequence in a microtidal (wave-dominated) estuary. (After Nichols, M. M., et al., 1991, Modern sediments and facies models for a microtidal coastal plain estuary, the James Estuary, Virginia: Jour. Sed. Petrology, v. 61. Fig. 12, p. 895, reproduced by permission of Society of Economic Paleontologists and Mineralogists, Tulsa, Okla.)

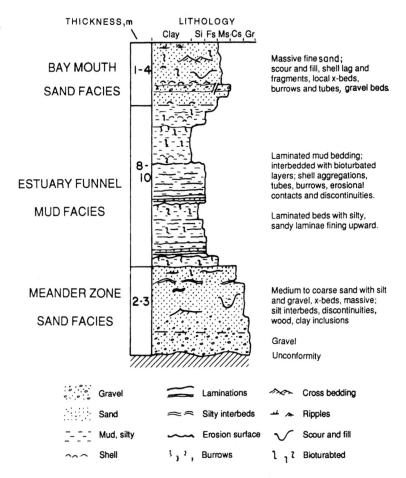

patterns in lagoons are much less affected by freshwater inflow than they are in estuaries, and many lagoons receive no freshwater discharge. Also, circulation with the open ocean is restricted by the presence of some type of barrier. Consequently, the principal movement of water within lagoons occurs in the form of tidal currents (which move in and out through the narrow inlets between barriers) and wind-forced waves.

On the basis of geomorphology and the nature of water exchange with the coastal ocean, Kjerfve and Magill (1989) identify three types of lagoons: (1) choked, (2) restricted, and (3) leaky (Fig. 11.38). **Choked lagoons** occur along coasts with high wave energy and significant alongshore drift (e.g., Coorong Lake, southern Australia). They are characterized by one or more long, narrow entrance channels; long residence times of water within the lagoon; and dominant water movement by wind forcing. Intense solar radiation coupled with inflow event can cause intermittent vertical stratification. **Restricted lagoons** commonly exhibit two or more entrance channels or inlets, have a well-defined tidal circulation, are strongly influenced by winds, and are generally vertically mixed (e.g., Lake Pontchartrain, Louisiana). **Leaky lagoons** typically

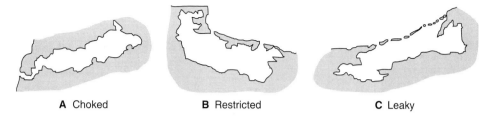

A Choked   **B** Restricted   **C** Leaky

FIGURE 11.38 Principal kinds of coastal lagoons (choked, restricted, leaky) based on the degree of water exchange with the adjacent coastal ocean. (From Kjerfve, B., and K. E. Magill, 1989, Geographic and hydrodynamic characteristics of shallow coastal lagoons: Marine Geology, v. 88. Fig. 2, p. 190, reproduced by permission.)

occur along coasts where tidal currents are a more important factor in sediment transport than are wind waves (e.g., Belize Lagoon, Belize). They may stretch along coasts for more than 100 km but commonly are no more than a few kilometers wide. They are characterized by wide tidal passes, efficient water exchange with the ocean, strong tidal currents, and sharp salinity and turbidity fronts.

Except within tidal channels that extend into the lagoon, lagoons are predominantly areas of low water energy. Tidal deltas commonly develop at the ends of these tidal inlets, both within the lagoon and on the ocean sides, and sandy sediment may also be deposited within the higher-energy tidal channels inside the lagoon. Otherwise, sedimentation within lagoons is dominated by deposition of silt and mud, although occasional high wave activity during storms can cause washover of sediment from the barrier.

Salinity within lagoons can range from hypersaline to essentially that of freshwater, depending upon the hydrologic conditions and the climate. Lagoons formed in arid or semiarid coastal areas, where little freshwater influx occurs, are commonly hypersaline, with salinities well above that of normal seawater. Lagoons in more humid regions may be characterized by brackish water. Salinity within lagoons may vary in response to seasonal rainfall and evaporation rates. Also, salinity at a particular time may not be uniform throughout a lagoon. Lagoons receiving considerable freshwater inflow commonly display distinct, lateral salinity zones (Fig. 11.39).

The deposits of lagoons may differ from those of estuaries in several ways. First, because many lagoons do not receive freshwater discharge from rivers, most or all the sediment in such lagoons is from marine sources. Lagoons are typically low-energy environments, although tidal currents move into lagoons through inlets between barriers, winds create some wave action along shorelines, storms provide occasional episodes of high-energy waves that wash over barriers into the lagoon, and prevailing winds may more or less continuously blow small amounts of sediment from barriers into the lagoon. Because of the dominance of low-energy conditions in lagoons, lagoonal deposits consist mainly of fine-grained sediments. Sandy sediments are confined principally to tidal deltas constructed at the mouths of the tidal inlets, some tidal channels that extend into the lagoon, washover lobes behind barriers, and some parts of the lagoonal shoreline (lagoonal beaches). Small amounts of sandy sediment blown from barriers may also be scattered throughout the lagoon. Sandy sediment in tidal channels is characterized by current ripples and internal small-scale cross-bedding that may dip in either a landward or a seaward direction. Most of the lagoonal bottom is covered with silty or muddy sediments, commonly extensively bioturbated, that may contain thin intercalations of sand brought in by storms or blown in by wind.

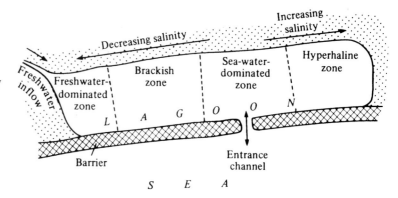

This sand is generally horizontally laminated, but it may display ripple cross-laminations. The faunas that inhabit lagoons are highly variable depending upon the salinity conditions of the lagoon, but they are generally characterized by low diversity. Lagoons with normal salinity show faunas similar to those of the open ocean, whereas brackish-water faunas dominate lagoons in front of river mouths. Hypersaline lagoons commonly contain few organisms because few species are adapted to such high salinities.

In areas where the availability of siliciclastic sediment is low and climatic conditions are favorable, sedimentation in lagoons is dominated by chemical and biochemical deposition. Under very arid conditions, lagoonal sedimentation may be characterized by deposition of evaporites, which are mainly gypsum but may include some halite and minor dolomites (e.g., lagoons in the Persian Gulf). Under less hypersaline conditions, carbonate deposition prevails, particularly in lagoons developed behind barrier reefs (e.g., Australia). Deposits in such lagoons may consist largely of carbonate muds and associated skeletal debris, although ooids may form in more agitated parts of the lagoon. Algal mats, commonly developed in the supratidal and shallow intertidal zone, may trap fine carbonate or siliciclastic mud to form stromatolites. Algal mats in the supratidal zone generally display mudcracks with curled margins. For additional details on the hydrology and sedimentology of lagoons, see Ward and Ashley (1989); carbonate sediments and environments are further discussed in Chapter 13.

Ancient Estuarine and Lagoonal Deposits

Both estuaries and lagoons are ephemeral features. Because they tend to fill with sediments in geologically short periods of time, the preservation potential of estuarine and lagoonal sediments is generally high. Nevertheless, relatively few estuarine deposits have been reported from the geologic record, possibly because they have not been widely recognized and distinguished from associated fluvial, deltaic, lagoonal, or shallow-marine deposits.

Estuarine deposits tend to have restricted faunal assemblages that include brackish-water species and that may be characterized by trace fossil assemblages reflecting brackish to stressed conditions (Reinson, 1992); however, no unique physical criterion exists for these deposits. Depending upon location within an estuary, estuarine deposits may consist almost entirely of cross-bedded sands, laminated or bioturbated muds, or combinations of sand and mud. Gradation from fluvial channel sands at the base of a vertical section through mixed fluvial-marine muds in the middle of the sec-

FIGURE 11.40 Composite stratigraphic section of the Cretaceous Peace River Formation, northern Alberta, Canada, showing estuarine deposits interbedded with marginal-marine and marine deposits in a transgressive sequence. A ravinement surface is a minor erosional surface. (After Leckie, D., and C. Singh, 1991, Estuarine deposits of the Albian Paddy Member (Peace River Formation) and the lowermost Shaftsbury Formation, Alberta, Canada: Jour. Sed. Petrology, v. 61. Fig. 2, p. 828, reproduced by permission of Society of Economic Paleontologists and Mineralogists, Tulsa, Okla.)

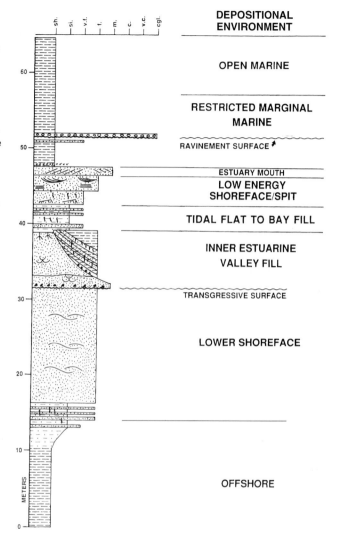

tion to marine (tidal) sands at the top suggests a transgressive estuarine deposit. In a transgressive succession, estuarine deposits may be interstratified with tidal deposits, beach shoreface sands, and open-marine shelf muds, as illustrated in Figure 11.40. Additional examples of ancient estuarine deposits are discussed by Brownridge and Moslow (1991) and Simpson (1991).

As discussed, lagoonal deposits may form in many settings, including parts of barrier-island complexes. Criteria that can be used to distinguish ancient lagoonal deposits from estuarine and other deposits include evidence for restricted circulation such as the presence of evaporites or anoxic facies (e.g., black shales), lack of strong tidal influence, slow rates of terrigenous sediment influx and dominantly fine-grained sediments, low faunal diversity, and extensive bioturbation (Davis, 1983). The St. Mary River Formation of southern Alberta, Canada, contains back-barrier deposits that illustrate some of the characteristics of lagoonal sediments (Fig. 11.41). This lagoonal suc-

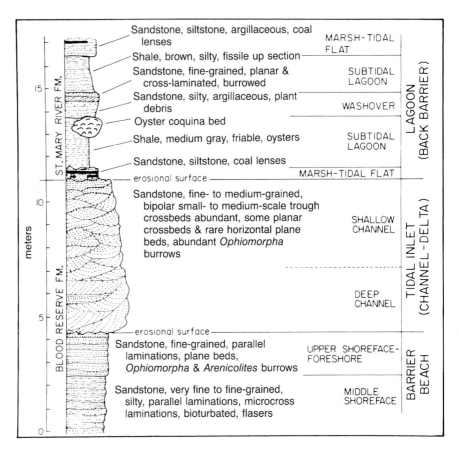

FIGURE 11.41   Composite stratigraphic section of Cretaceous formations in southern Alberta, Canada. Back-barrier lagoonal deposits of the St. Mary River Formation overlie tidal-inlet and tidal-delta deposits of the Blood Reserve Formation. (From Reinson, G. E., 1984, Barrier island and associated strand-plain systems, *in* R. G. Walker (ed.), Facies models, 2nd ed.: Geoscience Canada Reprint Ser. 1. Fig. 16, p. 128, reprinted by permission of Geological Association of Canada.)

cession begins with a basal section of sandstones, siltstones, and coals that represent marsh–tidal-flat deposits (Young and Reinson, 1975). These deposits lie on the eroded surface of tidal-inlet deposits of the underlying Blood Reserve Formation. The basal coal-bearing beds are succeeded upward by gray shales containing oysters, which are brackish-water faunas, and disseminated carbonaceous materials and imprints of plant remains. These deposits are interpreted as subtidal lagoonal sediments. They are overlain by fine-grained, planar and cross-laminated sandstones, which are probably washover deposits from the barrier beach. More subtidal shales lie above these washover sands, and the succession is capped by sandstones and siltstones with coal lenses of probable marsh–tidal-flat origin. Overall, the Blood Reserve and St. Mary River formations appear to represent a progradational barrier-island complex with the lagoonal deposits of the St. Mary River Formation forming the topmost part of the regressive succession.

11.5 TIDAL-FLAT SYSTEMS

Introduction

Tidal flats form primarily on mesotidal and macrotidal coasts (Fig. 11.42) where strong wave activity is absent. They develop either along open coasts of low relief and relatively low wave energy or behind barriers on high-energy coasts where protection is afforded from waves by barrier islands, spits, reefs, and other structures. Thus, they occur within estuaries, bays, the backshores of barrier-island complexes, and deltas, as well as along open coasts.

Tidal flats are marshy and muddy to sandy areas partially uncovered by the rise and fall of tides. They constitute almost featureless plains dissected by a network of tidal channels and creeks that are largely exposed during low tide (Fig. 11.43). As tide level rises, flood-tide waters move into the channels until at high tide the channels are overtopped and water spreads over and inundates the adjacent shallow flats. Ebb tide again exposes the channels and intervening flats. In temperate regions, salt marshes commonly cover the upper parts of tidal flats, and muds and silts accumulate near high-water level. At the same time, mixed mud and sand are deposited in the mid-tidal–flat region, and sands accumulate in channels and on the lower parts of the tidal flat. In arid to semiarid regions, tidal flats may become desiccated and marked by the formation of mudcracks and the growth of gypsum and halite crystals in muds. The surface of tidal flats in subarctic regions may be marked by surficial scars, caused by ice floes and ice-pushed boulders, and the presence of ice-rafted pebbles and cobbles. Modern tidal flats are primarily sites of siliciclastic deposition; however, carbonate sediments and, in a few areas, evaporites accumulate on some modern tidal flats such as those in the Bahamas, the Persian Gulf, Florida Bay, and the western coast of Australia.

Much of what is known about ancient tidal-flat sediments comes from research on modern tidal flats. Modern tidal flats have been studied intensively in many parts of the world since the 1950s, particularly in Germany; the North Sea coastline of the Netherlands; England; the Bay of Fundy, Nova Scotia; the Yellow Sea of Korea; and the Gulf of California. Important monographs on tidal sediments include those of de Boer,

FIGURE 11.42 Global classification of coastlines by tidal range. Microtidal = 0–2 m tidal range; mesotidal = 2–4 m; macrotidal > 4 m. (From Klein, G. deV., 1985, Intertidal flats and intertidal sand bodies, *in* R. A. Davis, Jr. (ed.), Coastal sedimentary environments, 2nd ed.: Springer-Verlag, New York. Fig. 3–1, p. 189. Redrawn from Davies, J. L., 1964, Zeitschrift für Geomorphologie, v. 8. Fig. 4, p. 136.)

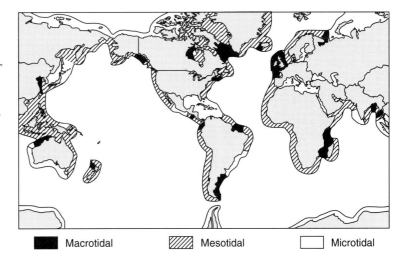

■ Macrotidal   ▨ Mesotidal   □ Microtidal

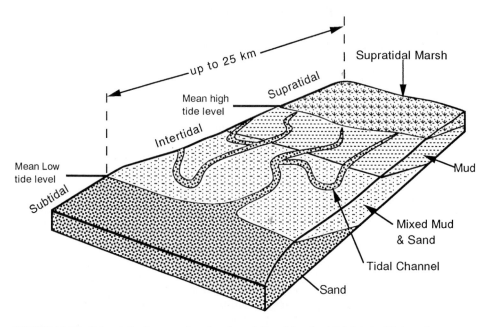

FIGURE 11.43   Schematic diagram showing the relationship of subtidal, intertidal, and supra-tidal zones in the tidal-flat environment. Note that mud is the dominant deposit in the upper part of the intertidal zone, mixed mud and sand predominate in the lower intertidal zone, and sand is deposited in the subtidal zone and in tidal channels. Muddy marsh deposits characterize the supratidal zone.

van Gelder, and Nio (1988); Ginsburg (1975); Klein (1976, 1977); Smith et al. (1991); Stride (1982); and Thompson (1968). Tidal environments and deposits have been summarized in shorter contributions by Boothroyd (1978); Evans (1965); Frey and Basan (1978); Klein (1970, 1985); Reineck (1972); Weimer, Howard, and Lindsay (1982); Shinn (1983); and Van Straaten (1961).

Oil and gas deposits have been discovered in both siliciclastic and carbonate tidal facies, and uranium is present in some sandy tidal facies. Therefore, tidal deposits have economic significance as well as general scientific interest.

Depositional Setting

Although tidal currents may operate in the ocean to depths of 2000 to 2500 m, the tidal-flat environment is confined to the shallow margin of the ocean. The vertical distance between the high- and low-tide line in most modern tidal environments commonly ranges from 1 to 4 m (mesotidal coasts), depending upon the locality, although tidal ranges up to 10 to 15 m or more (macrotidal coasts) occur in some localities, such as the Bay of Fundy. The total width of tidal flats may range from a few kilometers to as much as 25 km. Topographic relief within the tidal-flat environment is generally rather small, except for tidal channels, and slopes of the tidal flat are gentle, although commonly irregular.

The tidal-flat environment is divided into three zones: subtidal, intertidal, and supratidal. The **subtidal zone** encompasses the part of the tidal flat that normally lies below mean low-tide level. It is inundated with water most of the time and is normally subjected to the highest tidal-current velocities. Tidal influence in this part of the environment is particularly important within tidal channels, where bedload transport and deposition are predominant, although this zone is also influenced to some extent by wave processes. The **intertidal zone** lies between mean high- and low-tide levels. It is subaerially exposed either once or twice each day, depending upon local wind and tide conditions, but it commonly does not support significant vegetation. Both bedload and suspension sedimentation take place in this zone. The **supratidal zone** lies above normal high-tide level but is incised by tidal channels and flooded by extreme tides. This part of the tidal flat is exposed to subaerial conditions most of the time but may be flooded by spring tides twice each month or by storm tides at irregular intervals. Sedimentation is dominantly from suspension. On some tidal flats, the supratidal zone is a salt-marsh environment incised by tidal channels. In arid or semiarid climates, it is commonly an environment of evaporite deposition and is often referred to as a **sabkha.** The relationships among the subtidal, intertidal, and supratidal portions of the tidal flat are illustrated schematically in Figure 11.43.

Sedimentation Processes

Physical Processes. Sedimentation on tidal flats takes place in response to both tidal processes and waves. Sedimentation in the channels of tidal flats is dominated by tidal currents, but wind-driven waves and the currents generated by these waves also play an important role in deposition on the flats between channels. Tidal currents move up the gentle slope of the tidal flat during flood tide and back down during ebb tide. The tidal velocities achieved during reversing tides are commonly asymmetrical, and the velocities of flood tides may differ significantly from those of ebb tides. Within the channels, tidal currents can reach velocities of 1.5 m/s or more, and velocities on the flats range from 30 to 50 cm/s (Reineck and Singh, 1980). These velocities are adequate to cause transport of sandy sediment and produce ripple and dune bedforms, crossbedding, and plane bedding.

Deposition in the subtidal zone takes place mainly by lateral accretion of sandy sediment in tidal channels and point bars (Weimer, Howard, and Lindsay, 1982). Tidal channels can be quite large, ranging in depth to 15 m or more. These channels migrate laterally, similarly to meandering-river systems. Shinn (1983) compares highly channeled tidal flats to river deltas "turned wrong side out." The sea rather than the land is the sediment source, and the tidal channels with their distributaries branching in a landward direction provide the pathways for sediment delivery to the tidal flat.

The intertidal zone is a zone of mixed lateral accretion and suspension deposition. Sand or muddy sand deposition is dominant within channels, but both fine sand and mud can be deposited on the broad flats between tidal channels. Bedload transport and deposition of sand by tidal currents takes place during the higher-velocity phases of a tidal cycle, either rising or falling tide, producing ripple or dune bedforms. These bedforms may be modified by wind-generated waves and currents. During periods of lower velocity or zero-velocity flow accompanying high- or low-water slack periods, suspension deposition of clay- or silt-size sediment takes place as discrete particles or aggregates of particles (Klein, 1977, 1985). Deposition of these muddy sediments over rippled, sandy sediments gives rise to a variety of lenticular and flaser bedding, which are important diagnostic characteristics of sediments in this environment. Wave activity tends to be greatest in the lower part of the intertidal zone, leading to reworking of sediment and resuspension of fine sediment, sorting of sands, and a predominance of sandy deposits. Wave energy and current energy are lowest in the upper part of the intertidal zone, which is consequently dominated by deposition of muddy sediment. Periodic storms can also affect deposition in the middle and high intertidal zones. Significant quantities of sediments may be carried in during storms from farther offshore and deposited in the intertidal zone, and storm waves may erode and reentrain previously deposited sediments and modify or destroy bedforms and sedimentary structures.

The supratidal zone is only slightly affected by tidal currents and marginally affected by waves. The common presence of salt-marsh vegetation and mangrove trees in this zone further dampens wave energy. Thus, the supratidal zone is the zone of lowest energy on the tidal flat. Deposits of this zone are mainly muds, and sedimentation rates are very low. Desiccated, cracked muds (mudcracks) are a diagnostic feature of the supratidal and upper intertidal zones. Stratification in the muds may be destroyed by root growth from salt-marsh vegetation, and peats may accumulate in the supratidal zone. Most of the mud deposited in the intertidal and supratidal zones originates in the subtidal zone, where it is stirred into suspension during storms and transported to the tidal flats by storm waves.

Chemical and Biologic Processes. The preceding discussion refers primarily to deposition on siliciclastic tidal flats. Much the same processes occur on carbonate tidal flats, with the difference that biochemical and biologic sedimentation processes are important in these settings in supplying carbonate sediment. Carbonate muds are generated mainly in the shallow subtidal zone, in part through biogenic processs (Chapter 7). Lime muds and sand-size skeletal fragments, also generated within the subtidal zone, are transported into the intertidal and supratidal zones by tidal currents and waves. In arid and semiarid climates, chemical precipitation of gypsum, anhydrite, and dolomite may occur in the supratidal and upper intertidal zones owing to strong evaporation.

Biogenic processes are important in other ways. Although tidal flats are a harsh environment for many organisms, owing to intermittent exposure, they are inhabited by a limited number of species of organisms such as gastropods and pelecypods, crustaceans, polychaete worms, foraminifers, diatoms, and blue-green algae (cyanobacteria). Many of these animals produce fecal pellets of mud that accumulate on the flats; they may also cause extensive bioturbation of sediment and generate burrows belonging to the *Skolithos* ichnofacies (Chapter 5). In general, bioturbation appears to be strongest in mudflats, weaker in mixed mud- and sandflats, and least in sandflats. Blue-green algae are particularly important agents in the supratidal and intertidal zones in trapping and binding fine sediment to produce stromatolitic bedding.

Characteristics of Tidal-Flat Sediments

General Morphology. The body of sediments that make up tidal flats may range from a few kilometers to a few tens of kilometers in width. A tidal flat is elongated parallel to the coast and along open coasts may extend for tens to hundreds of kilometers, cut across here and there by major tidal channels and river estuaries. The tidal flats of bays and barrier-island systems are more restricted in size, and plan-view shape depends upon the shape of the bayshore. As mentioned, the upper surfaces of tidal flats are characterized by low relief, except for tidal channels. In cross section, tidal deposits form a gross wedge-shaped body tapering shoreward.

Lithofacies. Sediments on siliciclastic tidal flats are composed primarily of mud and sand. Muds predominate in the supratidal and upper intertidal zones, and the deposits

FIGURE 11.44 Distribution of sediments on the intertidal flats of Jade Bay. (From Reineck, H. E., and I. B. Singh, 1980, Depositional sedimentary environments. Fig. 591, p. 432, reprinted by permission of Springer-Verlag, Heidelberg. Originally after S. Gadow, 1970, 1. Sedimente und chemismus, *in* H. E. Reineck (ed.), Das Watt, Ablagerungs- und Lebensraum: W. Kramer, Frankfurt a. M.)

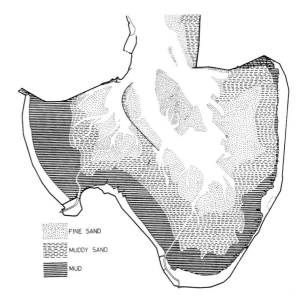

FINE SAND

MUDDY SAND

MUD

of supratidal marshes are further characterized by abundant plant debris, which may eventually form peat. Mixed mud and sand characterize the middle part of the intertidal zone, and sand dominates the lower intertidal zone as well as the channel and bar deposits of the shallow subtidal zone. Muds may be deposited between channels in the subtidal zone below wave base. The general areal distribution of siliciclastic tidal-flat facies in a modern bay tidal flat is illustrated in plan view in Figure 11.44. The proportion of mud and sand in modern tidal flats varies considerably. Some tidal flats are dominated by mud, whereas others are sand-dominated flats. Figure 11.45 illustrates, in both plan view and cross-sectional view, the typical facies of siliciclastic-dominated tidal flats. The distribution of facies on carbonate tidal flats is similar. Carbonate tidal flats are dominated by muddy deposits in low-energy areas of the flat and by carbonate sands composed of skeletal fragments, intraclasts, or ooids in channels and other high-energy areas. Evaporite minerals such as gypsum, anhydrite, and occasionally halite are present on all modern arid tidal flats (Shinn, 1983), commonly in association with dolomite and other carbonate minerals.

Sedimentary Structures. The predominant types of sedimentary structures in tidal-flat sediments vary in different parts of the tidal flat (Fig. 11.45). Channel sands are characterized by dunes and internal cross-bedding that may show bimodal directions of foreset dip. Reversing tides during an asymmetrical tidal cycle can cause erosion of ripple crests followed by redeposition during the next tidal cycle, producing **reactivation surfaces** (Fig. 11.46). Alternatively, suspended mud can be deposited over eroded ripple crests during slack-water stage to form **mud drapes**. Sandy and muddy sediments on the mixed flats are characterized by small-scale ripple cross-stratification, flaser bedding, wavy bedding, lenticular bedding, and, more rarely, finely laminated bedding (Fig. 11.45). Tidal currents too weak to produce ripples may deposit thin sand layers from suspension that alternate with mud laminae. Such beds that show cyclic changes in thickness owing to neap-spring-tide variations in tidal current velocity are **tidal rythmites** (e.g., Smith et al., 1991). Some tidal-flat sediments display **herringbone cross-stratification** in which cross-laminated sediments deposited during flood tide

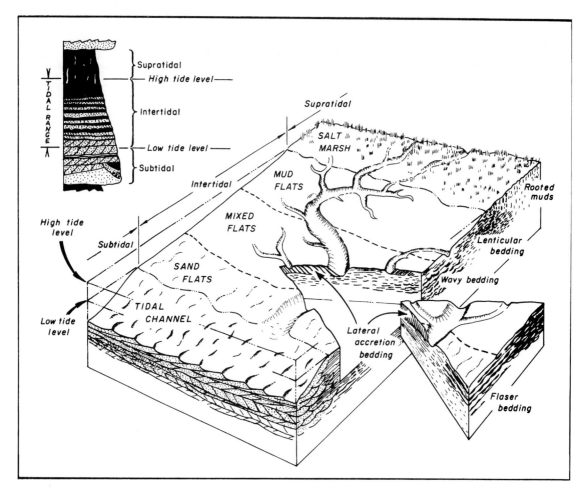

FIGURE 11.45 Schematic diagram of a typical siliciclastic tidal flat. The tidal flat fines toward the high-tide level, passing gradationally from sandflats, through fixed flats, to mudflats and salt marshes. An example of the upward-fining succession produced by tidal-flat progradation is shown in the upper left corner. (From Dalrymple, R. W., 1992, Tidal depositional systems, *in* R. G. Walker and N. P. James (eds.), Facies models: Geol. Assoc. Canada. Fig. 12, p. 201, reproduced by permission.)

dip in the opposite direction to those formed almost immediately afterward during ebb tide (Fig. 11.47).

Deposits on the mudflats are characterized by thick mud layers separated by thin sand laminae, which in arid climates may be disrupted by growth of gypsum, anhydrite, or halite crystals and development of mudcracks. Anhydrite is typically characterized by chickenwire texture (Chapter 8), and distortion of gypsum and anhydrite layering is common. Muddy deposits of supratidal marshes may shown some thin layering, but they are commonly bioturbated and disturbed by root growth. In addition to mudcracks, other structures indicating subaerial exposure may be present, including raindrop imprints, hail marks, and foam marks. Tidal flats in mid- to high latitudes may contain ice-rafted debris and soft-sediment deformation structures produced by

FIGURE 11.46 Reactivation surface developed owing to alternation of a dominant tidal phase (constructional event) with a subordinate phase (destructional event). A. Dunes develop by dominate-phase tidal flow and produce internal avalanche cross-stratification. B. Reversed tidal flow during subordinate phase destroys sharp asymmetry of dunes, producing a subdued, asymmetrical profile and a rounded reactivation surface (R). C. During the next constructional event, dunes build on the reactivation surface (R) and develop a superimposed set of avalanche cross-stratification. D. Destruction of the dune profile occurs during reversed, subordinate tidal flow and produces a second reactivation surface (R). (From Klein, G. deV., 1970, Depositional and dispersal dynamics of intertidal sand bars: Jour. Sed. Petrology, v. 40. Fig. 28, p. 1118, reprinted by permission of Society of Economic Paleontologists and Mineralogists, Tulsa, Okla.)

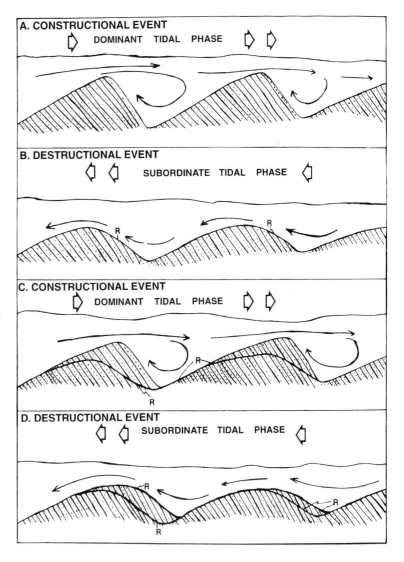

grounding of ice blocks. Sedimentary structures on carbonate tidal flats are generally similar to those formed on siliciclastic tidal flats; however, dunes, cross-stratification, and other current-generated structures tend to be less abundant, and mudcracked, algal stromatolites are more characteristic of the upper intertidal zone and supratidal zone.

Bioturbation structures are prevalent in tidal-flat sediments. They are most abundant in sediments of the muddy upper intertidal and supratidal zones and least abundant in sediments of the sandy lower intertidal zone and subtidal zone. Trace fossils produced by organic activity include not only burrows *(Skolithos)* and various kinds of feeding and resting traces of organisms living in the intertidal zone but also the tracks of birds, land animals, and insects.

Vertical Successions. Transgression and regression cause deposits of laterally adjacent tidal-flat environments to become superimposed, generating characteristic successions

FIGURE 11.47 Progradational sequence of tidal-flat deposits. Based on Middle Member, Wood Canyon Formation (Late Precambrian), Nevada. Diagram also shows interpretation of dominant sediment transport processes and depositional environments. (From Klein, G. deV., 1977, Clastic tidal facies. Fig. 76, p. 85, reprinted by permission of IHRDC Publications, Boston.)

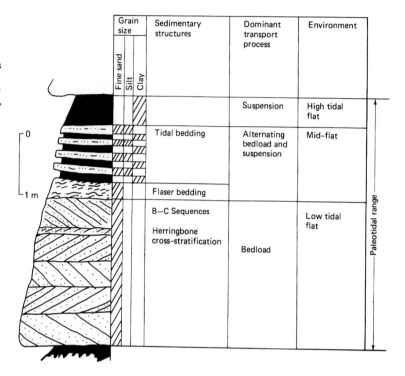

Grain size			Sedimentary structures	Dominant transport process	Environment	
Fine sand	Silt	Clay				
				Suspension	High tidal flat	
			Tidal bedding	Alternating bedload and suspension	Mid-flat	
			Flaser bedding			
			B—C Sequences		Low tidal flat	
			Herringbone cross-stratification	Bedload		

Paleotidal range

of vertical facies. Progradation produces a generalized fining-upward succession that begins with subtidal and lower intertidal cross-bedded sands, followed upward by mixed sand and mud in the middle intertidal zone, and mud and peat in the upper intertidal and supratidal zones. A typical vertical regressive (progradational) succession developed on a siliciclastic tidal flat is illustrated schematically in Figure 11.47 (also see Fig. 11.45). Transgression presumably generates a coarsening-upward succession that displays the same general facies but in reverse order; however, trangression may rework and destroy intertidal deposits. Similar patterns of subtidal, intertidal, and supratidal carbonate facies could be expected to develop on coasts characterized by carbonate tidal flats. On a local scale, lateral channel migration may also produce small-scale coarsening-upward vertical successions.

Ancient Tidal-Flat Sediments

Tidal-flat deposits have several distinctive characteristics that help to differentiate them from sediments of most other environments; however, their overall characteristics are similar to those of estuarine deposits. The most important criteria for recognition of ancient tidal-flat deposits are commonly regarded to include the following:

1. a bimodal direction of current-formed cross-bedding resulting from reversing tidal currents
2. the occurrence of facies that reflect repeated, small-scale alternations in sediment transport conditions (tidal rhythmites) and the joint occurrence of large-scale (channel) and small-scale (sand- and mudflat) structural units in superposition or juxtaposition

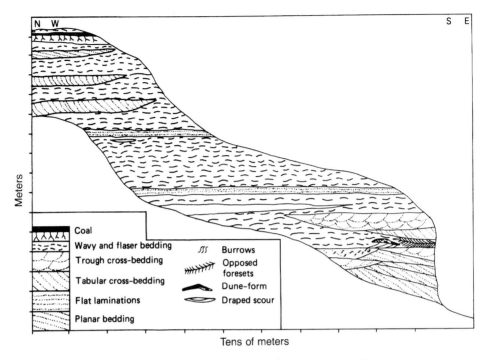

FIGURE 11.48 Tidal-flat deposits in exposure of the Jurassic Lower Coal Series of Bornholm, Denmark. (From Sellwood, B. W., 1972, Tidal-flat sedimentation in the Lower Jurassic of Born-hold, Denmark: Palaeogeography, Palaeoclimatology, Palaeoecology, v. 11. Fig. 3, p. 97, reprinted by permission of Elsevier Science Publishers, Amsterdam.)

3. the presence of abundant reactivation surfaces and flaser bedding

4. a high frequency of erosional contacts and abrupt facies changes

Other supporting criteria include the typical vertical succession of facies discussed above, the high degree of bioturbation of many tidal-flat sediments, the presence of mudcracked stromatolites, and other evidence of subaerial exposure such as raindrop imprints, hail marks, and animal or bird tracks. Nio and Yang (1991) and Terwindt (1988) further discuss these criteria.

Tidal-flat deposits have been reported from stratigraphic units of virtually all ages, from Precambrian to Holocene. Numerous examples are cited in the bibliography given by Reineck (1972), and by de Boer, van Gelder, and Nio (1988); Klein (1977); Smith et al. (1991); and Weimer, Howard, and Lindsay (1982). Several examples of both ancient siliciclastic and carbonate tidal-flat deposits are discussed in the compendium volume on tidal deposits edited by Ginsburg (1975).

Two preceding figures (Figs. 11.28 and 11.41) illustrating barrier and lagoonal deposits also include tidal deposits. The upper part of the Blood Reserve Formation of Alberta, shown in Figure 11.41, contains siliciclastic tidal-channel and tidal-delta deposits composed of fine- to medium-grained, trough cross-bedded sandstones. The generalized lagoonal succession from Carboniferous rocks of eastern Kentucky, shown in Figure 11.28, includes tidal-flat siltstones with sandstone flasers and rippled and cross-bedded tidal-channel deposits interbedded with other back-barrier deposits.

Sellwood (1972, 1975) describes a succession of tidal deposits from the Jurassic Lower Coal Series on the island of Bornholm near the mouth of the Baltic Sea (Fig.

11.48). This succession is about 15 m thick and has a basal cross-bedded sandstone unit about 5 m thick. The basal unit consists of fine- to medium-grained sandstone that displays large-scale tabular to trough cross-bedding, herringbone cross-stratification, and planar stratification. The herringbone cross-stratification appears to be the preserved remains of dunes with reactivation surfaces and, together with planar bedding, suggests deposition under upper-flow-regime conditions. A transition upward in the section to bedding characteristics indicative of small ripples suggests change in flow conditions to lower-flow regime. Burrows and shallow sand-filled channels are present in some of the beds. This basal unit probably reflects deposition in the tidal-dominated, low-tidal-flat environment (Fig. 11.47) where energy conditions are generally high. The succession lying above the cross-bedded unit is characterized by fine-grained sands that show wave and current ripples draped with clay. Flaser and lenticular bedding are prevalent. The characteristics of this succession correlate well with facies of the mid-flat environment (Fig. 11.47). The succession is capped by coal beds up to 30 cm thick and having root traces extending down into underlying flaser bedding. These marsh-type deposits represent the high-tidal-flat environment.

FURTHER READINGS

Deltaic Systems

Broussard, M. L. (ed.), 1975, Deltas: Models for exploration: Houston Geol. Society, 555 p.

Colella, A., and D. B. Prior (eds.), 1990, Coarse-grained deltas: Internat. Assoc. Sedimentologists Spec. Pub. 10, Blackwell, Oxford, 357 p.

Coleman, J. M., 1981, Deltas: Processes of deposition and models for exploration, 2nd ed.: Burgess Publishing Co., Minneapolis, Minn., 124 p.

Coleman, J. M., and D. B. Prior, 1980, Deltaic sand bodies: Am. Assoc. Petroleum Geologists Ed. Course Notes 15, Tulsa, Okla., 171 p.

Morgan, J. P. (ed.), 1970, Deltaic sedimentation—Modern and ancient: Soc. Econ. Paleontologists and Mineralogists Spec. Pub. 15, Tulsa, Okla., 312 p.

Nemec, W., and R. J. Steel (eds.), 1988, Fan deltas: Sedimentology and tectonic settings: Blackie, Glasgow and London, 444 p.

Shirley, M. L., and J. A. Ragsdale (eds.), 1966, Deltas in their geologic framework: Houston Geol. Society, 251 p.

Whateley, M. K. G., and K. T. Pickering (eds.), 1989, Deltas—Sites and traps for fossil fuels: Geol. Soc. Spec. Pub. 41, Blackwell, Oxford, 360 p.

Beach and Barrier-Island Systems

Davis, R. A., Jr. (ed.), 1985, Coastal sedimentary environments, 2nd ed.: Springer-Verlag, New York, 716 p.

Davis, R. A., Jr., and R. L. Ethington (eds.), 1976, Beach and nearshore sedimentation: Soc. Econ. Paleontologists and Mineralogists Spec. Pub. 24, Tulsa, Okla., 187 p.

Fisher, J. S., and R. Dolan (eds.), 1977, Beach processes and coastal hydrodynamics: Dowden, Hutchinson and Ross, Stroudsburg, Pa., 382 p.

Hardisty, J., 1990, Beaches—Form and process: Unwin Hyman, London, 324 p.

Komar, P. D., 1976, Beach processes and sedimentation: Prentice-Hall, Englewood Cliffs, N.J., 429 p.

Leatherman, S. P. (ed.), 1979, Barrier islands from the Gulf of St. Lawrence to the Gulf of Mexico: Academic Press, New York, 325 p.

Oertel, G. F., and S. P. Leatherman (eds.), 1985, Barrier islands: Marine Geology, v. 63, p. 1–396.

Pilkey, O. H., 1983, The beaches are moving: Duke University Press, Durham, N.C., 336 p.

Schwartz, M. L. (ed.), 1972, Spits and bars: Benchmark Papers in Geology, v. 3, Dowden, Hutchinson and Ross, Stroudsburg, Pa., 452 p.

Schwartz, M. L. (ed.), 1973, Barrier islands: Benchmark Papers in Geology, v. 9. Dowden, Hutchinson and Ross, Stroudsburg, Pa., 451 p.

Estuarine and Lagoonal Systems

Ashley, G. M. (ed.), 1988, The hydrodynamics and sedimentology of a back-barrier lagoon—salt marsh system, Great Sound, New Jersey: Marine Geology, v. 82, p. 1–132.

Barnes, R. S. K., 1980, Coastal lagoons: Cambridge University Press, Cambridge, 106 p.

Cronin, L. E. (ed.), 1975, Estuarine research, v. 2: Geology and engineering: Academic Press, New York, 587 p.

Dyer, K. R., 1986, Coastal and estuarine sediment dynamics: John Wiley & Sons, Chichester, 342 p.

Kjerfve, B. (ed.), 1978, Estuarine transport processes: University of South Carolina Press, 331 p.

Lauff, G. H. (ed.), 1967, Estuaries: Am. Assoc. for the Advancement of Science Spec. Pub. 83, 757 p.

Nelson, B. W. (ed.), 1972, Environmental framework of coastal plain estuaries: Geol. Soc. America Mem. 133, 619 p.

Officer, C. B. (chm.), 1977, Estuaries, geophysics, and the environment: National Academy of Sciences, Washington, D.C., 127 p.

Olaussen, E., and I. Cato (eds.), 1980, Chemistry and biogeochemistry of estuaries: John Wiley & Sons, New York, 452 p.

Ward, L. G., and G. M. Ashley, 1989, Physical processes and sedimentology of siliciclastic-domimated lagoonal systems: Marine Geology, v. 88, p. 181–364.

Wiley, M. (ed.), 1977, Estuarine processes, v. II: Circulation, sediments, and transfer of material in the estuary: Academic Press, New York, 428 p.

Tidal-Flat Systems

de Boer, P. L., A. van Gelder, and S. D. Nio (eds.), 1988, Tide-influenced sedimentary environments and facies: D. Reidel, Dordrecht, 530 p.

Ginsburg, R. N., 1975, Tidal deposits: A casebook of Recent examples and fossil counterparts: Springer-Verlag, New York, 428 p.

Klein, G. deV., 1977, Clastic tidal facies: Continuing Education Publication Co., Champaign, Ill., 149 p.

Klein, G. deV. (ed.), 1976, Holocene tidal sedimentation: Benchmark Papers in Geology, v. 5, Dowden, Hutchinson and Ross, Stroudsburg, Pa., 423 p.

Open University Course Team, 1989, Waves, tides and shallow-water processes: Pergamon, Oxford, 187 p.

Smith, D. G., G. E. Reison, B. A. Zaitlin, and R. A. Rahmani (eds.), 1991, Clastic tidal sedimentology: Canadian Soc. Petroleum Geologists Mem. 16, 307 p.

Stride, A. H. (ed.), 1982, Offshore tidal sands: Chapman and Hall, London, 222 p.

12

Siliciclastic Marine Environments

12.1 INTRODUCTION

The marine environment is that part of the ocean lying seaward of the zone dominated by shoreline processes. Water depth in the marine realm ranges from a few meters to more than 10,000 m. The salinity of seawater in the open ocean averages about 35o/oo, although higher or lower salinities can occur locally in restricted bodies of the ocean. Marine life forms are characterized by generally high diversity and large populations, and most are low-tolerance organisms adapted to conditions of normal salinity. The energy of the bottom water lying immediately above the ocean floor is generally low, except on the shallow continental shelf, which is affected by a variety of tidal processes and wind- and storm-wave activity, and on some parts of the deeper ocean floor that are swept by bottom currents.

The major subdivisions of the oceanic realm are the **continental margin** and the **ocean basin.** These in turn can be further subdivided as shown in Figure 12.1. The **continental shelf** extends seaward from the shoreline at a gentle slope of about 1° to a point where a perceptible increase in rate of slope, the **shelf break,** takes place. The shelf break occurs in the modern ocean at an average distance from shore of about 75 km, although the distance ranges from a few tens of meters to more than 1000 km. Average water depth at the shelf break is about 130 m. The **continental slope** descends from the shelf break to the deep seafloor with a typical slope of about 4°. On passive, or divergent, continental margins, the foot of the continental slope merges with the **continental rise,** which is a gently sloping surface created by coalescing submarine fans at the base of the slope. The continental rise passes gradually into the floor of the ocean basin. Parts of the deep ocean floor consist of nearly flat areas called **abyssal plains,** which are covered by sediment. Other parts of the ocean floor are characterized by the presence of volcanic hills that rise above the seafloor to elevations ranging from a few hundred meters to more than 1000 m. The central part of the major ocean basins

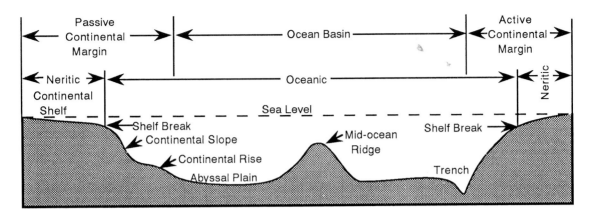

FIGURE 12.1 Schematic cross-sectional profile of the marine environment. Not to scale.

is occupied by a gigantic **mid-ocean ridge** that may protrude more than 2.5 km above the seafloor. On active, or convergent, margins, the continental slope may descend into a **deep-sea trench,** and the continental rise is absent. On the basis of water depth, we divide the ocean into two major zones: the neritic zone and the oceanic zone. The shallow **neritic zone** extends from the shoreline to the shelf break. The **oceanic zone** extends from shelf break to shelf break and encompasses the deeper part of the ocean.

In this chapter we consider marine environments characterized by transport and deposition of siliciclastic sediment. Carbonate marine environments are discussed in Chapter 13.

12.2 THE NERITIC (SHELF) ENVIRONMENT

The neritic zone encompasses the shallow-water areas of the ocean lying shoreward of the shelf break. Although the shelf break on modern shelves occurs at an average depth of about 130 m, as indicated, it may be located on some shelves at depths as shallow as 18 m or as deep as 915 m (Bouma et al., 1982). In the modern ocean, the shallow-marine environment occupies mainly the continental shelf area around the margin of the continents, forming what is referred to as a **pericontinental,** or marginal, sea (Heckel, 1972). At various times in the geologic past, broad, shallow **epicontinental,** or epeiric, seas occupied extensive areas within the continents (Fig. 12.2), somewhat like the present-day Hudson Bay area of the North American Arctic region. The following discussion of the neritic environment is focused primarily on the continental shelf environment because we can draw on the modern continental shelf environment as a model. Readers should keep in mind, however, that many of the shallow-marine deposits preserved in the geologic record may have been deposited in broad epicontinental seas, for which we may have no truly representative modern analogs. We may assume that similar sedimentologic processes operated on continental shelves and in epicontinental seaways, but, in fact, differences exist between these two environments. For example, epicontinental seas received sediments from nearly all sides, whereas continental shelves receive sediments from only one side. Furthermore, the wave and current regimes in epicontinental seas may have been different from those on shelves. In addition, modern continental shelves may not provide a good analog for ancient

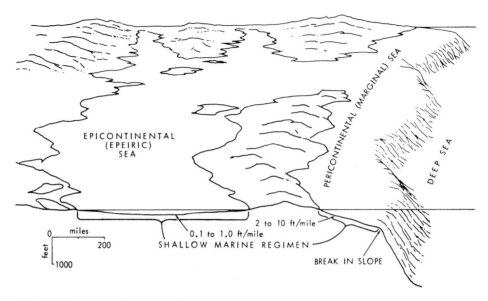

FIGURE 12.2   Schematic diagram illustrating the difference between pericontinental (continental shelf) and epicontinental shallow-marine environments. (From Heckel, P. H., 1972, Recognition of ancient shallow marine environments, *in* J. K. Rigby and W. K. Hamblin (eds.), Recognition of ancient sedimentary environments: Soc. Econ. Paleontologists and Mineralogists Spec. Pub. 16. Fig. 1, p. 227, reprinted by permission of SEPM, Tulsa, Okla.)

marginal seas because rapid rise of sea level following the final episode of Pleistocene glaciation has stranded coarse sediment in deeper parts of the shelves, creating conditions of sediment-water disequilibrium. Thus, sediment grain size on some parts of modern shelves is not consistent with present water depth, energy conditions, and sedimentation processes on the shelves.

Both siliciclastic and carbonate sediments can accumulate in the marine-shelf environment, although most modern continental shelves are covered by siliciclastic sediments. Carbonate sediments (Chapter 13) are restricted to a few shelves, mainly (but not exclusively) in tropical areas.

Physiography and Depositional Setting

The siliciclastic shelf environment is bounded by various coastal environments on the landward side and by the continental slope on the seaward side. It can be divided into the shallow **inner shelf**, which is dominated by tidal, wind-driven, and storm-wave processes, and the deeper **outer shelf**, which may be affected by intruding major ocean currents such as the Gulf Stream System (Fig. 12.3). The outer shelf may be affected also by density currents generated by temperature-salinity differences or suspended sediment differences in water bodies. The boundary between the inner and outer shelves is not well defined, and its position fluctuates with changing sea level. In fact, during greatly lowered sea level, the normal inner shelf is subaerially exposed, and the outer shelf may be exposed or covered only by very shallow water.

The width of shelves varies according to their plate-tectonic setting (Shepard, 1973). Shelves along the fore-arc region of convergent continental margins tend to be very narrow. By contrast, broad shelves and platforms occur in the back-arc basins of

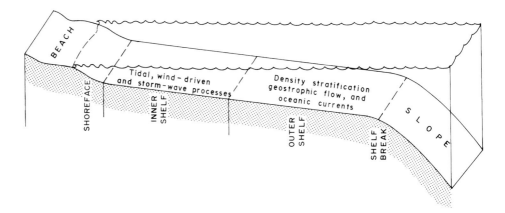

FIGURE 12.3   Subdivisions of the continental shelf. (From Galloway, W. E., and D. K. Hobday, 1983, Terrigenous clastic depositional systems. Fig. 7.1, p. 144, reprinted by permission of Springer-Verlag, Heidelberg.)

convergent margins, on divergent or trailing-edge continental margins, and on cratonic downwarps that open to the sea. Although shelves are fundamentally low-relief plat-forms, the shelf surface can vary considerably. It may be relatively smooth or covered by a variety of small- to large-scale bedforms. It may also contain banks, islands, or shoals near its offshore edge (Bouma et al., 1982). Shepard (1977) divides the present continental shelves into five major categories on the basis mainly of depositional set-ting. **Glaciated shelves** occur at high latitudes where glaciers spread from landmasses onto the continental shelf. They are characterized by ice-scoured troughs and glacially deposited sediments. **Shelves with elongate sand ridges** occur along some coasts, such as the U.S. Atlantic coast. These subparallel sand ridges and associated bedforms are composed of Pleistocene sediment now being reworked and modified within the mod-ern environment. **Shelves off large deltas** are low-relief platforms that build seaward as delta progradation occurs. **Shelves with coral reefs** occur in tropical waters and locally in subtropical waters but are very limited in the modern ocean. Coral reefs (Chapter 13) are typically developed along the shelf edge or shelf break, but they can occur also as patches on the inner shelf. **Shelves bordered by rocky banks and islands** are characterized by rocky elevations along their outer edges. These elevations may be either islands or rocks banks; within the elevated areas they may contain basins or channels partly or completely filled with sediment.

Depositional Processes

Ideas concerning transport and depositional processes on continental shelves continue to change, and research into both modern shelf processes and ancient shallow-marine systems is very active. R. G. Walker (1984) traces the history of shallow-marine studies beginning with early ideas of a "graded shelf" (Johnson, 1919), which was believed to display progressive decrease in grain size from coarse at the shoreline to very fine at the shelf edge, in response to presumed decrease in water energy seaward. The graded-shelf concept, at least for modern shelves, was eventually abandoned when it became obvious with additional study that the grain-size distribution on many modern shelves is patchy or irregular. The concept of **relict** shelf sediments was introduced to explain such irregular distribution patterns as the presence of coarse sands and gravels in deep

water. Relict sediments are deposits that are not in equilibrium with present hydrodynamic conditions. They were deposited on the shelf by fluvial or glacial processes during low stands of sea level. (Not all ancient shelves may have been affected by such
processes.) Relict sediments were initially believed to remain on the shelf floor without significant reworking as they were inundated by rising sea level (Shepard, 1932;
Emery, 1968). It was subsequently recognized, however, that some so-called relict sediments had likely been reworked to some extent during sea-level rise. This reworking
caused the sediments to be brought into partial or complete equilibrium with present
shelf processes and conditions. Such reworked sediment, having partly relict and
partly modern characteristics, was called **palimpsest** (Swift, Stanley, and Curray,
1971).

These authors also identified and discussed four different types of shelf currents
that operate on the shelf to rework and transport sediment:

1. tidal currents
2. storm-generated currents
3. intruding ocean currents, such as the Gulf Stream System
4. density currents

Shallow-marine systems have subsequently been divided on the basis of dominant
shelf processes into three main types:

1. **tide-dominated shelves** (about 17 percent of the world's shelves)
2. **storm-dominated shelves** (about 80 percent of the world's shelves)
3. **shelves dominated by intruding ocean currents** (about 3 percent of the world's
   shelves)

No modern shelves are dominated by density current processes. Changes in sea level
also play an extremely important role in the deposition and preservation of sediments
on continental shelves (e.g., MacDonald, 1991; Walker and James, 1992; Wilgus et al.,
1988).

Tidal Processes. Tides are generated by the gravitational attraction of the moon and
sun for Earth in conjunction with the rotation of Earth. Tidal influence is mainfested at
any given coastal locality by daily rise and fall of the sea over an average range of
about 1 to 4 m on open coasts, but tidal range may exceed 15 m in some enclosed
basins (e.g., the Bay of Fundy, Nova Scotia). Some localities (e.g., the U.S. Atlantic
coast) experience **diurnal tides,** characterized by two highs (of equal height) and two
lows (of equal height) each day. Others (e.g., parts of the Gulf of Mexico coast) have
semidiurnal tides, distinguished by one high and one low each day, and still others
(e.g., the U.S. Pacific coast) have **mixed tides**—two highs (of unequal height) and two
lows (of unequal height) each day. Tidal currents on continental shelves are propagated as a large wave or tidal bulge generated in deep ocean basins (Fox, 1983). In
major ocean basins, this tidal bulge rotates around a central point of no tidal movement called an **amphidromic point.** The tidal wave follows an elliptical path that is
almost circular in the open ocean. In more restricted areas, the ellipse is strongly elongated, forming a narrow, rectilinear pattern.

The vertical rise and fall of the tides are accompanied by horizontal movements
of water (Howarth, 1982; The Open University Team, 1989) that we refer to as tidal
currents. The currents generated on the shelf by tides are bidirectional but asymmetrical with respect to velocity. That is, flood-tide and ebb-tide velocities are commonly
different. Asymmetrical currents may result in net sediment transport in the direction
of the stronger current. Tidal-current velocity decreases with water depth; thus, tidal-

current transport is most important in shallow water. Tidal-current velocities ranging up to about 2 m/s have been measured in some enclosed basins—the Bay of Fundy, for example. Tidal currents on some shelves, such as those around the British Isles, have velocities that may exceed 1.5 m/s; however, tidal velocities on most shelves are less than about 1 m/s. Even so, close to the seabed, many tidal currents are strong enough to rework and transport significant quantities of sand and possibly gravel. On the other hand, tidal-current velocities on some shelves are so low that they are below the threshold velocities required for sediment entrainment and transport. Much of the movement of sediment by tidal currents occurs when tidal currents are aided by wave action. The orbital motion of waves may be sufficient to lift grains off the seafloor, which are then transported some distance by currents too weak to move the grains unaided (Komar, 1976; The Open University Team, 1989). An outstanding example of a modern shelf dominated by strong tidal currents is the North Sea, which lies between the United Kingdom and the coasts of Denmark and Norway.

Storm-generated Waves and Currents. Tidal currents play only a minor role in sediment transport and deposition on storm (wave)-dominated coasts. Instead, sediment movement is caused mainly by waves and by storm-influenced currents. **Waves,** as previously discussed, generate a circular orbital motion of water particles. This orbital motion is translated downward to a nearly horizontal to-and-fro motion as shoaling orbital waves feel bottom, leading to formation of oscillation ripples. Bottom-water motion generated by orbital waves can stir up bottom sediment and place it in suspension, where it can be transported by relatively weak, unidirectional currents, as mentioned. Also, the to-and-fro motion of water along the bottom may be asymmetrical in velocity, causing some net sediment transport that tends to be shoreward in the case of fair-weather waves. Fair-weather waves can entrain sediment only to depths of about 10 to 15 m; however, storm waves of much longer wave length are capable of disturbing sediment on the outer shelf to depths of 200 m or more. In these deeper-water areas, storm waves can stir up and rework bottom sediment, but they probably produce relatively little net transport of sand. By contrast, storm waves acting on the shoreline erode sediment from the beach and deposit it seaward on the shoreface and shelf.

Wind-forced currents are unidirectional currents generated by wind shear stress as wind blows across the water surface, gradually putting into motion deeper and deeper layers of water. Deeper layers of water are deflected by the Coriolis force, so that their direction of movement diverges from that of surface layers. The Coriolis force is generated by Earth's rotation, causing moving objects to be deflected to the right in the Northern Hemisphere and to the left in the Southern Hemisphere. If the velocity and duration of wind are great enough, water movement may extend to the seabed with enough velocity to transport sediments. Strong winds commonly create wind-forced currents that flow parallel to shore and therefore do not provide much offshore sediment transport. If, however, currents moving along the shoreline are deflected landward owing to the Coriolis force, an onshore pile-up of water takes place (e.g., the Oregon coast during winter). Piling up of water onshore creates an elevation of the water surface—a **coastal set-up**—of perhaps a meter or two. This set-up can apparently be enhanced by very low atmospheric pressure. The different water levels at the coast and offshore result in a hydrostatic pressure difference on the ocean floor that drives a bottom flow seaward (relaxation flow, Fig. 12.4). As the bottom water flows seaward, it is deflected laterally to form a **geostrophic current.** This current initially moves obliquely offshore but subsequently veers around, owing to the Coriolis force, to assume a direction roughly parallel to the bathymetric contours or isobaths, that is, roughly parallel to the shoreline.

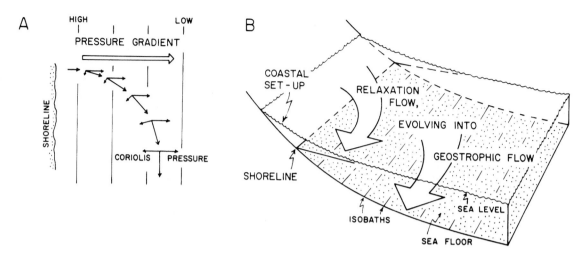

FIGURE 12.4   Seaward flow of water owing to coastal set-up (storm surge). Bottom water flows seaward following pile-up of water along the coast owing to storm waves but is deflected to the right (Northern Hemisphere) by the Coriolis force to form a geostrophic current that flows roughly parallel to isobaths. (From Walker, R. G., 1984, Shelf and shallow marine sands, *in* R. G. Walker (ed.), Facies models, 2nd ed.: Geoscience Canada Reprint Ser. 1, Fig. 1, p. 142, reprinted by permission of Geological Association of Canada.)

These geostrophic flows can achieve velocities at water depths of 10 to 20 m of as much as 60 cm/s (Walker and Plint, 1992). Flows of this magnitude may not be capable of transporting much sandy sediment unless they are accompanied by strong wave-driven oscillatory motion at the bed. As mentioned, oscillatory wave motion can provide the shear stress needed to lift grains off the bottom, which are then transported by the geostrophic currents (Snedden, Nummedal, and Amos, 1988). Unidirectional and oscillatory currents operating together are called **combined flows.** Tropical storms and hurricanes accompanied by strong winds that blow directly onshore can generate higher coastal set-ups than those generated by seasonal storms, creating much stronger seaward-flowing currents (sometimes referred to as storm-surge–ebb). Current-flow velocities of as much as 2 m/s have been reported on shelves during some tropical storms (Morton, 1988), and currents reaching velocities of 2 m/s have also been recorded flowing down submarine canyons (Hubbard, 1992). Sediment movement is obliquely offshore with some sand moving from the beach shoreface into offshore settings. Note again, however, that such transport by geostrophic currents tends to be mostly parallel to isobaths and not directly seaward. Therefore, sediment may not be transported by such currents to any great distance outward onto the shelf.

Ocean Currents. Major, semipermanent ocean currents intrude onto some shelves with sufficient bottom velocity to transport sandy sediment. About 3 percent of modern shelves are dominated by these ocean currents, which operate most effectively on the outer shelf. Modern examples include the northwestern Gulf of Mexico, which is affected by the Gulf Stream System; shelves swept by the Panama and North Equatorial Current off the northeast coast of South America; Taiwan Strait between Taiwan and mainland China, which is intruded by a branch of the Kuroshio Current flowing north from the Philippines; and the outer shelf of southern Africa, which is crossed by the southward-flowing Agulhas Current of the western Indian Ocean. These currents commonly contribute little if any new sediment to the shelf, but they are capable of transporting significant volumes of fine sediment along the shelf. Some achieve bottom velocities great enough to transport sandy sediment and create sand waves and other bedforms—for example, the Kuroshio Current (Boggs, Wang, and Lewis, 1979) and the Agulhas Current (Fig. 12.5; Fleming, 1980).

Density Currents. Density currents are created by density differences within water masses owing to variations in temperature, salinity, or suspended sediment. Density currents are important mainly in transport of fine suspended sediment. High concentrations of suspended sediments near river mouths may create dilute density currents that move along the bottom. Conversely, plumes of warm river water carrying suspended terrigenous clay may override denser seawater and transport the clay for some distance across the shelf before mixing and flocculation cause the clay to settle. Clay deposition may be aided by filter-feeding organisms that ingest mud from the water column and deposit it as fecal pellets. Transport of sediment within the near-bottom nepheloid layer, described in Chapter 3, is also related to water density. In arid climates, excessive nearshore evaporation may generate dense brines that flow seaward along the bottom as an underflow. Overall, density currents are not significant agents of shelf transport, and no modern shelf is dominated by such processes.

Effects of Sea-Level Change. Because geographic environments shift rapidly and change their form during sea-level fluctuations, sea-level changes constitute an important, and sensitive, depositional variable on shelves. They can affect both erosional and depositional processes and thus the kinds of sediments deposited on shelves. Among other things, sea-level changes are an important factor in establishing the stratigraphic architecture of shelf sediments, a topic that we explore in detail in Chapter 14. The increased importance that sedimentologists and stratigraphers now place on sea-level changes can perhaps be judged by the fact that eight full-length, English-language books on sea-level change were published between 1987 and 1994 (see "Further Readings" at the end of this chapter).

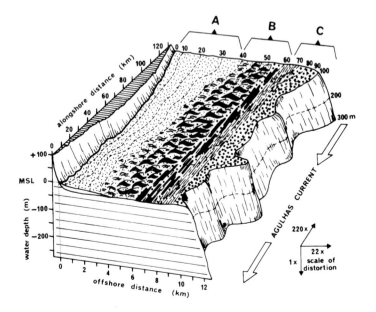

FIGURE 12.5 Sediment transport by the Agulhas Current off the southeastern tip of Africa. Sand waves migrate under the influence of the Agulhas Current. The stippled pattern indicates coarse lag deposits; black streaks indicate sand ribbons. Sand-wave fields are up to 20 km long and 10 km wide, and individual sand waves are up to 17 m high. (From Fleming. B.W., 1980, Sand transport and bedforms on the continental shelf between Durban and Port Elizabeth (southeast Africa continental margin): Sed. Geology, v. 26. Fig. 15, p. 194, reprinted by permission of Elsevier Science Publishers, Amsterdam.)

Biologic Activities. The modern continental shelves are among the most densely populated of all depositional environments, and the geologic record suggests that ancient epeiric seas were also inhabited by large populations of organisms. The shelf floor is habitat for a high diversity of invertebrate organisms, including molluscs, echinoderms, corals, sponges, worms, and arthropods. Both infauna and vagrant and sessile epifauna are represented. Organisms are most abundant in lower-energy areas of the shelf, and the greatest populations occur on the inner shelf just below wave base (Howard, 1978).

Organisms on siliciclastic-dominated shelves are particularly important as agents of bioturbation. Both type and abundance of bioturbation structures vary with sediment type and water depth. As discussed in Chapter 5, many burrows in the nearshore high-energy zone are escape structures that tend to be predominantly vertical. The burrow style changes to oblique or horizontal feeding structures with deepening of water across the shelf. In general, muddy sediments of the shelf are more highly bioturbated than are sandy sediments, and physical sedimentary structures in these sediments may be almost completely obliterated by bioturbation. By contrast, only a few species of organisms can survive in the very high-energy nearshore shelf and beach zone. Therefore, sandy sediments of the beach-shelf transition zone are dominated by physical structures such as cross-bedding rather than bioturbation structures. Nonetheless, some some bioturbation structures may be present if they escape destruction by reworking. Sandy layers deposited in deeper water on the shelf may be bioturbated to some degree in their upper part. Bioturbation structures associated with various water depths on the shelf are illustrated in Figure 12.6. In addition to their importance as bioturbation agents, some organisms produce fecal pellets from muddy sediment; these pellets may become hardened and coherent enough to behave as sand grains. Organisms with shells or other fossilizable hard parts also leave remains that may be preserved to become part of the sediment record.

Shelf Sediments

Source. Sediments on siliciclastic shelves of the modern ocean may include modern, relict, and palimpsest types. **Modern sediments** are deposits that are in equilibrium with the hydrodynamic conditions now existing on the shelf. They were transported onto the shelf and deposited by processes currently operating on the shelf, or they were generated in place by shelf processes (chemical/biochemical sediments). By far, the greatest volume of modern sediments are **siliciclastic muds and sands** contributed to the shelf by river outflow and shoreline erosion, together with some fine dust blown onto the shelf by winds. On shelves that border volcanic arcs, such as the Japan continental shelf, materials contributed by volcanism may be an important component of the shelf sediment (Boggs, 1984). Under the category of modern sediments, Emery (1968) also includes **biogenic sediments** composed mainly of carbonate shells and tests; **authigenic sediment,** which consists mainly of glauconite and phosphorite; and **residual sediments,** produced by *in situ* submarine weathering of bedrock.

Relict sediments were derived from sources on land and transported onto the shelves by rivers or glaciers during lowered sea level. Therefore, they may include significant quantities of gravels as well as sands. Emery (1968) estimates that about 70 percent of the area of the modern shelves is covered by relict sediment. In making this estimate, he did not take into account certain parts of the shelves that have been extensively reworked by physical and biological processes operating today. These reworked relict sediments are the palimpsest sediments of Swift, Stanley, and Curray (1971)

FIGURE 12.6 Bioturbation structures at different water depths on the shelf. (From Reineck, H. E., and I. B. Singh, 1980, Depositional sedimentary environments. Fig. 554, p. 402, reprinted by permission of Springer-Verlag, Heidelberg. Originally after H. E. Reineck, J. Dörjes, S. Dadow, and G. Hertweck, 1968, Sedimentologie, fauenzonierung und faziesabfolge vor der Ostküste der inneren Deutschen Bucht: Senckenbergiana Lethaea, v. 49, p. 261–309.)

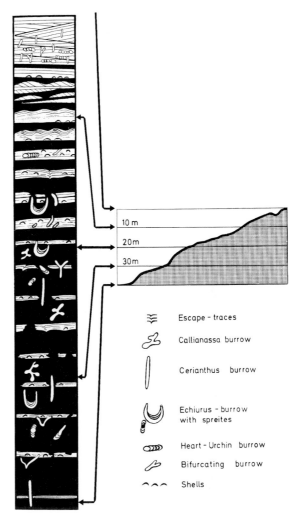

10 m
20 m
30 m

≋ Escape - traces

ᔓ Callianassa burrow

❘ Cerianthus burrow

∪ Echiurus - burrow
with spreites

⬭ Heart - Urchin burrow

ᔓ Bifurcating burrow

∿∿ Shells

mentioned above. The abundance of palimpsest sediment relative to relict sediment is not well established.

General Characteristics and Distribution. The sediments of modern siliciclastic shelves consist largely of muds and sands, although large patches of relict gravels occur in many areas. The areal distribution of muds and sands varies markedly on different shelves. On some shelves, nearshore sands grade seaward through a transition zone of mixed sand and mud to deeper-water muds, in essentially the classic pattern of seaward-decreasing grain size visualized by Johnson (1919) in his concept of a graded shelf. On some other shelves, coastal sands grade directly offshore into Pleistocene relict sands, or, alternatively, muds may cover the shelf floor right up to the shoreline (Reineck and Singh, 1980). Modern shelves with their mixture of modern and relict sediment may not be perfect analogs for ancient continental shelves and epeiric seaways. We know, for example, that some types of modern shelf sediment bodies—such as transgressive, shoreface-detached sand ridges—have not yet been reported from the ancient geologic record (Walker, 1984). In any case, the types of sediments and sedimentary structures found on modern shelves are related both to Pleistocene depositional patterns and to the dominant sedimentation processes now operating on the shelves.

Sediments of Tide-dominated Shelves. As discussed, tide-dominated shelves are distinguished by the presence of tidal currents with velocities ranging from about 50 to more than 150 cm/s. Modern examples include the North Sea; the Korea Bay of the Yellow Sea; the Gulf of Cambay, India; the shelf around the British Isles; Georges Bank in the outer part of the Gulf of Maine; and the northern Australia shelf. Tide-dominated shelves are characterized particularly by the presence of sand bodies of various types and dimensions. Large **sand waves (dunes)** a few meters to more than 20 m in height with wave lengths of tens to hundreds of meters typically occur in fields that may cover areas of 15,000 km^2 or more. Sand waves may have symmetrical cross-sectional shapes if produced by tidal currents with equal ebb and flood peak speeds; however, asymmetrical shapes caused by unequal ebb and flood velocities are more common (Belderson, Johnson, and Kenyon, 1982). **Tidal sand ridges** or **sand ribbons** up to 40 m high, 5 km wide, and 60 km long, and covering areas up to 5000 km^2, have been reported from the North Sea shelf (Swift, 1975). These ridges are oriented nearly parallel to tidal flow with steep sides that face in the direction of regional sediment transport. They have been referred to as "shoal retreat massifs." They are believed to have formed initially by shoreline detachment during Holocene transgression, but they are currently maintained by tidal flows. The faces of these ridges dip about 5°; therefore, migration of the ridges does not produce high-angle (angle-of-repose) cross-beds. In addition to sand waves, ridges, and ribbons, tide-dominated shelves include **sand sheets, sand patches** (Fig. 12.7), and **gravel sheets,** characterized by small-scale bedforms, and patches of bioturbated muds in areas sheltered from tidal currents and waves (Stride et al., 1982).

Because most of the shelf is constantly covered by water, the characteristics of sand waves, sand ridges, and other bedforms on modern shelves must be studied largely by indirect methods. Small-scale bedforms can be observed and photographed by divers or by remote-controlled cameras. Larger bedforms are investigated by sonar bottom-profiling and side-scan sonar techniques (e.g., Belderson et al., 1972; Belderson, Johnson, and Kenyon, 1982). Small-scale, internal sedimentary structures can be studied in cores of bottom sediment, and sub-bottom seismic profiling methods, described in Chapter 15, may be used to study some large-scale features such as bedding. None of these methods allows detailed examination of modern shelf structures. The idealized distribution of bedforms along the sediment transport path on tide-dominated shelves is illustrated in Figure 12.8. At high tidal velocities, of about 150 cm/s, the seafloor may be eroded, leaving furrows and gravel waves. With progressively diminishing velocity farther down the transport path, eroded sediments are deposited to form flow-parallel sand ribbons, large dunes, small dunes, a rippled sand sheet, and finally sand patches. Sand ridges may form in the dune belt if enough sand is present.

Most tidal shelf sands are characterized by cross-bedding. Small-scale cross-bedding and ripple cross-lamination, produced by migration of ripples and small dunes, and large-scale cross-bedding generated by migration of dunes and sand ridges are both common. Foreset dip directions of cross-lamination may be bidirectional or unidirectional depending upon tidal influence. Plane beds also develop under some upper-flow-regime flow conditions. Physical structures thus tend to dominate tidal shelf sands, which typically display fewer bioturbation structures than do muddy

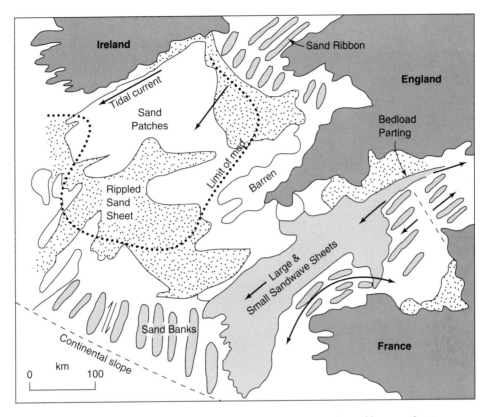

FIGURE 12.7 Tide-generated bedforms, including sand sheets, patches, ribbons, and waves (dunes), on the shallow floor of the modern Celtic Sea and the western part of the English Channel. (After Johnson, M. A., et al., 1982, Sand transport, *in* A. H. Stride (ed.), Offshore tidal sands: Chapman and Hall, London. Fig. 4.4, p. 66, as redrawn by Banerjee, 1991. Reproduced by permission.)

shelf sediments (deposited under other conditions), which are commonly highly bioturbated with few physical structures except possibly planar lamination (Fig. 12.6).

Sediments of Storm-dominated Shelves. Storm-dominated shelves predominate on most of the world's coasts. They are characterized by low tidal current velocities (commonly <25 cm/s), and fair-weather wave base is normally shallow (<~10 m). Because of these characteristics, little coarse sediment movement occurs on these shelves except during intense storm conditions. Examples of storm-dominated shelves include the Atlantic shelf off the eastern coast of the United States, the Pacific shelf off Oregon and Washington, and the Bering Sea. Sedimentation patterns on storm-dominated shelves may be quite complex, depending particularly upon the extent to which the shelves are mantled by relict sediments or modern sediments. Shelves with abundant relict sediment, such as the Atlantic, are characterized particularly by sand bodies. These bodies take several different forms. Shoal retreat massifs are large sand bodies similar to those described in the North Sea (Fig. 12.7). They formed during Holocene transgression but are reworked and maintained by storm activity. Linear sand ridges, up to 10 m high, 2 to 3 km wide, a few tens of kilometers long, and spaced 2 to 7 km apart, occur between or superimposed on retreat massifs. These ridges have been inter-

FIGURE 12.8 Idealized sequence of bedforms developed along a sediment transport path on a tide-dominated shelf. Maximum spring-tide current velocities associated with each bedform type are shown along the edge of the diagram. Sand ridges may form in the dune belt if sufficient sand is present. (After Belderson, R. H., M. A. Johnson, and N. H. Kenyon, 1982, Bedforms, *in* A. H. Stride (ed.), Offshore tidal sands: Chapman and Hall, London. Fig. 3.1, p. 28, reproduced by permission.)

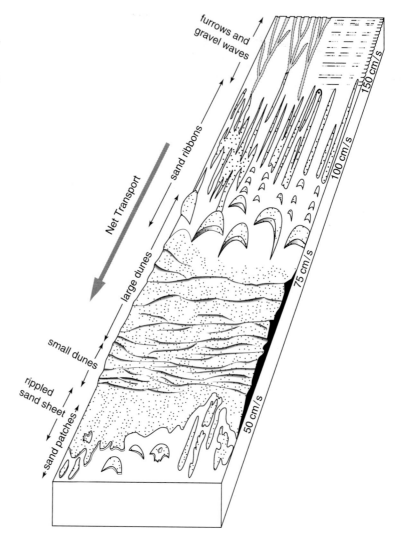

preted both as originating from relict barriers and as arising from modern shelf processes (Swift, 1975). Lower-relief sand bodies, characterized by ripple and dune bedforms, are also common. These bodies presumably form by a combination of oscillatory and unidirectional flow arising from storm activity (Walker, 1984a). Shelves with a greater component of modern vs. relict sediments, such as the Pacific shelf off Oregon and Washington, are typically characterized by less relief and a greater proportion of finer-grained sediments (muds) than are Atlantic-type shelves. Although sediment on these shelves may display a general trend of seaward fining, extensive "windows" occur where relict sands or gravels show through a discontinuous blanket of muddy modern sediment. Also, mixing of relict sands and modern muds may take place in some areas. Muds are typically thoroughly bioturbated.

Coarse-grained "storm layers" and hummocky cross-stratification are sedimentary structures that appear to be especially characteristic of storm-dominated shelf sediments. "Storm layers" are commonly thin layers consisting of concentrations of

Shallow-Marine Storm Beds

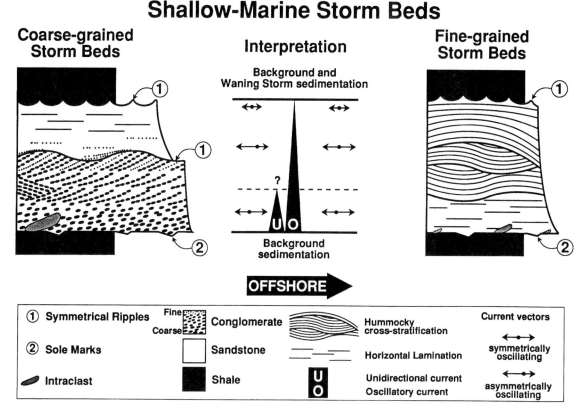

FIGURE 12.9 Schematic comparison of idealized coarse-grained storm beds and fine-grained, hummocky cross-stratified beds on storm-dominated shelves. The lengths of current vectors are proportional to the strength of the current in a given direction rather than the duration. (From Cheel, R. J., and D. A. Leckie, 1992, Coarse-grained storm beds of the Upper Cretaceous Chungo Member (Wapiabi Formation), southern Alberta, Canada: Jour. Sed. Petrology, v. 62. Fig. 14, p. 943, reproduced by permission of Society of Economic Paleontologists and Mineralogists, Tulsa, Okla.)

coarser grains interlayered or embedded in finer-grained muds (Fig. 12.9). The coarser material typically consists of coarse silt, fine sand, shell fragments, or, less commonly, gravel. The layers characteristically show vertical size grading. The exact origin of these layers is controversial (see review by Cheel and Leckie, 1992). Cheel and Leckie (1992) suggest that they form by a two-stage process: (1) transport from a beach by offshore-directed, storm-generated, combined (geostrophic) flows, followed by (2) reworking and selective sorting of bed material by asymmetrical oscillatory currents generated by shoaling swell waves propagating onshore (Fig. 12.9). Storm layers, also called tempestites, are described in considerable detail in a monograph by Aigner (1985). They are best developed on the inner shelf, but they have been found as far as 40 km from the coast (Reineck and Singh, 1980).

Hummocky cross-stratification has been identified in few, if any, modern shelf sediments, but it has been described in numerous ancient shelf sediments ranging in age from Precambrian to Pleistocene. It is thought to be particularly characteristic of shelf sediments, although it has been described also in shoreface (beach) and some lake sediments. As discussed in Chapter 5, it consists of curving, gently dipping laminae, both convex-up (hummocks) and concave-up (swales), that intersect at a low angle (Fig. 12.9). It is commonly interbedded with bioturbated mudstones. Most workers appear to agree that hummocky cross-stratification forms as a result of storm waves acting in some manner below fair-weather wave base; however, the exact mechanism of formation remains controversial (see review in Duke, Arnott, and Cheel, 1991). Duke, Arnott, and Cheel (1991) suggest the following sequence of events in the formation of hummocky cross-stratification on the inner shelf. Initially, waves interact with relatively weak, offshore-directed, coast-oblique geostrophic currents (resulting from coastal set-up) to scour the muddy substrate during rising phases of the storm. Coastal sand, moving in intermittent suspension and as bedload under the combined bottom flow, is transported to offshore localities. Sand accumulates rapidly as the storm current wanes, and flat lamination, much of it draping low-relief scours, is formed under oscillatory-dominant combined flow. Large ripples begin to form (and migrate) on the still-aggrading substrate, and anisotropic hummocky cross-stratification (asymmetrical dip directions and different dip angles) develops; however, much of the sand is emplaced and reworked by storm and swell waves as the bottom current subsides. Thus, in the final stage, isotropic hummocky cross-stratification (more or less symmetrical dip directions and uniform dip angles) is formed under very strongly oscillatory-dominant combined flow and purely oscillatory flow. This sequence of events is illustrated in Figure 12.10. According to this interpretation, hummocky cross-stratification thus originates by a combination of unidirectional and oscillatory flow related to storm activity.

Storm-generated deposits appear to have low preservation potential. Storm flows may increase gradually in velocity to about 2 m/s but then decrease almost as gradually. The waning flow tends to erase structures made by increasing storm-flow velocity. If storm-generated structures do survive waning flow, they may not survive bioturbation (Dott, 1983; Gagan, Johnston, and Carter, 1988). Gagan, Johnston, and Carter report that one year after Cyclone Winifred generated storm deposits on the Great Barrier Reef shelf, Australia, bioturbation had completely destroyed deposits in water more than 30 m deep. Structures in shallower water were preserved, apparently because subsequent

FIGURE 12.10 Possible sequence of events leading to formation of hummocky cross-stratification in inner shelf sands; 2D = two-dimensional. (From Duke, W. L., R. W. C. Arnott, and R. J. Cheel, 1991, Shelf sandstones and hummocky cross-stratification: New insights on a stormy debate: Geology, v. 19. Fig. 5, p. 628. Published by the Geological Society of America, Boulder, Co.)

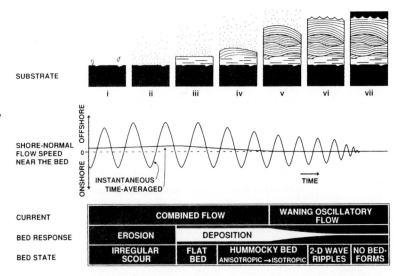

sediment buried them faster than the rate of bioturbation. Owing to bioturbation, many storm-generated deposits in the ancient record may not be recognizable as storm deposits. Thus, the preservation of storm-dominated deposits may be the exception rather than the rule.

Sediments on Shelves Dominated by Intruding Ocean Currents. These shelves make up only a very small percentage of the world's shelves. Sediment on such shelves is largely relict, but it is commonly reworked by intruding currents to form sand waves, sand ribbons, and coarse sand and gravel lag deposits. The best-studied example of this type is the southeastern shelf of South America which is intruded by the Agulhas Current of the Indian Ocean. Sand waves up to 17 m high with wave lengths up to 700 m occur in sand-wave fields up to 10 km wide and 20 km long (Fig. 12.5). Taiwan Strait between Taiwan and China is another broad shelf invaded by ocean currents that create extensive sand-wave fields (Boggs, 1974).

Ancient Siliciclastic Shelf Sediments

Although recognition of ancient shelf sediments is aided by study of modern continental shelves, modern shelves are not necessarily good analogs of ancient shelves. The prevalence of relict sediments on modern shelves may be atypical, and certain features of modern shelves such as shoal retreat massifs have not been recognized in the ancient rock record. Conversely, some structures believed to be diagnostic of ancient shelf sediments, notably storm-generated hummocky cross-stratification, have apparently not been recognized in modern shelf environments. Thus, no definitive model for shelf sediments combining data from modern environments and study of ancient rocks has yet been formulated, although tentative models have been proposed. In general, ancient shelf sediments appear to be distinguished by the following:

1. tabular shape
2. extensive lateral dimensions (1000s of km^2) and great thickness (100s of m)
3. moderate compositional maturity of sands with quartz dominating over feldspars and rock fragments
4. generally well-developed, even, laterally extensive bedding
5. the presence of storm beds in some shelf deposits
6. wide diversity and abundance of normal marine, fossil organisms
7. diagnostic associations of trace fossils

More specific characteristics are related to deposition under tide-dominated or storm-dominated conditions. The deposits of ancient tide-dominated shelves are characterized particularly by cross-bedded sandstone. Paleocurrents are mainly unimodal, apparently owing to the existence of ancient regional net sediment transport paths, although bipolar cross-stratification is present locally (Dalrymple, 1992). Reactivation surfaces are abundant. Ancient storm-dominated shelf deposits likely contain a greater proportion of mud than do tide-dominated deposits, and hummocky cross-stratification and storm layers are common (but not in modern deposits!). Walker and Plint (1992) suggest two main kinds of ancient storm-dominated shelf deposits:

1. sheetlike or patchy bodies of interbedded hummocky cross-stratified sandstone and bioturbated mudstone, representing aggrading offshore marine environments
2. long, narrow sand to conglomerate bodies surrounded by finer sediment

These narrow bodies are oriented more or less parallel to the shoreline and tend to be under- and overlain by regionally extensive erosion surfaces. Aggrading hummocky

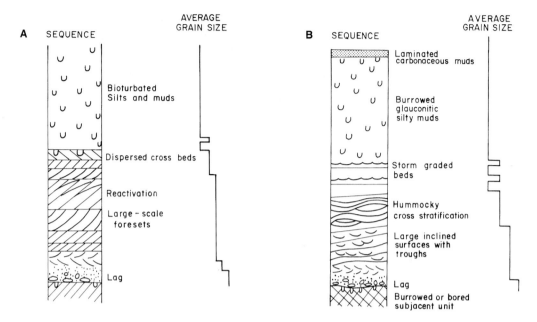

FIGURE 12.11   Schematic diagrams illustrating typical fining-upward transgressive shelf successions. A. Tide-dominated shelf. B. Storm-dominated shelf. (After Galloway, W. E., and D. K. Hobday, 1983, Terrigenous clastic depositional systems. Fig. 7.14, p. 159, Fig. 7.15, p. 160, reprinted by permission of Springer-Verlag, Heidelberg.)

cross-stratified sandstones are characteristic of highstand (transgressive) conditions, and the long, narrow sandstone/conglomerate bodies probably form under lowstand, or regressive, conditions.

Several kinds of vertical successions may thus be generated in shelf sediments, depending upon whether deposition takes place during transgression or regression and depending upon the dominant type of shelf processes operating during deposition. It is difficult to generalize about these successions except to say that transgression tends to produce fining-upward successions that may begin with coarse lag deposits, and regression produces coarsening-upward successions. Some idealized vertical shelf successions produced under different postulated sedimentation conditions are illustrated in Figures 12.11 and 12.12. Both tide-dominated and storm-dominated transgressive successions are shown in Figure 12.11 to illustrate differences in facies that can develop under these different shelf conditions. These successions should be considered only as working models. Actual transgressive and regressive successions may differ markedly in detail from these idealized profiles.

Ancient shelf deposits are known in stratigraphic units of all ages and from all continents. They are probably the most extensively preserved rocks in the geologic record. Examples of a storm-dominated shelf deposit (Fig. 12.13) and a tide-dominated shelf deposit (Fig. 12.14) are discussed below to illustrate some of the characteristics of ancient shelf sediments.

The Lower Cretaceous Grayson Formation of Texas (Fig. 12.13), described by Hobday and Morton (1984), displays many of the features commonly attributed to storm deposition. The basal part of this section consists of thick graded sandstones with well-developed hummocky cross-stratification overlain by parallel laminae, sepa-

FIGURE 12.12   Idealized coarsening-upward regressive shelf successions on a storm-dominated shelf. (After Galloway, W. E., and D. K. Hobday, 1983, Terrigenous clastic depositional systems. Fig. 7.17, p. 162, reprinted by permission of Springer-Verlag, Heidelberg.)

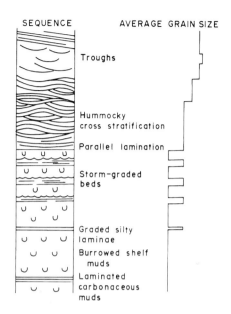

SEQUENCE   AVERAGE GRAIN SIZE

Troughs

Hummocky
cross stratification

Parallel lamination

Storm-graded
beds

Graded silty
laminae

Burrowed shelf
muds

Laminated
carbonaceous
muds

rated by subordinate siltstones. The upper few centimeters of the sandstones are bioturbated, causing ripple forms and cross-stratification to be almost entirely obliterated in some units. A thin siltstone unit separates the hummocky cross-laminated basal sandstone unit from an overlying thick, lenticular sandstone with scoured upper surface, interpreted as a channel-mouth bar deposit. This channel sand is overlain by mudstones with plant fragments and a sandstone unit with large wave ripples and concentrations of oysters and mussels that may have formed owing to storm swells. The upper part of the section above the wave-rippled sand consists of siltstones and mudstones with interbeds of thinner graded sandstones and layers with shell concentrations, capped at the top by a coarsening-upward mudstone-sandstone succession with a lignite bed above. The graded sandstones are sharp based with sole markings and elongate fossils oriented offshore. The associated mudstones intervening between successive sandstone beds contain some burrows but are not highly bioturbated. The uppermost sandstone of this succession is highly bioturbated with scattered molluscs.

Hobday and Morton (1984) interpret the hummocky cross-stratified sandstones and the wave-rippled sandstone in the base of the section as deposits formed above storm wave base at water depths less than about 30 m. The hummocky cross-stratification was produced by interaction of unidirectional currents and storm waves. The sharp-based, graded sandstones are the deposits of wind-forced currents. The mudstones separating the graded sandstone units appear to be suspension sediments deposited during fair-weather conditions at such a rapid rate that little bioturbation occurred. The uppermost mudstone, sandstone, and lignite units represent a change in sedimentation conditions from transgressive shelf deposition to regressive sedimentation along a prograding delta front.

Banerjee (1991) describes Cretaceous sandstones of southern Alberta, Canada, which he interprets as tidal sand sheet facies. Figure 12.14 is an example of one of these tidal sandstone bodies, the (lowermost) Sunkay Member of the Lower Cretaceous Alberta Group. This sandstone unit, approximately 10 m thick, rests with a sharp erosional base on nonmarine shales of the Beaver Mines Formation and is overlain by black marine shales. The Sunkay Member consists mainly of sandstone, with a thin lag gravel

FIGURE 12.13 Vertical section
through 50-m-thick unit of the
Lower Cretaceous Grayson Formation of Texas near Lake Texoma.
This section includes many depositional characteristics attributed to
storm-shelf deposition. (After Hobday, D. K., and R. A. Morton, 1984,
Lower Cretaceous shelf storm
deposits, northeast Texas, *in* R. W.
Tillman and C. T. Siemers (eds.),
Siliciclastic shelf sediments: Soc.
Econ. Paleontologists and Mineralogists Spec. Pub. 34. Fig. 3, p. 208,
reprinted by permission of SEPM,
Tulsa, Okla.)

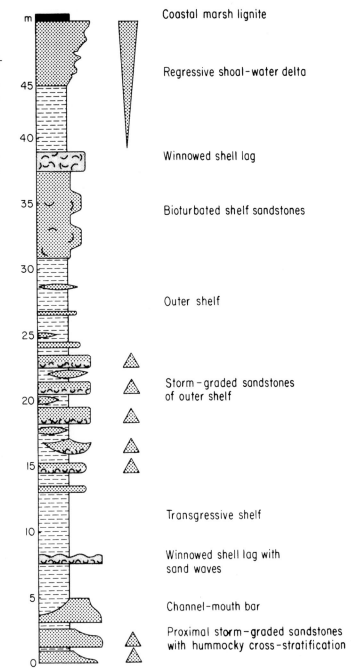

Coastal marsh lignite

Regressive shoal-water delta

Winnowed shell lag

Bioturbated shelf sandstones

Outer shelf

Storm-graded sandstones
of outer shelf

Transgressive shelf

Winnowed shell lag with
sand waves

Channel-mouth bar

Proximal storm-graded sandstones
with hummocky cross-stratification

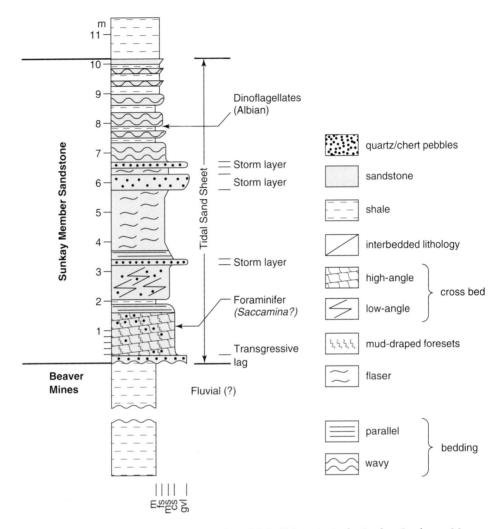

FIGURE 12.14 Vertical succession of sandy, tidal shelf deposits in the Sunkay Sandstone Member of the Lower Cretaceous Alberta Group, southern Alberta, Canada. Symbols in the grain-size scale are gvl = gravel, cs = coarse sand, ms = medium sand, fs = fine sand, and m = mud (silt-clay). (After Banerjee, I., 1991, Tidal sand sheets of the Late Albian Joli Fou-Kiowa-Skull Creek marine transgression, Western Interior Seaway of North America, *in* R. G. Smith et al. (eds.), Clastic tidal sedimentology: Canadian Soc. Petroleum Geologists Mem. 16. Fig. 9, p. 343.)

deposit at the base and a few thin interbedded conglomerate layers, interpreted as storm layers. The succession as a whole fines upward. The basal sandstone units are cross-bedded; the uppermost ones are characterized by flaser and wavy bedding. Foraminifers and dinoflagellates are present in some layers. Bioturbation is not common.

Banerjee interprets these sandstones as tidal sand sheets. Tidal interpretation is based in part on the presence of the following:

1. cross-stratification (>20° dip and sets 30–40 cm thick), with thin shale drapes adhering to foreset surfaces, in the lower part of the section
2. flaser, wavy, and lenticular bedding in the upper part

Transgression resulted in deepening of water at this site, causing deposition of black marine shales on top of the tidal sandsones.

12.3   THE OCEANIC (DEEP-WATER) ENVIRONMENT

In the discussion of depositional environments to this point, I have focused on the continental, marginal-marine, and shallow-marine environments because much of the preserved sedimentary record was deposited in these environments. In terms of size of the environmental setting, however, these nonmarine to shallow-water environments actually cover a much smaller area of Earth's surface than do deep-water environments. By far the largest portion of Earth's surface lies seaward of the continental shelf in water deeper than about 200 m. Approximately 65 percent of Earth's surface is occupied by the continental slope, the continental rise, deep-sea trenches, and the deep ocean floor. Even so, most textbooks that discuss sedimentary environments typically give only modest coverage to oceanic environments. This bias is probably so because, as a whole, deep-water sediments are much more poorly represented in the exposed rock record than are shallow-water sediments. Deep-water deposits are less abundant than shallow-water deposits in the exposed rock record because sedimentation rates overall are slower in deeper water; thus, the sediment record is thinner. [An exception is the submarine fan environment near the base of the slope where sedimentation rates from turbidity currents can exceed 10 m/1000 yr and turbidite sediments can achieve thicknesses of thousands of meters (e.g., Bouma, Normark, and Barnes, 1985).] Also, part of the sediment record of the deep seafloor may have been destroyed by subduction in trenches, and those deep-water sediments that have escaped subduction have required extensive faulting and uplift to bring them above sea level where they can be viewed. Deep-water sediments other than turbidites have not been studied as thoroughly as shallow-water sediments—perhaps in part because deep-water sediments have less economic potential for petroleum. Owing to the advent of seafloor spreading and global plate tectonics concepts, however, the deep seafloor has taken on enormous significance for geologists. Consequently, intensive research efforts have been focused on the continental margins and deep seafloor since the early 1960s. Also, the continuing need to add to our fossil fuel reserves is pushing petroleum exploration into deeper and deeper water, and the possibility of mining manganese nodules and metalliferous muds from the seafloor is also causing increased economic interest in the deep ocean.

Deep-sea research has been particularly stimulated by the Deep Sea Drilling Program (which began in 1968) and the Ocean Drilling Program, discussed in Chapter 1. Since initiation of these programs, several hundred holes have been drilled by DSDP and ODP teams throughout the ocean basins of the world to an average depth below seafloor of about 300 m (~1000 ft) and to maximum depths exceeding 1000 m. In addition to deep coring by DSDP and ODP, many thousands of shallow piston cores have been collected from the seafloor throughout the ocean by marine geologists from major oceanographic institutions of the world. Also, hundreds of thousands of kilometers of seismic profiling lines (Chapter 15) have been run in criss-cross patterns across the ocean floor in an attempt to unravel the sub-bottom structure of the ocean. Although much of this research has been aimed at understanding the larger-scale features of the ocean basins that illuminate the origin and evolutionary history of the ocean basins along plate tectonics lines, many data on sedimentary facies and sedimentary environments have also been collected. Much additional new information on ocean circulation and sediment transport systems has also been generated by oceanographers who study ocean-bottom currents and bottom-water masses. Thus, a significant increase in understanding of the ocean basins and the deep ocean floor has come about since the 1950s.

We shall concentrate our discussion here on the fundamental processes of sediment transport and deposition on continental slopes and the deep ocean floor and the principal types of facies developed in these environments. Additional details of the structure, stratigraphy, and sediment characteristics of continental margins and ocean basins are available in several symposium volumes and other monographs, including those of Bally (1981); Bouma, Moore, and Coleman (1978); Bouma, Normark, and Barnes (1985); Cook and Enos (1977); Cook, Hine, and Mullins (1983); Curray et al. (1977); Doyle and Pilkey (1979); Hay (1974); Hsü and Jenkyns (1974); Lisitzin (1972); Osborne (1991); Pickering, Hiscott, and Hein (1989); Saxov and Nieuwenhuis (1982); Siebold and Berger (1993); Siemers, Tillman, and Williamson (1981); Stanley and Moore (1983); Warme, Douglas, and Winterer (1981); and Weimer and Link (1991).

Depositional Setting

Continental Slope. The continental slope extends from the shelf break, which occurs at an average depth of about 130 m in the modern ocean, to the deep seafloor (Fig. 12.15). The lower boundary is typically located at water depths ranging from about 1500 to 4000 m, but locally in deep trenches it may extend to depths exceeding 10,000 m (Bouma, 1979). Continental slopes are comparatively narrow (10–100 km wide), and they dip seaward much more steeply than the shelf. The average inclination of modern continental slopes is about 4°, although slopes may range from less than 2° off major deltas to more than 45° off some coral islands.

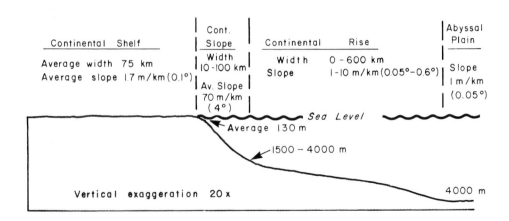

FIGURE 12.15 Principal elements of the continental margin. (After Drake, C. L., and C. A. Burk, 1974, Geological significance of continental margins, *in* C. A. Burk and C. L. Drake (eds.), The geology of continental margins. Fig. 9, p. 8, reprinted by permission of Springer-Verlag, Heidelberg. Modified by H. E. Cook, M. E. Field, and J. V. Gardner, 1982, Characteristics of sediments on modern and ancient continental slopes, *in* P. A. Scholle and D. Spearing (eds.), Sandstone depositional environments: Am. Assoc. Petroleum Geologists Mem. 31. Fig. 1, p. 329, reprinted by permission of AAPG, Tulsa, Okla.)

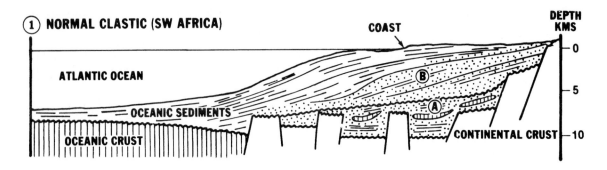

① NORMAL CLASTIC (SW AFRICA)

COAST

DEPTH KMS

ATLANTIC OCEAN

OCEANIC SEDIMENTS

OCEANIC CRUST

Ⓑ

Ⓐ

CONTINENTAL CRUST

0

5

10

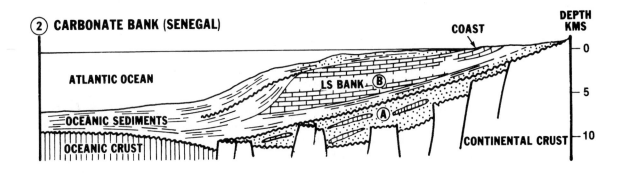

② CARBONATE BANK (SENEGAL)

COAST

DEPTH KMS

ATLANTIC OCEAN

OCEANIC SEDIMENTS

OCEANIC CRUST

LS BANK   Ⓑ

Ⓐ

CONTINENTAL CRUST

0

5

10

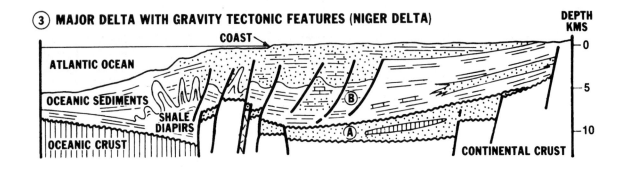

③ MAJOR DELTA WITH GRAVITY TECTONIC FEATURES (NIGER DELTA)

COAST

DEPTH KMS

ATLANTIC OCEAN

OCEANIC SEDIMENTS

SHALE DIAPIRS

OCEANIC CRUST

Ⓑ

Ⓐ

CONTINENTAL CRUST

0

5

10

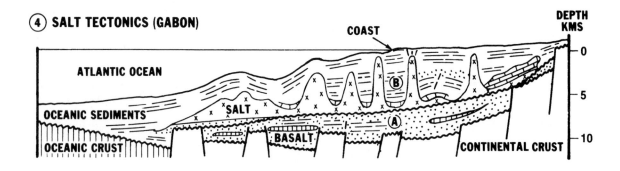

④ SALT TECTONICS (GABON)

COAST

DEPTH KMS

ATLANTIC OCEAN

OCEANIC SEDIMENTS

SALT

BASALT

OCEANIC CRUST

Ⓑ

Ⓐ

CONTINENTAL CRUST

0

5

10

FIGURE 12.16 **(Opposite)** Schematic examples of four principal kinds of passive continental margins, classified on the basis of sediment characteristics: (1) normal clastic, (2) carbonate bank, (3) major delta, and (4) salt tectonics. A = nonmarine sediments; B = marine sediments. (From Kingston, D. R., C. P. Dishroon, and P. A. Williams, 1983, Global basin classification system: Am. Assoc. Petroleum Geologists Bull., v. 67. Fig. 6, p. 2180, reproduced by permission of AAPG, Tulsa, Okla.)

The origin and internal structure of continental slopes are not of primary interest here; however, a brief description of differences in the characteristics of continental slopes on passive (Atlantic-type) and active (Pacific-type) continental margins is pertinent to succeeding discussion. On the basis of the most important controls on sedimentation, Kingston, Dishroon, and Williams (1983) recognize four types of passive margins (Fig. 12.16):

1. normal siliciclastic margins
2. margins with a carbonate bank (platform)
3. margins dominated by a large delta with major gravity-controlled tectonic features
4. margins dominated by salt tectonics (flowage of salt to produce salt domes and diapirs)

More than one of these passive-margin types may be present within a given geographic area. Active continental margins may be characterized by the presence of a fore-arc region only or by both fore-arc and back-arc regions, as, for example, the Japan margin (Fig. 12.17). Sedimentation can take place in both back-arc and fore-arc basins, on back-arc and fore-arc slopes, and in the fore-arc trench.

Continental slopes may have a smooth, slightly convex surface morphology, such as that found on passive siliciclastic margins (Fig. 12.16, Part 1), or they may be irregular on a small to very large scale (e.g., Fig. 12.16, Part 4). Active-margin slopes tend to be particularly irregular. For example, the Pacific slope off Japan, which descends to a depth of about 7000 m into the Japan Trench, is characterized by the presence of structural terraces and basins together with anticlinal welts and fault-bounded ridges arranged in an en echelon pattern roughly parallel to the Japan coast (Boggs, 1984). These ridges and folds form prominent structural "dams" behind which sediments are

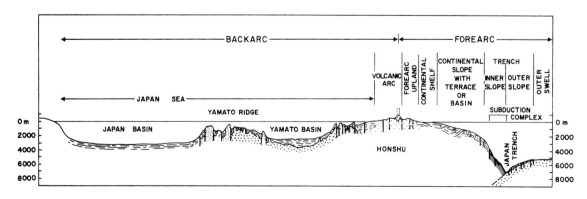

FIGURE 12.17 Schematic representation of an active continental margin (Japan), showing both the fore-arc and the back-arc characteristics of the margin. (From Boggs, S., Jr., 1984, Quaternary sedimentation in the Japan arc-trench system: Geol. Soc. America Bull., v. 95. Fig. 2, p. 670, reproduced by permission.)

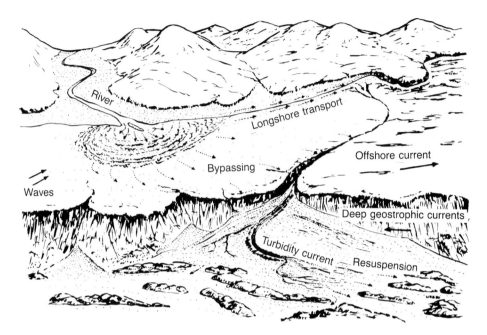

FIGURE 12.18 Transport and redistribution of sediment on a continental margin such as the west coast of North America. Coarser sediment input by rivers moves laterally along the coast by longshore transport until intercepted by a submarine canyon. Fine sediment bypasses the shelf and comes to rest on the continental slope or in deeper water. (From Seibold, E., and W. H. Berger, 1982, The sea floor. Fig. 4.1, p. 79, reprinted by permission of Springer-Verlag, Heidelberg. Redrawn from D. G. Moore, 1969, Reflecting profiling studies of the California continental borderland: Structure and Quaternary turbidite basins: Geol. Soc. America Spec. Paper 107, Fig. 5, p. B–11.)

ponded. In general, structural barriers on highly irregular slopes can inhibit movement of bottom sediment across the slope and create catchment basins for sediment.

Modern continental slopes are gashed to varying degrees by submarine canyons oriented approximately normal to the shelf break (Fig. 12.18), which provide accessways for turbidity currents moving across the slope. Most submarine canyons have their heads near the slope break and do not cross the shelf; however, a few major canyons on modern shelves extend onto the shelf and head up very close to shore. Some large canyons also extend seaward beyond the base of the slope to form deep-sea channels that may meander over the nearly flat ocean floor for hundreds of kilometers. The Toyama Deep Sea Channel in the Japan Sea, for example, winds its way across the seafloor from the mouth of the Toyama Trough for approximately 500 km before emptying onto the Japan Sea abyssal plain.

The origin of submarine canyons has been debated since the early part of the twentieth century (Pickering, Hiscott, and Hein, 1989, p. 134–136). Although downcutting by rivers that extended across the shelf during periods of lowered sea level may have initiated the formation of some canyons on the shelf, turbidity currents are the main agents of canyon cutting on the slope and deeper seafloor. Canyon development may be initiated by local slope failure (slumping), followed by headward growth of erosional scars. Turbidity currents are erosive in their initial stages and thus can deepen and lengthen the incipient canyons over time—aided by further slumping on the upper part of the slope. The locations and shapes of some submarine canyons may have been influenced by the presence of faults and folds (Green, Clarke, and Kennedy, 1991).

Continental Rise and Deep Ocean Basin. The continental rise and deep ocean basin encompass that part of the ocean lying below the base of the continental slope. Together, they make up about 80 percent of the total ocean seafloor. The deeper part of the ocean seaward of the continental slope is divided into two principal physiographic components:

1. the deep **ocean floor,** which is characterized by the presence of abyssal plains, abyssal hills (volcanic hills <1 km high), and seamounts (volcanic peaks >1 km high)
2. the **oceanic ridges**

Off passive continental margins, a **continental rise** (Fig. 12.1) is present at the base of the slope. The continental rise is a gently sloping surface that leads gradually onto the deep ocean floor and is built in part from submarine fans extending seaward from the foot of the slope. It commonly has little relief other than that resulting from the presence of incised submarine canyons and protruding seamounts. Continental rises are generally absent on convergent or active margins where subduction is taking place, such as along much of the Pacific margin. On margins of this type, a long, arcuate **deep-sea trench** commonly lies at the foot of the continental slope, and the rise is absent. Trenches in less active subduction zones, such as along Oregon-Washington, may be filled with sediment. **Abyssal plains** are extensive, nearly flat areas punctuated here and there by seamounts. Some abyssal plains are also cut by deep-sea channels, as mentioned. **Mid-ocean ridges** extend across some 60,000 km of the modern ocean and overall make up about 30 to 35 percent of the area of the ocean. Mid-ocean ridges are particularly prominent in the Atlantic, where they rise about 2.5 km above the abyssal plains on either side. Rocks on these ridges are predominantly volcanic, and the ridges are cut by numerous transverse fracture zones along which significant lateral displacements may be apparent. Ridges play a crucial role in the seafloor spreading process, but they are not particularly active areas of sedimentation. They do have a very important effect on circulation of deep bottom currents in the ocean and thus have an indirect effect on sedimentation in the deep ocean.

Transport and Depositional Processes to and within Deep Water

Transport across the Shelf. Most sediment deposited in deeper water originates on the shelf and must make its way across the shelf (Fig. 12.18) to get to deeper-water environments. As described under shelf transport, fine sediment can move across the shelf as freshwater surface plumes. Clay and fine silt are separated from sands in the surf zone during river outflow. The fine sediment may move outward, as a freshwater plume over the denser seawater, across the shelf, and into deeper water as far as 100 km offshore (Reineck and Singh, 1980) before mixing and flocculation cause the clay particles to settle. Movement of fine sediment to great distances off the shelf also takes place by nepheloid transport, which involves intermittent deposition, resuspension, and seaward migration of fine sediment in near-bottom turbid layers. Storm waves resuspend fine bottom sediment from the shelf floor, creating a turbid near-bottom layer of suspended sediment that may reach several tens of meters in thickness, particularly on the outer shelf. This suspended sediment can then be carried across and off the shelf owing to seaward-directed, wind-driven flow or the seasonal production of dense, cool, saline bottom waters that flow offshore under the influence of gravity (McGrail and Carnes, 1983).

Coarser sediment has a more difficult time crossing the shelf to deeper water. Where submarine canyons cut the shelf and head up close to shore, sandy sediment may move laterally along the coast by longshore transport into the heads of these canyons (Fig. 12.18) and then down the canyons to deeper water by sediment gravity-flow processes. If submarine canyons are absent on the shelf, sandy sediments probably cannot cross the barrier created by deeper water of the outer shelf, although tidal currents, turbidity currents, or seaward-directed geostrophic currents may move sands some distance seaward onto the inner shelf, as discussed. Therefore, during periods of high sea level such as the present, sands tend to be trapped in the nearshore zone and prevented from moving across and off the shelf to deeper water. By contrast, during low stands of the sea, rivers can flow across the subaerially exposed shelf and dump coarse sediment into the heads of submarine canyons that originate near the shelf break. Thus, sand transport onto the continental slope, and subsequent retransport by turbidity currents to deeper water, is commonly much greater during periods of low sea level.

Transport in Deeper Water beyond the Shelf. Processes capable of transporting sediment further into deeper water beyond the shelf edge can be grouped into the following categories:

1. suspension transport by near-surface water and by wind
2. near-bottom nepheloid-layer transport
3. tidal-current transport in submarine canyons
4. sediment gravity flows and other mass-transport processes, particularly in submarine canyons
5. transport by geostrophic contour currents
6. transport by floating ice

In addition, sedimentation in deep ocean basins occurs by the rain of dead pelagic organisms from near-surface waters and by the air fall and submarine settling of pyroclastic particles generated by explosive volcanism within and outside the ocean basin. Also, some sediment can be retransported off oceanic ridges and seamounts after accumulating there by pelagic rain or volcanic processes.

Suspension, Nepheloid, and Wind Transport. Where continental shelves are narrow, freshwater surface plumes carrying fine sediment across the shelf can move it considerable distances into deeper water before the sediment settles, as mentioned. Nepheloid-layer transport of fine-grained sediment is important in deeper water as well as on the shelf. Fine-grained sediment is injected into the water column owing to erosion of the seabed by bottom currents (to be discussed) and possibly by turbidity currents. It is then transported seaward by these currents while still in suspension. Because fine sediment can remain in suspension in nepheloid layers for periods ranging from weeks to months, perhaps even years, it travels much farther than does coarser bottom sediment, which may also be eroded and transported by bottom currents. Nepheloid layers in the modern ocean extend seaward for hundreds of kilometers and to water depths of 6000 m or more. Winds can also transport fine suspended dust particles seaward, where they settle out over the ocean hundreds of kilometers from shore. In fact, wind transport may be the primary mechanism by which siliciclastic pelagic sediment is transported to the deep ocean.

Tidal Currents. Tidal currents measured in submarine canyons at depths exceeding 1000 m may be capable of transporting silt and fine sand (Shepard, 1979; Pickering, Hiscott, and Hein 1989, p. 143–146). Two types of currents have been detected in submarine valleys:

1. ordinary tidal currents that rarely exceed 50 cm/s and that flow alternately up and down the valley in response to tidal reversal
2. occasional surges of strong downcurrent flow with velocities up to 100 cm/s

Shepard (1979) interprets the surges as low-velocity turbidity currents. There are few data as yet on the quantitative importance of net downcanyon transport owing to tidal currents; however, surge currents of the magnitude measured by Shepard are certainly capable of transporting fine sediment seaward.

Turbidity Currents and Other Mass-Transport Processes. Catastrophic or surge-type, high-velocity turbidity currents generated on the shelf or upper slope are probably the single most important mechanism for transporting sands and gravels to deeper water through submarine channels. On passive margins and in back-arc basins, the deposits of these flows spread out from the mouths of the canyons onto the deep seafloor to form deep-sea fans (Fig. 12.18), and they contribute in part to building of the continental rise. In the fore-arc region of active margins, submarine canyons discharge turbidites into fore-arc basins on the slope or into deep-sea trenches where they may spread out along the canyon axis. Downcanyon grain flow of beach sands swept into the heads of nearshore submarine canyons during storms and submarine debris flows may also be locally important. In addition to these processes, other mass-transport processes—such as creep, gliding (sliding), and slumping—appear to be responsible for large-scale en masse sediment transport on oversteepened continental slopes and ridge slopes. Some of these slump masses can be of enormous size—up to 300 m thick and 100 km long. The importance of these mass-transport processes as agents of slope modification and basin filling is recognized in a 1982 volume entitled *Marine Slides and other Mass Movements* edited by Saxov and Nieuwenhuis. See also Bouma, Normark, and Barnes (1985) and, for an updated look at turbidity-current processes, Normark and Piper (1991).

Contour Currents. Density differences in surface ocean water owing to temperature or salinity variations create vertical circulation of water masses in the ocean commonly referred to as **thermohaline circulation.** Circulation is initiated primarily at high latitudes as cold surface waters sink toward the bottom, forming deep water masses that flow along the ocean floor as bottom currents. The path of these bottom currents is influenced by the position of oceanic ridges and rises and other topographic features such as narrow passages through fracture zones. Owing to density stratification of ocean water, bottom currents adjacent to continental margins tend to flow parallel to depth contours or isobaths and thus are often called **contour currents.** The movement of these currents is also affected by the Coriolis force which likewise tends to deflect them into paths parallel to depth contours; thus, they are sometimes also called geostrophic contour currents. Because contour currents are best developed in areas of steep topography where the bottom topography extends through the greatest thickness of stratified water column (Kennett, 1982), they are particularly important on the continental slope and rise. Photographs of the deep seafloor have revealed the presence of current ripples in some areas and suspended sediment clouds and seafloor erosional features in others, both of which suggest that some contour currents can achieve velocities on or near the seafloor great enough to erode the seabed and transport sediment.

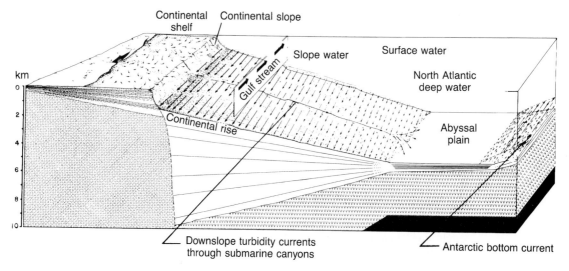

FIGURE 12.19 Schematic diagram illustrating the shaping of the continental rise of eastern North America by contour currents. (After Heezen, B. C., C. D. Hollister, and W. F. Ruddiman, 1966, Shaping the continental rise by deep geostrophic contour currents: Science, v. 152. Fig. 4, p. 507, reprinted by permission of American Association for the Advancement of Science, Washington, D.C. Copyright 1966 by the AAAS.)

Evidence is now available (e.g., Hollister and Nowell, 1991) that suggests that the speed of these bottom currents may be accelerated in some parts of the ocean to velocities on the order of 40 cm/s, owing perhaps to the superimposed influence of large-scale, wind-driven circulation at the ocean's surface. That is, eddy kinetic energy may be transmitted from the surface of the ocean to the deep seafloor. Resulting motions near the seafloor are so energetic that they have been referred to as "abyssal storms" or "benthic storms," particularly because huge amounts of fine sediments are stirred up and transported by these energetic pulses. Contour currents are believed to have had a particularly important role in shaping and modifying continental rises, such as those off the eastern coast of North America (Fig. 12.19).

Floating Ice. During glacial episodes of the Pleistocene when sea level was low and many land areas were covered by ice, rafting of sediment of all sizes into deeper water by icebergs was a particularly important transport process. Ice transport is still going on today on a more limited scale at high latitudes in the Arctic and Antarctic regions (e.g., Kempema, Reimnitz, and Barnes, 1988). Melting of the floating ice dumps sediments of mixed sizes, commonly referred to as glacial-marine sediment (Chapter 10), onto the shelf and the deep ocean floor. The overall quantitative significance of iceberg transport through geologic time has probably not been significant, but locally and at certain times it was quite important.

Pelagic Rain. Calcareous- and siliceous-shelled planktonic organisms settle through the ocean water column to the seafloor upon death, a process called **pelagic rain.** The geographic distribution of these organisms in surface waters is affected by nutrients and prevailing ocean currents. After the tiny shells settle onto the ocean floor, they may be retransported by turbidity currents or contour currents. These skeletal materials form extensive biogenic deposits, or **oozes,** in some areas of the modern ocean floor.

They are also important contributors to ancient deep ocean sediments, particularly in Jurassic and younger rocks, forming thick deposits of chalk, radiolarian chert, and diatomite (Chapter 8).

Explosive Volcanism. Volcanism within and along the margins of marine basins may contribute important quantities of sediment to both the shelf and deeper water, particularly near volcanic arcs. Volcanic ash, lapilli, and bombs can be ejected both subaerially and subaqueously. Coarse material ejected subaerially tends to be deposited by air fall close to the eruption column on all sides of the vent. If strong prevailing winds are blowing during eruption, fine ash will be carried downwind for considerable distances before settling. Pyroclastic particles ejected beneath the sea as well as air-fall particles can be dispersed still more widely within the ocean basin by various transport processes. Pumice may even be dispersed to some extent by floating on the ocean surface.

Principal Kinds of Modern Deep-Sea Sediments

Little agreement exists regarding the classification of deep-sea sediments. Suggested classifications range from those that are largely genetic (Shepard, 1973; Berger, 1974) to those that are largely descriptive (Dean, Leinen, and Stow, 1985; Pickering, Hiscott, and Hein, 1989, p. 43). No entirely satisfactory scheme that gives due regard to both genesis and descriptive properties of all kinds of deep-sea sediments has yet been devised. Two broad classes of deep-sea sediment, **terrigenous** and **pelagic,** are often mentioned; however, these two terms are difficult to define precisely. Terrigenous deposits include gravel, sand, and mud derived from land and transported within the more proximate parts of the deep ocean by a variety of processes (e.g., turbidity currents, contour currents, ice rafting). Some pelagic deposits (clays) are also derived from land but are deposited by slow settling in the more distal parts of the ocean; others consist of pelagic biogenic remains that rain down from near-surface waters. A third minor category of deeper-water sediments are shallow-water carbonate sediments (Chapter 13) that have been retransported from the shelf into deeper water **(allochthonous deep-water carbonates).** Additional details of the siliciclastic deep-sea sediments shown in Table 12.1 are discussed below.

TABLE 12.1 Principal kinds of deep-sea sediments

Terrigenous siliciclastic deposits
> **Hemipelagic mud**—mixtures of terrigenous mud and biogenic remains; deposited from nepheloid plumes and by suspension settling and pelagic rain-out
> **Turbidites**—graded gravel/sand/mud; deposited by turbidity currents
> **Contourites**—sandy or muddy sediments deposited and/or reworked by contour currents
> **Glacial-marine sediments**—Gravel, sand, and mud deposited by ice rafting
> **Slump and slide deposits**—Terrigenous or pelagic deposits emplaced downslope by mass-wasting processes

Pelagic deposits
> **Pelagic clay**—>2/3 siliciclastic clay; deposited by suspension settling and authigenic formation of clay minerals
> **Oozes**—>2/3 planktonic biogenic remains; deposited by pelagic rain-out
>> **Calcareous**—dominantly $CaCO_3$ biogenic remains
>> **Siliceous**—dominantly SiO_2 biogenic remains

Allochthonous deep-sea carbonates
> Shallow-water carbonates emplaced downslope by storms or sediment gravity flows

---

Terrigenous Sediments

A wide variety of deep-sea siliciclastic sediments are grouped under this heading, including, as shown in Table 12.1, hemipelagic muds, turbidites and other sediment gravity-flow deposits, contourites, glacial-marine sediments, and slumps and slides. These deposits are all composed mainly of siliciclastic materials, which may range in size from gravel to clay. Some deposits exhibit well-developed bedding and may or may not display vertical size grading. Others have a disorganized fabric with poorly developed bedding. For convenience of discussion, these terrigenous deposits are grouped below under mainly genetic subheadings that reflect their principal mode of transport and deposition.

Hemipelagic Muds. Hemipelagic ("half-pelagic") deposits are difficult to define precisely. According to Stow and Piper (1984a), they are muddy deposits that contain more than 5 percent biogenic remains and a terrigenous component of more than 40 percent silt, although some geologists may find this definition too restrictive. They are deposited under very low current velocities (e.g., by suspension settling), and they are probably the principal kind of sediment deposited on continental slopes and in fore-arc basins on slope. Much of the compendium volume on fine-grained sediments and deep-water processes and facies edited by Stow and Piper (1984b) is devoted to discussion of hemipelagic sediments.

Hemipelagic muds range in color from gray to green and, more rarely, to reddish brown. Textures range from clay to silty, sandy clay. They are commonly composed of fine terrigenous quartz, feldspar, micas, and clay minerals and/or volcanogenic sediments such as ash, fine pumice, and palagonite. Volcanic ash or glass may be intermixed with other hemipelagic sediment or concentrated into distinct **tephra** layers that range in thickness from less than 1 cm to more than 25 cm (e.g., Boggs, 1984). Granules and pebbles of pumice may occur as isolated fragments, pockets, or distinct layers associated with hemipelagic muds. Hemipelagic muds may also contain the remains of siliceous organisms, particularly diatoms, and calcareous organisms such as foraminfers and nannofossils, as well as fine lime muds swept off carbonate platforms into deeper water.

Hemipelagic muds are poorly laminated to massive and generally are moderately to highly bioturbated. In general, they are deposited closer to shore than are pelagic muds. They are widely distributed on continental slopes of volcanic arcs, such as those of the Western Pacific. They also occur in back-arc basins, on inner trench walls, and on the tops of some rises. Hemipelagic muds are probably deposited mainly from nepheloid layers and plumes. Deposition may be aided in some settings by planktonic organisms that aggregate fine sediment into fecal pellets that settle rapidly.

Turbidites. The general characteristics of turbidites are described in Chapter 5 of this book. Turbidites may occur in the lower reaches of submarine canyons and farther seaward in deep-sea channels, but most are deposited in broad, cone-shaped fans. These turbidite fans, or submarine fans, spread outward on the seafloor from the mouths of canyons. Where submarine canyons are closely spaced along the slope, the fans at the base of the slope may coalesce to build a broad, gently sloping continental rise. On active margins where a trench is present, turbidites commonly occur on the trench floor throughout the length of the trench owing to deposition from turbidity currents flowing longitudinally through the trench. Most turbidites are composed of sands, silty sands, or gravelly sands interbedded with pelagic clays. They are commonly characterized by normal size grading and may or may not display complete Bouma sequences

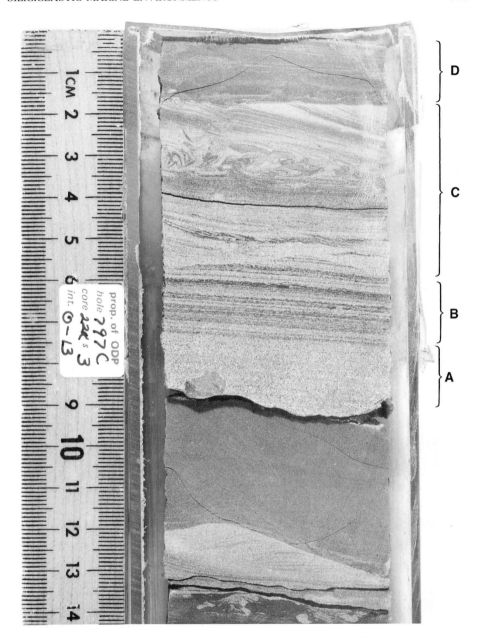

FIGURE 12.20 Graded volcaniclastic turbidite with Bouma divisions marked, from an Ocean Drilling Program (ODP) Leg 127 core of Miocene sediment in the Japan Sea back-arc basin. (Photograph courtesy of Ocean Drilling Program, Texas A & M University.)

(see Fig. 3.28). Figure 12.20 is an example of a sediment core composed of turbidites. As discussed in Chapter 3, many turbidites lack either the basal (A unit) or upper (D–E unit) part of the Bouma sequence, or both. Sole markings—including flute casts, groove casts, and load casts—are common on the base of many turbidite sequences. In addition to sandy turbidites, mud turbidites occur on many parts of the modern ocean floor. These turbidites are composed of normally graded silt and clay that may be either laminated or massive and that commonly lack extensive bioturbation. Turbidites are widely distributed in the modern ocean on passive margins and in both the back-arc and the fore-arc regions of active margins.

    Because the main depositional environments of turbidites are submarine fans, and most modern turbidites occur in such fans, geologists have displayed considerable interest in developing a general model for submarine fan sedimentation. An early fan model by Normark (1978), which divides fans into upper-fan, mid-fan, and lower-fan segments (Fig. 12.21), has been widely quoted and republished. Other earlier fan models include those of Walker (1978) and Mutti (1985). Mutti proposed three stages of fan

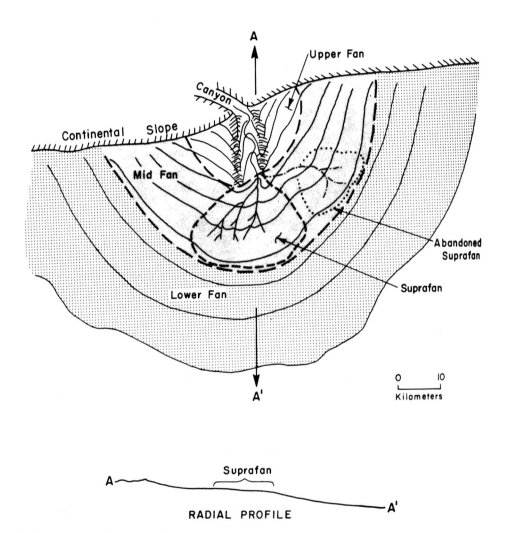

FIGURE 12.21   Schematic representation of submarine fan model proposed by Normark. This model emphasizes active and abandoned depositional lobes called suprafans. (From Normark, R. W., 1978, Fan valleys, channels, and depositional lobes on modern submarine fans: Characters for recognition of sandy turbidite environments: Am. Assoc. Petroleum Geologists Bull., v. 62. Fig. 1, p. 914, reprinted by permission of AAPG, Tulsa, Okla.)

FIGURE 12.22 Three types of submarine fans related (right) to three proposed stages of fan evolution during rising sea level. (After Mutti, E., 1985, Turbidite systems and their relation to depositional sequences, *in* G. G. Zuffa (ed.), Provenance of arenites: D. Reidel Publishing Co., Dordrecht. Fig. 2, p. 69, and Fig. 4, p. 73. Reprinted by permission of Kluwer Academic Publishers. As modified slightly by R. G. Walker, 1992; reproduced by permission.)

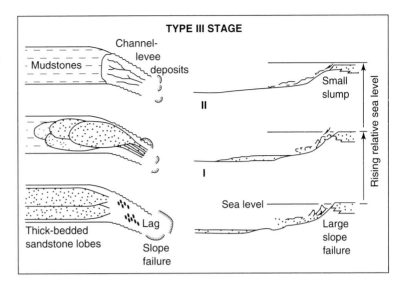

evolution that he relates to sea-level change (Fig. 12.22). Through the 1980s and into the 1990s, extensive research has been carried out on modern submarine fans through use of sidescan sonar and seismic profiling (Chapter 15). These studies have revealed previously unknown details of the surface morphology and internal stratigraphy of fans that cast doubt on the validity of these earlier sedimentologic fan models. In a review and synopsis of new information uncovered by these techniques, Walker (1992) recognizes four large-scale facies associations common to modern fans:

1. extensive channel-levee systems
2. continuous, unchannelized, sheetlike deposits (overbank or basin plain)
3. mass-transport complexes
4. slumps and debris flows associated with levee failure

Clearly, submarine fans are built by a combination of processes rather than by deposition from turbidity currents alone.

Modern submarine fans occur in many parts of the world ocean. The largest of these fans include the Amazon fan in the South Atlantic (330,000 km^2 surface area; Manley and Flood, 1988); the Rhone fan (70,000 km^2) in the Mediterranean Sea (Droz and Bellaiche, 1985); the Indus fan (1,100,000 km^2) off the coast of Pakistan and India (Kolla and Coumes, 1985); the Laurential fan (300,000 km^2) off the Grand Banks of New Foundland (Piper, Stow, and Normark, 1984); and the Mississippi fan (300,000 km^2) in the Gulf of Mexico (Weimer, 1989). Figure 12.23 shows the general shape and size of the Mississippi fan.

Attempts to relate the characteristics of modern fans to ancient turbidite systems are complicated by problems of scale (many features on modern fans are much larger than corresponding features in ancient fan systems) and by the nature of the features themselves (e.g., the abundance of channel-levee systems and the scarcity of depositional lobes in modern fans). For example, the presence of thickening- and thinning-upward successions in ancient turbidites has commonly been attributed respectively to progradation of fan lobes and gradual channel filling and abandonment. This interpretation is difficult to reconcile with the scarcity of lobes in modern fans (Walker, 1992).

FIGURE 12.23 Location and general configuration of the youngest lobe of the Plio-Pleistocene Mississippi fan, as determined from seismic stratigraphic information (Chapter 15). Note the submarine fan channel, which extends more than 500 km seaward from the shelf break. (After Weimer, P., 1989, Sequence stratigraphy of the Mississippi Fan (Plio-Pleistocene), Gulf of Mexico: Geo-Marine Letters, v. 9. © Springer-Verlag. Fig. 82, p. 264; as redrawn by Walker, R. G., 1992, Facies, facies models and modern stratigraphic methods, *in* R. G. Walker and N. P. James (eds.), Facies models: Response to sea level change: Geol. Assoc. Canada, Fig. 26.)

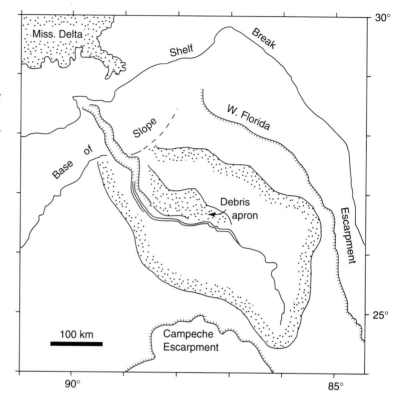

Thus, it appears that no noncontroversial model for turbidite fans, based on modern fans, is available to help us interpret ancient fan deposits. Those readers who wish to further pursue the subjects of turbidites and submarine fans will find the literature on turbidites voluminous. A good place to start is the compendium volumes of Bouma, Normark, and Barnes (1985); Osborne (1991); Stow and Piper (1984b); and Weimer and Link (1991). Also see the textbooks on deep-marine environments by Kennett (1982) and Pickering, Hiscott, and Hein (1989).

Contourites. Coring of sediments on continental rises has shown that in addition to turbidites, rise sediments also include both sandy and muddy sediments interpreted to be deposited by contour currents. Sandy contourites occur either as thin, irregular layers (<1–5 cm) or as thicker layers (5–25 cm), which are either structureless or thoroughly bioturbated or have some primary horizontal and cross-lamination preserved (Stow, 1986). The grain size is mainly fine sand to coarse silt, and sorting is poor to moderate. Contourites may have either sharp or gradational bed contacts, and vertical size grading may or may not be present. The composition can include both terrigenous constituents and biogenic components. Muddy contourites are silty to sandy clays. They tend to be homogeneous, or structureless, and thoroughly bioturbated, but some muddy contourites may display irregular lamination and lensing. Muddy contourites may be difficult to distinguish from hemipelagites. Sandy and muddy contourites commonly occur interbedded in vertical successions (Fig. 12.24). See Pickering, Hiscott, and Hein (1989, p. 219–245) for extended discussion of contourites.

FIGURE 12.24 Core of contourite sediments from the North American Atlantic continental rise off New England in 4746-m water. Note intercalations of dark mudstone and coarser, cross-laminated siltstone. The core is about 13 cm long. (From Heezen, B. C., and C. D. Hollister, 1971, The face of the deep. Fig. 9.66, p. 398, reproduced by permission of Oxford University Press, New York. Photograph courtesy of C. D. Hollister.)

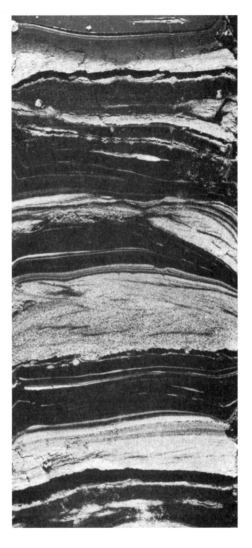

As discussed under a preceding section "Contour Currents," recent work has shown that modern contour currents develop velocities high enough to stir up and transport huge quantities of fine sediments during what is called abyssal or benthic "storms" (Hollister and Nowell, 1991). It thus appears that transport and deposition of sediment by contour currents may be even more important than previously believed. In the modern ocean, contourites appear to be particularly well developed on the continental rise off eastern North America (Fig. 12.24).

Glacial-marine Sediments. Sediments ice-rafted to deep water are typically poorly sorted gravelly sands or gravelly muds that show crude to well-developed stratification. The coarse fraction may include angular, faceted, and striated pebbles. Significant areas of the modern ocean floor at high latitudes are covered by these glacial-marine sediments, particularly the subpolar North Atlantic, the circum-Antarctic, some parts of the Arctic Ocean, the North Pacific, and the Norwegian Sea (Fig. 12.25). Glacial-marine deposits are discussed in further detail under "Glacial Systems" in Chapter 10.

Slump and Slide Deposits. These deposits consist of previously sedimented pelagic or terrigenous deposits that have been emplaced downslope owing to mass-movement processes. During the transport process, consistency of the slump masses is disturbed, resulting in faulted, contorted, and chaotic bedding and internal structure. Studies of the ocean floor with sidescan sonar and bottom and sub-bottom accoustical profiling show that slump and slide deposits are particularly common on continental slopes with high rates of deposition, such as off the Mississippi and Rhone deltas, and on slopes with glacial-marine deposits. Nardin et al. (1979) studied mass-flow deposits in the Santa Cruz Basin, California Borderland, using seismic reflection profiling, and they identified two main types of gravity deposits based on accoustical signatures on reflection records:

1. slides, in which failure took place elastically and only minor internal deformation of strata occurred
2. various kinds of failed masses, which were emplaced plastically and are characterized by different degrees of internal deformation

Much more study is needed to better understand the characteristics and volumetric importance of slump and slide deposits; however, emplacement of sediment along some modern lower continental slopes by mass-transport processes appears to be the dominant sediment transport process.

Pelagic Sediments

The term **pelagic sediment** has been defined in different ways, but it is generally taken to mean sediment deposited far from land influence by slow settling of particles suspended in the water column. Pelagic sediments may be composed dominantly of clay-size particles of terrigenous or volcanogenic origin, or they may contain significant amounts of silt- to sand-size planktonic biogenic remains. **Pelagic clays** are siliciclastic muds that contain clay minerals, zeolites, iron oxides, and windblown dust or ash. They are commonly red to red-brown owing to oxidation by oxygen-bearing deep waters in areas of very slow sedimentation. These clays cover vast areas of the deeper parts of the ocean below about 4500 m (Fig. 12.25). Pelagic sediments that contain significant quantities of biogenic remains are called oozes. Little agreement exists with regard to the amount of biogenic remains required to qualify a sediment as an ooze. In Table 12.1, I suggest that oozes have more than two-thirds biogenic components. Oozes composed predominantly of $CaCO_3$ tests are **calcareous oozes;** those composed mainly of siliceous tests are **siliceous oozes.** Calcareous oozes are dominated by the tests of foraminifers and nannofossils such as coccoliths, but they may include somewhat larger fossils such as petropods, which are planktonic molluscs. Calcareous oozes are widespread in the modern deep ocean at depths shallower than about 4500 m, the calcium carbonate compensation depth (Chapter 8), particularly in the Atlantic Ocean (Fig. 12.25). In deeper ocean basins such as the Pacific, they may occur on the shallow tops of ridges and rises. Lithified equivalents of calcareous oozes are called chalks (limestones).

Siliceous oozes are particularly abundant in the modern ocean at high latitudes in a belt more than 200 km wide stretching across the ocean (Fig. 12.25). They occur also in some equatorial regions of upwelling where nutrients are abundant and productivity of siliceous organisms is high. Siliceous oozes are composed primarily of the

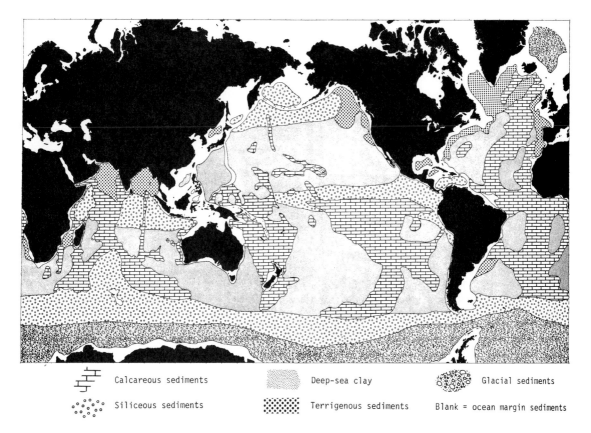

FIGURE 12.25 Distribution and dominant types of deep-sea sediments in the modern ocean. (From Davies, T. A., and D. S. Gorsline, 1976, Oceanic sediments and sedimentary processes, *in* J. P. Riley and R. Chester (eds.), Chemical oceanography, v. 5, 2nd ed. Fig. 24.7, p. 26, reprinted by permission of Academic Press, Orlando, Fla.)

remains of diatoms and radiolarians but may include other siliceous organisms such as silicoflagellates and sponge spicules. Diatom oozes occur mainly in high-latitude areas and along some continental margins, whereas radiolarian oozes are more characteristic of equatorial areas. Planktonic sediments that settle onto steep slopes on seamounts, ridges, and so on, may be retransported to adjacent basins by turbidity currents or by slumping and sliding. Excellent discussions of pelagic environments and sediments are given in Jenkyns (1986); Kennett (1982); Scholle, Arthur, and Ekdale (1983); and Stow and Piper (1984b).

Ancient Deep-Sea Sediments

As mentioned, deep-sea sedimentary deposits other than turbidites are not as abundant in the rock record as shallow-water sediments because the potential for preservation and uplift of these sediments above sea level is much lower. Nonetheless, they are known from stratigraphic units of all ages. Typically, deep-sea sedimentary rocks consist predominantly of pelagic and hemipelagic shales, turbidite sandstones and conglomerates, bedded cherts (recrystallized siliceous oozes), chalks and marls (lithified, clayey, pelagic calcareous oozes), limestone breccias (slope deposits), and carbonate turbidites (Chapter 13). Except for turbidites and carbonate breccias, which may be very coarse-grained deposits, they are distinguished in general by their fine grain size. Other than turbidites, most deep-sea deposits do not show vertical facies successions that change upward in any fixed order.

Physical sedimentary structures consist predominantly of thin, horizontal laminations, although rippled bedding and graded bedding are common in turbidites, and cross-lamination occurs in some contourites. The bedding of many deep-sea deposits is well developed, even, and laterally persistent. Deep-water turbidites commonly display repetitive, well-bedded successions of thin, graded units that are often referred to as **rhythmites.** Such successions are also called **flysch** facies. Colors of deep-water sediments are typically dark gray to black. Red pelagic shales are much rarer. Muds may be well bioturbated or essentially nonbioturbated, and they are commonly characterized by distinctive deep-water trace-fossil associations. Fine-grained deep-water sediments are characterized also by the presence of much greater concentrations of planktonic organisms than occur in shallow-water sediments. These organisms include diatoms, radiolarians, foraminifers, coccoliths, and, in older rocks, graptolites and ammonites. Deep-water sedimentary rocks occur in extensive tabular- or blanket-shaped deposits and may be underlain by ocean crustal rocks such as submarine basalts and ophiolite assemblages consisting of serpentized peridotite, dunite, gabbros, sheeted dikes, and pillow lavas.

The Cretaceous Rosario Group in Southern California (Nilsen and Abbot, 1981) provides a good example of an ancient submarine fan succession that illustrates some of the characteristics of deep-water siliciclastic deposits that include coarse-grained sediments. The Rosario Group in the San Diego area consists of the Point Loma Formation and the overlying Cabrillo Formation (Fig. 12.26). The basal Point Loma Formation is composed of a thick, massive mudstone unit with thin interbeds of graded siltstone and very fine-grained sandstone. The mudstones are thoroughly bioturbated, are rich in carbonaceous materials, and are inferred to be probable slope and basin-plain deposits. Overlying the mudstones is a distinctive unit composed of interbedded turbidite sandstones and mudstones, forming a series of thickening- and coarsening-upward successions. The thicker and coarser upper sandstone beds are locally scoured and channeled, whereas the thinner and finer lower beds are planar and more highly bioturbated. Abundant carbonaceous matter, mica, and megafossil fragments are present in this facies. Nilsen and Abbot (1981) interpret this unit as the deposits of outer-fan sandstone lobes. The uppermost part of the Point Loma Formation consists dominantly of well-defined, repetitive, thinning- and fining-upward sandstone successions with some mudstone interbeds. The thicker lower sandstones are locally pebbly and may contain dish structures, parallel laminae, wood fragments, and mudstone rip-up clasts. They range from nongraded to normally graded to reversely graded. The thinner upper sandstone beds contain abundant parallel laminae, convolute laminae, ripple markings, flute and groove casts, mica flakes, carbonaceous matter, and bioturbation structures. These deposits are interpreted as largely channel-fill sandstones deposited in a middle-fan environment. The basal part of the overlying Cabrillo Formation consists of thick beds of conglomerate with some interbedded sandstone. These coarse-grained sediments are inner-fan channel deposits that prograded over the middle-fan sandstone channel deposits of the Point Loma Formation. The inner-fan conglomerates are overlain by a retrogradational succession of mid-fan channel sandstones. The inferred paleogeographic setting in which the Rosario Group was deposited is shown in Figure 12.27. Lateral progradation and retrogradation of the turbidite fan generated the vertical succession of facies illustrated in Figure 12.26.

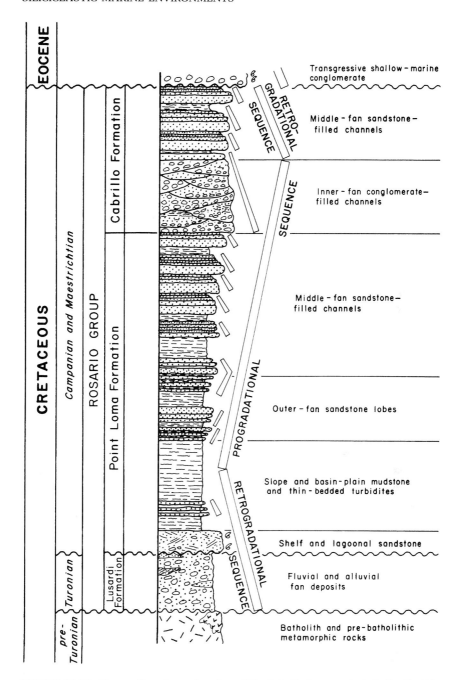

FIGURE 12.26 Composite columnal section of the Late Cretaceous strata in the San Diego area, California, showing progradational and retrogradational cycles of submarine fan deposition. (From Nilsen, T. H., and P. A. Abbott, 1981, Paleogeography and sedimentology of Upper Cretaceous turbidites, San Diego, California: Am. Assoc. Petroleum Geologists Bull., v. 65. Fig. 19, p. 1280, reprinted by permission of AAPG, Tulsa, Okla.)

FIGURE 12.27 Paleogeographic map of the San Diego area, California, in Late Cretaceous time, showing the inferred depositional setting of the submarine fan and associated deposits shown in Figure 12.26. (From Nilsen, T. H., and P. A. Abbott, 1981, Paleogeography and sedimentology of Upper Cretaceous turbidites, San Diego, California: Am. Assoc. Petroleum Geologists Bull., v. 65. Fig. 20, p. 1281, reprinted by permission of AAPG, Tulsa, Okla.)

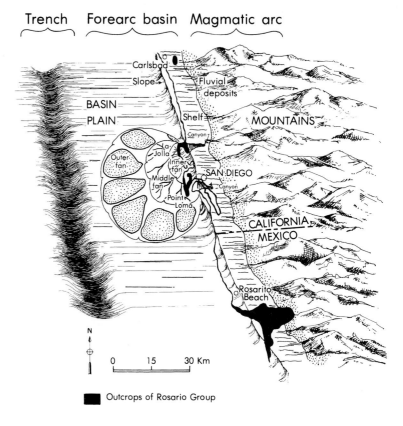

The Miocene Monterey Formation of California provides a well-known example of an ancient deep-water deposit that consists largely of lithified planktonic biogenic remains and pelagic clays. The Monterey Formation lies on the Early Miocene Rincon Shale (Fig. 12.28), which is massive to locally laminated, clay-rich shale. The Monterey can be divided into five major lithofacies or informal members (Isaacs, 1984). The **lower calcareous-siliceous member** consists mainly of irregularly laminated to massive and thick-bedded strata composed predominantly of lithified clayey, planktonic calcareous (foraminifers, coccoliths) and siliceous (mainly diatoms) organic remains. This unit is succeeded upward by the **carbonaceous (and phosphatic) marl member,** composed of organic-rich, thick shale units (to 15 m) with interbeds of diatomite or diatomaceous shale and calcareous rocks. The Carbonaceous marl member is gradational upward to the much thinner **transitional marl-siliceous member** and the **upper calcareous-siliceous member.** These two units are made up of various proportions of calcareous porcellanite, chert, shale, calcareous diatomaceous shale, and diatomite. The thick, uppermost **clayey-siliceous member** is composed of meter-thick units of regularly laminated, clay-poor siliceous rock (chert) interbedded with meter-thick units of massive, clay-rich siliceous rock.

The Monterey Formation rocks are pelagic/hemipelagic deposits laid down mainly in upper-slope to basinal environments (Pisciotto and Garrison, 1981). They were deposited under conditions of variable but generally high planktonic productivity, for both calcareous and siliceous organisms, in subsurface water masses conducive

FIGURE 12.28 Generalized lithostratigraphic column showing the principal hemipelagic/pelagic members of the Miocene Monterey Formation in the Santa Barbara coastal area, California. Ma = million of years before the present. (After Isaacs, C. M., 1984, Hemipelagic deposits in a Miocene basin, California: Toward a model of lithologic ariation and sequence, *in* D. A. V. Stow and D. J. W. Piper (eds.), Fine-grained sediments: Deep-water processes and facies: Blackwell Scientific Publications. Fig. 9, p. 285, reproduced by permission.)

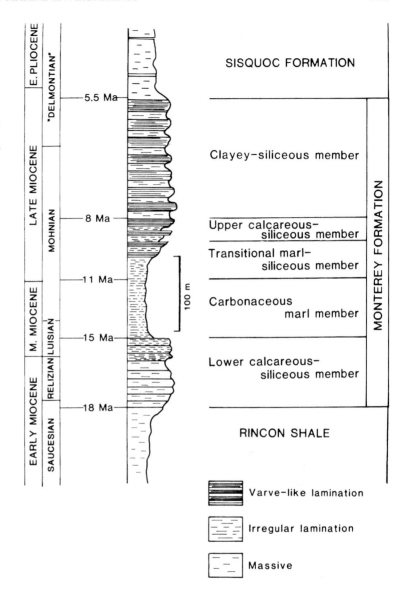

to preservation of organic-rich sediments. Subsequent diagenesis during burial converted the siliceous organic remains to diatomites and cherts, and the foraminifers and coccoliths to lithified carbonate rock (chalk).

FURTHER READINGS

Siliciclastic Shelf Systems

Aigner, T., 1985, Storm depositional systems: Springer-Verlag, Berlin, 174 p.

Aigner, T., and R. H. Dott (eds.), 1990, Processes and patterns in epeiric basins: Sed. Geology, v. 69, p. 165–313.

De Boer, P. L., A. van Gelder, and S. D. Nio (eds.), 1987, Tide-influenced sedimentary environments and facies: D. Reidel Publishing Co., Dordrecht, 530 p.

Greenwood, B., and R. A. Davis, Jr., 1984, Hydrodynamics and sedimentation in wave-dominated coastal environments: Elsevier, New York, 473 p.

MacDonald, D. I. M. (ed.), 1991, Sedimentation, tectonics and eustasy: Sea-level changes at active margins: Internat. Assoc. Sedimentologists Spec. Pub. 12, Blackwell, Oxford, 518 p.

Moslow, T. F., 1985, Depositional models of shelf and shoreline sandstones: Am. Assoc. Petroleum Geologists Ed. Course Notes Series 27, Tulsa, Okla.

Nittrouer, C. A. (ed.), 1981, Sedimentary dynamics of continental shelves: Elsevier, Amsterdam, 449 p.

Smith, G. G., G. E. Reinson, B. A. Zaitlin, and R. A. Rahmani, 1991, Clastic tidal sedimentology: Canadian Soc. Petroleum Geologists Mem. 16, 387 p.

Stride, A. H. (ed.), Offshore tidal sands: Processes and deposits: Chapman and Hall, London, 222 p.

Swift, D. J. P., G. F. Oertel, R. W. Tillman, and J. A. Thorne (eds.), 1991, Shelf sand and sandstone bodies: Internat. Assoc. Sedimentologists Spec. Pub. 14, Blackwell, Oxford, 532 p.

The Open University Team, 1989, Waves, tides and shallow-water processes: Pergamon, Oxford, 187 p.

Tillman, R. W., and C. T. Siemers (eds.), 1984, Siliciclastic shelf sediments: Soc. Econ. Paleontologists and Mineralogists Spec. Pub. 34, Tulsa, Okla., 268 p.

Tillman, R. W., D. J. P. Swift, and R. G. Walker (eds.), 1985, Shelf sands and sandstone reservoirs: Soc. Econ. Paleontologists and Mineralogists Short Course Notes No. 13, 708 p.

Wilgus, C. K., B. S. Hastings, C. G. St. C. Kendall, H. W. Posamentier, C. A. Ross, and J. C. Van Wagoner (eds.), 1988, Sea-level changes: An integrated approach: Soc. Econ. Paleontologists and Mineralogists Spec. Pub. 42, 407 p.

Continental Slope and Deep-Sea Systems

Bouma, A. H., W. R. Normark, and N. E. Barnes (eds.), 1985, Submarine fans and related turbidite systems: Springer-Verlag, New York, 351 p.

Dott, R. H., and R. H. Shaver (eds.), 1974, Modern and ancient geosynclinal sedimentation: Soc. Econ. Paleontologists and Mineralogists Spec. Pub. 19, Tulsa, Okla., 380 p.

Doyle, L. J., and D. H. Pilkey (eds.), 1979, Geology of continental slopes: Soc. Econ. Paleontologists and Mineralogists Spec. Pub. 27, Tulsa, Okla., 374 p.

Hay, W. W. (ed.), 1974, Studies in paleooceanography: Soc. Econ. Paleontologists and Mineralogists Spec. Pub. 20, Tulsa, Okla., 218 p.

Heezen, B. C., and Hollister, C. D., 1971, The face of the deep: Oxford University Press, New York, 659 p.

Hsü, K. J., and H. C. Jenkyns (eds.), 1974, Pelagic sedimentation on land and under the sea: Internat. Assoc. Sedimentologists Spec. Pub. 1, Blackwell, London, 447 p.

Lisitzin, A. P., 1972, Sedimentation in the world ocean: Soc. Econ. Paleontologists and Mineralogists Spec. Pub. 17, Tulsa, Okla., 218 p.

Middleton, G. V., and A. H. Bouma, (eds.), 1973, Turbidites and deep water sedimentation: Soc. Econ. Paleontologists and Mineralogists, Pacific Section Short Course, Anaheim, Calif., 157 p.

Nowell, A. R. M. (ed.), 1991, Deep ocean transport: Marine Geology, v. 89, no. 3/4, p. 275–460.

Osborne, R. H. (ed.), 1991, From shoreline to abyss: Contributions in marine geology in honor of Francis Parker Shepard: SEPM (Society for Sedimentary Geology) Spec. Pub. 46, 320 p.

Pickering, K. T., R. N. Hiscott, and F. J. Hein, 1989, Deep marine environments: Clastic sedimentation and tectonics: Unwin Hyman, London, 416 p.

Saxov, S., and J. K. Nieuwenhuis (eds.), 1982, Marine slides and other mass movements: Plenum, New York, 353 p.

Stanley, D. J., and G. T. Moore (eds.), 1983, The shelf-break: Critical interface on continental margins: Soc. Econ. Paleontologists and Mineralogists Spec. Pub. 33, Tulsa, Okla., 467 p.

Stow, D. A. V., and D. J. W. Piper (eds.), 1984, Fine-grained sediments: Deep-water processes and facies: The Geological Society, Blackwell, Oxford, 659 p.

Warme, J. E., R. G. Douglas, and E. L. Winterer, 1981, The Deep Sea Drilling Project: A decade of progress: Soc. Econ. Paleontologists and Mineralogists Spec. Pub. 32, Tulsa, Okla., 564 p.

Weimer, P., and M. H. Link (eds.), 1991, Seismic facies and sedimentary processes of submarine fans and turbidite systems: Springer-Verlag, New York, 447 p.

13

Carbonate Marine Environments

13.1 INTRODUCTION

Carbonate deposits constitute the dominant sediment cover on some modern continental shelves. These shelves are located primarily at low latitudes in clear, shallow, tropical to subtropical seas (Fig. 13.1) where little terrigenous siliciclastic detritus is introduced. Tropical carbonate-producing environments with low terrigenous input are found today on some shelves attached to the mainland, such as Florida Bay and western Australia. They occur also in a few smaller shelf areas that surround oceanic islands where terrigenous influx is extremely low. The Bahama Platform and the narrow shelves around Pacific atolls are examples. Carbonate sediments also form on some higher-latitude (30°–60°), cool-water shelves, where they consist predominantly of shell remains (Lees and Buller, 1972; Farrow, Allen, and Akpan, 1984). Temperate carbonate environments include the shelf off southern Australia between 32° and 40°+ south latitude, portions of the northwest European shelf, and the Orkney shelf off northeast Scotland.

The relatively minor importance of modern carbonate deposition is decidedly atypical of many geologic periods of the past when widespread deposition of carbonate sediments characterized sedimentation in broad epeiric seas hundreds of kilometers wide. During the middle Paleozoic, for example, carbonate deposition prevailed in shallow inland seas that spread over much of the continental interior of North America. In spite of the small areal extent of modern shelf carbonate environments, carbonate-dominated shelves nonetheless provide outstanding "laboratories" for studying the mechanisms of carbonate sedimentation. Much of what we now understand about carbonate textures and the basic processes of carbonate deposition has come from study of modern carbonate environments. On the other hand, we must turn to the ancient rock record itself for insight into the environmental conditions that typified carbonate-dominated epeiric seas.

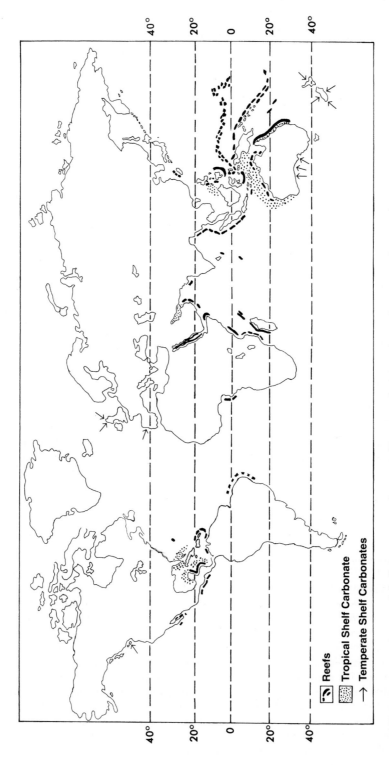

FIGURE 13.1 Distribution of shallow-marine carbonate sediments and reefs in the modern ocean. (From Wilson, J. L., 1975, Carbonate facies in geologic history. Fig. 1.1, p. 2, reprinted by permission of Springer-Verlag, Heidelberg.)

462

Geologists have been greatly interested in carbonate rocks for well over one hundred years. In fact, the science of microscopic petrography was initiated by an English geologist named Henry Clifton Sorby, who began his petrographic studies about 1851 with the study of limestones. Other historically interesting early studies of carbonates include investigation of carbonate sediments in the Bahamas by Black (1933) and Cayeux's (1935) classic publication on the carbonate rocks of France. Modern study of carbonate sediments and depositional processes is generally regarded to have begun in the 1950s with the publications of Newall et al. (1951), Illing (1954), and Ginsburg (1956) dealing with modern carbonate sediments in the Bahamas and Florida Bay. Since that time, the pace of research on carbonates has accelerated at an astounding rate, and dozens of books and hundreds of research papers have been devoted to both modern and ancient carbonate sediments and depositional processes. Many of these books tend to be narrowly focused on special aspects of carbonate sedimentology, but several with good general coverage of carbonate environments have been published. Probably the most comprehensive of these is the symposium volume *Carbonate Depositional Environments* (Scholle, Bebout, and Moore, 1983). Others include Bathurst (1975); Cook, Hine, and Mullins, (1983); Crevello et al. (1989); Frost, Weiss, and Saunders (1977); Hardie (1977); Laporte (1974); Logan et al. (1970, 1974); Milliman (1974); Scoffin (1987); Tucker et al. (1990); Tucker and Wright (1990); and Wilson (1975). Tropical shelf carbonate depositional environments have been summarized in shorter papers by Cook (1983), Enos (1983), Hine (1983), James and Kendall (1992), James and Borque (1992), Sellwood (1986), and Wilson and Jordan (1983). Temperate carbonates are discussed by Farrow, Allen, and Akpan (1984); Lees and Buller (1972); and Nelson (1988).

13.2 CARBONATE SHELF (NONREEF) ENVIRONMENTS

Depositional Setting

Carbonate sediments are deposited primarily on shallow-marine shelf platforms, including, in the geologic past, broad epeiric platforms covered by shallow water. The geologic record suggests that carbonate platforms can occur on the margins of cratonic blocks, in intracratonic basins, across the tops of major offshore banks, and on localized positive features on wide shelves (Wilson and Jordan, 1983). Carbonate environments may be present also in some parts of marginal-marine environments such as beaches, lagoons, and tidal flats. With respect to the nature of the platform edge, three basic types of carbonate platforms are recognized (Fig. 13.2): (1) rimmed carbonate shelves; (2) unrimmed carbonate platforms, including open shelves and carbonate ramps; and (3) isolated carbonate platforms (Harris, Moore, and Wilson, 1985; James and Kendall, 1992; Read, 1982, 1985).

Rimmed carbonate shelves are shallow platforms marked at their outer edges (margins) by a pronounced break in slope into deeper water. They have a nearly continuous rim or barrier along the platform edge. This barrier consists of either a reef buildup or a skeletal/ooid sand shoal that absorbs wave action and may restrict water circulation, creating a low-energy shelf environment, sometimes called a "lagoon," landward of the shelf-edge barrier. The lagoon commonly grades landward into a low-energy tidal-flat environment rather than a high-energy beach zone. An **unrimmed platform** has no pronounced marginal barrier. Unrimmed platforms occur today on the leeward side of large tropical banks and in all cool-water carbonate settings (James and Kendall, 1992). A **ramp** is a gently sloping (<~1°) unrimmed platform on which shallow-

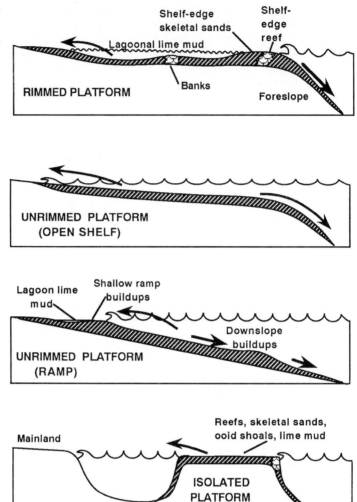

water deposits pass downslope with only a slight break in slope into deeper-water facies. The break in slope on a ramp is not marked by a pronounced reef trend, but discontinuous sand shoals may be present along the shelf edge where water energy is high. Water circulation across an unrimmed platform may be adequate to allow development of a moderately high-energy beach zone alongshore in addition to formation of skeletal or ooid-pellet sand shoals along the shelf edge. Thus, unrimmed carbonate platforms are affected by much the same physical processes as siliciclastic shelves. **Isolated platforms** (Bahama type) are shallow-water platforms tens to hundreds of kilometers wide, commonly located offshore from shallow continental shelves, surrounded by deep water that may range from several hundreds of meters to a few kilometers deep. The platforms may have gently sloping, ramplike margins or more steeply sloping margins resembling those of rimmed shelves.

Carbonate workers subdivide the carbonate shelf environment into inner, middle, and outer shelves (Fig. 13.3). The **inner shelf** includes nearshore shallow-water to subaerial environments such as beaches and tidal flats, which are environments not considered part of the shelf in our previous discussion of siliciclastic shelves. The **middle shelf** (the "lagoon" mentioned above) encompasses the shallow, subtidal zone lying between the nearshore areas and the shelf break. The **outer shelf** is a very narrow zone that constitutes the shelf break and that may encompass reef buildups or carbonate sand shoals along the shelf edge. Wilson (1975) recognized nine "facies belts" in carbonate environments extending seaward from the supratidal zone to the slope and basin. The shelf encompasses facies belts 4 through 9 (Fig. 13.3). Note that the above subdivisions of the shelf differ from those commonly used for siliciclastic shelves (Fig. 12.3). Furthermore, the term basin as used by carbonate sedimentologists simply signifies a deepening of water beyond the shelf edge to below normal wave base. The term does not necessarily imply that the carbonate basin environment is analogous to the

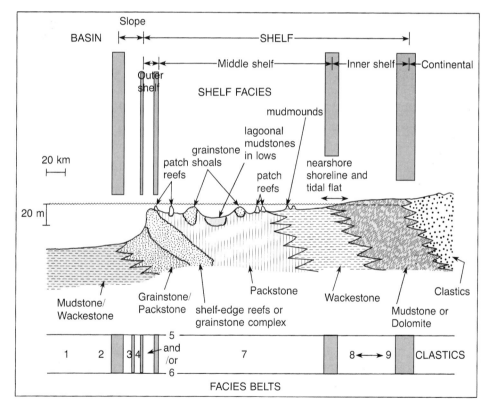

FIGURE 13.3 Schematic profile of the carbonate shelf environment. Note subdivisions of the shelf into major subdivisions, and contrast this subdivision with that of siliciclastic shelves (Fig. 12.3). This figure shows also the approximate positions of Wilson's (1975) carbonate facies belts: 1, basin; 2, open-sea shelf; 3, deep-shelf margin; 4, foreslope; 5, organic buildups; 6, winnowed platform edge (sands); 7, open-circulation shelf; 8, restricted-circulation shelf and tidal flats; and 9, evaporite sabhkas-salinas. (After Wilson, J. L., and C. Jordan, 1983, Middle shelf environment, *in* P. A. Scholle, D. G. Bebout, and C. H. Moore (eds.), Carbonate depositional environments: Am. Assoc. Petroleum Geologists Mem. 33. Fig. 1a, p. 298, reprinted by permission of AAPG, Tulsa, Okla.)

deep oceanic environment. In fact, carbonate basins generally lie well above abyssal depth, commonly less than a few hundred meters below sea level.

In contrast to most siliciclastic shelves, many carbonate platforms, particularly rimmed platforms, are characterized by some kind of topographic buildup at the shelf margin of the outer shelf. This buildup may be caused by the presence of organic reefs or banks, lime sand shoals, or small islands that create a barrier to incoming waves. This outer barrier is commonly dissected by a network of tidal channels that allow high-velocity tidal currents to flow through onto the shelf. Water depth may be only a few meters over this buildup, but depth increases over the middle shelf (Fig. 13.3) to perhaps several tens of meters. The outer shelf is the highest-energy zone of the shelf. Much of the middle shelf is commonly below fair-weather wave base. Water energy is thus low over most of the middle shelf except over patch reefs, localized banks, or shoals and along the shoreline of some carbonate ramp platforms. The elevation and lateral continuity of the shelf-edge carbonate barrier control water circulation over the entire shelf. The effect of this barrier on water circulation, coupled with the width of the shelf, strongly influences the type and distribution of carbonate facies that develop on the shelf. If a well-developed barrier is present, or if the shelf is very wide, water circulation on the shelf is restricted (Wilson and Jordan, 1983). This is so because, on wide shelves, water energy is expended in friction with the bottom, leading to poor water circulation. Restricted water circulation leads to development of salinity conditions that deviate from normal (~35o/oo). Salinities may rise well above normal in arid or semiarid climates where evaporation rates are high, or they may fall below normal in areas affected by considerable freshwater runoff. Variations in salinity affect the diversity and numbers of organisms living on shelves; the organisms, in turn, strongly affect carbonate deposition owing to the extremely important role they play in carbonate sedimentation processes (Chapter 7). The inner shelf is especially characterized by restricted conditions.

Although carbonate environments extend from the supratidal zone to deeper basins off the shelf, the shallow platform basin that constitutes the middle and outer shelves is the primary site of carbonate production. James (1984a) refers to this platform as the "subtidal carbonate factory" (Fig. 13.4). The sediments produced in this carbonate factory are deposited mainly on the shelf; however, some sediments are eventually transported landward onto tidal flats and beaches and into subtidal settings. Others are transported seaward off the shelf onto the slope and into the deeper basin. Little carbonate sediment is generated in the deeper-water basin environment off the shelf except for fallout of calcium carbonate–secreting plankton from near-surface waters.

FIGURE 13.4 The main areas of marine carbonate production. Most carbonates accumulate in water less than 30 m deep—the "subtidal carbonate factory." (From James, N. P., 1984, Introduction to carbonate facies models, *in* R. G. Walker (ed.), Facies models, 2nd ed.: Geoscience Canada Reprint Ser. 1. Fig. 2, p. 210, reprinted by permission of Geological Association of Canada.)

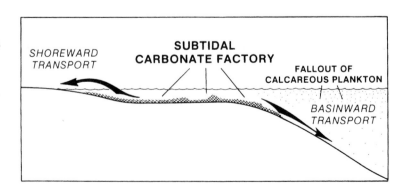

Sedimentation Processes

Unlike the deposition of siliciclastic shelf sediments—a process controlled primarily by physical processes—carbonate deposition is controlled by a combination of physical, chemical, biochemical, and biological processes (Chapter 7). Also in contrast to siliciclastic deposition, deposition of carbonate sediments is largely an autochthonous process in the sense that carbonate sediments are generated primarily within the same basin in which they are deposited. Siliciclastic sediments are largely allochthonous materials derived from extrabasinal sources. Carbonate muds and carbonate grains such as ooids, pellets, and fossils are formed mainly through chemical and biochemical processes, aided secondarily by physical processes such as water agitation. After formation, they commonly undergo some physical transport before final deposition.

Chemical and Biochemical Processes. The principal chemical and biological/biochemical controls on carbonate deposition are discussed in Chapter 7; they are reviewed only briefly here. Although the solubility of calcium carbonate is controlled by pH, temperature, and carbon dioxide content of seawater, the actual importance of chemical (inorganic) precipitation of calcium carbonate in the modern and ancient oceans is not definitely known (e.g., Shinn et al., 1989); that is, we do not know what percentage of modern (or ancient) carbonate sediment was deposited owing to inorganic processes. Carbonate deposition brought about by organisms capable of extracting calcium carbonate from the seawater to build their shells or skeletal structures may be a more important process in the modern ocean than are purely inorganic processes. Organisms also contribute to the formation of carbonate sediment through their feeding and bioturbation activities, which cause breakdown of skeletal fragments and other carbonate materials and generate various kinds of trace fossils. It is worth stressing that the organisms primarily responsible for carbonate production in the modern ocean are not necessarily the same as those that were major carbonate formers in the past. Figure 13.5 shows the relative importance of some major groups of organisms as carbonate formers during Phanerozoic time. Note from this figure that the principal carbonate formers have changed somewhat with time. For example, crinoids, bryozoans, and bra-

FIGURE 13.5 The approximate diversity, abundance, and relative importance of various calcareous marine organisms as sediment producers. P, Paleozoic; M, Mesozoic; C, Cenozoic. (Modified from Wilkinson, B. H., 1979, Biomineralization, paleooceanography, and evolution of calcareous marine organisms: Geology, v. 7, Fig. 1, p. 526. Published by Geological Society of America, Boulder, Co.)

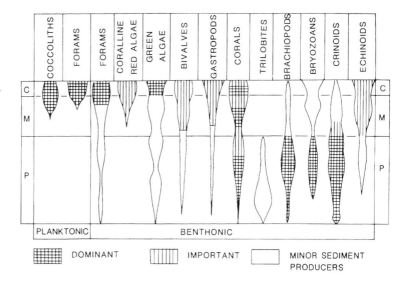

chiopods were more important during the Paleozoic than during the Cenozoic, whereas coccoliths, planktonic foraminifers, coralline algae, and green algae were particularly important carbonate formers during the Cenozoic.

Physical Processes. Physical processes are important primarily in the reworking and transport of carbonate materials on the shelf, but they also aid in the production of carbonate sediments. Circulation of water onto the shelf brings fresh nutrients, necessary for organic growth, from deeper water. Breaking of waves against reef barriers on the outer shelf increases oxygen content in the water by interaction with the atmosphere and decreases CO_2 owing to decreased water pressure. Thus, modern reefs are best developed in wave-agitated zones, and biogenic production of carbonate sediment in general is stimulated by strong water movement. On the other hand, strong waves crashing on the reef front cause breakdown of reef rock, producing sand- and gravel-size bioclasts that subsequently undergo transport both seaward and landward from the reef.

Agitated water is important to the formation of ooids, and currents aid in the generation and preservation of grapestones and hardened fecal pellets by submarine accretion and cementation. Waves and currents also winnow fine carbonate mud from coarser sediment and transport this mud off the shelf platform or into sheltered or protected areas of the shelf. Depending upon water energy, the coarser sediment itself may either remain as a winnowed lag deposit, forming sand- or gravel-covered flats, or be transported and deposited to create wave-formed bars and shoals, beaches, spits, or tidal deltas and bars. Wave- and current-transported and winnowed carbonate sand deposits are particularly common along the outer edge of the shelf platform, where water energy is highest. In resuspension and transport of sediment, storms are as important on carbonate shelves as they are on siliciclastic shelves. For example, most transport of sediment from the subtidal shelf into the intertidal (tidal-flat) environment is accomplished by storms. Absence of wave and current activity on the shelf leads to stagnant circulation, with consequent deviations from normal salinity and possibly anoxic conditions. Such restricted environments constitute unfavorable habitats for many normal marine organisms.

Sediment Characteristics

The deposition of carbonate sediments is favored in moderately shallow, warm water with low terrigenous siliciclastic input. Although carbonates form predominantly in warm-water settings, they can accumulate also in some cool-water, higher-latitude environments. In these cool-water environments, the carbonate sediment is composed almost entirely of the skeletal remains of organisms. Cool-water assemblages of organic remains are referred to as **foramol** assemblages (Lees and Buller, 1972; Jones and Desrochers, 1992), named for the dominance of foraminifers and molluscs. They are composed of benthic (bottom-dwelling) foraminifers, molluscs, barnacles, bryozoans, and calcareous red algae. By contrast, warm-water (>18°C) assemblages of organisms, called **chlorozoan** assemblages (named from chlorophyta plus zoantharia corals), are dominated by hermatypic corals (corals that live primarily in the photic zone) and calcareous green algae in addition to foramol components. Cool-water carbonates make important contributions to the deposits of some modern shelves (e.g., Farrow, Allen, and Akpan, 1984; Nelson, 1988) and may likewise have been important in ancient shelf environments.

Warm-water carbonates may include, in addition to skeletal remains, substantial amounts of ooids, aggregate grains, peloids, and lime mud. Table 13.1 provides a more

TABLE 13.1   Modern warm- and cool-water marine organisms and their counterparts in the fossil record

Modern, warm-water	Modern, cool-water	Ancient counterpart	Sedimentary aspect
Corals	Absent	Corals, stromatoporoids, stromatolites, coralline sponges, rudist bivalves	Large components of reefs and biogenic mounds
Bivalves, red algae, echinoderms	Bivalves, red algae, brachiopods, echinoderms, barnacles	Red algae, brachiopods, cephalopods, trilobites	Remain whole or break apart into several pieces to form sand- and gravel-size particles
Gastropods, benthic foraminifera	Gastropods, benthic foraminifera	Gastropods, benthic foraminifera	Whole skeletons that form sand- and gravel-size particles
Green (codiacean) and red algae	Red algae, bryozoans	Phylloid algae, crinoids and other echinoderms, bryozoans	Spontaneously disintegrate upon death to form many sand-size particles
Ooids, peloids	Absent	Ooids, peloids	Concentrically laminated or micritic sand-size particles
Planktonic foraminifera, coccoliths, pteropods	Planktonic foraminifera, coccoliths, pteropods	Planktonic foraminifera, coccoliths (post-Jurassic), styliolinids	Medium sand-size and smaller particles in basinal deposits
Encrusting foraminifera, red algae, bryozoans	Encrusting foraminifera, red algae, bryozoans, serpulid worms	Red algae, renalcids, encrusting foraminifera, bryozoans	Encrust on or inside hard substrates; build up thick deposits or fall off upon death to form sand grains
Dasyclad green algae	Absent	Dasyclad green algae	Spontaneously disintegrate upon death to form lime mud
Cyanobacteria and other calcimicrobes	Cyanobacteria and other calcimicrobes	Cyanobacteria and other calcimicrobes (especially pre-Ordovician)	Trap, bind, and precipitate fine-grained sediments to form mats and stromatolites or thrombolites

Source: James, N. P., and A. C. Kendall, 1992, Introduction to carbonate and evaporite facies models, *in* R. G. Walker and N. P. James (eds.), Facies models—Response to sea level change: Geol. Assoc. Canada. Table 2, p. 269.

complete listing of modern warm-water and cool-water organisms and their ancient counterparts. This table also suggests the manner in which these organisms contribute to the makeup of carbonate sediment. Notice the extremely important roles that organisms play in the formation of carbonate sediment.

As stated, carbonate sediment accumulates primarily in shallow-water settings; however, some carbonates are deposited in deeper water on foreslopes and in basins. Most carbonate sediment deposited in deeper water results from the fallout of calcareous plankton—foraminifers, green algae (coccoliths), and tiny gastropods (Fig. 13.4). These pelagic calcareous organisms evolved mainly in Jurassic and post-Jurassic time; therefore, deeper-water pelagic carbonates are not important in older rocks. In addition to pelagic carbonate, some shallow-water carbonate sediment may be swept off carbonate platforms into deeper water by storm waves or be transported by sediment gravity-flow processes (e.g., turbidity currents).

Modern Shallow Platform Environments

The plan-view setting of carbonate platforms typical of many modern rimmed carbonate shelves is illustrated in Figure 13.6. This figure graphically depicts the different subenvironments of the platform environment, each of which is characterized by specific environmental conditions and carbonate facies.

The outer shelf is commonly the highest-energy environment of the shelf. It is characterized by the development of lime sand or gravel sheets and shoals landward of platform margin reefs. These deposits are lag deposits composed of skeletal fragments of shallow-water organisms such as corals, calcareous algae, and bryozoa and/or ooids and peloids. Much of the skeletal material is derived from adjacent organic reefs and is commonly abraded and rounded by wave action. Sand shoal deposits are generally well sorted and cross-bedded. Facies of ancient carbonate rocks representative of this zone are mainly bioclastic and oolitic grainstones or packstones (some lime-mud matrix).

The middle shelf is a zone of generally low water energy, particularly on rimmed shelves. Much of the shelf floor is below wave base; however, local patch reefs, shoals, or banks extend above wave base. Water circulation ranges from open to partially restricted, depending upon the conditions of the shelf. Typically, the outer part of the middle shelf is characterized by open circulation, whereas the inner part of the shelf experiences somewhat more restricted circulation. This is a zone of high carbonate production of skeletal sands, lime muds, peloids, and grapestones. Organisms that populate the shelf are typically normal marine low-tolerance forms, but more tolerant forms may inhabit restricted areas of the shelf. Typical shelf fauna include brachiopods, pelecypods, gastropods, crinoids, echinoids, calcareous algae, bryozoans, benthic foraminifers, and, in older rocks, ammonites and nautiloids. Bioturbation is commonly extensive. Owing to the generally low water energy over the middle shelf, the sediments are typically poorly winnowed, with a high ratio of micrite to skeletal fragments and other grains. This condition may not be true of some carbonate ramps where higher-energy conditions exist across the platform and carbonate sands may be

FIGURE 13.6 Schematic plan view of a modern, rimmed, tropical carbonate platform showing the relative positions of reefs, lime-sand shoals, the lagoon, and tidal flats. Note that tidal flats occur both adjacent to land and in the lee of lime-sand shoals. (Based in part on James, N. P., 1984, Shallowing-upward sequences in carbonate rocks, *in* R. G. Walker (ed.), Facies models, 2nd ed.: Geoscience Canada Reprint Ser. 1. Fig. 5, p. 215, reprinted by permission of Geological Association of Canada.)

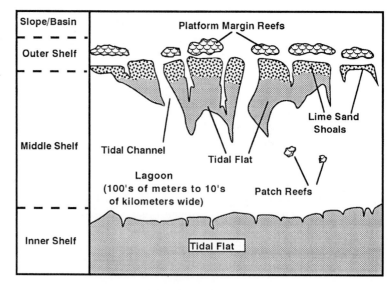

developed over much of the shelf. Ancient facies of the middle shelf zone are represented mainly by mudstones (micrites) and skeletal or pelleted wackestones (biomicrites, biopelmicrites, pelmicrites), although accumulations of better-winnowed grainstones or packstones may occur also. Sediments of this facies are very common in the geologic record.

The inner shelf in most carbonate environments is a low-energy, tidal-flat environment (Fig. 13.6) in which predominantly fine-grained sediments accumulate; however, on some ramp platforms, a higher-energy nearshore zone may be present where carbonate beaches or lime-sand shoals develop. The distinguishing physical characteristics of tidal-flat environments are described in Chapter 11. Carbonate sediments formed in this environment are sometimes referred to as **peritidal carbonates** (e.g., Pratt and James, 1992). Carbonate sediments deposited in the subtidal and lower intertidal portions of tidal flats are similar to those formed on the inner part of the middle shelf in that they consist generally of pelleted mudstones and some skeletal wackestones that are commonly burrowed and bioturbated. Where salinities are near normal, tidal ponds, creeks, and other water bodies support a restricted but prolific population of gastropods and foraminifers (James, 1984b). These gastropods graze on blue-green algae (cyanobacteria) and keep it cropped to the point that algal mats cannot flourish. In hypersaline areas where gastropods cannot survive, prolific growth of algal mats can occur.

The middle and upper parts of the intertidal zone are characterized by thin, graded, lime-mud layers, probably deposited by storm waves, and extensive development of algal (cyanobacteria) mats (James, 1984b; Pratt and James, 1992). Both mud layers and algal mats are disrupted by well-developed desiccation cracks. Bedding tends to be irregular, with algal mat layers alternating with graded storm layers. Trapping of fine lime mud by algal mats and repeated renewal and growth of these mats generate finely laminated stromatolites. Parts of the algal mats may rot away during burial, leaving irregular, subhorizontal, elongated cavities called "laminoid fenestrae," or fenestral porosity. Mud crusts formed on intermittently exposed parts of mudflats and tidal channels may be broken off and reworked by incoming tidal currents or storm waves to form intraclasts. Carbonate sediments of the supratidal zone are similar to those of the upper intertidal zone, except that long periods of subaerial exposure cause storm-deposited layers to form lithified surface crusts several centimeters in thickness and fractured into irregular mudcracks. Sediments of the supratidal zone commonly contain some dolomite in association with calcite and aragonite. In hypersaline environments, gypsum, anhydrite, and halite may be present also. Crystallization of evaporite minerals within the sediments can cause doming up of desiccation polygons to form "tepee" structures.

In summary, tidal-flat carbonates are characterized particularly by irregularly to evenly laminated algal mats and associated fenestrae; irregularly bedded, thin, graded storm layers; and desiccation cracks. Other desiccation features, such as raindrop imprints and animal or bird tracks, may be present. Ancient carbonate facies of the inner shelf are represented by pelleted mudstones (micrites), nearly homogeneous mudstones, fenestral (having elongated, subhorizontal pores) laminated mudstones, intraclast-rich wackestones (in channels), stromatolites, and in some deposits nodular anhydrite or gypsum. A good description of inner shelf deposits is provided by Enos (1983). Because of their distinctive characteristics, carbonate tidal flat deposits in the rock record are relatively easily identified.

With the exception of mud chips or intraclasts concentrated in tidal channels, coarse-grained carbonate sediments are relatively rare in the tidal-flat environment. On some carbonate ramp platforms where an outer barrier is absent, a high-energy sand-shoal beach facies may develop along the inner platform edge in lieu of the tidal-flat deposits of low-energy shelves. Carbonate grains on these beaches consist of skeletal fragments, ooids, pellets, and possibly intraclasts ripped up from layers of previously deposited, submarine, cemented lime mud. Inden and Moore (1983) discuss carbonate beaches, including both barrier beaches and mainland beaches, and indicate that sediments of the lower shoreface are commonly coarse-grained, poorly sorted carbonate sands that have a mud matrix and that are characterized by trough cross-bedding. Sediments of the upper shoreface are much better-sorted carbonate sands and gravels that display planar cross-bedding with foresets that dip mainly seaward at angles less than about 15°. Two modern examples of mainland carbonate beaches are present on the Trucial coast of the Persian Gulf (Inden and Moore, 1983).

Examples of Modern Carbonate Platforms

Modern carbonate shelves of both the ramp and the rimmed types have now been studied in considerable detail. Examples of unrimmed shelves or carbonate ramps include the eastern Gulf of Mexico off the Florida coast; the Yucatan Shelf, Mexico, in the southern part of the Gulf of Mexico; and the Persian Gulf. Examples of rimmed shelves include Florida Bay, the Bahama Platform (an isolated platform), the Belize Shelf in the western Caribbean off Guatemala, and the Great Barrier Reef area of Australia. Other important deposits of carbonate sediments in Australian waters occur along the western coast. The characteristics of several of these platforms are summarized by Jones and Desrochers (1992), Sellwood (1986), and Wilson and Jordan (1983). Sediment facies maps of three well-known carbonate shelves show some of the facies-distribution patterns of these types of shelves. Figure 13.7 illustrates an open shelf or carbonate ramp (the West Florida Shelf), and Figure 13.8 shows a rimmed shelf (in South Florida Bay). The best-studied example of a modern isolated platform is the Bahama Platform, which is also rimmed (Fig. 13.9). Note the general progression

FIGURE 13.7 Example of an open-shelf or carbonate ramp—the West Florida Shelf in the eastern Gulf of Mexico. (From Sellwood, B. W., 1978, Shallow-water carbonate environments, *in* H. G. Reading (ed.), Sedimentary environments and facies. Fig. 10.17, p. 276, reprinted by permission of Elsevier Science Publishers, Amsterdam. Originally after R. N. Ginsburg and N. P. James, 1974, Holocene carbonate sediments of continental shelves, *in* C. A. Burk and C. L. Drake (eds.), The geology of continental margins. Fig. 6, p. 140, Springer-Verlag, New York.)

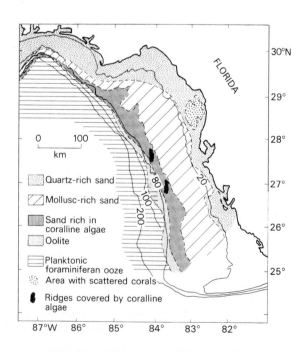

FIGURE 13.8 Example of a rimmed shelf in South Florida Bay. (From Sellwood, B. W., 1978, Shallow-water carbonate environments, *in* H. G. Reading (ed.), Sedimentary environments and facies. Fig. 10.21A, p. 281, reprinted by permission of Elsevier Science Publishers, Amsterdam. Originally after R. N. Ginsburg and N. P. James, 1974, Holocene carbonate sediments of continental shelves, *in* C. A. Burk and C. L. Drake (eds.), The geology of continental margins. Fig. 23, p. 150, Springer-Verlag, New York.)

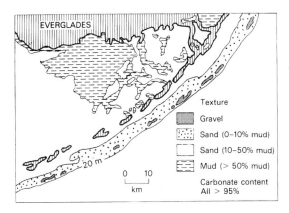

FIGURE 13.9 Sediment distribution on an isolated carbonate platform, the Great Bahama Banks. (From Sellwood, B. W., 1978, Shallow-water carbonate environments, *in* H. G. Reading (ed.), Sedimentary environments and facies. Fig. 10.21C, p. 281, reprinted by permission of Elsevier Science Publishers, Amsterdam. Originally after E. G. Purdy, 1963, Recent calcium carbonate facies of the Great Bahama Banks: Jour. Geology, v. 71, Fig. 1, p. 473.)

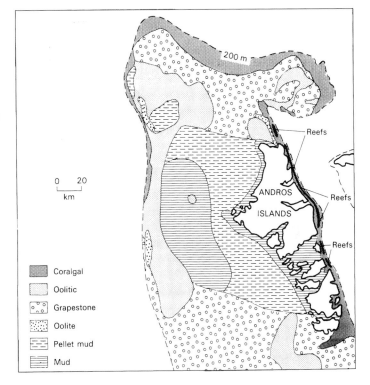

of facies on the rimmed shelves from reef buildups and shelf-edge sands on the outer shelf to carbonate muds and muddy carbonate sands on the middle and inner shelves. By contrast, most of the carbonate ramp in Figure 13.20 is covered by carbonate sand deposits mixed, on this shelf, with some terrigenous quartz sands.

Ancient Carbonate Shelf Successions

Carbonate shelf deposits are abundant in the geologic record in stratigraphic successions ranging in age from middle Precambrian to Holocene. Carbonate deposition was particularly prevalent during the middle Paleozoic, late Mesozoic, and Tertiary, especially in broad epeiric seas as opposed to the small marginal-platform (continental shelf) occurrences in the modern ocean. As indicated, shelf carbonate facies are characterized by distinctive suites of largely normal marine organisms and carbonate textures that are generally muddy, although lithofacies types range from lime mudstones, wackestones, grainstones, and packstones to stromatolitic boundstones and patch-reef boundstones. Shelf carbonates occur in distinctive shallowing-upward successions in which tidal-flat deposits are particularly diagnostic units. Bedding of shelf carbonates is variable, with lens- or wedge-shaped layers common, although some shelf carbonate beds may be laterally extensive. Shelf carbonates are commonly interbedded with thin shale beds. Sedimentary structures include cross-bedding in lime-sand units, extensive bioturbation structures and burrows, and flaser and nodular bedding.

The rock record shows that carbonate rocks are deposited in cyclic successions ranging from a few tens of meters to hundreds of meters thick. Many successions begin with a high-energy carbonate sand or conglomerate unit followed upward progressively in the depositional succession by sediments deposited in the lower-energy, subtidal, open-marine shelf; intertidal zone; supratidal zone; and possible nonmarine environment. Such a succession is basically regressive (progradational); however, because rates of carbonate sedimentation commonly exceed rates of basin subsidence or sea-level rise, sediments also build upward toward sea level. Sediment is thus deposited in progressively shallower water as the sediment surface accretes toward sea level, generating shallowing-upward successions (James, 1984b). Figure 13.10 shows the hypothetical succession of vertical facies that develop on a typical low-energy rimmed shelf as tidal-flat deposits build outward and upward over low-energy, subti-

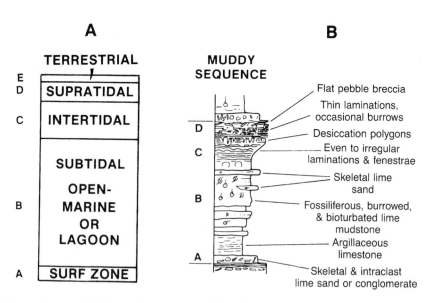

FIGURE 13.10 Hypothetical shallowing-upward sequence on a low-energy, intertidal tropical carbonate shelf. A. Subdivisions of the shallowing-upward model for carbonates. B. Sequence with a low-energy tidal-flat unit developed on a low-energy subtidal unit. (Modified from James, N. P., 1984, Shallowing-upward sequences in carbonate rocks, *in* R. G. Walker (ed.), Facies models, 2nd ed.: Geoscience Canada Reprint Ser. 1. Fig. 2, p. 214, Fig. 9, p. 218, reprinted by permission of Geological Association of Canada.)

FIGURE 13.11 Hypothetical shallowing-upward carbonate-shelf sequence formed under conditions where a high-energy intertidal beach developed adjacent to a low-energy subtidal environment. Letters refer to the subdivisions of the shallowing-upward models shown in Figure 13.10. Calcrete in sediment is cemented by calcium carbonate under subaerial conditions. (After James, N. P., 1984, Shallowing-upward sequence in carbonate rocks, *in* R. G. Walker (ed.), Facies models, 2nd ed.: Geoscience Canada Reprint Ser. 1. Fig. 17, p. 223, reprinted by permission of Geological Association of Canada.)

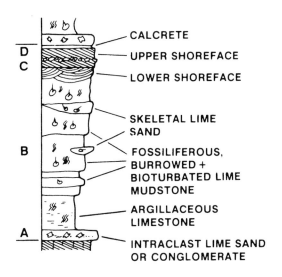

CALCRETE

UPPER SHOREFACE

LOWER SHOREFACE

SKELETAL LIME SAND

FOSSILIFEROUS, BURROWED + BIOTURBATED LIME MUDSTONE

ARGILLACEOUS LIMESTONE

INTRACLAST LIME SAND OR CONGLOMERATE

dal, open-shelf deposits. Figure 13.11 illustrates the hypothetical shallowing-upward succession that would form on a high-energy, nonrimmed, ramp-type shelf where a mainland beach occurs shoreward of the subtidal zone. These two successions can be considered generalized models for vertical successions fomed on rimmed and non-rimmed carbonate shelves; however, considerable deviation from these norms can occur depending upon exact depositional conditions. For example, the presence of shoaling areas within the subtidal area of rimmed shelves or extensive evaporite deposition in the supratidal zone can produce successions that differ in detail from these models.

Repetition of large-scale shallowing-upward successions may be largely the result of repeated episodes of rapid sea-level rise, flooding the carbonate platform, followed by periods of standstill during which shallowing-upward successions develop (Wilkinson, 1982). Osleger and Read (1991) suggest that meter-scale cyclicity is the result of Milankovich-forced sea-level oscillations (see discussion of stratigraphic cycles in Chapter 15), with a cyclicity on the order of 20,000 to 40,000 years. The development of small-scale successions may possibly be influenced also by the supply of carbonate sediment from the subtidal shelf, which is progressively reduced in area as intertidal sediments build out onto the shelf. When the area of the subtidal shelf becomes too small to supply sediments, deposition ceases until basin subsidence again creates a subtidal platform deep enough to supply sediments and begin the cycle anew (Wilkinson, 1982).

The Lower Cretaceous Edwards Formation of central Texas is a carbonate deposit that illustrates many of the characteristic features of carbonate shelf deposits. This formation was deposited on an extensive carbonate platform characterized by carbonate skeletal sands and lime muds. Reefs composed of rudistid clams developed along the platform margin and some parts of the inner shelf (Fig. 13.12). Over the Llano Uplift of central Texas, the Edwards Limestone contains tidal-flat mudstones, foraminiferal grainstones, and other skeletal carbonates. The inner shelf grades seaward over the middle shelf, where a series of high-energy grainstone bars or banks developed not far offshore (Wilson and Jordan, 1983). The banks are composed of oolitic grainstone with skeletal wackestones to the north in front of the bank. Rudistid reefs surrounded by wackestone occur farther offshore. The inner part of the middle shelf behind the banks

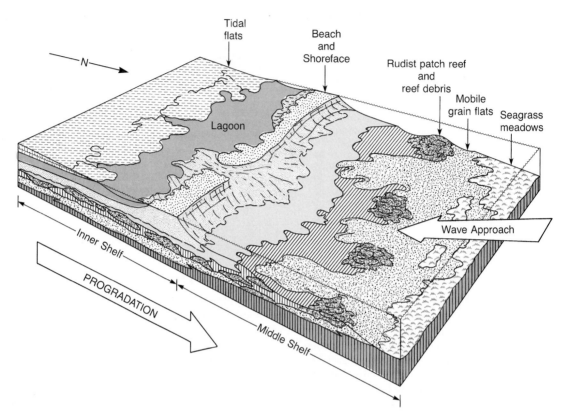

FIGURE 13.12   Depositional model of the Lower Cretaceous Edwards Limestone of Texas. The model shows progradational inner- and middle-shelf facies. (After Kerr, R. S., 1977, Facies, diagenesis, and porosity development in a Lower Cretaceous bank complex, Edwards Limestone, north-central Texas, *in* D. G. Bebout and R. G. Loucks (eds.), Cretaceous carbonates of Texas and Mexico, applications to subsurface exploration: Tex. Bur. Econ. Geology Report. Inv. 89. Fig. 9, p. 223, reprinted by permission of University of Texas, Austin.)

is characterized by scattered rudistid patch reefs; reef-derived skeletal sands; and muddy, pelleted skeletal packstones. These facies grade landward to carbonate sands formed in the beach environment, restricted lagoonal lime muds, and supratidal dolomites. A complete vertical succession of the Edwards Limestones thus grades upward from oolitic, skeletal, reef-derived sands deposited near the shelf margin through middle-shelf pelleted, skeletal packstones to inner-shelf bioclastic beach sands, lagoonal lime muds, and supratidal dolomitic muds. For some additional examples of ancient carbonate platform successions, see Crevello et al. (1989) and Tucker et al. (1990).

13.3   ORGANIC REEF ENVIRONMENTS

As mentioned, the outer shelf of many rimmed platforms is characterized by the presence of nearly continuous carbonate reefs that constitute an effective barrier to wave movement across the shelf. Reefs may be developed also as fringing masses along the shoreline or as isolated patches within the inner shelf (Figs. 13.3 and 13.6). Reefs constitute a unique depositional environment that differs greatly from other parts of the shelf environments. They have been studied intensively for years; however, discussion of reefs has long been plagued by confusion over the precise meaning of the term reef. Carbonate workers have been unable to agree on whether to restrict use of the term **reef** to carbonate buildups or bioherms that have a rigid organic framework or core, built of colonial organisms, or to extend the definition to include carbonate buildups of other types that do not have a rigid-framework core. The word **bioherm** is a nonspecific term used for lenslike bodies of organic origin that are enclosed in rocks of different lithology or character; a bioherm may or may not have a rigid internal organic framework. It carries no connotation of the internal structure or composition of the lens. Likewise, Wilson (1975) uses the term **carbonate buildup** for a body of locally formed, laterally restricted, carbonate sediment that possesses topographic relief—without regard to the internal makeup of the buildup.

Dunham (1970) attempted to solve the nomenclature dilemma by proposing two types of reefs:

1. **ecologic reefs,** which are rigid, wave-resistant topographic structures produced by actively building and sediment-binding organisms
2. **stratigraphic reefs,** characterized simply as thick, laterally restricted masses of pure or largely pure carbonate rock

The reef problem was reviewed again by Heckel (1974), who proposed his own definition of reef in still another effort to generate an acceptable, single meaning for the term. Longman (1981, p. 10) modified Heckel's definition slightly to arrive at the definition of a reef as "any biologically influenced buildup of carbonate sediment which affected deposition in adjacent areas (and thus differed to some degree from surrounding sediments), and stood topographically higher than surrounding sediments during deposition." He suggest using the term **reef complex** for "the specific type of reef having a significant rigid organic framework which generally forms in high wave energy, shallow water environments, as well as the genetically related facies associated with the framework." We have not likely heard the last word in this nomenclature wrangle, but this brief review of the nomenclature problem should at least provide readers with an understanding of the generally accepted meaning of reef, as well as some alternate usages of the term.

Modern Reefs and Reef Environments

Depositional Setting. Most modern reefs occur in shallow water. The most striking of these occurrences are the linear reefs located along platform margins, commonly called **barrier reefs.** These reefs are more or less laterally continuous, and the reef trend may extend for hundreds of kilometers—as, for example, the Great Barrier Reef of Australia which runs for some 1900 km along the eastern shelf of Australia. In a few modern localities where shelves are very narrow, linear reefs are located hard up against the shoreline, with no intervening lagoon, and thus are called **fringing reefs.** Isolated, doughnut-shaped reefs called **atolls** occur around the tops of some Pacific seamounts. These reefs form an outer wave-resistant barrier that encloses a shallow lagoon. Small, isolated reef masses commonly referred to as **patch reefs, pinnacle reefs,** or **table reefs** occur along some shelf margins or scattered on the middle shelf. In addition to these shallow-water reefs, reefs or reeflike carbonate buildups occur also in deeper water. For example, organically produced mounds 100 m long and 50 m high have been reported from the Straits of Florida in water 600 to 700 m deep (Neuman, Kofoed, and Keller, 1977). These mounds are composed of mud-cemented remains of various types of deep-water organisms such as crinoids, ahermatypic hexacorals, and sponges.

Reef Organisms. We tend to think of all reefs as coral reefs; however, many organisms in addition to corals can contribute to the formation of reefs. These organisms include blue-green algae (cyanobacteria), coralline red algae, green algae, encrusting foraminifera, encrusting bryozoa, sponges, and molluscs (Heckel, 1974; James and Macintyre, 1985). In the geologic past, reef-building organisms also included some now extinct groups such as the archaeocyathids, stromatoporoids, fenestellid bryozoans, and rudistid clams. Nonetheless, corals are certainly dominant constituents of modern reefs, and two types of corals are recognized. The principal corals in shallow-water reefs are hermatypic (zoanthellae) hexacorals. Hermatypic corals carry out a symbiotic relationship with several kinds of unicellular organisms, mainly algae, referred to collectively as zooxanthellae. These algae live in or between the living cells of the corals and aid them in gaining energy by producing photosynthetic products (Cowen, 1988). They may also facilitate the process of secreting calcium carbonate by removing CO_2 from the tissues during photosynthesis. Because the zooxanthellae require sunlit waters, hermatypic corals are restricted to living in very shallow water. Ahermatypic (azooxanthellae) corals lack the symbiotic relationship and are not restricted to shallow water. They are the principal organisms today that form carbonate buildups in deeper water. They range in their distribution from shallow water to water depths exceeding 2000 m (Stanley and Cairns, 1988). The growth forms of reef-building organisms are closely related to the water energy over the reef. Organisms that live in low-energy parts of the reef tend to have delicate, branching or platelike forms. Those living in higher-energy zones of the reef develop hemispherical, encrusting, or tabular forms that are better able to withstand strong wave action (Fig. 13.13).

Reef Deposits. We cannot discuss all types of modern reefs and reef facies here; however, we will examine the zoning and facies development of high-energy, platform margin reefs as a general model for high-energy reef environments. Figure 13.14 illus-

FIGURE 13.13 Growth forms of reef-building organisms and the principal types of environments in which they occur. (From James, N. P., 1983, Reef environment, *in* P. A. Scholle, D. G. Bebout, and C. H. Moore (eds.), Carbonate depositional environments: Am. Assoc. Petroleum Geologists Mem. 33. Fig. 59, p. 374, reprinted by permission of AAPG, Tulsa, Okla.)

GROWTH FORM		ENVIRONMENT	
		Wave Energy	Sedimentation
	Delicate, branching	low	high
	Thin, delicate, plate-like	low	low
	Globular, bulbous, columnar	moderate	high
	Robust, dendroid, branching	mod-high	moderate
	Hemispherical, domal irregular, massive	mod-high	low
	Encrusting	intense	low
	Tabular	moderate	low

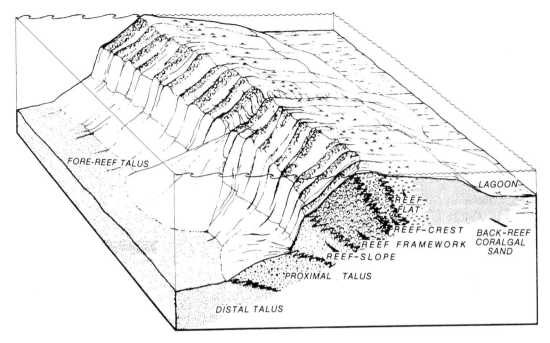

FIGURE 13.14 Idealized facies in a typical modern, mature coral reef with a well-developed reef framework. (From Longman, M. W., 1981, A process approach to recognizing facies of reef complexes, *in* D. F. Toomey (ed.), European fossil reef models: Soc. Econ. Paleontologists and Mineralogists Spec. Pub. 30. Fig. 10, p. 23, reprinted by permission of SEPM, Tulsa, Okla.)

trates schematically the principal facies subdivisions of a platform margin reef. Note that the reef consists of a central core, the **reef framework,** which grades seaward into the **reef slope,** and a loose accumulation of reef debris called the **fore-reef talus.** The nearly flat uppermost, shallowest part of the reef is called the **reef flat,** which grades landward into **back-reef coral-algal sands** and **subtidal lagoonal deposits.** James (1983) subdivides reefs physiographically into **fore-reef, reef-front, reef-crest, reef-flat,** and **back-reef zones** (Fig. 13.15). Figure 13.15 shows also the types of carbonate materials formed in different zones of the reef. The words rudstone, floatstone, bafflestone, bindstone, and framestone in Figure 13.15 are terms used by Embry and Klovan (1971) as modifications of Dunham's (1962) limestone classification (see Table 7.3B). Floatstone and rudstone are unbound carbonate grains, more than 10 percent of which are more than 2 mm in size; floatstones are mud supported; and rudstones are grain supported. Bafflestones are carbonate components bound together at the time of deposition by stalked organisms that trapped sediment by acting as baffles. Bindstones were bound during deposition by encrusting and binding organisms such as encrusting foraminifers and bryozoans, and framestones were bound by organisms such as corals, which build a rigid framework structure.

The water energy, dominant sedimentation processes, types of organisms, percentage of framework components, and grain size and sorting of sediment vary in each zone of the reef. Table 13.2 summarizes these characteristics, following the physiographic subdivisions of reefs illustrated in Figure 13.14. Water energy is highest on the reef crest, which also contains the highest percentage of framework constituents. As water energy decreases toward both the fore reef and the back reef, the percentage of

FIGURE 13.15 Cross-sectional view of a hypothetical zoned marginal reef illustrating the major zones, the principal types of limestones produced, and the growth forms of organisms in different parts of the reef. (From James, N. P., 1984, Reefs, *in* R. G. Walker (ed.), Facies models, 2nd ed.: Geoscience Canada Reprint Ser. 1. Fig. 9, p. 233, reprinted by permission of Geological Association of Canada.)

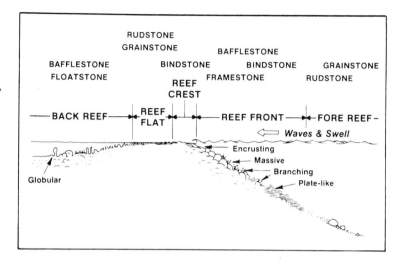

framework constituents also decreases. Note that overall the framework component of reefs is commonly much smaller than the volume of nonframework constituents. Longman (1981) compares the structure of reefs to that of an apple, which has a central core, or framework, surrounded by the much larger edible fruit. The nonframework fraction of reefs consists of organisms such as echinoderms, green algae, and molluscs that do not build framework structures, together with bioclasts broken from the reef by wave activity and, in lower-energy zones of the reef, some lime mud. The fore-reef talus slope and back-reef coral-algal sand zone are made up entirely of nonframework constituents that consist principally of reef-derived bioclasts. Relatively few organisms live in these zones.

Low-Energy Reef Facies. The facies of modern, high-energy, platform margin–type reefs thus consist fundamentally of a central framework core composed largely of corals and coralline algae; the core grades seaward through a zone of rubbly fore-reef talus to deeper-water lime muds or shales and landward through back-reef coral-algal sands to finer-grained lagoonal deposits. This model serves reasonably well for high-energy reefs developed in most settings; however, some reefs form under much lower-energy conditions. Low-energy reefs do not develop the characteristic zoning of high-energy reefs (Fig. 13.16) and tend to be circular to elliptical in plan view. Organisms growing on such reefs are dominated by the more delicate, branching forms (Fig. 13.13). Some low-energy reefs do not contain the typical reef structure described above but are constructed simply of carbonate sands and muds built by organisms that are very similar to reef-type organisms in composition (James, 1984c). Other low-energy buildups are composed largely of nonreef-type organisms. They consist of mound-shaped piles of skeletal fragments and/or bioclastic lime muds rich in skeletal organisms, with minor amounts of organic boundstone. These structures are called reef mounds or simply mounds. James and Bourque (1992) group mounds into three principal types:

1. **microbial mounds,** composed of calcimicrobes (organisms such as cyanobacteria or true algae that can either calcify directly or induce calcification or trap and bind sediment), stromatolites, and thrombolites
2. **skeletal mounds,** composed of the remains of organisms that trapped or baffled mud
3. **mud mounds,** formed by accumulations of mud plus various amounts of fossils

TABLE 13.2 Principal depositional processes and facies characteristics in modern reef complexes

Facies	Process of sedimentation and controls on organisms	Types of organisms likely to be preserved	Grain size	Sorting	Amount of framework (%)	Typical depth (m)	Dominant rock type
Lagoon	Low energy, much burrowing, sporadic currents and turbidity, possible terrigenous influx	Molluscs, echinoids, miliolids, forams, ostracods	Mud mixed with coarse skeletal debris	Poor	0	5–30	Wackestone
Back-reef sand	Sporadic storms and currents across reef, saltation, gravity sliding	Halimeda, miliolids, minor red algae, sparse finger corals	Coarse	Moderate to good	0	1–10	Grainstone
Reef flat	Sporadic storms, good current circulation, winnowing of mud	Finger corals, red and green algae, larger (benthic) forams, head corals	Coarse–very coarse	Moderate	0–10	1–3	Grainstone, rudstone, scattered corals
Reef crest	High wave energy, constant turbulence, good water circulation	Wave-resistant corals and algae	Very coarse	Moderate to good	0–80	0–2	Grainstone (minor bindstone)
Reef framework	Good water circulation, high wave energy—sporadic at greater depths	Abundant corals, algae, molluscs, echinoderms, forams	Framework and sand	Poor; mud in some cavities	20–80	1–30	Framestone
Reef slope	Limited light, sporadic turbulence, gravity transport of reef debris	Soft corals, flattened coral plates, sponges	Mixed	Poor	5–40	20–50	Packstone, bindstone, bafflestone
Proximal talus	Sporadic turbulence, gravity transport, little light, unstable substrate	Few living organisms	Medium to coarse	Poor to good	0	40–100	Grainstone, packstone, rudstone
Distal talus	Quiet water, no light, gravity sliding of sediments	Planktic forams	Fine	Moderate to good	0	100–200	Packstone

Source: Modified slightly from Longman, M. W., 1981. A process approach to recognizing facies of reef complexes, in D. F. Toomey (ed.), European fossil reef models: Soc. Econ. Paleontologists and Mineralogists Spec. Pub. 30. Table 3, p. 24, reprinted by permission of SEPM, Tulsa, Okla.

481

FIGURE 13.16 Schematic diagram illustrating variations in reef zonation in response to different energy conditions ranging from calm water to rough water. (From James, N. P., 1984, Reefs, *in* R. G. Walker (ed.), Facies models, 2nd ed.: Geoscience Canada Reprint Ser. 1. Fig. 10, p. 234, reprinted by permission of Geological Association of Canada.)

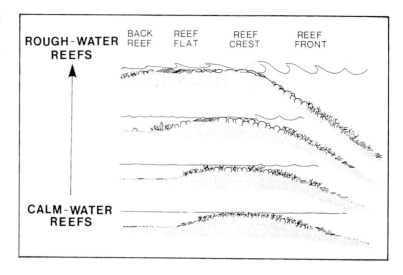

Ancient Reefs

Reef Deposits. Reefs as they appear in the fossil record differ in some important respects from modern reefs. First, we commonly see ancient reefs only in vertical exposures. We observe a two-dimensional limestone body composed of different components formed at different times (James, 1983). Thus, we cannot detect all of the facies zones displayed by modern reefs. Ancient reefs can commonly be divided into only three main facies:

1. the **reef core** consisting of the massive unbedded framework of the reef, composed of reef-building organisms cemented with a matrix of lime mud
2. the **reef-flank facies** consisting of bedded, commonly poorly sorted lime conglomerates (talus breccias) and/or lime sands that thin and dip away from the core
3. the **inter-reef facies**, composed of fine-grained, subtidal lime muds or possibly siliciclastic muds

The nature of the flank and inter-reef facies exposed in outcrop obviously may differ depending upon whether the vertical exposure cuts through the reef to expose a cross section from fore reef to back reef or a longitudinal section running parallel to the reef crest. An additional factor that can further complicate recognition of ancient reef facies is diagenesis, which may cause selective dolomitization or solution that can obliterate or destroy parts of the reef complex (e.g., Schroeder and Purser, 1986).

The carbonate buildups in the Cretaceous deposits of northern Mexico provide a good example that illustrates the general characteristics of ancient reef complexes. The generalized biofacies and lithofacies across the Golden Lane "atoll" in Mexico are shown in Figure 13.17 (Wilson, 1975). The reef core stands tens of meters above the deeper carbonate facies and consists of rudistid clams, colonial corals, stromatoporoids, and encrusting algae. The core facies grades shoreward through reef-derived oolitic-biogenic grainstones to back-reef (inner-reef) micrites, foraminiferal grainstones, and bioturbated wackestones with a fauna indicating restricted circulation. This facies grades farther shoreward into an even more restricted facies containing evaporites. In a basinward direction, reef-core facies grade through a reef-flank facies

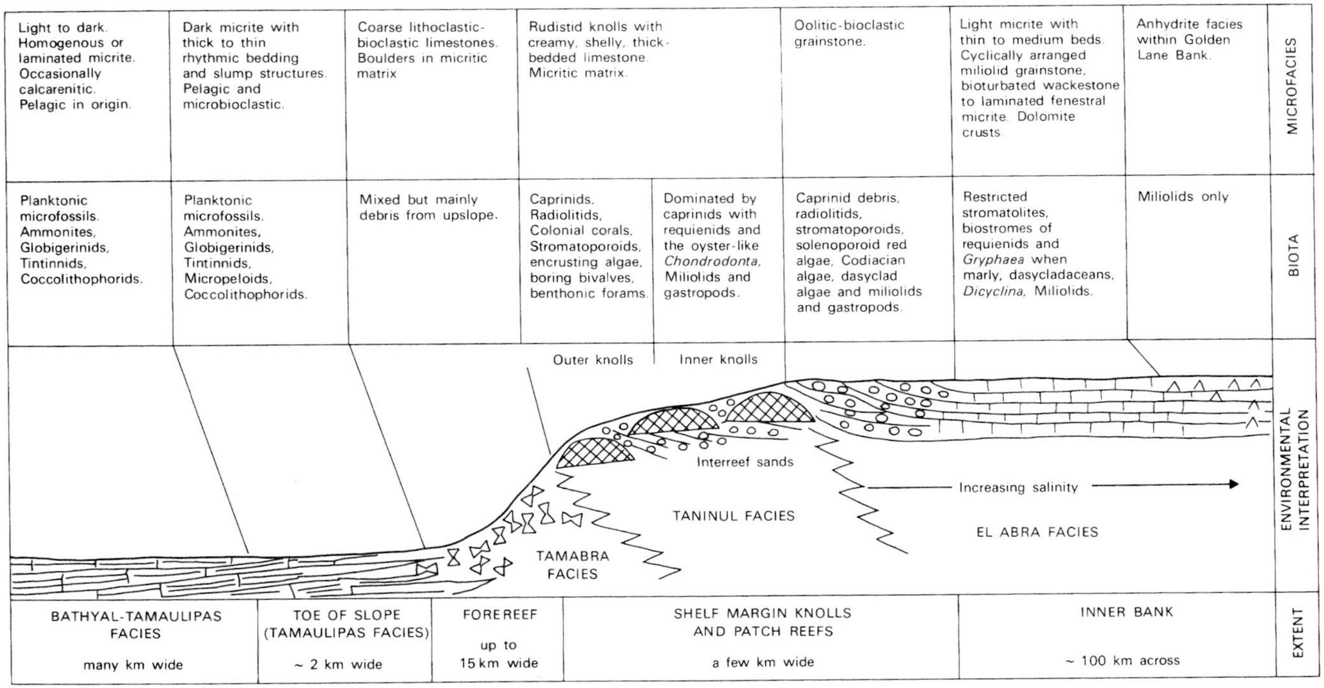
MICROFACIES
Light to dark. Homogenous or laminated micrite. Occasionally calcarenitic. Pelagic in origin.
Dark micrite with thick to thin rhythmic bedding and slump structures. Pelagic and microbioclastic.
Coarse lithoclastic-bioclastic limestones. Boulders in micritic matrix
Rudistid knolls with creamy, shelly, thick-bedded limestone. Micritic matrix.
Oolitic-bioclastic grainstone.
Light micrite with thin to medium beds. Cyclically arranged miliolid grainstone, bioturbated wackestone to laminated fenestral micrite. Dolomite crusts.
Anhydrite facies within Golden Lane Bank.
BIOTA
Planktonic microfossils. Ammonites, Globigerinids, Tintinnids, Coccolithophorids.
Planktonic microfossils. Ammonites, Globigerinids, Tintinnids, Micropeloids, Coccolithophorids.
Mixed but mainly debris from upslope.
Caprinids, Radiolitids, Colonial corals, Stromatoporoids, encrusting algae, boring bivalves, benthonic forams.
Dominated by caprinids with requienids and the oyster-like Chondrodonta, Miliolids and gastropods.
Caprinid debris, radiolitids, stromatoporoids, solenoporoid red algae, Codiacian algae, dasyclad algae and miliolids and gastropods.
Restricted stromatolites, biostromes of requienids and Gryphaea when marly, dasycladaceans, Dicyclina, Miliolids.
Miliolids only
ENVIRONMENTAL INTERPRETATION
Outer knolls
Inner knolls
Interreef sands
Increasing salinity
TANINUL FACIES
EL ABRA FACIES
TAMABRA FACIES
EXTENT
BATHYAL-TAMAULIPAS FACIES
many km wide
TOE OF SLOPE (TAMAULIPAS FACIES)
~ 2 km wide
FOREREEF
up to 15 km wide
SHELF MARGIN KNOLLS AND PATCH REEFS
a few km wide
INNER BANK
~ 100 km across

FIGURE 13.17 Generalized facies of Middle Cretaceous sediments across a large carbonate buildup in central Mexico. (From Wilson, J. L., 1975, Carbonate facies in geologic history. Fig. XI. 3, p. 323, reprinted by permission of Springer-Verlag, Heidelberg. As modified by B. W. Sellwood, 1978, Shallow-water carbonate sediments. Fig. 10.42, p. 307, Elsevier, New York.)

consisting of coarse intraclastic-biogenic boulders embedded in micrite. These in turn grade into more basinward facies composed mainly of micritic limestones with a fauna of pelagic organisms.

Depending upon their age, ancient reefs may also differ markedly from modern reefs in terms of the dominant reef-building organisms that make up the reef. The hermatypic corals that dominate modern coral reefs first appeared in the Mesozoic and thus are not components of older reefs. Older reefs are dominated by other kinds of frame-building organisms, such as tabulate corals, stromatoporoids, hydrozoans, sponges, encrusting bryozoa, coralline algae, and blue-green algae. Fagerstrom (1987) and Stanley and Fagerstrom (1988) provide additional information about reef organisms and the evolution of reef ecosystems through time.

Occurrence of Ancient Reefs. Reefs of some type are found in carbonate rocks of all ages. Although carbonate-secreting organisms were not present during the Precambrian, carbonate buildups composed of stromatolites have been reported from various localities in North America, Europe, Africa, and Australia. Phanerozoic (post-Precambrian) rocks of all ages contain reefs composed of calcium carbonate–secreting organisms. Many of these reefs were built by framework-constructing or encrusting organisms. Reef development was not uniform throughout geologic time, however, and reefs are much more abundant in some parts of the rock record than in others. Figure 13.18 graphically depicts the distribution of reefs through time and also shows the major kinds of organisms that were responsible for reef building at different times. Readers are referred to Heckel (1974), James (1983), and James and Macintyre (1985) for detailed discussions of the distribution and characteristics of reefs formed during each period of geologic time. Many other examples of ancient reefs are described in Frost, Weiss, and Saunders (1977); Geldsetzer, James, and Tebbutt (1988); Laporte (1974); Schroeder and Purser (1986); Toomey (1981); and Wilson (1975).

13.4 SLOPE/BASIN CARBONATES

Although we tend to think of carbonate sediments as strictly shallow-water deposits, deeper-water carbonates have been identified in several areas of the modern ocean, such as the slope and adjacent basin floor around the Bahama Platform. They have also been reported from many Phanerozoic-age stratigraphic successions. As shown in Figure 13.4, carbonate sediments are generated primarily on the shelf. No source of carbonate sediments exists within deep-water except that provided by the rain of calcareous pelagic organisms. Therefore, with the exception of calcareous oozes, carbonate sediments in deep water are derived from the shelf by transport processes that include storm waves, turbidity currents, debris and grain flows, slumping, sliding, and rock falls. Carbonate sediment deposited on the slope and basin by these processes generally consists of bioclastic debris and limestone blocks derived from the talus slopes off reef fronts. Also, sediments may be transported downslope from carbonate sand shoals or lime-mud deposits on the platform margin. Summaries of carbonate slope sedimentation and deposits are given by Conglio and Dix (1992), Cook and Mullins (1983), Enos and Moore (1983), McIlreath and James (1984), and Mullins and Cook (1986). Modern examples of carbonate slopes have been reported from the northern Bahamas–Florida region as well as from Belize, Jamaica, Grand Cayman, the northeast Australian coast, and several atolls in the Pacific and Indian oceans (Conglio and Dix,

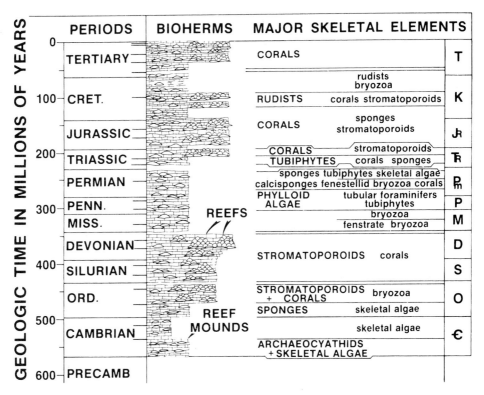

FIGURE 13.18 Distribution of reefs through the Phanerozoic showing times when there were only reef mounds, times when both reef mounds and reefs existed, and times when no reefs or bioherms were formed (gaps in the lithic column). (From James, N. P., 1983, Reef environment, *in* P. A. Scholle, D. G. Bebout, and C. H. Moore (eds.), Carbonate depositional environments: Am. Assoc. Petroleum Geologists Mem. 33, Fig. 61, p. 375, reprinted by permission of AAPG, Tulsa, Okla.)

1992). Modern slope carbonates consist mainly of pure carbonates, in contrast to many ancient slope deposits that include a large percentage of terrigenous clastic rocks.

McIlreath (1984) divides slope carbonates into five principal groups. **Hemipelagic carbonate sediments,** sometimes called "periplatform ooze," settle from the water column more or less constantly. They are augmented by episodic contributions of fine material swept off the shelf during storms or by warm, sediment-rich waters that float off the shelf over cooler basinal waters owing to tidal exchange. **Periplatform talus** consists of a debris apron of limestone blocks, skeletons of reef organisms, and lime sands and muds that accumulate directly seaward of reefs or platform-edge lime shoals. They accumulate owing mainly to rock fall and grain flow from shallow water. **Carbonate breccias** and **conglomerates** are coarse breccias derived from shallow water, or the slope, or both. They consist of carbonate clasts of various sizes, some of enormous dimensions, that commonly have a matrix of lime mud, lime sand, or argillaceous lime mud. The internal fabric of these deposits may range from chaotic to imbricated, horizontal, vertical, or even-graded. These deposits are transported on the slope by submarine debris flows or other mass-transport mechanisms. **Graded calcarenites** (calcareous turbidites) are the carbonate equivalents of siliciclastic turbidites. They commonly have sharp, planar bases, but sole markings are rare. Complete Bouma sequences (Fig. 3.28) may occur, but more commonly only the A division (and in some deposits the B and C divisions) is present. Graded calcarenites are deposited both on the slope and on the deeper seafloor at the base of the slope. These carbonate turbidites are derived mainly from loose carbonate sand and gravel accumulations near the platform margin, but they may also be derived from finer-grained sediment lower on the slope. **Nongraded calcarenites** (grainstones and packstones) are massive to cross-bedded and ripple-marked lime-sand deposits that generally have sharp bases and lenticular to irregular geometry. They contain various amounts of lime mud and may range from clean grainstones to fine- to coarse-grained wackestones. They are possibly formed by some type of grain-flow mechanism. Alternatively, they may be previously deposited sediments that were reworked and winnowed by bottom contour currents.

The distribution of carbonate facies that develops on carbonate slopes and adjacent basins depends upon the character of the shelf-margin sediments that serve as a source for slope carbonates, the nature of the platform edge, and the relief between the platform and basin. Margins characterized by a gentle transition from platform to slope are called "depositional margins." Figure 13.19 schematically depicts a depositional margin capped by lime-sand shoals and illustrates the distribution of facies on the slope and the hypothetical succession of deposits formed in the adjacent basin. By contrast, Figure 13.20 illustrates the facies and deposition succession developed on a "by-pass margin" with a very abrupt transition from platform to slope owing to the presence of a cliff-fronted reef. Slope to basin deep-water carbonate facies thus range from hemipelagic lime muds to relatively clean lime sands to graded turbidites to chaotic carbonate conglomerates and breccias to pelagic calcareous oozes. For discussion of other slope/basin facies models and additional amplification of slope/basin facies, see Conglio and Dix (1992) and Crevello and Harris (1985).

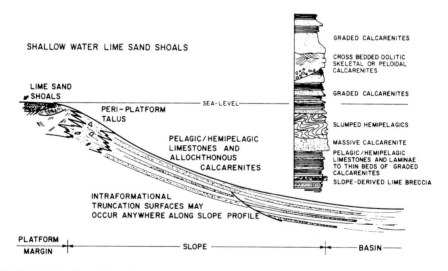

FIGURE 13.19 Depositional margin characterized by shallow-water lime-sand shoals. Schematic model shows the hypothetical sequence of deposits within the adjacent basin. Calcarenite = grainstone or packstone. (From McIlreath, I. A., and N. P. James, 1984, Carbonate slopes, *in* R. G. Walker (ed.), Facies models, 2nd ed.: Geoscience Canada Reprint Ser. 1. Fig. 13, p. 252, reprinted by permission of Geological Association of Canada.)

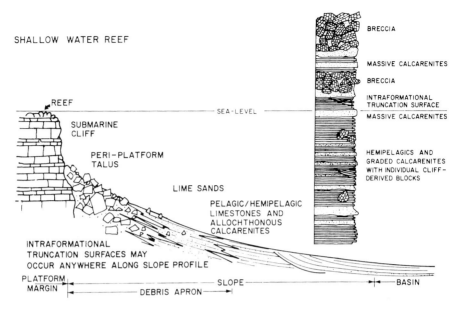

FIGURE 13.20 By-pass carbonate margin characterized by shallow-water reefs. Schematic model shows a hypothetical sequence of deposits formed in the adjacent basin. Calcarenite = packstone or grainstone. (From McIlreath, I. A., and N. P. James, 1984, Carbonate slopes, *in* R. G. Walker (ed.), Facies models, 2nd ed.: Geoscience Canada Reprint Ser. 1. Fig. 14, p. 253, reprinted by permission of Geological Association of Canada.)

13.5   MIXED CARBONATE-SILICICLASTIC SYSTEMS

To avoid possible confusion, the carbonate depositional environments described in the preceding sections are discussed as if only carbonate sediments are deposited in these systems. In fact, mixed carbonate and siliciclastic sediments are present in many stratigraphic successions. Such deposits are referred to as mixed carbonate-siliciclastic successions or carbonate-clastic transitional successions. Mixtures of carbonate and siliciclastic sediments can occur owing to laterals facies mixing (spatial variations in environments) that produces lateral interfingering of carbonate and clastic sediment. Mixtures may also occur as a result of sea-level change and/or variations in sediment supply, which cause vertical variations in the stratigraphic succession of facies (temporal variability) (Budd and Harris, 1990). Thus, siliciclastic facies may occur in a lateral interfingering relationship (see Chapter 14) with carbonate facies or as distinct interbeds within carbonate successions. Carbonate-siliciclastic transitions are known to be present in a variety of environments, including coastal and inner shelf, middle and outer shelf (including reef), and the slope to basin environments (Doyle and Roberts, 1988; Lomando and Harris, 1991). They may occur in temperate as well as tropical shelf environments (e.g., Haywick, Carter, and Henderson, 1992).

FURTHER READINGS

Bathurst, R. G. C., 1975, Carbonate sediments and their diagenesis: Elsevier, Amsterdam, 658 p.

Bhattacharyya, A., and G. M. Friedman (eds.), 1983, Modern carbonate environments: Benchmark Papers in Geology, v. 74. Dowden, Hutchinson and Ross, Stroudsburg, Pa., 376 p.

Cook, H. E., and P. Enos (eds.), 1977, Deep-water carbonate environments: Soc. Econ. Paleontologists and Mineralogists Spec. Pub. 25, Tulsa, Okla., 336 p.

Cook, H. E., A. C. Hine, and H. T. Mullins (ed.), 1983, Platform margin and deep water carbonates: Soc. Econ. Paleontologists and Mineralogists Short Course Notes No. 12, Tulsa, Okla.

Crevello, P. D., and P. M. Harris (eds.), 1985, Deep-water carbonates: Buildups, turbidites, debris flows and chalks: Soc. Econ. Paleontologists and Mineralogists Core Workshop No. 6, 527 p.

Crevello, P. D., J. L. Wilson, J. F. Sarg, and J. F. Read (eds.), 1989, Controls on carbonate platform and basin development: Soc. Econ. Paleontologists and Mineralogists Spec. Pub. 44, 405 p.

Doyle, L. J., and H. H. Roberts (eds.), 1988, Carbonate-clastic transitions: Elsevier, Amsterdam, 304 p.

Fagerstrom, J. A., 1987, The evolution of reef communities: John Wiley & Sons, New York, 600 p.

Frost, S. H., M. P. Weiss, and J. B. Saunders (eds.), 1977, Reefs and related carbonates—Ecology and sedimentology: Am. Assoc. Petroleum Geologists Studies in Geology 4, Tulsa, Okla., 421 p.

Geldsetzer, H. H. J., N. P. James, and G. E. Tebbutt (eds.), 1988, Reefs, Canada and adjacent areas: Canadian Soc. Petroleum Geologists Mem. 13, 775 p.

Laporte, L. F. (ed.), 1974, Reefs in time and space: Soc. Econ. Paleontologists and Mineralogists Spec. Pub. 18, Tulsa, Okla., 256 p.

Lomando, A. J., and P. M. Harris (eds.), 1991, Mixed carbonate-clastic siliciclastic successions: Soc. Econ. Paleontologists and Mineralogists Core Workshop No. 15, 568 p.

Scholle, P. A., D. G. Bebout, and C. H. Moore, 1983, Carbonate depositional environments: Am. Assoc. Petroleum Geologists Mem. 33, Tulsa, Okla., 708 p.

Schroeder, J. H., and B. H. Purser, 1986, Reef diagenesis: Springer-Verlag, Berlin, 455 p.

Scoffin, T. P., 1987, An introduction to carbonate sediments and rocks: Blackie, Glasgow, 274 p.

Toomey, D. F. (ed.), 1981, European fossil reef models: Soc. Econ. Paleontologists and Mineralogists Spec. Pub. 30, Tulsa, Okla., 546 p.

Tucker, M. E., J. L. Wilson, P. D. Crevello, J. R. Sarg, and J. F. Read, 1990, Carbonate platforms—Facies, successions and evolution: Internat. Assoc. Sedimentologists Spec. Pub. 9, Blackwell, Oxford, 328 p.

Tucker, M. E., and V. P. Wright, 1990, Carbonate sedimentology: Blackwell, Oxford, 482 p.

Stanley, G. D., Jr., and J. A. Fagerstrom (eds.), 1988, Ancient reef ecosystems: Palaios, v. 3, p. 111–254.

Wilson, J. L., 1975, Carbonate facies in geologic history: Springer-Verlag, Berlin, 471 p.

PART 6

PRINCIPLES OF STRATIGRAPHY AND BASIN ANALYSIS

Entrenched meanders cutting through well-stratified Permian sedimentary rocks at the Goosenecks of the San Juan River, Utah.

The emphasis in the preceding parts of this book has been on sedimentary processes, the environments in which these processes take place, and the properties of sedimentary rocks generated in these environments. In this part, we focus on a different aspect of sedimentary rocks. Our concern here is not so much sedimentary processes and detailed rock properties but rather the larger-scale vertical and lateral relationships between units of sedimentary rock that are defined on the basis of lithologic or physical properties, paleontological characteristics, geophysical properties, age relationships, and geographic position and distribution. It is the study of these characteristics of layered rocks that encompasses the discipline of **stratigraphy.** Understanding the principles and terminology of stratigraphy is essential to geologic study of sedimentary rocks because stratigraphy provides the framework within which systematic sedimentologic studies can be carried out. It allows the geologist to bring together the details of sediment composition, texture, structure, and other features into an environmental and temporal synthesis from which we can interpret the broader aspects of Earth history.

Prior to the 1960s, the discipline of stratigraphy was concerned particularly with stratigraphic nomenclature; the more classical concepts of lithostratigraphic, biostratigraphic, and chronostratigraphic successions in given areas; and correlation of these successions between areas. **Lithostratigraphy** deals with the lithology or physical properties of strata and their organization into units based on lithologic character. **Biostratigraphy** is the study of rock units on the basis of the fossils they contain. **Chronostratigraphy** (*chrono* = time) deals with the ages of strata and their time relations. This classical approach to stratigraphy is exemplified in Weller's (1960) textbook, *Stratigraphic Principles and Practice.* These established principles are still the backbone of stratigraphy; however, today's students must go beyond these basic principles. They must also have a thorough understanding of depositional systems and must be able to apply stratigraphic and sedimentological principles to interpretation of strata within the context of global plate tectonics. This means, among other things, becoming familiar with comparatively new branches of stratigraphy that have developed since the early 1960s as new concepts and methods of studying sedimentary rocks and other rocks by remote sensing techniques have unfolded. For example, the concept of depositional sequences, which are packages of strata bounded by unconformities, has gained particular prominence since the late 1970s. This concept has now become so important that we refer to study of sequences as **sequence stratigraphy.** Two new offshoots of stratigraphy that have made particularly important contributions to our understanding of the physical stratigraphic relationships, ages, and environmental significance of subsurface strata and oceanic sediments are **magnetostratigraphy,** which deals with stratigraphic relationships based on the magnetic properties of sedimentary rocks and layered volcanic rocks, and **seismic stratigraphy,** which is the study of stratigraphic and depositional facies as interpreted from seismic data. We explore all of these concepts in the next few chapters.

14

Lithostratigraphy

14.1 INTRODUCTION

Lithostratigraphy deals with the study and organization of strata on the basis of their lithologic characteristics. The term **lithology** is used by geologists in two different but related ways. Strictly speaking, it refers to the study and description of the physical character of rocks, particularly in hand specimens and outcrops (Bates and Jackson, 1980). It is used also to refer to these physical characteristics: Rock type, color, mineral composition, and grain size are all lithologic characteristics. For example, we may refer to the lithology of a particular stratigraphic unit as sandstone, shale, limestones, and so forth. Thus, lithostratigraphic units are rock units defined or delineated on the basis of their physical properties, and lithostratigraphy deals with the study of the stratigraphic relationships among strata that can be identified on the basis of lithology.

In this chapter, we begin study of stratigraphic principles by briefly discussing the nature of lithostratigraphic units, followed by an explanation of the various types of contacts that separate these units. We then explore the important concepts of sedimentary facies and depositional sequences. The essentials of stratigraphic nomenclature and classification as they apply to lithostratigraphic units are discussed next, including examination of the North American Code of Stratigraphic Nomenclature. Finally, correlation of lithostratigraphic units is explained, and the various methods of correlation are described.

14.2 TYPES OF LITHOSTRATIGRAPHIC UNITS

Lithostratigraphic units are bodies of sedimentary, extrusive igneous, metasedimentary, or metavolcanic rock distinguished on the basis of lithologic characteristics. A lithostratigraphic unit generally conforms to the **law of superposition,** which states that in any succession of strata, not subsequently disturbed or overturned since deposition, younger rocks lie above older rocks. Lithostratigraphic units are also commonly stratified and tabular in form. They are recognized and defined on the basis of observable rock characteristics. Boundaries between different units may be placed at clearly identifiable or distinguished contacts or may be drawn arbitrarily within a zone of gradation. Definition of lithostratigraphic units is based on a **stratotype** (a designated type unit), or type section, consisting of readily accessible rocks, where possible, in natural outcrops, excavations, mines, or bore holes. Lithostratigraphic units are defined strictly on the basis of lithic criteria as determined by descriptions of actual rock materials. They carry no connotation of age. They cannot be defined on the basis of paleontologic criteria, and they are independent of time concepts. They may be established in subsurface sections as well as in rock units exposed at the surface, but they must be established on the basis of lithic characteristics and not on geophysical properties or other criteria. Geophysical criteria, described in Chapter 15, may be used to aid in fixing boundaries of subsurface lithostratigraphic units, but the units cannot be defined exclusively on the basis of remotely sensed physical properties.

Wheeler and Mallory (1956) introduced the term **lithosome** to refer to masses of rock of essentially uniform character and having intertonguing relationships with adjacent masses of different lithology. Thus, we speak of shale lithosomes, limestone lithosomes, sand-shale lithosomes, and so forth. Krumbein and Sloss (1963) explain the meaning of lithosomes by asking readers to imagine the body of rock that would emerge if it were possible to preserve a single rock type, such as sandstone, and dissolve away all other rock types. The resulting sandstone body would thus appear as a roughly tabular mass with intricately shaped boundaries. These irregular boundaries would represent its surfaces of contact with erosion surfaces and with other rock masses of differing constitution above, below, and to the sides. Lithosomes have no specified size limits and may range in gross shape from thin, sheetlike or blanketlike units to thick prisms or narrow, elongated shoestrings.

Of course, stratigraphic units of a single lithology rarely exist as isolated bodies. They are commonly in contact with other rock bodies of different lithology. An important part of lithostratigraphy is identifying and understanding the nature of contacts between vertically superposed or laterally adjacent bodies. Another important aspect is the identification of single lithosomes, groups of lithosomes, or subdivisions of lithosomes that are so distinctive that they form lithostratigraphic units that can be distinguished from other units that may lie above, below, or adjacent.

The fundamental lithostratigraphic unit of this type is the formation. A **formation** is a lithologically distinctive stratigraphic unit that is large enough in scale to be mappable at the surface or traceable in the subsurface. It may encompass a single lithosome, or part of an intertonguing lithosome, and thus consist of a single lithology. Alternatively, a formation can be composed of two or more lithosomes and thus may include rocks of different lithology. Some formations may be divided into smaller stratigraphic units called **members,** which, in turn, may be divided into smaller distinctive units called **beds.** Beds are the smallest formal lithostratigraphic units. Formations having some kind of stratigraphic unity can be combined to form **groups,** and groups can be combined to form **supergroups.** All formal lithostratigraphic units are given names that are derived from some geographic feature in the area where they are studied.

Subdivision of thick units of strata into smaller lithostratigraphic units such as formations is essential for tracing and correlation of strata both in outcrop and in the subsurface. We will come back to discussion of these formal stratigraphic units near the end of this chapter. First, however, we examine the nature of contacts between stratigraphic units and the lateral and vertical facies relationships that characterize strata.

14.3 STRATIGRAPHIC CONTACTS

Different lithologic units are separated from each other by **contacts,** which are plane or irregular surfaces between different types of rocks. Vertically superposed strata are said to be either conformable or unconformable depending upon continuity of deposition. Conformable strata are characterized by unbroken depositional assemblages, generally deposited in parallel order, in which layers are formed one above the other by more or less uninterrupted deposition. The surface that separates conformable strata is a **conformity,** that is, a surface that separates younger strata from older rocks but along which there is no physical evidence of nondeposition. A conformable contact indicates that no significant break or hiatus in deposition has occurred. A **hiatus** is a break or interruption in the continuity of the geologic record. It represents periods of geologic time (short or long) for which there are no sediments or strata.

Contacts between strata that do not succeed underlying rocks in immediate order of age, or that do not fit together with them as part of a continuous whole, are called unconformities. Thus, an **unconformity** is a surface of erosion or nondeposition, separating younger strata from older rocks, that represents a significant hiatus. Unconformities indicate a lack of continuity in deposition and correspond to periods of nondeposition, weathering, or erosion, either subaerial or subaqueous, prior to deposition of younger beds. Unconformities thus represent a substantial break in the geologic record that may correspond to periods of erosion or nondeposition lasting millions or even hundreds of millions of years.

Contacts are also present between laterally adjacent lithostratigraphic units. These contacts are formed between rock units of equivalent age that developed different lithologies owing to different conditions in the depositional environment. Excluded from discussion here are contacts between laterally adjacent bodies that arise from postdepositional faulting. Contacts between laterally adjacent bodies may be gradational, where one rock type changes gradually into another, or they may be intertonguing, that is, pinching or wedging out within another formation (see Fig. 13.4).

Contacts between Conformable Strata

Contacts between conformable strata may be either abrupt or gradational. **Abrupt contacts** occur as a result of sudden, distinct changes in lithology. Most abrupt contacts coincide with primary depositional bedding planes that formed as a result of changes in local depositional conditions, as discussed in Chapter 5. In general, bedding planes represent minor interruptions in depositional conditions. Such minor depositional breaks, involving only short hiatuses in sedimentation with little or no erosion before deposition is resumed, are called **diastems.** Abrupt contacts may be caused also by postdepositional chemical alteration of beds, producing changes in color owing to oxidation or reduction of iron-bearing minerals, changes in grain size owing to recrystallization or dolomitization, or changes in resistance to weathering owing to cementation by silica or carbonate minerals.

Conformable contacts are said to be **gradational** if the change from one lithology to another is less marked than abrupt contacts, reflecting gradual change in depositional conditions with time. Gradational contacts may be of either the progressive

FIGURE 14.1 Types of gradational vertical contacts. A. Progressive gradual. B. Intercalated.

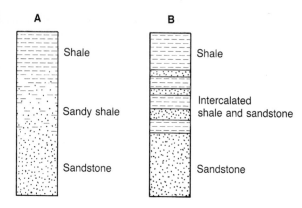

gradual type or the intercalated type. **Progressive gradual contacts** occur where one lithology grades into another by progressive, more or less uniform changes in grain size, mineral composition, or other physical characteristics. Examples include sandstone units that become progressively finer grained upward until they change to mudstones (Fig. 14.1A) or quartz-rich sandstones that become progressively enriched upward in lithic fragments until they change to lithic arenites. **Intercalated contacts** are gradational contacts that occur owing to an increasing number of interbeds of another lithology that appear upward in the section (Figure 14.1B).

Unconformable Contacts

Four types of unconformable contacts (unconformities) are recognized: (1) angular unconformity, (2) disconformity, (3) paraconformity, and (4) nonconformity. Unconformities are recognized on the basis of the presence or absence of an angular relationship between the unconformable strata, the presence or absence of a marked erosional surface separating these strata, and the nature of the rocks underlying the surface of unconformity. The first three types of unconformities occur between bodies of sedimentary rock. The last type occurs between sedimentary rock and metamorphic or igneous rock.

Angular Unconformity. An angular unconformity is a type of unconformity in which younger sediments rest upon the eroded surface of tilted or folded older rocks; that is, the older rocks dip at a different, commonly steeper, angle than do the younger rocks (Fig. 14.2A). Angular unconformities may be confined to limited geographic areas (local unconformities) or may extend for tens or even hundreds of kilometers (regional unconformities). Some angular unconformities are clearly visible in a single outcrop (Fig. 14.3). By contrast, regional unconformities between stratigraphic units of very low dip may not be apparent in a single outcrop and may require detailed mapping over a large area before they can be identified.

Disconformity. An unconformity surface above and below which the bedding planes are essentially parallel and in which the contact between younger and older beds is marked by a visible, irregular, or uneven erosional surface is a disconformity (Fig. 14.2B). Disconformities are most easily recognized by this erosional surface, which may be channeled and which may have relief ranging up to tens of meters. Disconformity surfaces, as well as angular unconformity surfaces, may be marked also by "fos-

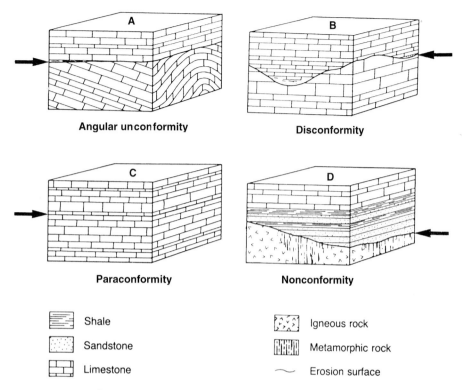

Angular unconformity

Disconformity

Paraconformity

Nonconformity

Shale

Sandstone

Limestone

Igneous rock

Metamorphic rock

Erosion surface

FIGURE 14.2 Four basic types of unconformities. Arrows point to unconformity surfaces. (After Dunbar, C. O., and J. Rodgers, 1957, Principles of stratigraphy: John Wiley & Sons, New York. Fig. 57, p. 117, reprinted by permission.)

sil" soil zones (paleosols) or may include lag-gravel deposits that lie immediately above the unconformable surface and that contain pebbles of the same lithology as the lithology of the underlying unit. Disconformities are presumed to form as a result of a significant period of erosion throughout which older rocks remained essentially horizontal during nearly vertical uplift and subsequent downwarping.

Paraconformity. A paraconformity is an obscure unconformity characterized by beds above and below the unconformity contact that are parallel and in which no erosional surface or other physical evidence of unconformity is discernible. The contact may even appear to be a simple bedding plane (Fig. 14.2C). Paraconformities are not easily recognized and must be identified on the basis of a gap in the rock record (owing to nondeposition or erosion) as determined from paleontologic evidence such as absence of faunal zones or abrupt faunal changes.

Nonconformity. An unconformity developed between sedimentary rock and older igneous or massive metamorphic rock that has been exposed to erosion prior to being covered by sediments is a nonconformity (Fig. 14.2D).

The presence of unconformities has considerable significance in sedimentological studies. Many stratigraphic successions are bounded by unconformities, indicating that these successions are incomplete records of past sedimentation. Not only do

FIGURE 14.3 Angular unconformity between steeply dipping Coaledo Formation (Eocene) sediments and overlying, horizontal Pleistocene deposits, southern Oregon coast. The nearly vertical dark stripe is a weathering stain.

unconformities show that some part of the stratigraphic record is missing, but they also indicate that an important geologic event took place during the time period (hiatus) represented by the unconformity—an episode of uplift and erosion or, less likely, an extended period of nondeposition.

Contacts between Laterally Adjacent Lithosomes

In the preceding discussion, we examined the kinds of stratigraphic contacts or boundaries that separate sedimentary units into distinct vertical lithologic successions. Stratigraphic units also have finite lateral boundaries. They do not extend indefinitely laterally but must eventually terminate, either abruptly as a result of erosion or more gradually by change to a different lithology. Lateral changes may be accompanied by progressive thinning of units to extinction—**pinch-outs** (Fig. 14.4A); lateral splitting of a lithologic unit into many thin units that pinch out independently—**intertonguing** (Fig. 14.4B); or **progressive lateral gradation,** similar to progressive vertical gradation (Fig. 14.4C).

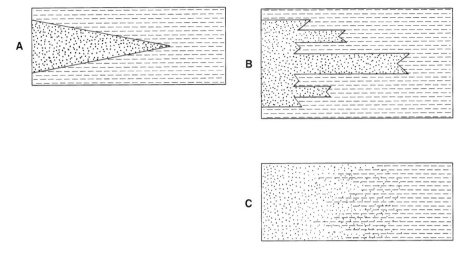

FIGURE 14.4   Lateral relationship of sedimentary units: A. Pinch-out. B. Intertonguing. C. Lateral gradation.

14.4   VERTICAL AND LATERAL SUCCESSIONS OF STRATA

Nature of Vertical Successions

As discussed, conformities and unconformities divide sedimentary rocks into vertical successions of beds, each characterized by a particular lithologic aspect. Different types of beds can succeed each other vertically in a great variety of ways, and distinctions can be drawn between rock units characterized by (1) lithologic uniformity, (2) lithologic heterogeneity, and (3) cyclic successions (Weller, 1960). Rock units that have complete lithologic uniformity are rare, although many beds may display a high degree of uniformity in color, grain size, composition, or resistance to weathering. Beds that are most likely to be uniform are fine-grained sediments deposited slowly under essentially uniform conditions in deeper water or coarser sediments that have been deposited rapidly by some type of mass sediment transport mechanism such as grain flow. Heterogeneous bodies of sedimentary strata are characterized by internal variations or irregularities in properties. Heterogeneous units may include strata such as extremely poorly sorted debris-flow deposits, as well as thick units broken internally by thinner beds characterized by differences in grain size or bedding features.

Cyclic Successions

Many stratigraphic successions display repetitions of strata that reflect a succession of related depositional processes and conditions that are repeated in the same order. Such repetitious events are referred to as **cyclic sedimentation** or rhythmic sedimentation. Cyclic sedimentation leads to the formation of vertical successions of sedimentary strata that display repetitive, orderly arrangement of different kinds of sediments. The term cyclic sediment has been used for a wide variety of repetitious strata, including such small-scale features as presumed annually deposited varves in glacial lakes as well as large-scale sediment cycles caused by long-period, recurring migration of depositional environments. Other common examples of repetitive deposits include rhythmically bedded turbidites, laminated evaporite deposits, limestone-shale rhythmic successions, coal cyclothems (repeated cycles involving coal deposition), black shale deposits, and chert deposits. Cyclic successions occur on all continents in essentially every stratigraphic system. They are produced by processes that range in geographic scope and duration from very local, short-term events—such as seasonal climatic changes that generate varves—to global changes in sea level that may involve entire geologic periods.

On the basis of the mechanisms that form cyclic deposits, two kinds of cyclic successions are recognized: (1) autocyclic and (2) allocyclic (Fig. 14.5). **Autocyclic successions** are controlled by processes that take place within the basin itself, and their beds show only limited stratigraphic continuity. Examples include nonperiodic storm

FIGURE 14.5 Schematic representation of autocyclic and allocyclic mechanisms that cause generation of rhythmic and cyclic successions. (After Einsele, G., W. Ricken, and A. Seilacher, 1991, Cycles and events in stratigraphy—Basic concepts and terms, *in* G. Einsele, W. Ricken, and A. Seilacher (eds.), Cycles and events in stratigraphy. Fig. 4, p. 9, reproduced by permission of Springer-Verlag, Berlin.)

a AUTOCYCLIC MECHANISMS

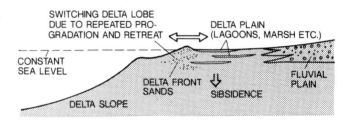

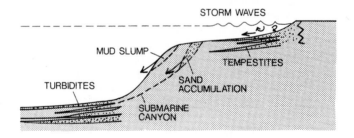

b ALLOCYCLIC MECHANISMS

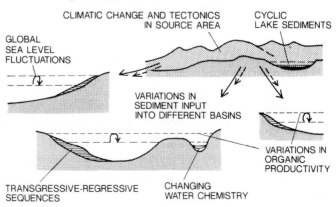

beds and turbidites. **Allocyclic successions** are caused mainly by variations external to the depositional basin, for example, climatic changes, tectonic movements in the source area and global sea-level fluctuations. Allocyclic successions may extend over long distances and perhaps even from one basin to another (Einsele, Ricken, and Seilacher, 1991a).

Some allocyclic successions appear to be related to changes in Earth's orbital parameters. Earth's axis precesses (the position of the rotational pole wobbles) with two predominant periods averaging 19,000 and 23,000 years; it undergoes change in axis inclination up to 3° (obliquity) in a cycle of about 41,000 years; and its orbit changes from almost circular to almost elliptical (eccentricity) in two cycles, one with a cycle ranging from 95,000 to 136,000 years, the other with a period of 413,000 years. These variations in Earth's orbital behavior produce periodic changes of climate, called **Milankovich cycles** (with periodicities ranging from about 20,000 to more than 400,000 years), which, in turn, influence depositional patterns. This process is sometimes referred to as **orbital forcing** (e.g., de Boer, 1991). Stratigraphic cycles are discussed in greater detail under "Sea-Level Analysis" in Chapter 15, particularly with respect to cyclic successions related to changes in sea level, and by Einsele and Seilacher (1982) and Einsele, Ricken, and Seilacher (1991b).

Sedimentary Facies

In preceding chapters dealing with depositional environments, I point out many examples of sediments of one type that grade laterally into sediments of a different type deposited in laterally contiguous parts of a given depositional setting. For example, sandy sediments of the beach shoreface may grade seaward to muddy sediments of the shallow inner shelf; delta-front sands and silts commonly grade seaward to prodelta muds; and shelf-edge skeletal or oolitic carbonate sands grade toward the open shelf to pelleted carbonate muds. I have already referred to such laterally equivalent bodies of sediment with distinctive characteristics as **facies.** Thus, a deposit may be characterized by shale facies, sandstone facies, limestone facies, and so forth. The concept of facies is so important in stratigraphy that a more detailed explanation of the meaning and significance of facies is necessary at this point.

The term facies was introduced into the geological literature by Nicolas Steno in 1669 (Teichert, 1958); however, modern scientific use of the term is credited to the Swiss geologist Amanz Gressly, who used the term in 1838 in his description of Upper Jurassic strata in the region of Solothurn in the Jura Mountains to describe marked changes in lithology and paleontology of these strata. Krumbein and Sloss (1963) maintain that Gressly intended to confine usage of the term to lateral changes within a stratigraphic unit, such as those illustrated in Figure 14.6. Other workers have interpreted Gressly's usage to include vertical changes in the character of rock units as well (Teichert, 1958). Subsequently, the term has been used with numerous meanings, many of which bear little resemblance to Gressly's original meaning. These various meanings have been summarized and discussed by Moore (1949), Teichert (1958), Weller (1958), Markevich (1960), and others. The extended meanings of facies have included referring to all strata of a particular type as a certain facies, such as referring to all redbeds as the "redbed facies," and even such nonstratigraphical usage as "metamorphic facies," "igneous facies," and "tectonic facies." Because of these rather loose and inconsistent usages of the term, the real meaning of facies has become considerably clouded.

A commonly accepted definition of facies in the United States is that of Moore (1949), who described facies as "any areally restricted part of a designated stratigraphi-

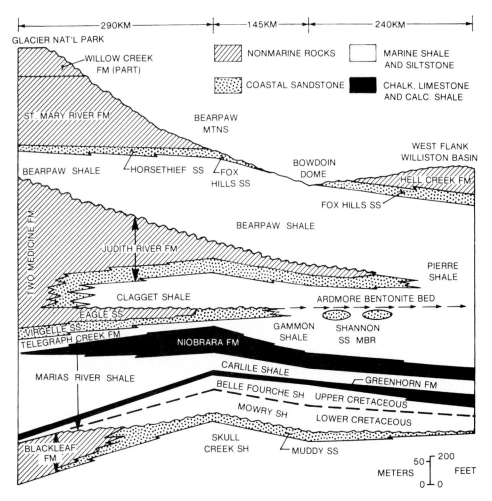

FIGURE 14.6   Facies relationships in Upper Cretaceous strata of the Rocky Mountains in Montana. (From Swift, D. J. P., and D. D. Rice, 1984. Sand bodies on muddy shelves: A model for sedimentation in the western interior Cretaceous seaway, North America, *in* R. W. Tillman and C. T. Siemers (eds.), Siliciclastic shelf sediments: Soc. Econ. Paleontologists and Mineralologists Spec. Pub. 34, Fig. 2, p. 45, reprinted by permission of SEPM, Tulsa, Okla.)

cal unit which exhibits characters significantly different from those of other parts of the unit." Facies comprise "one or any two or more different sorts of deposits which are partly or wholly equivalent in age and which occur side by side or in somewhat close neighborhood." According to Moore's definition, facies are restricted in areal extent, but the same facies could be found at different levels within the same stratigraphic unit. A different usage of facies that is closer to that of Gressly usage—and to that of many European geologists—is to consider facies simply as stratigraphic units distinguished by lithological, structural, and organic aspects detectable in the field. The areal distribution of facies thus designated may not be well known (Blatt, Middleton, and Murray, 1980), in contrast to the restricted areal distribution required by Moore's definition. See also the discussion by Walker (1992). Regardless of the exact definition followed in defining facies, it is now common practice to designate facies identified on the basis of lithologic characteristics as **lithofacies** and facies distinguished by paleontologic characteristics (fossil content) without regard to lithologic character as **biofacies.**

An important objective of facies studies is to ultimately make environmental interpretations from the facies. Thus, some geologists designate facies on the basis of assumed depositional environment and speak of "continental facies," "fluvial facies," "delta facies," and so on. Such generic usage involves subjective judgments that may not always be justified. It is better to make the usage of facies purely descriptive and objective and then make subjective interpretations of environment on the basis of these descriptive facies (Hallam, 1981).

Walther's Law of Succession of Facies

Relationship of Lateral and Vertical Facies. It is implicit in the concept of facies that different facies represent different depositional environments. As laterally contiguous environments in a given region shift with time in response to shifting shorelines or other geologic conditions, facies boundaries also shift so that eventually the deposits of one environment may lie above those of another environment. This deceptively simple idea embodies one of the single most important concepts in stratigraphy—the concept that a direct environmental relationship exists between lateral facies and vertically stacked or superimposed successions of strata. This concept was first formally stated by Johannes Walther in 1894 and is now called the **law of the correlation (or succession) of facies,** or simply **Walther's Law.** This law has often been misstated as "the same facies sequences are seen laterally as vertically." The correct statement of the law as translated by Middleton (1973, p. 979) is

> The various deposits of the same facies-area and similarly the sum of the rocks of different facies-areas are formed beside each other in space, though in a cross-section we see them lying on top of each other. . . . It is a basic statement of far-reaching significance that only those facies and facies-areas can be superimposed primarily which can be observed beside each other at the present time. (Walther, 1894)

Walther's Law is thus interpreted to mean that facies that occur in **conformable** vertical successions of strata also occurred in laterally adjacent environments. Middleton (1973) is careful to point out that the law states not that vertical successions always reproduce the horizontal succession of environments, but only that those facies can be superimposed that can now be seen developing side by side. For example, the beach and barrier-island environmental setting discussed in Chapter 11 may include several laterally adjacent environments such as beach, back-barrier lagoon, marsh, tidal flat, tidal channel, and tidal delta. Depending upon the manner in which these lateral environments shift with time, the vertical successions produced by deposition in a particular barrier-island setting might consist only of beach sands overlain by lagoonal muds and capped by marsh peats. The entire lateral succession of deposits formed in the contiguous environments may not be preserved, but those deposits that are preserved in the vertical succession originally occurred side by side.

Transgressions and Regressions. The principles embodied in Walther's Law are illustrated in Figure 14.7. The block diagram in this figure shows the lateral succession of environments typical of a clastic-dominated shelf undergoing transgression. As discussed in Chapter 11, **transgression** refers to movement of a shoreline in a landward direction, also called **retrogradation.** The column at the left side of the block in Figure 14.7 shows how sediments from different, laterally adjacent environments have

FIGURE 14.7   Schematic representation of Walther's Law. Transgression results in lateral shifts of environments and resulting facies, producing the vertical succession of facies shown in the column.

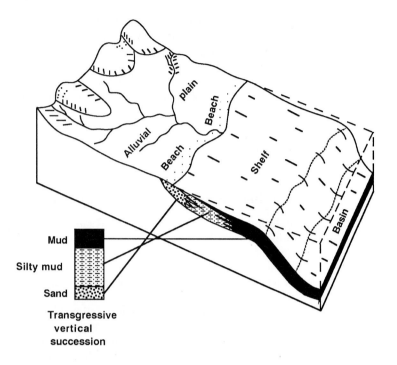

Mud

Silty mud

Sand

**Transgressive
vertical
succession**

become superimposed vertically as the shoreline advanced landward, resulting in sediments from more distant offshore environments being deposited progressively on top of sediments previously laid down nearer to shore. Transgression of shorelines thus produces vertical successions of sedimentary units in which deeper-water (commonly fine-grained) sediments are superimposed on coarser-grained nearshore sediments, creating fining-upward successions of strata. Transgressions occur during a relative rise in sea level when the influx of terrigenous sediments from land sources is low enough (Fig. 14.8A) to allow deeper-water marine sediments to encroach landward over nearshore deposits (coastal encroachment). Transgression will not occur during rising sea level if the influx of terrigenous sediments is so high that outbuilding of the shoreline takes place.

Under some conditions, shorelines and environments may shift in a seaward direction. Seaward movement of a shoreline is called **regression,** or **progradation.** Regression leads also to vertical superposition of contiguous lateral facies, but, in this case, nearshore (commonly coarse-grained) sediments become progressively stacked on top of finer-grained, deeper-water sediments, leading to coarsening-upward successions. Regression may occur during a relative rise in sea level or during static sea level if the influx of terrigenous clastics is high (Fig. 14.8B), or it may occur during a relative fall in sea level (Fig. 14.9). Transgression followed by regression tends to produce a wedge of sediments in which deeper-water sediments are deposited on top of shallower-water sediments in the basal part of the wedge, and shallower-water sediments are deposited on top of deeper-water sediments in the top part of the wedge (Fig. 14.10). Note the marked coastal onlap illustrated in Figure 14.10. The initial depositional surface at the base of a transgressive succession is commonly an unconformity. The bounding surface at the base of a regressive succession can also be an unconfor-

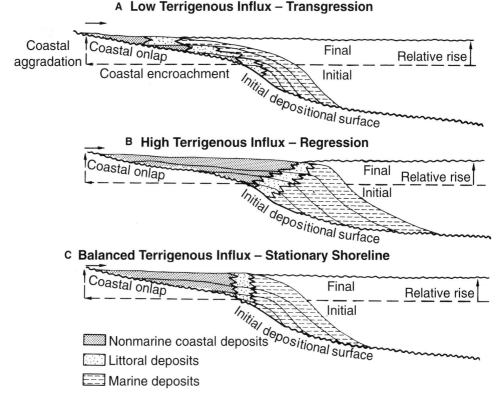

FIGURE 14.8 A relative rise in sea level can produce (A) transgression, (B) regression, or (C) a stationary shoreline, depending upon the rate of terrigenous influx. (After Vail, P. R., R. M. Mitchum, Jr., and S. Thompson, III, 1977. Seismic stratigraphy and global change of sea level. Part 3: Relative changes of sea level from coastal onlap, *in* C. E. Payton (ed.), Seismic stratigraphy—Applications to hydrocarbon exploration: Am. Assoc. Petroleum Geologists Mem. 26. Fig. 3, p. 66, reprinted by permission of AAPG, Tulsa, Okla.)

mity if regression occurs as a result of a relative fall in sea level accompanied by erosion.

Transgressions and regressions thus lead to deposition of vertically stacked successions of fining-upward and coarsening-upward deposits. Under at least two types of depositional conditions, transgression or regression may not occur for a geologically significant period of time. If during a relative rise in sea level the shoreline is stationary for a long period of time as a result of a balance in terrigenous influx, so that neither sediment outbuilding (progradation) nor coastal encroachment (retrogradation) occurs, lateral facies do not become vertically superimposed (Fig. 14.8C). During a relative standstill of the sea when relative sea level is neither rising nor falling and terrigenous influx is sufficiently high, a type of deposition called **coastal toplap** occurs (Vail, Mitchum, and Thompson, 1977a). Shelf sediments cannot build above effective wave base and aggrade because of the standstill of sea level, so onlap cannot be produced. Instead, each unit of strata laps out in a landward direction, but the successive terminations lie progressively seaward (Fig. 14.11).

FIGURE 14.9 Rapid fall in relative sea level indicated by downward shift in coastal onlap. A and B represent two different depositional sequences (see Fig. 14.12), and the numbers indicate relative ages of beds: 1 = oldest, 9 = youngest. (After Vail, P. R., R. M. Mitchum, Jr., and S. Thompson, III, 1977, Seismic stratigraphy and global change of sea level. Part 3: Relative changes of sea level from coastal onlap, *in* C. E. Payton (ed.), Seismic stratigraphy—Applications to hydrocarbon exploration: Am. Assoc. Petroleum Geologists Mem. 26. Fig. 8, p. 72, reprinted by permission of AAPG, Tulsa, Okla.)

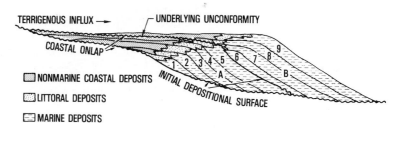

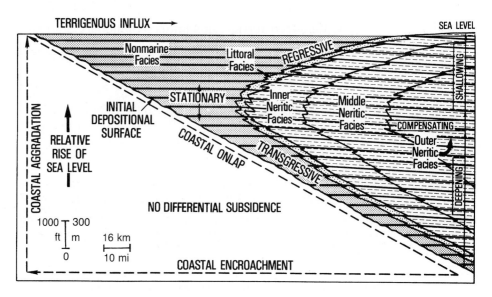

FIGURE 14.10 Coastal onlap owing to marine transgression and regression. During relative rise in sea level, littoral facies may be transgressive, stationary, or regressive. Neritic (shallow-shelf) facies may be deepening, shallowing, or compensating (maintaining a given depth). (From Vail, P. R., R. M. Mitchum, Jr., and S. Thompson, III, 1977, Seismic stratigraphy and global change of sea level. Part 3: Relative changes of sea level from coastal onlap, *in* C. E. Payton (ed.), Seismic stratigraphy—Applications to hydrocarbon exploration: Am. Assoc. Petroleum Geologists Mem. 26. Fig. 4, p. 67, reprinted by permission of AAPG, Tulsa, Okla.)

Practical Application of Walther's Law. To return to Walther's Law, this law not only offers a rational explanation for vertical successions of facies but also has practical application in the study of ancient sedimentary facies and environments. For example, utilization of Walther's Law can aid in understanding depositional environments of strata in a vertical succession. Only a limited number of depositional successions or associations have been found in the study of modern and ancient sediments. To illustrate, the environments recognized on an inner marine shelf undergoing regression

FIGURE 14.11 Coastal toplap. Coastal toplap indicates relative standstill of sea level. During a relative standstill of sea level, no relative rise of base level occurs; therefore, nonmarine coastal and/or littoral deposits cannot aggrade, and no onlap is produced. Instead, sediment bypassing takes place, producing toplap. (From Vail, P. R., R. M. Mitchum, Jr., and S. Thompson, III, 1977, Seismic stratigraphy and global change of sea level. Part 3: Relative changes of sea level from coastal onlap, *in* C. E. Payton (ed.), Seismic stratigraphy—Applications to hydrocarbon exploration: Am. Assoc. Petroleum Geologists Mem. 26. Fig. 6, p. 70, reprinted by permission of AAPG, Tulsa, Okla.)

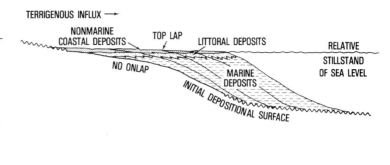

include tidal-flat, lagoon-bay, littoral (beach), wave-zone, shoreface, and the neritic zone below wave base. Sediments deposited in each of these environments have distinctive properties of grain size, sorting, sedimentary structures, and geometry that characterize that particular environment. Knowing that this lateral succession of environments and facies exists in a marine environment undergoing regression, geologists studying ancient regressive marine sediments can interpret the depositional environment of various parts of the vertical succession with greater confidence, and they can also recognize the absence of facies from a particular environment that may have been removed by erosion.

As an additional example of the application of Walther's Law, it is possible to study the vertical succession of beds in an outcrop section or well boring located along the edge of a basin and predict from this succession what the lateral succession of facies will be as a particular stratigraphic unit is followed into the basin. Such predictions are of particular importance to petroleum geologists, who must try to determine possible facies changes in petroleum reservoir beds that extend from outcrop areas or drilled regions into prospective undrilled regions.

Effects of Climate and Sea Level on Sedimentation Patterns

Preceding discussion shows that both the rate of influx of terrigenous clastic sediments and the change in relative sea level exert controls on sedimentation patterns in coastal areas and on the continental shelf. In turn, terrigenous influx is itself influenced by tectonism and climatic conditions. Tectonism produces changes in elevation of sediment source areas and thus affects rates of erosion, which generally increase with increase in land elevation. Also, source areas at higher elevations, and with steeper slopes, tend to shed coarser sediment than do those at lower elevations. Climate regulates sediment influx by controlling rates of weathering and erosion, sediment transport conditions, and sedimentation mechanisms. For example, in a given geographic area, significantly greater terrigenous influx will occur during periods of heavy rain (e.g., during the winter rainy season), when erosion rates are accelerated and stream transport is increased, than during dry periods. On a shorter time scale, more sediment, and coarser sediment, may be eroded and transported during a single unusually large, high-velocity flood that occurs only once every hundred years than during all the smaller floods that may have occurred during the preceding hundred years. [See Clifton (1988) for discussion of the sedimentologic consequences of large convulsive geologic events.] Thus, the rates of sediment influx and the grain sizes of sediments delivered to coastal areas from continents have varied throughout geologic time in response to these variables of tectonism and climate.

Changes in sea level also affect sedimentation patterns in coastal areas. Changes in sea level that are worldwide and that affect sea level on all continents essentially simultaneously are called **eustatic sea-level changes.** Changes of sea level that affect only local areas are referred to as **relative sea-level changes.** Relative sea-level changes may involve some global eustatic change but are also affected by local tectonic uplift or downwarping of the basin floor and sediment aggradation (buildup). Local tectonics and rates of sedimentation have little or no effect on worldwide sea levels.

Eustatic sea-level changes have been attributed to a variety of causes, all of which can be lumped under changes in volume of water and changes in volume of the ocean basins (Table 14.1). The most important changes in water volume are tied to continental glaciation. Sea level drops during glacial stages, when seawater is locked up on land as ice, and rises during interglacial stages as continental ice sheets melt. Water may also be tied up on land in lakes, reservoirs, and groundwater aquifers. Finally, fluctuations in ocean temperature (Table 14.1) may produce small variations in sea level. Changes in volume of an ocean basin may be brought about by a variety of causes. Sediment infill of an ocean basin, for example, would cause sea level to rise. Changes in the volume of the mid-ocean ridge system may be another cause. Changes in volume of mid-ocean ridges occur as a result of variations in rates of seafloor spreading. An increase in rates of seafloor spreading causes an increase in volume of mid-ocean ridges and a consequent rise in sea level, and a decrease in spreading rates generates a decrease in ridge volume and a corresponding fall in worldwide sea level. Pitman (1978) suggests, for example, that change in the rate of seafloor spreading from

TABLE 14.1 Postulated mechanisms of sea-level change

Mechanisms	Time scale (yr)	Order of magnitude
1. Ocean steric (thermohaline) volume changes		
Shallow (0–500 m)	0.1–100	0–1 m
Deep (500–4000 m)	10–10,000	0.01–10 m
2. Glacial accretion and wastage		
Mountain glaciers	10–100	0.1–1 m
Greenland Ice Sheet	100–100,000	0.1–10 m
East Antarctic Ice Sheet	1,000–100,000	10–100 m
West Antarctic Ice Sheet	100–10,000	1–10 m
3. Liquid water on land		
Groundwater aquifers	100–100,000	0.1–10 m
Lakes and reservoirs	100–100,000	0.01–0.1 m
4. Crustal deformation		
Lithosphere formation and subduction	$100{,}000–10^8$	1–100 m
Glacial isostatic rebound	100–10,000	0.1–10 m
Continental collision	$100{,}000–10^8$	10–100 m
Seafloor and continental epirogeny	$100{,}000–10^8$	10–100 m
Sedimentation	$10{,}000–10^8$	1–100 m

Source: Revelle, 1990.

2 cm/yr to 6 cm/yr in the modern ocean could produce a rise in sea level of more than 100 m over a period of 70 million years. Correspondingly, a decrease in spreading rate back to 2 cm/yr for the next 70 million years would cause sea level to drop by more than 100 m. Other possible ways of changing the volume of an ocean basin include glacial isostatic rebound, upwarping or downwarping of the seafloor, and continental collision (Table 14.1). Isostatic rebound after glaciation involves gradual rise in a land surface, which has been depressed owing to weight of the ice, after weight of the ice has been removed by melting. Changes in sea level and the methods that stratigraphers use to determine the magnitude of sea-level changes from the stratigraphic record are further discussed under "Sea-Level Analysis" in Chapter 15.

14.5 NOMENCLATURE AND CLASSIFICATION OF LITHOSTRATIGRAPHIC UNITS

To bring order to strata and to understand to the fullest extent the geologic history recorded in these strata, it is necessary to have a formal system for defining, classifying, and naming geologic units. Such a stratigraphic procedure promotes systematic study of the physical properties and successional relationships of sedimentary strata and is essential for correct interpretation of depositional environments and other aspects of Earth history. The need for systematic organization of strata was recognized as early as the latter half of the eighteenth century by European scientists such as Johann Gottlob Lehman, Giovanni Arduino, and Georg Christian Füchsel, who made early attempts to organize strata on the basis of relative age (Krumbein and Sloss, 1963). The gradual evolution of these efforts to organize and classify strata continued through the eighteenth and nineteenth centuries and eventually culminated in formulation of the internationally used Geologic Time Scale and the Geologic (Stratigraphic) Column. This evolution is one of the more fascinating chapters in the history of stratigraphic study. Succinct summaries of these early efforts at stratigraphic classification are given by Weller (1960), Krumbein and Sloss (1963), and Dunbar and Rogers (1967).

Development of the Stratigraphic Code

Local study of rock strata requires subdivision of the Stratigraphic Column into smaller units that are systematically arranged on the basis of inherent properties and attributes. The purpose of stratigraphic classification is thus to promote understanding of the geometry and successions of rock bodies. To ensure uniform usage of stratigraphic nomenclature and classification, attempts have been underway for several decades to adopt a code of stratigraphic nomenclature that formulates views on stratigraphic principles and practices designed to promote standardized classification and formal nomenclature of rock materials. In the United States, such codes have been drafted by the Committee on Stratigraphic Nomenclature, 1933, and its successors, the American Commission on Stratigraphic Nomenclature, 1961, and the North American Commission on Stratigraphic Nomenclature, 1983. The Code of Stratigraphic Nomenclature, published by the American Commission on Stratigraphic Nomenclature in 1961 and revised slightly in 1970, standardized terminology and practices used in stratigraphy in the United States at that time and was widely accepted by North American geologists. New concepts and techniques, particularly the concept of global plate tectonics, have developed in the past few decades. These developments have revolutionized the earth sciences and have necessitated revision of the 1961 code. The International Stratigraphic Guide, published by the International Subcommission on Stratigraphic Classification in 1976 (Hedberg, 1976), provided a comprehensive treatment of stratigraphic classification, terminology, and procedures from an international point of view. In order to incorporate new concepts and techniques and to recognize the contribution of international stratigraphic organizations, the North American Commission on Stratigraphic Nomenclature published a new North American Stratigraphic Code in May 1983. For the convenience of readers, this code is reproduced in full in Appendix B.

Major Types of Stratigraphic Units

The 1983 North American Code of Stratigraphic Nomenclature is a new code—not a revision of the 1961 code. Some categories of stratigraphic units included in the 1961 code have disappeared; others are new. In general, the new code has been prepared to be as consistent as possible with the International Stratigraphic Guide. The various categories of stratigraphic units recognized by the code are summarized in Table 14.2. Note that some stratigraphic units (e.g., lithostratigraphic units, biostratigraphic units) are based on observable characteristics of rocks. Such units are identified in the field on the basis of physical or biological properties that can be measured (e.g., grain size),

TABLE 14.2   Categories of stratigraphic units defined by the 1983 North American Stratigraphic Code

---

Material categories based on content or physical limits (composition, texture, fabric, structure, color, fossil content)

   Lithostratigraphic units—conform to the law of superposition and are distinguished on the basis of lithic characteristics and lithostratigraphic position

   Lithodemic units—consist of predominately intrusive, highly metamorphosed, or intensely deformed rock that generallly does not conform to the law of superposition

   Magnetopolarity units—bodies of rock identified by remnant magnetic polarity

   Biostratigraphic units—bodies of rock defined and characterized by their fossil content

   Pedostratigraphic units—consist of one or more pedologic (soil) horizons developed in one or more lithic units now buried by a formally defined lithostratigraphic or allostratigraphic unit or units

   Allostratigraphic units—mappable stratiform (in the form of a layer) bodies defined and identified on the basis of bounding discontinuities

Categories expressing or related to geologic age

   Material categories to define temporal spans (stratigraphic units that serve as standards for recognizing and isolating materials of a particular age)

      Chronostratigraphic units—bodies of rock established to serve as the material reference for all rocks formed during the same spans of time

      Polarity-chronostratigraphic units—divisions of geologic time distinguished on the basis of the record of magnetopolarity as embodied in polarity-chronostratigraphic units

   Temporal (nonmaterial) categories—(not material units but conceptual units, i.e., divisions of time)

      Geochronologic units—divisions of time distinguished on the basis of the rock record as expressed by chronostratigraphic units

      Polarity-chronologic units—divisions of geologic time distinguished on the basis of the record of magnetopolarity as embodied in polarity-chronostratigraphic units

      Diachronic units—comprise the unequal spans of time represented by one or more specific diachronous rock bodies, which are bodies with one or two bounding surfaces that are not time synchronous and thus "transgress" time

      Geochronometric units—isochronous units (units having equal time duration) that are direct divisions of geologic time expressed in years

---

Source: North American Commission on Stratigraphic Nomenclature, 1983, North American Stratigraphic Code: Am. Assoc. Petroleum Geologists Bull., v. 67, reprinted by permission of AAPG, Tulsa, Okla.

sensed by instruments (e.g., magnetic polarity), or described (e.g., sedimentary structures, kinds of fossils). Others are related to geologic ages of rocks. Stratigraphic units having time significance may be actual units of rock (e.g., chronostratigraphic units) that formed during particular time intervals, or they may simply be divisions of time (e.g., geochronologic units) and not actual rock units.

Categories and ranks of all stratigraphic units defined in the 1983 code are shown in Table 14.3. Procedures and requirements for defining formal stratigraphic units are set forth in detail in the code (Appendix B). These procedures include requirements for picking a name, designating a stratotype or type section, describing the units, specifying the boundaries between units, and publishing appropriate descriptions of the units in a recognized scientific medium. Our immediate concern

TABLE 14.3 Categories and ranks of stratigraphic units defined in the 1983 North American Stratigraphic Code

A   Material Units

Lithostratigraphic	Lithodemic	Magnetopolarity	Biostratigraphic	Pedostratigraphic	Allostratigraphic
Supergroup	Supersuite				
Group	Suite (Complex)	Polarity Superzone			Allogroup
Formation	Lithodeme	Polarity zone	Biozone (Interval, Assemblage, or Abundance)	Geosol	Alloformation
Member (or Lens, or Tongue)		Polarity Subzone	Subbiozone		Allomember
Bed(s) or Flow(s)			Realism		

B   Temporal and Related Chronostratigraphic Units

Chrono-stratigraphic	Geochronologic Geochronometric	Polarity chrono-stratigraphic	Polarity chronologic	Diachronic
Eonothem	Eon	Polarity Superchronozone	Polarity Superchron	
Erathem (Supersystem)	Era (Superperiod)			
System (Subsystem) Series	Period (Subperiod) Epoch	Polarity Chronozone	Polarity Chron	Episode Phase
Stage (Substage) Chronozone	Age (Subage) Chron	Polarity Subchronozone	Polarity Subchron	(Diachron) Span Cline

Source: North American Commission on Stratigraphic Nomenclature, 1983. North American Stratigraphic Code: Am. Assoc. Petroleum Gelogists Bull., v. 67, Table 2, p. 852, reprinted by permission of AAPG, Tulsa, Okla.

Note: Fundamental units are italicized.

here is with subdivision and nomenclature of lithostratigraphic units. Other types of stratigraphic units are described in subsequent parts of the text.

Formal Lithostratigraphic Units

The concept of formations and other formal lithostratigraphic units is briefly introduced in Section 14.2 above. In terms of size, the hierarchy of lithostratigraphic units in descending order is supergroup, group, formation, member, and bed (Table 14.4). Although a **formation** is not the largest lithostratigraphic unit, it is nonetheless the fundamental unit of lithostratigraphic classification. **All other lithostratigraphic units are defined as either assemblages or subdivisions of formations.** Note from Table 14.4 that a formation is defined strictly on the basis of lithology. Formations may be defined on the basis of a single lithic type, repetitions of two or more lithic types, or extreme lithic heterogeneity where such heterogeneity constitutes a form of unity when compared to adjacent units. For example, a formation might be composed entirely of shale, entirely of sandstone, or of an intimate mixture of sandstone and shale beds that is distinctive because of the mixed lithology. Boundaries of formations, as with all lithostratigraphic units, are placed at the position of lithic change. Boundaries between different formations may, therefore, occur both vertically and laterally. That is, a formation may be located above or below another formation or be positioned laterally adjacent to another formation where lateral facies changes occur. Illustrations of different types of formation boundaries are given in Figure 2, Appendix B. A formation must be of sufficient areal extent and thickness to be mappable at the scale of mapping commonly used in the region where it occurs.

Formal lithostratigraphic units are assigned names that consist of a geographic name combined with the appropriate rank (formation, member, etc.) or an appropriate lithic term, such as limestone, or both. Formation names thus consist of a geographic

TABLE 14.4 Hierarchy of lithostratigraphic units

Supergroup—a formal assemblage of related or superposed groups or of groups and formations.

Group—Consists of assemblages of formations, but groups need not be composed entirely of named formations.

Formation—a body of rock, identified by lithic characteristics and stratigraphic position, that is prevailingly but not necessarily tabular and is mappable at Earth's surface and traceable in the subsurface. Must be of sufficient areal extent to be mappable at the scale of mapping commonly used in the region where it occurs. **The fundamental lithostratigraphic unit**—formations are grouped to form higher-rank lithostratigraphic units and are divided to form lower-rank lithographic units.

Member—the formal lithostratigraphic unit next in rank below a formation and always part of some formation. A formation need not be divided entirely into members. A member may extend laterally from one formation to another.

Lens (or lentil)—a geographically restricted member that terminates on all sides within a formation.

Tongue—a wedge-shaped member that extends beyond the main boundary of a formation or that wedges or pinches out within another formation.

Bed—distinctive subdivisions of a member; the smallest formal lithostratigraphic unit of sedimentary rock. Members commonly are not divided entirely into beds.

Flow—the smallest formal lithostratigraphic unit of volcanic rock.

Source: North American Commission on Stratigraphic Nomenclature, 1983, North American Stratigraphic Code: Am. Assoc. Petroleum Geologists Bull., v. 67, reprinted by permission of AAPG, Tulsa Okla.

name followed by either the word formation or a lithic designation. For example, a particular formation might be called the Otter Point Formation (geographic name only) or the Eureka Quartzite (geographic name plus lithic designation). The names of members include a geographic name and the word member, or the name may have an intervening lithic designation such as Eau Claire Sandstone Member. A group name combines a geographic name with the word group, as in Arbuckle Group. The first letters of all words used in formal names of lithostratigraphic units are capitalized.

The North American Stratigraphic Code of 1983 recognizes that some lithostratigraphic bodies are bounded, top and bottom, by discontinuities (unconformities or diastems). The code introduces the name **allostratigraphic unit** for such mappable stratiform bodies of sedimentary rock that are defined on the basis of bounding, laterally traceable discontinuities rather than on the basis of lithologic change. The International Stratigraphic Guide (Hedberg, 1976) proposes the term **synthem** for unconformity-bounded units.

Informal names may be used for lithostratigraphic units when there is insufficient need, insufficient information, or an inappropriate basis to justify designation as a formal unit (Hedberg, 1976). Informal names may be applied to such units as oil sands, coal beds, mineralized zones, quarry beds, and key or marker beds. Informal names are not capitalized. Examples of informally designated names are "shaley zone," "coal-bearing zone," "pebbly beds," and "siliceous-shale member."

14.6 DEPOSITIONAL SEQUENCES AND SEQUENCE STRATIGRAPHY

Definition

The term sequence is often used informally by geologists to refer to any grouping or succession of strata. Sequence is also used in a more restricted sense to identify distinctive stratigraphic units that are commonly bounded by unconformities (i.e., similar to allostratigraphic units). Sloss (1963) considered sequences to be major rock-stratigraphic units of interregional scope that are separated and delimited by interregional unconformities. He recognized and named six major sequences on the North American craton, each separated by demonstrable regional unconformities that can be traced from the Cordilleran region of western North America to the Appalachian Basin in the east. Each succession or sequence represents a major cycle of transgression and regression, that is, advance and retreat of shorelines. Recognition of the sequences is based on physical relationships among rock units, although Sloss indicates that the sequences also have time-stratigraphic significance.

The sequence concept was subsequently extended and redefined by Mitchum, Vail, and Thompson (1977). These authors define a **depositional sequence as** "a stratigraphic unit composed of a relatively conformable succession of genetically related strata and bounded at its top and base by unconformities or their correlative conformities." Sequences as thus defined differ from Sloss's sequences in that they may be much smaller rock units (a few tens of meters to as much as a thousand meters; Wilson, 1992). Also, because they are bounded by interregional unconformities and their equivalent conformities, they may be traceable over major areas of ocean basins as well as continents. Distinct, related groups of depositional sequences superposed one on another are designated by Mitchum, Vail, and Thompson (1977) as supersequences. These supersequences are of the same general order of magnitude as Sloss's original sequences. The basic concept of depositional sequences is illustrated in Figure 14.12.

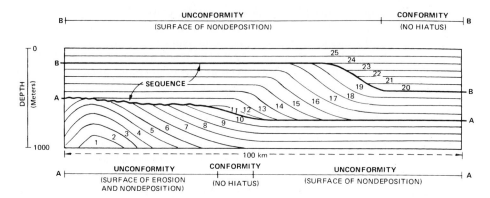

FIGURE 14.12   Illustration of the concept of depositional sequences. A depositional sequence is composed of relatively conformable, genetically related strata bounded at its base (A) and top (B) by unconformities that pass laterally to correlative conformities. Individual units of strata 1 through 25 are traced by following stratification surfaces; they are assumed to be conformable where successive strata are present. Where units of strata are missing, hiatuses are present. (From Mitchum, R. M., Jr., P. R. Vail, and S. Thompson, III, 1977, Seismic stratigraphy and global change of sea level. Part 2: The depositional sequence as a basic unit for stratigraphic analysis, *in* C. E. Payton (ed.), Seismic stratigraphy—Applications to hydrocarbon exploration: Am. Assoc. Petroleum Geologists Mem. 26. Fig. 1, p. 54, reprinted by permission of AAPG, Tulsa, Okla.)

Because sequences are defined on the basis of physical relationships of the strata—that is, bounded at the top and base by unconformities or their correlative conformities—they are not primarily dependent for recognition upon determination of rock types, fossils, or depositional processes.

Time Significance of Sequences

Depositional sequences have time-stratigraphic significance in the sense that all strata within a sequence were deposited during a given broad interval of time, although the age range of individual strata within the sequence may differ from place to place. Surfaces separating depositional sequences may be either unconformities or, in the case of correlative conformities, stratal surfaces or bedding planes (Fig. 14.12). The hiatus represented by unconformities may range from millions to hundreds of millions of years. On the other hand, the physical surfaces that separate groups of strata or individual beds and laminae within sequences were produced during a relatively short period of time and are essentially synchronous.

Internal Relationships

The strata that make up a depositional sequence may be either **concordant,** that is, essentially parallel to the sequence boundary, or **discordant,** lacking parallelism with respect to the sequence boundaries. Concordant relations can occur at either the upper or the lower boundary of a sequence and may be expressed as parallelism to an initially horizontal, inclined, or uneven surface (Fig. 14.13). Discordance is the most important physical criterion used in determining sequence boundaries. Two main types of discordance are recognized depending upon the manner in which the strata terminate against the sequence boundaries. **Truncation** is the lateral termination of strata owing to being cut off from their original depositional limits by erosion. Truncation occurs at the upper boundary of a sequence and may be of either local or regional extent. **Lapout** is the lateral termination of strata against a boundary at their original depositional limit. Lapout relationships are further divided into two types, baselap and toplap, depending upon the specific nature of the discordant relationship with the upper or lower sequence boundary.

Baselap occurs at the lower boundary of a depositional sequence and may itself be of two types. **Onlap** is baselap in which an initially horizontal or inclined stratum terminates against a surface of greater inclination. **Downlap** is baselap in which an initially inclined stratum terminates downdip against an initially horizontal or inclined surface. Onlap and downlap indicate nondepositional hiatuses (Fig. 14.13) and not erosional breaks in deposition.

Toplap is lapout at the upper boundary of a depositional sequence, for example, the lateral termination updip of the foreset beds of a deltaic complex. Toplap is also evidence of a nondepositional hiatus. Mitchum, Vail, and Sangree (1977) suggest that toplap results from a depositional base level, such as sea level, being too low to permit the strata to extend farther updip, thus allowing sedimentary bypassing and possibly

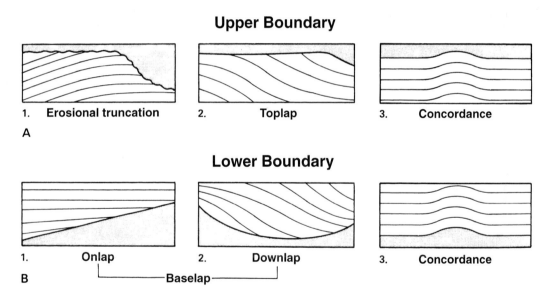

FIGURE 14.13 Relations of strata to the (A) upper boundary and (B) lower boundary of a depositional sequence. A1. Erosional truncation: strata terminate against the upper boundary mainly owing to erosion. A2. Toplap: initally inclined strata terminate against an upper boundary created mainly as a result of nondeposition (e.g., foreset strata terminating against an overlying horizontal surface where no erosion or deposition took place). A3. Top concordance: strata at the top of a sequence do *not* terminate against an upper boundary. B1. Onlap: strata terminate updip against an inclined surface. B2. Downlap: initally inclined strata terminate downdip progressively against initally horizontal or inclined surface. B3. Base-concordance: strata at base of a sequence do *not* terminate against lower boundary. (From Mitchum, R. M., Jr., P. R. Vail, and S. Thompson, III, 1977, Seismic stratigraphy and global change of sea levels. Part 2: The depositional sequence as a basic unit for stratigraphic analysis, *in* C. E. Payton (ed.), Seismic stratigraphy—Applications to hydrocarbon exploration: Am. Assoc. Petroleum Geologists Mem. 26. Fig. 2, p. 58, reprinted by permission of AAPG, Tulsa, Okla.)

FIGURE 14.14 Terminology for relations that define unconformable boundaries of a depositional sequence. (After Mitchum, R. M., Jr., P. R. Vail, and J. B. Sangree, 1977, Seismic stratigraphy and global change of sea level. Part 6: Stratigraphic interpretation of seismic reflection patterns in depositional sequences, *in* C. E. Payton (ed.), Seismic stratigraphy—Applications to hydrocarbon exploration: Am. Assoc. Petroleum Geologists Mem. 26, Fig. 1, p. 118, reprinted by permission of AAPG, Tulsa, Okla.)

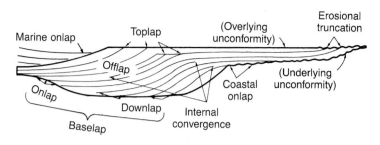

minor erosion to occur above base level while prograding strata are deposited below base level.

Toplap and baselap relations of the strata in a depositional sequence should not be confused with the foreset bedding of cross-laminated units which form parts of beds rather than sequences. Figure 14.14 diagrammatically illustrates in a regional setting the relationships of strata in depositional sequences to sequence boundaries.

Identification of Depositional Sequences

Depositional sequences can be identified both in outcrop sections and in subsurface sections by searching for the presence of unconformities (erosional surfaces, truncation of strata, missing strata). Subsurface identification depends largely upon the use of instrumental well logs, such as electric logs that measure resistivity of rock units, and seismic data (Chapter 15) to locate and trace unconformities, truncations, and lapout relationships. In fact, the sequence concept was developed initially from seismic data. Sequence boundaries are identified on the basis of physical stratigraphic relationships, but where fossil information is available, as from well-core data, the geologic ages of the sequences are also determined. An example of sequences defined in a subsurface section by correlation of strata using electric log data is given in Figure 14.15 (also see Van Wagoner et al., 1990).

Stratigraphic Rank of Sequences

With respect to the Code of Stratigraphic Nomenclature (Appendix B), a sequence is a kind of informal stratigraphic unit, not a formal stratigraphic unit. As originally defined by Sloss (1963), sequences were considered to be "rock-stratigraphic units of higher rank than group, megagroup, or supergroup"; however, as redefined by Mitchum, Vail, and Sangree (1977), depositional sequences are an order of magnitude smaller than Sloss's sequences. Thus, a depositional sequence might include a single formation, groups of formations, or, on a smaller scale, subdivisions of formations. Sequences are similar to but not exactly the same as allostratigraphic units. Allostratigraphic units are descriptive formal units, defined by the code, that are bounded by discontinuities, irrespective of how the discontinuities formed. Seqences are theoretical, informal, interpretive units recognized on the basis of bounding unconformities and correlative conformities.

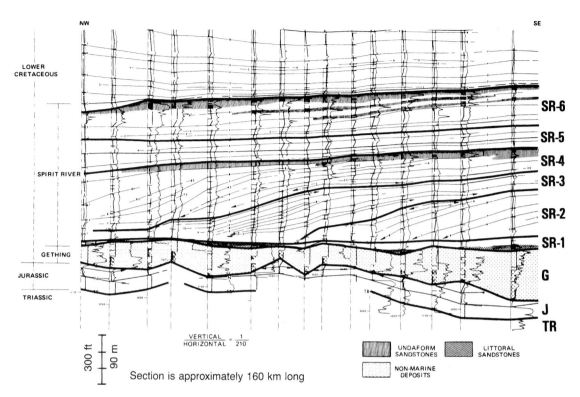

FIGURE 14.15 Sequences (SR-1 through SR-6) defined in the subsurface on the basis of well-log marker correlation. (From Mitchum, R. M., Jr., P. R. Vail, and S. Thompson, III, 1977, Seismic stratigraphy and global change in sea level. Part 6: Stratigraphic interpretation of seismic reflection patterns in depositional sequences, *in* C. E. Payton (ed.), Seismic stratigraphy—Applications to hydrocarbon exploration: Am. Assoc. Petroleum Geologists Mem. 26. Fig. 3, p. 60, reprinted by permission of AAPG, Tulsa, Okla.)

Sequence Stratigraphy

The sequence concept proposed by Mitchum, Vail, and Sangree (1977), which is a largely theoretical concept tied very closely to the assumption of eustatic sea-level changes, has been honed in a series of subsequent papers by Vail, Mitchum, and coworkers at Exxon Production Research Company in Houston (e.g., Posamentier, Jervey, and Vail, 1988; Van Wagoner et al., 1988; Van Wagoner et al., 1990). This honing process has been accompanied by a corresponding increase in terminology related to sequences, for example, depositional system, systems tract, parasequence, and parasequence sets. Table 14.5 explains some of these terms and shows the hierarchy of sequence-stratigraphic units.

Sequences are composed of **depositional systems,** which are major three-dimensional assemblages of lithofacies enclosed within sequence boundaries. Two types of sequence boundaries are recognized in coastal deposits, where rocks are subject to subaerial exposure (Van Wagoner et al., 1988).

Type 1 sequence boundaries are the result of subaerial exposure and accompanying subaerial erosion associated with stream rejuvenation and downcutting, a basin-

TABLE 14.5   Hierarchy of sequence-stratigraphic units

Depositional Sequence—genetically related strata bounded by surfaces of erosion or nondeposition or their correlative conformities. Two kinds of sequence boundaries are recognized:

Type 1 sequence boundary—characterized by subaerial exposure and concurrent erosion associated with stream rejuvenation, a basinward shift of facies, a downward shift in coastal onlap, and onlap of overlying strata.

Type 2 sequence boundary—marked by subaerial exposure and a downward shift in coastal onlap landward of the depositional-shoreline break; however, it lacks both subaerial erosion associated with stream rejuvenation and a basinward shift in facies.

Stratal units within sequences include:

Depositional System—a three-dimensional assemblage of lithofacies, genetically linked by active (modern) or inferred (ancient) processes and environments (e.g., fluvial, deltaic, barrier-island).

Systems Tract—a subdivision of a depositional system. Four kinds are recognized: (1) **lowstand**—lies directly on a type 1 sequence boundary; (2) **highstand**—the upper system tract in either a type 1 or a type 2 sequence; (3) **shelf-margin**—the lowermost system tract associated with a type 2 sequence boundary; and (4) **transgressive**—the middle systems track of both type 1 and 2 sequences.

Parasequence Set—a succession of genetically related parasequences that form a distinctive stacking pattern that is bounded, in many cases, by major marine-flooding surfaces and their correlative surfaces.

Parasequence—a relatively conformable succession of genetically related beds or bedsets (within a parasequence set) bounded by marine flooding surfaces or their correlative surfaces.

Marine-flooding Surfaces—a surface that separates younger from older strata, across which there is evidence of an abrupt increase in water depth.

Source: Van Wagoner et al., 1988; Van Wagoner et al., 1990.

ward shift in facies, a downward shift in coastal onlap, and onlap of overlying strata. As a result of such a basinward shift of facies, nonmarine or very shallow-marine rocks (e.g., braided-stream or estuarine sandstones) above a sequence boundary may directly overlie deeper-water marine rocks (e.g., lower shoreface sandstone or shelf mudstone) below a boundary—with no intervening rocks deposited in intermediate depositional environments (Fig. 14.16A). A type 1 sequence boundary forms when the rate of eustatic sea-level fall exceeds the rate of basin subsidence at the depositional-shoreline break, producing a relative fall in sea level at that position. The depositional-shoreline break is a position on the shelf, landward of which the depositional surface is at or near base level, commonly sea level, and seaward of which the depositional surface is below base level (Fig. 14.16). (*Note:* A sequence bounded by type 1 boundaries is called a type 1 sequence.)

Type 2 sequence boundaries are also associated with subaerial exposure and are marked by a downward shift in coastal onlap landward at the depositional-shoreline break. They differ from type 1 boundaries in that they lack both subaerial erosion associated with stream rejuvenation and a basinward shift in facies. Onlap of overlying strata landward of the depositional-shoreline break also marks a type 2 sequence boundary (Fig. 14.16B). Type 2 sequence boundaries are generated when the rate of eustatic sea-level fall is less than the rate of basin subsidence at the depositional-shoreline break; thus, no relative fall in sea level occurs at this shoreline position (Van Wagoner et al., 1988).

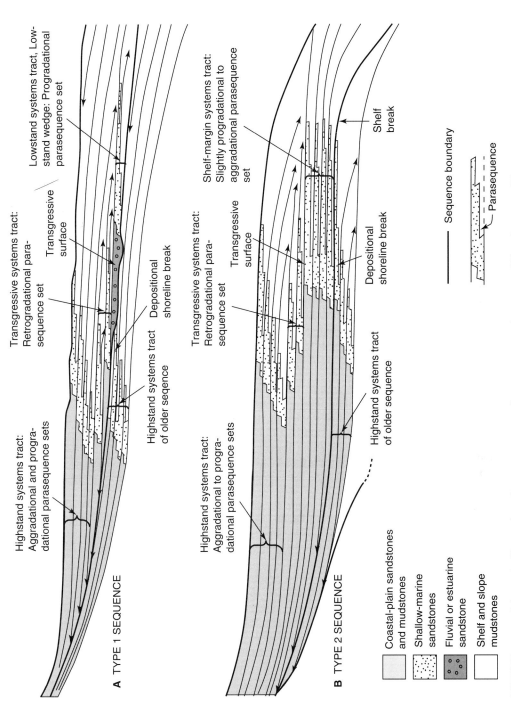

FIGURE 14.16 Schematic illustration of systems tracts and parasequences in a type 1 (A) and a type 2 (B) sequence. The arrows show truncation of strata against some kind of bounding surface (e.g., a transgressive surface or a sequence boundary). (After Van Wagoner, J. C., et al., 1990, Siliciclastic sequence stratigraphy in well logs, cores, and outcrops: Am. Assoc. Petroleum Geologists Methods in Exploration Series 7. Fig. 20A, p. 28, and Fig. 20B, p. 29, reproduced by permission of AAPG, Tulsa, Okla.)

Highstand systems tract: Aggradational and progradational parasequence sets

Transgressive systems tract: Retrogradational parasequence set

Transgressive surface

Lowstand systems tract, Lowstand wedge: Progradational parasequence set

Highstand systems tract of older seqence

Depositional shoreline break

A TYPE 1 SEQUENCE

Highstand systems tract: Aggradational to progradational parasequence sets

Transgressive systems tract: Retrogradational parasequence set

Transgressive surface

Shelf-margin systems tract: Slightly progradational to aggradational parasequence set

Shelf break

Depositional shoreline break

Highstand systems tract of older sequence

B TYPE 2 SEQUENCE

Coastal-plain sandstones and mudstones

Shallow-marine sandstones

Fluvial or estuarine sandstone

Shelf and slope mudstones

Sequence boundary

Parasequence

Depositional systems are composed of successions of genetically related strata deposited under particular environmental conditions—such as fluvial, deltaic, and shelf—associated with different parts of a cycle of eustatic rise and fall. These associated depositional systems are called **systems tracts** (Table 14.5). Systems tracts form during different parts of a cycle of base-level change that goes from emergence to submergence (maximum water depth) and back to emergence. They may lie immediately above or below a sequence boundary or lie in the middle of a sequence; four types are recognized depending upon the sea-level conditions under which they form.

Highstand systems tracts (Fig. 14.16) lie immediately below either a type 1 or a type 2 sequence boundary and form during the late part of a sea-level rise, or during a sea-level standstill or during the early part of a sea-level fall. They form under conditions that allow progradation, and fluvial sediments characterize the latter part of highstand systems tracts. They are terminated by the unconformity produced by the next eustatic sea-level fall.

Lowstand systems tracts lie above a type 1 sequence boundary. They may form during a time of rapid eustatic sea-level fall or during late sea-level fall or early rise. During a rapid eustatic fall, basin-floor fans develop when sea level drops below the shelf edge, causing the shelf to become subaerially exposed. Rivers become incised and bypass the shelf to feed sediments directly onto the slope, where they may be transported downslope by sediment gravity flows and deposited as basin-floor submarine fans. Once the rate of eustatic sea-level fall begins to slow, subsidence at the shelf edge may again exceed the rate of fall so that a relative rise in sea level is achieved. Under these conditions, deltaic sediments may form and fill previously incised river valleys. As relative sea level continues to rise, deposition from sediment gravity flows generates prograding submarine fans on the slope. Deposition of prograding lowstand-fan systems can continue as long as the rate of sediment supply exceeds the rate of relative sea-level rise. Deposition terminates when the rate of sediment supply cannot keep up with the combined rates of eustatic sea-level rise and basin subsidence.

Shelf-margin systems tracts are the lowermost systems tract associated with a type 2 sequence boundary; that is, the base of the systems tract is a type 2 boundary. These systems tracts form when the rate of eustatic sea-level fall is lower than that associated with the lowstand systems tract. Aggradation (vertical sediment accumulation), rather than progradation, takes place during deposition of shelf-margin systems tracts.

Transgressive systems tracts lie in the middle of either a type 1 or a type 2 sequence (Fig. 14.16) and form during a rapid eustatic sea-level rise. This rapid rise floods the shelf, preventing rivers from incising. Under these conditions, little fluvial sediment is delivered to the shelf, and marine sediments build in a landward direction.

Systems tracts are made up of smaller depositional units called **parasequence sets,** which range in thickness from a few tens of meters to a few hundreds of meters (Wilson, 1992). Parasequence sets, in turn, are composed of **parasequences** (Table 14.5). Parasequences are bounded by marine flooding surfaces, surfaces that separate younger strata from older, across which there is evidence of an abrupt increase in water depth. They are the smallest facies units of depositional sequences and range in thickness from about 10 to 100 m. Parasequence boundaries are thus sharp boundaries created by sea-level change. A parasequence is deposited during a single episode of submergence. It begins when rising sea level reaches its maximum extent and sediment starts to accumulate. After sea level is stabilized at this maximum level, coastal sediments are deposited to push the shoreline seaward, creating a flat-topped body of sediment whose surface is governed by the position of mean sea level. Thus, siliciclastic parasequences are progradational, shallowing- and (generally) coarsening-upward depositional units. Carbonate parasequences are commonly aggradational (vertical accumulation of sediment) and also shallow upward.

The sequence concept produced a revolution in stratigraphy during the 1980s and early 1990s and gave rise to the term **sequence stratigraphy,** defined as the study of rock relationships within a chronostratigraphic (time-stratigraphic) framework of repetitive, genetically related strata bounded by surfaces of erosion or nondeposition or their correlative conformities (Van Wagoner et al., 1988). Numerous geologists have embraced the sequence stratigraphy concept, and it is currently being widely applied to a variety of stratigraphic successions, particularly within the petroleum industry. Since the initial concept was proposed, numerous papers on sequence stratigraphy have been presented at scientific meetings and published in the geological literature. The concept has not, however, been without its detractors. It has been criticized, among other things, for being largely theoretical and introduced without specific, worked-out examples; for inadequate attention to problems of scale; and for an overemphasis on eustatic sea-level cycles without due regard to the effects of local tectonism and sediment supply (e.g., Walker, 1990; Miall, 1991). Also, it is difficult to apply the sequence concept to nonmarine and deep-sea sediments, which may not be subdivided into well-defined unconformity-bounded units. Nonetheless, the sequence concept has certainly revived interest in stratigraphy and has produced many exciting new ideas. The concept of sequence stratigraphy emerged initially from study of seismic stratigraphic principles (seismic stratigraphy), which we will explore further in Chapter 15.

Galloway (1989) proposed an alternative method for defining sequences based on the concept of repetitive episodes of progradation punctuated by periods of transgression and flooding of depositional surfaces. Galloway's sequences are bounded by features formed during maximum submergence, in contrast to the Exxon sequences which are bounded by features formed during maximum emergence. At this time, the Exxon scheme for defining sequences appears to be more widely understood and used.

14.7 CORRELATION OF LITHOSTRATIGRAPHIC UNITS

Introduction

In the simplest sense, stratigraphic correlation is the **demonstration of equivalency of stratigraphic units.** Correlation is a fundamental part of stratigraphy, and much of the effort by stratigraphers that has gone into creating formal stratigraphic units has been aimed at finding practical and reliable methods of correlating these units from one area to another. Without correlation, treatment of stratigraphy on anything but a purely local level would be impossible.

The concept of correlation goes back to the very roots of stratigraphy. The fundamental principles of correlation have been presented in numerous early textbooks on geology and stratigraphy; especially interesting reviews of these general principles are given in Dunbar and Rodgers (1957), Weller (1960), and Krumbein and Sloss (1963). The continued strong interest in correlation is demonstrated by more recent publication of several books and articles dealing with correlation, particularly statistical methods of correlation (e.g., Agterberg, 1990; Cubitt and Reyment, 1982; Mann, 1981; Merriam, 1981).

The fundamental concepts of stratigraphic correlation were already firmly established by the 1950s and 1960s. These basic principles are still important today; however, the emergence of new concepts and more advanced analytical tools has changed our perception of correlation to some degree, as well as adding new methods for correlation. The development of the field of magnetostratigraphy since the late 1950s, for example, has provided an extremely important new tool for global time-stratigraphic correlation on the basis of magnetic polarity events. Also, rapid advances in computer technology and availability and the application of computer-assisted statistical methods to stratigraphic problems have added a new quantitative dimension to the field of stratigraphic correlation. I will attempt in this section to bring out some of these new developments, along with discussion of the more "classical" concepts of stratigraphic correlation.

Definition of Correlation

In spite of the fact that the concept of correlation goes back to the early history of stratigraphy, disagreement has persisted over the exact meaning of the term. Historically, two points of view have prevailed. One view rigidly restricts the meaning of correlation to demonstration of time equivalency, that is, to demonstration that two bodies of rock were deposited during the same period of time (Dunbar and Rodgers, 1957; Rodgers, 1959). From this point of view, establishing the equivalence of two lithostratigraphic units on the basis of lithologic similarity does not constitute correlation. A broader interpretation of correlation allows that equivalency may be expressed in lithologic, paleontologic, or chronologic terms (Krumbein and Sloss, 1963). In other words, two bodies of rock can be correlated as belonging to the same lithostratigraphic or biostratigraphic unit even though these units may be of different ages. It is clear, from a pragmatic point of view, that most geologists today accept the broader view of correlation. Petroleum geologists, for example, routinely correlate subsurface formations on the basis of lithology of the formations, the specific "signatures" recorded within the formations by instrumental well logs, or the reflection characteristics on seismic records. The 1983 North American Stratigraphic Code (Appendix B) recognizes three principal three kinds of correlation:

1. **lithocorrelation,** which links units of similar lithology and stratigraphic position
2. **biocorrelation,** which expresses similarity of fossil content and biostratigraphic position
3. **chronocorrelation,** which expresses correspondence in age and chronostratigraphic position

Even though our concern in this chapter is correlation based on lithology, it is important to clarify the relationship between chronocorrelation and lithocorrelation. Chronocorrelation can be established by any method that allows matching of strata by age equivalence. Correlation of units defined by lithology may also yield chronostratigraphic correlation on a local scale, but when traced regionally many lithostratigraphic units transgress time boundaries. Stratigraphic units deposited during major transgressions and regressions are notably time-transgressive. Perhaps the most famous North American example of a time-transgressive formation is the Cambrian Tapeats Sandstone in the Grand Canyon region. This sandstone is all Early Cambrian in age at the west end of the canyon and all Middle Cambrian in age at the eastern end (Fig. 14.17). Thus, the Tapeats Sandstone, which can be traced continuously through the canyon region, correlates from one end of the canyon to the other as a lithostratigraphic unit but not as a chronostratigraphic unit. The important point stressed here is that the boundaries defined by criteria used to establish time correlation of stratigraphic units may not be the same as those defined by criteria used to establish lithologic correla-

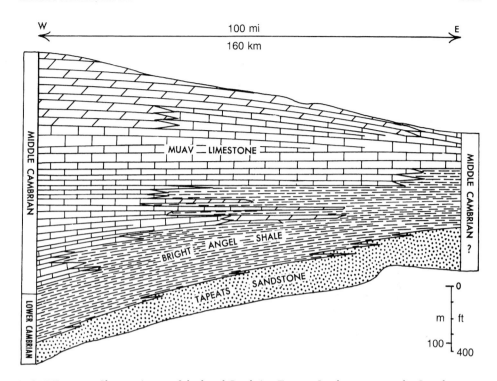

FIGURE 14.17 Changes in age of the basal Cambrian Tapeats Sandstone across the Grand Canyon region. (From Clark, T. H., and C. W. Stern, 1968, Geological evolution of North America, 2nd ed. Fig. 7.10, p. 138, reprinted by permission of John Wiley & Sons, Inc. Originally from E. D. McKee, 1954, Cambrian history of the Grand Canyon region. Part 1. Stratigraphy and ecology of the Grand Canyon Cambrian: Carnegie Inst. Washington Pub. 563, Washington, D.C.)

tion. Because of this fact, different methods of correlation (lithocorrelation, biocorrelation, chronocorrelation) may yield different results when applied to the same stratigraphic succession.

Another point that requires some clarification is the difference between matching of stratigraphic units and correlation of these units. Matching has been defined simply as correspondence of serial data without regard to stratigraphic units (Schwarzacher, 1975; Shaw, 1982). For example, two rock units identified in stratigraphic sections at different localities as having essentially identical lithology (e.g., two black shales) can be matched on the basis of lithology; however, these units may have neither time equivalence nor lithostratigraphic equivalence. Physical tracing of the units between the localities may show that one unit lies stratigraphically above the other. Matching by lithologic characteristics in this particular case does not constitute demonstration of equivalence. Shaw (1982) states that the process of correlation is the demonstration of geometric relationships between rocks, fossils, or successions of geologic data for interpretation and inclusion in facies models, paleontologic reconstructions, or structural models. The object of correlation is to establish equivalency of stratigraphic units between geographically separated parts of a geologic unit. Implicit in this definition is the concept that correlation is made between stratigraphic units, that is, lithostratigraphic units, biostratigraphic units, or chronostratigraphic units. The difference

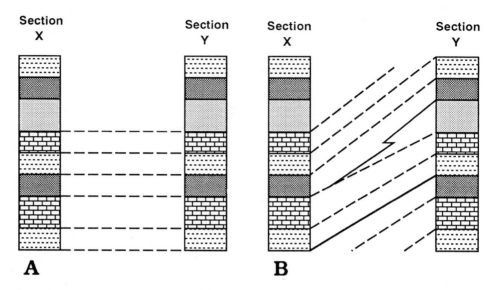

FIGURE 14.18   Illustration of the difference between matching and correlation. A. Apparent correlation achieved by matching of similar-appearing strata. B. Actual lithocorrelation.(After Shaw, A. B., 1964, Time in stratigraphy. Fig. 30.1, p. 214, McGraw-Hill, New York.)

between correlation and matching is illustrated in Figure 14.18. Figure 14.18A shows two stratigraphic sections that appear to be perfectly matched. The actual lithocorrelation is shown in Figure 14.18B. The tie lines in Figure 14.18A do not constitute correlation because they do not encompass equivalent lithostratigraphic units.

Correlation can be regarded as either direct (formal) or indirect (informal) (Shaw, 1982). **Direct correlation** can be established physically and unequivocally. Physical tracing of continuous stratigraphic units is the only unequivocal method of showing correspondence of a lithic unit in one locality to that in another. **Indirect correlation** can be established by numerous methods such as visual comparison of instrumental well logs, polarity reversal records, or fossil assemblages; however, such comparisons have different degrees of reliability and can never be totally unequivocal. Indirect correlation made on the basis of a single unique physical or biological attribute that is deemed to be both necessary and sufficient to establish equivalence is called monothetic (one essential element) correlation. Alternatively, demonstration of equivalence may be determined statistically on the basis of the greatest number of shared characteristics, when no single characteristic is essential or adequate for correlation (polythetic correlation—more than one essential element). Polythetic correlation commonly requires systematic measures involving statistical applications rather than simple visual comparisons. The differences between matching, formal correlation, and indirect correlations are illustrated in Figure 14.19.

Lithocorrelation

We turn now to the methods used for correlating strata on the basis of lithology. Methods of biocorrelation and chronocorrelation are discussed in appropriate sections of Chapters 17 and 18.

FIGURE 14.19 The relationship of formal (direct) correlation, indirect correlation, and matching. (From Shaw, B. R., The correlation of geologic sequences, *in* J. M. Cubitt and R. A. Reyment (eds.), Quantitative stratigraphic correlation. Fig. 2, p. 11, © 1982 by John Wiley & Sons, Ltd., Chichester, England.)

Correlation	Formal	Physical tracing of stratigraphic units		
	Indirect	Arbitrary	Systematic	
		Visual comparisons	Monothetic	Polythetic
			Numeric equivalence	Statistical equivalence
Matching		Comparisons of nonstratigraphic units		

Continuous Lateral Tracing of Lithostratigraphic Units

Direct, continuous tracing of a lithostratigraphic unit from one locality to another is the only correlation method that can establish the equivalence of such a unit without doubt. This correlation method can be applied only where strata are continuously or nearly continuously exposed. The most straightforward way of tracing lithostratigraphic units laterally is by walking out the beds. A geologist who traces a stratigraphic unit continuously from one locality to another by walking along the top of a particular bed can be quite confident that correlation has been established. Thus, the application of field boots and a bit of physical effort yields the satisfaction of achieving a virtually unequivocal correlation. Another useful, but somewhat more equivocal, method of tracing stratigraphic units laterally is to follow the beds on aerial photographs. In areas where surface exposures are abundant and visibility is little hampered by soil or vegetation cover, lateral tracing of thick, distinctive stratigraphic units on aerial photographs can be done rapidly and effectively. This method is limited to tracing of distinctive beds that are thick enough to show up on photographs of a suitable scale (e.g., Fig. 14.20).

FIGURE 14.20 Evenly bedded, laterally extensive Permo-Pennsylvanian formations exposed along the Colorado River at Dead Horse Point, near Moab, Utah.

Although physical tracing of beds is the only unequivocal method of correlation, it is not without limitations. The most serious of these is the fact that in most areas where geologists do mapping, beds cannot be traced continuously for more than a very short distance before encountering areas covered by soil or vegetation, structural complications (faults), or erosional terminations, as across a large valley. In fact, it is often impossible to trace a given stratigraphic unit more than a few hundred meters before the unit is lost for one of these reasons. An additional problem may arise if the beds being traced pinch out or merge with others laterally, a very common occurrence in nonmarine strata. In such a case, tracing of an individual bed or bedding plane will be impossible. Therefore, in practice, geologists commonly trace a gross lithostratigraphic unit (e.g., a member or a formation) consisting of beds of like character, rather than trying to trace individual beds.

Lithologic Similarity and Stratigraphic Position

Lithologic Similarity. Geologists working in areas where direct lateral tracing of beds is not possible must depend for correlation of lithostratigraphic units upon methods that involve matching strata from one area to another on the basis of lithologic similarity and stratigraphic position. Because matching of strata does not necessarily indicate correlation, correlation by lithologic similarity has varying degrees of reliability. The success of such correlation depends upon the distinctiveness of the lithologic attributes used for correlation, the nature of the stratigraphic succession, and the presence or absence of lithologic changes from one area to another. Facies changes that take place in lithostratigraphic units between two areas under study obviously complicate the problem of lithologic correlation.

Lithologic similarity can be established on the basis of a variety of rock properties. These include gross lithology (e.g., sandstone, shale, or limestone), color, heavy mineral assemblages or other distinctive mineral assemblages, primary sedimentary structures such as bedding and cross-lamination, and even thickness and weathering characteristics. The greater the number of properties that can be used to establish a match between strata, the stronger the likelihood of a reliable match. A single property such as color or thickness may change laterally within a given stratigraphic unit, but a suite of distinctive lithologic properties is less likely to change. I caution again that matching of strata on the basis of lithology is not a guarantee that correlation has been established. Strata with very similar lithologic characteristics can form in similar depositional environments widely separated in time or space. It may be quite possible, for example, to obtain an excellent lithologic match between a clean, well-sorted, cross-bedded, eolian sandstone unit of Triassic age and a virtually lithologically identical sandstone of Jurassic age, yet these sandstones do not correlate as either lithostratigraphic or chronostratigraphic units. Correlation on the basis of lithologic identity is particularly difficult between cyclic successions, such as the Pennsylvanian cyclothems of the U.S. mid-continent region. Very similar-appearing successions of units can be repeated over and over in the stratigraphic section owing to the fact that similar environmental conditions can reappear in a region time after time during repeated transgressive-regressive cycles of deposition.

The most reliable lithologic correlations are made when it is possible to match not just one or two distinctive beds or rock types but a succession of several distinctive units. For example, the Triassic and Jurassic formations of the Colorado Plateau in the western United States consist of a highly distinctive succession of largely nonmarine red to green siltstone and mudstone units (the Moenkopi, Chinle, Kayenta,

FIGURE 14.21 Correlation of strata between two localities on the Colorado Plateau based on similar lithology of distinctive stratigraphic units. (From Mintz, L. W., 1981, Historical geology, The science of a dynamic Earth, 3d ed. Fig. 10.1, p.241, reprinted by permission of Charles E. Merrill Publishing Co., Columbus, OH.)

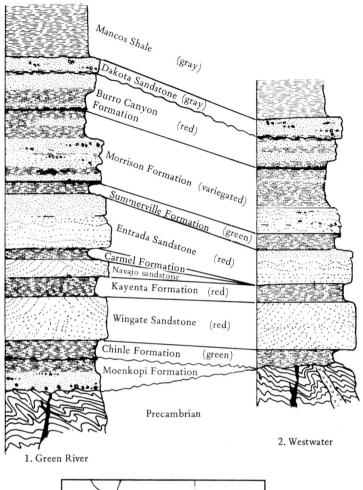

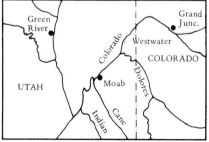

Summerville, Morrison formations) interstratified with red to white, cross-bedded eolian (?) sandstones (the Wingate, Navajo, Entrada formations). This succession of formations is so distinctive that it can be recognized and correlated lithologically with considerable confidence over wide areas on the Colorado Plateau (Fig. 14.21). In some cases, it may be possible to improve the reliability of correlation by applying statistical and computer-assisted techniques. These quantitative methods can provide a probability measure of whether a proposed correlation is valid or invalid (Agterberg, 1990).

Stratigraphic Position in a Succession. The preceding illustration points out the importance of position in a stratigraphic succession when correlating units by lithologic identity. Several of the Colorado Plateau formations are lithologically similar, but because they occur in a succession of strata distinctive enough to be correlated from one area to another, individual formations can be correlated also by their position in this succession. Another way in which position in a stratigraphic succession is important has to do with establishing correlation of strata by relation to some highly distinctive and easily correlated unit or units. Such distinctive beds serve as control units for correlation of other strata above and below. For example, a thin, ash-fall unit or bentonite bed may be present and easily recognized throughout a particular region. If it is the only such bed in the stratigraphic succession in the region, and thus cannot be confused with any other bed, it can serve as a **key bed,** or **marker bed,** to which other strata are related. Strata immediately above or below this control unit can be correlated with a reasonable degree of confidence with strata that are in a similar stratigraphic position with respect to the control unit in other areas. If two or more marker beds are present in a succession, this gives even greater reliability to correlation of units that lie between the marker beds. Obviously, correlation becomes more equivocal with increasing stratigraphic separation above or below the control units.

Correlation by Instrumental Well Logs

Well logs are simply curves sketched on paper charts that are produced from data obtained from measurements in well bores. These traces record variations in such rock properties as electrical resistivity, transmissibility of sound waves, or adsorption and emission of nuclear radiation in the rocks surrounding a bore hole. These variations are a reflection of changes in gross lithology, mineralogy, fluid content, porosity, and so forth, in the subsurface formations. Thus, correlation by use of well logs is not based totally on lithology. Nonetheless, most of the rock properties measured by well logs are closely related to lithology.

As discussed briefly in Chapter 1, well logs are obtained by the following procedure. After an exploratory well is drilled by a petroleum company, the well is logged before being completed as an oil or gas producer or abandoned as a dry hole. The logging procedure begins with lowering an instrument called a **sonde** to the bottom of the well bore. The sonde may be designed to measure the electrical resistivity of a rock unit, natural or induced gamma radiation emitted by the unit, the velocity of sound waves passing through the rock, or other rock properties. As the sonde is slowly withdrawn from the bore hole through a succession of stratigraphic units, it continuously measures the particular property of the rock that it is designed to analyze, and it electrically transmits this information to a digital tape and display unit located in a logging truck at the surface.

One common type of well log is the **electric log,** or resistivity log, which records resistivity of rock units as the sonde passes up the bore hole in contact with the wall of the hole. Resistivity is affected by the lithology of the rock units and the amount and nature of pore fluids in the rock. For example, a marine shale that has its pore spaces filled with saline formation water will have a much lower electrical resistivity (higher conductivity) than does a porous sandstone or limestone filled with oil or gas. With experience in a given geological province, petroleum geologists can recognize the particular signatures represented by the traces on the log and can relate these signatures to particular types of lithostratigraphic units or to a specific formation. Lithology cannot be read directly from such logs, but the characteristics of the log traces are a reflection of lithology (and fluid content). Other types of logging tools measure the natural gamma radiation in rock units **(gamma ray logs)** or measure the velocity with which a sound signal passes through rock units **(sonic logs).** In addition to their usefulness in correlation, sonic logs can also be used to determine the porosity of subsurface formations owing to the fact that sound waves are slowed in their passage through rocks by the presence of fluid-filled pores.

Other types of well logs are also in general use (e.g., geochemical, formation microscanner, magnetic susceptibility), but all have the common characteristic that they consist of electrically produced signatures or traces that represent some particular property of subsurface lithostratigraphic units that is related in some way to lithology, fluid content, bed thickness, or other properties. One common type of log data display consists of two different types of traces that are arranged on either side of a central column that represents the well bore. This central column is calibrated in feet (or meters) to show depth below the surface. Figure 14.22 illustrates a section of a well log showing a sonic curve opposite a gamma ray curve. The curve shapes generated by a particular lithostratigraphic unit are not unique, but a trained, experienced well-log analyst can learn to recognize the signature of a particular formation or succession of formations and can match up the signatures in logs from one area to those from nearby wells. See Asquith (1982) and Rider (1986) for additional description of well logs and log interpretation.

Characteristically, the well-log curves of adjacent wells are very similar, but the degree of similarity decreases in more distant wells. By working with a series of closely spaced wells, however, a geologist can carry a correlation across an entire sedimentary basin, even when pinch-outs or facies changes occur. In fact, one of the reasons why petroleum geologists find correlation of well logs so useful in petroleum exploration is that correlation permits recognition of pinch-outs and facies changes that may be potential traps for oil and gas. Figure 14.23 is an example of correlation by electric (resistivity) logs across a portion of a basin. Geologists often add lithologic information obtained from drill cores or cuttings to the well logs; however, I stress again that correlation by well logs is not necessarily correlation based entirely on lithologic identity because the shapes of the curves can represent a variety of rock properties, including porosity and fluid content. Correlation by well logs is actually based more on the position of each unit in a succession of units represented on the logs rather than on the character of any individual unit reflected in the curves. Correlation by well logs is thus the approximate subsurface equivalent of correlation of surface sections by position in the succession.

Correlation by instrumental well logs can be a laborious process involving large numbers of logs; it is also subject to considerable subjectivity owing to the similarity of the log curves or traces in different parts of a logged stratigraphic section. Differences between stratigraphic units may be manifested only by very subtle differences in the digital plots and can be difficult to discern visually. The availability of computers and sophisticated statistical techniques now makes it possible to apply automated approaches to stratigraphic correlation of well logs, removing some of the subjectivity in correlation. These approaches involve use of digital tapes to segment the logs for use in computational systems. These systems then provide a statistical match for correlation purposes. Details of automated well-log correlation are given by Shaw and Cubitt (1978), Griffiths (1982), and Olea (1988).

FIGURE 14.22 A section of a sonic–gamma ray log. The gamma ray curve is shown on the left. The sonic log (interval transit time log) is on the right. Depths (in feet) below the surface are shown in the central column. Four distinct correlatable units are indicated.

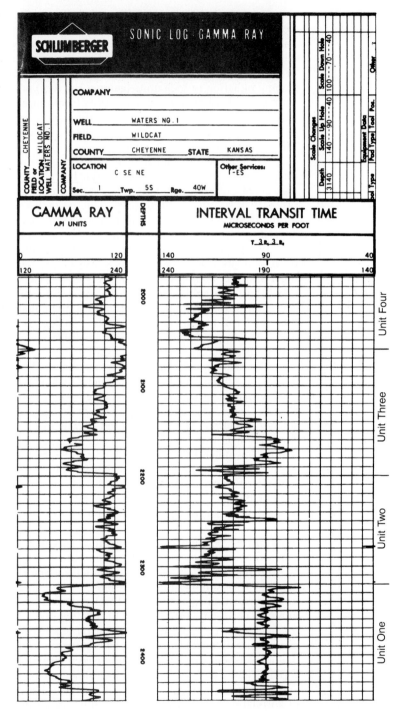

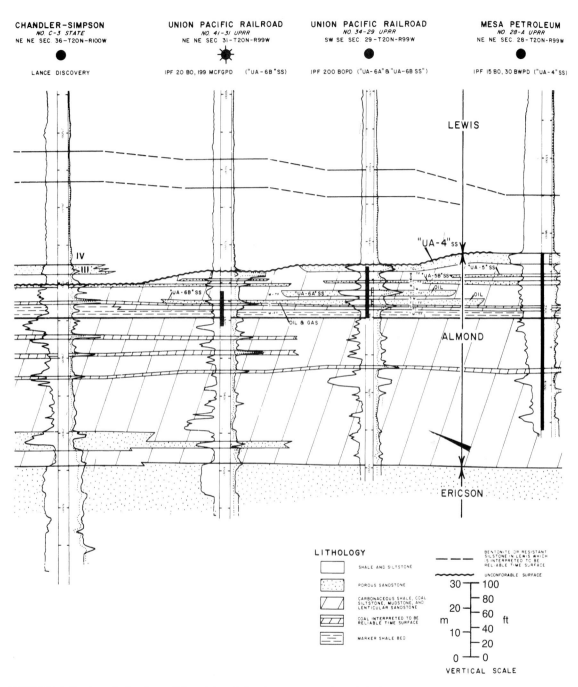

FIGURE 14.23 Correlation by well logs (electric logs) across a portion of the West Desert Springs oil field in Wyoming. Lithologic information has been added to the well logs, but correlation is based on the "signatures" of the well-log curves. (After Weimer, R. J., J. D. Howard, and D. R. Lindsay, 1982, Tidal flats and associated tidal channels, *in* P. A. Scholle and D. Spearing (eds.), Sandstone depositional environments: Am. Assoc. Petroleum Geologists Mem. 31. Fig. 44, p. 224, 225, reprinted by permission of AAPG, Tulsa, Okla.)

FURTHER READINGS

Agterberg, F. P., 1990, Automated stratigraphic correlation: Elsevier, Amsterdam, 424 p.

Childs, O. E., 1983, Correlation of stratigraphic units of North America, COSUNA, 1977–83: Am. Assoc. Petroleum Geologists and U.S. Geological Survey, 49 p.

Conklin, B. M., and J. W. Conklin, 1984, Stratigraphy: Foundations and concepts: Benchmark Papers in Geology, v. 82, Van Nostrand Reinhold, New York, 363 p.

Cubitt, J. M., and R. A. Reyment (eds.), 1982, Quantitative stratigraphic correlation: John Wiley & Sons, New York, 301 p.

Donovan, D. T., 1966, Stratigraphy—An introduction to principles: John Wiley & Sons, New York, 199 p.

Dunbar, C. O., and J. Rodgers, 1957, Principles of stratigraphy: John Wiley & Sons, New York, 356 p.

Einsele, G., and A. Seilacher (eds.), 1982, Cyclic and event stratification: Springer-Verlag, Berlin, 536 p.

Einsele, G., W. Ricken, and A. Seilacher (eds.), 1991b, Cycles and events in stratigraphy: Springer-Verlag, Berlin, 955 p.

Gill, D., and D. F. Merriam (eds.), 1979, Geomathematical and petrophysical studies in sedimentology: Pergamon Press, New York, 267 p.

Hallam, A., 1981, Facies interpretation and the stratigraphic record: W. H. Freeman, San Francisco, 287 p.

Hedberg, H. D., 1976, International Stratigraphic Guide: A guide to stratification classification, terminology, and procedure: John Wiley & Sons, New York, 200 p.

James, D. P., and D. A. Leckie (eds.), 1988, Sequences, stratigraphy, sedimentology: Surface and subsurface: Canadian Soc. Petroleum Geologists Mem. 15, 586 p.

Krumbein, W. C., and L. L. Sloss, 1963, Stratigraphy and sedimentation: W. H. Freeman, San Francisco, 660 p.

Merriam, D. F. (ed.), 1981, Computer applications in the earth sciences—An update of the 70s: Plenum Press, New York, 385 p.

North American Commission on Stratigraphic Nomenclature, 1983, North American Stratigraphic Code: American Assoc. Petroleum Geologists Bull., v. 67, p. 841–875.

Schwartzacher, W., 1975, Sedimentation models and quantitative stratigraphy: Developments in sedimentology 19: Elsevier, Amsterdam, 382 p.

Van Wagoner, J. C., R. M. Mitchum, K. M. Campion, and V. D. Rahmanian, 1990, Siliciclastic sequence stratigraphy in well logs, cores, and outcrops: Am. Assoc. Petroleum Geologists Methods in Exploration Series No. 7, Tulsa, Okla., 55 p.

Visher, G. S., 1984, Exploration stratigraphy: Pennwell, Tulsa, Okla., 334 p.

Weller, J. M., 1960, Stratigraphic principles and practice: Harper and Brothers, New York, 725 p.

Wilgus, C. K., B. S. Hastings, C. G. St. C. Kendall, H. W. Posamentier, C. A. Ross, and J. C. Van Wagoner (eds.), 1988, Sea-level changes: An integrated approach: Soc. Econ. Paleontologists and Mineralogists Spec. Pub. 42, 407 p. (Many of the papers in this volume explore the relationship of sea-level changes to sequence stratigraphy.)

15

Seismic Stratigraphy

15.1 INTRODUCTION

Seismology is the study of earthquakes and the structure of Earth on the basis of the characteristics of seismic waves. Although the broad subject of seismology lies outside the scope of this book, some aspects of seismology have a very important application to stratigraphy. The emphasis of this chapter is on what is commonly referred to as exploration seismology and more specifically on application of the techniques of exploration seismology to stratigraphic study. Exploration seismology deals with the use of artificially generated seismic waves to obtain information about the geologic structure, stratigraphic characteristics, and distributions of rock types. The techniques of exploration seismology were developed initially to locate structural traps for petroleum deposits, and they are still used extensively for that purpose. Seismic methods are now being used also for many other scientific purposes to increase knowledge of Earth's structure and stratigraphy.

In this chapter, we examine the application of seismic methods to stratigraphic problems. Seismic methods can be used to delineate rock bodies with distinctive geometries and internal structures, which can be thought of as constituting "seismic facies." Study of the lateral and vertical variations in seismic facies patterns is also used as a basis for interpreting lithology, depositional environments, and geologic history of subsurface stratigraphic units. **Seismic stratigraphy** is thus the study of seismic data for the purpose of extracting stratigraphic information. Seismic stratigraphy is a relatively new science, born in the early 1960s. Owing to its wide applicability to subsurface study both on land and at sea, where other types of stratigraphic data are few, it has already achieved an important position alongside the more traditional branches of stratigraphy.

15.2   EARLY DEVELOPMENT OF SEISMIC METHODS

The use of seismic methods for obtaining information about subsurface rocks and structures involves the natural or artificial propagation of seismic (elastic) waves. These waves pass downward into Earth until they encounter a discontinuity and are reflected back to the surface, where they can be picked up by detectors. Seismic waves travel at velocities ranging from less than 2 km/s (some sediments) to more than 8 km/s (some ultramafic rocks), depending upon the kinds of rocks through which they pass and their depth below Earth's surface (Dobrin, 1976). If we know the velocity with which seismic waves travel through a particular kind of rock and we can time their passage downward to a reflector and back to the surface, we can calculate the depth to the reflecting horizon. This principle forms the basis for application of seismic methods to geologic study.

Much of the theory of elasticity and propagation of seismic waves through rock materials that constitutes the theoretical basis for seismology was developed in the early part of the nineteenth century. An English seismologist, Robert Mallet, was the first scientist to measure the velocity of seismic waves in subsurface materials. He initiated experimental seismology in 1848 by measuring the speed of seismic waves through comparatively near-surface materials. He used black powder as an energy source to create a disturbance in the rocks and the surface of a bowl of mercury as the detector for the arriving seismic waves. The possibility of using seismic techniques to define the characteristics of subsurface rocks was apparently first put forward by a scientist named Milne in 1898. Two other interesting applications of the principles of seismology were experimented with in the early part of the twentieth century. These were a method for detecting icebergs—after the sinking of the *Titanic* by an iceberg in 1912—and the use of mechanical seismographs to detect the position of large enemy guns during World War I.

15.3   PRINCIPLES OF REFLECTION SEISMIC METHODS

On-Land Surveying

Practical application of seismology to the detection of rock structures began immediately after the end of World War I in both the United States and Europe, especially Germany and England. The first applications were in petroleum exploration in Germany and the Gulf Coast region of the United States, particularly in exploration for petroleum traps associated with salt domes. These early exploration efforts used the refraction seismic method for determining the structure of subsurface formations. This method is based on the principle that artificially generated seismic waves (produced in the early years of exploration by explosives) are refracted or bent at discontinuity surfaces as they travel downward below the surface. The waves then travel along these discontinuities before being refracted back to the surface, where their arrival is picked up by detectors placed at various distances away from the explosion (shot) point. The time that elapses during passage of the seismic waves downward to the discontinuity and back to the surface is used to compute the depth to the discontinuity.

Although several shallow petroleum deposits in salt domes were discovered during the early years of seismic exploration by the refraction method, this method did not work well for deeper structures because of the excessive distances required between shot points and detectors. Therefore, it was soon largely supplanted in petroleum exploration by the **reflection** seismic method. In the reflection method, waves created by an explosion are reflected back to the surface directly from subsurface rock interfaces without being refracted and by traveling laterally along discontinuity surfaces. Therefore, detectors can be located at relatively short distances from the shot points, and reflection seismic techniques can be used for delineating very deep structures. After introduction of the reflection method about 1930, it quickly became the primary tool in the petroleum industry for locating buried anticlines and other structural traps.

A very brief summary of basic principles of reflection seismology is given here. Additional details of the physical principles upon which reflection methods are based, as well as the development history of the reflection seismic method, are provided by Anstey (1982), Dobrin (1976), and Sheriff and Geldart (1982).

As mentioned, the reflection seismic method for delineating the structure of subsurface rock units is based on the principle that elastic or seismic waves travel at known velocities through rock materials. These velocities vary with the type of rock (typical average velocities: shale = 3.6 km/s; sandstone = 4.2 km/s; limestone = 5.0 km/s; Christie-Blick, Mountain, and Miller, 1990). Where the subsurface lithology is known relatively well from drill-hole information, it is possible to make accurate calculations of the time required for a seismic signal to travel from the surface to a given depth and then be reflected back to the surface. The reflection technique involves first generating elastic waves at the surface at a point source, originally called a shot point because explosives were first used to create the seismic waves. Seismic detectors, called geophones, are laid out in arrays extending outward from the shot point. Seismic waves reflected back from subsurface discontinuities are picked up by these detectors and fed electronically to a recording device. The principal discontinuities that reflect seismic waves are bedding planes and unconformities. By multiplying one-half of the travel time elapsed from initiation of the elastic waves at the point source to their arrival at the detector by the velocity of travel, geophysicists can accurately calculate depths to the discontinuities. This procedure thus allows the subsurface position of the discontinuities to be determined. The data obtained in this manner can then be displayed as seismic sections or profiles that depict the structure of the major rock units as they appear in cross section. Alternatively, the data may be used to prepare structure contour maps on the tops of particular reflecting horizons.

In practice, the technique for on-land reflection seismic profiling involves the following steps:

1. selecting locations for shot points and emplacing the energy source, either explosive or other devices
2. laying out and burying the geophones in a predetermined array and connecting the geophones to the recording equipment by long cables
3. triggering the energy source
4. digitizing and recording on magnetic tape or disk the seismic signals picked up by the detectors
5. computer processing of the tapes and preparation of visual analog displays of the seismograms

As mentioned, in the early stages of seismic exploration, seismic waves were generated by explosives placed in shallow shot holes drilled through the near-surface weathered zone to bedrock. Nonexplosive energy sources located on the surface are now also in common use. These nonexplosive energy sources include vibratory devices that produce continuous vibrations at the surface or devices that drop heavy weights onto a metal plate placed on the ground surface.

The general principles of on-land reflection seismic "shooting" are illustrated in the interesting old diagram (Fig. 15.1) from Nettleton's 1940 *Geophysical Prospecting for Oil.* The equipment and techniques for surveying locations, shooting, and recording and processing seismic data have changed and significantly improved since the 1940s; however, the basic principles illustrated in this figure still apply. As seismic waves pass downward and outward from the point energy source through the subsurface formations, they are reflected from successively deeper formations back to the surface where they are picked up by the electronic detectors. The signal from the detectors is then amplified, filtered to remove excess "noise," digitized, and fed to a recording truck to be recorded on magnetic tape or disk.

The data recorded on the magnetic tape or disk must then be presented in visual form for monitoring and interpretation. Prior to the use of magnetic tapes or disks for recording, the visual seismic records, or seismograms, were mechanically produced, wiggly-trace records such as that shown on the right in Figure 15.1. Photographic or dry-paper recording methods are now used for visual display, and several modes of displaying the amplitude of arriving seismic waves against arrival time are in use. A common type of display, called a **variable-density mode** display, is generated by a technique by which light intensity is varied to display differences in wave amplitude (height) by producing alternating light and dark areas on film or paper, thereby accentuating the amplitude of waves from a particular reflecting surface. For example, all wave traces having an amplitude greater than a given value are shaded black; traces with lower amplitudes are unshaded. Thus, a strong reflection event will show up as a black line on the record, as illustrated in Figure 15.2.

Marine Seismic Surveying

Early seismic surveys were carried out on land; however, reflection seismic methods can be used also in the ocean or on lakes. Some marine operations in very shallow water began in the late 1920s and 1930s, but extensive marine seismic surveys did not get underway until about the middle 1940s. Marine seismic operations employ the same principles as those used on land, but they differ in the speed at which they take place and in the specific details of the shooting and detection processes. Sound sources and detectors are towed behind the survey ship, which can operate at a speed of 6 knots or more on a continuous 24-h/day basis. In the early years of marine operations, a half-pound block of TNT was tossed over the ship's side every 3 minutes to provide a continuous seismic record. This method was potentially dangerous and also very damaging to fish and other ocean life. It has now been largely replaced by techniques that use acoustic sources such as airguns, which produce sound energy by releasing highly compressed air.

Early marine operations were severely hampered by problems of accurately locating the shot point and detector positions, and the operations had to be carried out within sight of land so that locations could be determined by land-based surveying methods. The development about 1949–50 of radio navigation methods made possible operations in the ocean away from land. Subsequent development of satellite navigation methods that "home in" on orbiting satellites to fix the position of ships at sea now allow the position of the survey ships in the open ocean to be accurately and continuously determined. Another significant development that came about as a result of

FIGURE 15.1 (Opposite)   Diagram illustrating the equipment and procedures used in seismic exploration in the 1940s. (From Nettleton, L. L., 1940, Geophysical prospecting for oil. Fig. 155, following p. 332, reprinted by permission of McGraw-Hill Book Company, New York.)

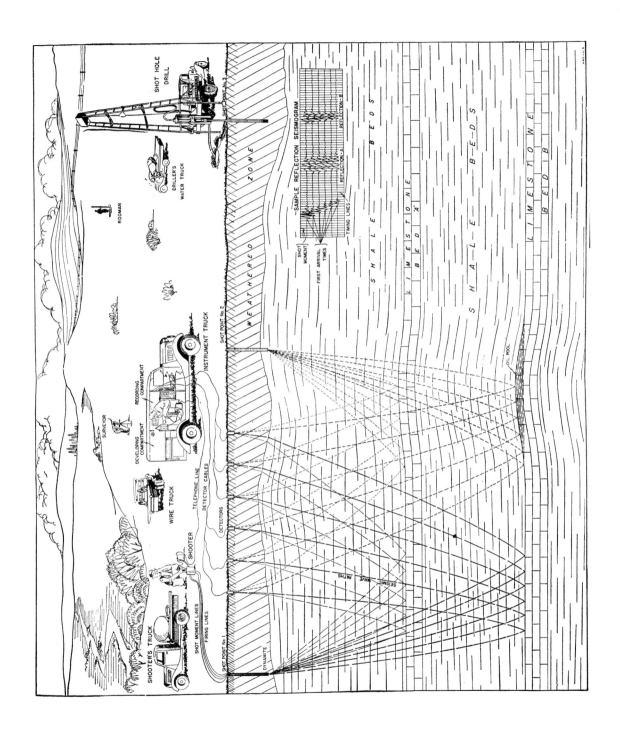

535

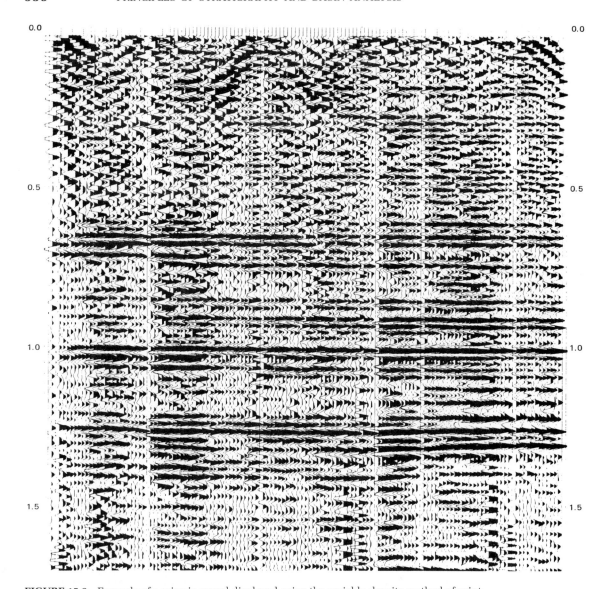

FIGURE 15.2 Example of a seismic record displayed using the variable-density method of printing. The vertical scale is given in two-way seismic-wave travel time (in seconds) rather than depth. (From Neidell, N. S., 1979, Stratigraphic modeling and interpretation: Geophysical principles and techniques: Am. Assoc. Petroleum Geologists Education Short Course Notes Ser. 13. Fig., p. 49, reprinted by permission of AAPG, Tulsa, Okla.)

new advancements during World War II was invention of the floating streamer cable, which allowed detectors (called hydrophones) to be towed in a floating cable behind the ship. Streamers may be up to several kilometers in length. These technical advances made possible rapid progress in marine seismic surveying methods in the years following. The general principles of marine seismic profiling are illustrated in Figure 15.3.

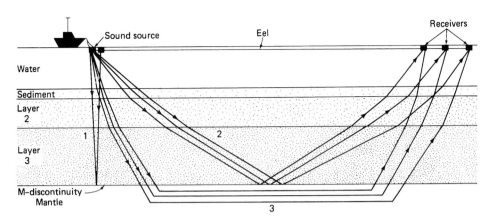

FIGURE 15.3 Diagram illustrating the principle of marine seismic surveying: 1, vertical incidence reflector, 2, wide-angle reflector, and 3, refracted waves. (From Kennett, J., Marine geology, © 1982, Fig. 2.14, p. 40. Reprinted by permission of Prentice-Hall, Englewood Cliffs, N.J.)

Subsequent progress in both marine and land-based seismic surveying techniques have included improvements in energy sources, including sources that do not require explosives; development of new detection equipment and procedures; and advances in treatment and analysis of seismic data. In particular, computer analysis of seismic data has allowed a quantum leap forward in filtering and enhancing seismic signals and in methods of displaying and interpreting seismic data.

15.4 APPLICATION OF REFLECTION SEISMIC METHODS TO STRATIGRAPHIC ANALYSIS

The science of seismic stratigraphy was developed largely by petroleum companies out of pragmatic necessity to locate petroleum deposits in deep, unexplored basins both on land and offshore. Geologists have not yet discovered a successful geochemical method for directly detecting the presence of oil or gas in deep subsurface formations, although some progress has been made in direct detection of hydrocarbon deposits by seismic methods. Therefore, the successful search for petroleum still requires that explorationists locate and drill petroleum traps such as anticlines and salt domes. Because most shallow petroleum traps were located and tested long ago during the earlier phases of petroleum exploration, petroleum companies have been forced to extend exploration efforts to deeper formations and to frontier basins, which are undrilled or sparsely drilled basins, on land and offshore. Inasmuch as successful oil search depends upon a knowledge of stratigraphic relationships as well as structural anomalies, and because poorly explored basins lack sufficient well control for stratigraphic analysis, new techniques had to be developed that would allow stratigraphic information to be extracted from seismic data. Thus, seismic stratigraphy was born in the 1960s as a tool that made possible the integration of stratigraphic concepts with geophysical data—that is, a geologic approach to stratigraphic interpretation of seismic data (Payton, 1977; Berg and Wolverton, 1985; Vail, 1987; Cross and Lessenger, 1988).

Seismic reflections are generated by physical surfaces in subsurface rocks. In the conventional structural application of seismic data, seismic reflections are used to identify and map the structural attitudes of subsurface sedimentary layers. By contrast, seismic stratigraphy uses seismic reflection correlation patterns to identify depositional sequences, to predict the lithology of seismic facies by interpreting depositional processes and environmental settings, and to analyze relative changes in sea level as recorded in the stratigraphic record of coastal regions. Seismic stratigraphy thus makes possible many types of stratigraphic interpretations, such as geologic time correlations, definition of genetic depositional units, and thickness and depositional environment of genetic units (Vail and Mitchum, 1977).

Parameters Used in Seismic Stratigraphic Interpretation

To accomplish the objective of interpreting stratigraphy and depositional facies from seismic data, geologists must identify characteristic features of seismic reflection records (seismograms) and relate these features to the geologic factors responsible for the reflections. An understanding of the factors that generate seismic reflections is therefore critical to the entire concept of seismic stratigraphy. Fundamentally, primary seismic reflections occur in response to the presence of significant density-velocity changes at either unconformity or bedding surfaces. Reflections are generated at unconformities because unconformities separate rocks having different structural attitudes or physical properties, particularly different lithologies. The density-velocity contrast along unconformities may be further enhanced if the rocks below the unconformity have been altered by weathering. Reflections are generated at bedding surfaces because, owing to lithologic or textural differences, a velocity-density contrast exists between some sedimentary beds; however, not every bedding surface will generate a seismic reflection. Also, a given reflection event identified on a seismic record may not necessarily be caused by reflection from a single surface, but it may represent the sum or average of reflections from several bedding surfaces, particularly if beds are thin.

The seismic records produced as a result of primary reflections from unconformities or bedding surfaces have distinctive characteristics that can be related to depositional features such as lithology, bed thickness and spacing, and continuity. The principal parameters that are useful in seismic stratigraphy for interpreting geologic information are reflection configuration, continuity, amplitude, and frequency; interval velocity; and external form and association of seismic facies units (Table 15.1).

Reflection Configuration

Reflection configuration refers to the gross stratification patterns identified on seismic records. Four basic types of configurations are recognized (Mitchum, Vail, and Sangree, 1977). **Parallel patterns,** including subparallel and wavy patterns (Fig. 15.4A and B), are generated by strata that were probably deposited at uniform rates on a uniformly subsiding shelf or in a stable basin setting. **Divergent configurations** are characterized by a wedge-shaped unit in which lateral thickening of the entire unit is caused by thickening of individual reflection subunits within the main unit (Fig. 15.4C). Divergent configurations are interpreted to signify lateral variations in rates of deposition or progressive tilting of the sedimentary surface during deposition. **Prograding reflection configurations** are reflection patterns generated by strata that were deposited by lateral outbuilding or progradation to form gently sloping depositional surfaces called clinoforms (next paragraph). As represented on seismic records, prograding reflection configurations may have a variety of patterns, including **sigmoid** (superposed S-shaped reflectors), **oblique,** or **hummocky** (Fig. 15.5). These stratal configura-

TABLE 15.1 Seismic reflection parameters commonly used in seismic stratigraphy, and the geologic significance of these parameters

Seismic facies parameters	Geologic interpretation
Reflection configuration	Bedding patterns Depositional processes Erosion and paleotopography Fluid contacts
Reflection continuity	Bedding continuity Depositional processes
Reflection amplitude	Velocity-density contrast Bed spacing Fluid content
Reflection frequency	Bed thickness Fluid content
Interval velocity	Estimation of lithology Estimation of porosity Fluid content
External form and areal association of seismic facies units	Gross depositional environment Sediment source Geologic setting

Source: Mitchum, R. M., Jr., P. R. Vail, and J. B. Sangree, 1977, Seismic stratigraphy and global change of sea level, Part 6: Stratigraphic interpretation of seismic reflection patterns in depositional sequences, *in* C. E. Payton (ed.), Seismic stratigraphy—Applications to hydrocarbon exploration: Am. Assoc. Petroleum Geologists Mem. 26. Table 2, p. 122, reprinted by permission of AAPG, Tulsa, Okla.

tions are all caused in some way by outbuilding of strata—commonly from shallow water into deeper water, as along the front of a delta—or by infilling of channels. Differences in configurations of the clinoforms represent variations in sediment supply or rates of basin subsidence, changes in sea level, or changes in water energy of the depositional environment or water depth. **Chaotic reflection** patterns (Fig. 15.6) are interpreted to represent a disordered arrangement of reflection surfaces owing to penecontemporaneous, soft-sediment deformation, or possibly to deposition of strata in a variable, high-energy environment. Some chaotic reflections may also be related to overpressured, or geopressured, zones in deep formations. Reflection-free areas on seismic records may represent homogeneous, nonstratified units such as igneous masses or thick salt deposits, or highly contorted or very steeply dipping strata.

The terms undaform, clinoform, and fondaform were introduced by Rich (1951) to describe depositional environments in relation to wave base (Fig. 15.7). The **undaform** is the more or less flat topographic surface that exists in an aqueous environment above wave base where bottom sediments are moved or stirred by waves and currents, particularly during storms. The **clinoform** is the sloping surface extending from wave base down to the generally flat floor, called the **fondaform,** of the water body.

Reflection Continuity

Reflection continuity depends upon the continuity of the density-velocity contrast along bedding surfaces or unconformities. It is closely associated with continuity of strata, and it provides information about depositional process and environment. Continuous reflectors (Fig. 15.8) characteristically indicate stratified deposits that are continuous over large areas, although this is not necessarily true of all widespread reflec-

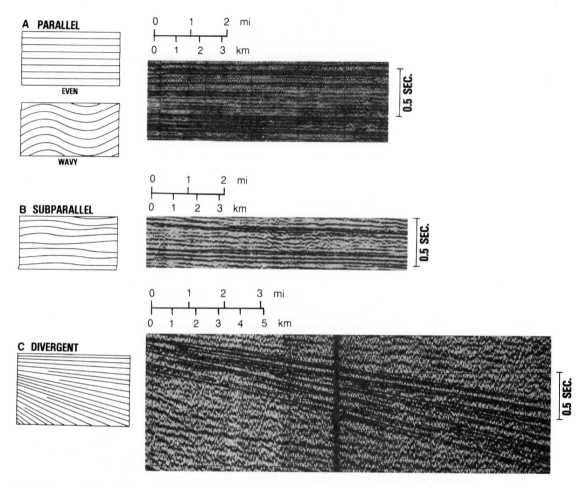

FIGURE 15.4   Principal types of seismic reflection configurations. A. Parallel (even or wavy). B. Subparallel. C. Divergent. (After Mitchum, R. M., Jr., P. R. Vail, and J. B. Sangree, 1977. Stratigraphic interpretation of seismic reflection patterns in depositional sequences, *in* C. E. Payton (ed.), Seismic stratigraphy—Applications to hydrocarbon exploration: Am. Assoc. Petroleum Geologists Mem. 26. Fig. 4, p. 123, Fig. 5, p. 124, reprinted by permission of AAPG, Tulsa, Okla.)

tors. In some depositional sequences, such reflectors may, for reasons that are still poorly understood, reflect isochronous (having the same age everywhere) horizons that cut across some stratal surfaces. In contrast to continuous reflectors, reflection patterns showing reflection terminations (Fig. 15.9) indicate stratigraphic relationships such as onlap, downlap, and toplap that occur in coastal regions in response to transgressions and regressions.

FIGURE 15.5 (Opposite)   Example of reflection patterns interpreted as prograding clinoforms. A. Sigmoid. B. Mostly tangential. C. Mostly parallel oblique. D. Complex sigmoid-oblique. E. Shingled. F. Hummocky clinoforms. (After Mitchum, R. M., Jr., P. R. Vail, and J. B. Sangree, 1977, Stratigraphic interpretation of seismic reflection patterns in depositional sequences, *in* C. E. Payton (ed.), Seismic stratigraphy—Applications to hydrocarbon exploration: Am. Assoc. Petroleum Geologists Mem. 26. Fig. 6, p. 125, Fig. 7, p. 126, Fig. 8, p. 127, reprinted by permission of AAPG, Tulsa, Okla.)

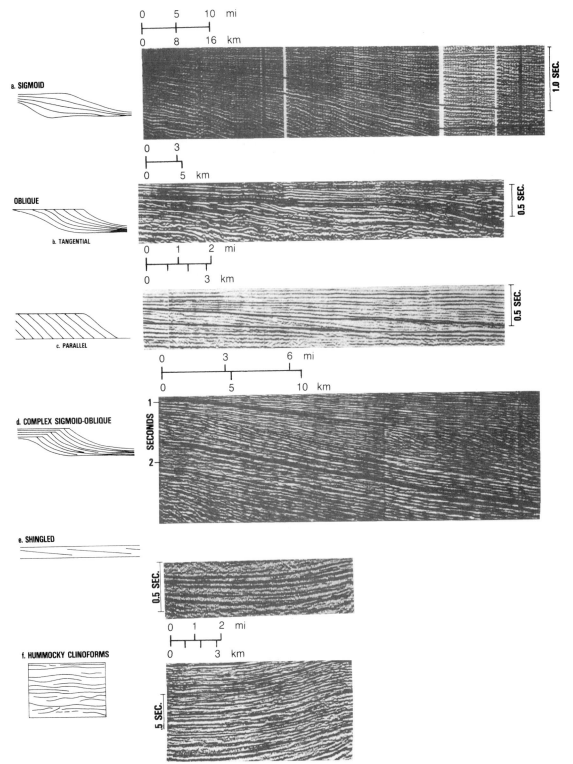

a. SIGMOID

OBLIQUE

b. TANGENTIAL

c. PARALLEL

d. COMPLEX SIGMOID-OBLIQUE

e. SHINGLED

f. HUMMOCKY CLINOFORMS

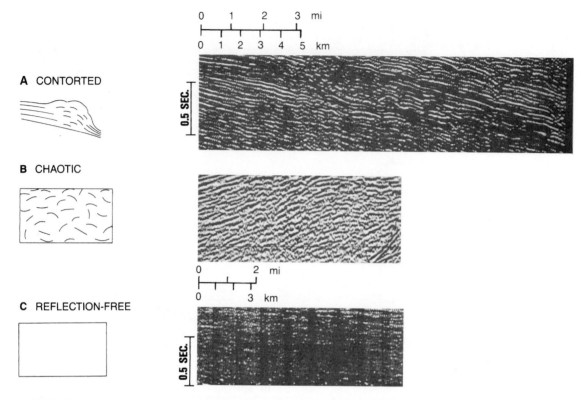

FIGURE 15.6   Examples of chaotic and reflection-free reflection patterns. A. Interpreted as the reflection from contorted stratal surfaces that are still recognizable after penecontemporaneous (soft-sediment) deformation. B. Reflections are so chaotic that no stratal pattern can be reliably interpreted. C. Largely reflection-free, where no or very few reflections occur in seismically homogenous strata. (After Mitchum, R. M., Jr., P. R. Vail, and J. B. Sangree, 1977, Stratigraphic interpretation of seismic relection patterns in depositional sequences, *in* C. E. Payton (ed.), Seismic stratigraphy—Applications to hydrocarbon exploration: Am. Assoc. Petroleum Geologists Mem. 26. Fig. 9, p. 128, Fig. 10, p. 129, reprinted by permission of AAPG, Tulsa, Okla.)

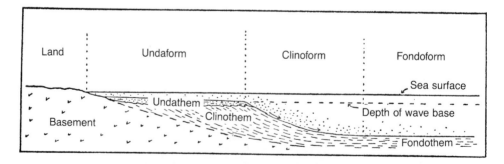

FIGURE 15.7   Sketch illustrating the meaning of the terms undaform, clinoform, and fondoform as used by Rich. (After Rich, J. L., 1951, Three critical environments of deposition and criteria for recognition of rocks deposited in each of them: Geol. Soc. America Bull., v. 62, Fig. 1, p. 3.)

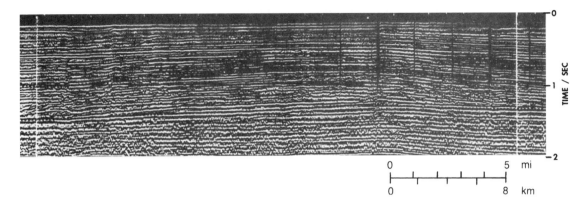

FIGURE 15.8 Illustration of broad, low-relief seismic mound facies. This example is from off-shore Africa. It shows mounded external form with reflections concordant at the top and down-lapping in opposite directions at the base. (From Sangree, J. B., and J. M. Widmier, 1977, Seismic stratigraphy and global change in sea level, Part 9: Seismic interpretation of clastic depositional facies, *in* C. E. Payton (ed.), Seismic stratigraphy—Applications to hydrocarbon exploration: Am. Assoc. Petroleum Geologists Mem. 26. Fig. 4, p. 172, reprinted by permission of AAPG, Tulsa, Okla.)

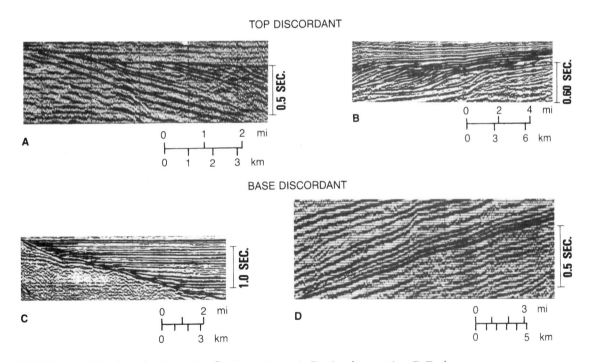

FIGURE 15.9 Top-discordant seismic reflection patterns: A. Erosional truncation. B. Toplap. Base-discordant seismic reflection patterns: C. Onlap. D. Downlap. (After Mitchum, R. M., Jr., P. R. Vail, and J. B. Sangree, 1977, Stratigrapic interpretation of seismic reflection patterns in depositional sequences, *in* C. E. Payton (ed.), Seismic stratigraphy—Applications to hydrocarbon exploration: Am. Assoc. Petroleum Geologists Mem. 26. Fig. 2, p. 119, Fig. 3, p. 120, reprinted by permission of AAPG, Tulsa, Okla.)

Reflection Amplitude and Frequency

Reflection amplitude has to do with seismic wave height and is a function of the energy of seismic waves. On a seismic record, amplitude is measured as the distance from the mid-position of a wave to the extreme position. Amplitude is thus equal to one-half the height of the wave above the adjacent trough (Fig. 15.10). The amplitude of reflected seismic waves is controlled principally by the velocity and density contrast along individual reflecting surfaces and increases with increasing velocity-density contrast. It is also affected by the spacing between reflecting surfaces. Where bed spacing is optimum, lower-energy responses are phased together constructively (constructive interference) to intensify or amplify the reflected energy and thus increase amplitude. For example, if bed thickness is less than the wave length of the seismic wave—for example, one-fourth of a wavelength—the reflections from the top and base of the bed can be phased together to give exceptionally large amplitudes. When beds are very thick, the reflections from the top and base of the beds are completely separate (Sheriff, 1980). Assuming a seismic source with a frequency of 20 hertz (explained below), the practical one-quarter wavelength limit means that a single sedimentary bed, such as a shale bed, thinner than 50 m cannot be resolved. If a sufficiently high-frequency signal is generated, resolution might be increased to allow single reflections from beds as thin as 10 m to be detected (Christie-Blick, Mountain, and Miller, 1990). Figure 15.11 shows the effect of bed thickness on amplitude in the case of a sandstone with a gradational base. Analysis of wave amplitudes allows geophysicists to calculate the thickness of beds, if a nearby contact with a layer thicker than 1 wavelength is present for calibration (Christie-Blick, Mountain, and Miller, 1990).

The amplitude of reflected seismic waves can be affected also by fluid content of sedimentary beds or by accumulations of gas in the beds. The presence of hydrocarbons in beds can produce a marked increase in amplitude of waves that shows up on seismic records as so-called "bright spots." These bright spots actually appear blacker than surrounding events, so the meaning of the name is not clear—perhaps it simply means that they stand out clearly on a seismic record. Bright-spot analysis was introduced in the petroleum industry in the early 1970s and is now used as a method for direct detection of hydrocarbon deposits.

Reflection frequency refers to the number of vibrations or oscillations of seismic waves per second. It is numerically equal to wave velocity divided by wave length. The frequency of a seismic wave is commonly expressed in hertz (Hz) or kilohertz (kHz). A hertz is a unit of frequency equal to one cycle per second; a kilohertz is 1000 hertz. The frequency of seismic waves affects both the depth of penetration of the

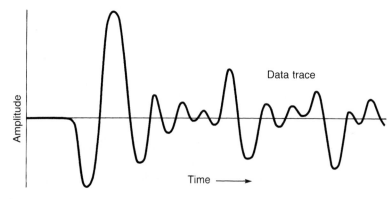

FIGURE 15.10 Schematic representation of the amplitude of seismic waves. The amplitude is the vertical distance above or below the mid-point line drawn through the wave traces. Time refers to arrival time of the waves at the seismic detector. (After Neidell, N. S., 1979, Stratigraphic modeling and interpretation: Geophysical principles and techniques: Am. Assoc. Petroleum Geologists Education Short Course Notes 13. Fig., p. 31, reprinted by permission of AAPG, Tulsa, Okla.)

FIGURE 15.11 Seismic response for a sand with a gradational base. The 9 m thickness is about ⅛ wave length. (After Neidell, N. S., and E. Poggiagliolmi, 1977, Stratigraphic modeling and interpretations—Geophysical principles and techniques, *in* C. E. Payton (ed.), Seismic stratigraphy—Applications to hydrocarbon exploration: Am. Assoc. Petroleum Geologists Mem. 26. Fig. 27, p. 413, reprinted by permission of AAPG, Tulsa, Okla.)

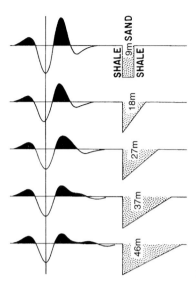

waves into the subsurface and the resolution of the seismic records, that is, the sharpness with which details of the seismograms can be distinguished. Lower frequencies give greater depth of penetration but less resolving power. The frequency of seismic waves is induced by the particular energy sound source used to create the waves. As the waves pass downward through subsurface formations and are reflected back to the surface, the initial induced frequency is attenuated by bed thickness, which controls the spacing of reflectors. Thus, attenuations of the initial induced frequency of seismic waves is related to bedding characteristics. Frequency is also affected by lateral changes in fluid content of beds—the presence of hydrocarbon accumulations, for example—and by lateral thickness changes in beds.

Interval (Seismic-Wave) Velocity

Interval velocity refers to the average velocity of seismic waves between reflectors. Seismic wave velocity is affected by several factors, especially porosity, density, external pressure, and pore (fluid) pressure. Porosity has a particularly significant effect on velocity, which increases as porosity decreases. Thus, because porosity commonly decreases with depth, velocity increases with depth. Velocity also increases with density of the rocks and with increasing overburden pressure. For example, the velocity for a typical sandstone increases from about 4 km/s at the surface to more than 5 km/s at a depth of 5000 m. Velocity decreases with increasing interstitial fluid pressure; the presence of gas at low saturations in the pore spaces of the rocks also causes a decrease in velocity.

Seismic velocity is of particular interest because of the possibility that different rock types, which are characterized by different densities, porosities, pore fluid pressures, and other characteristics, can be differentiated on the basis of seismic velocity. Seismic velocity is calculated from signal travel times and reflection amplitudes. Inverse models are then used to deduce the geologic causes of the seismic response—one cause being rock lithology. Figure 15.12 shows, for example, that the velocity of seismic waves is lower in siliciclastic rocks such as sandstones than in carbonate rocks and salt. Thus, where velocity contrasts in different rocks are large, velocity can be

FIGURE 15.12 Characteristic velocity-depth relations for terrigenous clastic sedimentary rocks, carbonate rocks, and salt. Younger rocks tend to have lower velocities than older rocks because they generally have higher porosities, are less cemented, and have undergone less deformation. (From Sheriff, R. E., 1976, Inferring stratigraphy from seismic data: Am. Assoc. Petroleum Geologists Bull., v. 60. Fig. 6, p. 533, reprinted by permission of AAPG, Tulsa, Okla.)

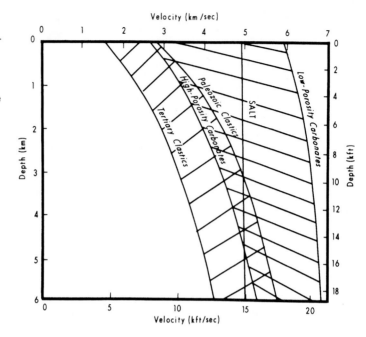

used as an indicator of gross lithology. Unfortunately, there is appreciable overlap in the velocities of seismic waves in various rock types, owing mainly to variations in porosity, so that velocity alone is not sufficient to unequivocally distinguish rock types. For example, the velocity of seismic waves in a low-porosity sandstone can be approximately the same as the velocity in a high-porosity carbonate rock. Therefore, to predict lithology, velocity information must be combined with other types of seismic data to allow interpretation of depositional processes and environmental settings. Nonetheless, velocity is one of the most critical factors in seismic data processing and interpretation.

External Form of Reflecting Units

The external form or geometry of stratigraphic bodies that generate seismic reflections can be interpreted from seismic data (Fig. 15.16). Thus, these data can be used to identify "seismic facies," which may be interpreted in terms of depositional environments of the lithologic analogs of these seismic facies. This procedure of interpreting the external form of stratigraphic packages from seismic data is part of the process of seismic facies analysis, which also provides information on sediment source and geologic setting, including major facies changes. Seismic facies analysis is an extremely important aspect of seismic stratigraphy and is discussed in greater detail in the following paragraphs.

Procedures in Seismic Stratigraphic Analysis

The significance of the seismic stratigraphic approach to study of subsurface sedimentary rocks lies in the fact that it permits geologists and geophysicists to interpret stratigraphic relationships and depositional processes as well as to use seismic data for conventional structural mapping. Interpretation is a subjective process, but when seismic stratigraphic analysis is pursued in a logical manner and interpretation is based upon analogy with established stratigraphic and depositional models that have been generated by other types of studies, seismic stratigraphic analysis becomes an extremely valuable tool. Seismic stratigraphy can thus provide insight into such stratigraphic and depositional factors as lithofacies changes, relief and topography of unconformities, paleobathymetry (depth relationships and topography of ancient oceans), geologic time correlations, depositional history, and subsidence and tilting history (burial history). The procedures for interpreting stratigraphy from seismic data involve three principal stages: (1) seismic sequence analysis, (2) seismic facies analysis, and (3) interpretation of depositional environments and lithofacies (Vail, 1987). Seismic stratigraphic analysis is applied also to interpretation of ancient sea-level changes.

Seismic Sequence Analysis

Seismic sequence analysis involves identification of major reflection "packages" that can be delineated by recognizing surfaces of discontinuity. We commonly think of discontinuities as unconformities, which are surfaces of erosion or nondeposition that represent major hiatuses, and we identify four different kinds of unconformities (Chapter 14). In seismic stratigraphic analysis, however, two kinds of discontinuities are recognized:

1. **erosional unconformity surfaces** that represent a significant hiatus owing to subaerial or subaqueous erosional truncation
2. unconformable surfaces called **downlap surfaces,** which are marine surface representing a hiatus but without evidence of erosion

A downlap surface marks the change from a retrogradational (transgressive) to an aggradational (stable shoreline position) parasequence set (Fig. 15.13) and is the surface of maximum flooding (Van Wagoner et al., 1988). Downlap surfaces commonly occur above condensed sections of sediments. A **condensed section** is a facies consisting of thin marine beds that accumulate at a very slow rate. Condensed sections develop owing to a rise in relative sea level, which results in a landward shift in sites of sediment accumulation and reduction in rate of sediment accumulation (starved sedimentation) further basinward. Downlap surfaces occur within a sequence and between bounding unconformities (Fig. 15.13).

The concept of depositional sequence is introduced in Chapter 14, where sequences are defined as stratigraphic units separated by unconformities or their correlative conformities (Mitchum, Vail, and Thompson, 1977). The mapping of unconformities is thus the key to seismic sequence analysis. Discontinuities are generally good reflectors and also commonly separate rock units having different dips, at least on a regional scale. Discontinuities may thus be recognized by interpreting systematic patterns of reflection terminations along the discontinuity surfaces. Two patterns, onlap and downlap (Chapter 14), occur above discontinuities. Three patterns—truncation, toplap, and apparent truncation (Fig. 15.13)—occur below discontinuities (Vail, 1987).

Seismic resolution is generally not adequate to delineate minor sedimentary sequences because the practical vertical resolution is on the order of 10 to 50 m. On the other hand, major depositional units or systems such as progradational delta-slope systems, carbonate shelf-margin systems, or marine offlap-onlap systems can be identified. Once basinwide correlation of depositional sequences has been made, these

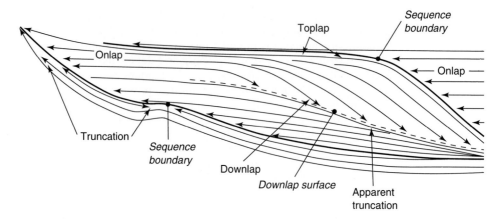

FIGURE 15.13   Diagram illustrating sequence boundaries (unconformities), downlap (maximum flooding) surfaces, and various kinds of reflection terminations. Apparent truncation refers to termination by depositional thinning. (From Vail, P. R., 1987, Seismic stratigraphic interpretation using sequence stratigraphy, *in* A. W. Bally (ed.), Seismic stratigraphy: Am. Assoc. Petroleum Geologists Studies in Geology 27. Fig. 1, p. 2, reproduced by permission of AAPG, Tulsa, Okla.)

sequences provide a first-order stratigraphic framework within which more detailed seismic facies studies can be carried out. Figure 15.14, a seismic section from the Beaufort Sea (north of Alaska), provides an example of sequences delineated by picking unconformities that separate different depositional units.

The procedure for carrying out seismic sequence analysis thus involves the following steps:

1. picking unconformities in a given area by recognizing reflection terminations along their surfaces
2. extending or extrapolating these boundaries over the complete section, including areas where the reflectors are conformable, to define the sequences completely
3. repeating the process of delineating sequence boundaries on seismic records from other parts of a basin or region and correlating the sequences throughout the seismic grid to produce a three-dimensional framework of successive stratified seismic sequences separated by unconformities or correlative conformities
4. mapping sequence units on the basis of thickness, geometry, orientation, or other features to see how each sequence relates to neighboring sequences

Details of sesimic sequence mapping and its application to geologic problems are given by Hubbard, Pape, and Roberts (1985a, 1985b); Vail (1987); and Vail et al. (1991).

Seismic Facies Analysis

Seismic facies analysis takes the interpretation process one step beyond seismic sequence analysis by examining within sequences and systems tracts (Chapter 14) smaller reflection units that may be the seismic response to lithofacies. In seismic facies analysis, the most common reflection characteristics used to distinguish one seismic facies from another are the geometry of reflections or reflection terminations with respect to the two unconformity surfaces bounding the sequence, the external geometry of the facies, and the internal configuration and character of the reflections. A **seismic facies unit** is a mappable, areally definable, three-dimensional unit com-

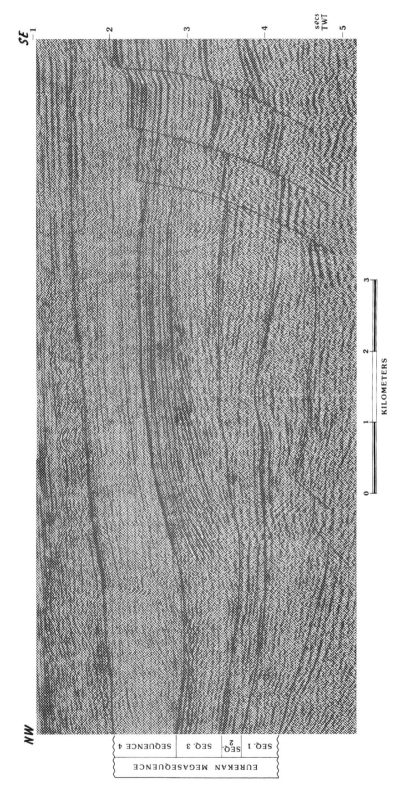

FIGURE 15.14 Depositional sequences as defined from seismic records. In this example from the Beaufort Sea, seismic sequence boundaries are shown by solid black lines. The three heavy, sloping lines in the right-hand part of the seismic record indicate large faults. The vertical scale on this record is given in seismic wave travel time rather than depth. (From Hubbard, R. J., J. Pape, and D. G. Roberts, 1985, Depositional sequence mapping as a technique to establish tectonic and stratigraphic framework and evaluate hydrocarbon potential on a passive continental margin, *in* O. R. Berg and R. G. Wolverton (eds.), Seismic stratigraphy II—An integrated approach: Am. Assoc. Petroleum Geologists Mem. 39, p. 84, reprinted by permission of AAPG, Tulsa, Okla.)

posed of seismic reflections whose characteristic reflection elements differ from those of adjacent units. It is considered to represent or express the gross lithologic aspect and stratification characteristics of the depositional unit that generates the reflections. Some of the more important seismic reflection patterns that constitute seismic facies are offlap, submarine onlap, submarine mounds, channel/overbank complexes, slumps, slope-front-fill, climbing toplap, and drape (Fig. 15.15).

Procedures for Interpreting Seismic Facies. The objective of seismic facies analysis is regional interpretation of lithology, depositional environments, and geologic history. Several distinct steps are involved in the interpretation process (Mitchum and Vail, 1977; Vail, 1987).

The first step is recognizing and delineating seismic facies units within each sequence on all of the seismic sections in the region being mapped. The most useful seismic parameters in seismic facies analysis are the following:

1. the geometry of reflections (Fig. 15.15) and reflection terminations
2. reflection configuration (parallel, divergent, sigmoid, or oblique)
3. three-dimensional form (Fig. 15.16)

Reflection terminations and configurations can be analyzed visually from two-dimensional seismic profiles. Determination of external three-dimensional geometry must be done by mapping based on many different seismic profiles. Each seismic facies unit is distinguished from adjacent units on the basis of these parameters.

After the distribution (geometry) and thickness of the reflection packages have been mapped, the next step in seismic facies analysis is to combine this information with any other distinctive seismic information, such as interval velocity, and any available nonseismic data, such as well and outcrop data, that shed light on the regional geology. Vail et al. (1991) provide an in-depth overview of these procedures.

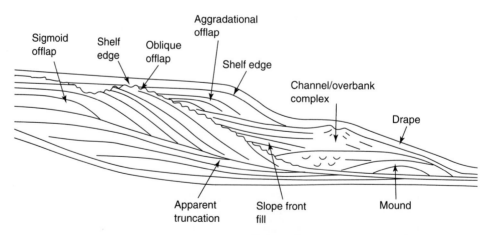

FIGURE 15.15 A simulated seismic section illustrating some common seismic facies patterns that can be identified from seismic records. (From Vail, P. R., 1987, Seismic stratigraphy interpretation using sequence stratigraphy, *in* A. W. Bally (ed.), Seismic stratigraphy, Am. Assoc. Petroleum Geologists Studies in Geology 27. Fig. 6, p. 6, reproduced by permission of AAPG, Tulsa, Okla.)

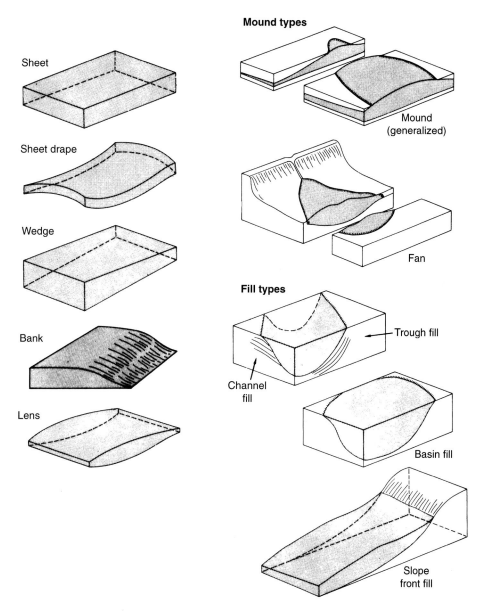

FIGURE 15.16 External form of some stratigraphic bodies as interpreted from seismic facies units. (From Mitchum, R. J., Jr., P. R. Vail, and J. B. Sangree, 1977, Stratigraphic interpretation of seismic reflection patterns in deposited sequences, *in* C. E. Payton (ed.), Seismic stratigraphy— Applications to hydrocarbon exploration: Am. Assoc. Petroleum Geologists Mem. 26. Fig. 12, p. 131, reprinted by permission of AAPG, Tulsa, Okla.)

Interpretation of Lithofacies and Depositional Environments

Once the objective aspects of delineating seismic sequences and facies have been completed, the final objective is to interpret the facies in terms of lithofacies, depositional environments, and paleobathymetry. For example, seismic facies that show prograding reflection characteristics commonly indicate deltaic deposits. The presence of reflection patterns showing laterally adjacent undaform (above wave base), clinoform (seaward sloping), and fondaform (flat basin floor) beds suggest changes in water depth from shelf to slope to deep basin. Parallel reflectors that extend over large areas suggest shelf deposits or possibly deeper-water deposits in a stable basin.

To a large extent, environmental interpretation of seismic facies is a process of elimination (Brown and Fisher, 1980). The interpreter may be able to immediately eliminate certain lithofacies or depositional environments because of obvious inconsistencies with available data or because of personal knowledge of the basin or region under study. Additional analysis involving study of relationships to other units, reflections characteristics, and other properties commonly allows further reduction of the remaining options until only one or two depositional or lithofacies models fit the available data. Lateral facies equivalents must be given special attention in the interpretation process, and the interpreter must be experienced in and have a good knowledge of depositional process and systems, that is, lithofacies composition, geometry, and spatial relationships. Even so, it may not always be possible to arrive at a final, unique interpretation, and the interpreter may have to settle for the best conclusion that can be made consistent with available data.

Figure 15.17 demonstrates the process of seismic facies interpretation by showing how the seismic reflection patterns depicted in Figure 15.15 are interpreted in terms of lithofacies and environmental setting. The first step in making such an interpretation is to learn as much as possible about the regional geology from well and outcrop control. Such nonseismic data can commonly show whether the sedimentary section consists of carbonates (and/or evaporites), siliciclastics, or mixed carbonates and siliciclastics. The seismic facies patterns can then be interpreted. For example, the lowstand systems tract (lying directly above the sequence boundary, i.e., the unconformity shown by the wiggly line) in Figure 15.15 is interpreted as mainly siliciclastic; the channel/overbank complex consists of marine silt and clay (mudstone); and the mound is composed of deep-water sand. The highstand systems tract (lying immediately below the sequence boundary) consists of carbonates, which are shelf, slope, or basin carbonates depending upon their relation to the offlap pattern (Vail, 1987).

Sea-Level Analysis

In addition to seismic sequence and facies analysis, the principles of seismic stratigraphy have also been applied to study and interpretation of ancient sea level and changes in sea level throughout geologic time. Studies of sea-level changes have special relevance with respect to analysis of cyclic sequences in the stratigraphic record. Sea-level changes through time have been studied particularly intensively by P. R. Vail and his associates at the Exxon research laboratory in Houston (Vail, Mitchum, and Thompson, 1977a, 1977b; Haq, Hardenbol, and Vail, 1988). These authors used seismic data and surface-outcrop data to integrate occurrences of coastal onlap, marine (deep-water) onlap, baselap, and toplap into a model that involves asymmetric cycle oscillations of relative sea level.

Vail and his group inferred changes in relative sea level by reference to coastal onlap charts. These charts were constructed by estimating from seismic profiles the magnitude of sea-level rise, as measured by coastal aggradation (the thickness of

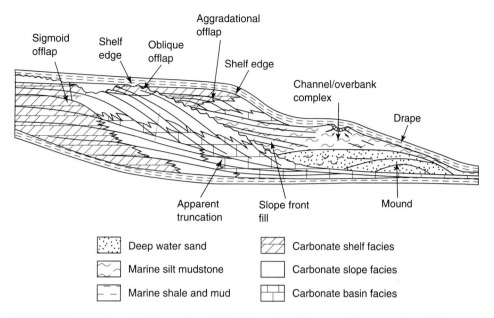

FIGURE 15.17 Schematic illustration of lithologic and environmental interpretation of the simulated seismic facies patterns shown in Figure 15.15. (From Vail, P. R., 1987, Seismic stratigraphic interpretation using sequence stratigraphy, *in* A. W. Bally (ed.), Seismic stratigraphy, Am. Assoc. Petroleum Geologists Studies in Geology 27. Fig. 9, p. 10, reproduced by permission of AAPG, Tulsa, Okla.)

coastal sediments deposited during sea-level rise). The amount of sea-level drop is determined by measuring the magnitude of downward shifts in coastal onlap, that is, the elevation (vertical) difference between the point of maximum coastal onlap reached at maximum sea level and the point of maximum sea-level fall, which is determined from the seismic records by the position where the next (younger) onlap unit lies above the unconformable surface produced during the sea-level fall (Vail, Mitchum, and Thompson, 1977a; Vail, Hardenbol, and Todd, 1984). The procedures used in constructing a relative coastal onlap chart from coastal and marine sequences are illustrated in Figure 15.18.

The first step involves analysis of sequences such as those shown as units A through E of Figure 15.18A. Sequence boundaries, areal distributions, and the presence or absence of coastal onlap and toplap are determined by tracing reflections on seismic profiles. Available age controls from well data are used to establish the geologic-time range of each sequence. An environmental analysis is also made from seismic and other available data to distinguish coastal facies from marine facies. The second step is to construct a chronostratigraphic (time-stratigraphic) chart of the sequences. Both stratal surfaces and unconformities give time-stratigraphic information. Because they are depositional surfaces, the seismic response to strata surfaces are assumed to be chronostratigraphic reflectors. In addition, because seismic reflectors are isochronous (have the same age everywhere), they can cross lithologic boundaries. That is, the seismic reflections from a given surface may extend laterally through a variety of lithofacies. Seismic reflectors may be traced continuously, for example, through a shelf system, over the shelf edge, and downward through an equivalent slope system. Unconformities are not isochronous surfaces; however, strata below an

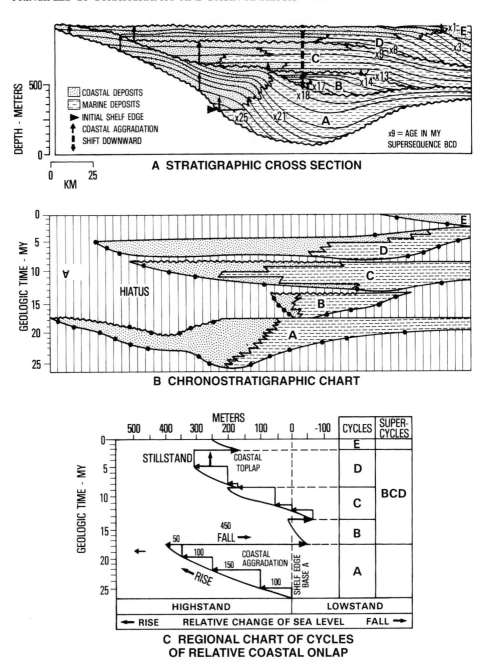

FIGURE 15.18 Diagram illustrating the procedure used by Exxon geologists for constructing a regional chart of cycles of relative coastal onlap. (After Vail, P. R., R. M. Mitchum, Jr., and S. Thompson, III, 1977, Seismic stratigraphic and global change of sea level, Part 3: Relative changes of sea level from coastal onlap, *in* C. E. Payton (ed.), Seismic stratigraphy—Applications to hydrocarbon exploration: Am. Assoc. Petroleum Geologists Mem. 26. Fig. 13, p. 78, reprinted by permission of AAPG, Tulsa, Okla.)

unconformity are older than strata above it. Therefore, strata between unconformities constitute time-stratigraphic units. After determining the ages of depositional sequences, such as those shown on the stratigraphic cross section in Figure 15.18A, from well-control or other information, workers plot the stratigraphic information against geologic time to construct a chronostratigraphic correlation chart (Fig. 15.18B). The final step in the procedure is to identify cycles of relative coastal onlap in each seismic sequence, measure the magnitude of aggradation and seaward downshifts in coastal onlap that result from relative rise and fall of sea level, and plot these changes and sea-level standstills against geologic time as shown in Figure 15.18C. Thus, the magnitude of coastal aggradation is a measure of a relative rise in sea level; a relative standstill is indicated by coastal toplap (Fig. 14.11); and seaward shifts in coastal onlap indicate a relative sea-level fall. The plots of relative coastal onlap are repeated for each sequence (cycles A though E of Fig. 15.18C) to complete the relative coastal onlap chart. Finally, the changes in relative coastal onlap are used as the basis of inferring changes in relative sea level. For simplicity, it assumed that there has been no subsidence of the margin and that the bed-thinning effects of compaction under deep burial have been considered (original thickness restored).

The time interval occupied by a relative rise and fall of sea level, as interpreted from a coastal onlap chart such as that shown in Figure 15.18C, constitutes a cycle of relative sea-level change in a region. Vail, Mitchum, and Thompson (1977b) correlated regional cycles and used this information to construct composite charts of global cycles of relative sea-level change. They interpreted these cycles to be worldwide and apparently to be controlled by absolute or eustatic sea-level changes. Figure 15.19 shows the chart by Vail, Mitchum, and Thompson (1977b); a slightly different chart constructed by Hallam (1984) from compilations of continental flooding is shown also for comparison. Upon publication, the Exxon coastal onlap charts, and the relative sea-level curves inferred from these charts, immediately generated lively interest, discussion, and controversy among geologists. They continue to be a focus of controversy. Before we discuss this controversy and the status of the Exxon sea-level curves, some additional explanation of stratigraphic cycles (Chapter 14) is necessary.

Orders of Stratigraphic Cycles

Five orders of stratigraphic cycles, with periodicities ranging from hundreds of millions to tens of thousands of years, have been defined (Table 15.2) and are described in the following paragraphs.

First-Order Cycles

Two first-order cycles, which last from 200 to 400 million years, are recognized in the Phanerozoic record (Fig. 15.20). First-order cycles appear to to reflect eustatic sea-level changes resulting from formation and breakup of supercontinents (e.g., the Permian Pangaea). Global sea level falls when continents join together because the volume of spreading ocean ridges is minimized and ocean basin volume is maximized owing to thermal subsidence. The reverse is true during supercontinent breakup, when new spreading ridges form and reduce ocean-basin volume.

Second-Order Cycles

These cycles have durations of 10 to 100 million years and correlate well with Sloss's (1963) North American cratonic sequences, as shown in Figure 15.20. They can appar-

FIGURE 15.19   Eustatic sea-level curves for Phanerozoic time. A. Hallam, 1984. B. Vail, Mitchum, and Thompson, 1977b. (From Hallam, A., 1984, Pre-Quaternary sea-level changes: Ann. Rev. Earth and Planetary Sciences, v. 12. Fig. 5, p. 220, reprinted by permission.)

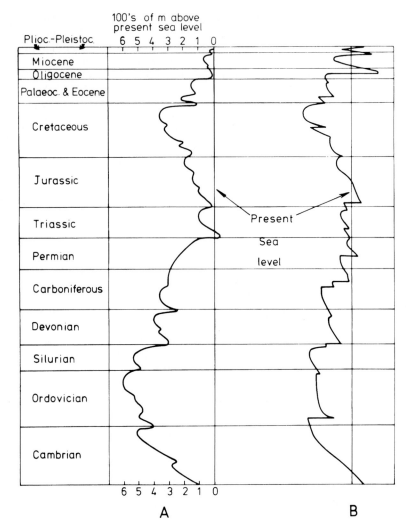

ently be correlated between continents. Second-order cycles reflect changes in the volume of oceanic ridges, which are related to changes in rates of seafloor spreading as previously discussed.

Third-Order Cycles

Third-order cycles span periods of time ranging from 1 to 10 million years but are commonly shorter than 3 million (Haq, Hardenbol, and Vail, 1988). They may be related to changes in spreading ridges or continental ice growth and decay; however, the control for these cycles is controversial (Plint et al., 1992). They represent the highest-frequency sea-level events portrayed on the Exxon curve. It may be possible to correlate third-order cycles between continents; however, global correlation of most third-order cycles has not been conclusively demonstrated.

TABLE 15.2 Stratigraphic cycles and their postulated causes

Type	Other terms	Duration, m.y.	Probable cause
First-order	—	200–400	Major eustatic cycles caused by formation and breakup of supercontinents
Second-order	Super cycle (Vail, Mitchum, and Thompson, 1977b); sequence (Sloss, 1963)	10–100	Eustatic cycles induced by volume changes in global mid-ocean spreading ridge system
Third-order	Mesothem (Ramsbottom, 1979); megacyclothem (Heckel, 1986)	1–10	Possibly produced by ridge changes and continental ice growth and decay
Fourth-order	Cyclothem (Wanless and Weller, 1932); major cycle (Heckel, 1986)	0.2–0.5	Milankovich glacioeustatic cycles, astronomical forcing
Fifth-order	Minor cycle (Heckel, 1986)	0.01–0.2	Milankovich glacioeustatic cycles, astronomical forcing

Source: Vail, P. R., R. M. Mitchum, Jr., and S. Thompson, III (1977); Miall (1990, p. 447).

Fourth- and Fifth-Order Cycles

Many thin cyclic deposits are present in the Phanerozoic sedimentary record that appear to record sea-level cycles with durations less than 500,000 years. Fourth-order cycles last from 200,000 to 500,000 years, and fifth-order cycles from about 10,000 to 200,000 years. These cycles are believed to be driven by changes in climate caused by cyclic changes in Earth's orbital parameters. These orbital cycles, called **Milankovich cycles** after the mathematician who calculated the climatic effects of the cycles, are related to the following:

1. changes in the eccentricity of Earth's orbit about the sun (400,000 and 100,000 years)
2. changes in the tilt of Earth's axis with respect to the plane in which it orbits the sun (41,000 years)
3. precession, or wobble, owing to the tilt axis sweeping out a cone (21,000 years)

These cyclic changes in Earth's tilt and orbit cause cyclic variations in the intensity and seasonal distribution of incoming solar radiation. These variations in solar intensity, in turn, result in alternate accumulation and melting of continental ice sheets (at least in the Quaternary), producing cycles of falling and rising sea level. It has also been suggested that climatic fluctuations related to Milankovich cycles might affect sea levels by controlling the volume of groundwater stored on continents (Hay and Leslie, 1990). As mentioned in Chapter 14, the postulated process of astronomically driven climate and sea-level cycles is often referred to as **orbital forcing.** Goodwin and Anderson (1985) coined the term **punctuated aggradational cycles** (PAC) for small-scale (1–5 m thick), shallowing-upward cycles separated by surfaces marked by abrupt changes to deeper facies (parasequences?) that, presumably, are the result of glacial eustasy driven by orbital perturbations.

The existence of Milankovich cycles has been studied particularly through examination of the oxygen isotope record in Quaternary deep-sea sediments. Fluctuations in oxygen isotope values can be correlated to eustatic sea-level changes arising from

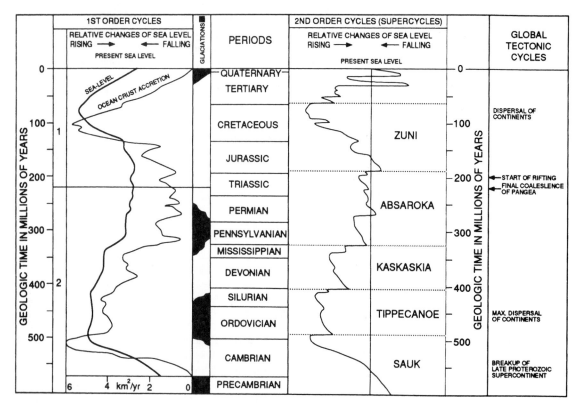

FIGURE 15.20   Illustration of first- and second-order global sea-level cycles. First-order cycles are correlated with ocean-crust accretion related to formation and breakup of continents. Second-order cycles are related to changes in rates of seafloor spreading. (From Plint et al., 1992, Control of sea-level changes, *in* R. G. Walker and N. P. James (eds.), Facies models—Response to sea level change: Geol. Assoc. Canada. Fig. 3, p. 18, reproduced by permission.)

episodes of glaciation and deglaciation, as described in Chapter 18. Some workers have proposed that Milankovich cycles can be recognized also in pre-Quaternary sediments; however, the models of glacioeustatic sea-level variation established on the basis of oxygen isotope studies of Quaternary deep-sea sediments may not be applicable to older rocks. These rocks were deposited at times when the distribution of continents, patterns of continental ice sheets, and possibly the Milankovich cycles themselves may have been different—nonetheless, climatic changes did occur.

The Reliability of Sea-Level Analysis from Seismic Data

Following publication of the coastal onlap curves of Vail, Mitchum, and Thompson in 1977, several workers (e.g., Brown and Fisher, 1980; Kerr, 1984; Miall, 1986) challenged the basic premise of Vail et al. that these curves primarily reflect eustatic changes in sea level, and they pointed out that tectonism (basin subsidence) and rates of sediment supply also affect these curves. Nonetheless, interest by the geologic community in sea-level analysis continued at a high level. The publication in 1988 of the SEPM Special Publication *Sea-Level Changes: An Integrated Approach,* edited by Wilgus et al., is an indication of this interest. This volume contains several articles on analysis of sea-level changes as well as articles dealing with sea-level changes and sequence stratigraphy. In this volume, Haq, Hardenbol, and Vail (1988) present a new generation of coastal onlap curves and corresponding eustatic sea-level curves for the Mesozoic and Cenozoic (originally published by the same authors in 1987) that have been the subject of considerable subsequent discussion and controversy (e.g., Christie-Blick, Mountain, and Miller, 1990; Miall, 1991a, 1992; Walker, 1990). Haq, Hardenbol, and Vail (1988) indicate that these new curves are based on well-log and outcrop data as well as seismic data and have greater resolution than that obtainable from seismic data alone. These highly detailed curves are not reproduced in full here; however, a highly simplified portion of the late Tertiary and Quaternary curve is shown in Figure 15.21 to illustrate the coastal onlap chart and show the nature of the second-order and third-order sea-level cycles interpreted from this chart. Figure 15.22 shows a more complete, but still simplified, part of the cycle chart for the Tertiary.

Christie-Blick, Mountain, and Miller (1990) provide a comprehensive critique of seismic stratigraphy and its usefulness in sea-level analysis and interpretation of the stratigraphic record. With respect to the Haq, Hardenbol, and Vail (1988) curves, Christie-Blick et al. (1990) are particularly concerned by what they call the uncritical interpretation of all second- and third-order boundaries as resulting from eustatic sea-level changes. They maintain that amplitudes of eustatic fluctuations cannot be inferred from seismic stratigraphic data alone because coastal aggradation (the vertical component of onlap) is primarily a result of basin subsidence, not sea-level rise. Furthermore, downward shifts in onlap reflect only the rate of sea-level fall in relation to the rate of basin subsidence. They suggest that the large component of basin subsidence cannot be easily or objectively removed to derive the smaller eustatic signal. Christie-Blick et al. point out further that another major limitation to the global onlap chart is the

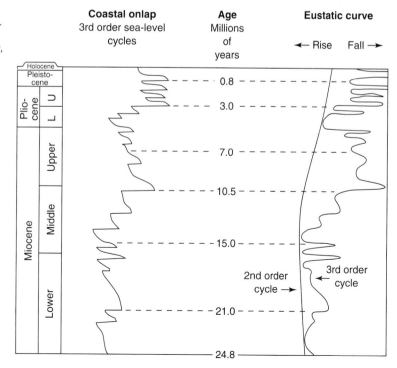

FIGURE 15.21 Coastal onlap and long-term (second-order) and short-term (third-order) sea-level curves for the late Tertiary and Quaternary, as determined by Haq et al., 1988. (Redrawn from Haq, B. U., J. Hardenbol, and P. R. Vail, 1988, Mesozoic and Cenozoic chronostratigraphy and cycles of sea-level change, *in* C. K. Wilgus et al. (eds.), Sea-level changes: An integrated approach: Soc. Econ. Paleontologists and Mineralogists Spec. Pub. 42. Fig. 14, p. 94, reproduced by permission of SEPM, Tulsa, Okla.)

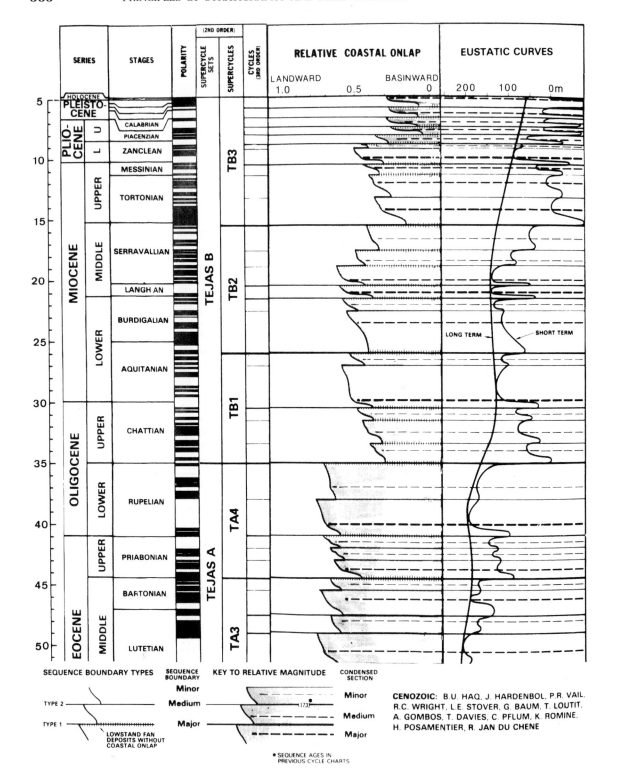

FIGURE 15.22 (Opposite) Simplified global sequence chart for part of the Tertiary and Quaternary. The polarity column shows the geomagnetic time scale (explained in Chapter 16); series and stages are discussed in Chapter 18. (After Haq, B. U., J. Hardenbol, and P. R. Vail, 1987, Chronology of fluctuating sea levels since the Triassic: Science, v. 235. Fig. 2, p. 1158. Copyright 1987 by the AAAS. As redrawn by Vail et al., 1991.)

uncertainties about the calibration of many boundaries to the geologic time scale. Another potential problem in interpretation arises from the operation of autocyclic mechanisms (Chapter 14), such as delta switching, that can generate small-scale cycles. Miall (1991a, 1992) also concludes that the implied precision of the Mesozoic–Cenozoic global cycle chart of Haq et al. (1988) is not justified. He maintains that the chart is not an independently tried and tested global standard and that the implied precision is unsupportable because it is greater than that of the best available chronostratigraphic techniques, such as those used to construct the global standard time scale (Chapter 18).

In spite of these criticisms, interest in sea-level analysis and seismic stratigraphy in general remains high. To quote Plint et al. (1992), "The Exxon sea level curve is here to stay, although it is likely to undergo progressive evolution and refinement as more data are gathered and better chronostratigraphic control becomes available." Over the next several years, geologists all over the world will surely be watching the outcome of the sea-level controversy. For application of the principles of sea-level analysis and seismic stratigraphy to various geologic systems, see the symposium volumes edited by MacDonald (1991); Nummedal, Pilkey, and Howard (1987); and Wilgus (1988).

15.5 CORRELATION BY SEISMIC EVENTS

As discussed in Section 15.2, seismic methods are predicated on the fact that elastic waves transmitted downward from a point source at the surface are reflected back to the surface from discontinuities. These discontinuities are either bedding planes or unconformities. Mitchum, Vail, and Sangree (1977) emphasize the fact that both bedding planes and unconformities have chronostratigraphic significance. The physical surfaces that separate individual beds, laminae, or groups of strata are essentially synchronous surfaces, in contrast to the boundaries of major lithostratigraphic units which may or may not transgress time boundaries. These authors assume that although the hiatus represented by an unconformity may not be of the same duration everywhere, unconformities nonetheless have time-stratigraphic significance because the rocks that lie above an unconformity are everywhere younger than those that lie below, as mentioned.

Because elastic waves are reflected from bedding planes and unconformities rather than from the boundaries between lithostratigraphic units (formation and other boundaries), seismic reflection patterns can be used for large-scale time-stratigraphic correlation throughout a region even where lithologic units are markedly diachronous (do not have the same age everywhere). I pointed out above, for example, that seismic reflectors may be traced continuously in some cases through a shelf system, over the shelf edge, and downward through an equivalent slope system. Figure 15.23 provides an example of time-stratigraphic correlations of regional scope based on lateral tracing of seismic reflection horizons. Inasmuch as seismic reflectors commonly do not coincide with the boundaries of lithologic units, seismic reflection patterns cannot be used for lithocorrelation.

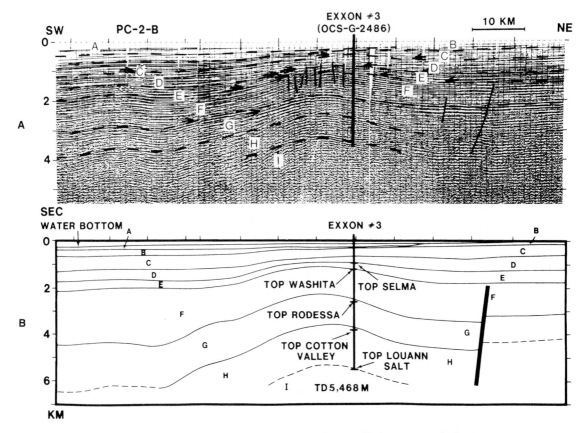

FIGURE 15.23   Correlation of stratigraphic units across the northeast Gulf of Mexico on the basis of seismic reflections. A. Original seismic section in which the vertical scale is calibrated in seconds. B. Time has been converted to depth, and the geologist's interpretation of the correlation of units A–I is shown more clearly. (From Addy, S. K., and R. T. Buffler, 1984, Seismic stratigraphy of shelf and slope, northwestern Gulf of Mexico: Am. Assoc. Petroleum Geologists Bull., v. 68. Fig. 2, p. 1786, reprinted by permission of AAPG, Tulsa, Okla.)

15.6   NOMENCLATURE AND CLASSIFICATION OF SEISMIC STRATIGRAPHIC UNITS

The terminology of seismic stratigraphic units used in this chapter was developed principally by geologists at the Exxon research laboratory in Houston. At this time, no formal system of nomenclature for seismic stratigraphic units, equivalent to that established for lithostratigraphic and magnetostratigraphic units, has been proposed by either the International Subcommission on Stratigraphic Nomenclature or the North American Commission on Stratigraphic Nomenclature, although the 1976 International Stratigraphic Guide does suggest that seismic units can be treated as informal zones.

The degree to which seismic reflections can be segregated into identifiable sequences and seismic facies (parasequences, parasequence sets, systems tracts) depends upon the resolving power of the seismic technique and the equipment in use. Packages of seismic reflections may represent stratigraphic intervals encompassing lithostratigraphic units as large as groups or supergroups or as small as very thick beds. Some seismic sequences probably represent packages of lithofacies containing several formations or groups of formations that together may make up the rocks of an entire geologic system, such as the Jurassic system. Seismic facies units, on the other hand, may include formations or lithostratigraphic units of lesser rank, such as members. I emphasize the point that there is no direct equivalent relationship between seismic sequences and facies and lithostratigraphic units such as groups, formations, and members. This lack of relationship is especially true because seismic units are generally time-stratigraphic units that may transgress the boundaries of lithologic units.

FURTHER READINGS

Anstey, N. A., 1982, Simple seismics: International Human Resources Development Corporation, Boston, 168 p.

Bally, A. W. (ed.), 1987, 1988, 1989, Atlas of seismic stratigraphy, Am. Assoc. Petroleum Geologists Studies in Geology 27, 3 volumes.

Brown, L. F., Jr., and W. L. Fisher, 1980, Seismic stratigraphic interpretation: Geophysical principles and techniques: Education Course Note Series 16, Am. Assoc. Petroleum Geologists, Tulsa, Okla., 56 p.

Einsele, G., W. Ricken, and A. Seilacher (eds.), 1991, Cycles and events in stratigraphy: Springer-Verlag, Berlin, 955 p.

Hardage, B. A. (ed.), 1987, Seismic stratigraphy: Handbook of geophysical exploration, v. 9: Geophysical Press, London, 432 p.

MacDonald, D. I. M., 1991, Sedimentation, tectonics and eustasy: Sea-level changes at active margins: Internat. Assoc. Sedimentologists Spec. Pub. 12, Blackwell, Oxford, 518 p.

Neidell, N. S., 1979, Stratigraphic modeling and interpretation: Geophysical principles and techniques: Education Course Note Series 13, Am. Assoc. Petroleum Geologists, Tulsa, Okla., 141 p.

Nummedal, D., O. H. Pilkey, and J. D. Howard (eds.), 1987, Sea-level fluctuation and coastal evolution: Soc. Econ. Paleontologists and Mineralogists Spec. Pub. 41, 267 p.

Payton, C. E. (ed.), 1977, Seismic stratigraphy—Applications to hydrocarbon exploration: Am. Assoc. Petroleum Geologists Mem. 26, Tulsa, Okla., 516 p.

Revelle, R. R. et al. (eds.), 1990, Sea level change: National Research Council Studies in Geophysics, National Academy Press, Washington, D.C., 234 p.

Sheriff, R. E., 1980, Seismic stratigraphy: International Human Resources Development Corporation, Boston, 227 p.

Sheriff, R. E., and L. P. Geldart, 1982, Exploration seismology: History, theory, and data acquisition: Cambridge University Press, Cambridge, 253 p.

Walker, R. G., and N. P. James (eds.), 1992, Facies models—Response to sea level change: Geol. Assoc. Canada, 409 p.

Weimer, P., and M. H. Link (eds.), 1991, Seismic facies and sedimentary processes of submarine fans and turbidite systems: Springer-Verlag, New York, 447 p.

Wilgus, C. K., B. S. Hastings, C. G. St. C. Kendall, H. W. Posamentier, C. A. Ross, and J. C. Van Wagoner (eds.), 1988, Sea-level changes: An integrated approach: Soc. Econ. Paleontologists and Mineralogists Spec. Pub. 42, 407 p.

16

Magnetostratigraphy

16.1 INTRODUCTION

Magnetic stratigraphy, or **magnetostratigraphy,** is a relatively new branch of stratigraphy developed largely since about the mid-1960s. The principles of magnetic stratigraphy were initially applied to the study of volcanic rocks and sediments younger than about 5 million years. Magnetic stratigraphic techniques have now been applied to much older rocks, and a detailed magnetic polarity time scale has been extended to the Jurassic and parts of the older record. Magnetic stratigraphy came about through the discovery that magnetic iron-rich minerals in igneous and sedimentary rocks can preserve the orientation or field direction of Earth's magnetic field at the time the rocks were formed. During the cooling of molten rock, iron-bearing minerals become magnetized in alignment with Earth's magnetic field as they cool through a critical temperature of about 500°C to 600°C (for magnetite)—the **Curie point.** As they approach this temperature, the influence of the magnetic field exerts itself, and the atomic-scale magnetic fields within the crystal lattices of the minerals begin to line up parallel to one another and to the direction of the magnetic lines of force around Earth. With further cooling, these atoms become locked into this orientation, and each mineral in essence becomes a small magnet, with polarity parallel to Earth's magnetic field. During deposition of sediments, small magnetic mineral grains are able to rotate in the loose, unconsolidated sediment on the depositional surface and thus align themselves mechanically with Earth's magnetic field. This (statistically) preferred orientation of magnetic minerals in igneous and sedimentary rocks imparts bulk magnetic properties to the rocks. These properties are retained for geologically long periods of time unless the rocks are again heated to near the Curie point. Therefore, this residual magnetism is called **remanent** magnetism. Because sediment grains can be disturbed by bioturbating organisms or by physical and chemical processes during burial and diagenesis, the magnetization of sedimentary rocks is less stable, as well as weaker, than that of volcanic lavas. The study of remanent magnetism in rocks of various ages to determine the intensity and direction of Earth's magnetic field in the geologic past is called **paleomagnetism.**

Remanent magnetism is measured by instruments called magnetometers. Early magnetometers were capable of making paleomagnetic measurements only in igneous rocks and highly magnetized iron-bearing red sediments. Modern superconducting magnetometers can measure the magnetism in much more weakly magnetic sediments, including carbonates. Remanent magnetism is complex and can include secondary magnetism caused by prolonged effects of Earth's present magnetic field or by chemical changes owing to alteration of one magnetic mineral to another. Demagnetization techniques are available for destroying this secondary magnetic effect in the laboratory so that the primary magnetization can be measured. It is this primary magnetic component, recording Earth's geomagnetic field at the time volcanic or sedimentary rocks formed, that is of interest in stratigraphic studies.

The significance of primary remanent magnetism for stratigraphic studies stems from the fact that Earth's magnetic field has not remained constant throughout geologic history but has experienced frequent reversals. The geomagnetic field is generated in some poorly understood way by the motion of highly conducting nickel-iron fluids in the outer part of Earth's core; this motion is assumed to be controlled by thermal convection and by the Coriolis force generated by Earth's rotation. Studies of the remanent magnetism in igneous and sedimentary rocks show that the dipole (main) component of Earth's magnetic field has reversed its polarity at irregular intervals from Precambrian time onward, apparently owing to instabilities in the outer-core convection. When Earth's magnetic field has the present orientation, it is said to have **normal** polarity. When this orientation changes 180°, it has **reversed** polarity. Figure 16.1 illustrates diagrammatically the magnetic lines of force around Earth during normal and reversed polarity epochs and shows what the orientation of a compass needle would be at points in the Northern and Southern hemispheres at such times. Reversals of Earth's magnetic field are recorded in sediments and igneous rocks by patterns of normal and reversed remanent magnetism. The direction of magnetization of a rock is defined by its **north-seeking magnetization.** If the north-seeking magnetization of rocks points toward Earth's present magnetic north pole, the rock is said to have **normal-polarity magnetization.** If the north-seeking magnetization points toward the present-day south magnetic pole, the rock has **reversed-polarity magnetization,** or reversed polarity. Thus, sedimentary and igneous rocks that display bulk remanent magnetic properties of the same magnetic polarity as the present magnetic field of Earth have **normal** polarity, whereas those that have the opposite magnetic orientation have **reverse** polarity.

These geomagnetic reversals are contemporaneous, worldwide phenomena. Thus, they provide unique stratigraphic markers in igneous and sedimentary rocks. The process of reversal is thought to take place over a period of 1000 to 10,000 years (e.g., Clement, Kent, and Opdyke, 1982). A decrease in intensity of the magnetic field by 60 to 80 percent occurs over a period of about 10,000 years preceding reversal. The actual reversal requires about 1000 to 2000 years, followed by a buildup of intensity for the next 10,000 years (Cox, 1969). The last unquestioned reversal of the magnetic field took place approximately 700,000 years ago, although a brief reversal or excursion of the field probably occurred about 20,000 years ago. In the early years of paleomagnetic study, intervals of reversed or normal polarity lasting 100,000 years or more were called epochs, and those having a duration of about 10,000 to 100,000 years were called events. Geomagnetic polarity reversals are now known to occur on a much broader spectrum of time scales ranging from less than 10,000 to greater than 10 mil-

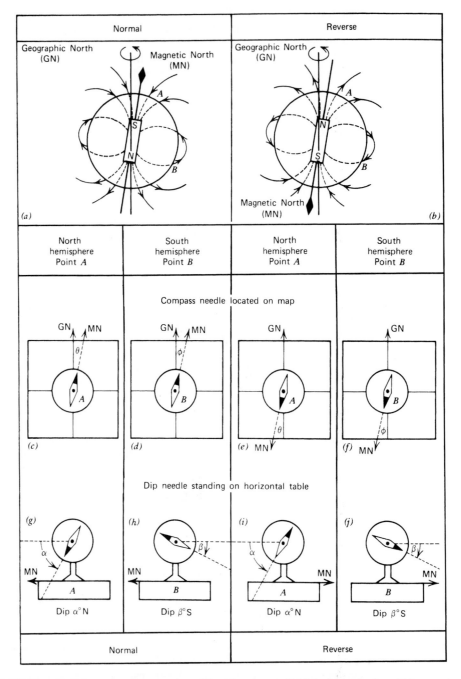

FIGURE 16.1   Schematic representation of Earth's magnetic field during episodes of (*a*) normal and (*b*) reversed polarity. The diagram also shows behavior during these episodes of a horizontal compass needle (*c–f*) and a vertical compass needle (*g–j*) at two different points on Earth. (From Wyllie, P. J., The way the Earth works, Fig. 9.1, p. 120, © 1976, John Wiley & Sons, Inc. Reprinted by permission of John Wiley & Sons, Inc., New York.)

lion years; however, geologists still use terminology equivalent to epochs and events although the terminology is now different (e.g., epochs are now referred to as chrons; Section 16.4). Magnetic stratigraphy in pre-Pleistocene rocks is based on these changes in polarity recorded in sediments or volcanic rocks that produce recognizable patterns of alternating polarity stratigraphic units that can be used for chronological and correlation purposes.

16.2 SAMPLING, MEASURING, AND DISPLAYING REMANENT MAGNETISM

To determine remanent magnetism in sedimentary rocks, geologists commonly must remove samples from the field for subsequent laboratory analysis, although techniques are being developed to measure remanent magnetism in well bores (e.g., Bouisset and Augustin, 1993). Three kinds of samples may be taken:

1. Samples cored with a portable drill. Cores taken by this method are commonly 2.5 cm in diameter and 6 to 12 cm long. Before cores are broken out of the rock on the outcrop, the orientation of the cores must be determined and marked on the sample. The inclination (dip) of the core axis is determined and the azimuth of the core axis (deviation from geographic north) is measured by use of a magnetic and/or sun compass.
2. Oriented block (hand) samples. These samples, which are broken from the outcrop with a hammer, are easier to obtain than core samples but present more difficulties with later orientation in the laboratory instruments.
3. Cores of lake- or ocean-bottom samples. Sediment cores, commonly obtained by piston-coring apparatuses, are assumed to penetrate the sediment vertically but are azimuthly unoriented.

In the laboratory, the remanent magnetism of a rock specimen is measured by means of a magnetometer. Several types of magnetometers (balanced fluxgate, cyrogenic, astatic, spinner) are in use (Hailwood, 1989; Butler, 1992). Three orthogonal components of magnetism are commonly measured; these three components are then combined to give the direction and intensity of the magnetic vector of the specimen. Because most rocks carry different components of magnetization that have been acquired at different times, the signature of secondary magnetism must be removed to reveal the primary remanent magnetism. This removal is accomplished by progressive demagnetization methods that may involve alternating field demagnetization, thermal demagnetization, and chemical demagnetization.

Once primary remanent magnetism has been measured, vector directions in paleomagnetism are described in terms of the following:

1. **inclination** (with respect to horizontal in the original bed at the collecting station)
2. **declination** (with respect to geographic north) (Fig. 16.2)

Inclination is called positive if downward directed and negative if upward directed. Positive inclination in the Northern Hemisphere indicates normal polarity; negative inclination means reversed polarity. Inclination directions are opposite relative to the polarity in the Southern Hemisphere (Fig. 16.1). Inclination and declination together define the geomagnetic field vector (**F** in Fig. 16.2). Inclination is a function of the latitude at which the rock specimens formed, and declination shows the deviation of the ancient paleomagnetic pole from the geographic pole. The polarity data are commonly

FIGURE 16.2 Description of the direction of the geomagnetic field. The total magnetic field vector **F** can be divided into a vertical component and a horizontal component. The angle D is the declination, the azimuthal angle between magnetic north and geographic north. The angle I is the inclination (dip), the angle between the horizontal and **F**.

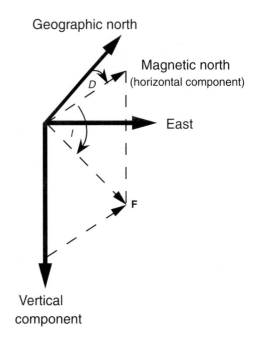

displayed graphically in polarity-reversal stratigraphic columns by plotting intervals of normal polarity in black and intervals of reversed polarity in white (e.g., Fig. 16.8). Dating of the polarity intervals by radiometric and biochronologic methods forms the basis of the magnetic polarity time scale, described in the next paragraph. Data may be plotted also as "virtual geomagnetic pole" (VGP) latitudes. The VGP represents the position of the effective north geomagnetic pole calculated from both inclination and declination information (Hailwood, 1989).

16.3 DEVELOPMENT OF THE MAGNETIC POLARITY TIME SCALE

The principle of developing a polarity time scale is illustrated in Figure 16.3. This figure shows three lava flows. The oldest erupted about 1.9 million years ago when Earth's magnetic field was normal. Thus, when it cooled, it acquired normal magnetic polarity (Fig. 16.3A). The second erupted about 1.5 million years ago during an episode of reversed magnetic polarity (Fig. 16.3B), and the youngest erupted 0.5 million years ago after the magnetic field had reversed back to normal polarity (Fig. 16.3C). Although the lavas have subsequently undergone weathering and erosion, each has retained its original magnetic polarity until the present time (Fig. 16.3D). By measuring the remanent magnetization in each lava flow and then determining its age by radiometric dating methods, geologists can construct a polarity time scale for these lavas, shown at the left of the diagram.

The concept of remanent magnetism is well known to today's students of geology; however, only a few studies of rock magnetism had been made prior to the 1960s. The basic principles of magnetostratigraphy were developed in the early and mid-1960s in the remarkably short time of about five years by two groups of scientists working independently and competetively—one group in northern California and one

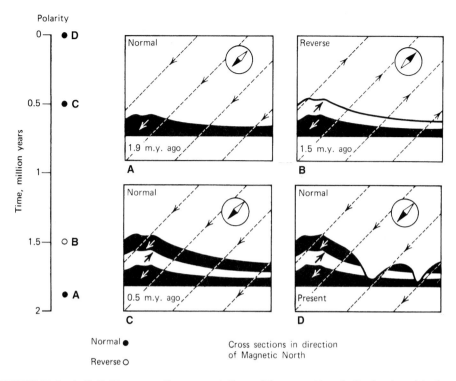

FIGURE 16.3 A, B, C. Diagrammatic representations of the magnetic polarity developed in three lava flows that erupted during the last 2 million years. Each of these lavas became magnetized at the time of eruption, with the direction of magnetization oriented parallel to the magnetic lines of force (dashed arrows) around Earth. D. Present time, showing retention of magnetic polarity. By measuring the radiometric ages of the lavas, geologists can construct a polarity time scale (left). (After Wyllie, P. J., The way the Earth works, Fig. 9.2, p. 122, © 1976, John Wiley & Sons, Inc. Reprinted by permission of John Wiley & Sons, Inc., New York.)

in Australia. The initial development of a magnetic polarity sequence by these groups of scientists is summarized by Cox (1973a), Glen (1982), McDougall (1977), and Watkins (1972).

The use of magnetic polarity reversals as a stratigraphic tool is based on identifying characteristic patterns of reversals. If the absolute age of each reversal can be established, a quantitative time scale for reversals can be set up. The first such polarity scales were achieved by measuring the ages and magnetic polarities of young volcanic rocks on land using potassium-argon (K-Ar) techniques to estimate the ages of the rocks (Cox, 1969). These polarity scales were initially developed only for the last 4.5 million years of geologic time (Fig. 16.4) because extension of the time scale was limited by lack of resolution of the K-Ar dating method. For ages greater than about 5 Ma, the typical value of ±2 percent for the precision of a K-Ar date is equivalent to ±0.1 m.y., which is longer than the duration of many of the shorter polarity intervals (Cox, 1969; Channell, 1982). The polarity time scale was subsequently extended to about 7 million years by use of stratigraphically related Icelandic lavas (McDougall et al., 1977). Note from Figure 16.4 that this original polarity time scale is subdivided into polarity "epochs," each named for a distinguished scientist who contributed to development of the field of geomagnetism; shorter "events" are named for localities where

FIGURE 16.4 The geomagnetic time scale for the last 4.5 million years as published in 1969. Each short horizontal line represents the magnetic polarity and K-Ar date of one volcanic cooling unit. The duration of events is based in part on paleomagnetic data from deep-sea sediments. (From Cox, A., 1969, Geomagnetic reversals: Science, v. 163. Fig. 4, p. 240, reprinted by permission of American Association for the Advancement of Science, Washington, D.C. Copyright 1969 by the AAAS.)

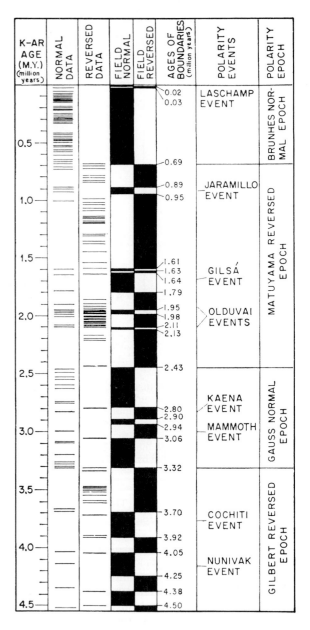

definitive study of the paleomagnetic characteristics of specific groups of rocks has been carried out.

In addition to study of the polarity of volcanic rocks on land, a second very important source of information about magnetic reversal sequences is provided by the linear anomaly patterns discovered in volcanic rocks of the ocean floor, particularly along mid-ocean ridges, and first interpreted by Vine and Matthews (1963). (Magnetic anomalies are significant deviations from Earth's magnetic background on either a local or a regional scale.) These linear "stripes" of normal- and reversed-polarity mag-

FIGURE 16.5 Linear magnetic anomaly patterns in the northeastern Pacific. Positive anomalies (normal polarity) are black. (After Mason, R. G., and A. D. Raff, 1961, Magnetic survey of the west coast of North America, 32°N. latitude to 42°N. latitude: Geol. Soc. America Bull., v. 72, Fig. 1, p. 1260; and A. D. Raff and R. G. Mason, 1961, Magnetic survey of the west coast of North America, 40°N. latitude to 52°N. latitude: Geol. Soc. America Bull., v. 72, Fig. 1, p. 1268; as modified by K. C. Conde, 1982, Plate tectonics and crustal evolution, 2nd ed., Fig. 4.22, p. 65, reprinted by permission of Pergamon Press, New York.)

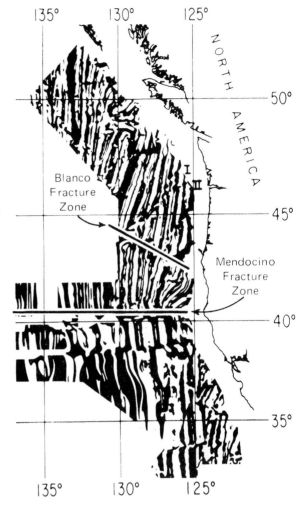

netic rocks (Fig. 16.5) are roughly parallel to ridge crests and are typically 5 to 50 km wide and hundreds of kilometers long. They were produced by reversals in Earth's magnetic field as successive flows of lava erupted along ridge crests and cooled below the Curie point. Previously magnetized volcanic rock was pushed or pulled aside from the ridges as new volcanic rock formed and became magnetized. Vine and Matthews (1963) hypothesized that the linear magnetic anomaly patterns on the ocean floor correlate with normal and reversed polarity intervals in the geomagnetic scale established on land, allowing the ages of the anomalies to be estimated. The fact that the magnetic anomalies are roughly symmetrical about spreading ridges was a critically important piece of evidence used in developing the concept of seafloor spreading.

The discovery of these linear magnetic anomalies on the seafloor provided the necessary tool for extending and calibrating the magnetic polarity time scale developed on land. Geophysicists assign numbers to particularly characteristic magnetic anomalies, beginning with number 1 at ridge axes, as illustrated in Figure 16.6. These ocean-floor magnetic anomalies do not in themselves determine an independent rever-

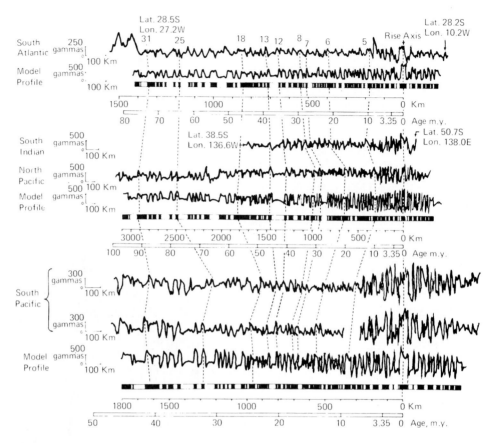

FIGURE 16.6   Magnetic profiles from the Atlantic, Indian, and Pacific ocean basins. Model profiles are also given from the South Atlantic, South Pacific, and North Pacific based on the normal (black) and reversely (white) magnetized bands beneath each model profile. Suggested correlations of anomalies are shown by dashed lines. Each time scale was constructed assuming an age of 3.35 million years for the end of the Gauss epoch. (From Heirtzler, J. R., G. O. Dickson, E. M. Herron, W. C. Pitman, III, and X. Le Pichon, Marine magnetic anomalies, geomagnetic field reversals, and motions of the ocean floor and continents: Jour. Geophysical Research, v. 73. Fig. 1, p. 2120. © 1968, American Geophysical Union, Washington, D.C.)

sal time scale because the ages of volcanic rocks are not usually known; however, once the anomalies are calibrated against known points on the radiometrically or biochronologically dated polarity scale, they provide a nearly continuous record of magnetic polarity intervals. The seafloor magnetic anomaly record is particularly valuable because it is continuous, unlike the on-land record, and may include polarity events that exist within gaps in the on-land radiometric data. The principal problem with the oceanic record is that it is very difficult to date directly. Paleontologic ages on the oldest sediments overlying marine magnetic anomalies are available where basement has been reached by DSDP or ODP drill holes, but large uncertainties are often associated with these age determinations. Figure 16.7 illustrates the method of dating seafloor anomalies by use of paleontologic data.

A polarity time scale for Mesozoic and Cenozoic oceanic events has been constructed by extrapolating ages on the basis of rates of seafloor spreading. Heirtzler et al.

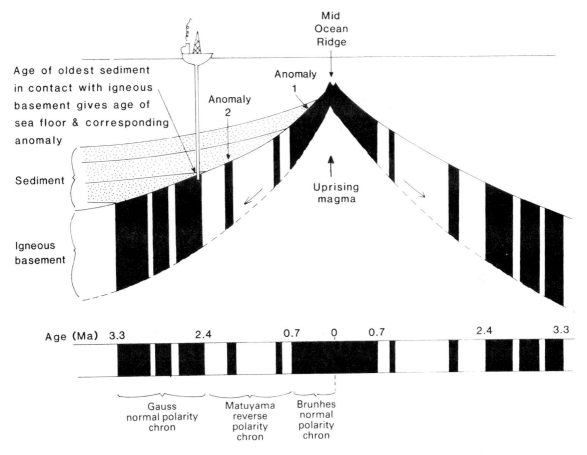

FIGURE 16.7 Schematic illustration of the principle by which the biostratigraphic age of sediment (determined by deep-sea drilling) overlying igneous basement can be used to date a particular marine magnetic anomaly. (From Hailwood, E. A., 1989, Magnetostratigraphy: Geological Soc. Spec. Report 19, Blackwell Scientific Publications. Fig. 17, p. 29, reproduced by permission.)

(1968) assumed that magnetic anomaly profiles above the ocean ridges and basins were manifestations of earlier reversals in the polarity of Earth's magnetic field. By further assuming a constant rate (1.9 cm/yr) of seafloor spreading since Late Cretaceous time in the South Atlantic, they assigned ages to the magnetic reversal time scale by extrapolation from a date of 3.35 m.y. for the older reversal boundary correlated to the Gauss-Gilbert magnetic polarity epoch (Figs. 16.4 and 16.6). The assumption of nearly constant spreading rate turned out to be surprisingly good, although subsequent work indicated that the calculated spreading rate was about 7 percent too low (LaBrecque, Kent, and Cande, 1977). Cande and Kent (1992) published the first new polarity time scale for Late Cretaceous and Cenozoic events since Heirtzler et al.'s 1968 scale. These authors suggest that the spreading rate of the world's ocean basins has not been constant since the late Cretaceous but has decreased (somewhat erratically) from about 70 mm/yr in Late Cretaceous time to the recent rate of about 32 mm/yr.

A third method of obtaining information on magnetic polarity is based on study of the record of polarity events in land sections of sedimentary rocks and in oceanic cores. Bulk remanent magnetism is produced in sediments by mechanical alignment of magnetic iron-bearing minerals during slow settling through water or shortly after deposition while the sediment is still highly water saturated. Study of polarity reversals in sediments has been hampered by several factors, including gaps in the stratigraphic record; variable rates of deposition; and chemical alteration (authigenesis) of magnetic iron-bearing minerals, causing secondary magnetism. Also, many sediments or sedimentary rocks are weakly magnetized; early magnetometers were not sensitive enough to measure their magnetic polarity. The development of modern superconducting magnetometers and improvements in laboratory techniques for removal of unstable secondary components of magnetic overprinting, as discussed, have alleviated some of these problems and have allowed extension of paleomagnetic studies to many types of sedimentary rocks.

The major advantage of paleomagnetic studies of sediments and sedimentary rocks is that even though gaps exist in the sedimentary record, sedimentary successions are stratigraphically far more continuous than most volcanic successions. Furthermore, ages of magnetic anomalies can be estimated on the basis of associated fossils. Conventional piston-coring techniques in the oceans have provided paleontologic control on reversal successions as far back as the early Miocene (McElhinney, 1978). Cores recovered during DSDP and ODP drilling have provided useful ages through the Cretaceous, although their usefulness is limited by incomplete recovery and by physical distortion of cores and disturbance of magnetism caused by rotary drilling. Owing to gaps in the stratigraphic record, many land-based sections of sedimentary rock are inappropriate for detailed magnetic polarity studies; however, some land sections provide good paleontologic calibration, allowing extension of the polarity time scale for the Tertiary and Mesozoic into the Paleozoic. For example, an essentially complete section of Middle Cretaceous to Paleocene calcareous, pelagic sediments exposed at Gubbio in the Umbrian Apennines of Italy (see Fig. 5.11) yielded well-defined reversal stratigraphy that can be tied to detailed foraminiferal biostratigraphy (Arthur and Fischer, 1977; Lowrie and Alvarez, 1977; Alvarez et al., 1977).

Land-section magnetostratigraphy can be correlated with reversal stratigraphy derived from oceanic anomalies, thereby making possible paleontological dating of the oceanic anomalies. This procedure allows the establishment of dated calibration points in the oceanic geomagnetic reversal time scale. Interpolation of ages of anomalies between these points is made by extrapolation, assuming constant rates of seafloor spreading. As many as 11 calibration points have now been established for the Late Cretaceous and Cenozoic (Lowrie and Alvarez, 1981; Cande and Kent, 1992); these data have allowed extension and dating of the magnetic polarity time scale into the Jurassic, well beyond 100 million years (Fig. 16.8). A new geomagnetic polarity time scale for the Late Cretaceous to Pleistocene part of the geologic record was published by Cande and Kent (1992). Note from Figure 16.8 that an extended period of normal polarity occurred during the Cretaceous. Such periods are commonly referred to as "quiet zones," in contrast to zones characterized by frequent reversals. Although extension of the oceanic polarity time scale beyond the Late Cretaceous was somewhat more difficult, it has now been extended into the Late Jurassic (~160 Ma), as shown in Figure 16.8 (e.g., Larson and Hilde, 1975; Lowrie and Ogg, 1986; Handschumacher et al., 1988).

A detailed magnetic polarity time scale for rocks older than the Jurassic has not yet been established because the continuous oceanic geomagnetic time scale cannot be extrapolated beyond the age of the oldest oceanic crust (about 150–160 m.y); however, magnetic reversals are known to occur in land sections in rocks at least as old as 1.5

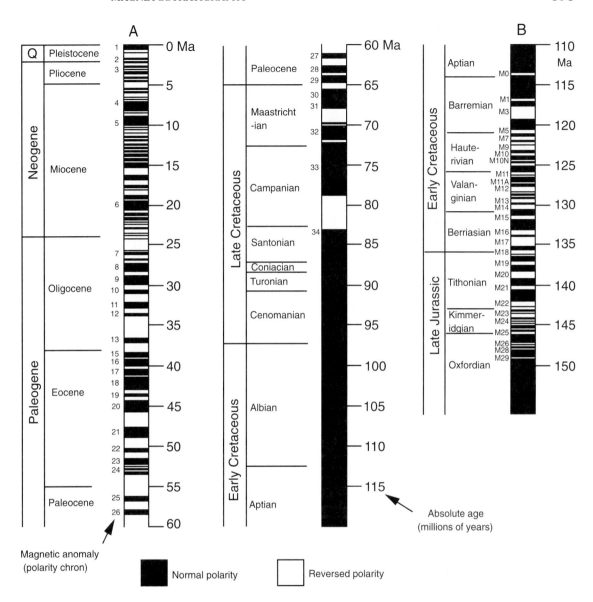

FIGURE 16.8 Magnetic polarity time scale for the last ~150 million years of geologic time. Note that the absolute age calibration for the Aptian Stage of the Early Cretaceous as shown by Cox (A) differs slightly from that shown by Lowrie and Ogg (B.) (A, redrawn from Cox, 1982, *in* W. B. Harland et al. (eds.), A geologic time scale. Fig. 4.6, p. 76–77. Reprinted with permission of Cambridge University Press. B, redrawn from Lowrie, W., and J. G. Ogg, 1986, A magnetic polarity time scale for the Early Cretaceous and Late Jurassic: Earth and Planetary Science Letters, v. 76. Fig. 3, p. 345.)

FIGURE 16.9 Major polarity bias magnetic anomalies (polarity chrons) for Phanerozoic time. Absolute age is shown to the left of the polarity bias column, with age limits of polarity superchrons shown in bold type. Some of the Paleozoic chrons are based on Soviet data, which have not been verified. (Data from Harland, W. B., et al., 1990; Ogg, in press.)

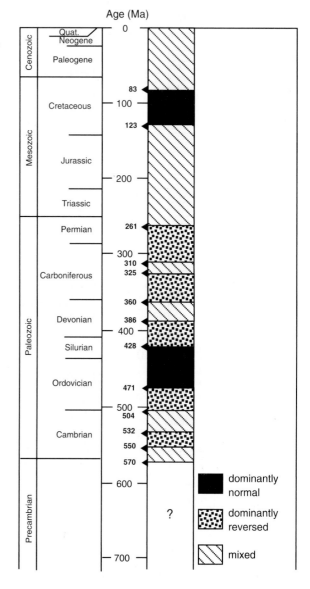

billion years (Conde, 1982). Although study of on-land stratigraphic sections from various parts of the world has provided some data on the paleomagnetic characteristics of early Mesozoic and older rocks, our knowledge of the polarity time scale for these rocks is much less refined than that for younger rocks. Figure 16.9 shows generalized polarity reversal patterns for the Phanerozoic; not all of these reversal patterns have been definitely verified (Ogg, in press). Note that throughout much of Paleozoic time, except from about middle Ordovician to middle Silurian time, the magnetic polarity was either dominantly reversed or mixed.

16.4 NOMENCLATURE AND CLASSIFICATION OF MAGNETOSTRATIGRAPHIC UNITS

Magnetostratigraphy, as we see from the preceding discussion, is the study of remanent magnetism in sedimentary and volcanic rocks. It is the element of stratigraphy that deals with the magnetic characteristics of rocks. In a broad sense, magnetostratigraphy encompasses all aspects of the study of remanent magnetism. Therefore, it includes study of magnetic susceptibility (the ratio of induced magnetism to the strength of the field causing magnetism); dipole-field position (variations in Earth's magnetic field intensity and pole positions); nondipole components (secular or short-term variations on the order of 5000–10,000 years in the direction and intensity of the Earth's magnetic field); and magnetic inclination (the angle at which magnetic field lines dip), as well as study of magnetic field reversals. Field reversals are of special interest for stratigraphy, however, because reversals in magnetic polarity have proven to be the most useful property for sytematically subdividing strata on the basis of their remanent-magnetic properties.

Although units of rock having uniform magnetic properties, such as a uniform direction of magnetic polarity, are not necessarily coincident with lithostratigraphic units, remanent magnetism is a physical attribute of rocks and may be used to characterize a body of rock and separate it stratigraphically from other bodies with different magnetic properties. As discussed, early workers in the field of magnetic stratigraphy set up a magnetic polarity scale based on informal "epochs" lasting more than about 100,000 years and "events" lasting between 10,000 and 100,000 years. Efforts have been underway since the early 1970s to develop a formal system of classification and nomenclature for stratigraphic bodies defined on the basis of magnetic properties. These efforts have been spearheaded by the International Subcommission on Stratigraphic Classification (ISSC) of the International Union of Geological Sciences (IUGS) and by the American Commission on Stratigraphic Nomenclature, which is now the North American Commission on Stratigraphic Nomenclature. The work of these two organizations has culminated in separate publications, each of which sets forth proposed terminology and procedures for establishing stratigraphic units based on magnetic polarity.

The IUGS International Subcommission on Stratigraphic Nomenclature and the IUGS/IAGA Subcommission on a Magnetic Polarity Time Scale (1979) published a supplementary chapter, "Magnetostratigraphy Polarity Units," to the ISSC International Stratigraphic Guide. This supplement presents the recommendations of the International Subcommission on Stratigraphic Classification for naming and defining formal stratigraphic units based on magnetic properties. Rock units characterized by any type of magnetic property (magnetic susceptibility, magnetic field intensity, direction of natural remanent magnetism) are classified under the general heading of **magnetostratigraphic units,** or magnetozones, which are "bodies of rock strata unified by similar magnetic characteristics which allow them to be differentiated from adjacent strata" (p. 579). Rock units defined specifically on the basis of their magnetic polarity are called **magnetostratigraphic polarity units,** which are defined as "bodies of rock in original sequences, unified by their magnetic polarity which allows them to be differentiated from adjacent strata" (p. 579).

Because magnetostratigraphic polarity units are of primary interest in stratigraphy, these units are formally subdivided into polarity superzones, polarity zones, and polarity subzones. The equivalent geochronologic time units are referred to as chrons or superchrons (Table 16.1). The **polarity zone** is the fundamental polarity unit for

TABLE 16.1 Nomenclature of magnetostratigraphic polarity units

Magnetostratigraphic polarity units	Description	Geochronologic (time) equivalent	Chronostratigraphic equivalent	Example	
				New name	Old name
Polarity superzone (10^6–10^7 years duration)	Magnetostratigraphic unit composed of two or more polarity zones	Chron (or superchron)	Chronozone (or superchronozone)		
Polarity zone (10^5–10^6 years duration)	Magnetostratigraphic unit distinguished by a single direction of magnetic polarization or by a distinctive alternation of normal and reversed polarities	Chron	Chronozone	Brunhes Normal Polarity Zone or Brunhes Normal Polarity Chron	Brunhes Normal Epoch
Polarity subzone (10^4–10^5 years duration)	Subdivision of a polarity zone	Chron (or subchron)	Chronozone (or subchronozone)	Jaramillo Polarity subzone or Jaramillo Polarity subchron	Jaramillo Event

Note: See Chapter 18 for an explanation of geochronologic and chronostratigraphic units subdivision of stratigraphic sections. A polarity zone may consist of strata with a single direction of polarization throughout; may be composed of an intricate alternation of normal and reversed units; or may be dominantly either normal or reversed but with minor subdivisions of the opposite polarity. A **polarity superzone** consists of two or more polarity zones, and a **polarity subzone** is a subdivision of a polarity zone. The principal polarity zone names now in use are the well-established names such as Brunhes, Matuyama, Gauss, and Gilbert, which are used for the last 5 million years of Earth's history. Historically, these units have been called epochs (Fig. 16.4); however, it is now recommended that these "epochs" be called the Brunhes, Matuyama, Gauss, and Gilbert polarity zones, or the Brunhes, Matuyama, Gauss, and Gilbert polarity chrons, if referring to time. Similarly, the so-called "events," such as the Jaramillo, Gilsá, and Olduvai, should now be referred to as the Jaramillo, Gilsá, and Olduvai polarity subzones (in rocks) or subchrons (for time).

In the 1983 North American Stratigraphic Code (Appendix B), the North American Commission on Stratigraphic Nomenclature follows approximately the same scheme of nomenclature for remanent-magnetic stratigraphic units as that proposed by the IUGS, and it defines magnetostratigraphic units and magnetopolarity units in roughly the same way. With regard to magnetopolarity units, the 1983 code further states that the upper and lower limits of a magnetostratigraphic unit are defined by boundaries marking a change in polarity and that such boundaries may represent either a depositional discontinuity **(polarity-reversal horizon)** or a magnetic field transition **(polarity transition-zone).** According to the code, a polarity-reversal horizon is either a single, clearly definable surface or a thin body of strata constituting a transitional interval across which a change in magnetic polarity is recorded. Polarity-reversal horizons describe transitional intervals of 1 m or less. If the change in polarity takes place over a stratigraphic interval greater than 1 m, the term polarity transition-zone should be used. Polarity-reversal horizons and polarity transition-zones provide the boundaries for polarity zones, although they may also be contained within a polarity zone, where they mark an internal change subsidiary in rank to those at its boundaries.

The North American Commission on Stratigraphic Nomenclature also follows the IUGS in considering the polarity zone as the fundamental unit of magnetostratigraphic classification, and it defines a polarity zone as "a unit of rock characterized by the polarity of its magnetic signature" (Appendix B, Article 46). Polarity zones may be grouped into polarity superzones or subdivided into polarity subzones.

Names, numbers, and letters are currently in use for designating magnetostratigraphic units. Both the IUGS and the North American Commission on Stratigraphic Nomenclature recommend that magnetostratigraphic units be given formally designated names, consisting preferably of a geographic name followed by the term polarity zone, polarity subzone, or polarity superzone, as appropriate. The name may be modified by including the words normal, reversed, or mixed, as in Deer Park Reversed Polarity Zone. These recommendations appear to be only rarely followed by magnetostratigraphic workers.

The IUGS guide and the North American Commission on Stratigraphic Nomenclature both recommend that a stratotype, or type section, be established for magnetostratigraphic units and that boundaries be defined in terms of recognizable lithostratigraphic or biostratigraphic units, in much the same manner as stratotypes are established for lithostratigraphic units. A special problem arises with selecting stratotypes for magnetopolarity units, however, because the best sequential record of reversals of Earth's magnetic field available for the past approximately 150 million years is that preserved in the linear seafloor-spreading anomalies described in Section 16.3.

These anomalies have been dated by extrapolation and interpolation from radiometric and paleontologic evidence in on-land sections. Because they are deduced from remotely sensed shipboard magnetometer records rather than by magnetic polarity determinations on outcropping or cored volcanic or sedimentary rocks (and because they are global events), it is not possible to designate any satisfactory type intervals or type boundaries for them, as required of conventional stratotypes. Instead, the standard reference for these ocean-floor magnetic anomalies is magnetometer profiles such as those shown in Figure 16.6.

16.5   APPLICATIONS OF MAGNETOSTRATIGRAPHY AND PALEOMAGNETISM

Correlation

The primary application of magnetostratigraphy lies in its use as a tool for global correlation of marine strata. Magnetostratigraphic correlation is particularly important where paleontologic or lithologic correlation is difficult. It has special significance for international correlation because geomagnetic reversals are contemporaneous, synchronous, worldwide phenomena. They have worldwide scope owing to the fact that reversals of Earth's magnetic field affect the magnetic field everywhere on Earth at the same time. Because the polarity time scale can be calibrated radiometrically or paleontologically, polarity events thus provide a precise tool for chronostratigraphic (time) correlation. The first significant application of magnetostratigraphic techniques to correlation and age determinations of rocks was correlation of linear ocean-floor magnetic anomalies to on-land sections of volcanic strata whose ages had been determined by radiometric methods. These correlation techniques were subsequently extended to cores of oceanic sediments.

Until very recently, correlation of sediment cores by use of magnetic polarity events had its greatest application in the study of marine sediments younger than about 6 to 7 million years. Correlation was previously restricted to very young rocks because the magnetic time scale had not been developed beyond about 7 Ma, and because most gravity and piston cores of ocean-floor sediment did not penetrate deeply enough to sample older sediments. As mentioned, the detailed geomagnetic time scale has subsequently been extended to about 150 to 160 Ma. Furthermore, deeper coring by use of hydraulic piston cores now makes it possible to obtain undisturbed cores of sediments as old as about middle Miocene. Longer cores obtained during the Deep Sea Drilling Program and Ocean Drilling Program by rotary coring methods have recovered rocks as old as the middle Jurassic; however, these rotary cores are commonly too badly disturbed to provide unambiguous paleomagnetic data. Because paleomagnetic methods have now been extended to correlation of on-land sections, this development opens up the possibility of even more extensive future use of paleomagnetic methods for correlating on-land stratigraphic sections. Magnetostratigraphy thus becomes an important tool for international correlation of older, on-land strata as the magnetic polarity time scale is extended farther back into geologic time.

Figure 16.10 provides an easily visualized example of paleomagnetic correlation in cores of young oceanic sediment. Beginning with the Brunhes Normal Epoch (polarity chron) at the top of the cores, the correlation can be carried downward on the basis of the patterns of reversed and normal polarity. With longer cores and older sediment, correlation becomes more difficult because the magnetostratigraphic record consists of many sets of reversals (Figs. 16.4 and 16.8) that may look very much alike. Correlation

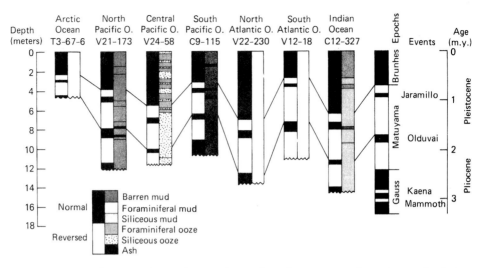

FIGURE 16.10 Paleomagnetic correlations of cores from the Arctic, Pacific, Indian, and Atlantic oceans. Cores have different lithologies and fossil assemblages. (From Opdyke, N., Paleomagnetism of deep-sea cores: Reviews Geophysics and Space Physics, v. 10. Fig. 20, p. 244, © 1972, American Geophysical Union, Washington, D.C.)

of these reversal patterns may require independent radiometric or paleontologic age evidence to first establish stratigraphic position. Paleomagnetic reversal patterns are particularly useful for correlating long distances across biogeographic boundaries where correlation by fossils, even planktonic fossils, may be difficult owing to the fact that different biogeographic provinces are marked by different fossil assemblages. Figure 16.10 illustrates that paleomagnetic correlation can be carried across the Arctic, Pacific, Indian, and Atlantic ocean basins, each of which is characterized by sediments composed of different lithologies with different fossil assemblages. In a similar manner, Figure 16.11 illustrates how on-land stratigraphic sections can be correlated on the basis of magnetic polarity reversal stratigraphy.

Geochronology

Although magnetostratigraphic sequence in itself does not normally provide unequivocal ages for geologic events preserved in strata, correlation of magnetic polarity zones or anomalies from areas where the ages of magnetic events have been established by radiometric methods or paleontologic data to areas where the ages of the strata are unknown, or poorly known, provides a means of estimating the ages of events in the new areas. Magnetostratigraphic geochronology may be particularly useful in determining ages within stratigraphic successions that are so nonfossiliferous that little biostratigraphic control exists as to their ages. For example, Heller and Tungsheng (1984) used magnetostratigraphic methods to work out the chronology of nonfossiliferous Chinese loess (eolian) deposits.

Magnetostratigraphic chronometry can also provide absolute ages for sediments that have been zoned by fossils and whose ages have been estimated from fossil data, as described by Hailwood (1989, p. 51). For example, McNeill et al. (1988) used magnetostratigraphy to refine the dating of Pliocene–Pleistocene sediments in a 91-m core of shallow-water carbonates from San Salvadore, Bahamas. Remanent magnetism in

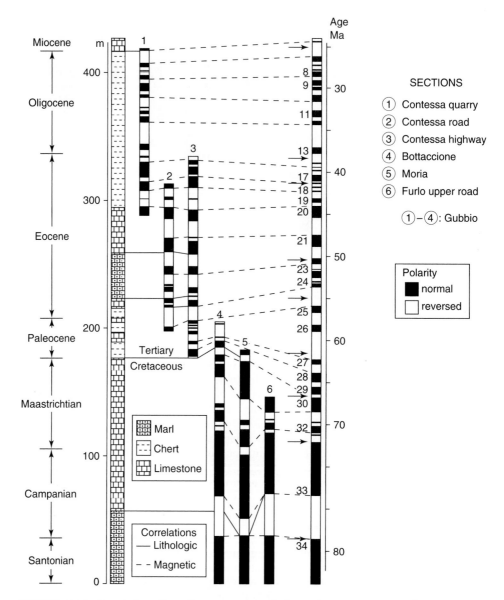

FIGURE 16.11 Correlation of Late Cretaceous through Cenozoic magnetostratigraphic sections in the Umbrian Apennines with the seafloor magnetic anomaly sequence. Age from biostratigraphic (foraminiferal) zonation (Miocene–Santonian) is given at the left of the column showing dominant lithology. Polarity zones in individual stratigraphic sections are correlated with each other and with the seafloor magnetic anomaly sequence shown by the polarity column at the right. Magnetic anomaly numbers and paleontologic calibration points (shown by the arrows) are noted at the left side of the seafloor polarity column. (Redrawn and modified from Lowrie, W., and W. Alvarez, 1981, One hundred million years of geomagnetic history: Geology, v. 9. Fig. 1, p. 393.)

these carbonate rocks is contained in magnetite that was precipitated with the aid of microbes at the time of carbonate deposition. Small samples (2 cm × 3 cm) were cut from the larger (8-cm diameter) core at approximately 0.5-m intervals. Remanent magnetism was measured in these samples with a superconducting magnetometer, which allows determination of very weak magnetic signals. The samples were then demagnetized by alternating field and thermal demagnetization techniques to isolate characteristic components of remanent magnetism. Because the main core was not oriented with respect to magnetic north, magnetic polarities were determined on the basis of magnetic inclination [relative down (positive) and up (negative) directions]; positive inclination in the Northern Hemisphere means normal polarity, and negative inclination means reversed polarity.

To match polarity zones determined in the core with the standard geomagnetic polarity time scale, McNeill et al. used a well-established biostratigraphic datum—the disappearance of Bowden assemblage molluscs and the coral *Stylophora affinis* shown by the dashed line in Figure 16.12—as the starting point for age determinations. On the basis of biostratigraphic age and the presence of this datum in a normal zone, they interpreted this datum to be in the upper part of the Gauss normal chron (Fig. 16.12). The age of this datum was thus estimated to be 2.6 to 2.7 Ma by reference to the standard polarity time scale. Once this calibration point had been established, the remaining polarity zones in the core could be matched to the standard geometric polarity time

FIGURE 16.12 Application of magnetostratigraphic techniques to determination of the age of Pliocene–Pleistocene carbonate sediments from San Salvador, Bahamas. The dashed line shows the stratigraphically highest appearance of Bowden equivalent molluscs (BEA) and the coral *Stylophora affinis* (S) in the core. Matching the measured magnetic polarities in the polarity column to the standard geomagnetic time scale places the age of the datum at 2.6–2.7 Ma. Ages of other parts of the core can be established by extrapolation. (After McNeill, D. F., et al., 1988, Magnetostratigraphic dating of shallow-water carbonates from San Salvador, Bahamas: Geology, v. 16. Fig. 4, p. 10.)

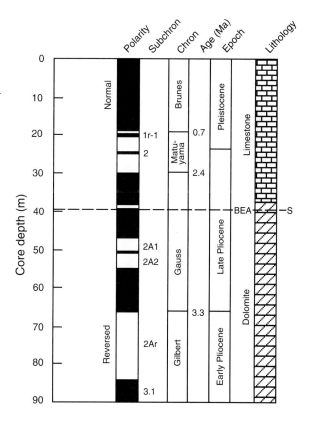

scale, allowing absolute ages at several points in the core sediments to be extrapolated from this scale. This study confirmed and refined the local timing of the disappearance of coral and molluscan species from the Bahamas as upper late Pliocene (between 2.6 and 2.7 Ma), and it established new time-stratigraphic markers for the Pliocene-Pleistocene of the Bahamas. This study showed also that the remanent magnetism in these carbonate sediments was not affected by original depositional conditions or subsequent diagenesis (replacement of limestones in the lower part of the section by dolomite).

Magnetostratigraphy is thus a useful, if somewhat limited, chronological tool. At the present time, the chronological framework is most accurate for the last 5 to 7 million years, but establishment of correlation points in older sequences of strata whose ages can be determined has extended the usefulness of geomagnetic chronometry to rocks as old as 100 million years or more.

Another example is the use of magnetostratigraphy of sediment cores as a tool for estimating the ages of volcanic eruptions that took place either on land or in the ocean. This is done by determining the ages of erupted ash that fell or was washed into the ocean and preserved as ash beds in oceanic sediments. By establishing the magnetic chronology from the reversal patterns in cores, geologists can determine the ages of ash layers in the cores by reference to this paleomagnetic time scale, a technique called **tephrochronology.** Figure 16.13 illustrates the principle; the ages of ash layers in piston cores in the Antarctic Ocean were estimated by comparing the position of the ash layers to polarity reversals in the cores.

A related application is in determination of rates of sedimentation for deep-sea sediments. Paleomagnetic correlation of deep-sea cores with rocks on land whose ages have been determined radiometrically allows absolute ages to be assigned to the boundaries between different geomagnetic events in the cores. The thickness of sediments between horizons within the cores whose ages are thus determined can then be used to calculate the sedimentation rate. For example, if we assume that 10 m of sediment were deposited in a given area of the ocean during the time represented in a core by the Matuyama Reversed Polarity chron extending from 2.4 Ma to 0.7 Ma, a time interval of 1.7 million years, the sedimentation rate for this area of the ocean can be calculated as 10 m/1.7 million yr = 5.8 m/m.y.

Paleoclimatology

Ages of sediment cores determined by paleomagnetic methods have also been used to study paleoclimate oscillations during the Quaternary and late Pliocene. For example, the magnetostratigraphy of deep-sea sediments provided a means of estimating the ages of ice-rafted debris in piston cores. It also furnished a method of studying the timing of siliceous ooze deposition, which reflects increased biologic productivity owing to increased oceanic upwelling and resulting increase in nutrients during cooler periods. Quantitative determinations of variations in microfossil assemblages, particularly planktonic foraminiferal assemblages, have also been studied in relation to climatic cycles. Some microfossil species are much more abundant during cooling trends, whereas others are more abundant during warming periods. Also, fluctuations in oxygen-isotope ratios in carbonate shells in deep-sea sediments are related to climate changes (Chapter 18). Use of paleomagnetic methods to estimate the ages of these climate-related biologic oscillations, fluctuations in oxygen isotope ratios, and variations in distribution of ice-rafted material in the oceans has added significantly to our knowledge of climatic fluctuations on land. It has also radically changed our ideas about the number of climatic cycles that occurred during the Quaternary. We now

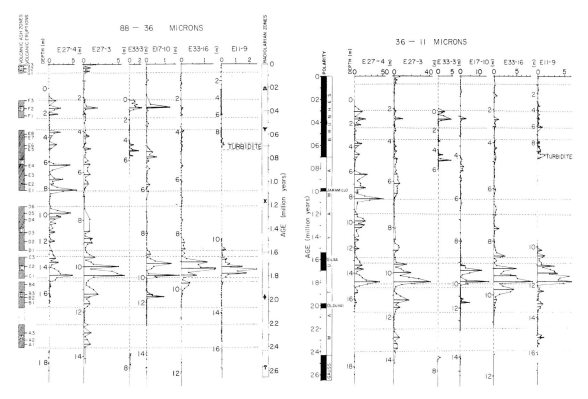

FIGURE 16.13 Use of magnetostratigraphy in tephrochronology of a suite of piston cores from the Antarctic Ocean. Periods of maxima on volcanic glass accumulation rates are labeled A to G, with members 1 to 8 indicating possible single eruptions or closely spaced series of eruptions. Black = normal polarity; clear = reversed polarity. (From Huang, T. C., N. D. Watkins, and D. M. Shaw, 1975, Atmospherically transported volcanic glass in deep-sea sediments: Volcanism in sub-Antarctic latitudes of the South Pacific during Pliocene and Pleistocene time: Geol. Soc. America Bull., v. 86. Fig. 3, p. 1307.)

know, for example, that many more cycles of cooling and warming took place than the four major glacial advances and retreats postulated from land-based studies.

Other Applications of Paleomagnetism

The study of paleomagnetism can be applied to a number of other geologic problems, not all of which, strictly speaking, are stratigraphic problems. These applications include dating of archaeological materials; tracing the source or provenance of these materials; study of magnetic fabrics in sedimentary, igneous, and metamorphic rocks; study of apparent polar-wandering paths of Earth through time; and paleogeographic and tectonic plate reconstruction (Butler, 1992; Khramov, 1987; Piper, 1987; Tarling, 1983; Van der Voo, Scotese, and Bonhommet, 1984; Van der Voo, 1993).

For example, the magnetic inclination of ancient magnetized rocks is now being used extensively as a tool to examine presumed movements of continental masses and smaller blocks. By measuring the remanent magnetic inclination and declination in ancient rocks, geologists can reconstruct the original geographic position of these rocks at the time they formed. These studies have shown not only that major continents have shifted their positions with time, but also that many smaller blocks of rock have moved from their original locations. That is, these blocks are now located in different latitudes from those in which they formed. Quite commonly the blocks have different lithologies and structural attitudes from those of adjoining areas. These exotic blocks are often called **suspect terranes,** referring to the probability that they are not now in the geographic positions in which they originally formed. There is growing evidence that large portions of many continental margins are made up of a collage of these suspect terranes, assembled by seafloor spreading and subduction processes over long periods of time from different parts of Earth.

FURTHER READINGS

Butler, R. F., 1992, Paleomagnetism: Blackwell, Boston, 319 p.

Conde, K. C., 1982, Plate tectonics and crustal evolution, 2nd ed.: Pergamon, New York, 310 p.

Cox, A. (ed.), 1983, Plate tectonics and geomagnetic reversals: W. H. Freeman, San Francisco, 702 p.

Glen, W., 1982, The road to Jaramillo: Critical years of the revolution in the earth sciences: Stanford University Press, 459 p.

Hailwood, E. A., 1989, Magnetostratigraphy: Geol. Soc. Spec. Report 19, Blackwell, Oxford, 84 p.

IUGS International Subcommission on Stratigraphic Classification and IUGS/IAGA Subcommission on a Magnetic Polarity Time Scale, 1979, Magnetostratigraphic polarity units—A supplementary chapter of the ISSC International Stratigraphic Guide: Geology, v. 7, p. 578–583.

Kennett, J. P. (ed.), 1980, Magnetic stratigraphy of sediments: Benchmark Papers in Geology 54, Dowden, Hutchinson and Ross, Stroudsburg, Pa., 438 p.

Khramov, A. N., 1987, Paleomagnetology: Springer-Verlag, Berlin, 308 p.

Piper, D. J. A., 1987, Paleomagnetism and the continental crust: The Open University Press, Milton Keynes, 434 p.

Tarling, D. H., 1983, Paleomagnetism: Principles and applications in geology, geophysics, and archaeology: Chapman and Hall, London, 379 p.

Van der Voo, R., 1993, Paleomagnetism of the Atlantic, Tethys, and Iapetus oceans: Cambridge University Press, New York.

Van der Voo, R., C. R. Scotese, and N. Bonhommet (eds.), 1984, Plate reconstruction from Paleozoic paleomagnetism: Geodynamics Series, v. 12, Amer. Geophys. Union, Washington, D.C., 136 p.

Wyllie, P. J., 1976, The way the Earth works: John Wiley & Sons, New York, 296 p.

17

Biostratigraphy

17.1 INTRODUCTION

The preceding three chapters focus on stratigraphic relations in sedimentary successions that exist owing to the physical characteristics of sedimentary rocks, either lithology or physical properties that can be remotely sensed by seismic or magnetic instrumentation. We turn now in the present chapter to examination of the exceedingly important role that fossil organisms play in stratigraphy. First, fossils provide an additional and highly useful method for subdividing sedimentary rocks into identifiable stratigraphic (biostratigraphic) units. In addition, they make possible the ordering and relative-age dating of strata and their correlation on both a continental and a global scale. The characterization and correlation of rock units on the basis of their fossil content is called **biostratigraphy.** Stratigraphy based on the paleontologic characteristics of sedimentary rocks is also referred to as **stratigraphic paleontology,** the study of fossils and their distributions in various geologic formations.

Separation of rock units on the basis of fossil content may or may not yield stratigraphic units whose boundaries coincide with the boundaries of lithic stratigraphic units. In fact, lithostratigraphic units such as formations commonly can be subdivided by distinctive fossil assemblages into several smaller biostratigraphic units. Indeed, one of the primary objectives of biostratigraphy is to make possible differentiation of strata into small-scale subunits or zones that can be dated and correlated over wide geographic areas, allowing interpretation of Earth history within a precise framework of geologic time. On the other hand, it is quite common for biologically defined stratigraphic units to span the boundaries of formally defined lithostratigraphic units. Some biostratigraphic units may thus include parts of two members or formations, for example, or even encompass two or more entire members or formations (Fig. 17.1).

The concept of biostratigraphy is based on the observation that organisms have undergone successive changes throughout geologic time. Thus, any unit of strata can

FIGURE 17.1 Relationship of biostratigraphic units to lithostratigraphic units. (Modified from Berg, R. R., C. A. Nelson, and W. C. Bell, 1956, Upper Cambrian rocks in southeastern Minnesota, in Lower Paleozoic of the upper Mississippi Valley: Geol. Soc. America Guidebook Ser., Minneapolis meeting, Field Trip 2.)

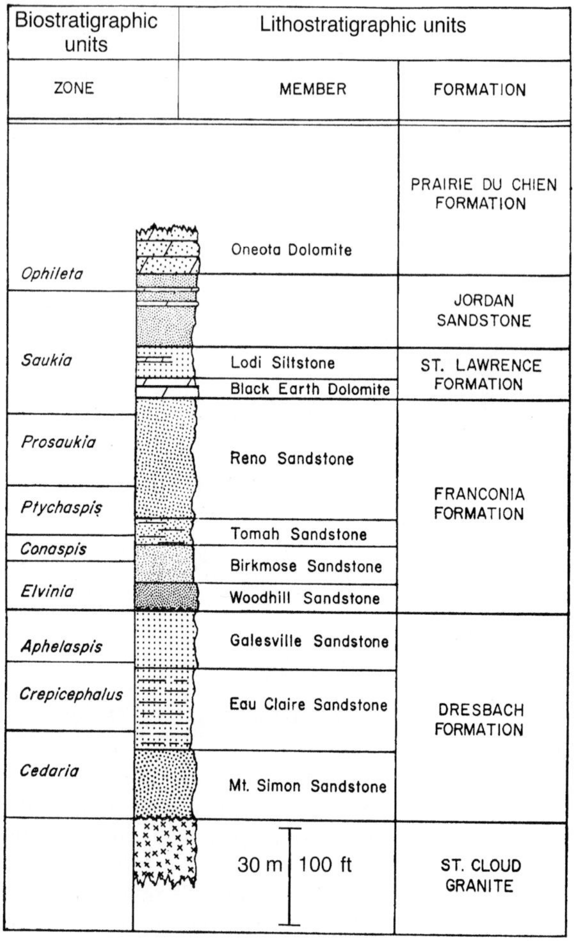

Biostratigraphic units	Lithostratigraphic units		
ZONE		MEMBER	FORMATION
			PRAIRIE DU CHIEN FORMATION
Ophileta		Oneota Dolomite	
			JORDAN SANDSTONE
Saukia		Lodi Siltstone	ST. LAWRENCE FORMATION
		Black Earth Dolomite	
Prosaukia		Reno Sandstone	
Ptychaspis			FRANCONIA FORMATION
		Tomah Sandstone	
Conaspis		Birkmose Sandstone	
Elvinia		Woodhill Sandstone	
Aphelaspis		Galesville Sandstone	
Crepicephalus		Eau Claire Sandstone	DRESBACH FORMATION
Cedaria		Mt. Simon Sandstone	
		30 m 100 ft	ST. CLOUD GRANITE

be dated and characterized by its fossil content. That is, any stratigraphic unit can be differentiated on the basis of its contained fossils from stratigraphically younger and older units. Biostratigraphy is obviously closely allied to paleontology, and a skilled biostratigrapher must also be a well-trained paleontologist. In fact, the application of biostratigraphy is for specialists who have intimate knowledge of large groups of organisms and their temporal and spatial distribution. Because stratigraphic paleontology is such a complex field, comprehensive treatment of this subject is beyond the scope of this book. The aim of this chapter is to introduce some very basic concepts and principles of biostratigraphy. Readers who wish more in-depth treatment of biostratigraphy should consult standard reference works on paleontology as well as more specialized, biostratigraphically oriented monographs such as Berry (1987), Cubitt and Reyment (1982), Dodd and Stanton (1981), Gradstein et al. (1985), and Kauffman and Hazel (1977).

We begin discussion of biostratigraphy by examining the concept that fossils constitute a valid basis for stratigraphic subdivision. As part of this examination, the origin and development of methods for biostratigraphic zonation are traced, and the stratigraphic procedures currently in use for classifying, naming, and describing biostratigraphic units are discussed. Organic evolution and the distribution of organisms in both time and space are explored next. We conclude the chapter with a discussion of the extremely important role that biostratigraphy plays in correlation of stratigraphic units. The use of fossils for calibrating the geologic time scale, called biochronology, is discussed in Chapter 18.

17.2   FOSSILS AS A BASIS FOR STRATIGRAPHIC SUBDIVISION

Principle of Faunal Succession

An English surveyor and civil engineer named William Smith, who worked in England and Wales in the late 1700s, is credited with discovering the fundamental principle of biostratigraphy. Previous workers had recognized that fossils are the remains of once-living organisms, and some workers had even suggested the possibility that certain species of marine shelled organisms had become extinct. Smith was evidently the first to utilize fossils as a practical tool for characterizing, subdividing, and correlating strata from one area to another. In his work as a surveyor and canal builder, he had discovered by about 1796 that the strata in and around Bath in Somerset and for some distance outward were always found in the same order of superposition—the order in which rocks are placed above one another. Furthermore, he noted that each layer in the stratigraphic succession was characterized by the same distinctive fossil assemblage wherever it was found throughout the region. Soon, Smith was able to assign any fossil-bearing rock to its proper superpositional interval by comparing its fossils with others whose stratigraphic position he knew from previous study. He thus discovered that fossil-bearing strata occur in a definite and determinable order. On the basis of Smith's discovery, we now know that rocks formed during any particular interval of geologic time can be recognized and distinguished by their fossil content from rocks formed during other time intervals. This concept has consequently become known as the **principle (law) of faunal succession.** Even without assigning names to fossils, Smith was successful in using them to establish a stratigraphic succession and to subdivide the rocks into mappable units by a combination of lithologic characteristics and fossil assemblages.

It is important to stress that Smith did not subdivide rock successions on the basis of fossils alone. His strata were first delineated and named according to their lithology. Then, their characteristic fossils were collected and studied. The use of fossils alone to subdivide thick, essentially lithologically homogeneous formations did not come about for another 15 years. The French scientist Georges Cuvier, a contemporary of Smith's, recognized the desirability of using fossils to subdivide rocks but did not attempt this process himself. Subdivision of rock successions on the basis of fossils was first carried out on sediments of Tertiary age in the early 1830s. Deshayes in France (1830), Bronn in Germany (1831), and Lyell in England (1833) all proposed subdivisions of Tertiary strata based on fossils (Hancock, 1977). Lyell's subdivisions are historically noteworthy. He split the Tertiary strata into four units on the basis of the proportions of living to extinct species in the rocks (Table 17.1). Thus, we see here for apparently the first time the use of fossils as an essential part of the definition of units of geologic time and the possibility of biostratigraphy freed from lithologic control.

TABLE 17.1 Lyell's subdivisions of the Tertiary

Name of subdivision	Extant species in the rocks (%)
Pliocene (more recent)	
Newer Pliocene	90
Older Pliocene	33–50
Miocene (less recent)	18
Eocene (dawn of Recent)	3.5

Concept of Stage

Although Smith's principle of faunal succession was to be the cornerstone for all subsequent biostratigraphy, his own work on biostratigraphic successions led to only vaguely defined time units. Lyell's subdivisions were similarly vague and, in any case, were confined to the Tertiary. A closer look at fossil successions was needed to refine their use in dating and correlation. This important step came with introduction of the concept of stage, credited to the French paleontologist Alcide d'Orbigny. About 1842, d'Orbigny came up with the idea of erecting major subdivisions of strata, each systematically following the other and each bearing a unique assemblage of fossils. Like Smith, d'Orbigny recognized that similarity of fossil assemblages was the key to correlating rock units, but he went a step further to propose that strata characterized by distinctive and unique fossil assemblages might include many formations (lithostratigraphic units) in one place or only a single formation or part of a formation in another place. He defined as **stages** groups of strata containing the same major fossil assemblages. He named these stages after geographic localities with particularly good sections of rock that bear the characteristic fossils on which the stages are based. Using the stage concept, he was able to divide the rocks of the Jurassic system into ten stages and the Cretaceous rocks into seven stages, each characterized strictly by its fossil fauna.

The boundaries of d'Orbigny's stages were defined at intervals marked by the last appearance, or disappearance, of distinctive assemblages of life forms and their replacement in the rock record by other assemblages. He conceived these stages as having worldwide extent and to be the result of repeated catastrophic destruction of life on Earth followed by new creations. His ideas on catastrophic destruction and special new creation, like those of Cuvier who preceded him, failed to gain lasting acceptance among geologists, and subsequent study has shown that his stages and their characteristic faunas are local rather than worldwide. Nonetheless, d'Orbigny's concept of stages as major bodies of strata characterized by large assemblages of fossils unique to that part of the total stratigraphic column was a significant and lasting contribution to the growing discipline of biostratigraphy. Somewhat different interpretations of the meaning of stage have been used by subsequent workers, but d'Orbigny's basic concept is still valid.

Concept of Zone

The stage concept of d'Orbigny permitted subdivision of strata into major successions on the basis of fossils. What they did not provide was a method by which fossiliferous strata could be divided into small-magnitude, clearly delimited units. Friedrich Quenstedt in Germany was particularly critical of d'Orbigny's stages because, according to him, d'Orbigny's method "centered around the acceptance, as the diagnostic faunal aggregate, of species of many strata in many localities, lumped together without enough regard for their precise stratigraphic ranges" (Berry, 1987, p. 125). Quenstedt maintained that only by extremely detailed study of strata on essentially a centimeter-by-centimeter basis could full understanding of the succession of faunas be developed. Quenstedt's own work did not bring this notion of detailed biostratigraphic subdivision into full fruition. It remained for his student, Albert Oppel, to expand, synthesize, and meld Quenstedt's ideas into the concept of the **zone.**

Oppel introduced the concept of zone in 1856 and thereby altered for all time the practice of biostratigraphy. Working with Jurassic rocks in various parts of Germany, he conceived the idea of small-scale units defined by the **stratigraphic ranges** of fossil species irrespective of lithology of the fossil-bearing beds. Oppel noted that the vertical ranges of some species were very short; that is, the species existed for only a very short time geologically. Others were quite long, but most were of some intermediate length. Oppel noted also that the assemblages of fossils that characterized the strata were made up of **overlapping** ranges of fossils. He defined his zones by exploring the vertical range of each separate species. Each zone was characterized by the joint occurrence of species not found together above or below this zone. Thus, the range of some species began at the base of a zone (the first appearance of a species), others ended at the top of a zone (the last appearance of a species), whereas still others ranged throughout the zone or even extended beyond it. Using species ranges, Oppel discovered that he could delineate the boundaries between small-scale rock units and distinguish a succession of unique fossil assemblages. Each of these assemblages was bounded at its base by the appearance of distinctive new species and at its top—that is, the base of the succeeding section—by the appearance of other new species. It is, however, the overlapping stratigraphic ranges of the species that make up the fossil assemblage that typifies a zone (Fig. 17.2). Because a zone represents the time between the appearance of species chosen as the base of the zone and the appearance of other species chosen as the base of the next succeeding zone, recognition of zones thus permits delineation of clear-cut, small-scale **time units.** Each of Oppel's zones was named after a particular fossil species, called an **index fossil,** or **index species,** which is but one fossil species in the assemblage of species that characterize the zone.

FIGURE 17.2 Diagrammatic illustration of an Oppel zone defined by the overlapping ranges of two or more taxa.

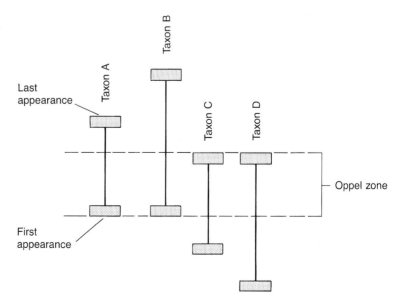

The concept of zone thus allowed subdivision of stages into two or more smaller, distinctive biostratigraphic units that could be recognized and correlated over long distances. Oppel was able, for example, to subdivide the Jurassic rocks of Western Europe into 33 zones. It should be noted that Oppel did not start with d'Orbigny's stages and subdivide them into zones. Instead, he delineated zones on the basis of fossil ranges and then combined the zones into stages, all of which did not necessarily fit into one of d'Orbigny's stages as then defined (Hancock, 1977).

Zones were slow to be adopted into stratigraphic practice, especially in the United States, but with minor modifications of Oppel's method they have now become the common denominator of biostratigraphic study. Zones have been extended to all parts of the fossil record, not just the Jurassic, and to all areas of the world. It is now recognized, however, that there are definite geographic limits beyond which most zones cannot be traced. The area within which a zone can be recognized is a **biogeographic province.** Because zones constitute the basic unit of biostratigraphic classification, considerable work has gone into efforts to standardize their usage. The current usage of zone and the different kinds of zones now recognized are described in the following section.

17.3   BIOSTRATIGRAPHIC UNITS

As the preceding discussion makes clear, a biostratigraphic unit is a body of rock strata characterized by its fossil content that distinguishes and differentiates it from adjacent strata. Furthermore, the zone, or **biozone,** is the fundamental biostratigraphic unit. Zones do not have any prescribed thickness or geographic extent. They may range in thickness from thin beds a few meters thick to units thousands of meters thick and in geographic extent from local units to those with nearly worldwide distribution. Recent attempts to standardize nomenclature and usage of zones have been made by the International Subcommission on Stratigraphic Classification (Hedberg, 1976) in the International Stratigraphic Guide and by the North American Commission on Stratigraphic Nomenclature (1983) in the North American Stratigraphic Code. The usage in this book follows that recommended in the North American Stratigraphic Code, but reference is made to usage in the International Stratigraphic Guide where appropriate.

Principal Categories of Zones

The North American Stratigraphic Code (Appendix B, Articles 49–52) subdivides biostratigraphic units into three principal kinds of biozones: interval zones, assemblage zones, and abundance zones. Each of these zones is distinguished by different criteria, as explained below.

An **interval zone,** or subzone, is the body of strata between two specific, documented lowest and/or highest occurrences of single taxa. Taxa is the plural of taxon, which is a general term used to signify a taxonomic group or entity, that is, a species or higher group. The boundaries of interval zones are defined by lowest and/or highest occurrences of single taxa, and three basic types of interval zones are recognized:

1. *The interval between the lowest and highest occurrences of a single taxon* (Fig. 17.3) constitutes the simplest type of interval zone in that it involves a single species or other taxon. This type of interval zone is called a **taxon range zone** in the International Stratigraphic Guide. It is the body of strata representing the total range of occurrence of specimens of a particular taxon, as opposed to the local range.

FIGURE 17.3 Principal kinds of interval zones as defined by the North American Stratigraphic Code and the International Stratigraphic Guide (ISG).

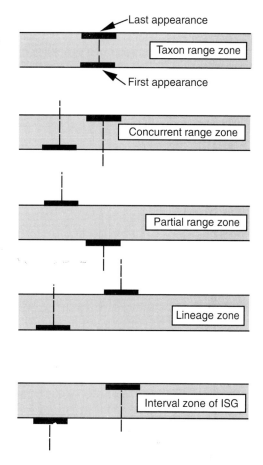

2. *The interval between the documented lowest occurrence of one taxon and the documented highest occurrence of another taxon* defines a type of interval zone that is called a concurrent range zone in some cases. For example, the International Stratigraphic Guide uses the term **concurrent range zone** for such an interval zone when the occurrence results in stratigraphic overlap of the taxa (Fig. 17.3). An example of the use of concurrent range zones for subdividing the Cretaceous (Maastrichtian) chalks of northwestern Europe is shown in Figure 17.4. When such occurrences do not result in stratigraphic overlap, the interval zone may be called a **partial range zone.**

3. *The interval between documented successive lowest occurrences or successive highest occurrences of two taxa* forms the final type of interval zone. When the interval is between successive lowest occurrences of unrelated taxa or between successive highest occurrences of either related or unrelated taxa, it corresponds to the **interval zone** of the International Stratigraphic Guide (Fig. 17.3). Note that the interval zone, defined in the International Stratigraphic Guide as an interval between two distinctive biostratigraphic horizons, is thus much more restrictive than the general definition of an interval zone given by the North American Stratigraphic Code. When the interval is between successive documented lowest occurrences within an evolutionary lineage (Fig. 17.3), it is the **lineage zone** of the International Stratigraphic Guide.

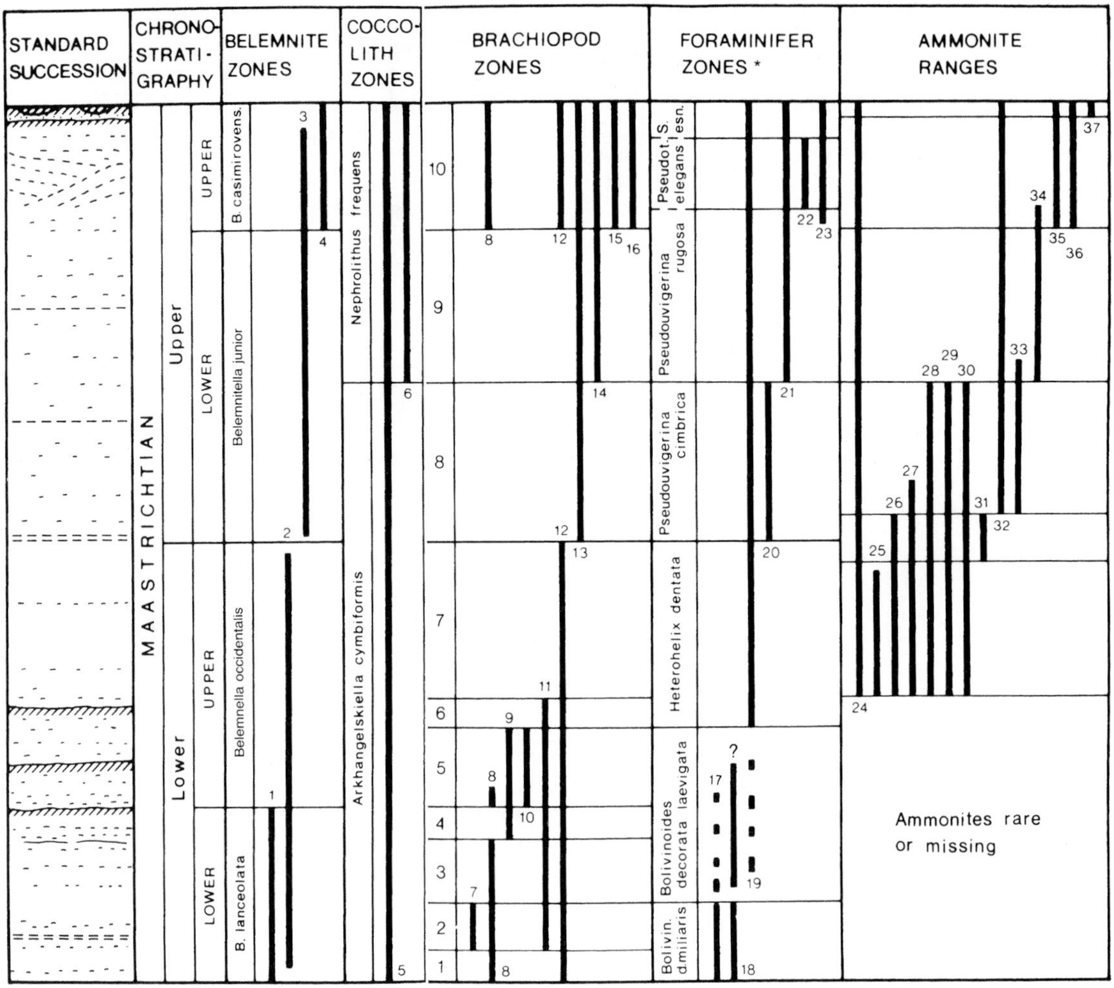

*S.esn = Stensioeina esnehensis; Pseudot. = Pseudotextularia

FIGURE 17.4 Subdivision of the Cretaceous (Maastrichtian) chalks in northwest Europe using concurrent range zones (biozones). Numbers beside range bars refer to individual species. (From Surlyk, F., and T. Birkelund, 1977, An integrated stratigraphical study of fossil assemblages from the Maastrichtian White Chalk of northwestern Europe, *in* E. G. Kauffman and J. E. Hazel (eds.), Concepts and methods of biostratigraphy. Fig. 13, p. 280. Copyright 1977, 1983 by Van Nostrand Reinhold. All rights reserved.)

An **assemblage zone** is defined by the North American Stratigraphic Code as a biozone characterized by the association of three or more taxa. An assemblage zone may consist of a geographically or stratigraphically restricted assemblage or may incorporate two or more contemporaneous assemblages with shared characterizing taxa (Appendix B, Fig. 5C), in which case it may be referred to as a **composite assemblage zone.** The code suggests that two concepts can be used to define assemblage zones. If the zone is characterized by taxa without regard to their range limits, it is called an **assemblage zone** (Fig. 17.5). If, on the other hand, it is characterized by more than two

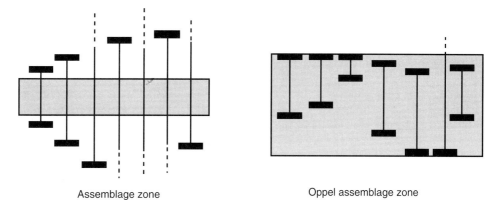

Assemblage zone   Oppel assemblage zone

FIGURE 17.5 Two kinds of assemblage zones as defined by the North American Stratigraphic Code.

taxa and is further characterized by having boundaries based on two or more documented first and/or last occurrences of the included characterizing taxa, it is called an **Oppel zone.** The inclusion of the Oppel zone as a subcategory of an assemblage zone differs from its usage in the International Stratigraphic Guide. In the guide, an Oppel zone is defined as a zone characterized by an association or aggregation of selected taxa of restricted and largely concurrent range, chosen as indicative of approximate contemporaneity. It is considered in the guide to be a subcategory of a concurrent range zone. These differences may cause readers some confusion, but that's the way it is.

An **abundance zone** is a biozone characterized by quantitatively distinctive maxima of related abundance of one or more taxa (Appendix B, Article 52). The abundance zone is equivalent to the **acme zone** of the International Stratigraphic Guide. The guide suggests that an acme zone represents the maximum development, which commonly means maximum abundance or frequency of occurrence, of some species, genus, or other taxon, but not its total range. Figure 17.6 illustrates subdivision of strata on the basis of abundance (acme) zones. Note that gaps may exist between zones defined on the basis of taxon abundance.

Rank of Biostratigraphic Units

The zone, or biozone, is the fundamental unit of biostratigraphic classification. Other biostratigraphic units are formed by either grouping or subdividing zones. The International Stratigraphic Guide suggests that some kinds of biozones, such as assemblage zones and Oppel zones, may be subdivided into subzones and/or grouped into superzones. The North American Stratigraphic Code provides that a biozone may be completely or partly divided into formally designated subbiozones (subzones) if such divisions serve a useful purpose.

Naming Biostratigraphic Units

The formal procedure for establishing and naming biostratigraphic units is outlined in the North American Stratigraphic Code (Appendix B). Each biozone is given a unique name, which is compound and which designates the kind of biozone, for example, the *Exus albus* assemblage zone or the *Rotalipora cushmani* taxon range zone. The name may be based on one or two characteristic and common taxa that are restricted to the

FIGURE 17.6 Schematic illustration of the abundance (acme) zones of three hypothetical fossil species. Note that each species reaches peak abundance (total number of individuals) at a particular time and then declines in abundance. Ages of the strata increase downward, and relative abundance increases to the right.

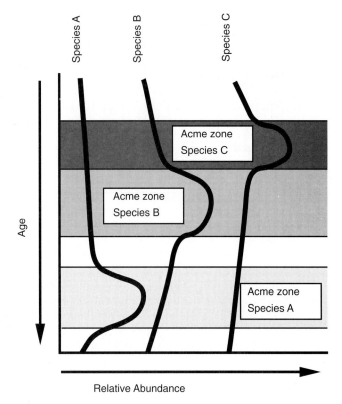

biozone or have their total stratigraphic overlap within the biozone. These names most commonly are those of genera or subgenera, binomial designations of species, or trinomial designations of subspecies.

17.4 THE BASIS FOR BIOSTRATIGRAPHIC ZONATION: CHANGES IN ORGANISMS THROUGH TIME

Evolution

The practicality of biostratigraphy as a tool for characterizing and correlating strata had been clearly established by Smith, d'Orbigny, and Oppel by the middle of the nineteenth century. None of these workers fully understood why fossil assemblages changed from one stratigraphic layer to another, although d'Orbigny apparently gave considerable thought to the origin of his stage boundaries. Charles Darwin provided the answer to this puzzle a few years after Oppel conceived the zone concept. In his monumental work on the origin of species published in 1859, Darwin demonstrated the existence of organic evolution and thereby greatly changed subsequent geological and philosophical thought, although his ideas were by no means accepted by all of his contemporaries.

Darwin was not the first to conceive the general idea of evolution, but previous workers had marshaled little supporting evidence for their ideas. By contrast, Darwin drew together data on the fossil records of extinct organisms, the results of selective breeding of domestic animals, observations on ecological adaptations and variations among living organisms, and details on comparative anatomy. He then wove these data into a powerful argument for organic evolution. Darwin pointed out that all organisms have high reproductive rates, yet populations of these organisms remain essentially constant over the long run. He explained this observation by suggesting that not all organisms of the same kind (species) are equally well equipped to survive, and therefore many individuals die before reproducing. Each individual of a species differs from other individuals as a result of variations that arise within an organism entirely by chance. Some of these chance variations may be an advantage to the organism in coping with its environment in its struggle for existence. Others may be a disadvantage. Successful variations help organisms survive and extend their environment and range. Unsuccessful variations result in extinction.

Darwin termed this process of weeding out the unfit and survival of the fittest **natural selection.** Furthermore, he proposed that these favorable variations are inheritable and can be transmitted from one generation to the next. Darwin's fundamental contribution to understanding evolution was thus recognition that natural selection was the process by which new species arise. New species appear because the composition of populations changes with time owing to the fact that those individuals that undergo favorable adaptations will stand a better chance of surviving and reproducing. He did not understand how variations arose or how these traits were passed on from one generation of organisms to the next. The concept of spontaneous changes in genes that we now call **mutations** was not known at the time Darwin published *The Origin of Species.*

Taxonomic Classification and Importance of Species

Organisms can be classified in a variety of ways, including habitat (planktonic, nektonic, benthonic) and environmental distribution (littoral, neritic, bathyl, etc.); however, **taxonomic classification** based on morphological and developmental similarities and presumed genetic relationships is most pertinent to recognition of evolution and biostratigraphic zonation. The basic system of taxonomic classification now in use was introduced in 1735 by the Swedish naturalist Linnaeus, who grouped organisms into a hierarchy of different categories based on the number of distinctive characteristics shared in common. Organisms in the lowest, or least inclusive, category have the greatest number of common characteristics; those in the next highest category have fewer common characteristics; and so on, until the highest, or most inclusive, category is reached. In the last category, organisms share only a very few common characteristics or traits. Linnaeus's system of classification, as modified by some later additions, is illustrated in Table 17.2.

The Linnaean system of taxonomic classification brought to light the fact that degrees of similarities among organisms differ at different levels of classification. Dif-

TABLE 17.2 Taxonomic systems for classifying organisms

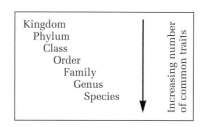

ferences among groups of organisms are greatest at the kingdom level and least at the species level. Species have thus become the fundamental entity of biostratigraphy. Biologists define species as a breeding community that preserves its genetic identity by its ability to exchange genes with other breeding communities. In other words, all members of a given species have the ability to interbreed, but they do not normally breed with members of a different species. Thus, a species constitutes a group of inter-breeding organisms that are reproductively isolated from other such groups. The criteria for identifying a biologic species are difficult to apply to fossil organisms. Therefore, fossil species are commonly characterized mainly on the basis of shell, or skeletal, morphology. Because the skeletal morphology of different members of the same species can be quite variable, determination of fossil species must be made by taxonomic specialists. Such determination may require quantitative measurements of shell parameters and computer analysis of measurement data to provide statistical rigor to fossil species identification.

Changes in Species through Time

The importance of species in biostratigraphic study lies in the fact that species do not remain immutable for all time. If environmental conditions remained absolutely constant through time, perhaps species would never change. The fact is that environments do change, and as they change, species also change, although environments do not directly cause species to change. Both gene mutation, or gene pool combinations, and shifting environmental conditions are essential to the evolution of species. Most species are well adjusted to their normal environments, but if an appropriate variation appears in a species just at the time when it is becoming inadaptive to a changing environment, the force of natural selection may preserve this novel variant (Shaw, 1964). Thus, species have evolved through time as a result of natural selection of those random, chance mutations that brought the species into better adjustment with changing environmental conditions.

All indications from the geologic record suggest that species variations are one-directional and nonreversible. Once a species has become extinct, it does not reappear in the fossil record. As members of a new species increase in numbers, they may eventually become abundant and widespread enough to show up in the geologic record as the **first appearance** of the species. When the species is no longer able to adjust to shifting environmental conditions, its members decrease in number and eventually disappear—the extinction, or **last appearance,** of the species. **Extinction** refers to the disappearance by death of every individual member of a species or higher taxonomic group so that the lineage no longer exists. Paleontologists recognize also that a species may experience pseudoextinction. **Pseudoextinction,** or phyletic extinction, refers to an evolutionary process whereby a species evolves into a different species. Thus, the original species becomes extinct, but the lineage continues in the daughter species. Some species exist for only a fraction of a geologic period. Others may persist for longer periods of time. Organisms that were abundant and geographically widespread and had relatively short ranges have the greatest time-stratigraphic utility, that is, the greatest usefulness for biostratigraphic study (Fig. 17.7).

Models and Rates of Evolution. There is currently considerable controversy among paleontologists concerning the "tempo" of change in organic evolution. Two principal points of view prevail. One view states that evolution proceeds mainly as a gradual change by slow, steady transformation of well-established lineages—**phyletic evolution,** or **gradualism.** The gradualist concept has been the traditional view of species

FIGURE 17.7 The most important macrofossil groups of marine invertebrate organisms for biostratigraphic zonation. The white columns show the time span of distribution, the black columns the time span in which the organisms are important as index fossils. (From Thenius, E., 1973, Fossils and the life of the past, Fig. 50, p. 79, reprinted by permission of Springer-Verlag, Heidelberg.)

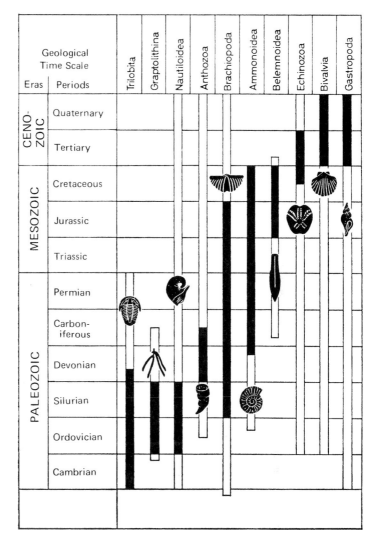

evolution. The second view holds that most species arise very rapidly from small populations of organisms that have become isolated from the parental range and then subsequently change very little after their successful origin. This latter view represents evolution by speciation or branching of lineages, the so-called **punctuated equilibria** model of Eldredge and Gould (1972). Differences in these two postulated modes of evolution are illustrated graphically in Figure 17.8. In the punctuational model, **speciation,** or branching of species, is viewed as a very rapid process, requiring only tens of thousands of years or possibly as little as a few hundred years (Stanley, 1979) after a population becomes reproductively isolated from the parent population. Although the duration of species from first appearance to extinction may be measured in millions of years (Table 17.3), species are believed by the punctuationalists to change morphologically very little and only very slowly after initial speciation. This concept is stated very succinctly by Eldredge and Gould (1977), who emphasize the importance of speciation (splitting) and claim "that most morphological differences between two species

FIGURE 17.8 Diagrammatic sketch of hypothetical phylogenies (lines of direct descent in a group of organisms) representing the punctuational model (A) and the gradualistic model (B). Note that some phyletic evolution is indicated in model A and that some speciation events in model B display accelerated evolution. (From Macroevolution, patterns and process, by S. M. Stanley. W. H. Freeman and Company. Fig. 2.4, p. 17, © 1979.)

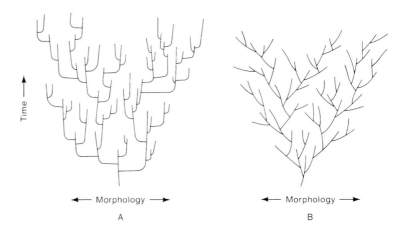

TABLE 17.3 Estimated mean species duration (in millions of years) for a variety of biological groups

Biological group	Estimated mean species duration (Ma)
Marine diatoms	25
Benthic foraminifers	20–30
Planktonic foraminifers	>20
Bryophytes	>20
Marine bivalves	11–14
Marine gastropods	10–14
Higher plants	8–>20
Ammonites	~5 (but with a mode in the 1–2 Ma range)
Freshwater fish	3
Graptolites	2–3
Beetles	>2
Snakes	>2
Mammals	~1–2
Trilobites	>1

Source: Stanley (1985).

appear in conjunction with the speciation process itself, whereas most of a species' history involves little further change, at least of a progressive nature."

Whether species evolution takes place by gradual evolutionary change or by punctuated speciation, or both, is still an unresolved issue. Both sides of the controversy continue to be aired in the paleontological literature (e.g., Levinton and Simon, 1980; Gingerich, 1985; Stanley, 1985; Gould and Eldredge, 1986; Levinton, 1988). The answer to this question has pragmatic as well as philosophical interest. From a practical point of view, the task of delineating the boundaries of species is more difficult if evolution occurs by phyletic gradualism because a chain of intermediate species is present in the geologic record and the boundaries between successive species are arbitrary. Thus, it is difficult to pick points in the evolutionary sequence at which distinct species boundaries are recognizable. If, on the other hand, evolution occurs by speciation (punctuated equilibrium), most morphological change presumably occurs at branch intersections in the evolutionary line (Fig. 17.8A), which represent discrete points in time. Thus, the task of picking species boundaries should theoretically be easier and less error prone if evolution occurs by speciation rather than by phyletic gradualism. On the other hand, the initial appearance of a new species in different provinces may show a time lag owing to lags in migration (see Sections 17.5 and 17.6), which makes identification of the first-appearance species boundary more difficult.

The practicality of identifying species boundaries, and of establishing the boundaries of biostratigraphic zones, is further complicated by problems involving the following:

1. sampling intervals (How small must they be to ensure that species boundaries are detected?)
2. changes in the fossil record induced by burial and the vagaries of preservation
3. constancy and rates of sedimentation (Smaller sampling intervals are required for sediments that accumulated very slowly vs. those that accumulated very rapidly.)
4. intermittent or punctuated patterns of sedimentation and erosion that yield an incomplete stratigraphic record, thus giving the appearance of punctuated speciation

A related question about evolution concerns the factors that control the rates at which evolution and extinction occur. Different groups of species are known to evolve at greatly different rates. Stanley (1985) indicates, for example, that marine bivalve groups evolve at a rate that yields only three or four species in 20 Ma. By contrast, mammalian families evolve at a rate that yields roughly 80 species in 20 Ma. What controls the rate of evolution? Is it chiefly environmental factors external to organisms that control organic change, or does change arise from some independent and internal dynamic control, some key biologic innovation, within organisms themselves (Schopf, 1977)? Most workers seem to agree that environment is the crucial factor controlling diversity of organisms, but the exact way in which environment regulates evolution is less clear, although competition appears to be the key element of the model. Kauffman (1977) maintains that the major controls on rates of evolution are externally manifest and include "population size, degree of isolation, rate of isolation, diversity of niches, variations in selective pressures, size and mobility of organisms and ecological controls on their distribution, trophic relationships, and breadth of ecological tolerances." Stanley (1989) suggests that rates of speciation and extinction also increase with increasing level of stereotypical behavior (increasing complexity of behavior). Increasing complexity of behavior constitutes an important isolating mechanism that favors speciation; however, it results also in a narrow dimension of niche breadth, which renders a species vulnerable to environmental change and at a greater risk of extinction than a less specialized species. Evolution and extinction rates decrease with dispersal ability, a force that opposes isolation (and speciation) but promotes survival.

Deterministic vs. Probabilistic Evolution. An interesting side issue to the problem of evolutionary controls relates to the question of whether or not such evolutionary events as adaptive radiation and periods of mass extinction are deterministic or probabilistic. That is to say, are evolutionary events explainable only in terms of causal factors, or are there statistical laws or generalizations that can explain these events on the basis of random variations or processes? Probabilistic evolutionary models are called **stochastic models.** Van Valen (1973, p. 1) asserted, for example, that "all groups for which data exist go extinct at a rate that is constant for a given group." Such statements should not be taken to mean that extinctions occur without cause. Extinction of a species may be the result of any number of specific causes—predation, aging, starvation, or genetic deformity, for example—and it may therefore be invalid to attribute the death of individuals to chance. If frequency of death is considered at population levels, however, it may be mathematically valid to describe the frequency as being governed by random stochastic processes. In other words, individuals in a given population of organisms may die owing to a number of different specific causes; however, the population as a whole will become extinct at a constant rate, depending upon its size, regardless of the specific causes of death of the individuals. Therefore, the observed frequency of death can be used to compute the probability of extinction if the population size is known (Raup, 1977). Thus, in the stochastic approach, the pattern of evolution as a whole is perceived to be a random process, although individual fluctuations in this pattern can be explained by cause and effect.

Raup (1977) states that "stochastic models serve to separate those features of the evolutionary record which are amenable to deterministic explanations from those where the search for a specific cause is not warranted." For example, the fairly rapid demise of the dinosaurs at the end of Cretaceous time can probably be explained by some specific environmental or catastrophic event such as dramatic climatic change resulting from meteorite impact; however, the gradual decline of conodonts and their eventual extinction at the end of the Triassic appears to be more difficult to attribute to a specific cause or causes. Raup (1977) points out, however, that, owing particularly to the questionable validity of some of the underlying assumptions used in the models, there are many pitfalls in application of stochastic models to evolutionary studies. Furthermore, not all geologists and paleontologists agree with the stochastic approach to evolution. Many take the view, for example, that the apparent correlation of episodes of massive extinction with key environmental events argues that these extinction waves are deterministic (Valentine, 1977a). Stanley (1979) also favors deterministic models, stating that "rapid radiation and decline of large clades (clusters of lineages) normally relates to nonrandom causal factors: adaptive innovations and the agents of species selection."

Mass Extinctions. The geologic record shows that many groups of organisms became extinct or suffered dramatic reductions in numbers and diversity at particular times in the geologic record. Examples of such intervals are the Late Cambrian, Late Ordovician, Late Devonian, Late Permian, Late Triassic, and Late Cretaceous (Table 17.4). As Table 17.4 indicates, extinctions affected both terrestrial and marine forms. Many of these mass extinction episodes are so dramatic that it is indeed difficult to accept the stochastic model as an explanation for the demise of these groups of organisms, and scientists feel compelled to seek specific causal factors to explain these extinction waves. The extinction of the dinosaurs, mentioned above, and ammonites at the end of the Cretaceous and the extinction of the fusulinid foraminifers at the end of the Permian are but two examples of abrupt disappearance of major groups of organisms. These dramatic extinctions have taxed the imagination of paleontologists and other geologists to provide an acceptable causal explanation. The Late Permian extinction phase has received particular attention because of the number of major groups affected and the sharpness of the change with which these groups disappeared from the geologic record between the Late Permian and the Triassic.

Mass extinctions are of enormous interest to geologists because of the questions they raise, among other things, about possible recurring catastrophic events in Earth's history. The 1980s and early 1990s saw explosive growth in research into the patterns, rates, causes, and consequences of extinction (Flessa, 1990) but little overall agreement about the causes of extinction. Theories about extinction fall into three groups:

1. catastrophic extinction
2. gradual extinction
3. stepwise extinction—extinction that occurs in a series of discrete steps in the vicinity of major stratigraphic boundaries, such as the Permian/Triassic boundary

TABLE 17.4 Major extinctions of organisms during the Phanerozoic

Extinction episode	Major animal groups strongly affected	Percentage of families extinct
Late Cretaceous	Ammonites*	
	Belemnites	
	Rudistid bivalves*	
	Corals	
	Echinoids	26
	Bryozoans	
	Sponges	
	Planktonic foraminifers	
	Dinosaurs*	
	Marine reptiles*	
Late Triassic	Ammonites	
	Brachiopods	
	Conodonts*	35
	Reptiles	
	Fish	
Late Permian	Ammonites	
	Rugose corals*	
	Trilobites*	
	Blastoids*	
	Inadunate, flexibiliate, and camerate crinoids*	50
	Productid brachiopods*	
	Fusulinid foraminifers*	
	Bryozoans	
	Reptiles	
Late Devonian	Corals	
	Stromatoporoids	
	Trilobites	
	Ammonoids	30
	Bryozoans	
	Brachiopods	
	Fish	
Late Ordovician	Trilobites	
	Brachiopods	24
	Crinoids	
	Echinoids	
Late Cambrian	Trilobites	
	Sponges	52
	Gastropods	

Source: Facies interpretation and the stratigraphic record, by A. Hallam. W. H. Freeman and Company. Table 10.1, p. 216, 217, © 1981.

*Last appearance of group.

These various theories have been expounded in numerous research papers as well as in several recent books (e.g., Clube, 1988; Donovan, 1989; Kauffman and Walliser, 1990; Larwood, 1988; Sharpton and Ward, 1990).

Proponents of the catastrophic theory, especially for the sharp Cretaceous/Tertiary boundary event, suggest that the impact of extraterrestrial objects (bolides) created major climatic change (global winter) by throwing up huge clouds of dust and/or generated acid rain, tsunamis, and wildfires that caused extinction of some taxonomic groups. Alternatively, intense explosive volcanic activity may have adversely affected climates through discharge of excessive gas clouds (greenhouse warming). Other geologists suggest that such extraterrestrial causes are not needed to explain most extinction events. Gradual, progressive changes in climate together with changes

TABLE 17.5 Possible cases of major extinction events in the late Precambrian and Phanerozoic

Extinction event	Suggested cause
Late Pleistocene	Postglacial warming plus predation by humans
Eocene to Oligocene	Stepwise extinction associated with severe cooling, glaciation, and changes of oceanographic circulation, driven by the development of the circum-Antarctic current
End of Cretaceous	Impact of extraterrestrial body (bolide), producing catastrophic environmental disturbance
Late Triassic	Possibly related to increased rainfall with implied regression
End of Permian	Gradual reduction in diversity produced by sustained period of cold climates, associated with widespread regression and reduction in area of warm, shallow seas
Late Devonian	Global cooling associated with (causing?) widespread anoxia of epeiric seas
Late Ordovician	Controlled by the growth and decay of the Gondwanan ice sheet followed by a sustained period of environmental stability associated with high sea level
Late Cambrian	Habitat reduction, probably in response to a rise in sea level, producing a reduction in number of component communities
Late Precambrian	Complex, including widespread regression, physical stress (restricted circulation and oxygen deficiency), and biological stress (increased predation, scavenging, and bioturbation)

Source: Donovan (1989, Table 0.1).

in sea level (e.g., lowering of sea level reduces habitats for shallow-water organisms) are adequate, they say, to account for extinctions. Still other geologists suggest that some extinctions occur in a stepwise fashion by a series of pulses—some before, some at, and some just after a major boundary. These extinctions are presumably the result of a succession of events, such as brief showers of comets superimposed on a background of progressive environmental deterioration.

More extended discussion of possible causes of extinction related to catastrophic events is given by Hallam (1989b), and principal extinction events and suggested probable causes are summarized in Table 17.5, after Donovan (1989). Readers should be aware, however, that not all researchers may agree with the suggested causes described in Table 17.5. In any event, worldwide extinctions of major groups of organisms, while very interesting, play only a limited role in biostratigraphy because they provide only a few correlation horizons. Changing local environmental conditions are probably a more significant cause of extinction of individual species, which form the most important basis for biostratigraphy.

17.5 DISTRIBUTION OF ORGANISMS IN SPACE: PALEOBIOGEOGRAPHY

When d'Orbigny introduced the concept of stage, he believed that the fossil assemblages upon which his stages were based had worldwide distribution. We now know that the fossil species and assemblages that characterize biostratigraphic units are not necessarily found everywhere that rocks of the appropriate ages occur. Few species are distributed throughout the entire world. Most, in fact, are restricted in their geographic range, although some fossil groups ranged widely throughout whole ecological realms at times in the geologic past. The region within which a particular group or groups of plants or animals is distributed is called a **biogeographic province.** Biogeographic provinces are separated by physical or climatic barriers. Land areas are barriers to marine organisms; open marine water is a barrier to land animals and plants; deep water is a barrier to shallow-water, shelf-dwelling organisms; cold water is a barrier to warm-water organisms; freshwater is a barrier to organisms adapted to saline marine condition; and so forth. A particular type of barrier may be impenetrable by one species of organism but not by another. For example, benthonic organisms that do not have a long-lived, juvenile planktonic larval stage find deep water a barrier to dispersal. By contrast, planktonic organisms, which live in near-surface waters in the ocean, are distributed widely throughout the oceans in both shallow and deep water.

Dispersal of Organisms

Paleontologists regard species as the fundamental biologic units in nature. They are the basic unit that undergoes evolution; the species niche is the basic functional unit in ecological interactions; and species are the fundamental units of biogeography, biostratigraphic zonation, and correlation. Because of their central importance in biostratigraphy, it is essential that we understand the factors that control the dispersal and distribution of species. Obviously, factors affecting the dispersal of land organisms and plants are different from those that control the dispersal of marine organisms. Also, the distribution of invertebrate marine organisms is controlled by different factors from those that control the distribution of vertebrate marine groups. Because of the overriding importance of marine invertebrates in biostratigraphic studies, we shall confine our discussion of dispersal here to invertebrate organisms in the marine setting, which consists of the pelagic realm (the water column) and the benthic realm (the seafloor, or bottom environment) (Fig. 17.9).

Marine invertebrate organisms can be divided into three fundamental types on the basis of habitat: plankton, nekton, and benthos (Table 17.6). **Plankton** are mainly microscopic-size organisms that live suspended at shallow depths within the water column (pelagic realm) and have very weak or limited ability to direct their own movements. They are distributed more or less passively by currents and wave action and may be dispersed widely into all types of open-ocean environments. Planktonic organisms are exceptionally useful fossils for biostratigraphic zonation and correlation because of their widespread distribution. They reflect the habitat of the pelagic realm and not the bottom environment into which they fall upon death; therefore, their presence in ancient marginal-marine sedimentary rocks is of limited value in environmental interpretation, although they are useful in some paleooceanographic applications (e.g., interpretation of water paleotemperature from oxygen isotopes). A few plankton such as graptolites do have some value as indicators of bottom environments. Graptolites were too fragile to survive in high-energy, shallow-water environments and are hence preserved mainly in the facies of quiet-water environments. Thus, they constitute "facies fossils."

Nekton also inhabit the pelagic realm and include all animals that are able to swim freely. Modern nekton are distributed in the ocean at depths ranging from the surface to thousands of meters and encompass many advanced groups of animals such as fish, whales, and mammals. Nekton are less abundant in the fossil record than

FIGURE 17.9 Subdivision of the marine environment into the pelagic (water column) and benthic (bottom) realms. The pelagic realm is inhabited by planktonic and nektonic organisms; benthonic organisms occupy bottom environments of the benthic realm.

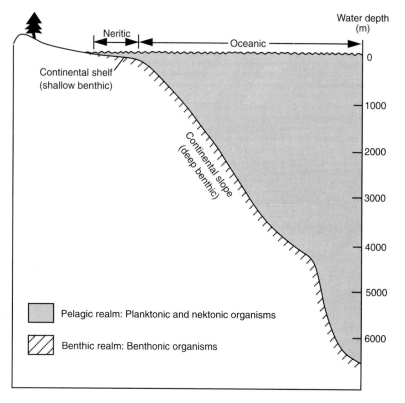

TABLE 17.6 Classification of organisms by habitat or life style

Classification	Description	Example
Planktonic	Organisms that live suspended in the upper water column and which have only a very weak or limited ability to direct their own movements	
Phytoplankton	Have the ability to carry on photosynthesis; primary food producers, or autotrophs	Diatoms, dinoflagellates coccolithophoridae
Zooplankton	Do not carry on photosynthesis and thus cannot produce their own food (heterotrophs); feed on phytoplankton	Foraminifers, radiolarians, graptolites
Meroplankton	Spend only their juvenile stage as plankton; later become free-swimming or bottom-dwelling organisms	Larvae of most benthonic organisms such as molluscs
Pseudoplankton	Organisms distributed by waves and currents as a result of attachment to floating seaweed, driftwood, etc.	Mussels, barnacles, etc.
Benthonic	Bottom-dwelling organisms that live either on or below the ocean floor	
Sessile benthos	Benthos that attach themselves to the substrate (epifauna)	Crinoids, oysters, brachiopods
Vagrant benthos	Benthos that either creep or swim over the bottom (epifauna) or burrow into the bottom (infauna)	Starfish, echinoids, crabs Clams, worms
Nektonic	Organisms able to swim freely and thus move about largely independently of waves and currents	Mobile cephalopods, fish, sharks

planktonic and benthonic organisms (below) and thus overall appear to have somewhat less value in biostratigraphic studies. Nektonic fossils include fish remains in some deep-sea clays, belemnites, and other mobile cephalopods, and probably conodonts. Conodonts are an interesting type of fossil, with considerable biostratigraphic significance, that occur in rocks ranging in age from Cambrian to Triassic. They are tiny, toothlike phosphate fossils whose origin remained an enigma until 1982. A complete specimen of the conodont-bearing animal has been found in Lower Carboniferous rocks of the Edinburgh district, Scotland (Briggs, Clarkson, and Aldridge, 1983). The specimen is an elongate, soft-bodied animal 40 mm long and 1.8 mm wide. The conodont apparatus occurs in the head or anterior region of the body and may have served as teeth or possibly some type of internal support.

Benthos are bottom-dwelling organisms that live either on or below the ocean floor (benthic realm). Benthos (benthonic organisms) with preservable hard parts are particularly important for environmental interpretation because their remains are commonly preserved in the same environment in which they lived. Because most benthos live in shallow water and have limited ability to move long distances along the bottom, they tend to be more provincial, and of somewhat less biostratigraphic significance, than plankton. Nonetheless, benthos can be dispersed outside their local environment because many benthonic species have a planktonic, juvenile larval stage during which they can be dispersed by currents. Some workers originally questioned the importance of larval transport as a mechanism for dispersal of shallow-water benthos over long distances, such as across ocean basins. It was initially believed that the duration of the larval stage was so short that the larvae would change to the adult phase while the organisms were still over deep water, causing them to perish when they settled to the bottom. Subsequent work (e.g., Scheltema, 1977) has demonstrated, however, that there are many shoal-water or continental shelf benthic invertebrates whose larvae have a pelagic stage lasting from six months to over one year. Scheltema refers to such larvae as **teleplanic,** or "far wandering." Such long-lived larval species could cross the modern Atlantic Ocean, for example, by way of the main surface currents. Many workers interpret this ability to become dispersed by pelagic mechanisms as being favorable for interpopulation migration and, therefore, gene flow. Valentine (1977b, p. 145) summarizes the importance of these long-lived larval forms as follows:

> Species with the more long-lived and hardy pelagic larvae have the greater chance to be widely dispersed after reproduction. . . . Therefore, species with such attributes would commonly be able to colonize habitats that lie at some distance from their parental ranges, and would usually be able to maintain gene flow to such outlying populations. Species with shorter planktonic development periods, smaller broods, or more restricted larval requirements would tend to colonize only localities that are fairly close to their parental regions. If a population became established at any considerable distance from others, gene exchange might be sporadic or lacking altogether, leading to divergence between the colonists and the parental population, and a reduction in their usefulness in correlation.
>
> For a species in a given locality, then, a geographic range exists for which colonization is essentially obligatory, as the region lies within the normal migratory range of the population; thus, by some standard time, occupation is virtually assured. This can be called the **local range.**

Barriers to Dispersal

Each species thus has a potential geographic range that is determined by its habitat requirements. Few species actually occur throughout their potential range. Their distribution is restricted owing either to the presence of barriers of some type that prevent their expansion into all areas of suitable habitat or because the species may not have had time to spread to all suitable areas, especially if barriers are present. At any given time, there are many regions in the world that could be colonized by species if they could reach them in appropriate numbers, but they are barred from reaching them by intervening inhospitable areas. Many species eventually find ways to broach narrow barriers and perhaps in time even to cross wider barriers. Once barriers are crossed, or barriers disappear, the migrant species may find itself in competition for environmental niches with similar species or similarly adapted species in the new province. In the face of this competition, either the indigenous species or the migrant species may become extinct. Alternatively, the less well-adapted species could evolve and become adapted to a different environmental niche. Once a barrier is surmounted, the colonizers typically expand their range at the new location until it is circumscribed by other barriers, filling out their new local range. The intruding species may subsequently broach still other barriers, hopping from one habitable region to another across barriers of varying difficulty of penetration, episodically expanding their total species range (Valentine, 1977a).

The broaching of barriers thus leads to expansion of the total range of a species, although in some cases it may lead to extinction of the species in the new region or to its evolution to a more adaptable species. On the other hand, if the opposite situation prevails and a barrier "suddenly" appears and divides a once-continuous area of suitable habitat, the result is the segregation of the species into different populations separated by the barrier. The separated populations would gradually evolve into different species, each with a more restricted geographic range than that of the parent species (Dodd and Stanton, 1981).

Numerous ecological factors can act as barriers to dispersal of organisms, all of which can be grouped under two major categories: (1) habitat failure—as when shelf habitats give way to deep-sea conditions or to land—and (2) temperature. Hallam (1981) suggests that the major controls on faunal provinciality are climate and plate movements. Dodd and Stanton (1981) indicate that the principal factors controlling geographic distribution of species are depth-elevation—that is, water depth and land elevation—and temperatures. These possible controls on provinciality are discussed in further detail in the following paragraphs.

Temperature. Temperature is a major barrier to migration of species, and it commonly affects larvae more than adult organisms. Because the distribution of worldwide temperatures is latitudinally controlled, temperature barriers are most important latitudinally, although seasonal and even diurnal temperature changes are also important. The boundaries of all modern biotic provinces are in part temperature controlled, and ancient biotic provinces were undoubtedly similarly controlled. Warm-water taxa are restricted primarily to the equatorial zone of the ocean because no other large parts of the ocean, either at the surface or at depth, are warm enough to sustain these tropical species. Cold-water taxa, on the other hand, can extend their range closer to the equatorial region by migrating down the bathymetric gradient into deeper and colder water, the phenomenon of **submergence,** if they are capable of adapting to greater depths. Also, if a polar species can manage to find a way of breaking through the temperature barrier and crossing the equatorial region, it can find suitable cold-water habitats at or near the surface in the higher latitudes of the other hemisphere.

Some species of organisms are adapted to a wide range of temperatures and may thus be distributed through a much wider range of temperature zones than less tolerant species. Nonetheless, even tolerant species are sensitive to temperature variations and do not occur throughout all temperature zones. It must be recognized also that marine temperature zones have changed throughout geologic time as world climatic zones have shifted in response to plate movements and episodes of glaciation. A given geographic region of the world may thus record a succession of colder- or warmer-water faunas through time in response to these shifting climatic conditions.

Geographic Barriers. The terms habitat failure, plate movements, and depth-elevation used above are all different ways of expressing the concept of geographic barriers. These geographic barriers arise out of the distribution pattern of landmasses and oceans and variations in water depths of the oceans. All marine organisms have limited water depths at which they can survive. Thus, water that is either too deep or too shallow can constitute a barrier to a particular species of organism. Landmasses constitute barriers to the dispersal of marine organisms, and the open ocean is a barrier to migration of land animals and plants from one continent to another. The most important factors influencing geographic barriers appear to be changes in sea level and changes in the nature and geographic distribution of landmasses and the ocean floor brought about by plate movements (discussed in a following section).

Sea-Level Changes. Causes of major cycles of sea-level change are discussed in Chapter 14. Fluctuations in sea level cause significant interruptions in biogeographic provinces owing to variations in water depths on the continental shelves. During a major drop in sea level, water is withdrawn from the continental shelves, exposing much of the inner shelf. The habitable area of shallow water is greatly reduced, leading to crowding and increased competition among shallow-water species that cannot move seaward into deeper water, and to probable extinction of less adaptable groups. During major rises in sea level, water depths on the outer continental shelf are increased, and the total area of shallow water along continental margins is also vastly increased owing to spread of the seas over the edges of the continents. The available environmental niches for shallow-water organisms are correspondingly increased, resulting in less competition among species for available space and food. These conditions lead to expansion of the local ranges of species as they move into favorable habitats, and also probably to rapid emergence of new species (speciation) as a result of adaptive radiation of groups that survived the preceding episode of lowered sea level.

Hallam (1981) states that an analysis of genera across the world indicates a clear inverse relationship between endemism and the area of continents covered by sea. **Endemism** is the tendency of species or other taxa to have a very limited geographic range, as contrasted with **pandemism,** which is the tendency of species to have worldwide distribution. He suggests that at times of low sea level and restriction of seas, faunal migrations between continental shelf areas would be rendered more difficult, resulting in less gene flow. Thus, there would be more local speciation among the dispersible organisms that occupied shallower-water habitats.

Plate Movements. Tectonism is the major factor controlling the distribution of landmasses and ocean basins. Major changes in the environmental framework of the marine realm occur as the geographic positions, configurations, and sizes of continents and ocean basins are changed by global plate tectonic processes. Plate movements can greatly affect topographic barriers by producing changes in oceanic widths and depths. As previously discussed, changes in rates of seafloor spreading may have a major effect on sea level. Plate movements can also alter latitudinal temperature gradients by shifting the geographic position of continents, and they can even affect the distribution patterns of major ocean currents. The creation or destruction of migration barriers may thus be tied closely to plate tectonics events.

Provinciality of species is greatest during times when plate motion has produced a maximum number of separate continents, such as at the present time (Dodd and

FIGURE 17.10 Schematic illustration of the relationships between crustal plates and continents as they affect the distribution of organisms. Table 17.7 shows the biogeographical implications for the continental shelf for each case. (After Valentine, J. W., 1971, Plate tectonics and shallow marine diversity and endemism, an actualistic model: Systematic Zoology, v. 20. Fig. 4, p. 261, reprinted by permission of Society of Systematic Zoology, Norman, Okla.)

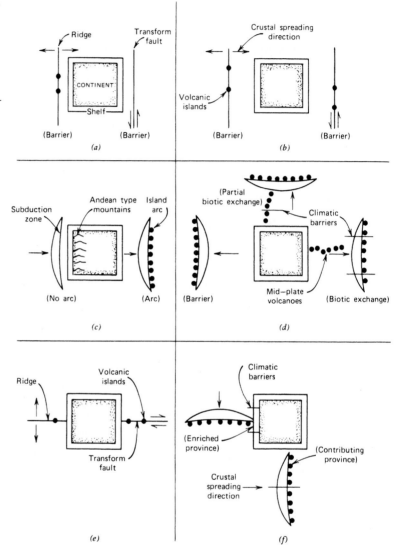

Stanton, 1981). Barriers are fewer and provinciality is less when plate motions have welded continents together, as at the end of the Paleozoic. Several generalities (Valentine, 1971) between plate tectonics and biogeography can be stated (Fig. 17.10 and Table 17.7):

1. When spreading ridges (e.g., the Mid-Atlantic Ridge) lie parallel to continents, they produce deep and ever-widening ocean basins and thus are barriers to shallow-marine species and terrestrial organisms.
2. Transform (strike-slip) faults that run parallel to continental margins are also usually associated with deep-water barriers.
3. Subduction zones (e.g., the Peru-Chili Trench) that are oriented parallel to and dipping toward the continent create deep-water barriers.

TABLE 17.7   The relationship of crustal plates to continental margins (Figure 17.10) and the effect of this relationship on biogeography

Geometry of relation	Character of margin	Distance	Biogeographic implications for continental shelf	Fig. 17.10
Parallel	Ridge or transform	Near	Barrier but with depauperate provincial outliers on isolated islands	a
		Far	Barrier	b
	Subduction zone	Near	Barrier if truly marginal with no island arc; source of rich biota and dispersal route in arc present	c
		Far	No effect unless intervening region bridged by midplate volcanoes, then source of rich biota and dispersal route if no climatic barriers intervene	d
High angle	Ridge or transform	Near	Little effect, with depauperate provincial outliers on isolated islands	e
		Far	Not a case	
	Subduction zone	Near	N–S shelf, E–W arc: arc system a source of rich biota for local province. E–W shelf, N–S arc: proximal province of arc system a source of rich biota for entire shelf	f
		Far	Not a case	

Source: After Valentine, J. W., 1971. Plate tectonics and shallow marine diversity and endemism, an actualistic model: Systematic Zoology, v. 20. Table 1, p. 262, reprinted by permission of Society of Systematic Zoology, Norman, Okla.

4. Subduction zones dipping away from continents (e.g., the subduction zone running from Burma to New Hebrides) may have island arcs that aid in breaking down barriers.
5. Mid-plate volcanoes (e.g., the Hawaiian Islands) help break down deep-water barriers.
6. Subduction zones, spreading ridges, and associated island arcs at high angles to continents (e.g., the Aleutian Islands) may provide migration pathways, thus breaking down barriers.

Three patterns of changing faunal distributions with time that are related to plate movements have been identified. **Convergence** refers to the phenomenon whereby the degree of resemblance of faunas in different regions increases from an earlier to a later time. **Divergence** refers to the reverse process, that is, decrease in degree of resemblance of faunas in different regions from an earlier to a later time. Table 17.8 lists several examples of this phenomenon, including examples of both nonmarine vertebrates and marine invertebrates. Note from this table that plate tectonics movements that cause closure of ocean basins have tended to produce convergence of shallow-water benthonic organisms. Distinctly different faunal provinces on either side of a wide ocean are gradually brought together by closure of the intervening ocean between contiguous continents, eventually producing a single faunal province. Opening of oceans has commonly produced divergence. Hallam (1981) identifies a third biogeographic

TABLE 17.8 Correlation of plate tectonic events and changes in faunal distribution patterns

Plate tectonic event	Convergence	Divergence
Closure of Proto-Atlantic (Ordovician, Silurian)	Trilobites, graptolites, corals, brachiopods, conodonts, anaspids, and thelodonts of the two continents flanking the Proto-Atlantic	
Closure of Urals Seaway	Post-Permian continental vertebrates of Eurasia	
Opening of Atlantic (Cretaceous, Tertiary)		Cretaceous bivalves and benthic foraminifers of Caribbean and Mediterranean; Upper Cretaceous ammonites of United States and W. Europe–N. Africa; post-Lower Eocene mammals of North America and Europe; Tertiary mammals of Africa and South America
Opening of Indian Ocean (Cretaceous)		Bivalves of East African and Indian shelves
Closure of Tethys (Late Cretaceous) (mid-Tertiary)	? Ammonites of Eurasia and Africa–Arabia; mammals of Eurasia and Africa	Molluscs, foraminifers, etc., of Indian Ocean and Mediterranean–Atlantic

Source: Facies interpretation and the stratigraphic record, by A. Hallam. W. H. Freeman and Company. Table 10.2, p. 237, © 1981.

pattern, which is the distributional changes of contiguous marine and terrestrial organisms that occurs when one group exhibits convergence and the other divergence. Creation of a land connection between two previously isolated areas of continents, for example, allows convergence of the terrestrial faunas to take place, while at the same time the land connection creates a barrier to marine organisms, causing divergence as a result of genetic isolation.

Other Barriers. Other, less important barriers than temperature and geographic barriers may also help to define the boundaries of biogeographic provinces. Salinity differences constitute an important boundary between freshwater and marine provinces; however, salinity is a relatively unimportant barrier within the marine realm itself. Marked salinity increases can occur in some small, restricted arms of the ocean where evaporation rates are high. Conversely, lower than normal salinities may ensue in some coastal areas where freshwater runoff is high. These salinity variations can control local communities of organisms but not the distribution of organisms on a provincial level. In the open ocean, salinity tends to be highest in the equatorial region where evaporation rates are at a maximum, and lowest in the middle latitudes where some dilution occurs as a result of freshwater runoff from the continents. Even so, the salinity in these regions varies only a few parts per thousand from the average ocean salinity (35o/oo), a variation not adequate to seriously affect the dispersal of organisms in the open ocean.

Currents aid in the dispersal of planktonic species and the larvae of benthonic species, but they help also, in some parts of the ocean, to maintain the temperature gradients that create barriers to dispersal. Thus, currents may act as either a barrier or an aid to dispersal. The long-term pattern of currents is itself affected by plate movements, as discussed.

17.6 COMBINED EFFECTS OF THE DISTRIBUTION OF ORGANISMS IN TIME AND SPACE

Eicher (1976) points out that both the environmental and the temporal records are important for interpretation of geologic history, where temporal relates to variations with time. If organisms throughout geologic time had been spread over the world and not confined to specific biogeographic provinces and environments, worldwide correlation of strata on the basis of fossils would be greatly facilitated, assuming that evolutionary changes have been simultaneous and worldwide. Under these conditions, however, fossils would provide little or no help in working out ancient depositional environments because more or less the same organisms would have lived in all environments. Conversely, if organisms were distributed in biogeographic provinces as they are today, but organic evolution never occurred, we would be able to interpret local ancient environments with great confidence because ancient sedimentary rocks would contain the same species as modern environments. By the same token, these species would be of no value in correlation and the unraveling of local chronologies because the same species would have existed throughout geologic time.

The real fossil record reflects the fact that both segregation into biogeographic provinces and organic evolution took place. Owing to organic evolution, we are able to correlate strata of a given age from one area to another and to work out the relative chronology of strata in a given area. Because many organisms were confined to biogeographic provinces in the past, however, we cannot always correlate time-equivalent strata from different environments because the organisms that existed in different biogeographic provinces during the same period of time were different. Thus, correlation between biogeographic provinces is difficult, and it is commonly not possible to make worldwide correlations. On the other hand, because different groups of organisms were confined to different provinces and different environments, the provinciality of ancient organisms provides an invaluable tool for interpreting ancient sedimentary environments.

On the debit side again, the provinciality of organisms creates special problems from the standpoint of determining the total vertical stratigraphic range of a species. A species may exist in one province for long periods of time before broaching a particular barrier and spreading into a nearby province. After migration into the new province, the species may die out in the old province while continuing to thrive for some time in the new region. Therefore, the **local vertical range** of a species in a given province, sometimes called the teil zone, may be much shorter than the **total range** of the species on a global scale. Paleontologists must be extremely careful about recognizing this possibility when using fossils for time correlations. This problem is demonstrated in Figure 17.11, which illustrates some of the factors that can affect the range of a species. This diagram shows that the range of a species is affected both by evolutionary changes and by the presence of barriers that can regulate the times of migration into and first appearance in nearby provinces.

17.7 BIOCORRELATION

Biostratigraphic units are observable, objective stratigraphic units identified on the basis of their fossil content. As such, they can be traced and matched from one locality to another just as lithostratigraphic units are traced. Biostratigraphic units may or may not have time significance. For example, assemblage zones and abundance zones may

FIGURE 17.11 Diagram illustrating the difference in local range and total range of a hypothetical species (F). Species F first appears in Province A and is restricted to Province A by a barrier. Later removal of the barrier allows migration to Province B, where the species persists for a time after it has died out in Province A. FAD = first appearance datum; LAD = last appearance datum.

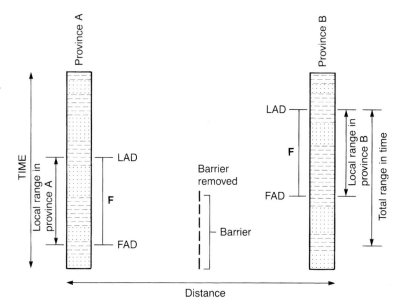

cross time lines (be diachronous) when traced laterally. On the other hand, interval zones, particularly those defined by first appearances of taxa, yield correlation lines that coincide in general with time lines. Biostratigraphic units may be correlated, irrespective of their time significance, using much the same principles employed in correlation of lithostratigraphic units—matching by identity and position in the stratigraphic sequence, for example. In this section, we will first examine correlation by assemblage zones and abundance zones, which can be correlated as biostratigraphic units even though they may not have time-stratigraphic significance. We will then discuss biocorrelation methods based on interval zones and other zones that yield time-stratigraphic correlations.

Correlation by Assemblage Zones

Assemblage zones, as opposed to Oppel assemblage zones, are based on distinctive groupings of three or more taxa without regard to their range limits (Fig. 17.5). They are defined by different successions of faunas or floras, and they succeed each other in a stratigraphic section without gaps or overlaps. Assemblage zones have particular significance as an indicator of environment, which may vary greatly regionally. Therefore, they tend to be of greatest value in local correlations. Nonetheless, some assemblage zones based on marine planktonic assemblages may be used for correlation over much wider areas. The principle of correlation by assemblage zones is illustrated graphically in the very simple example shown in Figure 17.12.

Shaw (1964) points out that the boundary between assemblage zones is inherently fuzzy because above and below the limits of this zone will be transition zones in which part of the characteristic fossil assemblage will be missing because it has not yet appeared or has already vanished. Therefore, there are practical limits to the accuracy that can be achieved by assemblage zone correlations. Part of the problem in correlation by assemblage zones stems from the fact that the number of fossil taxa that a biostratigrapher must work with is so large that it is difficult to visually assimilate the

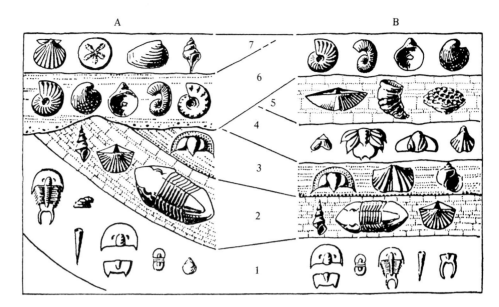

FIGURE 17.12 Generalized diagram illustrating the principle of correlation by fossil assemblages. (From Moore, R. C., C. G. Lalicker, and A. G. Fischer, Invertebrate fossils, © 1952. Fig. 1.3, p. 8, reprinted by permission of McGraw-Hill Book Company, New York.)

data and draw meaningful zone boundaries (Fig. 17.13). To overcome this problem, earlier workers tended to reduce the number of taxa whose distributions would be studied, or they tried to make composites of the samples. A more recent solution to this problem is to apply the techniques of multivariate statistical analysis to recognition and delineation of assemblage zones. These quantitative techniques provide a rational statistical basis for delineating zones based on large numbers of taxa without taking the decision making out of the hands of the biostratigrapher. Details of these multivariate techniques are given in Hazel (1977), Brower (1981), and Gradstein et al. (1985).

Correlation by Abundance Zones

As mentioned, abundance zones, or acme zones, are defined by the quantitatively distinctive maxima of relative abundance of one or more species, genus, or other taxon rather than by the range of the taxon. They represent a time or times when a particular taxon was at the peak, or acme, of its development with respect to numbers of individuals. Some biostratigraphers previously used abundance zones for time-stratigraphic correlation under the assumption that there is a time in the history of every taxon when it reaches its maximum abundance and that this abundance peak occurs everywhere at the same time. The current prevailing opinion among biostratigraphers is that most abundance zones are unreliable and unsatisfactory for time-stratigraphic correlation. This opinion is based on the apparent fact that not all species achieve a maximum abundance, or that if they do, this peak is not necessarily recorded by layers of abundant specimens. Furthermore, peak abundances that are recorded in the stratigraphic record may be related to favorable local ecological conditions that can occur at different times in different areas and that may persist in one area much longer than in another. Maximum abundance may thus represent local, sporadically favorable envi-

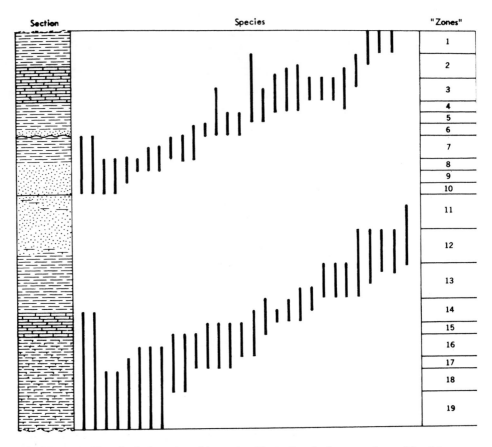

FIGURE 17.13 Hypothetical stratigraphic section illustrating the large numbers of fossil taxa that may be involved in correlation by assemblage zones. Vertical black lines represent the composite ranges of the species found at various local sections. The column at the right shows one interpretation that could be drawn from these fossil data. (From Hazel, J. E., 1977, Use of certain multivariate and other techniques in assemblage zonal biostratigraphy: Examples utilizing Cambrian, Cretaceous, and Tertiary benthic invertebrate, *in* G. G. Kauffman and J. E. Hazel (eds.), Concepts and methods of biostratigraphy. Fig. 1, p. 189. Copyright 1977, 1983 by Van Nostrand Reinhold. All rights reserved.)

ronments, suddenly unfavorable environments that caused mass mortality, or mechanical concentrations of the shells of organisms after death. Some of the problems of correlating by abundance zones are illustrated in Figure 17.14. In short, abundance zones may be used for biostratigraphic correlation, but they do not provide a reliable means of time-stratigraphic correlation. Although they are sometimes used locally for correlation within provinces, biostratigraphers usually prefer correlations based on assemblage zones or interval zones.

Chronocorrelation by Fossils

Chronostratigraphic correlation is the matching up of stratigraphic units on the basis of time equivalence. Establishing the time equivalence of strata is the backbone of global

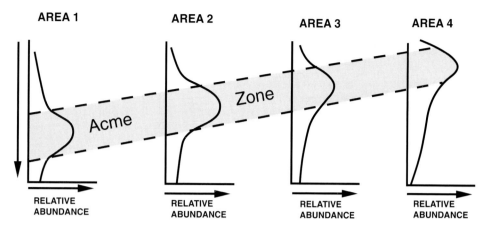

FIGURE 17.14 Schematic diagram illustrating why correlation by abundance (acme) zones may not yield a true time correlation. The same species may achieve its maximum abundance at different times in different localities. Age of strata increases downward; relative abundance increases to the right.

stratigraphy and is considered by most stratigraphers to be the most important type of correlation. Methods for establishing time-stratigraphic correlation fall into two broad general categories: biological and physical/chemical. As mentioned, time-stratigraphic correlation by biological methods is based mainly on use of concurrent range zones and other interval zones. Biological correlation methods also include statistical treatment of range-zone data and correlation by biogeographical acme zones, which are biological events related to climate fluctuations. A variety of physical and chemical methods are available for chronostratigraphic correlation and are discussed in the following chapter. Logically, this discussion of chronostratigraphic correlation by fossils also belongs in the next chapter; however, I am including it here to keep all material relating to correlation by fossils in a single unit. The discussion of biocorrelation that follows represents a very general introduction to this subject. For more rigorous treatment of biocorrelation, see Gradstein et al. (1985) and Guex (1991).

Correlation by Biologic Interval Zones

Interval zones are biozones that constitute the strata that fall between the highest and/or lowest occurrence of taxa. Several kinds of interval zones are recognized, including those formed by overlapping ranges of taxa. Figure 17.3 illustrates several ways that the first and last appearances of taxa may be used to define interval zones:

1. the interval between the first and last appearance of a single taxon (taxon range zone)
2. the interval between the first appearances or last appearances of two different taxa
3. the interval between the first appearance of one taxon and the last appearance of another
4. intervals defined by overlapping range zones (concurrent range zones)

These different interval zones have varying degrees of usefulness in time-stratigraphic correlation, as described below.

FIGURE 17.15   Correlation between two hypothetical sections based on interval zones. Note that several types of interval zones are used here for correlation. For example, Zone 1 is defined by the total vertical range of Species A; Zone 2 is an interval zone defined by the last appearance of Species A and the first appearance of Species B; Zone 4 is formed by the overlapping ranges of Species B and C; and so forth.

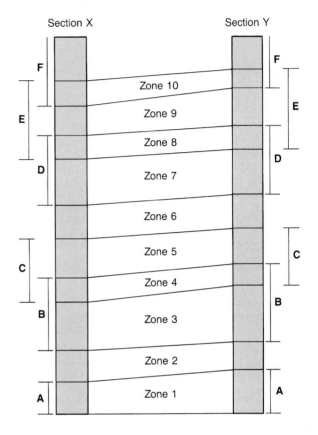

Taxon Range Zones. Taxon range zones may be very useful for time correlation if the taxa upon which they are based have very short stratigraphic ranges. They are of little value if the taxa range through an entire geologic period or several periods. Correlation by taxon range zone is often referred to as correlation by **index fossils.** Index fossils are considered to be those taxa that have very short stratigraphic ranges, were geographically widespread, were abundant enough to show up in the stratigraphic record, and are easily identifiable. Unfortunately, the the term index fossil has also been used in other ways and can have other connotations. Therefore, it is less confusing when speaking of correlation based on the entire range of a taxon to refer to it simply as correlation by (taxon) range zone. Correlation by taxon range zone is illustrated diagrammatically in Zone 1 of Figure 17.15.

Other Interval Zones. When individual taxon range zones are very long and correlation by taxon range zone is thus not suitable, much finer-scale correlation is possible by use of other types of interval zones. Interval zones defined by the first (stratigraphically lowest) appearance of two taxa, for example, are particularly useful in time-stratigraphic correlation because they are based on evolutionary changes, along phyletic lineages, that tend to occur very rapidly. Thus, the interval between the first documented appearance of two taxa may represent a very short span of time, and the age of the strata in this interval may be nearly synchronous throughout their extent. Interval zones defined on the last (stratigraphically highest) appearances of taxa are commonly considered to have less time significance than those based on first appearances because extinctions of taxa commonly do not occur with the same suddenness that new species appear through phyletic evolution.

Figure 17.15 illustrates some of the various methods that can be used for correlating between two stratigraphic sections on the basis of interval zones. Note from this illustration that interval zones can be identified that represent much shorter spans of time than that represented by the range zones of most individual taxa. Corrrelation can be made also between stratigraphic sections simply on the basis of first or last appearances of specific taxa, without correlating entire zones. In other words, a correlation line can be drawn from the stratigraphic position represented by the first appearance of a particular taxon, called the first appearance datum or FAD, to the FAD of the same taxon in another stratigraphic section. Similarily, correlation can be made between the last appearance datums, LADs, of a given taxon in different stratigraphic sections. FADs and LADs are further discussed in the following chapter in Section 18.4.

Graphic Method for Correlating by Taxon Range Zone. Although interval zones can be used to define units of strata deposited during relatively short periods of time, they do not necessarily yield precise time-stratigraphic correlations. Organisms may migrate laterally and appear in other areas at somewhat later times than their true first appearance (Fig. 17.11), or they may migrate out of a local area before their final extinction elsewhere. These variables of behavior make the boundaries between interval zones inherently "fuzzy." The exact boundary between biozones can never be known because such boundaries are determined empirically. Additional collecting in a new area always holds the possibility of extending the known range of previously defined species or taxa, because they may have appeared earlier or persisted longer in the new area than in originally defined areas. One way to minimize the problem of fuzzy zonal boundaries is to to treat range data statistically, utilizing the first and last appearances of all the species present in a stratigraphic section rather than the ranges of just one or two species. A. B. Shaw in 1964 was the first to propose a graphical method for establishing time equivalence of strata in two stratigraphic sections by plotting first and last appearances of all the species in one section against the first and last appearances of the same species in another section. This method is now widely used by stratigraphers for detailed time-stratigraphic correlation between stratigraphic sections, particularly local sections.

Shaw's method, as further elaborated by Miller (1977), involves first selecting a single stratigraphic section as a reference section to which other sections can be compared and correlated. This reference section should be the thickest section available, should be free of faulting or other structural complications, and should contain a large and varied fossil content. The reference section is measured and sampled as completely as possible, and the first and last appearances of all species are documented in terms of their positions in the stratigraphic section above an arbitrarily chosen reference point, that is, number of meters above the base of the measured section. The species ranges recorded by the first and last appearances in this local reference section may not be the true (total) ranges for all of the species; however, this fact does not preclude using them to help establish correlation, as we shall see. A second stratigraphic section is then chosen to be compared with the reference section, and the first and last appearances of the same species, and of any other species, are determined in this section.

From two such stratigraphic sections (Fig. 17.16), a graph is constructed in which distance above the base of the reference section, say, section A, is indicated on the horizontal axis and distance above the base of the second measured section, section B, is plotted on the vertical axis (Fig. 17.17). The first and last last appearances of

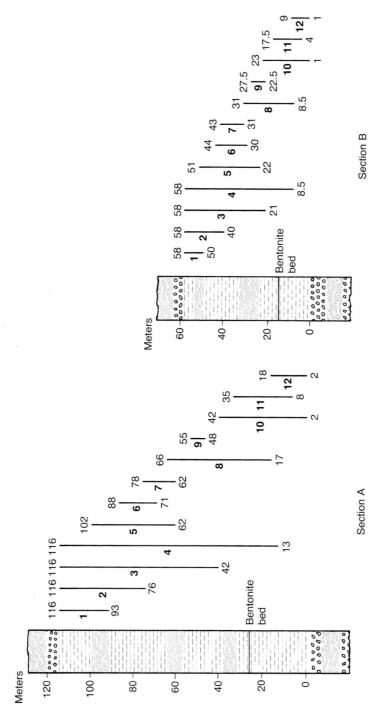

FIGURE 17.16 Two stratigraphic sections with ranges of fossil species (Species 1 through 12) graphed in meters above the base of the section. Sections A and B contain identical fossils with identical time spans; however, Section B represents only half the rate of sediment accumulation. Use of these fossil ranges in A. B. Shaw's (1964) graphic correlation method is illustrated in Figure 17.17. (After Eicher, D. L., Geologic time, 2nd ed., © 1976, Fig. 5.8, p. 112.

FIGURE 17.17 Ilustration of A. B. Shaw's (1964) graphic correlation method using the data shown in Figure 17.16. (After Eicher, D. L., Geologic time, 2nd ed., © 1976, Fig. 5.9, p. 113. Reprinted by permission of Prentice-Hall, Englewood Cliffs, N.J.)

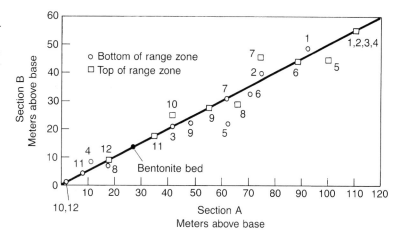

each species in the reference section can then be plotted against the first and last appearances of the same species in the second measured section. In Figure 17.16, for example, Species 1 first appears in reference Section A at 93 m above the base of the section and in measured Section B at about 47 m above the base. A single point can be plotted on the graph to represent these values. Similarly, additional points are plotted to represent the first and last appearances of all the species in the two sections. This procedure yields a series of points that tend to cluster around a straight line (Fig. 17.17). This line can be drawn visually to yield a "best-fit" line, or it can be drawn by use of statistical regression methods. The x and y coordinates of any point on this line provide a precise time-stratigraphic correlation between the two sections. In Figure 17.17, for example, the bed at 60 m in Section A correlates with the bed at 30 m in Section B, and the bed at 100 m in Section A correlates with the bed at about 49 m in Section B.

First and last appearances of species represented by points that plot well off the best-fit line in Figure 17.17 indicate species that appear in or disappear from Section A at distinctly different times than in Section B. Either such species are environmentally controlled (facies dependent), or their migration between Sections A and B was impeded by biogeographic barriers causing them to appear in the two sections at different times.

This graphic correlation method can take advantage of physical events, such as ash falls, or stable isotopic events that have time-stratigraphic significance, to verify the position of the best-fit line. For example, ash falls occur over wide geographic areas almost instantaneously. Their presence in two stratigraphic sections constitutes a precise time marker (that can be dated by radiometric methods) that provides a very reliable point for the best-fit line and should fall exactly on this line.

In addition to its usefulness in correlating between two stratigraphic sections, the graphic correlation method also provides a powerful tool for evaluating differences in rates of sedimentation between two sections or the presence of a hiatus in a section. The slope of the best-fit line indicates the relative rates of sedimentation between the areas. If an abrupt change occurs in this slope (Fig. 17.18), this change suggests a sudden relative increase or decrease in sedimentation rates in the sections. The change in slope at about 75 m in Figure 17.18, for example, indicates an increase in the rate of sedimentation in Section A compared to that in Section B. The presence of a hiatus in deposition in one section shows up as a horizontal line segment in the best-fit curve (Fig. 17.19).

FIGURE 17.18 Increase in rate of sedimentation in Section A compared to that in Section B shown by a "dogleg" in the correlation line determined by A. B. Shaw's (1964) graphic correlation method. (After Eicher, D. L., Geologic time, 2nd ed., © 1976, Fig. 5.10, p. 113. Reprinted by permission of Prentice-Hall, Englewood Cliffs, N.J.)

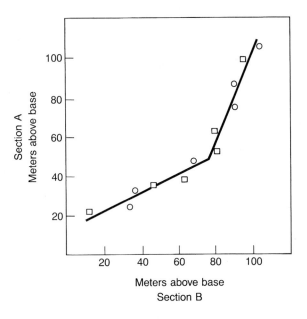

Not only can the graphic correlation method be used for correlating between any two local sections, but it also can be expanded by correlating one section after another to compile what Shaw (1964) refers to as a **composite standard reference section** that can be used for regional (and possibly global) time-stratigraphic correlations. In any reference section chosen for study, the ranges of some fossil species will be at their total stratigraphic maximum. Others fossils will have incomplete ranges owing to the environmental or biogeographical factors described above or to accidents of preservation. The purpose of creating a composite standard reference section is to establish the

FIGURE 17.19 A hiatus in deposition in Section A shows up as a horizontal line in the graphic correlation plot, offsetting the correlation line. (After Eicher, D. L., Geologic time, 2nd ed., © 1976, Fig. 5.11, p. 114. Reprinted by permission of Prentice-Hall, Englewood Cliffs, N.J.)

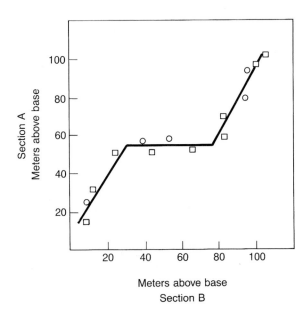

total composite standard range of each species or taxon by compounding information from other correlatable sections into the composite standard reference section. The tops and bases of the stratigraphic ranges of each taxon are adjusted in the standard composite reference section by correlating with other sections until a point is reached where the tops have been adjusted upward as high as they occur in any correlatable section and the bases downward as low as they occur. These adjusted bases and tops thus represent, as nearly as it is possible to determine, the times of speciation (evolutionary first appearance) and global extinction (last appearance anywhere on Earth) of the taxon. Once the total composite standard ranges of each taxon has been determined and the composite standard reference section established, it may be possible to make time-stratigraphic correlations on a regional to global scale.

Correlation by Biogeographical Acme Zones

Under the heading of "Biocorrelation," I discussed correlation by fossil abundance (acme) zones and pointed out that acme zones are unreliable for time-stratigraphic correlation because they are affected by environmental conditions and other factors that

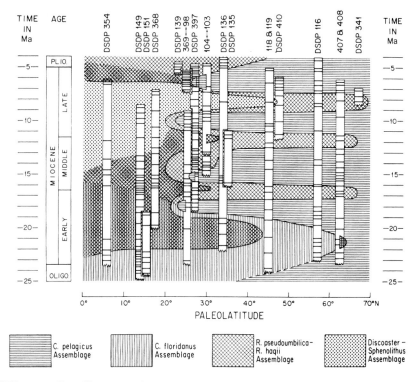

FIGURE 17.20 Use of biogeographical acme zones as a means of time correlation. Cycles of latitudinal shifts of calcareous nannoplankton assemblages in the North Atlantic Ocean during the Miocene are interpreted in response to major fluctuations in climate. The major shifts of relatively warmer, mid-latitude assemblages into higher latitudes can be used for the refinement of the biochronological scale in the higher latitudes from which marker, low-latitude taxa are normally excluded. (From Haq, B. V., and T. R. Worsley, Biochronology—Biologic events in time resolution, their potential and limitations, in G. S. Odin (ed.), Numerical dating in stratigraphy, © 1982 by John Wiley & Sons, Ltd. Fig. 4, p. 27, reprinted by permission of John Wiley & Sons, Ltd.)

can cause them to be diachronous. A different approach to the use of acme zones yields correlations that have time-stratigraphic significance; this approach is correlation based on the maximum abundance of a taxon that results from geographical shifts of an environmentally sensitive fossil assemblage (Haq and Worsley, 1982). Owing to latitudinally related temperature differences in the ocean, some species or other taxa are restricted to biogeographic provinces that are defined by latitude. Thus, low-latitude taxa are ecologically excluded from high latitudes, and vice versa; however, changes in climate can allow shifts of these taxa into a different biogeographic province. During major glacial stages, for example, high-latitude taxa can expand into lower latitudes, and during warming trends between major glacial stages low-latitude taxa can expand into higher latitudes. From a geochronological point of view, the

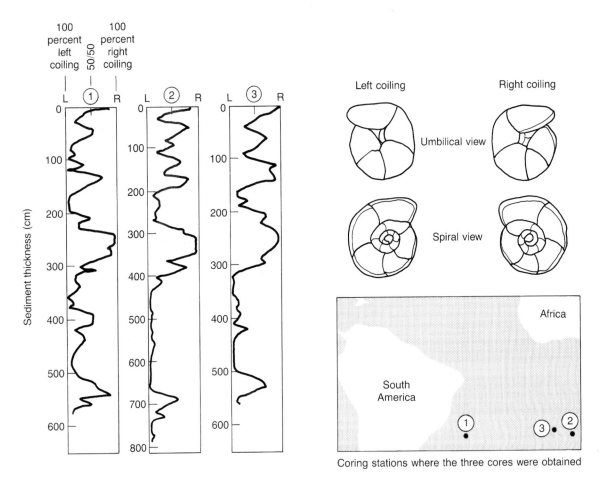

FIGURE 17.21   Biogeographical acme zone correlation based on coiling ratios of foraminifers. Correlation is based on coiling ratios of *Globorotalia truncatulinoides* in three South Atlantic Ocean cores. The depositional time represented by the cores is about 1.5 million years. (After Eicher, D. L., Geologic time, 2nd ed., © 1976, Fig. 5.12, p. 115. Reprinted by permission of Prentice-Hall, Englewood Cliffs, N.J. Original data from D. B. Ericson and G. Woolin, 1968, Pleistocene climates and chronology in deep-sea sediments: Science, v. 162, p. 1227–1234.

spreading out of certain planktonic species in response to major climatic fluctuations is essentially isochronous.

Climate-related shifts in planktonic taxa at specific times thus provides biogeographical acme events that can be correlated from one area to another. In each core or outcrop section studied, climatic curves are constructed on the basis of percentages of warm-climate to cool-climate taxa or relative abundance of a particular taxon. These curves can then be used to identify episodes of warming and cooling that can be correlated from one section to another. Figure 17.20, constructed from this type of information, illustrates how climatically controlled latitudinal shifts in calcareous nannoplankton assemblages in the North Atlantic during Miocene time has been used for chronostratigraphic correlation in DSDP cores.

A related approach is time-stratigraphic correlation based on the coiling ratios of planktonic foraminifera, as described by Eicher (1976). The multi-chambered shells of some foraminifera are known to coil in one direction when the species lives in areas of warm water and in the opposite direction when it lives in areas of cold water. The foraminifer *Globorotalia truncatulinoides,* for example, has dominantly right-handed coils in warm water and left-handed coils in cold water. Figure 17.21 shows that during times of glacial cooling of the ocean in the Pleistocene, predominantly right-coiled populations of *Globorotalia truncatulinoides* were replaced in middle and low latitudes by dominantly left-coiled populations. These changes in coiling ratios of foraminiferal species provide a means of correlating short-term fluctuations of climatic change in the Pleistocene that are essentially synchronous throughout at least a part of an ocean basin.

The major drawback to these correlation methods based on biologic response to climate fluctuations is that their use is restricted mainly to correlating sediments deposited during the Quaternary and Late Tertiary, when several episodes of cooling and warming in the world ocean took place. Nonetheless, they provide a useful supplement to correlation methods based on oxygen isotopes (Chapter 18), which also involve climate fluctuations in the Late Tertiary and Quaternary.

FURTHER READINGS

Berry, W. B. N., 1987, Growth of a prehistoric time scale—Based on organic evolution: Blackwell, Palo Alto, 202 p.

Clube, S. V. M. (ed.), 1989, Catastrophes and evolution: Astronomical foundations: Cambridge University Press, 239 p.

Dodd, R. J., and R. J. Stanton, Jr., 1981, Paleoecology: Concepts and applications: John Wiley & Sons, New York, 559 p.

Donovan, S. K. (ed.), 1989, Mass extinctions: Processes and evidence: Columbia University Press, New York, 266 p.

Gradstein, F. M., E. P. Agterberg, J. C. Brower, and W. S. Schwarzacher, 1985, Quantitative stratigraphy: D. Reidel Publishing, Dordrecht, 598 p.

Guex, J., 1991, Biochronological correlations: Springer-Verlag, Berlin, 252 p.

Kauffman, E. G., and O. H. Walliser (eds.), 1990, Extinction events in Earth history: Springer-Verlag, Berlin, 432 p.

Larwood, G. P. (ed.), 1988, Extinctions and survival in the fossil record: Oxford University Press, 365 p.

Levinton, J., 1988, Genetics, paleontology, and macroevolution: Cambridge University Press, 637 p.

Nitecki, M. H. (ed.), 1984, Extinctions: University of Chicago Press, 354 p.

Sharpton, V. L., and P. D. Ward (eds.), 1990, Global catastrophes in Earth history: Geol. Soc. America Spec. Paper 247, 631 p.

Stanley, S. M., 1979, Macroevolution: Pattern and process: W. H. Freeman, San Francisco, 332 p.

18

Chronostratigraphy and Geologic Time

18.1 INTRODUCTION

The stratigraphic units described in the preceding chapters are rock units distinguished by lithology, magnetic characteristics, seismic reflection characteristics, or fossil content. As such, they are observable or measurable material reference units that depict the descriptive stratigraphic features of a region. Definition of these units allows the vertical and lateral relationships between rock units to be recognized and provides a means of correlating the units from one area to another. As Krumbein and Sloss (1963) point out, however, descriptive stratigraphic units do not lend themselves to interpretation of the local stratigraphic column in terms of Earth history. To interpret Earth history requires that stratigraphic units be related to geologic time; that is, the ages of rock units must be known.

Geologists relate stratigraphic units to time through the use of **geologic time units.** In this chapter, we examine the concept of geologic time units and explore the relationship of time units to other types of stratigraphic units. We will also see how geologic time units are used to create the Geologic Time Scale, and we will discuss methods of calibrating the time scale.

18.2 GEOLOGIC TIME UNITS

Geologic time units are conceptual units rather than actual rock units, although most geologic time units are based on rock units. In fact, we recognize two distinct types of formal stratigraphic units that can be distinguished by geologic age: (1) units, called **stratotypes,** based on actual rock sections, and (2) units independent of reference rock sections (Fig. 18.1). Ideally, the reference rock bodies for geologic time units are **isochronous units.** That is, they are rock units formed during the same span of time

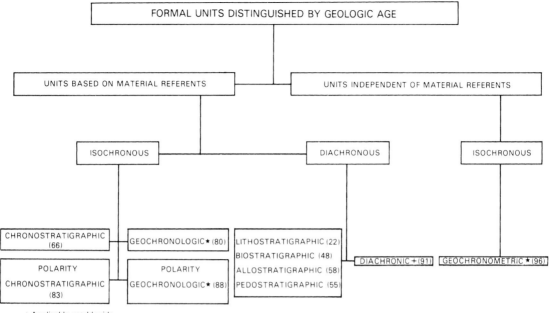

FIGURE 18.1 Major types of geologic time units and their relation to the kinds of rock-unit referents on which most are based. (From North American Commission on Stratigraphic Nomenclature, 1983, North American Stratigraphic Code: Am. Assoc. Petroleum Geologists Bull., v. 67, Fig. 1, p. 850, reprinted by permission of AAPG, Tulsa, Okla.)

and everywhere bounded by synchronous surfaces, which are surfaces on which every point has the same age. The North American Stratigraphic Code also recognizes diachronous rock bodies, which are not everywhere bounded by surfaces having the same age; however, diachronous stratigraphic units are not commonly selected as reference units for geologic time studies.

The International Stratigraphic Guide (Hedberg, 1976) recognizes two fundamental types of isochronous geologic time units: chronostratigraphic units and geochronologic units. **Chronostratigraphic units** are tangible bodies of rock that are selected by geologists to serve as reference sections, or material referents, for all rocks formed during the same interval of time. **Geochronologic units,** by contrast, are divisions of time distinguished on the basis of the rock record as expressed by chronostratigraphic units. They are not in themselves stratigraphic units. If the distinction between these two types of units seems somewhat confusing, the following illustration may help to clarify the difference. Chronostratigraphic units have been likened to the sand that flows through an hourglass during a certain period of time. By contrast, corresponding geochronologic units can be compared to the interval of time during which the sand flows (Hedberg, 1976). The duration of the flow measures a certain interval of time, such as an hour, but the sand itself cannot be said to be an hour.

Traditional internationally accepted chronostratigraphic units were previously based primarily on the time spans of lithostratigraphic or biostratigraphic units. We now also formally recognize (as chronostratigraphic units) polarity chronostratigraphic

TABLE 18.1   Geologic time units

Chronostratigraphic Unit—an isochronous body of rock that serves as the material reference for all rocks formed during the same spans of time; it is always based on a material reference unit, or stratotype, which is a biostratigraphic, lithostratigraphic, or magnetopolarity unit.

   Eonothem—the highest ranking chronostratigraphic unit; three recognized: **Phanerozoic,** encompassing the Paleozoic, Mesozoic, and Cenozoic erathems, and the **Proterozoic** and **Archean,** which together make up the Precambrian (Harrison and Peterman, 1980).

   Erathems—subdivisions of an eonothem; none in the Precambrian; the Phanerozoic erathems, names originally chosen to reflect major changes in the development of life on Earth, are the Paleozoic ("old life"), Mesozoic ("intermediate life"), and Cenozoic ("recent life").

   System—the primary chronostratigraphic unit of worldwide major rank (e.g., Permian System, Jurassic System); can be subdivided into subsystems or grouped into supersystems but most commonly are divided completely into units of the next lower rank (series).

   Series—a subdivision of a system; systems are divided into two to six series (commonly three); generally take their name from the system by adding the appropriate adjective "Lower," "Middle," or "Upper" to the system name (e.g., Lower Jurassic Series, Middle Jurassic Series, Upper Jurassic Series); useful for chronostratigraphic correlation within provinces; many can be recognized worldwide.

   Stage—smaller scope and rank than series; very useful for intraregional and intracontinental classification and correlation; many stages also recognized worldwide; may be subdivided into substages.

   Chronozone—the smallest chronostratigraphic unit; its boundaries may be independent of those of ranked stratigraphic units.

Geochronologic Unit—a division of time distinguished on the basis of the rock record as expressed by chronostratigraphic units; not an actual rock unit, but corresponds to the interval of time during which an established chronostratigraphic unit was deposited or formed; thus, the beginning of a geochronologic unit corresponds to the time of deposition of the bottom of the chronostratigraphic unit upon which it is based, and the ending corresponds to the time of deposition of the top of the reference unit; the hierarchy of geochronologic units and their corresponding geochronostratigraphic units are:

Geochronologic Unit	Corresponding Geochronostratigraphic Unit
Eon	Eonothem
Era	Erathem
Period	System
Epoch	Series
Age	Stage
Chron	Chronozone

Geochronometric Units—direct divisions of geologic time with arbitrarily chosen age boundaries; they are not based on the time span of designated chronostratigraphic stratotypes; a chronometric time scale is commonly used for Precambrian rocks, which cannot be subdivided into globally recognized chronostratigraphic units; ages are generally expressed in millions of years before the present (Ma) but may be expressed also in thousands of years (Ka) or billions of years (Ga).

Source: North American Stratigraphic Code (1983) and International Stratigraphic Guide (Hedberg, 1976).

units, which are geologic time units based on the remanent magnetic fields in rocks (Chapter 16).

The characteristics and hierarchical rankings of geologic time units are briefly described in Table 18.1. Chronostratigraphic units are discussed first in this table because they are the reference stratigraphic sections upon which time units (geochronologic) units are based. Chronostratigraphic units are themselves based upon biostratigraphic, lithostratigraphic, or magnetopolarity units (Table 16.1). The fundamental chronostratigraphic unit is the **system;** higher-ranking units are groupings of systems, and lower-ranking units are subdivisions of systems. In principle, chronostratigraphic units have worldwide extent and can be recognized throughout the world. In practice, worldwide use of chronostratigraphic units depends upon the

FIGURE 18.2 Proposed chronometric time scale for the Precambrian of the United States and Mexico. (After Harrison, J. E., and Z. E. Peterman, 1980, North American Commission on Stratigraphic Nomenclature Note 52—A preliminary proposal for a chronometric time scale for the Precambrian of the United States and Mexico: Geol. Soc. America Bull., v. 91, Fig. 1, p. 378. Subsequent revisions, J. E. Harrison, personal communication, 1985.)

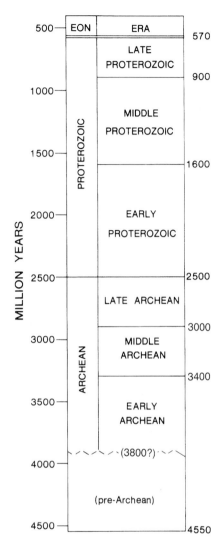

extent to which the time-diagnostic features that characterize these units can be recognized worldwide.

Geochronologic units represent the interval of time during which a correspondingly ranked chronostratigraphic unit was deposited. The fundamental geochronologic unit is thus the **period**—the time equivalent of a system. Names for periods and lower-ranked geochronologic units are identical with those for their corresponding chronostratigraphic units. For example, the Jurassic Period is the time during which the Jurassic System of rocks was deposited. Periods are divided into epochs. Epochs represent the time during which a series was deposited. They take their name from the period by adding the adjective Early, Middle, and Late (e.g., Early Jurassic Epoch, Middle Jurassic Epoch, Late Jurassic Epoch). Note from Table 18.1 the different usage of Lower, Middle, and Upper for subdivision of series—because series are rock units, not units of time. Most names for eons and eras are the same as the names of the corresponding eonothems and erathems.

Geochronometric units are pure time units. They are not based on the time spans of designated chronostratigraphic stratotypes but are simply time divisions of an appropriate magnitude or scale, with arbitrarily chosen boundaries. A geochronometric time scale has been developed for Precambrian rocks (Fig. 18.2) because these rocks have not yet proven generally susceptible to analysis and subdivision by superposition or by application of other lithologic or biologic principles that we commonly use in subdividing the Phanerozoic rocks. Thus, the Archean Eon (oldest Precambrian time) and the Proterozoic Eon (youngest Precambrian time), which have been tentatively formalized for the North American Precambrian, are geochronometric units, with the boundary between them chosen arbitrarily at 2.5 billion years.

18.3   THE GEOLOGIC TIME SCALE

Purpose and Scope

Classifying rocks on the basis of time involves systematic organization of strata into named units, each corresponding to specific intervals of geologic time. These units provide a basis for time correlation and a reference system for recording and systematizing specific events in the geologic history of Earth. Thus, the ultimate aim of creating a standardized geologic time scale is to establish a hierarchy of chronostratigraphic units of international scope that can serve as a standard reference to which the ages of rocks everywhere in the world can be related. Establishing the relative ordering of events in Earth's history is the main contribution that geology makes to our understanding of time.

A standard geologic time scale should (Harland, 1978):

1. express any age in any place
2. express broad and general ages as well as detailed and particular ages
3. be understandable, clear, and unambiguous
4. be independent of opinion and therefore have some objective reference that is accessible
5. be stable, that is, not subject to frequent change
6. be agreed to and used internationally in all languages

Development of the Geologic Time Scale

Chronostratigraphic Scale. Geologists have been working for more than 200 years to develop a systematic scheme for a global time-stratigraphic classification of rock units. This slow process has evolved through two fundamental stages of development:

1. determining time-stratigraphic relationships from local stratigraphic sections by applying the principle of superposition, supplemented by fossil control and, more recently, radiometric ages
2. using these local stratigraphic sections as a basis for establishing a composite international chronostratigraphic scale, which serves as the material reference for constructing a standardized international geologic time scale

An international chronostratigraphic scale ideally consists of a systematically arranged hierarchy of named and defined chronostratigraphic units that encompass the entire stratigraphic succession or geologic column without gaps or overlaps. This scale serves as a standard framework for expressing the ages of all rock strata and their positions with respect to Earth history. (In practice, the boundaries of chronostratigraphic units are defined in stratotypes; however, rocks of all ages may not be present within the boundaries of these stratotypes.) The procedure of naming units of geologic time to represent spans of time during which well-described units of rock were deposited or formed has long been standard practice in the study of Phanerozoic rocks. It has not yet been possible to extend this practice to classification of Precambrian rocks owing to the generally unfossiliferous character and structural complexity of these rocks.

Because the fundamental tool that stratigraphers use for putting events in Earth's history in chronologic order is the principle of positional relationships, or relative ages of rocks, the geologic time scale is thus based on actual sections of rocks. As mentioned, the ideal material reference for the geologic time scale should be a single stratigraphic section containing a complete succession of strata that represent all geologic time, with no gaps in the stratigraphic record. Unfortunately, there is nowhere on Earth a continuous record of strata representing all intervals of geologic time. The geologic systems in use today that constitute the reference sections for the geologic time scale were not proposed by any one geologist or group of geologists in any one locality; nor were they developed as a result of any specific organized approach. They developed gradually over a long period of time, without a specific plan, through the efforts of many different geologists working independently in various parts of the world, particularly Europe.

The first attempts at placing rocks into some type of systematic succession began as early as about 1750. The historical development of the geologic time scale, which eventually culminated in the establishment and naming of the geologic systems shown in Table 18.2, has been traced by numerous workers (e.g., Dunbar and Rodgers, 1957; Krumbein and Sloss, 1963; Berry, 1987; and Eicher, 1976). The geologic time table evolved slowly over nearly two centuries, during which many different schemes for subdividing the rock record were proposed. Many of these schemes were ultimately rejected; however, by the early to middle part of the twentieth century, the system names shown in Table 18.2 had been adopted more or less internationally. As this table suggests, the criteria used by various workers to define the systems were not the same in all cases. Some systems were established on the basis of distinctive lithologies and the presence of major bounding discontinuities. Other system boundaries were defined on the basis of fossils. Many of the system boundaries have subsequently been defined by distinctive fossils, in some cases in localities other than the original type locality, making it possible to recognize the geologic systems worldwide.

The geologic systems listed in Table 18.2 form a fundamental part of the conventional hierarchy of chronostratigraphic units that constitutes the basis for the geologic time scale. These systems are grouped into erathems (Paleozoic, Mesozoic, Cenozoic), and each system, in turn, can be subdivided into smaller units (series, stage). Although the systems are accepted by the international geologic community as the basic reference sections for the geologic time scale, considerable controversy still exists regarding the exact placement of system boundaries and the subdivision of some systems. For example, some systems have been subdivided into units considered by one group of geologists to be series and by another group to be stages. Therefore, the chronostratigraphic units that form the material referents for the geologic time scale are still in a state of flux, and revision of the boundaries of these units continues to the present time.

Table 18.3 shows the complete hierarchy of chronostratigraphic units in general use throughout most of the world. This chronostratigraphic and geochronometric scale was compiled by Salvador (1985) as part of the COSUNA (Correlation of Stratigraphic Units of North America) project. Some of the provincial stage names commonly used

TABLE 18.2   The internationally accepted geologic systems and their type localities

System name	Type locality	Named or proposed by	Date proposed	Remarks
Cambrian	Western Wales	Adam Sedgwick	1835	Defined mainly on lithology
Ordovician	Western Wales	Charles Lapworth	1879	Set up as an intermediate unit between the Cambrian and Silurian to resolve boundary dispute; boundary defined by fossils
Silurian	Western Wales	Roderick I. Murchison	1835	Defined by lithology and fossils
Devonian	Devonshire, southern England	Roderick I. Murchison and Adam Sedgwick	1840	Boundaries based mainly on fossils
Carboniferous	Central England	William Conybeare and William Phillips	1822	Named for lithologically distinctive, coal-bearing strata but recognizable by distinctive fossils
Mississippian	Mississippi Valley, United States	Alexander Winchell	1870	The Mississippian and Pennsylvanian are subdivisions of the Carboniferous; not used outside the United States
Pennsylvanian	Pennsylvania, United States	Henry S. Williams	1891	
Permian	Province of Perm, Russia	Roderick I. Murchison	1841	Identified by distinctive fossils
Triassic	Southern Germany	Frederick von Alberti	1843	Defined lithologically on the basis of a distinctive threefold division of strata; also defined by fossils
Jurassic	Jura Mountains, northern Switzerland	Alexander von Humboldt	1795	Defined originally on the basis of lithology
Cretaceous	Paris Basin	Omalius d'Halloy	1822	Defined initially on the basis of strata composed of distinctive chalk beds
Tertiary	Italy	Giovanni Arduino	1760	Originally defined by lithology; redefined with type section in France on the basis of distinctive fossils
Quaternary	France	Jules Desnoyers	1829	Defined by lithology, including some unconsolidated sediment

in North America are also shown; however, there now appears to be a general movement among North American stratigraphers to abandon these provincial stage names and adopt the European (global) stages as standards for North America. Stratigraphers in Europe and many other parts of the world have for many years subdivided the Ter-

TABLE 18.3 (p. 633–635)   Nomenclature of chronostratigraphic units generally used throughout the world. Some North American stage names are also shown, together with a chronometric scale.

Source: Salvador, A., 1985, Chronostratigraphic and geochronometric scales in COSUNA stratigraphic nomenclature charts of the United States: Am. Assoc. Petroleum Geologists Bull., v. 69. Figs. 1–3, p. 182–184, reprinted by permission of AAPG, Tulsa, Okla.

ERATHEM	SYSTEMS	SERIES / STAGES			NORTH AMERICAN CHRONOSTRATIGRAPHIC UNITS SERIES / STAGES	NUMERICAL TIME SCALE (Ma)
GLOBAL CHRONOSTRATIGRAPHIC UNITS						

ERATHEM	SYSTEMS	SERIES		STAGES	NORTH AMERICAN SERIES / STAGES	NUMERICAL TIME SCALE (Ma)
C E N O Z O I C	QUATERNARY	HOLOCENE / PLEISTOCENE			NORTH AMERICAN PLEISTOCENE GLACIAL STAGES ONLY WHEN APPLICABLE AND NECESSARY	0.01 / 1.7 to 2.8
	T E R T I A R Y — N E O G E N E	PLIOCENE		PIACENZIAN		4.6 / 5.3 — 5
				ZANCLEAN		
		M I O C E N E	UPPER	MESSINIAN		6.7
				TORTONIAN		10.8 — 10
			MIDDLE	SERRAVALLIAN		
				LANGHIAN		15.4 — 15 / 17
			LOWER	BURDIGALIAN		— 20
				AQUITANIAN		23 / 25 — 25
	T E R T I A R Y — P A L E O G E N E	OLIGOCENE	UPPER	CHATTIAN	PACIFIC AREA STAGES	— 30
					OR	33
			LOWER	RUPELIAN	MAMMALIAN STAGES ONLY	— 35
		EOCENE	UPPER	PRIABONIAN	WHEN APPLICABLE	38 — 40 / 41
				BARTONIAN	AND NECESSARY	— 45 / 45
			MIDDLE	LUTETIAN		
			LOWER	YPRESIAN		50 — 50 / 55 — 55
		PALEOCENE	UPPER	THANETIAN		— 60 / 62
			LOWER	DANIAN		— 65 / 67

633

Table 18.3   (continued)

GLOBAL CHRONOSTRATIGRAPHIC UNITS			NORTH AMERICAN CHRONOSTRATIGRAPHIC UNITS	NUMERICAL TIME SCALE (Ma)	
ERATHEM	SYSTEMS	SERIES / STAGES	SERIES / STAGES		
				67	65
		MAASTRICHTIAN			70
		CAMPANIAN		72	
		UPPER — SANTONIAN		80	80
		CONIACIAN		85	
		TURONIAN		90 / 92	90
		CENOMANIAN	SAME AS GLOBAL	100	100
	CRETACEOUS	ALBIAN		108	110
		APTIAN			
		LOWER — BARREMIAN		115	120
		HAUTERIVIAN		125	
		VALANGINIAN		130	130
M E S O Z O I C		BERRIASIAN		135	
		TITHONIAN		140	140
		UPPER — KIMMERIDGIAN		145	150
		OXFORDIAN		155	
		CALLOVIAN		160	160
		MIDDLE — BATHONIAN	SAME AS GLOBAL	165	
	JURASSIC	BAJOCIAN		170	170
		AALENIAN		175	
		TOARCIAN		180	180
		LOWER — PLIENSBACHIAN		185	
		SINEMURIAN		190	190
		HETTANGIAN		195	
				200	200
		RHAETIAN			210
		UPPER — NORIAN		215	
		CARNIAN	SAME AS GLOBAL	220	220
	TRIASSIC			230	230
		MIDDLE — LADINIAN			
		ANISIAN		240	240
		LOWER — SCYTHIAN		245	
				250	250

Table 18.3 (*continued*)

ERATHEM	SYSTEMS	GLOBAL CHRONOSTRATIGRAPHIC UNITS — SERIES / STAGES			NORTH AMERICAN CHRONOSTRATIGRAPHIC UNITS — SERIES / STAGES		NUMERICAL TIME SCALE (Ma)
PALEOZOIC	PERMIAN	UPPER		TATARIAN		OCHOAN	250 / 250
				KAZANIAN		GUADALUPIAN	255 / 260
				KUNGURIAN			270 / 270
		LOWER		ARTINSKIAN		LEONARDIAN	275 / 280
				SAKMARIAN		WOLFCAMPIAN	285
				ASSELIAN			290 / 290
	CARBONIFEROUS	UPPER	STEPHANIAN	GZHELIAN	PENNSYLVANIAN SUB-SYSTEM	VIRGILIAN	300
				KASIMOVIAN		MISSOURIAN	
			WESTPHALIAN	MOSCOVIAN		DESMOINESIAN / ATOKAN	310 / 310 / 315
		MIDDLE		BASHKIRIAN		MORROWAN	320
			"NAMURIAN"				330 / 330
				SERPUKHOVIAN	MISSISSIPPIAN SUB-SYSTEM	CHESTERIAN	340
		LOWER	VISEAN			MERAMECIAN	350
						OSAGEAN	355
			TOURNAISIAN			KINDERHOOKIAN	360 / 365
	DEVONIAN	UPPER	FAMENNIAN		CHAUTAUQUAN	CONEWANGOAN	370
						CASSADAGAN	
			FRASNIAN		SENECAN	CHEMUNGIAN / FINGERLAKESIAN	380 / 380 / 385
		MIDDLE	GIVETIAN		ERIAN		
			EIFELIAN				390 / 390
		LOWER	EMSIAN		ULSTERIAN	ESOPUSIAN	395
			SIEGENIAN			DEERPARKIAN	400 / 400
			GEDINNIAN			HELDERBERGIAN	405
	SILURIAN	UPPER	PRIDOLIAN		CAYUGAN		410
			LUDLOVIAN		NIAGARAN	LOCKPORTIAN	415
		LOWER	WENLOCKIAN			CLIFTONIAN / CLINTONIAN	420 / 420
			LLANDOVERIAN		ALEXANDRIAN		425
	ORDOVICIAN	UPPER	ASHGILLIAN		CINCINNATIAN	RICHMONDIAN	430
			CARADOCIAN			MAYSVILLIAN	440
						EDENIAN	450 / 455
					SHERMANIAN KIRKFIELDIAN ROCKLANDIAN	BLACKRIVERIAN	460 / 460
		MIDDLE	LLANDEILIAN			CHAZYAN	470
			LLANVIRNIAN		CHAMPLAINIAN	WHITEROCKIAN	475 / 480
			ARENIGIAN				485
		LOWER	TREMADOCIAN		CANADIAN		490 / 490
							500 / 500
	CAMBRIAN	UPPER				TREMPEALEAUAN	
						FRANCONIAN	510
						DRESBACHIAN	515
		MIDDLE					520 / 530 / 540 / 540
		LOWER					550 / 560 / 570 / 570

635

tiary into two subsystems, the **Paleogene** and the **Neogene,** with the top of the Oligocene Series as the dividing boundary between the two. Geologists in North America have now also adopted this practice. They have likewise adopted the European usage of the **Carboniferous** as a system name, but with subdivision in North America into the **Mississippian** and **Pennsylvanian** subsystems. Other versions of the chronostratigraphic scale exist (e.g., Harland et al., 1990) and may differ somewhat from this scale, particularly in naming of series and stages, as well as in ages of series and stage boundaries. The geologic community has not yet achieved the ideal of a truly international chronostratigraphic scale that is accepted and used by all geologists worldwide.

Geochronologic Scale. Table 18.3 is a chronostratigraphic scale with units and boundaries based on physical divisions of the rock record, but it is not in itself a time scale. To function as a geologic time scale for expressing the age of a rock unit or a geologic event, the chronostratigraphic scale must be converted to a geochronologic scale consisting of units that represent intervals of time rather than bodies of rock that formed during a specified time interval. The geologic time scale is derived from the chronostratigraphic scale by substituting for chronostratigraphic units the corresponding geochronologic units (Table 18.1). Thus, the geologic time scale is expressed in eras, periods, epochs, and ages rather than erathems, systems, series, and stages. The subdivision boundaries of the geologic time scale are calibrated in absolute ages; however, the geologic time scale differs from a true geochronometric scale, which is based purely on time without regard to the rock record (e.g., Fig. 18.2). By contrast, the subdivisions of the Phanerozoic time scale are of unequal length, owing to the fact that they are based on chronostratigraphic units that were deposited during unequal intervals of time.

The geologic time scale has been in existence for several decades, and during that time it has continued to evolve, with refinements being made particularly in subdivision of the epochs and ages and absolute-age calibration of the boundaries between periods, epochs, and ages. Figure 18.3 shows a version of the geologic time scale published in North America by the Geological Society of America as part of the efforts involved in publication of the 27 synthesis volumes of the *Decade of North American Geology,* or *DNAG* (Palmer, 1983). This DNAG time scale is subdivided into ages based on the European stages, with boundaries between ages calibrated in absolute time. Absolute ages are given in millions of years before the present (Ma), where the present refers to 1950. Methods for absolute age calibration of the geologic time scale are discussed below. Note that the magnetic polarity scale for the most recent approximately 170 million years is also included in the time scale. Note also the use of a geochronometric scale for the Precambrian, with the dividing boundary between the Archean and the Proterozoic set arbitrarily at 2500 million years, as discussed.

The traditional procedure followed in establishing the geologic time scale has been first to define and describe the chronostratigraphic reference units (system, series, stage) and then derive the corresponding geochronologic unit (period, epoch, age) from these already established rock units. In other words, geologic time intervals were conceived as being the time equivalents of rock referent units already defined. Harland (1978) and Harland et al. (1982) suggest that this procedure be reversed. They recommend that the geochronologic units be established first by selecting well-defined reference points in stratigraphic sections. **Each reference point represents a specific time of deposition. Pairs of such points thus define the intervening time span. The period, for example, is first defined in this way by picking upper and lower reference points that can be calibrated by absolute age. The system in turn becomes the rock formed during that particular defined period.** Current stratigraphic practice is to establish the bound-

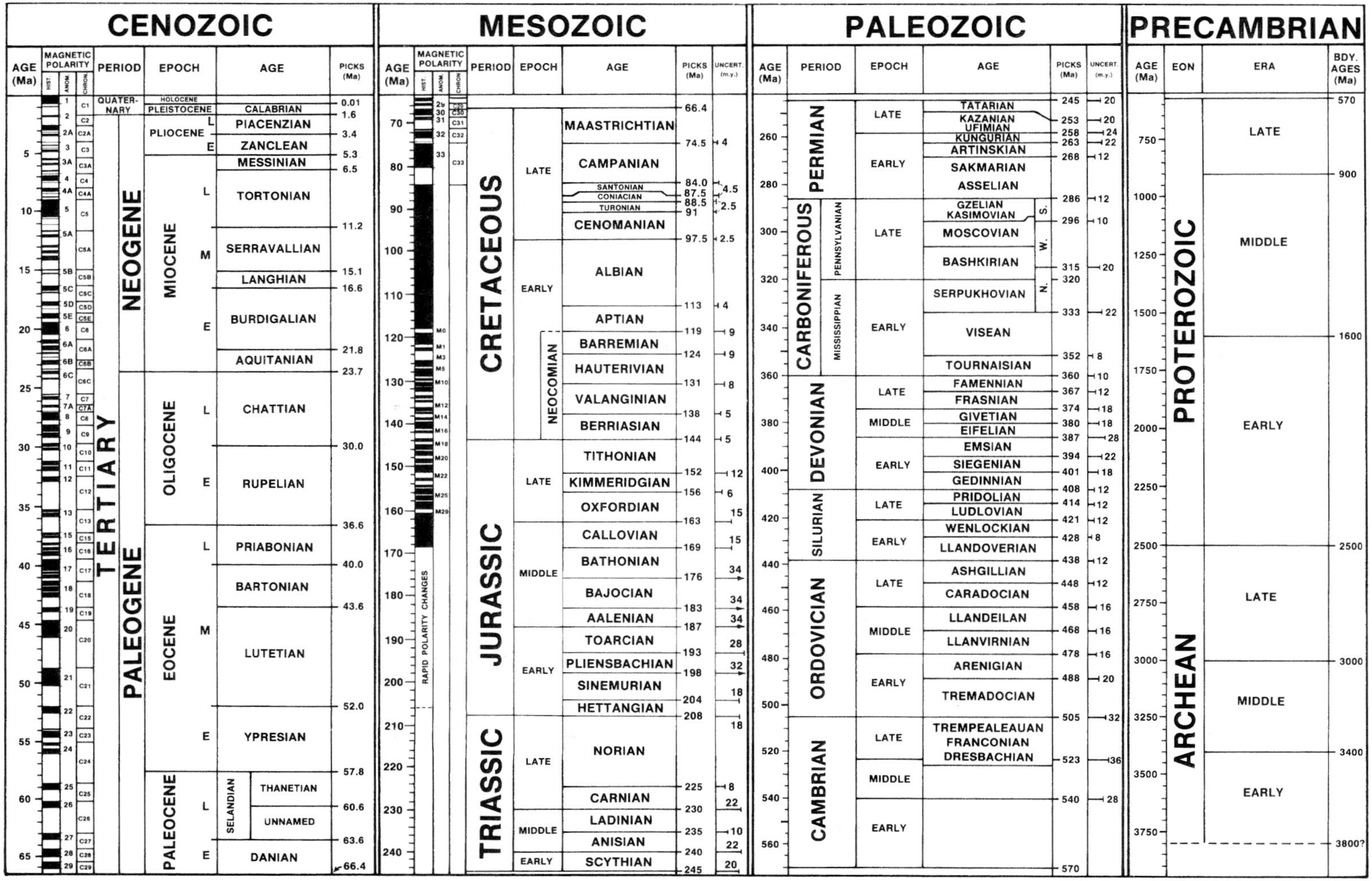

FIGURE 18.3 Geologic time table calibrated in absolute ages. The magnetic polarity scale for the last 170 million years is also shown. (From Palmer, A. R. (comp.), 1983, The decade of North American geology 1983 geologic time scale: Geology, v. 11, p. 504.)

aries of time intervals (geochronologic units) first, on the basis of distinctive datable horizons, before establishing the corresponding rock (chronostratigraphic) unit. Thus, stratigraphers and other geologists throughout the world are actively working together to achieve the ultimate goal of an international geologic time scale by establishing global boundary stratotypes and end points.

18.4 CALIBRATING THE GEOLOGIC TIME SCALE

The geologic time scale has evolved slowly over a long period of time. To develop the scale to its present level of usefulness for fixing the position in time of a particular rock unit or geologic event, two types of information had to be available to stratigraphers:

1. some method of arranging rocks in an orderly succession on the basis of their relative position in time, or relative ages
2. a method of determining the ages of the boundaries between rock units on the basis of their absolute position in time with respect to some fixed time horizon, for example, the present

Placing strata in stratigraphic order in terms of their relative ages has been the guiding principle used by stratigraphers in constructing the geologic time scale. Relative ordering was determined by applying the principle of superposition, aided by use of fossils. The principle of superposition means simply that in a normal succession of strata which have not been tectonically overturned since deposition, the youngest strata are on top and the ages of the strata increase with depth. Application of this simple but highly important principle was of key importance in building the early chronostratigraphic scale and is still useful for determining the relative ages of strata in any local stratigraphic succession.

Owing to the irreversible nature of evolution, fossil organisms succeed each other in the stratigraphic record in an orderly fashion—the principle of faunal succession. This organic continuum makes possible the determination of relative ages of strata because each fossiliferous unit of rock in a stratigraphic succession contains distinctive fossils that distinguish it from younger strata above and older strata below. These distinctive fossil assemblages also allow the correlation of strata from one area to another, as discussed in Chapter 17, making possible determination of relative ages of strata outside the boundaries of original areas of study. Most of the divisions in the current global chronostratigraphic scale are based on fossils, and early efforts to create an international chronostratigraphic scale before methods of absolute-age determinations were developed would have been impossible without the use of fossils.

Fortunately, methods are now available not only for determining the relative ages of strata but also for fixing within reasonable limits of uncertainty the absolute ages of some strata. Development of these methods of absolute-age estimation have made it possible to place approximate absolute ages on boundaries of the chronostratigraphic scale initially established by relative age-determination methods. Absolute age data can also be used for determining ages of poorly fossiliferous Precambrian rocks that cannot be placed in stratigraphic order by relative-age determination methods. The principal method for determining the absolute ages of rocks is based on decay of radioactive isotopes of elements in minerals. Other methods of determining the absolute passage of geologic time include counting the following: lake-sediment varves, which are presumed to represent annual sediment accumulations; growth increments in the shells of some invertebrate organisms; growth rings in trees; and Milankovich climate cycles in sediments (Chapter 15). These alternative methods are useful only for marking the passage of short periods of time in local and regional areas and are not of importance in calibrating the geologic time scale, except possibly some parts of the Pleistocene and Pliocene.

Thus, the major tools for finding ages of sediments to calibrate the geologic time scale are relative-age determinations by use of fossils—biochronology—and absolute age estimates based on isotopic decay—radiochronology. These tools may be used both for calibrating the chronostratigraphic scale directly and for calibrating the succession of reversals of Earth's magnetic field; this succession constitutes the magnetostratigraphic time scale discussed in Chapter 16. We shall now discuss each of these dating methods, beginning with biochronology.

Calibrating the Geologic Time Scale by Fossils: Biochronology

Biochronology is the organization of geologic time according to the irreversible process of evolution in the organic continuum (Chapter 17). Useful fossil horizons are more widespread and abundant in Phanerozoic rocks than are horizons whose ages can be estimated by radiochronology. Furthermore, biologic events can commonly be correlated in time more precisely than can radiometric data in all but Cenozoic rocks. Because of these factors, fossils have conventionally provided the most readily available tool for dating and long-distance correlations of Phanerozoic rocks. It is necessary, however, to make a clear distinction between biochronology and biostratigraphy. Biostratigraphy (Chapter 17) aims simply at recognizing the distinctive fossils that characterize a known stratigraphic level in a sedimentary section without regard to the inherent time significance of the fossils. For example, William Smith was able to use fossils very effectively for identifying and correlating strata even though he had little or no idea of the time relationships or time significance of the fossils. Biochronology, on the other hand, is concerned with the recognition of fossils as having ages that fall at known points in the span of evolutionary time, as measured by fossils of a reference biostratigraphic section. Therefore, by establishing identifiable horizons in reference sections based on fossils, biochronology provides a tool both for international correlation and for worldwide age determination.

The aim of biochronology is to make possible correlation and dating of the geologic record beyond the limits of local stratigraphic sections. To do this most effectively, stratigraphers use features or events in the paleontologic record that are widespread and easily identifiable and that occurred during short periods of geologic time. These events are considered to be biochronologic **datum events** because they mark a particular short period of time in the geologic past. The datum events most commonly used are the immigrations (first appearances) and extinctions (last appearances) of a fossil species or taxon. The first appearance of a species as a result of immigration from another area commonly occurs very rapidly after its initial appearance owing to evolution from its ancestral morphotype. The first appearance is so rapid, in fact, that geologically speaking we consider speciation and immigration as essentially synchronous events. Extinction of a taxon may also occur very rapidly, although commonly not as rapidly as speciation.

Stratigraphers speak of the first and last appearances of a taxon as the **first appearance datum (FAD)** and the **last appearance datum (LAD).** These FADs and LADs are not totally synchronous owing to the fact that even though immigrations and extinctions can take place quite rapidly, they are not actually instantaneous events. Some planktonic species have been reported to spread worldwide in 100 to 1000 years (see Fig. 18.8); however, bioturbation of sediment after deposition can mix fossils through a zone several centimeters thick, and accidents in preservation as well as bias in collection and analytical methods can combine to create uncertainties in the age of the FADs and LADs that can amount to thousands of years. Nevertheless, the duration of the FADs of many planktonic species may be as little as 10,000 years; that is, the ages of the first appearance datum of a species will not vary by more than 10,000 years in different parts of the world (Berggren and Van Couvering, 1978). The error caused by an age discrepancy of this magnitude becomes insignificant when applied to estimation of the ages of rocks that are millions to hundreds of millions of years old. Thus, the FADs and LADs of many fossil species can be considered essentially synchronous for the utilitarian purposes of biochronology.

FADs and LADs are the most easily utilized and communicated types of fossil information upon which to base biochronology, and they can be used over great distances within the range of the defining taxa. Therefore, they have come to dominate global biochronological subdivision. The procedure for establishing the biochronology of any fossil group based on FADs and LADs involves the following steps (described by Haq and Worsley, 1982) and is illustrated graphically in Figure 18.4:

1. Identify and locate in local biostratigraphic units the FADs and LADs of distinctive fossil taxa that have wide geographic distribution.
2. If possible, assign ages to these events by direct or indirect calibration through radiochronology or magnetostratigraphy. If ages can be assigned to any two events, the sedimentation rates for strata between these events can be calculated by dividing the age difference between the two by the thickness of sediment separating them. The sedimentation rates can then be used to calculate the approximate age of each event enclosed within the dated succession (Fig. 18.4).

FIGURE 18.4 Schematic illustration of the application of biochronolgy to age calibration of a local stratigraphic section. The ages of the FAD for species A and the LAD for species D are established by radiometric dating of some closely associated physical feature (e.g., an ash bed). The FAD for species B and the LAD for species C cannot be dated radiometrically; however, the ages can be calculated from the sedimentation rate determined between FAD (A) and LAD (D). This rate (3 m/Ma) can then be used to determine the age difference between FAD (A) and FAD (B) (3 m/Ma × 15 m = 5 Ma) and between LAD (D) and LAD (C) (3 m/Ma × 10 m = 3 Ma).

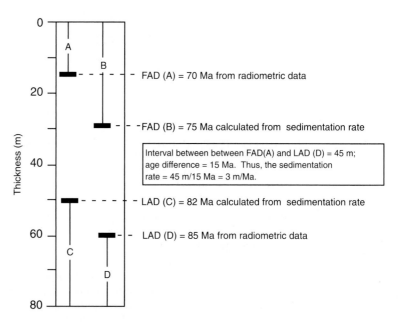

FIGURE 18.5 An example of biochronological dating by use of nannofossil datum events correlated with magnetic polarity events. (After Gartner, S., 1977, Calcareous nannofossil biostratigraphy and revised zonation of the Pleistocene: Marine Micropaleontology, v. 2. Fig. 5, p. 12, reprinted by permission of Elsevier Science Publishers, Amsterdam.)

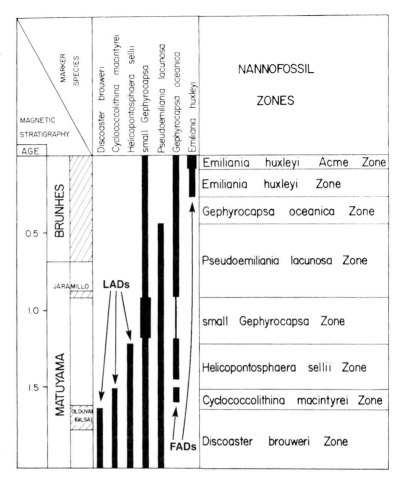

3. If radiometric or magnetostratigraphic calibration of FADs or LADs in the local section cannot be accomplished, then the ages of the datum levels must be found in a different way.

Under these conditions, ages of the FADs and LADs are estimated on the basis of their stratigraphic position with respect to calibrated datum levels of other fossil groups that also occur in the sedimentary succession and whose ages have been found by studying one or more successions elsewhere. An example of biochronologic calibration is illustrated in Figure 18.5, which shows the use of calcareous nannoplankton to establish a biochronology for the Pleistocene by direct correlation with magnetostratigraphic units.

Calibrating by Absolute Ages: Radiochronology

General Principles

Fossils provide an exceptionally useful tool for determining relative ages of rocks and geologic events; however, prior to the beginning of the twentieth century, scientists had no accurate method for determining the absolute passage of geologic time. Many early attempts were made to estimate the age of Earth, but with highly inconsistent results. For example, in 1897 Lord Kelvin, an outstanding physicist of the nineteenth century, used an assumed rate of cooling of a presumed molten Earth as a basis for estimating Earth's age. He used this assumed rate of cooling, together with estimates of the time during which the sun would remain hot without a major internal energy source, to calculate an age for Earth of 20 to 40 million years. Two years later, a geologist named John Joly, on the basis of the present salinity of ocean water and an assumed annual rate of delivery of salts to the ocean by rivers, calculated the age of the oceans to be 90 million years. Between about 1860 and 1910, several geologists calculated the age of Earth by dividing the estimated thickness of the total stratigraphic record by assumed rates of sedimentation. These calculations yielded ages for Earth ranging from 3 million years to 1584 million years, although most calculations gave ages less than 100 million years.

We now know that the results of these early calculations are several orders of magnitude too small and that the methods used to obtain them were based on faulty assumptions and incorrect measurements (e.g., ignoring heat generated by decay of radioactive minerals; incorrect sedimentation rates). The breakthrough in developing an accurate method for determining absolute ages of rocks began about 1896 with the discovery by the French physicist Antoine Henri Becquerel that uranium has the ability to spontaneously emit rays that cause fogging of photographic plates in total darkness—a property he called **radioactivity.** This discovery led B. B. Boltwood, an American chemist at Yale University, to suggest in 1905 that radioactive breakdown of uranium leads ultimately to the production of lead. From chemical analyses of uranium minerals from many parts of the world, Boltwood was able to show by about 1907 that uranium minerals from older rocks contain more radiogenically produced lead than do uranium minerals from younger rocks. Using rough estimates of the decay rate of uranium to lead, he obtained ages based on uranium/lead ratios that ranged from 340 million years for Carboniferous rocks to 1640 million years for Precambrian rocks.

Significant improvements have been made since Boltwood's time in the analytical techniques used for measuring radiometric ages, and the accuracy and precision of the ages have improved correspondingly; however, the basic principles of estimating absolute ages remain the same. That is, the age of a radiogenic mineral is usually calculated from the measured ratio of parent radionuclide to daughter product in the mineral, using the known decay rate of the parent material. The decay rate is measured in the laboratory by special counters and is commonly expressed as the half-life of the radioactive isotope (i.e., the time required for one-half of the parent material to decay to daughter product).

The equation for calculating radiometric age is

$$\log \frac{N}{N_o} = -0.3y \qquad\qquad (18.1)$$

where N = the total number of atoms of an element (e.g., uranium) present in any given amount of the element, N_o = the original number of radioactive atoms in the sample of parent material, and y = the number of half-lives that have elapsed since the mineral containing the radioactive element formed (Faure, 1986, Chapter 4). N can be measured directly with a mass spectrometer; N_o is deduced from the amount of daughter product present. Once y has been calculated from formula 18.1, multiplying y by the half-life of the radioactive element (Table 18.4) gives the age in years.

TABLE 18.4 Principal methods of radiometric age determination

Parent nuclide	Daughter nuclide	Half-life (years)	Approximate useful dating range (years B.P.)	Materials commonly dated
Carbon-14	*Nitrogen-14	5,730	**<~40,000	Wood, peat, CaCO$_3$ shells, charcoal
Protactinium-231 (daughter nuclide of uranium-235)	*Actinium-227	33,000	<150,000	Deep-sea sediment, aragonite corals
Thorium-230 (daughter nuclide of uranium-238/234)	*Radium-226	77,000	<250,000	Deep-sea sediment, aragonite corals
Uranium-238	Lead-206	4,510 million	>5 million	Monazite, zircon, uraninite, pitchblende
Uranium-238	Spontaneous fission tracks	—	**<~65 million	Volcanic glass, zircon, apatite
Uranium-235	Lead-207	713 million	>60 million	Monazite, zircon, uraninite, pitchblende
Potasssium-40	Argon-40	1,300 million	>~100,000	Muscovite, biotite, hornblende, glauconite, sanidine, whole volcanic rock
Rubidium-87	Strontium-87	47,000 million	>5 million	Muscovite, biotite, lepidolite, microcline, glauconite, whole metamorphic rock

*Not used in calculating radiometric ages

**Can be used for dating older rocks under favorable circumstances

Some of the most useful radionuclides for estimating absolute ages and the minerals, rocks, and organic materials most suitable for age determination are shown in Table 18.4. Additional, more specialized dating methods (e.g., argon-40/argon-39, amino-acid racemization method, obsidian hydration method) are available also (Faure, 1986; Geyh and Schleicher, 1990). Details of radiochronologic methods and discussions of errors and uncertainties in radiometric age determinations are available in several published volumes (e.g., Bowen, 1988; Easterbrook, 1988; Geyh and Schleicher, 1990; Mahaney, 1984; McDougall and Harrison, 1988; Odin, 1982a). I shall confine discussion here to the application of radiochronologic methods to calibration of the geologic time scale.

Radiometric Methods

Although radiochronologic methods can be applied to a variety of rock materials and organic substances (Table 18.4), they have limited application to the direct estimation of ages of sedimentary rocks. Most of the potentially usable minerals in sedimentary rocks are terrigenous minerals that when analyzed yield the age of the parent source rock, not the time of deposition of the sedimentary rock, although a few marine minerals such as glauconite can be used for direct dating of sedimentary rocks. Therefore, much of the geologic time scale has been calibrated by indirect methods of estimating ages of sedimentary rocks on the basis of their relationship to igneous or metamorphic rocks whose ages can be estimated by radiochronology. The types of rocks that are most useful for isotopic calibration of the geologic time scale are described in Table 18.5. We will now examine in greater detail the most common methods used to find ages of the sedimentary rocks of the international chronostratigraphic scale.

Finding Ages of Sedimentary Rocks by Analyzing Interbedded "Contemporaneous" Volcanic Rocks. Lava flows and pyroclastic deposits such as ash falls can be incorporated very quickly into an accumulating sedimentary succession without significantly interrupting the sedimentation process. Volcanic materials may be erupted onto "soft" unconsolidated sediment and then buried during subsequent, continued sedimentation, leading to a succession of interbedded sedimentary rocks and volcanic rocks that are essentially contemporaneous in age. Thus, estimates of the ages of such associated volcanic rocks also establish the ages of contemporaneous sedimentary rocks.

Ages of whole volcanic rock can be estimated relatively easily by the potassium-argon method, and ages of minerals in these rocks can be determined by either potassium-argon or rubidium-strontium methods. Potassium-argon methods can be used to study rocks ranging in age from about 50,000 years to the age of Earth, and rubidium-strontium methods are useful for studying rocks older than about 5 million years (Table 18.4). Volcanic rocks that occur in association with nearly contemporaneous sedimentary rocks whose ages can also be determined by fossils provide extremely useful reference points for calibration. In fact, establishing the absolute ages of fossiliferous sedimentary rocks by association with contemporaneous volcanic flows whose ages can be radiometrically estimated has probably been the single most important method of calibrating the geologic time scale.

For this method to work, the contemporaneity of the interbedded volcanic and sedimentary rocks must first be established. If a pyroclastic flow such as an ash fall or a lava flow erupts over an older, exposed sedimentary rock surface where erosion is taking place or sedimentation is inactive, the flow is not contemporaneous with the underlying sedimentary rock. The age calculated for such a flow indicates only that the

TABLE 18.5   Categories of rocks most useful for geochronologic calibration of the geologic time scale

Type of rock	Stratigraphic relationship	Reliability of age data
Volcanic rock (lava flows and ash falls)	Interbedded with "contemporaneous" sedimentary rocks	Give actual ages of sedimentary rocks in close stratigraphic proximity above and below volcanic layers
Plutonic igneous rocks	Intrude (cut across) sedimentary rocks	Give minimum ages for the rocks they intrude
	Lie unconformably beneath sedimentary rocks	Give maximum ages for overlying sedimentary rocks
Metamorphosed sedimentary rocks	Constitute the rocks whose ages are being determined	Give minimum ages for metamorphosed sedimentary rocks
	Lie unconformably beneath nonmetamorphosed sedimentary rocks	Give maximum ages for the overlying nonmetamorphosed sedimentary rocks
Sedimentary rocks containing contemporary organic remains (fossils, wood)		Give actual ages of sedimentary rocks
Sedimentary rocks containing authigenic minerals such as glauconite		Give minimum ages for sedimentary rocks

rock below the flow is older and the rock above younger than the flow. A geologist can establish contemporaneity by determining if fossils in sedimentary layers above and below the flow belong to the same biostratigraphic zone or by looking, along the basal contact of the flow unit, for physical evidence that may show that the underlying sediment was still soft at the time of the volcanic eruption, for example, ash-fall material mixed by bioturbation into underlying sediment, mixing of soft sediment into the base of a submarine lava flow, or other such relationships.

Bracketed Ages from Associated Igneous or Metamorphic Rocks. The radiometric ages of igneous rock that are not contemporaneous with associated sedimentary rocks can be used to estimate the ages of associated sedimentary rocks if two or more igneous bodies "bracket" the sedimentary unit. In this case, the age of the sedimentary unit can be established only as lying between those of the bracketing igneous bodies. The sedimentary unit will be older than an igneous body that intrudes it, but younger than an igneous body upon which it rests unconformably (Fig. 18.6A). For example, a sedimentary succession deposited on the eroded, weathered surface of a granite batholith may subsequently be intruded by a dike or a sill. The sedimentary unit is obviously younger than the batholith but older than the dike or the sill. Unfortunately, there is no way to determine how much younger or older unless other evidence is available. Because erosional and depositional processes are relatively slow, the time represented by a bracketed age may be so long as to be of relatively little use in cali-

FIGURE 18.6 Determining the ages of sedimentary rocks indirectly by (A) bracketing between two igneous bodies and (B) bracketing between regionally metamorphosed sedimentary rocks and an intrusive igneous body.

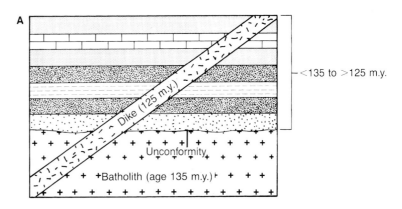

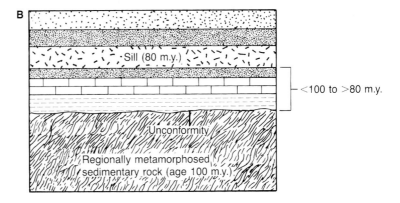

brating the geologic time scale. Only a few points on the time scale have been calibrated by this method.

Metamorphic minerals that develop in sedimentary rocks owing to regional or contact metamorphism can be studied also to provide a method of bracketing the ages of sedimentary rocks (Fig. 18.6B). The radiometric age of metamorphic minerals gives a minimum age for the metamorphosed sediment; that is, the metamorphosed sedimentary rocks are older than the time of metamorphism. If a succession of metamorphic rocks is overlain unconformably by nonmetamorphosed sedimentary rocks, the nonmetamorphosed rocks are obviously younger than the age of metamorphism.

Direct Radiochronology of Sedimentary Rocks

The calibration methods discussed above allow the estimation of ages of sedimentary rocks only through their association in some manner with igneous or metamorphic rocks whose ages can be determined by radiometric methods. Clearly, the uncertainties involved in finding ages of sedimentary rocks by these indirect methods could be avoided if ages could be estimated directly. As mentioned, terrigenous minerals in sedimentary rocks are not useful for radiochronology because they yield ages for the parent rocks, not the time of deposition of the sediment. The only materials in sedimentary rocks that can be used for direct radiochronology are organic remains (wood, calcium carbonate fossils, and other such remains) that were deposited with the sediment and authigenic minerals that formed in the sediment while still on the seafloor or shortly after burial. The principal methods that have been used for direct radiochronology of sedimentary rocks are as follows:

1. the carbon-14 technique for organic materials
2. the potassium-argon and rubidium-strontium techniques for glauconites
3. the thorium-230 technique for ocean-floor sediments
4. the thorium-230/protactinium-231 technique for fossils and sediment

A short discussion of the advantages and disadvantages of each of these methods follows. For a description of other possible direct dating methods, such as amino-acid racemization and other methods based on radioactive disequilibrium of uranium, thorium, and protactinium, see Geyh and Schleicher (1990).

Carbon-14 Method. The carbon-14 method can be applied to the radiochronology of materials such as wood, peat, charcoal, bone, leaves, and the $CaCO_3$ shells of marine organisms. The method has been used extensively for estimating ages of archaeological materials, but it has had limited application in geology owing to the very short useful age range of the method. Carbon-14 decays rapidly with a half-life of only 5730 years. Consequently, the carbon-14 method commonly can be used only for materials less than about 40,000 years old—older materials contain too little carbon-14 to be determined by standard analytical methods. Special techniques that make use of mass spectrometers that allow analysis of smaller amounts of carbon-14, or special proportional counters with high counting efficiencies (e.g., Bowen, 1988), make it possible to extend the usable ages to as much as 60,000 to 70,000 years (Stuivers, Robinson, and Yang, 1979). These special methods are very expensive and have not been widely used in the past. Also, they are exceptionally subject to systematic error because of contamination of samples with young carbon.

The carbon-14 method has been used successfully for such applications as estimating ages of very young sediment in cores of deep-sea sediment and unraveling recent glacial history by analyzing wood in glacial deposits. Its extremely short range renders the method of little value in calibrating the geologic time scale except for very recent Quaternary events.

Potassium-40/Argon-40 and Rubidium-87/Strontium-87 Radiochronology of Glauconites. The term **glauconite** is used loosely for a group of green clay minerals all of which are complex potassium-aluminum-iron silicates that commonly occur in sediments as small rounded grains or pellets. The term **glaucony** is also used as a "facies" name for these green pellets, especially by European geologists. The origin of glauconite is still not thoroughly understood; however, it appears to form authigenically on the seafloor by alteration of substrate materials such as skeletal debris, the fecal pellets or coprolites of organisms, and various types of mineral grains, particularly micas. The glauconization process requires exchange with seawater; therefore, authigenic growth of glauconite grains must take place within the top few centimeters of muddy sediment or the top few meters of coarse sandy sediment in order for such exchange to occur (Odin and Dodson, 1982).

Radioactive ^{40}K is incorporated into the glauconite grains as they evolve by alteration processes on the seafloor. When the glauconite grains are fully formed, they theoretically become closed systems with respect to gain or loss of potassium or argon; that is, no additional radioactive potassium is taken into the grains, and the ^{40}Ar that forms by gradual decay of potassium remains trapped within the glauconite grains. Measurement of the $^{40}K/^{40}Ar$ ratio in the glauconite grains thus allows the age of the grains to be estimated. The half-life of potassium-40 is 1300 million years; therefore, it is theoretically possible to apply the K-Ar method to radiochronology of rocks ranging in age from about 50,000 years to the age of Earth. As we shall see, however, uncertainties in the radiochronology of glauconites reduce the usable range of K-Ar ages for sedimentary rocks.

The wide distribution of glauconite in sediments of all ages and its undoubted early authigenic origin gives glauconite significant potential for estimating ages of sedimentary rocks. Glauconite has been studied more intensively for direct age determinations of sedimentary rocks than has any other sedimentary mineral; however, owing to concern that heating of glauconite during burial may cause loss of argon, considerable difference of opinion has existed regarding the reliability of K-Ar glauconite ages. Comparison of glauconite ages with ages obtained by other radiometric methods has led several workers to suggest that, owing to argon loss, glauconite ages are commonly 10 to 20 percent too young. On the other hand, calculated glauconite ages may be too old in some cases owing to the presence of inherited radiogenic argon that was already in sediment at the time the glauconite grains formed. More recent work has shown that some of the uncertainties in estimating ages of glauconites can be removed by using only glauconite grains that contain more than about 7 percent K_2O; such grains appear to be less likely to contain significant amounts of inherited argon. Nonetheless, uncertainties in glauconite ages may range from thousands of years to almost a million years (Fig. 18.7).

Obviously, the formation of glauconite grains and their closure to loss of argon do not occur simultaneously with deposition of the enclosing sediment. Glauconite grains, therefore, must yield a slightly younger age than the sediment in which they occur, even if uncertainties about inherited or lost argon are not a problem. Odin and Dodson (1982) suggest that the time required for glauconites to evolve and become closed systems may range up to 25,000 years or more. Thus, in relation to biostratigraphic zonation, the glauconite K-Ar ages are closer to those of fossils in

FIGURE 18.7 Stratigraphic and genetic uncertainties related to the use of major types of chronometers for dating sedimentary rocks. The best chronometers are those that plot nearest to the *x-y* intercept in the figure. (From Odin, G. S., Introduction: Uncertainties in evaluating the numerical time scale, *in* G. S. Odin (ed.), Numerical dating in stratigraphy, Fig. 3, p. 10, © 1982, John Wiley & Sons, Ltd. Reprinted by permission of John Wiley & Sons, Ltd. Chichester, England.)

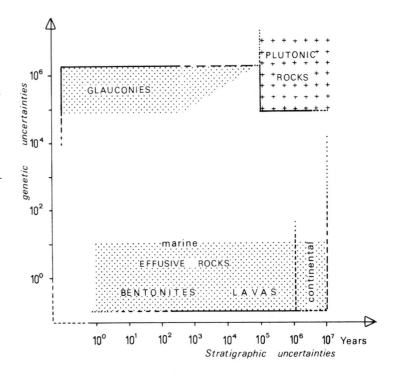

the horizon immediately above the glauconites than to the fossils deposited with the glauconites.

The ages of glauconites can also be estimated by the rubidium-strontium method (Table 18.4). Radioactive rubidium (^{87}Rb) is incorporated into glauconites as they form, along with potassium-40. Rubidium-87 decays to strontium-87 (^{87}Sr), with a half-life of 47,000 million years. This long half-life limits the use of the rubidium-strontium method to radiochronology of rocks older than about 5 million years. Like K-Ar ages for glauconites, the ages determined by the Rb-Sr method are commonly believed to be 10 to 20 percent too young and are thus minimum ages for the glauconites. Also, there appear to be somewhat greater analytical problems and uncertainties associated with the radiochronology of glauconites by the Rb-Sr method, and it has not been as widely used as the K-Ar method. Details of the Rb-Sr method as applied to the radiochronology of sedimentary rocks are given by Clauer (1982).

Estimating Ages of Sedimentary Rocks by Use of Other Authigenic Minerals. In addition to glauconite, several other authigenic minerals have been used in direct radiochronology of sedimentary rocks by the K-Ar and Rb-Sr methods. These minerals include other types of clay minerals such as illite, montmorillonite, and chlorite; zeolites; carbonate minerals; and siliceous minerals such as chert and opal. Owing to uncertainties about their origin—that is, authigenic or detrital—and time of closure to seawater interactions, none of the clay minerals except glauconite have so far proven to yield reliable ages. Zeolites, carbonate minerals, and siliceous minerals have been used for direct radiochronology of sedimentary rocks with some success, but the overall usefulness and reliability of methods based on these minerals have not yet been adequately investigated.

Thorium-230 and Thorium-230/Protactinium-231 Methods for Estimating Ages of Recent Sediments. Uranium-238 decays through several intermediate daughter products, including uranium-234, to thorium-230. Uranium-238 is fairly soluble in seawater and is present in detectable amounts in seawater. By contrast, the thorium-230 daughter product precipitates quickly from seawater by adsorption onto sediment or inclusion in certain authigenic minerals and becomes incorporated into accumulating sediment on the seafloor. Thorium-230 is an unstable isotope and itself decays with a half-life of 75,000 years to still another unstable daughter product, radium-226. Owing to this fairly rapid decay of ^{230}Th, cores of sediment taken from the ocean floor exhibit a measurable decrease in ^{230}Th content with increasing depth in the cores. Assuming that sedimentation rates and the rates of precipitation of ^{230}Th have remained fairly constant through time, the concentration of ^{230}Th should decrease exponentially with depth. The ages of the sediments at various depths in a core can be calculated by comparing the amount of remaining ^{230}Th at any depth to the amount in the top layer of the core (surface sediment). This method can be applied to the dating of sediments younger than about 250,000 years old, which makes it useful for bridging the gap between maximum carbon-14 ages and minimum K-Ar ages.

Protactinium-231 is the unstable daughter product of uranium-235 and itself decays with a half-life of about 34,000 years to actinium-227. Protactinium-231, like thorium-230, precipitates quickly from seawater and becomes incorporated into sediment along with thorium-230. Because protactinium-231 decays about twice as rapidly as thorium-230, the ^{231}Pa/^{230}Th ratio in the sediments changes with time. Thus, in a sediment core, this ratio is largest in the surface layer of the core and decreases progressively with depth in the core. The age of the sediment at any depth in the core is determined by comparing the ^{231}Pa/^{230}Th ratio at that depth to the ratio in the surface sediment. The reliability of the ages determined by this method rests on the assumption that protactinium-231 and thorium-230 are produced everywhere in the ocean at a constant rate and that the starting ratio of these two isotopes in surface sediment is constant throughout the ocean.

An alternative method for calculating ages of sediment based on protactinium-231 and thorium-230 involves measuring the ratio of these daughter products to their parent isotopes in the skeletons of marine invertebrates such as corals. Dissolved uranium-238 and uranium-235 in seawater are incorporated into corals as they grow, whereas seawater contains no appreciable protactinium-231 and thorium-231, owing to the rapid precipitation of these daughter products. Therefore, any protactinium-231 or thorium-230 present in corals results from decay of the parent uranium isotopes within the corals. The ratio of parent isotope to daughter product decreases systematically with time, providing a method for dating the corals. These ratios approach an equilibrium value with increasing passage of time, owing to the fact that the daughter products themselves continue to decay. Thorium-230 reaches a steady state after about 250,000 years, and protactinium-231 after about 150,000 years. Thus, these methods can be used only for radiochronology of rocks younger than these ages. Owing to the fact that corals and other skeletal materials tend to recrystallize with burial and diagenesis, the ^{231}Pa-^{230}Th method has severe limitations. Recrystallization may open the initially closed system and allow escape of the daughter or parent isotopes. Therefore, this method cannot be applied to estimating ages of skeletal materials that have undergone recrystallization.

Summary

Radiochronology of sedimentary rocks whose relative positions in the stratigraphic column are already established can be accomplished by several methods. The choice of method depends upon the age of the rocks and the types of materials present in them. In general, calibration of the time scale by estimating ages of volcanic rocks associated with essentially contemporaneous sedimentary rocks that can be easily correlated by marine fossils is the most useful and reliable approach. Radiochronology of sedimentary glauconites or bracketing the ages of sedimentary rocks from associated plutonic intrusive rocks may also yield usable ages—the only ages available in some cases. Therefore, different methods may have to be applied to estimating ages of rocks in each geologic system. Details of the methods used for estimating ages of boundaries between and within the different systems are given in Odin (1982a), Harland et al. (1990), and Snelling (1985a).

Figure 18.3 shows the calibration of the Decade of North American Geology 1983 Geologic Time Scale based on absolute ages obtained from a number of different sources (Palmer, 1983). Readers should be aware, however, that other published geologic time scales have slightly different values for some of these boundaries (e.g., Odin, Cury et al., 1982; Harland et al., 1990; Snelling 1985b), indicating differences in opinion about the ages of the boundaries. Calibration of the geologic time table has changed steadily over the years as radiochronologic methods have improved and more absolute ages have become available. Although the ages now used to calibrate the major boundaries of the geologic time scale are unlikely to undergo major revision in the future, it is safe to assume that refinements in these ages will continue for some time.

18.5   CHRONOCORRELATION

Chronostratigraphic units are extremely important in stratigraphy because they form the basis for provincial to global correlation of strata on the basis of age equivalence. We have already established that chronostratigraphic correlation is correlation that expresses correspondence in age and chronostratigraphic position of stratigraphic units. To many geologists, correlation on the basis of age equivalence is by far the most important type of correlation, and, in fact, it is the only type of correlation possible on a truly global basis. Methods of establishing the age equivalence of strata by magnetostratigraphic, seismic, and biologic techniques have already been discussed (Chapters 15, 16, and 17). Several other methods of time-stratigraphic correlation are also in common use, including correlation by short-term depositional events, correlation based on transgressive-regressive events, correlation by stable isotope events, and correlation by absolute ages. These methods are discussed below.

Event Correlation and Event Stratigraphy

Event correlation constitutes part of what has come to be known as **event stratigraphy.** Event stratigraphy focuses on the specific events that generate a stratigraphic unit or succession rather than on the physical or biological characteristics of the unit. For example, a eustatic rise in sea level can affect sedimentation patterns worldwide (Chapter 15). As a result of this event, sedimentary facies are generated in a variety of environments in various parts of the world. These facies may not be equivalent in terms of their physical characteristics; however, they are equivalent in the sense that they were produced as a result of the same event. Thus, they are chronological equivalents.

Events can be considered to have different scales depending upon their duration (Fig. 18.8), intensity, and geologic effect. Some convulsive events are extraordinarily energetic, occur quickly, and have regional influence [e.g., explosive volcanic eruptions, impact of large extraterrestrial bodies (bolides), great earthquakes, catastrophic floods, large violent storms, large tsunamis]. These events may produce widespread effects, including mass extinctions. Because of their magnitude, the deposits of such events may form important parts of the geologic record; in fact, the stratigraphic record tends to overemphasize extraordinary perturbations (Seilacher, 1992). On the other hand, the products of a particular event may not be well enough preserved in the geologic record to be recognized as an event marker (Clifton, 1988), and synchroneity of event deposits from one region to another may not be easily recognized. Other events

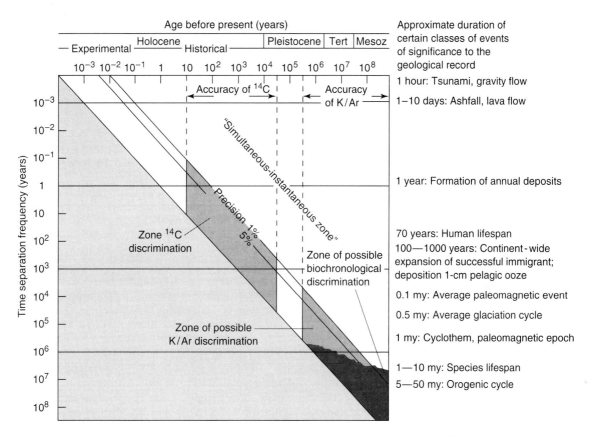

FIGURE 18.8 Resolving power of geochronologic systems in the Cenozoic on the basis of ^{14}C and K-Ar absolute age discrimination and biochronological discrimination. The vertical axis shows the duration of events ranging from hours to hundreds of millions of years, and the horizontal axis shows age before the present ranging from hours to hundreds of millions of years. Note that ^{14}C dating can resolve events that range in age from tens of years to less than 100,000 years and that are years to tens of thousands of years apart. K-Ar dating can resolve events that are older than 100,000 years and that are separated by at least 10,000 years. Biochronology is most effective in resolving events that are older than about 1 million years and that are spaced at least 1 million years apart. (After Berggren, W. A., and J. A. Van Couvering, 1978, Biochronology, *in* G. V. Cohee, M. F. Glaessner, and H. D. Hedberg (eds.), The geologic time scale: Am. Assoc. Petroleum Geologists Studies in Geology 6. Fig. 1, p. 43, reprinted by permission of AAPG, Tulsa, Okla.)

occur more slowly and produce important stratigraphic successions that may be well preserved and recognized over large areas—for example, rise and fall of sea level that generates a transgressive-regressive stratigraphic succession.

Correlation by Short-Term Depositional Events. Correlation made on the basis of such short-term geologic event markers is called **event correlation.** Some events produce key beds, or marker beds, that can be traced in outcrop or subsurface sections for long distances. These marker beds are useful for time-stratigraphic correlation, as well as for lithostratigraphic correlation, if they were deposited as a result of a geologic events that took place essentially "instantaneously." The most striking short-term depositional event is ash fall from volcanic eruptions, which can take place in 1 to 10 days (Fig. 18.8). Beds formed from ash falls are called ash layers, tephra layers, bentonite beds (if the ash alters to bentonite clays), or tuff layers. The ash fall from a single eruption may produce ash layers several centimeters in thickness that can cover thousands to hundreds of thousands of square kilometers. For example, ash from the eruption of Mt. Mazama in southeastern Oregon about 6500 to 7000 years ago—an eruption that subsequently led to the formation of the Crater Lake caldera—was carried northeastward by winds and deposited as far away as Saskatchewan and Manitoba, Canada. Ash from the May 1980 eruption of Mt. St. Helens also spread over thousands of square kilometers east and north of Mt. St. Helens in Washington and Idaho. Other historic examples of widespread ash falls include the 1932 eruption of Quizapú in Chile—an eruption that distributed volcanic ash eastward for 1500 km across South America and into the Atlantic Ocean—and the eruption of Perbuatan Volcano at Krakatoa Island, Indonesia, in 1883—an eruption that spread volcanic dust around the world.

Tephra layers make extremely useful reference points in stratigraphic sections. They provide a means for reliable time-stratigraphic correlation if they are of sufficient lateral and vertical extent and if they can be identified as the product of a particular volcanic eruption. Identification of individual ash layers or bentonite beds can often be made on the basis of petrographic characteristics—types of mineral grains, rock fragments, glass shards, or other components—or trace-element composition. Ages of these layers may be determined also by radiometric methods, allowing the layers to be identified and correlated by contemporaneous age. Tephra layers are particularly useful in correlating across marine basins, and it may even be possible to correlate ash layers in marine basins to well-dated lava flows or ash layers on land, thereby extending marine correlations onto land.

Turbidity currents constitute another type of "instantaneous" geologic event that can produce thin, widespread deposits. Turbidites may have chronostratigraphic significance if a particular turbidite bed, or succession of beds, can be differentiated from other turbidite units and traced laterally. Unfortunately, most turbidites commonly consist of rhythmic or cyclic successions of units that have very similar appearance and are very difficult to differentiate. Thus, in practice, the usefulness of turbidites in time-stratigraphic correlation is rather limited. Figure 18.9 shows an example of a distinctive tuff bed or tephra layer that provides a chronostratigraphic marker horizon that can be recognized and correlated in several wells that penetrated deep-water conglomerates, graded turbidite sandstones, and mudstone in the Ventura Basin, California. Turbidite units and other deep-water deposits can be correlated by their position in the succession with respect to the tuff horizon. Without this widespread marker bed, the turbidite units could not otherwise be correlated.

Other types of "catastrophic" short-term geologic events include dust storms that produce fine-grained loess deposits on land or silt-sand layers in marine basins. Storms at sea can stir up and transport sediment on the continent shelf to produce thin

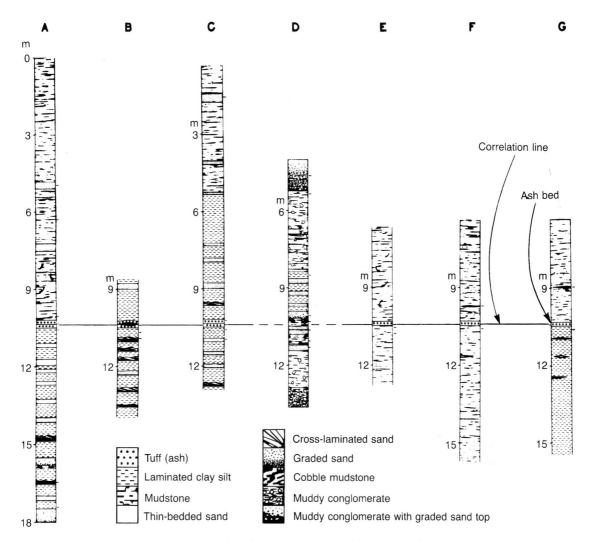

FIGURE 18.9   The Late Pliocene Bailey Ash in the Ventura Basin, California, provides a chrono-stratigraphic marker bed that can be correlated through outcrop sections (A through G) across the basin, allowing mudstone and sandstone beds above the marker unit to be correlated in relation to the marker. The ash bed is missing in Section D. (After Hsü, K. J., K. Kelts, and J. W. Valentine, 1980, Resedimented facies in Ventura Basin, California, and model of longitudinal transport of turbidity currents: Am. Assoc. Petroleum Geologists Bull., v. 64. Fig. 3, p. 1038, 1039, reprinted by permission of AAPG, Tulsa, Okla.)

"storm layers" of sand or silt, as discussed in a preceding chapter. Even the formation of microtektite or micrometeorite horizons in the deep sea can have time-stratigraphic significance (Glass and Zwart, 1979).

   Slower, noncatastrophic depositional conditions also may generate thin, distinctive, widespread stratigraphic marker beds under some depositional conditions. Deposition of these beds does not necessarily take place "instantaneously." Nevertheless, they can be used for time-stratigraphic correlation if they formed as a result of deposition that took place over a large part of a basin during a relatively short period of time under essentially uniform depositional conditions. For example, a thin, widespread limestone bed within a dominantly shale or silt succession implies deposition of the limestone under conditions that were in effect essentially simultaneous throughout a geologic province. Such a thin limestone bed within a succession of nonmarine clastic units may represent a brief incursion of marine conditions into a nonmarine environment or the temporary ponding of freshwater to form a large, shallow lake. Thin limestone units in a thick succession of marine clastic deposits may indicate shelf carbonate deposition during brief periods when clastic detritus was temporarily trapped in estuaries or deltaic environments and thus prevented from escaping onto the shelf. By contrast, thin interbeds of sand, clay, or silt in a thick carbonate or evaporite succession may represent temporary incursions of clastic detritus into a carbonate or evaporite basin. Such incursions may be due to a sudden increase in the supply of detritus as a result of tectonic events, periodic flooding on land, or deposition by windstorms or turbidity currents. Widespread, thin, continuous evaporite beds may also have time-stratigraphic significance because they appear to represent nearly simultaneous deposition throughout a large evaporite basin.

Event Correlation Based on Transgressive-Regressive Events. A different approach to event correlation is represented by local correlation based on position within a transgressive-regressive succession or cycle (Ager, 1981). According to Ager, event correlation in this case is based on the correlation of corresponding peaks of symmetric sedimentary cycles that are presumed to be synchronous. The events represented in this type of correlation are the result of transgressions and regressions that may represent either worldwide, simultaneous eustatic changes in sea level or more local changes owing to uplift, subsidence, or fluctuation in sediment supply.

The principle of correlation based on transgressive-regressive events is illustrated in Figure 18.10. The deposits formed during any transgressive-regressive cycle contain one particular time plane that represents the time of maximum inundation by the sea, that is, the time at which water depth was greatest at any particular locality. Rocks lying stratigraphically below this time plane were deposited during transgression, and those above during regression. This time plane can be identified by use of fossil data to determine depth zonation and maximum water depth at various localities, as illustrated in Figure 18.10. The position of the time plane can be established also from lithologic evidence by determining in the vertical stratigraphic section at each locality the position within the section where the rocks are symmetrically distributed with respect to the most basinward facies present. A surface connecting the most basinward rocks in each of the vertical sections defines the approximate position of the time plane and thus the time-stratigraphic correlation between the sections. Figure 18.11 further illustrates the method. Note from this illustration how time-equivalent points on the cycle are related, resulting in a correlation in which glauconitic clays at the east end of the succession are equated to laminated beds at the west end. Correlation is expressed, as Ager (1981) puts it, in terms of degrees of "marineness." Correlation in this manner can be considered to be a part of sequence stratigraphy (Chapter 15).

Correlation by Stable Isotope Events

Variations in the relative abundance of certain stable, nonradioactive isotopes in marine sediments and fossils can be used as a tool for chronostratigraphic correlation of marine sediments. Geochemical evidence shows that the isotopic composition of oxygen, carbon, sulfur, and strontium in the ocean has undergone large fluctuations, or

FIGURE 18.10 Time correlation by position in a transgressive-regressive cycle. The line connecting points of deepest-water conditions is a time line. (After Israelski, M. C., 1949, Oscillation chart: Am. Assoc. Petroleum Geologists Bull.,v. 33. Fig. 3, p. 98, reprinted by permission of AAPG, Tulsa, Okla.)

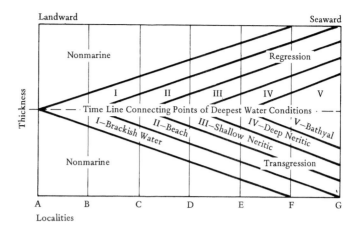

"excursions," in the geologic past—fluctuations that have been recorded in marine sediments. Because the mixing time in the oceans is about 1000 years or less, marine isotopic excursions are considered to be essentially isochronous throughout the world. Variations in isotopic composition of sediments or fossils allow geochemists to construct **isotopic composition curves** that can be used as stratigraphic markers for correlation purposes. To be useful for correlation, fluctuations in isotopic composition must be recognizable on a global scale and must be of sufficiently short duration to show up as a shift on isotopic composition curves. Also, stratigraphers must be able to fix the relative stratigraphic position of these fluctuations in relation to biostratigraphic, pale-

FIGURE 18.11 Transgressive-regressive cycle sedimentation and event correlation in the Eocene of the Isle of Wight in southern England. (From Ager, D. V., The nature of the stratigraphical record, 2nd ed., Fig. 7.2, p. 70, © 1981. Reprinted by permission of Macmillan, London and Basingstoke.)

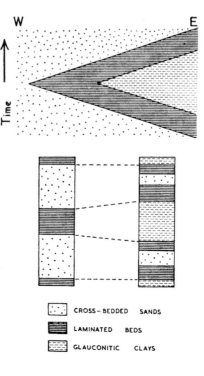

TABLE 18.6   Abundances and natural isotope ranges of carbon, oxygen, and sulfur stable isotopes

Element	Atomic number	Isotope	Geochemical abundances (ppm)	Relative isotopic abundances (%)	Range of natural variations (‰)
C	6	12	230	98.99	
		13*		1.108	90
		14		10^{-12}	
O	8	16	470,000	99.759	
		17		0.037	
		18*		0.203	110
S	16	32	470	95.0	
		33		0.76	
		34*		4.22	140
		36		0.014	

Source: Odin, G. S., M. Renard, and C. V. Grazzini, Geochemical events as a means of correlation, *in* G. S. Odin (ed.), Numerical dating in stratigraphy, Table 2, p. 55, © 1982, John Wiley & Sons, Ltd. Reprinted by permission of John Wiley & Sons, Ltd., Chichester, England.

omagnetic, or radiometric scales. Of the various potentially useful isotopes, oxygen isotopes seem most nearly to meet these requirements and have proven to be particularly useful for chronostratigraphic correlation of Quaternary and Late Tertiary sediments. Carbon, sulfur, and strontium isotopes are also useful for correlating rocks of certain ages.

Oxygen Isotopes. The natural isotopes of oxygen are shown in Table 18.6. Most of the oxygen in the oceans occurs as oxygen-16. Oxygen-18 is much rarer (about 0.2 percent of total oxygen), but it is present in measurable amounts. The ratio of $^{18}O/^{16}O$ in the ocean at any given time in the past is built into contemporaneous marine carbonate minerals and the calcium carbonate shells of marine organisms as a permanent record of the isotopic composition of the ocean at those times. Fluctuations in oxygen isotope ratios in the ocean with time thus show up in the geologic record as fluctuations in the isotopic ratios of these marine carbonates and fossils. Classification of deep-sea sediments on the basis of oxygen isotope ratios in the shells of calcareous marine organisms, particularly foraminifers, has given rise to a new stratigraphy for Quaternary sediments. This stratigraphic method is commonly referred to as **oxygen isotope stratigraphy.** It was first used by Emiliani (1955), who studied the isotopic composition of foraminifers in deep-sea cores and used oxygen isotope ratios to subdivide the core sediments. Oxygen isotope stratigraphy has now developed into a major tool for correlating Quaternary and late Tertiary marine successions, as explained below.

The $^{18}O/^{16}O$ ratio in biogenic marine carbonates reflects both the temperature and the $^{18}O/^{16}O$ ratio of the water in which these carbonates formed. The relationship of ocean paleotemperature *(T)* to oxygen isotopic composition has been shown by Shackleton (1967) to be

$$T \, (°C) = 16.9 - 4.38(\delta_c - \delta_w) + 0.10(\delta_c - \delta_w)^2$$

where δ_c = the equilibrium oxygen isotopic composition of calcite and δ_w = the oxygen isotopic composition of the water from which the calcite was precipitated. The δ_c and δ_w notations refer not to the actual oxygen isotopic abundances in calcite and water but to the per mil (parts per thousand) deviation of the $^{18}O/^{16}O$ ratio in calcite and water from that of an arbitrary standard. A commonly used standard for oxygen isotopes in the past was the University of Chicago PDB standard, where PDB refers to a particular fossil belemnite from the Pee Dee Formation of South Carolina. More commonly now, the isotope composition of ocean water (Standard Mean Ocean Water, or SMOW) is used as a standard (e.g., Coplen et al., 1983). The per mil deviation from the standard, referred to as $\delta^{18}O$, is expressed by the relationship

$$\delta^{18}O = \frac{[(^{18}O/^{16}O) \text{ sample} - (^{18}O/^{16}O) \text{ standard}]}{(^{18}O/^{16}O) \text{ standard}} \times 1000$$

Oxygen isotope stratigraphy is based on the fact that $\delta^{18}O$ values in biogenic marine carbonates reflect both the temperature and the isotopic composition of the water from which the calcite precipitates. These factors are both, in turn, a function of the climate. When water evaporates at the surface of the ocean, the lighter ^{16}O isotopes are preferentially removed in the water vapor, leaving the heavier ^{18}O in the ocean. This isotopic fractionation process thus causes water vapor to be depleted of ^{18}O with respect to the seawater from which it evaporates. When water vapor condenses to form rain or snow, the water containing heavy oxygen will tend to precipitate first, leaving the remaining vapor depleted of ^{18}O compared to the initial vapor. Thus, the precipitation that falls near the coast and runs back quickly to the ocean will contain heavier oxygen than that which falls in the interior of continents or in polar regions, where it returns more slowly to the ocean. There is a correlation also between the air temperature and the $^{18}O/^{16}O$ ratio of precipitates: The colder the air, the lighter the rain or snow (Odin, Renard, and Grazzini, 1982). For example, the overall average oxygen isotope composition of seawater is −0.28o/oo (per mil); however, the precipitation that falls in the crests of the Greenland Ice Sheet is about −35o/oo, and in relatively inaccessible parts of the Antarctic Ice Sheet it is as negative as −58o/oo.

The ^{18}O-depleted moisture that falls in polar regions is locked up as ice on land and is thus prevented from quickly returning to the ocean. Owing to this retention of light-oxygen water in the ice caps, the ocean becomes progressively enriched in ^{18}O as ^{18}O-depleted ice caps build up during a glacial stage. Marine carbonates that precipitate in the ocean during a glacial stage, particularly biogenic carbonates such as foraminifers, will be enriched in ^{18}O relative to those that precipitate during times when the climate is warmer and ice caps are absent, or are much smaller, on land. Changes in the $\delta^{18}O$ content of biogenic marine calcite thus reflect changes in the volumes of ice on land.

Decrease in temperature of the seawater in which biogenic calcite precipitates also causes an increase in the $\delta^{18}O$ values that are built into the calcite. Thus, during glacial periods both decrease in temperature of ocean water and changes in isotopic composition of ocean water owing to buildup of ice caps on the continents combine to cause an increase in the $\delta^{18}O$ content of biogenic calcites. Conversely, melting of polar ice caps, with consequent return of light-oxygen water to the oceans, and increase in ocean temperature will be reflected in a decrease in $\delta^{18}O$ values in marine biogenic carbonates.

Different kinds of marine organisms tend to incorporate somewhat different ratios of oxygen isotopes into their shells (fractionate oxygen isotopes to different degrees), as indicated in Figure 18.12. Therefore, to evaluate changes in oxygen isotopes in the ocean as a function of time requires that we analyze the same kind of fossil organism in rocks of different ages. Planktonic foraminifers are the most common fossil used in oxygen isotope studies of this kind.

Study of oxygen isotopes in sediment cores of late Tertiary to Quaternary age shows the presence of numerous $\delta^{18}O$ maxima and minima. Figure 18.13 shows an

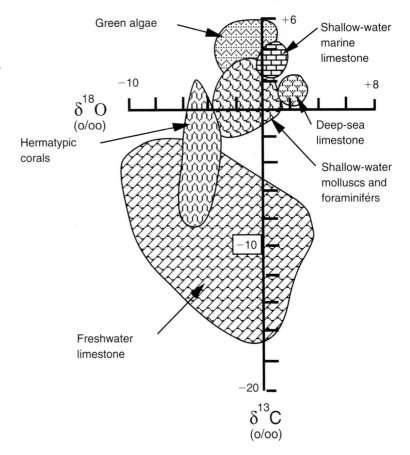

FIGURE 18.12 Distribution of $\delta^{18}O$ and $\delta^{13}C$ values in various types of marine carbonates. (After Milliman, J. D., 1974, Marine carbonates. Fig. 19, p. 33, reprinted by permission of Springer-Verlag, Heidelberg.)

Green algae

Shallow-water marine limestone

$\delta^{18}O$ (o/oo)

−10

+6

+8

Hermatypic corals

Deep-sea limestone

Shallow-water molluscs and foraminifers

−10

Freshwater limestone

−20

$\delta^{13}C$ (o/oo)

example of oxygen isotope values in three Pleistocene cores from widespread localities in the Pacific and Atlantic oceans (Wei, 1993). The numbers on the isotope curves are oxygen isotope stage numbers (e.g., Kennett, 1982; Ruddiman et al., 1989; Raymo et al., 1989). Where available, the magnetostratigraphic chrons are shown also. Once the isotope stages in a core have been identified and numbered, they can be correlated to the same isotope stages in other cores across the world ocean. These isotopic events appear to be related to the Milankovich orbital-climate cycles discussed in Chapter 15. Thus, they provide a record of fourth- and fifth-order cycles that are presumably driven by changes in climate related to changes in Earth's orbital parameters. These orbital changes cause fluctuations in the intensity of solar radiation reaching Earth at different latitudes, which, in turn, result in alternate accumulation and melting of continental ice sheets, producing fall and rise of sea level.

Carbon Isotopes. Carbon-12 and carbon-13 are the nonradioactive isotopes of carbon. Carbon-12 is much more abundant than carbon-13 and makes up most of the carbon in

FIGURE 18.13 (Opposite) Oxygen-isotope stratigraphy of Pleistocene cores from DSDP sites in the Pacific and Atlantic oceans. Note the numbered isotope stages, which can be correlated from one core to another. M/B refers to the Matuyama/Brunhes polarity chrons. (After Wei, W., 1993, Calibration of Upper Pliocene–Lower Pleistocene nannofossil events with oxygen isotope stratigraphy: Paleooceanography, v. 8. Fig. 4, p. 91, published by the American Geophysical Union.)

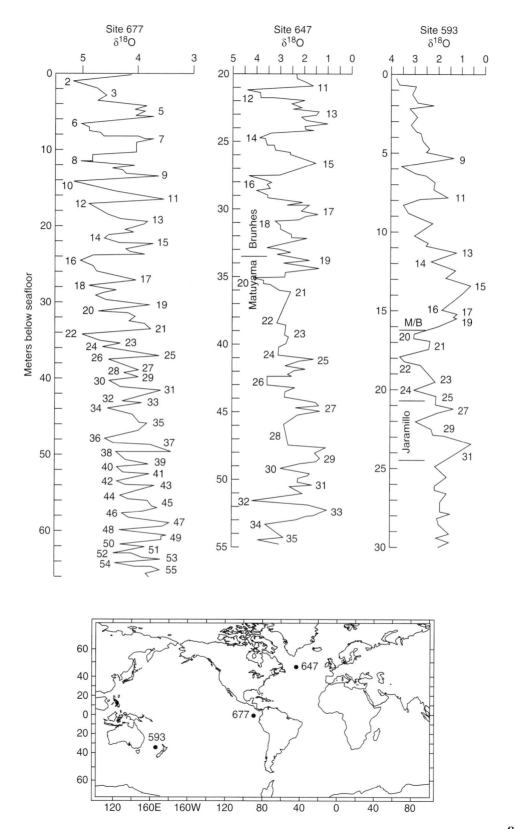

seawater (Table 18.6). The isotopic $^{13}C/^{12}C$ ratio can be expressed in terms of per mil deviation ($\delta^{13}C$) from the PDB belemnite standard, just as oxygen isotope ratios are expressed. The $\delta^{13}C$ values in marine carbonates reflect the $^{13}C/^{12}C$ ratio of CO_2 dissolved in deep ocean water; this ratio is, in turn, a reflection of the source of carbon in the CO_2. Carbon dioxide dissolves in the ocean by interchange with the atmosphere, and it is generated also by the decay of organic matter that originates both in the ocean and on land. Organisms preferentially incorporate light carbon (^{12}C); therefore, carbon dioxide derived from decaying organic matter is sharply depleted of ^{13}C compared to that derived from the atmosphere. Thus, water runoff from the continents (where soil organic matter is abundant) brings organic-rich waters with low $^{13}C/^{12}C$ ratios into the ocean, significantly lowering the $\delta^{13}C$ content of surface ocean waters near the continents. (Note from Fig. 18.12 the low $\delta^{13}C$ values in freshwater carbonates deposited in lakes.)

Another factor that influences the $\delta^{13}C$ content of ocean water, and thus the $\delta^{13}C$ content in the shells of marine organisms that live in these waters, is the residence time of deep-water masses in the ocean. Carbon-13 is depleted in deep-water masses that have long residence times near the ocean bottom, owing to oxidation of low $\delta^{13}C$ marine organic matter that sinks from the surface. Oxidation of this low $\delta^{13}C$ organic matter leads to production of low $\delta^{13}C$ dissolved bicarbonate (HCO_3^-). Respiration by bottom-dwelling organisms also apparently causes a decrease in $\delta^{13}C$ of deep bottom waters (Kennett, 1982). If low $\delta^{13}C$ bottom water is later circulated to the surface in some manner, carbonate-secreting organisms will built this low $\delta^{13}C$ isotope ratio into their shells.

Because the $\delta^{13}C$ in the calcareous shells of marine organisms is a function of the $\delta^{13}C$ content of the waters in which they live, changes in the $\delta^{13}C$ content of fossil marine organisms indicate changes in ocean water masses. Abrupt decreases in the $\delta^{13}C$ in fossil marine calcareous organisms may reflect changes in deep ocean paleocirculation and upwelling patterns that caused low $\delta^{13}C$ deep waters to spread upward and outward into other parts of the ocean. Or such decrease may reflect changes in surface circulation patterns that brought low $\delta^{13}C$ surface ocean waters from continental margins into deeper basins. Significant increases in the total biomass produced on the continents during any particular geologic time interval could cause an increase in runoff of low $\delta^{13}C$ to the oceans, an increase that may be reflected also as an episode of low $\delta^{13}C$ surface water. Increased rates of erosion of organic-rich sediments, such as dark shales and limestones, on land could produce much the same effect of increasing runoff of low $\delta^{13}C$ to the ocean. These abrupt changes in circulation patterns or organic carbon runoff from the continents may have affected the area of a single ocean basin, such as the Pacific, or in some cases the entire ocean. On the other hand, increased rates of sediment burial in the ocean may have the opposite effect of removing fine organic matter containing low $\delta^{13}C$ from interaction with seawater. This would have the effect of increasing $\delta^{13}C$ in ocean water. Vincent and Berger (1985) suggest, for example, that both carbonate $\delta^{13}C$ values and burial of organic carbon in continental margin sediments are related and that these were accomplished by lowered levels of atmospheric CO_2 and consequent climatic cooling.

Inasmuch as changes in the $\delta^{13}C$ content of the ocean are reflected in the $\delta^{13}C$ content of marine calcareous organisms, these isotopic "excursions" can be used for correlation. For example, a distinct decrease in $\delta^{13}C$, of apparent global extent, has been detected in marine sediments of Miocene age at about 5.9 to 6.2 Ma, probably reflecting some change in deep ocean circulation and upwelling patterns (Kennett, 1982, p. 87). Other events that reflect a marked decrease in $\delta^{13}C$ of ocean water are recorded near the Eocene–Oligocene boundary and the Paleocene–Eocene boundary. A similar decrease near the Cretaceous–Tertiary boundary apparently reflects an injection of freshwater from the Arctic (Odin, Renard, and Grazzini, 1982). On the other hand, several increases in $\delta^{13}C$ took place in the Middle Miocene between about 12 and 18 Ma (Fig. 18.14) that appear to be associated with exceptionally rapid burial of organic carbon and lowered levels of atmospheric CO_2 (e.g., Woodruff and Savin, 1991). All of these carbon isotopic excursions are essentially synchronous events that can be correlated over wide areas of the ocean in DSDP and ODP cores.

Sulfur Isotopes. Sulfur has four stable isotopes (Table 18.6); sulfur-32 is the most abundant, followed by sulfur-34. The $^{34}S/^{32}S$ ratio is used in most stratigraphic studies involving sulfur isotopes and is expressed in terms of $\delta^{34}S$, which is per mil deviation of the $^{34}S/^{32}S$ ratio relative to a meteorite standard—troilite (an FeS mineral) from the Canyon Diablo meteorite. Figure 18.15 shows the $\delta^{34}S$ values in various materials relative to the Canyon Diablo meteorite.

The major means of sulfur isotope fractionation in the oceans is by bacterial reduction of sulfate (SO_4^{2-}) in seawater to sulfides (H_2S, HS^-, HSO_4^-). Bacterial reduction of dissolved seawater sulfate at the sediment-seawater interface causes isotopic fractionation of the sulfate, resulting in enrichment of the remaining seawater sulfate in $\delta^{34}S$ by about +20o/oo and depletion in the reduced sulfide by about −9o/oo (Schopf, 1980). Precipitation of evaporites from dissolved marine sulfates introduces an additional fractionation (~+1.65o/oo), causing the $\delta^{34}S$ of evaporites to be higher than that of dissolved marine sulfates. Other minor factors that can influence the $\delta^{34}S$ content of seawater include oxidation of bacterial H_2S, producing sulfates depleted of ^{34}S relative to original sulfates, and local emanations of sulfate or sulfide through volcanic activity.

Marine sulfates in the present ocean have a mean $\delta^{34}S$ of about +21o/oo; however, the $\delta^{34}S$ of ancient marine evaporites ranges from about +10 to +30 (Fig. 18.15). On the basis of sulfur isotope ratios in ancient evaporite deposits, it appears that the sulfur isotope ratios in the surface waters of the world ocean have undergone major changes, or excursions, at various times (Fig. 18.16). These major excursions, called "catastrophic chemical events" by Holser (1977), are characterized by sharp rises in $\delta^{34}S$ in the surface waters of the world ocean and by greater "overshoots" locally. Three major events of sharply increased $\delta^{34}S$ have been recognized and named for the evaporite formations in which the $\delta^{34}S$ increase is most strongly manifested: the **Yudomski event** in very late Precambrian time (about 635 Ma), the **Souris event** in the Late Devonian (approximately 370 Ma), and the **Rot event** in Early to Middle Triassic (approximately 240 Ma). In addition to these excursions of increased $\delta^{34}S$, reverse events occurred also during the Late Permian and Late Paleogene (Fig. 18.16). Many of these sulfur isotope excursions appear to have affected the ocean worldwide.

Chemical events characterized by sharply increased $\delta^{34}S$ may be caused by catastrophic mixing of deep ^{34}S-rich brines with surface waters. Brines generated by evaporite deposition are stored in deep basins. Underneath the brines, bacterial reduction of sulfates to form pyrite builds up a store of brine heavy in ^{34}S sulfate. Catastrophic mixing of these ^{34}S-rich brines with surface waters, owing to destruction of the storage basin by tectonism, causes a sharp rise in the $\delta^{34}S$ of surface ocean waters and consequently in the evaporite deposits formed from these surface waters (e.g., Holser, 1977). Gradual decrease in the $\delta^{34}S$ of surface ocean waters with time after a catastrophic event is attributed to on-land erosion of dominantly sulfide materials into the ocean in an amount that exceeds evaporite deposition (Holser, 1977; Claypool et al., 1980). The

FIGURE 18.14 $\delta^{13}C$ values (upper dashed line) in cores from three DSDP sites in the Pacific and Atlantic oceans. The numbers show specific carbon isotope events that have been identified in each core, which can be correlated from core to core. Several oxygen isotope events, identified by letters in the lower solid line, are shown also. Ages of the carbon and oxygen isotope events can be determined from the age scale at the bottom of the diagram. (After Woodward, F., and S. M. Savin, 1991, Mid-Miocene isotope stratigraphy in the deep sea: High resolution correlations, paleoclimate cycles, and sediment preservation: Paleooceanography, v. 6. Fig. 13, p. 772, published by the American Geophysical Union.)

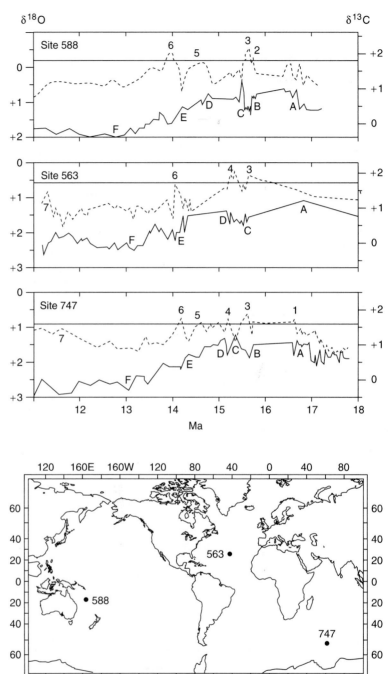

FIGURE 18.15 δ^{34}S values of marine sulfates shown relative to that of the Canyon Diablo meteorite. Values for seawater sulfate, rainwater sulfate, and sedimentary sulfides are shown for comparison. (Data from Degens, 1965.)

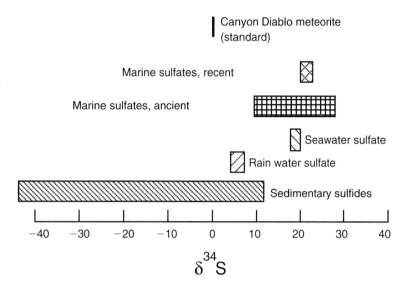

cause of sharp reverse δ^{34}S events, such as the Late Paleogene event shown in Figure 18.16, is not fully understood.

In any case, these sulfur isotope excursions constitute a sulfur isotope age curve because each catastrophic chemical event occurred within a very short interval of geologic time. Each major event thus represents a synchronous stratigraphic marker that can be correlated in evaporite deposits from one area to another. Some of the events can be correlated on a global basis. Thus, they provide an important method for international chronostratigraphic correlation of evaporite deposits, which commonly cannot be correlated by other means because they do not contain fossils or other datable materials.

Strontium Isotopes. The principal isotopes of strontium are ^{87}Sr and ^{86}Sr. The ratio of ^{87}Sr/^{86}Sr in the oceans is controlled by the chemical erosion of the following (Faure, 1986):

1. young volcanic rocks on continents and in the ocean basins
2. Precambrian sialic rocks of the continental crust
3. marine carbonate rocks of Phanerozoic age

Because the ^{87}Sr/^{86}Sr ratio in each of these source rocks is different, upon weathering they furnish different ratios of ^{87}Sr/^{86}Sr to the ocean. Weathering of marine carbonates is a particularly important source of strontium. The strontium isotope ratio of ocean water is constant throughout the ocean at any given time but has varied through time owing to variations in strontium contributed by weathering processes. Strontium is removed from the oceans by coprecipitation with calcium in calcium carbonate minerals. Therefore, analysis of marine carbonates of various ages allows the strontium isotope composition of the ocean through time to be determined.

Figure 18.17 shows variations in the ^{87}Sr/^{86}Sr ratio of Phanerozoic-age marine carbonates, reflecting variations in these isotope ratios in ocean water since Precambrian time owing to changing proportions of strontium contributed to the ocean from different sources. Note a general decrease in ^{87}Sr/^{86}Sr ratios of the ocean from Precam-

FIGURE 18.16 Sulfur isotopic ratios in sulfates collected from Phanerozoic formations. Dotted figures are for European samples, half-black figures are for North American samples, and open figures are for samples from other parts of the world. (From Odin, G. S., M. Renard, and C. V. Grazzini, Geochemical events as a means of correlation, *in* G. S. Odin (ed.), Numerical dating in stratigraphy. Fig. 11, p. 59, © 1982, John Wiley & Sons, Ltd. Reprinted by permission of John Wiley & Sons, Ltd., Chichester, England.)

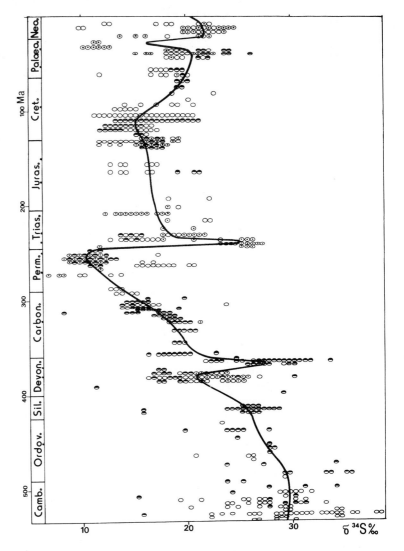

brian time until the Jurassic, with several pronounced excursions of lowered ratios (e.g., Ordovician, Devonian, Mississippian, Permian). A general increase in the ratio since Jurassic time is evident from Figure 18.17.

The major $^{87}Sr/^{86}Sr$ excursions in the Paleozoic and Mesozoic shown in Figure 18.17 involve too much time to be of much value in chronocorrelation of small-scale stratigraphic units, although the $^{87}Sr/^{86}Sr$ curve can be used as a geochronometer for estimating the ages of marine sedimentary rocks (Faure, 1982). On the other hand, correlation by strontium isotopes has been applied with very good results to some Tertiary formations. For example, Depaola and Finger (1988) use strontium isotopes to correlate marine sediments of the Miocene Monterey Formation of California at resolutions comparable to those of biostratigraphic methods. By correlating the $^{87}Sr/^{86}Sr$ ratios in these sediments with the strontium isotope vs. time curve for seawater derived from DSDP coreholes in the southwestern and central Pacific, these authors determined ages with resolutions ranging from less than 0.1 to 2.5 Ma.

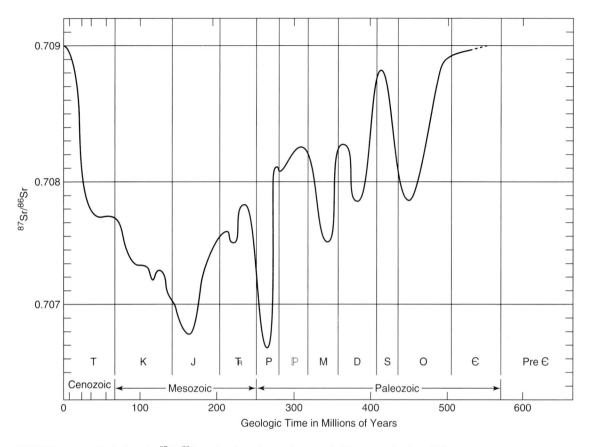

FIGURE 18.17 Variations in $^{87}Sr/^{86}Sr$ ratio of marine carbonates in Phanerozoic time. (After Faure, G., Principles of isotope geology, 2nd ed., Fig. 11.2, p. 188, © 1986, John Wiley & Sons. Reprinted by permission of John Wiley & Sons.)

Problems with Isotopic Chronocorrelation

The field of stable isotope geochemistry is very complex; many questions remain with regard to the validity of observed oxygen, carbon, sulfur, and strontium isotope excursions in the geologic record and use of these excursions in chronocorrelation. A discussion of these problems and the details of correlation by stable isotopes is outside the scope of this book. Additional information on this subject is available in numerous publications such as Claypool et al. (1980); Veizer, Holser, and Wilgus (1980); Arthur et al. (1983); Holland (1984); Holland and Trendall (1984); Holser (1984); Faure (1986); Hoefs (1987); and Williams, Lerche, and Full (1988).

Correlation by Absolute Ages

As discussed, radiometric dating of igneous and sedimentary rocks provides a basis for detailed calibration of the geologic time scale. In spite of the usefulness of radiochronology for that purpose, radiometrically determined age data are still not widely used for chronostratigraphic correlation of sedimentary rocks. We might logically assume that absolute-age data should constitute the primary basis for chronostratigraphic correlation; however, many sedimentary rocks do not contain datable materials. Furthermore, the precision of radiometric ages obtained from sedimentary rocks that can be dated is often not good enough for correlation purposes.

Carbon-14 and uranium-238 (^{230}Th/^{231}Pa) disequilibrium ages can be used for correlating very young rocks; however, geologists have been less successful in correlating older sedimentary rocks by absolute age owing to uncertainties in the direct-dating methods used for older rocks. Figure 18.7 shows, for example, that uncertainties in glauconite ages can range from thousands of years to as much as a million years. An error of a million years is statistically small in the dating of rocks hundreds of millions of years old, but it is nonetheless a large error for correlation purposes.

On the other hand, radiometric ages provide virtually the only available tool for correlating most Precambrian rocks. Lithologic and structural complexity and lack of fossils in these rocks preclude their correlation by most other methods. Absolute age data are also a valuable asset in stratigraphic studies of volcanic rocks and have proven particularly useful in correlating Cenozoic volcanic successions.

FURTHER READINGS

Arthur, M. A., T. F. Anderson, I. R. Kaplan, J. Veizer, and L. S. Land, 1983, Stable isotopes in sedimentary geology: Soc. Econ. Paleontologists and Mineralogists Short Course No. 10, Tulsa, Okla.

Bowen, R., 1988, Isotopes in the earth sciences: Elsevier, London, 647 p.

Clifton, H. E. (ed.), 1988, Sedimentological consequences of convulsive geologic events: Geol. Soc. America Spec. Paper 229, 157 p.

Easterbrook, D. J., (ed.), 1988, Dating Quaternary sediments: Geol. Soc. America Spec. Paper 227, 165 p.

Geyh, M. A., and H. Schleicher, 1990, Absolute age determination: Springer-Verlag, Berlin, 503 p.

Harland, W. B., R. L. Armstrong, A. V. Cox, L. E. Craig, A. G. Smith, and D. G. Smith, 1990, A geologic time scale 1989, 2nd ed.: Cambridge University Press, Cambridge, 263 p.

Hedberg, H. D. (ed.), 1976, International Stratigraphic Guide: A Guide to Stratigraphic Classification, Terminology, and Procedure: International Subcommission on Stratigraphic Classification of IUGS Commission on Stratigraphy: John Wiley & Sons, New York, 200 p.

Mahaney, W. C. (ed.), 1984, Quaternary dating methods: Developments in paleontology and stratigraphy, v. 7. Elsevier, Amsterdam, 431 p.

North American Stratigraphic Commission on Stratigraphic Nomenclature, 1983, North American Stratigraphic Code: Am. Assoc. Petroleum Geologists Bull., v. 67, p. 841–875.

Odin, G. S. (ed.), 1982, Numerical dating in stratigraphy: John Wiley & Sons, New York, Parts I and II, 1040 p.

Snelling, N. J. (ed.), 1985, The chronology of the geological record: Geol. Soc. Mem. 10, Blackwell, Oxford, 343 p.

Williams, D. F., I. Lerche, and W. E. Full, 1988, Isotope chronostratigraphy: Theory and methods: Academic Press, San Diego, 345 p.

19

Basin Analysis, Tectonics, and Sedimentation

19.1 INTRODUCTION

Preceding chapters of this book introduce the fundamental principles of sedimentology and stratigraphy. I have reviewed in those chapters the basic processes that generate sedimentary rocks; discussed the physical, chemical, and biological properties of these rocks; and described the various environments in which they accumulate. Further, I have emphasized those properties of sedimentary rocks that can be used to subdivide strata into meaningful stratigraphic units on the basis of lithology, seismic reflection characteristics, magnetic polarity, fossil content, and age.

Geologists study sedimentary rocks to develop a critical understanding of their geologic history or to evaluate their economic potential. Effective study requires utilization of all the sedimentological and stratigraphic principles discussed in this book. Because most sedimentary rocks were deposited in basins, it is common to refer to such detailed, integrated study as **basin analysis.** The concept of basin analysis, which goes back to the 1940s, is attributed to one of geology's most famous sedimentologists, Francis Pettijohn (Kleinspehn and Paola, 1988). Numerous research papers and several books that describe the principles of basin analysis have appeared in print since that time, particularly during the late 1970s to 1990s. A look at some of this literature reveals rather different points of view about what constitutes basin analysis. For example, Potter and Pettijohn (1977, p. 1) state that "consideration of the sedimentary basin as a whole provides a truly unified approach to the study of sediments." Miall (1990) suggests that basin analysis is the study of strata that play host to valuable resources of fuels and metals and that its most important product is the documentation of the subsidence history and paleogeographic evolution of a sedimentary basin. Allen and Allen (1990) refer to basin analysis as the integrated study of sedimentary basins as geodynamical entities, and they suggest that sedimentary basins contain in their geometry, tectonic evolution, and stratigraphic history valuable clues to the way in which the lithosphere deforms. Klein (1987) remarks that current research in basin analysis focuses on interdisciplinary integration of geodynamics, mathematical and physical modeling, and sedimentary geology. The common theme that emerges from all of these points of view is that basin analysis is an integrated program of study that involves application of sedimentologic, stratigraphic, and tectonic principles to develop a full understanding of the rocks that fill sedimentary basins for the purpose of interpreting their geologic history and evaluating their economic importance.

During the earlier years of geologic study from about 1860 to 1960, most geologists regarded marine basins as mainly linear troughs, called **geosynclines,** in which great thicknesses of predominantly shallow-marine deposits accumulated owing to continued subsidence of the geosynclines. With the emergence of the plate tectonics concept in the late 1950s and early 1960s, geological thinking shifted away from geosynclines. Today, geologists recognize that there are many kinds of basins and many different mechanisms by which basins form. Under the general rubric of basin analysis, geologists are concerned about the global tectonic controls that create basins and the geologic controls (e.g., sea-level changes, sediment supply, basin subsidence) that govern basin filling. Some of these factors are discussed in preceding chapters.

In this final chapter, attention is focused particularly on the different kinds of basins that we now recognize, the processes that form these basins, the processes that bring about filling of basins and the nature of the fills, and the relationship of tectonics (including source-area characteristics) to sedimentation patterns. This chapter rounds out the text coverage by providing an overall perspective regarding the integrated nature of sedimentologic, environmental, and stratigraphic analysis.

19.2   BASIN CLASSIFICATION AND ORIGIN

We now recognize that the origin of sedimentary basins is related in some way to crustal movements and plate tectonics processes. Sedimentary basins are commonly classified in terms of the following (Dickinson, 1974; Miall, 1990, p. 501):

1. the type of crust on which the basins rest
2. the position of the basins with respect to plate margins
3. for basins lying close to a plate margin, the type of plate interactions occurring during sedimentation

Several tectonic classifications for basins have been proposed (e.g., Bally and Snelson, 1980; Kingston, Dishroon, and Williams, 1983; Mitchell and Reading, 1986; Klein, 1987; Ingersoll, 1988). The classification of Mitchell and Reading (1986), used here, is straightforward and reasonably representative of the various classifications. Mitchell and Reading, suggest that we group tectonic settings for sediment accumulation into six basic types, listed in Table 19.1 and illustrated diagrammatically in Figure 19.1.

Interior Basins, Intracontinental Rifts, and Aulacogens

Interior basins (Fig. 19.1A) are relatively large, commonly ovate downwarps within the interiors of more or less stable cratonic shields. Some interior basins are filled with marine siliciclastic, carbonate, or evaporite sediment deposited from epicontinental seas; others contain nonmarine sediments. Various mechanisms have been suggested to form these basins:

TABLE 19.1 Principal kinds of depositional basins

Interior basins, intracontinental rifts, and aulacogens (including thermally related rifts and collision-related rifts) **Passive or rifted continental margins** (Atlantic-type margins comprising a shelf, slope, and rise) **Oceanic basins and rises** (the deep ocean floor and smaller basins associated with mid-ocean spreading ridges and rises, but excluding transform fault–related settings) **Subduction-related settings** (convergent plate boundaries characterized by a trench, arc-trench gap, and volcanic arc) **Strike-slip/transform fault–related settings** **Collision-related settings** (resulting from closure of an oceanic or marginal basin)

Source: Mitchell and Reading (1986).

1. subsidence owing to the presence of ancient underlying rift systems
2. cooling and subsidence following a thermal event, such as intrusion of dense material in the mantle
3. mantle phase changes
4. mantle hot spots
5. shallow subduction

The Michigan Basin in the upper midwestern part of the United States is an example of a nearly ovate interior basin (Fig. 19.2). The history of the Michigan Basin began in late Precambrian time with rifting across the mid-continent region, forming an initial rift basin. Rifting was followed by thermal subsidence, leading to initiation of a sag basin in Late Cambrian time. Major basin subsidence began in Early Ordovician time, creating the present basin configuration by the end of Early Ordovician time. Additional structural deformation occurred in Late Mississippian and Pennsylvanian time. The basin is filled with about 4800 m of mostly marine sedimentary rock (carbonates, sandstones, shale, evaporites, coal), ranging in age from Cambrian to Pennsylvanian (Catacosinos, Harrison, and Daniels, 1990). A small area of nonmarine Jurassic rocks is present in the center of the basin. Most of the basin is overlain by Quaternary glacial deposits. Other examples of interior basins are the Illinois and Williston basins in the United States, the Paris Basin in France, the Paraná Basin in Brazil, and the Carpentaria Basin in Australia (Leighton et al., 1990).

Rifts (Fig. 19.1B) are narrow, fault-bounded valleys that range in size from grabens a few kilometers wide to gigantic rifts such as the East African Rift System, which is nearly 3000 km long and 30 to 40 km wide. Rifts result from a thermal event that causes spreading and rifting. The East African Rift System (Fig. 19.3) is an example of a young rift zone. Different stages in the development of the rift are illustrated in Figure 19.3B. The East African Rift is filled mainly with volcanic rocks; however, a great variety of sedimentary environments can exist within rifts, ranging from nonmarine (fluvial, lacustrine, desert) to marginal marine (delta, estuarine, tidal flat) and marine (shelf, submarine fan). Thus, the deposits of rift basins can include conglomerates, sandstones, shales, turbidites, coals, evaporites, and carbonates.

Many ancient rift systems were associated with newly rifted continental plates, such as that developed between North America and Europe during Triassic time as a result of rifting accompanying development of the Mid-Atlantic Ridge. Many such rift systems eventually evolved into continental margin basins (discussed below).

Aulacogens are special kinds of rifts, often referred to as failed rifts, that are presumably the failed arms of a triple-point junction (Fig. 19.1C). One arm of a trilete

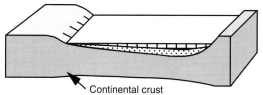

A. Interior basin

Continental crust

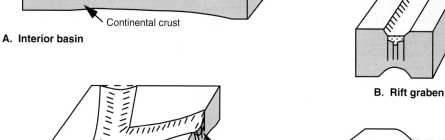

B. Rift graben

C. Aulacogen

Oceanic crust

Failed rift (Aulacogen)

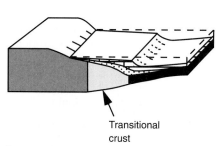

D. Passive-margin basin

Transitional crust

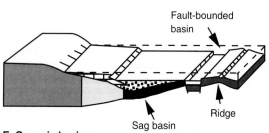

E. Oceanic basins

Fault-bounded basin

Sag basin

Ridge

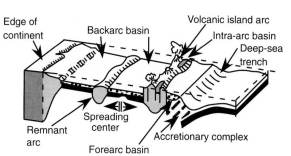

F. Subduction-related basin

Edge of continent

Backarc basin

Volcanic island arc

Intra-arc basin

Deep-sea trench

Remnant arc

Spreading center

Forearc basin

Accretionary complex

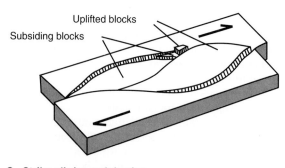

G. Strike-slip/wrench basins

Uplifted blocks

Subsiding blocks

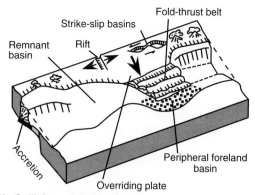

H. Collision-related basins

Fold-thrust belt

Strike-slip basins

Rift

Remnant basin

Accretion

Overriding plate

Peripheral foreland basin

FIGURE 19.1 (Opposite)   Schematic representation of major kinds of tectonically formed basins. (After Dickinson and Yarborough, 1976; Kingston, Dishroon, and Williams, 1983; Mitchell and Reading, 1986; and Einsele, 1992.)

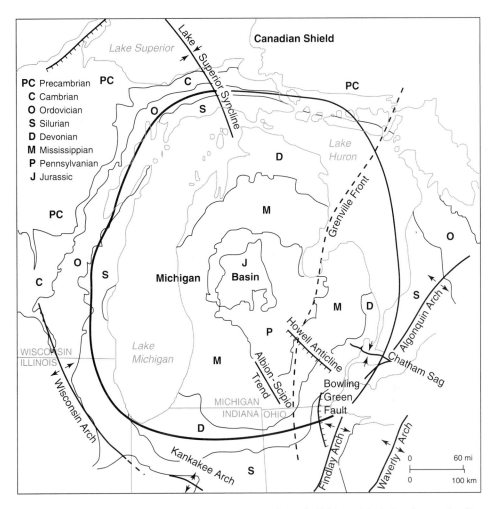

FIGURE 19.2   The Michigan Basin, showing structural trends (folds and faults) and ages of sedimentary rocks that crop out at the surface. The approximate limits of the basin are shown by the heavy solid line. (From Catacosinos, P. A., et al., 1990, Structure, stratigraphy and petroleum geology of the Michigan Basin, *in* M. W. Leighton et al. (eds.), Interior cratonic basins: Am. Assoc. Petroleum Geologists Mem. 51. Fig. 30.1, p. 562, reproduced by permission of AAPG, Tulsa, Okla.)

spreading rift system is believed to stop spreading after a few million years. The rifts making up the other two arms continue to spread, eventually causing separation of the continent and development of an ocean. The long, narrow troughs that make up the arms of aulacogens extend into continental cratons at a high angle from fold belts. Deposition of thick sequences of sediment can take place in aulacogens over long periods of time. These deposits may include nonmarine (e.g., alluvial-fan) sediments,

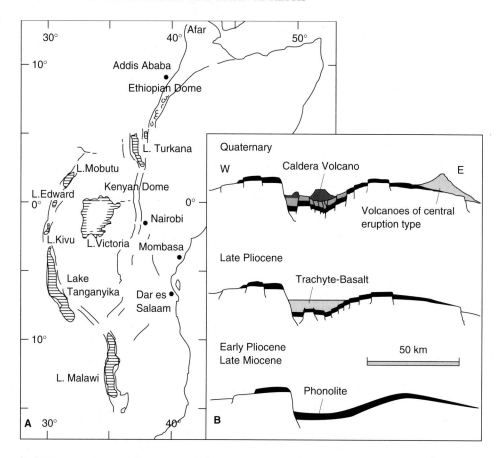

FIGURE 19.3 Map (A) showing the surface configuration of the East African Rift Zone and cross sections and (B) illustrating stages in evolution of the rift from late Miocene through the Quaternary. The rift is floored by volcanic rocks and volcaniclastic detritus. (From Einsele, G., 1992, Sedimentary basins. Fig. 12.4, p. 434, reprinted by permission of Springer-Verlag, Berlin.)

marine shelf deposits, and deeper-water facies such as turbidites. Examples of aulacogens include the Reelfoot Rift of late Paleozoic age in which the Mississippi River flows, the Amazon rift in which the Amazon River flows, the Benue Trough of Cretaceous age in which the Niger River is located, and the aulacogen north of the Black and Caspian seas on the Russian platform (Fig. 19.4).

Passive (Rifted) Continental Margins

Passive continental margins, also called Atlantic-type margins, are characterized by the presence of a sedimentary prism (Fig. 19.1D) and experience little or no seismic or volcanic activity. They are created as a result of rifting caused by spreading from mid-ocean ridges, but they lack the distinctive island arcs, trenches, and associated basins that characterize margins undergoing active subduction. Sediments accumulate in several parts of passive-margin settings—shelf, slope, and continental rise at the foot of the slope (see Fig. 12.16). Sediments deposited in this setting can include shallow neritic sands, muds, carbonates, and evaporites on the shelf; hemipelagic muds on the

FIGURE 19.4   Aulacogen north of the Black and Caspian seas on the Russian platform (After Burke, K., 1977, Aulacogens and continental breakup. Reproduced, with permission, from the Annual Review of Earth and Planetary Sciences, Volume 5, © 1977 by Annual Reviews, Inc.)

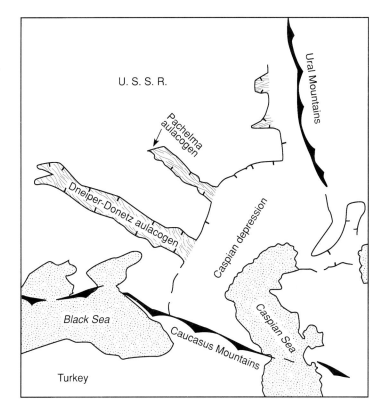

slope; and turbidites on continental rises. Thick prisms of sediment may accumulate owing to long-continued subsidence, which may be caused by deep crustal metamorphism (causing increase in density of lower crustal rocks), crustal stretching and thinning, and sediment loading.

As mentioned, rift basins may evolve into continental margin basins. A good example is the rift basins located off the eastern United States and the southeastern Canada coast (Blake Plateau Basin, Carolina Tough, Baltimore Canyon Trough, Georges Bank Basin, Nova Scotian Basin) which were created in late Triassic to early Jurassic time by rifting accompanying the breakup of Pangaea (Manspeizer, 1988). Some of these basins were isolated from the sea and accumulated thick deposits of arkosic clastic sediments and lacustrine deposits, intercalated with basic volcanic rocks. Others, with some connections to the sea, accumulated deposits ranging from evaporites to deltaic sediments, turbidites, and black shales. Figure 19.5 shows some of the sediments in the Baltimore Canyon Trough. Other examples of passive-margin basins include the Campos Basin, Brazil; the northwest shelf of Australia; and the sedimentary basins of Gabon on the west coast of Africa (Edwards and Santogrossi, 1990). Some passive-margin basins are prolific producers of petroleum and natural gas.

Oceanic Basins and Rises

Oceanic basins (Fig. 19.1E) occur in various parts of the deep ocean floor. These basins are created by rifting and subsidence accompanying opening of an ocean owing to con-

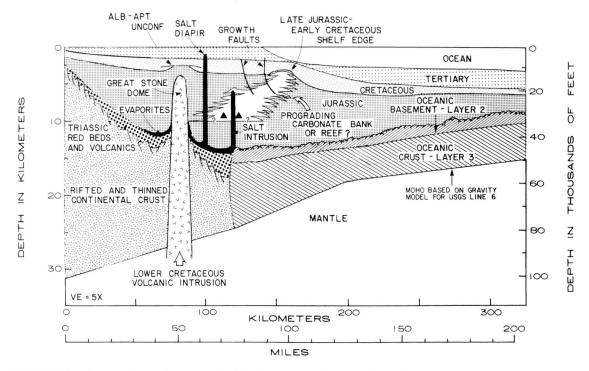

FIGURE 19.5 Interpretive section across the Atlantic continental margin of North America in the vicinity of Baltimore Canyon trough. Based on geophysical data and drill-hole information. (After Grow, J. A., 1981, Structure of the Atlantic margin of the United States, *in* Geology of passive continental margins: Am. Assoc. Petroleum Geologists Ed. Course Note Ser. No. 19. Fig. 13, p. 3–20, reprinted by permission of AAPG, Tulsa, Okla.)

tinental rifting. Oceanic basins may include ocean-floor sag basins as well as fault-bounded basins associated with ridge systems. Sediments that accumulate in these basins are mainly pelagic clays, biogenic oozes, and turbidites. Sediments deposited in oceanic basins adjacent to active margins may eventually be subducted into a trench and consumed during an episode of ocean closing. Alternatively, they may be off-scraped in trenches during subduction to become part of a subduction (accretionary) complex (Fig. 19.1F).

Subduction-related Settings

Subduction-related settings (Fig. 19.1F) are features of seismically active continental margins, such as the modern Pacific Ocean margin. These settings are characterized by a deep-sea trench, an active volcanic arc, and an arc-trench gap separating the two. The most important depositional sites in subduction-related settings are the deep-sea trench; the fore-arc basins that lie within the arc-trench gap; and the back-arc, or marginal, basins that lie behind the volcanic arc in some arc-trench systems (e.g., the modern Japan Sea back-arc basin; see Fig. 12.17). Subduction-related settings may occur also along a continental-margin arc (not shown in Fig. 19.1) rather than an oceanic arc. In these continental-margin settings, so-called retro-arc basins (intermontane basins within an arc orogen) may lie on continental crust behind fold-thrust belts. Sediments deposited in subduction-related basins are mainly siliciclastic deposits derived mainly from volcanic sources in the volcanic arc. These deposits include sands and muds deposited on the shelf and muds and turbidites deposited in deeper water in slope, basin, and trench settings. Sediments in the trench may include terrigenous deposits transported by turbidity currents from land, together with sediments scraped from a subducting oceanic plate—forming an accretionary complex (Fig. 19.1F).

Examples of modern trench–fore-arc sedimentation sites include the Sundra, Japan, and the Aleutian arc-trench systems (Leggett, 1982). A good example of an ancient fore-arc basin is the Great Valley fore-arc basin in California (Ingersoll, 1982). The Taranaka Basin, New Zealand, and the Magdalena Basin, Columbia, both petroleum producers, are additional examples of active-margin basins (Biddle, 1991).

Strike-Slip/Transform-Fault–related Settings

Strike-slip–related basins occur along ocean spreading ridges, along the transform boundaries between some major crustal plates, on continental margins, and within continents on continental crust. Movement along strike-slip faults can produce a variety of pull-apart basins, only one kind of which is illustrated in Figure 19.1G. Most basins formed by strike-slip faulting are small (a few tens of kilometers across) and may show evidence of significant local syndepositional relief, such as the presence of fault-flank conglomerate wedges. Because strike-slip basins occur in a variety of settings, they may be filled with either marine or nonmarine sediments, depending upon the setting. Sediments in many of these basins tend to be quite thick, owing to high sedimentation rates that result from rapid stripping of adjacent elevated highlands, and may be marked by numerous localized facies changes.

The Los Angeles Basin, California, is an example of a strike-slip basin formed in a transform-margin setting. It originated as a depositional basin in mid-Miocene to early Pliocene time owing to crustal extension associated with strike-slip deformation and was subsequently shortened by additional strike-slip movement (Biddle, 1991). The basin is characterized by the presence of several sediment-filled grabens bounded by major faults (Fig. 19.6). Largely Neogene-age sediment totaling more than 5000 m in thickness fills the basin (Blake, 1991). These sediments include conglomerates, sandstones, silty shale, diatomaceous and phosphatic shale, and volcaniclastic detritus, deposited in environments ranging from fluvial to shallow marine to deep marine. The Los Angeles Basin is a major petroleum-producing basin.

Collision-related Settings

Collision-related basins are formed as a result of closing of an ocean basin and consequent collision between continents or active arc systems, or both. Collision can generate compressional forces, resulting in development of fold-thrust belts and associated foreland basins along the collision suture belt. Collision may also give rise to strike-slip movement and the creation of strike-slip basins. Owing to the irregular shapes of continents and island arcs, and the fact that landmasses tend to approach each other obliquely during collision, portions of an old ocean basin may remain unclosed after collision occurs. These surviving embayments are called remnant basins. Figure 19.1H illustrates some of the basins that may be generated as a result of plate collision. Foreland basins may be isolated from the ocean and receive only nonmarine gravels, sands, and muds, or they may have an oceanic connection and contain carbonates, evaporites,

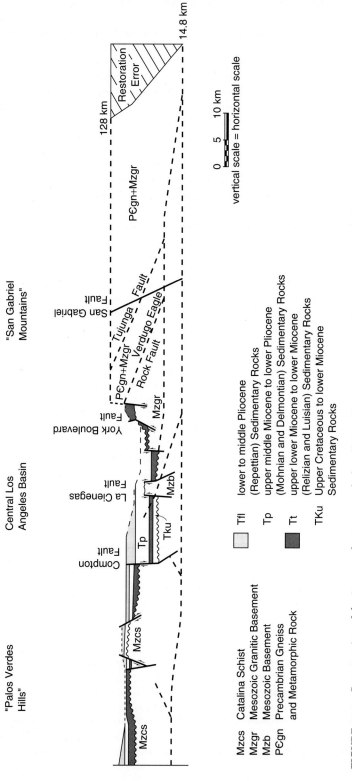

FIGURE 19.6 Cross section of the Los Angeles Basin with the effects of late-stage tectonic shortening removed. Note the presence of numerous faults that have created down-dropped blocks (grabens) to form the central basin. Restoration error refers to the amount of shortening that occurred owing to late-stage compressional forces and that is restored in the cross section. [From Biddle, K. T., 1991, The Los Angeles Basin—An overview, *in* K. T. Biddle (ed.), Active margin basins, Am. Assoc. Petroleum Geologists Mem. 52. Fig. 14, p. 20, reproduced by permission of AAPG, Tulsa, Okla.]

and/or turbidites. Remnant basins are characterized especially by turbidite sedimentation.

The Marathon Basin, Texas, provides an example of Pennsylvanian-age sedimentation in a remnant basin adjacent to a fore-arc basin (Fig. 19.7). Structural weaknesses developed in this region in the Late Precambrian/Early Cambrian and were reactivated in the late Paleozoic as reverse faults in response to compressional stresses (Wuellner, Lohtonen, and James, 1986). An early phase of sedimentation filled part of the fore-arc basin with volcaniclastic detritus. Subsequently, sediments of the Tesnus Formation accumulated in the fore-arc and remnant basin. Later deposition of the Dimple Limestone and Haymond Formation (not shown in Fig. 19.7) generated a total of more than 3400 m of Pennsylvanian sediment in the basin. Sediments include sandstones, shales, and limestones, deposited in environments ranging from shelf/platform to submarine fan (turbidite) settings.

Other examples of basins developed in collision-related settings include foreland basins of western Taiwan; foreland basins of the Alpennines and eastern Pyrenees; the Magallanes Basin at the southern tip of South America; basins in the northwestern Himalayas; and various basins in the Appalachians, Rocky Mountains, and western Canada. See Allen and Homewood (1986) and Macqueen and Leckie (1992) for details of these and other foreland basins.

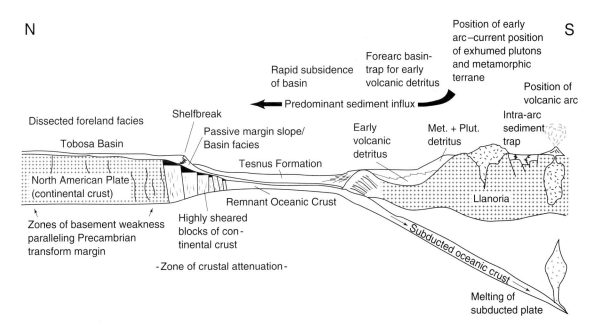

FIGURE 19.7 Cross-sectional diagram showing the remnant basin and associated basins that existed in the Marathon Basin, Texas, during deposition of Carboniferous sedimentary rocks. (From Wuellner, E. E., L. R. Lehtonen, and W. C. James, 1986, Sedimentary tectonic development of the Marathon and Val Verde basins, West Texas, U.S.A.: A Permo-Carboniferous migrating foredeep, *in* P. A. Allen and P. Homewood (eds.), Foreland basins: Internat. Assoc. Sedimentologists Spec. Pub. 8, Fig. 5, p. 354, reprinted by permission of Blackwell Scientific Publications, Oxford.)

19.3   SEDIMENTARY BASIN FILL

The preceding discussion focused on the structural characteristics of sedimentary basins and the tectonic processes that create these basins. The particular concern of basin analysis is, however, with the sediments that fill the basins. This concern encompasses the processes that produce the filling, the characteristics of the resulting sediments and sedimentary rocks, and the genetic and economic significance of these rocks. The fundamental processes that generate sediments (weathering/erosion) and bring about their transport and deposition; the physical, chemical, and biological properties of these rocks; the depositional environments in which they form; and their stratigraphic significance are discussed in preceding chapters of this book. The factors that control or affect these depositional processes and sediment characteristics, discussed also in appropriate parts of the book, include the following:

1. the lithology of the parent rocks (e.g., granite, metamorphic rocks) present in the sediment source area, which controls the composition of sediment derived from these rocks
2. the relief, slope, and climate of the source area, which control the rate of sediment denudation and thus the rate at which sediment is delivered to depositional basins
3. the rate of basin subsidence (owing to tectonism and sediment loading) together with rates of sea-level rise or fall

Some of the tectonic processes that may cause basin subsidence are discussed under "Interior Basins" above. The weight of sediments deposited in a basin (sediment loading) also affects basin subsidence. These two factors, together with sea level, control the available space in which sediments can accumulate (accommodation) as well as affecting sediment transport and deposition. Thus, owing to subsidence, thousands of kilometers of sediments may accumulate in shallow-water basins.

The purpose of basin analysis is to interpret basin fills to better understand sediment provenance (source), paleogeography, depositional environments, and geologic history and to evaluate the economic potential of basin sediments. As suggested by Klein (1987), interpretation of basin fills has evolved with increasing sophistication since the 1960s. Basin analysis incorporates the interpretive basis of sedimentology (sedimentary processes); stratigraphy (spatial and temporal relations of sedimentary rock bodies); facies and depositional systems (organized response of sedimentary products and processes into sequences and rock bodies of a contemporaneous or time-transgressive nature); paleooceanography, paleogeography, and paleoclimatology; sea-level analysis; and petrographic mineralogy (interpretation of sediment source). Further, biostratigraphy provides a means of establishing a temporal framework for correlating time-equivalent facies and systems and to constrain timing of specific events, and radiochronology also allows the dating of specific sedimentological events and stratigraphic boundaries. Recent research in sedimentary geology and basin analysis has focused particularly on analysis of sedimentary facies, cyclic subsidence events, changes in sea level, ocean circulation patterns, paleoclimates, and life history.

19.4   TECHNIQUES OF BASIN ANALYSIS

Analysis of the characteristics of sediments and sedimentary rocks that fill basins, and interpretation of these characteristics in terms of sediment and basin history, demands a variety of sedimentological and stratigraphic techniques. These techniques require the acquisition of data through outcrop studies and subsurface methods that can include deep drilling, magnetic polarity studies, and geophysical exploration. These data are then commonly displayed for study in the form of maps and stratigraphic cross sections, possibly using computer-assisted techniques. In this section, we take a brief look at the more common techniques of basin analysis.

Preparation of Stratigraphic Maps and Cross Sections

Structure-Contour Maps. It is often desirable in basin studies to determine the regional structural attitude of the rocks as well as the presence of local structural features such as anticlines and faults. Structure-contour maps are prepared for this purpose. These maps provide information about a basin's shape and orientation and the basin-fill geometry. Structure-contour maps are prepared by drawing lines on a map through points of equal elevation above or below some datum—commonly mean sea level. Elevations are typically determined on the top of a particular formation or key bed at a number of control points. Elevation data may be obtained through outcrop study and/or subsurface interpretation of mechanical or lithologic well logs. After control points are plotted on a map base, a suitable contour interval is selected and structure-contour lines are drawn by hand or by use of computers. Figure 19.8 shows an example of a structure-contour map.

Structure contours may also be prepared on the top of prominent subsurface reflectors from seismic data (Chapter 15). Depth to a particular reflector may be plotted initially as two-way travel time. Thus, the initial map shows contour lines of equal travel time. If the seismic-wave velocity can be determined from well information, the travel times can be converted to actual depths, allowing maps to be redrawn in terms of actual elevations on the reflecting horizon.

Structure-contour maps can reveal the locations of subbasins or depositional centers within a major basin as well as axes of uplift (anticlines, domes). Structural features may be related also to syndepositional topography. Thus, analysis of these maps can provide clues to local paleogeography and facies patterns. Structural maps are useful also in economic assessment (e.g., petroleum exploration) of basins.

FIGURE 19.8 Schematic illustration of a structure-contour map drawn on the top of a formation. The contour interval is 20 m. The negative contour values indicate that this formation is located below sea level and is thus a buried (subsurface) formation. Note the presence of a syncline, dome, anticline, and fault.

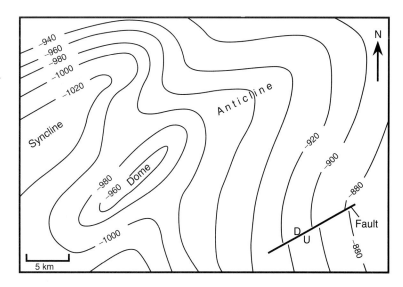

Isopach Maps. Isopachs are contour lines of equal thickness. An isopach map is a map that shows by means of contour lines the thickness of a given formation or rock unit. The thickness of sediment in a basin is determined by the rate of supply of sediment and the accommodation space in the basin, which in turn is a function of basin geometry and rate of basin subsidence. Abnormally thick parts of a stratigraphic unit suggest the presence of major depositional centers in a basin (basin lows), whereas abnormally thin parts of the unit suggest predepositional highs—or possibly areas of postdepositional erosion. Isopach maps thus provide information about the geometry of the basin immediately prior to and during sedimentation. Furthermore, analysis of a succession of isopach maps in a basin can provide information about changes in the structure of a basin through time.

To construct an isopach map, the thickness of a formation or other stratigraphic unit must be determined from outcrop measurements and/or subsurface well-log data at numerous control points. The thickness of the unit at each of these control points is entered on a base map, and the map is contoured in the same way that a structure-contour map is prepared. Figure 19.9 is an example.

Paleogeologic Maps. Paleogeologic maps are maps that display the areal geology either below or above a given stratigraphic unit. Imagine, for example, that we could strip off the rocks that make up a particular formation (and all the rocks above that formation) to reveal the rocks beneath—the rocks on which the formation was deposited. We could then construct a geologic map on top of these underlying formations. Such a map has been referred to as a **subcrop map** (Krumbein and Sloss, 1963). In a similar manner, the rocks above a formation or rock body may also be mapped. This kind of map, looked at as though from below, is called a **worm's eye view map** or supercrop map. Subcrop and worm's eye maps are commonly constructed at unconformity boundaries; however, they can be constructed at the top and base of any distinctive rock unit, whether or not an unconformity is present. The main purpose of such maps is to illustrate paleodrainage patterns, pattern of basin fill, shifting shorelines, or gradual burial of a preexisting erosional topography (Miall, 1990). To construct paleogeologic maps requires identification of the stratigraphic units that lie immediately below (or above) a given formation or other stratigraphic unit at numerous control points. Such data are gathered from outcrops or from subsurface well logs.

FIGURE 19.9   Example of an isopach map of a hypothetical formation drawn at a contour interval of 40 m. Note that the formation thickens to more than 240 m in the basin low (depositional center), thins over a basin high, and thins to zero along the northwest and north sides of the map.

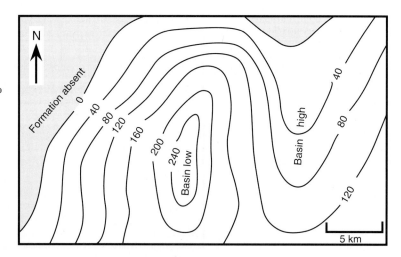

Lithofacies Maps. Facies maps depict variation in lithologic or biologic characteristics of a stratigraphic unit. The most common kinds of facies maps, and the only kind discussed here, are lithofacies maps, which depict some aspect of composition or texture. Maps based on faunal characteristics are biofacies maps. Many kinds of lithofacies maps are in use. Some are plotted as ratios of specific lithologic units (e.g., the ratio of siliciclastic to nonsiliciclastic components) or as isopachs of such units [e.g., sandstone isopach (or isolith) maps, limestone isopach (or isolith) maps]. Others examine the relative abundance and distribution of three end-member components (e.g., sandstone, shale, limestone). Two kinds of lithofacies maps are discussed here to illustrate the method: (1) clastic-ratio maps and (2) three-component lithofacies maps. Additional examples are given in Krumbein and Sloss (1963) and Miall (1990).

   Clastic-ratio maps are maps that show contours of equal clastic ratio, which is defined as the ratio of total cumulative thickness of siliciclastic deposits to the thickness of nonsiliciclastic deposits, for example,

$$\frac{(\text{conglomerate} + \text{sandstone} + \text{shale})}{(\text{limestone} + \text{dolomite} + \text{evaporite} + \text{coal})}$$

Values are computed for a number of control points, from outcrop or subsurface data, and plotted on a map. The map is then contoured in the same manner as that described for isopach maps. An example is shown in Figure 19.10. Clastic-ratio maps are useful for showing the relationship of lithologic units along the margin of a basin in which both siliciclastic and nonsiliciclastic deposits accumulated. Such maps provide limited information also about the location of the siliciclastic sediment source.

   Three-component lithofacies maps show by means of patterns or colors the relative abundance, within a formation or other stratigraphic unit, of three principal lithofacies components. Figure 19.11 shows an example of such a map based on the relative thickness of sandstone, shale, and limestone. A ternary diagram (inset in Fig. 19.11) is drawn using the three lithofacies components as end members. The triangle is subdivided into fields, each of which is indicated by a suitable pattern or color. The thickness of each end-member component is measured at as many control points in outcrop or in the subsurface as practical. The relative values (normalized to 100 percent if necessary) at each control point are plotted on the ternary diagram. The appropriate pat-

FIGURE 19.10 Example of a clastic-ratio (clastic/nonclastic) map. The progressive increase in the ratio from southeast to northwest across the map indicates a progressively increasing percentage of siliciclastic components in the stratigraphic section toward the northwest. Thus, the source of the sediments must have been located somewhere to the northwest. The small arrows show the probable direction of sediment transport.

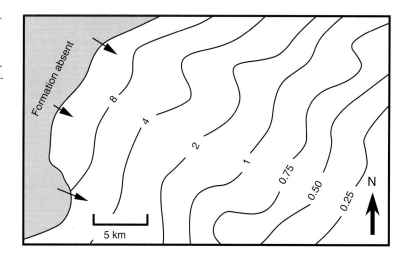

FIGURE 19.11 Hypothetical example of a three-component (sandstone, shale, limestone) lithofacies map. Siliciclastic sediments are predominant in the northwest portion of the map, whereas the stratigraphic section in the southeast part of the map is dominated by carbonate rocks. Note that not all lithofacies shown in the lithofacies triangle (e.g., sandstone) are actually present in the mapped area.

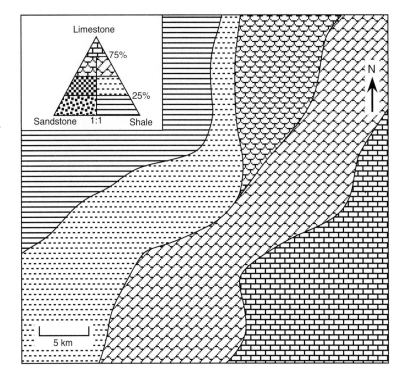

tern or color for that control point is then plotted on the map. When all points have been plotted with a pattern or color, lines are drawn on the map between different patterns or colors to show the geographic distribution of each subfacies present in the mapped area. If more than three lithofacies are present in the stratigraphic section, the additional lithofacies must be omitted (and the remaining three lithofacies normalized to 100 percent) or combined with other lithofacies to yield a total of three lithofacies. For example, if a conglomerate facies is present, it could be combined with the sandstone facies, or an evaporite facies might be combined with the limestone facies. Three-component lithofacies maps provide a convenient means of visualizing the relative importance of each lithofacies throughout a geographic area. Like clastic-ratio maps, however, they provide only a very rough guide to depositional environments and sediment-source locations.

Computer-generated Maps. All of the maps discussed above can be drawn by hand, and geologists have been drawing them this way for many years; however, construction of such maps is now being done more and more by computer. Computer application is particularly prevalent in the petroleum industry, where basin analysis is a commonplace procedure. Computers are able to handle large quantities of data, such as stratigraphic and structural data obtained from well records, and they allow these data to be easily manipulated for a variety of statistical and mapping purposes. Appropriate base maps are stored in the computer, and the locations of outcrop sections and subsurface wells can be easily plotted on these maps. Lithologic, structural, and stratigraphic data obtained from study of outcrop and subsurface sections are likewise stored in the computer. Selected data (e.g., thickness of lithologic units, structural elevations) can be retrieved as needed and added to the base maps, which can then be contoured by the computer by using appropriate software and special printers to draw the maps. Thus, any of the maps described above can be generated by computer, as well as other kinds of maps such as **trend-surface maps.** Trend-surface analysis allows separation of map data into two components: regional trends and local fluctuations. The regional trend is mathematically subtracted by the computer, leaving residuals, which correspond to local variations. For example, the regional structural trend might be extracted from a structure-contour map to more clearly reveal the nature of local structural anomalies. Computer-generated maps are not necessarily better or more accurate than hand-drawn maps. Their principal advantage is in the ease and rapidity with which they can be drawn. See Robinson (1982) and Jones, Hamilton, and Johnson (1986) for additional details.

Stratigraphic Cross Sections. Another important technique in basin analysis is preparation of stratigraphic cross sections. Stratigraphic cross sections are used extensively for correlation purposes, structural interpretation, and study of the details of facies changes that may have environmental or economic significance. Cross sections may be drawn to illustrate local features of a basin, often in conjunction with preparation of lithofacies maps, or they may depict major stratigraphic successions across an entire basin. The information needed to prepare stratigraphic cross sections is obtained by study of outcrop sections, subsurface lithologic logs, or mechanical well logs (petrophysical logs). Most stratigraphic cross sections depict in two dimensions the lithologic and/or structural characteristics of a particular stratigraphic unit, or units, across a given geographic region. Several examples of such cross sections are given in preceding chapters of this book. See, for example, Figures 14.6 and 14.21. Stratigraphic information may be presented also in a **fence diagram.** These diagrams attempt to give a three-dimensional view of the stratigraphy of an area or region (Fig. 19.12). Thus, they have the advantage of giving the reader a better regional perspective on the stratigraphic relationships. On the other hand, they are more difficult to construct than conventional two-dimensional diagrams, and parts of the section are hidden by the fences in front.

Paleocurrent Analysis and Paleocurrent Maps

Paleocurrent analysis is a technique used to determine the flow direction of ancient currents that transported sediment into and within a depositional basin, which reflects the local or regional paleoslope (see Chapter 5, Section 5.6). By inference, paleocurrent analysis also reveals the direction in which the sediment source area, or areas, lay. Further, it aids in understanding the geometry and trend of lithologic units and in interpretation of depositional environments. Paleocurrent analysis is accomplished by measuring the orientation of directional features such as sedimentary structures (e.g., flute casts, ripple marks, cross-beds) or the long-axis orientation of pebbles. Numerous orientation measurements must be made within a given stratigraphic unit to obtain a statistically reliable paleocurrent trend. Grain-size trends, lithologic characteristics, and sediment thickness may also have directional significance when mapped, as previously discussed. An example of a paleocurrent map, constructed mainly on the basis of cross-bedding orientation, is shown in Figure 19.13. Note that the average (statistical) paleoflow direction indicated by this map is from the northwest to southeast, suggesting that the sediment source area lay somewhere to the northwest. For more detailed discussion of the application of paleocurrent analysis to interpretation of sediment-dispersal patterns and basin-filling mechanisms, see Potter and Pettijohn (1977, Chapter 8).

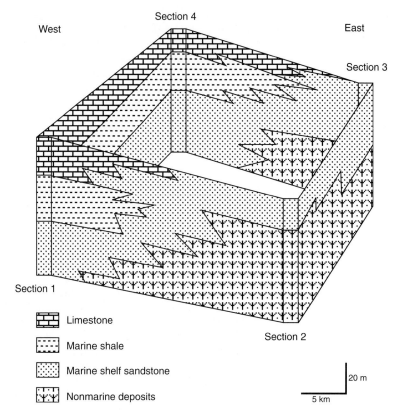

FIGURE 19.12   Schematic illustration of a fence diagram showing intertonguing facies relationships between marine and nonmarine deposits. Sections 1–4 are hypothetical measured stratigraphic sections.

Siliciclastic Petrofacies (Provenance) Studies

The composition of siliciclastic sediments that fill sedimentary basins is determined to a large extent by the lithology of the source rocks that furnished sediment to the basin, as well as by the climate and weathering conditions of the source area. Therefore, analysis of the particle composition of siliciclastic mineral assemblages (and rock fragments) provides a method of working backward to understand the nature of the source area. We commonly refer to such study as provenance study, where provenance is considered to include the following:

1. the lithology of the source rocks
2. the tectonic setting of the source area
3. the climate, relief, and slope of the source area

Provenance studies provide important information about the paleoclimatology and paleogeography of the basin setting.

The lithology of the source rocks is interpreted on the basis of the kinds of minerals and rock fragments present in siliciclastic sedimentary rocks, particularly in sandstones and conglomerates. For example, the presence of abundant alkali feldspars suggests derivation from granitic source rocks; abundant volcanic rock fragments suggest derivation from volcanic source rocks; and so on. The siliciclastic mineralogy also provides information about the tectonic setting (continental block, magmatic arc, colli-

FIGURE 19.13 Example of the use of paleocurrent data to locate source areas, Brandywine gravel of Maryland. The contours show modal grain size (in mm). (From Potter, P. E., and F. J. Pettijohn, 1977, Paleocurrents and basin analysis. Fig. 8–9, p. 282, reprinted by permission of Springer-Verlag, Berlin.)

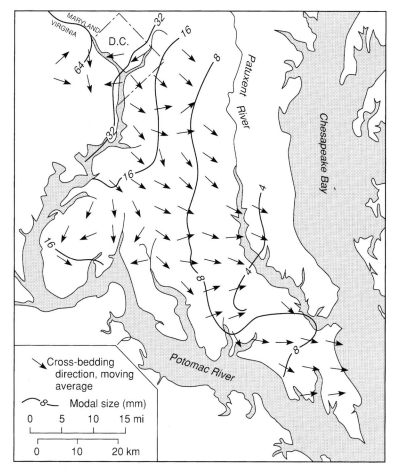

sion orogen)—sediment derived from continental block settings is likely to be enriched in quartz and alkali feldspars; sediment derived from a magmatic arc is dominated by volcanic rock fragments and plagioclase feldspars; and sediment stripped from highlands along orogenic collision belts is characterized particularly by an abundance of sedimentary and metasedimentary rocks fragments. Climate, relief, and slope of the source area are more difficult to interpret from siliciclastic mineral assemblages, but some clues are provided by quartz/feldspar ratios and by degree of weathering of feldspars. The essentials of provenance analysis are discussed in greater detail in Chapter 6, Section 6.5. Zuffa (1985) and Morton, Todd and Haughton (1991) provide a broader and more detailed view of provenance analysis.

Geophysical Studies

Geophysical investigations, including both seismic and paleomagnetic studies of various kinds, play an important role in basin analysis. **Seismic techniques** are used to document the regional structural trends and overall basin geometry as well as to identify local structural features such as anticlines and faults that may provide traps for hydrocarbons. These seismic techniques are described in Chapter 15, which also discusses the application of seismic methods to stratigraphic interpretation, including identification of seismic sequences and facies as well as correlation of stratal reflectors. The most widespread application of **paleomagnetism** in basin analysis is the study of magnetic polarity reversals as a tool for correlation and other purposes, as described in Chapter 16. Airborne or ground-based magnetic and gravity surveys are sometimes conducted also to provide information about basin structure in addition to that provided by seismic data. Gravity surveys determine the presence of gravity anomalies (abnormally high or low gravity values), which can be related to the presence of dense (e.g., basalt) or light (e.g., thick sediment) rock bodies in the subsurface. These magnetic and gravity surveys furnish data that can be used to construct maps showing large-scale characteristics of basins (e.g., major basin highs and lows); however, these maps commonly do not have sufficient resolution to identify small-scale structural features, that is, features such as small anticlines or faults that might constitute petroleum traps.

19.5   APPLICATIONS OF BASIN ANALYSIS

Interpreting Geologic History

As pointed out, basin analysis is an integrated program of study that involves application of sedimentologic, stratigraphic, and tectonic principles to develop a full understanding of the rocks that fill sedimentary basins and to use this information to interpret the geologic history and evaluate the economic importance of these rocks. Thus, one major goal of basin analysis is simply to develop a better understanding of Earth history as recorded in particular depositional basins. Through analysis of sedimentary textures, structures, particle and chemical composition, fossils, and the stratigraphic characteristics of sedimentary rocks (as revealed by physical, biological, paleomagnetic, and seismic-reflection characteristics), geologists are able to interpret the important tectonic and sedimentologic events that transpired to generate and fill a particular sedimentary basin. Thus, these various kinds of basin studies, commonly involving preparation of appropriate maps and stratigraphic sections, allow geologists to interpret past tectonic, climatic, and sedimentologic events and conditions (including source-area characterization and interpretation of depositional environments) to reconstruct the paleogeography and paleogeology of Earth during specific times in the past.

An example of the application of basin-analysis principles to interpretation of the geologic history of a particular region is provided by Clifton, Hunter, and Gardner (1988). These authors examined the influence of eustatic, tectonic, and sedimentological influences on the generation of transgressive and regressive depositional cycles in the Merced Formation near San Francisco, California. The Merced Formation consists of approximately 2000 m of shallow-marine and coastal nonmarine sediments of late Cenozoic age.

To work out the history of the Merced Formation, Clifton et al. first defined the stratigraphic boundaries (base and top) of the formation through field study and reference to previously published information. Ages of different parts of the formation were established by a combination of paleontologic data and radiometric dating of intercalated ash beds. Various depositional facies within the Merced were then characterized on the basis of contained fossils, physical and biogenic sedimentary structures, and textural characteristics (Table 19.2). Ten facies types were identified, which (from the base of the formation upward) were determined to represent ten different depositional

TABLE 19.2 Characteristics and inferred origin of sedimentary facies in sea-cliff exposures of the Merced Formation, San Francisco, California

Inferred origin	Texture	Biota	Physical structures	Biogenic structures
Outer shelf	Sandy silt	Molluscs (*in situ* and scattered shells)	None	Intense bioturbation
Mid-shelf	Sandy silt	Molluscs	Shell lags, parallel lamination, hummocky cross-stratification, ripple laminatiion	Sharp-topped bioturbated intervals, between sets of laminae
Inner shelf	Very fine sand, scattered small pebbles near top	Molluscs, echinoids	Shell lags, parallel lamination, hummocky cross-stratificaion	Locally intense bioturbation
Nearshore	Fine to coarse sand gravel	Molluscs, echinoids	Lenticular gravel and sand beds, high- and low-angle cross-bedding, parallel lamination	Vertical burrows, *Macaronichnus* (near top)
Foreshore	Fine to coarse sand, gravel fining-upward trend	Molluscs, echinoids	Parallel lamination, heavy mineral layers near top	*Macaronichnus* (near base)
Embayment	Mud, sand, gravel, some fining-upward cycles	Molluscs, ostracods	Fine sand-silt-clay lamination, ripples and ripple bedding (sand), cross-bedding (sand and gravel), shell lags	Bioturbation, root-rhizome structures
Backshore	Fine sand	None	Parallel lamination, low-angle cross-bedding, climbing adhesion ripple bedding, ripple lamination	Root-rhizome structures, vertebrate footprints, vertical tubes
Eolian dune	Fine sand	None	Medium- and large-scale and medium- and high-angle cross-bedding	Local mottling, vertical tubes
Alluvial	Gravel and pebbly sand	None	Indistinct stratification, lenticular bedding, medium- and small-scale trough cross-bedding	None
Pond/swamp/ marsh	Mud, peat, or lignite	Freshwater diatoms, insect wings, terrestrial vertebrate bones	Flat-bedded (mud)	Root structures, burrows and intrastratal trails

Source: Clifton, Hunter, and Gardner, 1988.

environments: pond/swamp marsh, alluvial, eolian dune, backshore, embayment, foreshore, nearshore, inner shelf, midshelf, and outer shelf.

In vertical succession, these facies define alternating episodes of transgression and regression that probably reflect Pleistocene eustatic sea-level fluctuations. The transgression-regression fluctuations were matched with reasonable confidence to a Pleistocene sea-level curve, determined from oxygen-isotope data. Finally, Clifton et al. were able to calculate that the Merced Formation was deposited under shallow-marine/coastal conditions in a setting undergoing subsidence at an average rate of 1 m/1000 years—a rapid rate that must have been influenced by tectonic deformation.

This example illustrates how geologists use an armory of different sedimentological and stratigraphic approaches (i.e., basin analysis) to determine the physical and biologic characteristics of a sedimentary formation, from which the sedimentological and tectonic history of the formation can be interpreted. Several additional examples of this approach may be examined in Kleinspehn and Paola (1988).

Economic Applications

A second goal of basin analysis is to utilize the principles and techniques described above to evaluate the economic importance of sedimentary rocks and identify economically exploitable deposits of minerals or fossil fuels. Basin analysis finds its greatest economic application in the fields of petroleum geology and, to a lesser extent, hydrogeology. In spite of the fact that petroleum geologists have been trying for many years to locate petroleum (hydrocarbon) accumulations by geochemical analysis of surficial rocks and soil overlying such deposits, no successful method has yet been developed for direct detection of hydrocarbon deposits. To find an oil or gas deposit, geologists must (1) carry out exploration in basins that have the right conditions for the formation and migration of hydrocarbons and (2) locate a suitable trap, such as a structural anticline, in which the hydrocarbons may have accumulated.

What, exactly, are the right conditions for formation and migration of hydrocarbons? Most petroleum geologists believe that petroleum originates from fine-size plant or animal organic matter through complex biochemical processes that take place during sediment burial. Because organic matter is preserved preferentially in fine-grained sediments, organic-rich shales are believed to be the principal source rocks for petroleum. Thus, the first condition for successful hydrocarbon exploration is to find a basin that contains suitable source rocks. Further, the source rocks should have been buried to depths where the temperature is adequate to "crack" oil and gas from the organic matter ($\sim$90°C–125°C) but not so high that the hydrocarbons are destroyed. After oil and gas are generated within the shale source rocks, they migrate in some (poorly understood) manner from the source rock into associated porous and permeable rocks called **reservoir rocks.** (Porosity refers to that percentage of the volume of a rock occupied by pore space; permeability refers to the ability of a rock to transmit a fluid.) Drilling experience has shown that the best reservoir rocks for hydrocarbons are sandstones and limestones. Roughly 55 percent of the world's oil and 75 percent of its natural gas occur in sandstone reservoirs; most of the remaining world reserves are in carbonate reservoirs.

Most of the pore spaces in porous sandstones and carbonate rocks in the subsurface are filled with water. When hydrocarbons migrate from source beds into these overlying or underlying (?) reservoir rocks, some of the pore water must be displaced to make room for the hydrocarbons. Because the hydrocarbons have a lower specific gravity than water, they tend to move updip to get above the water (oil floats on water). Thus, by progressive displacement of pore water, hydrocarbons will gradually move up the regional dip of a sedimentary basin. This migration will continue until the oil reaches a trap or, if no trap is encountered, until it is expelled at the surface as an oil seep.

A hydrocarbon **trap** is some kind of structural or stratigraphic feature of a rock into which oil or gas can easily migrate but from which it is difficult to escape. The three most common kinds of petroleum traps are anticlines, faults, and stratigraphic pinch-outs (see Fig. 8.26). Oil or gas will rise to the structurally highest position in an anticline, fault trap, or stratigraphic pinch-out, where it is prevented from further movement by the presence of an impermeable rock layer (e.g., shale or evaporite) that lies above the reservoir bed and seals the trap. Thus, the search for a petroleum deposit requires that a trap be located and drilled in a basin that contains suitable source and reservoir rocks. The trap, when drilled, may or may not contain petroleum. Only about 10 to 15 percent of the traps drilled in new exploration areas commonly yield commercial quantities of petroleum. Structural traps (e.g., anticlines and faults) are located primarily by reflection seismic techniques (Chapter 15). Stratigraphic traps are located through detailed analysis of stratigraphic information obtained through study of stratigraphic cross sections (based on subsurface lithologic logs and well logs) and seismic cross sections.

Basin analysis to the petroleum geologist thus means locating within depositional basins suitable source rocks, reservoir rocks, and traps. To do this successfully calls into play most of the principles of sedimentology and stratigraphy, as well as the various techniques for analyzing sedimentary successions and presenting data, discussed in this book. Petroleum geologists must have a thorough understanding of the physical, chemical, and biological characteristics of sedimentary rocks; must understand depositional environments and depositional systems; must be well grounded in all aspects of stratigraphy; must have a working knowledge of the principles of geophysics and structural geology; and must have a basic understanding of the principles involved in the flow of fluids through porous, subsurface rocks. No single geologist is normally capable of carrying out the many complex studies required to develop a major, successful petroleum "play." Basin assessment is commonly carried out by teams of workers, which may include stratigraphers, sedimentologists, sedimentary petrologists, paleontologists, geophysicists, and hydrologists. These teams of scientists work together to develop an understanding of the sedimentary, stratigraphic, and structural factors that control the accumulation of oil and gas in drilled basins where commercial hydrocarbon deposits have already been found. They seek then to extend the knowledge gained from such studies to exploration for new hydrocarbon deposits in untested basins.

The application of basin-assessment techniques to developing a petroleum play is described by Allen and Allen (1990). Before new exploratory wells (Fig. 19.14) are drilled in an untested basin, petroleum geologists must first work out the stratigraphic characteristics of the basin to determine the kinds of potential source rocks and reservoir rocks that are present and the thickness and characteristics of these rocks. They then collect and analyze samples of potential source rocks (organic-bearing shales) to see if enough organic matter (kerogen; Chapter 8), and the right kind of kerogen, is present to generate economically significant quantities of petroleum or natural gas. Some kinds of kerogen generate liquid petroleum; others generate gas. Through outcrop study, and possibly drilling of small-diameter holes for information purposes, they determine which potential reservoir rocks have adequate porosity and permeability to make them targets for drilling.

If suitable source and reservoir rocks are present in the basin, the next step is to locate one or more traps that are large enough to hold a commercial quantity of oil or gas. As mentioned above, traps are located mainly by seismic prospecting and detailed stratigraphic analysis. An essential component of a trap is that reservoir rocks in the trap must be covered or capped by some kind of nonpermeable rock such as shale or evaporites to prevent oil in the reservoir rock from escaping upward. Finally, geologists study the fluid-flow patterns in the subsurface rocks of the basin to see if fluids are flowing with adequate pressure to move petroleum into a well bore that penetrates the reservoir rock. If all the requisite conditions of source rock, reservoir rocks, and trap are present, the well is drilled.

FIGURE 19.14   An exploratory gas well (Chevron Can Sup Waterton) drilling in the northwest corner of the Western Canada Sedimentary Basin, Alberta, Canada. The well tested Mississippian and Devonian carbonate formations and encountered gas below a depth of 4000 m. (Photograph courtesy of Sidney Smith.)

19.6   RELATIONSHIP OF TECTONICS AND SEDIMENTATION

I have made numerous references throughout this book regarding the effects of tectonics on sedimentation patterns and stratigraphic characteristics. Before closing the book, I wish to add a short note to reiterate and emphasize the importance of this relationship. Plate tectonics provides a first-order control on sedimentation through its influence on the sediment source area. As mentioned, siliciclastic sediments are derived from source rocks located in three kinds of tectonic settings—continental blocks, collision orogens, and magmatic arcs—each characterized by particular kinds (lithologies) of source rocks. Thus, tectonics determines the initial mineralogy and chemical composition of sediments eroded from tectonic highlands. The relief and slope of these highlands (a function of tectonics), combined with climatic conditions, further determine the rate of sediment denudation and thus the rate at which sediments are delivered to depositional basins. The kinds of basins in which sediments accumulate are directly related to tectonic processes. For example, some basins form as a result of tectonic processes that produce faulting; others are sag basins created by crustal cooling and subsidence or other tectonic processes. In any case, tectonic forces control the size, shape, and location of the basins. Tectonic processes, together with sediment loading, further determine the rate of basin subsidence and thus the space available (accommodation) for sediment accumulation. As discussed in Chapter 15, tectonics (e.g., rates of seafloor spreading) appears to have played a major role throughout geologic time in controlling rise and fall of sea level. In turn, changes in sea level (together with rates of sediment delivery and basin subsidence) have a major impact on sediment characteristics as expressed in depositional systems, depositional sequences, and system tracts (Chapter 14).

In summary, tectonics affects the composition of siliciclastic sedimentary rocks by controlling the lithology of the source rocks. Further, it influences sediment denudation rates and the accommodation space for sediments—and thus sedimentation rates and sediment thickness—and it controls the location and size of depositional basins. Finally, tectonics exerts a major control on sea level, which affects sedimentary processes and environments and, in turn, sediment characteristics (e.g., textures, bedding, sedimentary structures) and stratigraphic relationships.

FURTHER READINGS

Allen, P. A., and J. R. Allen, 1990, Basin analysis—Principles and applications: Blackwell, Oxford, 451 p.

Allen, P. A., and P. Homewood (eds.), 1986, Foreland basins: Blackwell, Oxford, 453 p.

Biddle, K. T. (ed.), 1991, Active margin basins: Am. Assoc. Petroleum Geologists Mem. 52, Tulsa, Okla., 324 p.

Edwards, J. D., and P. A. Santogrossi (eds.), 1990, Divergent/passive margin basins: Am. Assoc. Petroleum Geologists Mem. 48, Tulsa, Okla., 252 p.

Einsele, G., 1992, Sedimentary basins: Springer-Verlag, Berlin, 628 p.

Foster, N. H., and E. A. Beaumont (comps.), 1987, Geologic basins I: Classification, modeling, and predictive stratigraphy: Am. Assoc. Petroleum Geologists Reprint Series No. 1, 457 p.

Katz, B. J. (ed.), 1990, Lacustrine basin exploration: Am. Assoc. Petroleum Geologists Mem. 50, 340 p.

Kent, P., M. H. P. Bott, D. P. McKenzie, and C. A. Williams (eds.), 1982, The evolution of sedimentary basins: Philos. Trans. Royal Soc. London, v. 305, p. 1–338.

Kleinspehn, K. L., and C. Paola (eds.), 1988, New perspectives in basin analysis: Springer-Verlag, New York, 453 p.

Leggett, J. K. (ed.), 1982, Trench-forearc geology: Sedimentation and tectonics on modern and ancient active plate margins: Geol. Soc. London Spec. Pub. 10, Blackwell, Oxford, 576 p.

Leighton, M. W., D. R. Kolata, D. F. Oltz, and J. J. Eidel (eds.), 1990, Interior cratonic basins: Am. Assoc. Petroleum Geologists, Mem. 51 Tulsa, Okla., 819 p.

Macqueen, R. W., and D. A. Leckie, 1992, Foreland basins and fold belds: Am. Assoc. Petroleum Geologists Mem. 55, Tulsa, Okla., 460 p.

Miall, A. D., 1990, Principles of sedimentary basin analysis, 2nd ed.: Springer-Verlag, New York, 668 p.

APPENDIX A
Principal Methods for Estimating Diagenetic Paleotemperatures

Method	Basis for technique	Useful temperature range	Remarks	References
Conodont color alteration	Color changes in conodonts from pale yellow (thermally unaltered) to brown to black, owing to carbon fixation within trace amounts of organic material in the conodonts as a result of increase in temperature; followed by change in color to white (extreme thermal alteration) caused by carbon loss and loss of water by crystallization and recrystallization.	<50°–>400°C Visually recognizable levels in the conodont color alteration index (CAI): 1 <50°–80°C (pale yellow) 1.5 50°–90° 2 60°–140° (bn/dk bn) 3 110°–200° 4 190°–300° 5 300°–400° (black)	Conodonts are marine microfossils composed of the mineral apatite, but also containing trace amounts of organic material. They range in age from Cambrian to Triassic and occur principally in carbonate rocks and shales.	Harris (1979) Epstein, Epstein, and Harris (1977)
Vitrinite reflectance	Light reflection from vitrinite (thermally altered organic grains) is measured quantitatively with a reflectance microscope. Reflection increases with increasing degree of thermal alteration (metamorphism).	Up to about 240°C; % vitrinite reflectance is related to minimum temperature by the scale: <0.48% <100°C 0.59 125 0.72 145 0.86 165 1.00 180 1.16 195 1.42 210 1.50 220 1.70 230 1.92 235 2.14 240	Vitrinite is structured or unstructured woody tissue plus tissue impregnations that occurs as disseminated grains in sediment and is a major constituent in coals (Chapter 8).	Bostock (1979) Diessel and Offler (1975) Staplin et al. (1982)

Method	Basis	Temperature range	Description	References
Analysis of graphitization levels in kerogen	The carbon atoms in kerogen become increasingly well ordered (structured) with increasing levels of thermal diagenesis—a process called graphitization. Increase in the degree of ordering, or level of graphitization, as determined by X-ray diffraction methods can thus be related to increase in diagenetic temperature.	Up to about 600°C	Kerogen is the disseminated organic matter of sedimentary rocks and is insoluble in nonoxidizing acids, bases, and organic solvents. The organic matter initially deposited in sediments is converted to kerogen during diagenesis by thermocatalytic processses. Kerogen occurs principally in shales.	Harrison (1979) Landis (1971)
Analysis of clay mineral assemblages	With increasing temperature and metamorphic grade, smectite clay minerals convert to illite through a mixed-layer illite/smectite series; chlorite appears, and kaolinite as well as K-feldspar disappears. Thus, the relative abundance of these clay minerals in rock determines the metamorphic grade and can be related to paleotemperatures (e.g., smectites occur at temperatures below about 100°C; mixed-layer clays are stable up to about 200°C; illite forms at temperatures above 200°C).	Up to about 300°C	Clay minerals are phyllosilicate minerals made up of two-dimensional layer structures. They belong to four major clay mineral groups distinguished by differences in layer structure and composition: kaolinite, smectite, illite, and chlorite. Clay minerals are most abundant in shales but also occur as a matrix in sandstones and in minor amounts in limestones.	Hoffman and Hower (1979) Dunoyer de Segonzac (1970)
Analysis of zeolite facies mineral assemblages	Temperature theoretically exerts a strong control on the types of zeolite minerals that occur together. For example, heulandite and analcite tend to occur at temperatures below about 100°–125°C, whereas laumontite and pumpellyite occur at temperatures between about 100°–125°C and 175°–200°C. Thus paleotemperature can be established roughly on the basis of the zeolite-facies minerals present in a rock.	Up to about 200°–250°C	Zeolite facies minerals include a large group of minerals that develop authigenically in volcaniclastic sediments through alteration of chemically reactive volcanic materials. They form in the overlapping temperature range of diagenesis and metamorphism and show a progression of mineral facies that is clearly a reflection of temperature (and pressure) of burial.	Ghent (1979) Coombs (1971) Merino (1975a, 1975b)

Principal Methods for Estimating Diagenetic Paleotemperatures

Method	Basis for technique	Useful temperature range	Remarks	References
Analysis of fluid inclusions	Recrystallization of minerals or formation of overgrowths on minerals during diagenesis may trap fluid as minute inclusions in the crystals. Fluid inclusions commonly consist of a liquid plus a bubble of gas. Presumably at the time of formation, the inclusion consisted of a single fluid phase which separated into two upon cooling. By reheating the mineral until the phase boundary between the liquid and gas can be seen to just disappear, the approximate temperature at which the inclusion formed can be established—taking into account an estimate of the pressure of formation.	25°–150°/200°C	Fluid inclusions are found in geodes, vugs, and veins in sediments; sedimentary ore deposits; carbonate and quartz cements in terrigenous sedimentary rocks; salt and sulfur deposits; petroleum reservoir rocks; and sphalerite (zinc-bearing ore mineral) in bituminous coal beds.	Roedder (1979) Roedder (1976)
Oxygen isotope ratio	The ratio of $^{18}O/^{16}O$ in two coexisting oxygen-bearing minerals (from the same specimen) such as quartz and illite is commonly different. The amount of this difference has been shown to be a function of the maximum temperature to which the rock containing the minerals has been heated during diagenesis. Therefore, the isotopic fractionation, or difference between the $^{18}O/^{16}O$ ratios of two minerals which have reached equilibrium with each other, can be used to calculate the maximum diagenetic temperature to which the rock was heated. For example, isotopic fractionation between quartz and illite pairs is greatest at low temperatures and decreases with increasing temperature.	Up to about 400°C	The fractionation factor α between two coexisting minerals is defined as $$\alpha_{A-B} = \frac{(^{18}O/^{16}O)A}{(^{18}O/^{16}O)B}$$ where A and B refer to two oxygen-containing minerals.	Eslinger, Savin, and Yeh (1979) Yeh and Savin (1977)

APPENDIX B

North American Stratigraphic Code[1]

NORTH AMERICAN COMMISSION ON STRATIGRAPHIC NOMENCLATURE

FOREWORD

This code of recommended procedures for classifying and naming stratigraphic and related units has been prepared during a four-year period, by and for North American earth scientists, under the auspices of the North American Commission on Stratigraphic Nomenclature. It represents the thought and work of scores of persons, and thousands of hours of writing and editing. Opportunities to participate in and review the work have been provided throughout its development, as cited in the Preamble, to a degree unprecedented during preparation of earlier codes.

Publication of the International Stratigraphic Guide in 1976 made evident some insufficiencies of the American Stratigraphic Codes of 1961 and 1970. The Commission considered whether to discard our codes, patch them over, or rewrite them fully, and chose the last. We believe it desirable to sponsor a code of stratigraphic practice for use in North America, for we can adapt to new methods and points of view more rapidly than a worldwide body. A timely example was the recognized need to develop modes of establishing formal nonstratiform (igneous and high-grade metamorphic) rock units, an objective which is met in this Code, but not yet in the Guide.

The ways in which this Code differs from earlier American codes are evident from the Contents. Some categories have disappeared and others are new, but this Code has evolved from earlier codes and from the International Stratigraphic Guide. Some new units have not yet stood the test of long practice, and conceivably may not, but they are introduced toward meeting recognized and defined needs of the profession. Take this Code, use it, but do not condemn it because it contains something new or not of direct interest to you. Innovations that prove unacceptable to the profession will expire without damage to other concepts and procedures, just as did the geologic-climate units of the 1961 Code.

This Code is necessarily somewhat innovative because of: (1) the decision to write a new code, rather than to revise the old; (2) the open invitation to members of the geologic profession to offer suggestions and ideas, both in writing and orally; and (3)

the progress in the earth sciences since completion of previous codes. This report strives to incorporate the strength and acceptance of established practice, with suggestions for meeting future needs perceived by our colleagues; its authors have attempted to bring together the good from the past, the lessons of the Guide, and carefully reasoned provisions for the immediate future.

Participants in preparation of this Code are listed in Appendix I, but many others helped with their suggestions and comments. Major contributions were made by the members, and especially the chairmen, of the named subcommittees and advisory groups under the guidance of the Code Committee, chaired by Steven S. Oriel, who also served as principal, but not sole, editor. Amidst the noteworthy contributions by many, those of James D. Aitken have been outstanding. The work was performed for and supported by the Commission, chaired by Malcolm P. Weiss from 1978 to 1982.

This Code is the product of a truly North American effort. Many former and current commissioners representing not only the ten organizational members of the North American Commission on Stratigraphic Nomenclature (Appendix II), but other institutions as well, generated the product. Endorsement by constituent organizations is anticipated, and scientific communication will be fostered if Canadian, United States, and Mexican scientists, editors, and administrators consult Code recommendations for guidance in scientific reports. The Commission will appreciate reports of formal adoption or endorsement of the Code, and asks that they be transmitted to the Chairman of the Commission (c/o American Association of Petroleum Geologists, Box 979, Tulsa, Oklahoma 74101, U.S.A.).

Any code necessarily represents but a stage in the evolution of scientific communication. Suggestions for future changes of, or additions to, the North American Stratigraphic Code are welcome. Suggested and adopted modifications will be announced to the profession, as in the past, by serial Notes and Reports published in the *Bulletin* of the American Association of Petroleum Geologists. Suggestions may be made to representatives of your association or agency who are current commissioners, or directly to the Commission itself. The Commission meets annually, during the national meetings of the Geological Society of America.

1982 NORTH AMERICAN COMMISSION
ON STRATIGRAPHIC NOMENCLATURE

[1]Manuscript received, December 20, 1982; accepted, January 21, 1983. Copies are available at $1.00 per copy postpaid. Order from American Association of Petroleum Geologists, Box 979, Tulsa, Oklahoma 74101.

[1]Reprinted by permission from American Association of Petroleum Geologists Bulletin, v. 67, no. 5 (May 1983), p 841–875.

CONTENTS

696

698

699

PERSPECTIVE

Codes of Stratigraphic Nomenclature prepared by the American Commission on Stratigraphic Nomenclature (ACSN, 1961) and its predecessor (Committee on Stratigraphic Nomenclature, 1933) have been used widely as a basis for stratigraphic terminology. Their formulation was a response to needs recognized during the past century by government surveys (both national and local) and by editors of scientific journals for uniform standards and common procedures in defining and classifying formal rock bodies, their fossils, and the time spans represented by them. The most recent Code (ACSN, 1970) is a slightly revised version of that published in 1961, incorporating some minor amendments adopted by the Commission between 1962 and 1969. The Codes have served the profession admirably and have been drawn upon heavily for codes and guides prepared in other parts of the world (ISSC, 1976, p. 104-106). The principles embodied by any code, however, reflect the state of knowledge at the time of its preparation, and even the most recent code is now in need of revision.

New concepts and techniques developed during the past two decades have revolutionized the earth sciences. Moreover, increasingly evident have been the limitations of previous codes in meeting some needs of Precambrian and Quaternary geology and in classification of plutonic, high-grade metamorphic, volcanic, and intensely deformed rock assemblages. In addition, the important contributions of numerous international stratigraphic organizations associated with both the International Union of Geological Sciences (IUGS) and UNESCO, including working groups of the International Geological Correlation Program (IGCP), merit recognition and incorporation into a North American code.

For these and other reasons, revision of the American Code has been undertaken by committees appointed by the North American Commission on Stratigraphic Nomenclature (NACSN). The Commission, founded as the American Commission on Stratigraphic Nomenclature in 1946 (ACSN, 1947), was renamed the NACSN in 1978 (Weiss, 1979b) to emphasize that delegates from ten organizations in Canada, the United States, and Mexico represent the geological profession throughout North America (Appendix II).

Although many past and current members of the Commission helped prepare this revision of the Code, the participation of all interested geologists has been sought (for example, Weiss, 1979a). Open forums were held at the national meetings of both the Geological Society of America at San Diego in November, 1979, and the American Association of Petroleum Geologists at Denver in June, 1980, at which comments and suggestions were offered by more than 150 geologists. The resulting draft of this report was printed, through the courtesy of the Canadian Society of Petroleum Geologists, on October 1, 1981, and additional comments were invited from the profession for a period of one year before submittal of this report to the Commission for adoption. More than 50 responses were received with sufficient suggestions for improvement to prompt moderate revision of the printed draft (NACSN, 1981). We are particularly indebted to Hollis D. Hedberg and Amos Salvador for their exhaustive and perceptive reviews of early drafts of this Code, as well as to those who responded to the request for comments. Participants in the preparation and revisions of this report, and conferees, are listed in Appendix I.

Some of the expenses incurred in the course of this work were defrayed by National Science Foundation Grant EAR 7919845, for which we express appreciation. Institutions represented by the participants have been especially generous in their support.

SCOPE

The North American Stratigraphic Code seeks to describe explicit practices for classifying and naming all formally defined geologic units. *Stratigraphic procedures* and principles, although developed initially to bring order to strata and the events recorded therein, are applicable to all earth materials, not solely to strata. They promote systematic and rigorous study of the composition, geometry, sequence, history, and genesis of rocks and unconsolidated materials. They provide the framework within which time and space relations among rock bodies that constitute the Earth are ordered systematically. Stratigraphic procedures are used not only to reconstruct the history of the Earth and of extra-terrestrial bodies, but also to define the distribution and geometry of some commodities needed by society. *Stratigraphic classification* systematically arranges and partitions bodies of rock or unconsolidated materials of the Earth's crust into units based on their inherent properties or attributes.

A *stratigraphic code* or guide is a formulation of current views on stratigraphic principles and procedures designed to promote standardized classification and formal nomenclature of rock materials. It provides the basis for formalization of the language used to denote rock units and their spatial and temporal relations. To be effective, a code must be widely accepted and used; geologic organizations and journals may adopt its recommendations for nomenclatural procedure. Because any code embodies only current concepts and principles, it should have the flexibility to provide for both changes and additions to improve its relevance to new scientific problems.

Any system of nomenclature must be sufficiently explicit to enable users to distinguish objects that are embraced in a class from those that are not. This stratigraphic code makes no attempt to systematize structural, petrographic, paleontologic, or physiographic terms. Terms from these other fields that are used as part of formal stratigraphic names should be sufficiently general as to be unaffected by revisions of precise petrographic or other classifications.

The objective of a system of classification is to promote unambiguous communication in a manner not so restrictive as to inhibit scientific progress. To minimize ambiguity, a code must promote recognition of the distinction between observable features (reproducible data) and inferences or interpretations. Moreover, it should be sufficiently adaptable and flexible to promote the further development of science.

Stratigraphic classification promotes understanding of the *geometry* and *sequence* of rock bodies. The development of stratigraphy as a science required formulation of the Law of Superposition to explain sequential stratal relations. Although superposition is not applicable to many igneous, metamorphic, and tectonic rock assemblages, other criteria (such as crosscutting relations and isotopic dating) can be used to determine sequential arrangements among rock bodies.

The term *stratigraphic unit* may be defined in several ways. Etymological emphasis requires that it be a stratum or assemblage of adjacent strata distinguished by any or several of the many properties that rocks may possess (ISSC, 1976, p. 13). The scope of stratigraphic classification and procedures, however, suggests a broader definition: a naturally occurring body of rock or rock material distinguished from adjoining rock on the basis of some stated property or properties. Commonly used properties include composition, texture, included fossils, magnetic signature, radioactivity, seismic velocity, and age. Sufficient care is required in defining the boundaries of a unit to enable others to distinguish the material body from those adjoining it. Units based on one property commonly do not coincide with those based on another and, therefore, distinctive terms are needed to identify the property used in defining each unit.

The adjective *stratigraphic* is used in two ways in the remainder of this report. In discussions of lithic (used here as synonymous with "lithologic") units, a conscious attempt is made to restrict the term to lithostratigraphic or layered rocks and sequences that obey the Law of Superposition. For nonstratiform rocks (of plutonic or tectonic origin, for example), the term *lithodemic* (see Article 27) is used. The adjective *stratigraphic* is also used in a broader sense to refer to those procedures derived from stratigraphy which are now applied to all classes of earth materials.

An assumption made in the material that follows is that the reader has some degree of familiarity with basic principles of stratigraphy as outlined, for example, by Dunbar and Rodgers (1957), Weller (1960), Shaw (1964), Matthews (1974), or the International Stratigraphic Guide (ISSC, 1976).

RELATION OF CODES TO INTERNATIONAL GUIDE

Publication of the International Stratigraphic Guide by the International Subcommission on Stratigraphic Classification (ISSC, 1976), which is being endorsed and adopted throughout the world, played a part in prompting examination of the American Stratigraphic Code and the decision to revise it.

The International Guide embodies principles and procedures that had been adopted by several national and regional stratigraphic committees and commissions. More than two decades of effort by H. D. Hedberg and other members of the Subcommission (ISSC, 1976, p. VI, 1, 3) developed the consensus required for preparation of the Guide. Although the Guide attempts to cover all kinds of rocks and the diverse ways of investigating them, it is necessarily incomplete. Mechanisms are needed to stimulate individual innovations toward promulgating new concepts, principles, and practices which subsequently may be found worthy of inclusion in later editions of the Guide. The flexibility of national and regional committees or commissions enables them to perform this function more readily than an international subcommission, even while they adopt the Guide as the international standard of stratigraphic classification.

A guiding principle in preparing this Code has been to make it as consistent as possible with the International Guide, which was endorsed by the ACSN in 1976, and at the same time to foster further innovations to meet the expanding and changing needs of earth scientists on the North American continent.

OVERVIEW

CATEGORIES RECOGNIZED

An attempt is made in this Code to strike a balance between serving the needs of those in evolving specialties and resisting the proliferation of categories of units. Consequently, more formal categories are recognized here than in previous codes or in the International Guide (ISSC, 1976). On the other hand, no special provision is made for formalizing certain kinds of units (deep oceanic, for example) which may be accommodated by available categories.

Four principal categories of units have previously been used widely in traditional stratigraphic work; these have been termed lithostratigraphic, biostratigraphic, chronostratigraphic, and geochronologic and are distinguished as follows:

1. A *lithostratigraphic unit* is a stratum or body of strata, generally but not invariably layered, generally but not invariably tabular, which conforms to the Law of Superposition and is distinguished and delimited on the basis of lithic characteristics and stratigraphic position. Example: Navajo Sandstone.

2. A *biostratigraphic unit* is a body of rock defined and characterized by its fossil content. Example: *Discoaster multiradiatus* Interval Zone.

3. A *chronostratigraphic unit* is a body of rock established to serve as the material reference for all rocks formed during the same span of time. Example: Devonian System. Each boundary of a chronostratigraphic unit is synchronous. Chronostratigraphy provides a means of organizing strata into units based on their age relations. A chronostratigraphic body also serves as the basis for defining the specific interval of geologic time, or geochronologic unit, represented by the referent.

4. A *geochronologic unit* is a division of time distinguished on the basis of the rock record preserved in a chronostratigraphic

unit. Example: Devonian Period.

The first two categories are comparable in that they consist of material units defined on the basis of content. The third category differs from the first two in that it serves primarily as the standard for recognizing and isolating materials of a specific age. The fourth, in contrast, is not a material, but rather a conceptual, unit; it is a division of time. Although a geochronologic unit is not a stratigraphic body, it is so intimately tied to chronostratigraphy that the two are discussed properly together.

Properties and procedures that may be used in distinguishing geologic units are both diverse and numerous (ISSC, 1976, p. 1, 96; Harland, 1977, p. 230), but all may be assigned to the following principal classes of categories used in stratigraphic classification (Table 1), which are discussed below:

I. Material categories based on content, inherent attributes, or physical limits,

II. Categories distinguished by geologic age:
    A. Material categories used to define temporal spans, and
    B. Temporal categories.

Table 1. Categories of Units Defined*

MATERIAL CATEGORIES BASED ON CONTENT OR PHYSICAL LIMITS

    Lithostratigraphic (22)
    Lithodemic (31)**
    Magnetopolarity (44)
    Biostratigraphic (48)
    Pedostratigraphic (55)
    Allostratigraphic (58)

CATEGORIES EXPRESSING OR RELATED TO GEOLOGIC AGE

    Material Categories Used to Define Temporal Spans
        Chronostratigraphic (66)
        Polarity-Chronostratigraphic (83)
    Temporal (Non-Material) Categories
        Geochronologic (80)
        Polarity-Chronologic (88)
        Diachronic (91)
        Geochronometric (96)

*Numbers in parentheses are the numbers of the Articles where units are defined.
**Italicized categories are those introduced or developed since publication of the previous code (ACSN, 1970).

Material Categories Based on Content or Physical Limits

The basic building blocks for most geologic work are rock bodies defined on the basis of composition and related lithic characteristics, or on their physical, chemical, or biologic content or properties. Emphasis is placed on the relative objectivity and reproducibility of data used in defining units within each category.

Foremost properties of rocks are composition, texture, fabric, structure, and color, which together are designated *lithic characteristics*. These serve as the basis for distinguishing and defining the most fundamental of all formal units. Such units based primarily on composition are divided into two categories (Henderson and others, 1980): lithostratigraphic (Article 22) and lithodemic (defined here in Article 31). A lithostratigraphic unit obeys the Law of Superposition, whereas a lithodemic unit does not. A *lithodemic unit* is a defined body of predominantly intrusive, highly metamorphosed, or intensely deformed rock that, because it is intrusive or has lost primary structure through metamorphism or tectonism, generally does not conform to the Law of Superposition.

Recognition during the past several decades that remanent magnetism in rocks records the Earth's past magnetic characteristics (Cox, Doell, and Dalrymple, 1963) provides a powerful new tool encompassed by magnetostratigraphy (McDougall, 1977; McElhinny, 1978). *Magnetostratigraphy* (Article 43) is the study of remanent magnetism in rocks; it is the record of the Earth's magnetic polarity (or field reversals), dipole-field-pole position (including apparent polar wander), the non-dipole component (secular variation), and field intensity. Polarity is of particular utility and is used to define a *magnetopolarity unit* (Article 44) as a body of rock identified by its remanent magnetic polarity (ACSN, 1976; ISSC, 1979). Empirical demonstration of uniform polarity does not necessarily have direct temporal connotations because the remanent magnetism need not be related to rock deposition or crystallization. Nevertheless, polarity is a physical attribute that may characterize a body of rock.

Biologic remains contained in, or forming, strata are uniquely important in stratigraphic practice. First, they provide the means of defining and recognizing material units based on fossil content (biostratigraphic units). Second, the irreversibility of organic evolution makes it possible to partition enclosing strata temporally. Third, biologic remains provide important data for the reconstruction of ancient environments of deposition.

Composition also is important in distinguishing pedostratigraphic units. A *pedostratigraphic unit* is a body of rock that consists of one or more pedologic horizons developed in one or more lithic units now buried by a formally defined lithostratigraphic or allostratigraphic unit or units. A pedostratigraphic unit is the part of a buried soil characterized by one or more clearly defined soil horizons containing pedogenically formed minerals and organic compounds. Pedostratigraphic terminology is discussed below and in Article 55.

Many upper Cenozoic, especially Quaternary, deposits are distinguished and delineated on the basis of content, for which lithostratigraphic classification is appropriate. However, others are delineated on the basis of criteria other than content. To facilitate the reconstruction of geologic history, some compositionally similar deposits in vertical sequence merit distinction as separate stratigraphic units because they are the products of different processes; others merit distinction because they are of demonstrably different ages. Lithostratigraphic classification of these units is impractical and a new approach, allostratigraphic classification, is introduced here and may prove applicable to older deposits as well. An *allostratigraphic unit* is a mappable stratiform body of sedimentary rock defined and identified on the basis of bounding discontinuities (Article 58 and related Remarks).

Geologic-Climate units, defined in the previous Code (ACSN, 1970, p. 31), are abandoned here because they proved to be of dubious utility. Inferences regarding climate are subjective and too tenuous a basis for the definition of formal geologic units. Such inferences commonly are based on deposits assigned more appropriately to lithostratigraphic or allostratigraphic units and may be expressed in terms of diachronic units (defined below).

Categories Expressing or Related to Geologic Age

Time is a single, irreversible continuum. Nevertheless, various categories of units are used to define intervals of geologic time, just as terms having different bases, such as Paleolithic, Renaissance, and Elizabethan, are used to designate specific periods of human history. Different temporal categories are established to express intervals of time distinguished in different ways.

Major objectives of stratigraphic classification are to provide a basis for systematic ordering of the time and space relations of rock bodies and to establish a time framework for the discussion of geologic history. For such purposes, units of geologic time traditionally have been named to represent the span of time during which a well-described sequence of rock, or a chronostratigraphic unit, was deposited ("time units based on material referents," Fig. 1). This procedure continues, to the exclusion of other possible approaches, to be standard practice in studies of Phanerozoic rocks. Despite admonitions in previous American codes and the International Stratigraphic Guide (ISSC, 1976, p. 81) that similar procedures should be applied to the Precambrian, no comparable chronostratigraphic units, or geochronologic units derived therefrom, proposed for the Precambrian have yet been accepted worldwide. Instead, the IUGS Subcommission on Precambrian Stratigraphy (Sims, 1979) and its Working Groups (Harrison and Peterman, 1980) recommend division of Precambrian time into *geochronometric units* having no material referents.

A distinction is made throughout this report between *isochronous* and *synchronous*, as urged by Cumming, Fuller, and Porter (1959, p. 730), although the terms have been used synonymously by many. *Isochronous* means of equal duration; *synchronous* means simultaneous, or occurring at the same time. Although two rock bodies of very different ages may be formed during equal durations of time, the term *isochronous* is not applied to them in the earth sciences. Rather, isochronous bodies are those bounded by synchronous surfaces and formed during the same span of time. *Isochron*, in contrast, is used for a line connecting points of equal age on a graph representing physical or chemical phenomena; the line represents the same or equal time. The adjective *diachronous* is applied either to a rock unit with one or two bounding surfaces which are not synchronous, or to a boundary which is not synchronous (which "transgresses time").

Two classes of time units based on material referents, or stratotypes, are recognized (Fig. 1). The first is that of the traditional and conceptually isochronous units, and includes *geochronologic units*, which are based on *chronostratigraphic units*, and *polarity-geochronologic units*. These isochronous units have worldwide applicability and may be used even in areas lacking a material record of the named span of time. The second class of time units, newly defined in this Code, consists of *diachronic units* (Article 91), which are based on rock bodies known to be diachronous. In contrast to isochronous units, a diachronic term is used only where a material referent is present; a diachronic unit is coextensive with the material body or bodies on which it is based.

A *chronostratigraphic unit*, as defined above and in Article 66, is a body of rock established to serve as the material reference for all rocks formed during the same span of time; its boundaries are synchronous. It is the referent for a *geochronologic unit*, as defined above and in Article 80. Internationally accepted and traditional chronostratigraphic units were based initially on the time spans of lithostratigraphic units, biostratigraphic units, or other features of the rock record that have specific durations. In sum, they form the Standard Global Chronostratigraphic Scale (ISSC, 1976, p. 76-81; Harland, 1978), consisting of established systems and series.

A *polarity-chronostratigraphic unit* is a body of rock that contains a primary magnetopolarity record imposed when the rock was deposited or crystallized (Article 83). It serves as a material standard or referent for a part of geologic time during which the Earth's magnetic field had a characteristic polarity or sequence of polarities; that is, for a *polarity-chronologic unit* (Article 88).

A *diachronic unit* comprises the unequal spans of time represented by one or more specific diachronous rock bodies (Article 91). Such bodies may be lithostratigraphic, biostratigraphic, pedostratigraphic, allostratigraphic, or an assemblage of such units. A diachronic unit is applicable only where its material referent is present.

A *geochronometric* (or chronometric) *unit* is an isochronous direct division of geologic time expressed in years (Article 96). It has no material referent.

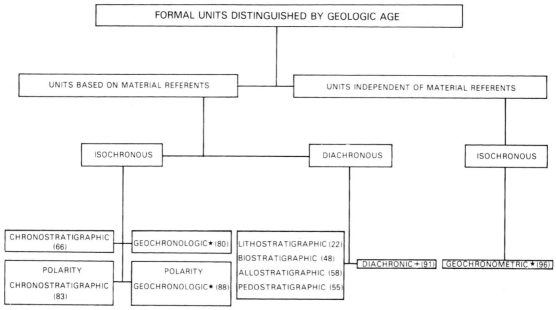

FORMAL UNITS DISTINGUISHED BY GEOLOGIC AGE

UNITS BASED ON MATERIAL REFERENTS

UNITS INDEPENDENT OF MATERIAL REFERENTS

ISOCHRONOUS

DIACHRONOUS

ISOCHRONOUS

CHRONOSTRATIGRAPHIC (66)

GEOCHRONOLOGIC ★ (80)

LITHOSTRATIGRAPHIC (22)
BIOSTRATIGRAPHIC (48)
ALLOSTRATIGRAPHIC (58)
PEDOSTRATIGRAPHIC (55)

POLARITY CHRONOSTRATIGRAPHIC (83)

POLARITY GEOCHRONOLOGIC ★ (88)

DIACHRONIC + (91)

GEOCHRONOMETRIC ★ (96)

★ Applicable world-wide.
+Applicable only where material referents are present.
( )Number of article in which defined.

FIG. 1.—Relation of geologic time units to the kinds of rock-unit referents on which most are based.

Pedostratigraphic Terms

The definition and nomenclature for pedostratigraphic[2] units in this Code differ from those for soil-stratigraphic units in the previous Code (ACSN, 1970, Article 18), by being more specific with regard to content, boundaries, and the basis for determining stratigraphic position.

The term "soil" has different meanings to the geologist, the soil scientist, the engineer, and the layman, and commonly has no stratigraphic significance. The term *paleosol* is currently used in North America for any soil that formed on a landscape of the past; it may be a buried soil, a relict soil, or an exhumed soil (Ruhe, 1965; Valentine and Dalrymple, 1976).

A *pedologic soil* is composed of one or more soil horizons.[3] A *soil horizon* is a layer within a pedologic soil that (1) is approximately parallel to the soil surface, (2) has distinctive physical, chemical, biological, and morphological properties that differ from those of adjacent, genetically related, soil horizons, and (3) is distinguished from other soil horizons by objective compositional properties that can be observed or measured in the field. The physical boundaries of buried pedologic horizons are objective traceable boundaries with stratigraphic significance. A buried pedologic soil provides the material basis for definition of a stratigraphic unit in pedostratigraphic classification (Article 55), but a buried pedologic soil may be somewhat more inclusive than a pedostratigraphic unit. A pedologic soil may contain both an 0-horizon and the entire C-horizon (Fig. 6), whereas the former is excluded and the latter need not be included in a pedostratigraphic unit.

The definition and nomenclature for pedostratigraphic units

in this Code differ from those of soil stratigraphic units proposed by the International Union for Quaternary Research and International Society of Soil Science (Parsons, 1981). The pedostratigraphic unit, geosol, also differs from the proposed INQUA-ISSS soil-stratigraphic unit, pedoderm, in several ways, the most important of which are: (1) a geosol may be in any part of the geologic column, whereas a pedoderm is a surficial soil; (2) a geosol is a buried soil, whereas a pedoderm may be a buried, relict, or exhumed soil; (3) the boundaries and stratigraphic position of a geosol are defined and delineated by criteria that differ from those for a pedoderm; and (4) a geosol may be either all or only a part of a buried soil, whereas a pedoderm is the entire soil.

The term *geosol*, as defined by Morrison (1967, p. 3), is a laterally traceable, mappable, geologic weathering profile that has a consistent stratigraphic position. The term is adopted and redefined here as the fundamental and only unit in formal pedostratigraphic classification (Article 56).

FORMAL AND INFORMAL UNITS

Although the emphasis in this Code is necessarily on formal categories of geologic units, informal nomenclature is highly useful in stratigraphic work.

Formally named units are those that are named in accordance with an established scheme of classification; the fact of formality is conveyed by capitalization of the initial letter of the *rank* or *unit* term (for example, Morrison Formation). Informal units, whose unit terms are ordinary nouns, are not protected by the stability provided by proper formalization and recommended classification procedures. Informal terms are devised for both economic and scientific reasons. Formalization is appropriate for those units requiring stability of nomenclature, particularly those likely to be extended far beyond the locality in which they were first recognized. Informal terms are appropriate for casually mentioned, innovative, and most economic units, those

---

[2]From Greek, *pedon*, ground or soil.
[3]As used in a geological sense, a *horizon* is a surface or line. In pedology, however, it is a body of material, and such usage is continued here.

defined by unconventional criteria, and those that may be too thin to map at usual scales.

Casually mentioned geologic units not defined in accordance with this Code are informal. For many of these, there may be insufficient need or information, or perhaps an inappropriate basis, for formal designations. Informal designations as beds or lithozones (the pebbly beds, the shaly zone, third coal) are appropriate for many such units.

Most economic units, such as aquifers, oil sands, coal beds, quarry layers, and ore-bearing "reefs," are informal, even though they may be named. Some such units, however, are so significant scientifically and economically that they merit formal recognition as beds, members, or formations.

Innovative approaches in regional stratigraphic studies have resulted in the recognition and definition of units best left as informal, at least for the time being. Units bounded by major regional unconformities on the North American craton were designated "sequences" (example: Sauk sequence) by Sloss (1963). Major unconformity-bounded units also were designated "synthems" by Chang (1975), who recommended that they be treated formally. Marker-defined units that are continuous from one lithofacies to another were designated "formats" by Forgotson (1957). The term "chronosome" was proposed by Schultz (1982) for rocks of diverse facies corresponding to geographic variations in sedimentation during an interval of deposition identified on the basis of bounding stratigraphic markers. Successions of faunal zones containing evolutionarily related forms, but bounded by non-evolutionary biotic discontinuities, were termed "biomeres" (Palmer, 1965). The foregoing are only a few selected examples to demonstrate how informality provides a continuing avenue for innovation.

The terms *magnafacies* and *parvafacies*, coined by Caster (1934) to emphasize the distinction between lithostratigraphic and chronostratigraphic units in sequences displaying marked facies variation, have remained informal despite their impact on clarifying the concepts involved.

Tephrochronologic studies provide examples of informal units too thin to map at conventional scales but yet invaluable for dating important geologic events. Although some such units are named for physiographic features and places where first recognized (e.g., Guaje pumice bed, where it is not mapped as the Guaje Member of the Bandelier Tuff), others bear the same name as the volcanic vent (e.g., Huckleberry Ridge ash bed of Izett and Wilcox, 1981).

Informal geologic units are designated by ordinary nouns, adjectives or geographic terms and lithic or unit-terms that are not capitalized (chalky formation or beds, St. Francis coal).

No geologic unit should be established and defined, whether formally or informally, unless its recognition serves a clear purpose.

CORRELATION

Correlation is a procedure for demonstrating correspondence between geographically separated parts of a geologic unit. The term is a general one having diverse meanings in different disciplines. Demonstration of temporal correspondence is one of the most important objectives of stratigraphy. The term "correlation" frequently is misused to express the idea that a unit has been identified or recognized.

Correlation is used in this Code as the demonstration of correspondence between two geologic units in both some defined property and relative stratigraphic position. Because correspondence may be based on various properties, three kinds of correlation are best distinguished by more specific terms. *Lithocorrelation* links units of similar lithology and stratigraphic position (or sequential or geometric relation, for lithodemic

---

[4]This article is modified slightly from a statement by the International Commission of Zoological Nomenclature (1964, p. 7-9).

units). *Biocorrelation* expresses similarity of fossil content and biostratigraphic position. *Chronocorrelation* expresses correspondence in age and in chronostratigraphic position.

Other terms that have been used for the similarity of content and stratal succession are homotaxy and chronotaxy. *Homotaxy* is the similarity in separate regions of the serial arrangement or succession of strata of comparable compositions or of included fossils. The term is derived from *homotaxis*, proposed by Huxley (1862, p. xlvi) to emphasize that similarity in succession does not prove age equivalence of comparable units. The term *chronotaxy* has been applied to similar stratigraphic sequences composed of units which are of equivalent age (Henbest, 1952, p. 310).

Criteria used for ascertaining temporal and other types of correspondence are diverse (ISSC, 1976, p. 86-93) and new criteria will emerge in the future. Evolving statistical tests, as well as isotopic and paleomagnetic techniques, complement the traditional paleontologic and lithologic procedures. Boundaries defined by one set of criteria need not correspond to those defined by others.

PART II. ARTICLES

INTRODUCTION

Article 1.—**Purpose.** This Code describes explicit stratigraphic procedures for classifying and naming geologic units accorded formal status. Such procedures, if widely adopted, assure consistent and uniform usage in classification and terminology and therefore promote unambiguous communication.

Article 2.—**Categories.** Categories of formal stratigraphic u-nits, though diverse, are of three classes (Table 1). The first class is of rock-material categories based on inherent attributes or content and stratigraphic position, and includes litho-stratigraphic, lithodemic, magnetopolarity, biostratigraphic, pe-dostratigraphic, and allostratigraphic units. The second class is of material categories used as standards for defining spans of ge-ologic time, and includes chronostratigraphic and polarity-chro-nostratigraphic units. The third class is of non-material temporal categories, and includes geochronologic, polarity-chronologic, geochronometric, and diachronic units.

GENERAL PROCEDURES

DEFINITION OF FORMAL UNITS

Article 3.—**Requirements for Formally Named Geologic Units.** Naming, establishing, revising, redefining, and abandoning formal geologic units require publication in a recognized scientific medium of a comprehensive statement which includes: (i) intent to designate or modify a formal unit; (ii) designation of category and rank of unit; (iii) selection and derivation of name; (iv) specification of stratotype (where applicable); (v) description of unit; (vi) definition of boundaries; (vii) historical background; (viii) dimensions, shape, and other regional aspects; (ix) geologic age; (x) correlations; and possibly (xi) genesis (where applicable). These requirements apply to subsurface and offshore, as well as exposed, units.

Article 4.—**Publication.**[4] "Publication in a recognized scientific medium" in conformance with this Code means that a work, when first issued, must (1) be reproduced in ink on paper or by some method that assures numerous identical copies and wide distribution; (2) be issued for the purpose of scientific, public, permanent record; and (3) be readily obtainable by purchase or free distribution.

Remarks. (a) **Inadequate publication.**—The following do not constitute publication within the meaning of the Code: (1) distribution of microfilms, microcards, or matter reproduced by similar methods; (2)

Table 2. Categories and Ranks of Units Defined in This Code*

A. Material Units

LITHOSTRATIGRAPHIC	LITHODEMIC	MAGNETOPOLARITY	BIOSTRATIGRAPHIC	PEDOSTRATIGRAPHIC	ALLOSTRATIGRAPHIC
Supergroup	Supersuite Complex				
Group	Suite	Polarity Superzone			Allogroup
Formation	*Lithodeme*	*Polarity zone*	*Biozone* (Interval, Assemblage or Abundance)	*Geosol*	*Alloformation*
Member (or Lens, or Tongue)		Polarity Subzone	Subbiozone		Allomember
Bed(s) or Flow(s)					

B. Temporal and Related Chronostratigraphic Units

CHRONO-STRATIGRAPHIC	GEOCHRONOLOGIC GEOCHRONOMETRIC	POLARITY CHRONO-STRATIGRAPHIC	POLARITY CHRONOLOGIC	DIACHRONIC
Eonothem	Eon	Polarity Superchronozone	Polarity Superchron	
Erathem (Supersystem)	Era (Superperiod)			
System (Subsystem)	*Period* (Subperiod)	*Polarity Chronozone*	*Polarity Chron*	*Episode*
Series	Epoch			Diachron Phase
Stage (Substage)	Age (Subage)	Polarity Subchronozone	Polarity Subchron	Span
Chronozone	Chron			Cline

*Fundamental units are italicized.

distribution to colleagues or students of a note, even if printed, in explanation of an accompanying illustration; (3) distribution of proof sheets; (4) open-file release; (5) theses, dissertations, and dissertation abstracts; (6) mention at a scientific or other meeting; (7) mention in an abstract, map explanation, or figure caption; (8) labeling of a rock specimen in a collection; (9) mere deposit of a document in a library; (10) anonymous publication; or (11) mention in the popular press or in a legal document.

(b). **Guidebooks.**—A guidebook with distribution limited to participants of a field excursion does not meet the test of availability. Some organizations publish and distribute widely large editions of serial guidebooks that include refereed regional papers; although these do meet the tests of scientific purpose and availability, and therefore constitute valid publication, other media are preferable.

Article 5.—**Intent and Utility.** To be valid, a new unit must serve a clear purpose and be duly proposed and duly described, and the intent to establish it must be specified. Casual mention of a unit, such as "the granite exposed near the Middleville schoolhouse," does not establish a new formal unit, nor does mere use in a table, columnar section, or map.

Remark. (a) **Demonstration of purpose served.**—The initial definition or revision of a named geologic unit constitutes, in essence, a proposal. As such, it lacks status until use by others demonstrates that a clear purpose has been served. A unit becomes established through repeated demonstration of its utility. The decision not to use a newly proposed or a newly revised term requires a full discussion of its unsuitability.

Article 6.—**Category and Rank.** The category and rank of a new or revised unit must be specified.

Remark. (a) **Need for specification.**—Many stratigraphic controversies have arisen from confusion or misinterpretation of the category of a unit (for example, lithostratigraphic vs. chronostratigraphic). Specification and unambiguous description of the category is of paramount importance. Selection and designation of an appropriate rank from the distinctive terminology developed for each category help serve this function (Table 2).

Article 7.—**Name.** The name of a formal geologic unit is compound. For most categories, the name of a unit should consist of a geographic name combined with an appropriate rank (Wasatch Formation) or descriptive term (Viola Limestone). Biostratigraphic units are designated by appropriate biologic forms (*Exus albus* Assemblage Biozone). Worldwide chronostratigraphic units bear long established and generally accepted names of diverse origins (Triassic System). The first letters of all words used in the names of formal geologic units are capitalized (except for the trivial species and subspecies terms in the name of a biostratigraphic unit).

Remarks. (a) **Appropriate geographic terms.**—Geographic names derived from permanent natural or artificial features at or near which the unit is present are preferable to those derived from impermanent features such as farms, schools, stores, churches, crossroads, and small communities. Appropriate names may be selected from those shown on topographic, state, provincial, county, forest service, hydrographic, or comparable maps, particularly those showing names approved by a national board for geographic names. The generic part of a geographic name, e.g., river, lake, village, should be omitted from new terms, unless required to distinguish between two otherwise identical names (e.g., Redstone Formation and Redstone River Formation). Two names should not be derived from the same geographic feature. A unit should not be named for the source of its components; for example, a deposit inferred to have been derived from the Keewatin glaciation center should not be designated the "Keewatin Till."

(b) **Duplication of names.**—Responsibility for avoiding duplication, either in use of the same name for different units (homonymy) or in use of different names for the same unit (synonymy), rests with the proposer. Although the same geographic term has been applied to different categories of units (example: the lithostratigraphic Word Formation and the chronostratigraphic Wordian Stage) now entrenched in the literature, the practice is undesirable. The extensive geologic nomenclature of North America, including not only names but also nomenclatural history of formal units, is recorded in compendia maintained by the Committee on Stratigraphic Nomenclature of the Geological Survey of Canada, Ottawa, Ontario; by the Geologic Names Committee of the United States Geological Survey, Reston, Virginia; by the Instituto de Geología, Ciudad Universitaria, México, D.F.; and by many state and provincial geological surveys. These organizations respond to inquiries regarding the availability of names, and some are prepared to reserve names for units likely to be defined in the next year or two.

(c) **Priority and preservation of established names.**—Stability of nomenclature is maintained by use of the rule of priority and by preservation of well-established names. Names should not be modified without explaining the need. Priority in publication is to be respected, but priority alone does not justify displacing a well-established name by one neither well-known nor commonly used; nor should an inadequately established name be preserved merely on the basis of priority. Redefinitions in precise terms are preferable to abandonment of the names of well-established units which may have been defined imprecisely but nonetheless in conformance with older and less stringent standards.

(d) **Differences of spelling and changes in name.**—The geographic component of a well-established stratigraphic name is not changed due to differences in spelling or changes in the name of a geographic feature. The name Bennett Shale, for example, used for more than half a century, need not be altered because the town is named Bennet. Nor should the Mauch Chunk Formation be changed because the town has been renamed Jim Thorpe. Disappearance of an impermanent geographic feature, such as a town, does not affect the name of an established geologic unit.

(e) **Names in different countries and different languages.**—For geologic units that cross local and international boundaries, a single name for each is preferable to several. Spelling of a geographic name commonly conforms to the usage of the country and linguistic group involved. Although geographic names are not translated (Cuchillo is not translated to Knife), lithologic or rank terms are (Edwards Limestone, Caliza Edwards; Formación La Casita, La Casita Formation).

Article 8.—**Stratotypes.** The designation of a unit or boundary stratotype (type section or type locality) is essential in the definition of most formal geologic units. Many kinds of units are best defined by reference to an accessible and specific sequence of rock that may be examined and studied by others. A stratotype is the standard (original or subsequently designated) for a named geologic unit or boundary and constitutes the basis for definition or recognition of that unit or boundary; therefore, it must be illustrative and representative of the concept of the unit or boundary being defined.

Remarks. (a) **Unit stratotypes.**—A unit stratotype is the type section for a stratiform deposit or the type area for a nonstratiform body that serves as the standard for definition and recognition of a geologic unit. The upper and lower limits of a unit stratotype are designated points in a specific sequence or locality and serve as the standards for definition and recognition of a stratigraphic unit's boundaries.

(b) **Boundary stratotype.**—A boundary stratotype is the type locality for the boundary reference point for a stratigraphic unit. Both boundary stratotypes for any unit need not be in the same section or region. Each boundary stratotype serves as the standard for definition and recognition of the base of a stratigraphic unit. The top of a unit may be defined by the boundary stratotype of the next higher stratigraphic unit.

(c) **Type locality.**—A type locality is the specified geographic locality where the stratotype of a formal unit or unit boundary was originally defined and named. A type area is the geographic territory encompassing the type locality. Before the concept of a stratotype was developed, only type localities and areas were designated for many geologic units which are now long- and well-established. Stratotypes, though now mandatory in defining most stratiform units, are impractical in definitions of many large nonstratiform rock bodies whose diverse major components may be best displayed at several reference localities.

(d) **Composite-stratotype.**—A composite-stratotype consists of several reference sections (which may include a type section) required to demonstrate the range or totality of a stratigraphic unit.

(e) **Reference sections.**—Reference sections may serve as invaluable standards in definitions or revisions of formal geologic units. For those well-established stratigraphic units for which a type section never was specified, a principal reference section (lectostratotype of ISSC, 1976, p. 26) may be designated. A principal reference section (neostratotype of ISSC, 1976, p. 26) also may be designated for those units or boundaries whose stratotypes have been destroyed, covered, or otherwise made inaccessible. Supplementary reference sections often are designated to illustrate the diversity or heterogeneity of a defined unit or some critical feature not evident or exposed in the stratotype. Once a unit or boundary stratotype section is designated, it is never abandoned or changed; however, if a stratotype proves inadequate, it may be supplemented by a principal reference section or by several reference sections that may constitute a composite-stratotype.

(f) **Stratotype descriptions.**—Stratotypes should be described both geographically and geologically. Sufficient geographic detail must be included to enable others to find the stratotype in the field, and may consist of maps and/or aerial photographs showing location and access, as well as appropriate coordinates or bearings. Geologic information should include thickness, descriptive criteria appropriate to the recognition of the unit and its boundaries, and discussion of the relation of the unit to other geologic units of the area. A carefully measured and described section provides the best foundation for definition of stratiform units. Graphic profiles, columnar sections, structure-sections, and photographs are useful supplements to a description; a geologic map of the area including the type locality is essential.

Article 9.—**Unit Description.** A unit proposed for formal status should be described and defined so clearly that any subsequent investigator can recognize that unit unequivocally. Distinguishing features that characterize a unit may include any or several of the following: composition, texture, primary structures, structural attitudes, biologic remains, readily apparent mineral composition (e.g., calcite vs. dolomite), geochemistry, geophysical properties (including magnetic signatures), geomorphic expression, unconformable or cross-cutting relations, and age. Although all distinguishing features pertinent to the unit category should be described sufficiently to characterize the unit, those not pertinent to the category (such as age and inferred genesis for lithostratigraphic units, or lithology for biostratigraphic units) should not be made part of the definition.

Article 10.—**Boundaries.** The criteria specified for the recognition of boundaries between adjoining geologic units are of paramount importance because they provide the basis for scientific reproducibility of results. Care is required in describing the criteria, which must be appropriate to the category of unit involved.

Remarks. (a) **Boundaries between intergradational units.**—Contacts between rocks of markedly contrasting composition are appropriate boundaries of lithic units, but some rocks grade into, or intertongue with, others of different lithology. Consequently, some boundaries are necessarily arbitrary as, for example, the top of the uppermost limestone in a sequence of interbedded limestone and shale. Such arbitrary boundaries commonly are diachronous.

(b) **Overlaps and gaps.**—The problem of overlaps and gaps between long-established adjacent chronostratigraphic units is being addressed by international IUGS and IGCP working groups appointed to deal with various parts of the geologic column. The procedure recommended by the Geological Society of London (George and others, 1969; Holland and others, 1978), of defining only the basal boundaries of chronostratigraphic units, has been widely adopted (e.g., McLaren, 1977) to resolve the problem. Such boundaries are defined by a carefully selected and agreed-upon boundary-stratotype (marker-point type section or "golden spike") which becomes the standard for the base of a chronostratigraphic unit. The concept of the mutual-boundary stratotype (ISSC, 1976, p. 84-86), based on the assumption of continuous deposition in selected sequences, also has been used to define chronostratigraphic units.

Although international chronostratigraphic units of series and higher rank are being redefined by IUGS and IGCP working groups, there may be a continuing need for some provincial series. Adoption of the basal boundary-stratotype concept is urged.

Article 11.—**Historical Background.** A proposal for a new name must include a nomenclatorial history of rocks assigned to the proposed unit, describing how they were treated previously and by whom (references), as well as such matters as priorities, possible synonymy, and other pertinent considerations. Consideration of the historical background of an older unit commonly provides the basis for justifying definition of a new unit.

Article 12.—**Dimensions and Regional Relations.** A perspective on the magnitude of a unit should be provided by such information as may be available on the geographic extent of a unit; observed ranges in thickness, composition, and geomorphic expression; relations to other kinds and ranks of stratigraphic units; correlations with other nearby sequences; and the bases for recognizing and extending the unit beyond the type locality. If the unit is not known anywhere but in an area of limited extent, informal designation is recommended.

Article 13.—**Age.** For most formal material geologic units, other than chronostratigraphic and polarity-chronostratigraphic, inferences regarding geologic age play no proper role in their definition. Nevertheless, the age, as well as the basis for its assignment, are important features of the unit and should be stated. For many lithodemic units, the age of the protolith should be distinguished from that of the metamorphism or deformation. If the basis for assigning an age is tenuous, a doubt should be expressed.

Remarks. (a) **Dating.**—The geochronologic ordering of the rock record, whether in terms of radioactive-decay rates or other processes, is generally called "dating." However, the use of the noun "date" to mean "isotopic age" is not recommended. Similarly, the term "absolute age" should be suppressed in favor of "isotopic age" for an age determined on the basis of isotopic ratios. The more inclusive term "numerical age" is recommended for all ages determined from isotopic ratios, fission tracks, and other quantifiable age-related phenomena.

(b) **Calibration**—The dating of chronostratigraphic boundaries in terms of numerical ages is a special form of dating for which the word "calibration" should be used. The geochronologic time-scale now in use has been developed mainly through such calibration of chronostratigraphic sequences.

(c) **Convention and abbreviations.**—The age of a stratigraphic unit or the time of a geologic event, as commonly determined by numerical dating or by reference to a calibrated time-scale, may be expressed in years before the present. The unit of time is the modern year as presently recognized worldwide. Recommended (but not mandatory) abbreviations for such ages are SI (International System of Units) multipliers coupled with "a" for annum: ka, Ma, and Ga[5] for kilo-annum (10^3 years), Mega-annum (10^6 years), and Giga-annum (10^9 years), respectively. Use of these terms after the age value follows the convention established in the field of C-14 dating. The "present" refers to 1950 AD, and such qualifiers as "ago" or "before the present" are omitted after the value because measurement of the duration from the present to the past is implicit in the designation. In contrast, the duration of a remote interval of geologic time, as a number of years, should not be expressed by the same symbols. Abbreviations for numbers of years, without reference to the present, are informal (e.g., y or yr for years; my, m.y., or m.yr. for millions of years; and so forth, as preference dictates). For example, boundaries of the Late Cretaceous Epoch currently are calibrated at 63 Ma and 96 Ma, but the interval of time represented by this epoch is 33 m.y.

(d) **Expression of "age" of lithodemic units.**—The adjectives "early," "middle," and "late" should be used with the appropriate geochronologic term to designate the age of lithodemic units. For example, a granite dated isotopically at 510 Ma should be referred to using the geochronologic term "Late Cambrian granite" rather than either the chronostratigraphic term "Upper Cambrian granite" or the more cumbersome designation "granite of Late Cambrian age."

Article 14.—**Correlation.** Information regarding spatial and temporal counterparts of a newly defined unit beyond the type area provides readers with an enlarged perspective. Discussions of criteria used in correlating a unit with those in other areas should make clear the distinction between data and inferences.

Article 15.—**Genesis.** Objective data are used to define and classify geologic units and to express their spatial and temporal relations. Although many of the categories defined in this Code (e.g., lithostratigraphic group, plutonic suite) have genetic connotations, inferences regarding geologic history or specific environments of formation may play no proper role in the definition of a unit. However, observations, as well as inferences, that bear on genesis are of great interest to readers and should be discussed.

Article 16.—**Subsurface and Subsea Units.** The foregoing procedures for establishing formal geologic units apply also to subsurface and offshore or subsea units. Complete lithologic and paleontologic descriptions or logs of the samples or cores are required in written or graphic form, or both. Boundaries and divisions, if any, of the unit should be indicated clearly with their depths from an established datum.

Remarks. (a) **Naming subsurface units.**—A subsurface unit may be named for the borehole (Eagle Mills Formation), oil field (Smackover Limestone), or mine which is intended to serve as the stratotype, or for a nearby geographic feature. The hole or mine should be located precisely, both with map and exact geographic coordinates, and identified fully (operator or company, farm or lease block, dates drilled or mined, surface elevation and total depth, etc).

(b) **Additional recommendations.**—Inclusion of appropriate borehole geophysical logs is urged. Moreover, rock and fossil samples and cores and all pertinent accompanying materials should be stored, and available for examination, at appropriate federal, state, provincial, university, or museum depositories. For offshore or subsea units (Clipperton Formation of Tracey and others, 1971, p. 22; Argo Salt of McIver, 1972, p. 57), the names of the project and vessel, depth of sea floor, and pertinent regional sampling and geophysical data should be added.

(c) **Seismostratigraphic units.**—High-resolution seismic methods now can delineate stratal geometry and continuity at a level of confidence not previously attainable. Accordingly, seismic surveys have come to be the principal adjunct of the drill in subsurface exploration. On the other hand, the method identifies rock types only broadly and by inference. Thus, formalization of units known only from seismic profiles is inappropriate. Once the stratigraphy is calibrated by drilling, the seismic method may provide objective well-to-well correlations.

REVISION AND ABANDONMENT OF FORMAL UNITS

Article 17.—**Requirements for Major Changes.** Formally defined and named geologic units may be redefined, revised, or abandoned, but revision and abandonment require as much justification as establishment of a new unit.

Remark. (a) **Distinction between redefinition and revision.**—Redefinition of a unit involves changing the view or emphasis on the content of the unit without changing the boundaries or rank, and differs only slightly from redescription. Neither redefinition nor redescription is considered revision. A redescription corrects an inadequate or inaccurate description, whereas a redefinition may change a descriptive (for example, lithologic) designation. Revision involves either minor changes in the definition of one or both boundaries or in the rank of a unit (normally, elevation to a higher rank). Correction of a misidentification of a unit outside its type area is neither redefinition nor revision.

---

[5]Note that the initial letters of Mega- and Giga- are capitalized, but that of kilo- is not, by SI convention.

Article 18.—**Redefinition.** A correction or change in the descriptive term applied to a stratigraphic or lithodemic unit is a redefinition which does not require a new geographic term.

Remarks. (a) **Change in lithic designation.**—Priority should not prevent more exact lithic designation if the original designation is not everywhere applicable; for example, the Niobrara Chalk changes gradually westward to a unit in which shale is prominent, for which the designation "Niobrara Shale" or "Formation" is more appropriate. Many carbonate formations originally designated "limestone" or "dolomite" are found to be geographically inconsistent as to prevailing rock type. The appropriate lithic term or "formation" is again preferable for such units.

(b) **Original lithic designation inappropriate.**—Restudy of some long-established lithostratigraphic units has shown that the original lithic designation was incorrect according to modern criteria; for example, some "shales" have the chemical and mineralogical composition of limestone, and some rocks described as felsic lavas now are understood to be welded tuffs. Such new knowledge is recognized by changing the lithic designation of the unit, while retaining the original geographic term. Similarly, changes in the classification of igneous rocks have resulted in recognition that rocks originally described as quartz monzonite now are more appropriately termed granite. Such lithic designations may be modernized when the new classification is widely adopted. If heterogeneous bodies of plutonic rock have been misleadingly identified with a single compositional term, such as "gabbro," the adoption of a neutral term, such as "intrusion" or "pluton," may be advisable.

Article 19.—**Revision.** Revision involves either minor changes in the definition of one or both boundaries of a unit, or in the unit's rank.

Remarks. (a) **Boundary change.**—Revision is justifiable if a minor change in boundary or content will make a unit more natural and useful. If revision modifies only a minor part of the content of a previously established unit, the original name may be retained.

(b) **Change in rank.**—Change in rank of a stratigraphic or temporal unit requires neither redefinition of its boundaries nor alteration of the geographic part of its name. A member may become a formation or vice versa, a formation may become a group or vice versa, and a lithodeme may become a suite or vice versa.

(c) **Examples of changes from area to area.**—The Conasauga Shale is recognized as a formation in Georgia and as a group in eastern Tennessee; the Osgood Formation, Laurel Limestone, and Waldron Shale in Indiana are classed as members of the Wayne Formation in a part of Tennessee; the Virgelle Sandstone is a formation in western Montana and a member of the Eagle Sandstone in central Montana; the Skull Creek Shale and the Newcastle Sandstone in North Dakota are members of the Ashville Formation in Manitoba.

(d) **Example of change in single area.**—The rank of a unit may be changed without changing its content. For example, the Madison Limestone of early work in Montana later became the Madison Group, containing several formations.

(e) **Retention of type section.**—When the rank of a geologic unit is changed, the original type section or type locality is retained for the newly ranked unit (see Article 22c).

(f) **Different geographic name for a unit and its parts.**—In changing the rank of a unit, the same name may not be applied both to the unit as a whole and to a part of it. For example, the Astoria Group should not contain an Astoria Sandstone, nor the Washington Formation, a Washington Sandstone Member.

(g) **Undesirable restriction.**—When a unit is divided into two or more of the same rank as the original, the original name should not be used for any of the divisions. Retention of the old name for one of the units precludes use of the name in a term of higher rank. Furthermore, in order to understand an author's meaning, a later reader would have to know about the modification and its date, and whether the author is following the original or the modified usage. For these reasons, the normal practice is to raise the rank of an established unit when units of the same rank are recognized and mapped within it.

Article 20.—**Abandonment.** An improperly defined or obsolete stratigraphic, lithodemic, or temporal unit may be formally abandoned, provided that (a) sufficient justification is presented to demonstrate a concern for nomenclatural stability, and (b) recommendations are made for the classification and nomenclature to be used in its place.

Remarks. (a) **Reasons for abandonment.**—A formally defined unit may be abandoned by the demonstration of synonymy or homonymy, of assignment to an improper category (for example, definition of a lithostratigraphic unit in a chronostratigraphic sense), or of other direct violations of a stratigraphic code or procedures prevailing at the time of the original definition. Disuse, or the lack of need or valid purpose for a unit, may be a basis for abandonment; so, too, may widespread misuse in diverse ways which compound confusion. A unit also may be abandoned if it proves impracticable, neither recognizable nor mappable elsewhere.

(b) **Abandoned names.**—A name for a lithostratigraphic or lithodemic unit, once applied and then abandoned, is available for some other unit only if the name was introduced casually, or if it has been published only once in the last several decades and is not in current usage, and if its reintroduction will cause no confusion. An explanation of the history of the name and of the new usage should be a part of the designation.

(c) **Obsolete names.**—Authors may refer to national and provincial records of stratigraphic names to determine whether a name is obsolete (see Article 7b).

(d) **Reference to abandoned names.**—When it is useful to refer to an obsolete or abandoned formal name, its status is made clear by some such term as "abandoned" or "obsolete," and by using a phrase such as "La Plata Sandstone of Cross (1898)". (The same phrase also is used to convey that a named unit has not yet been adopted for usage by the organization involved.)

(e) **Reinstatement.**—A name abandoned for reasons that seem valid at the time, but which subsequently are found to be erroneous, may be reinstated. Example: the Washakie Formation, defined in 1869, was abandoned in 1918 and reinstated in 1973.

CODE AMENDMENT

Article 21.—**Procedure for Amendment.** Additions to, or changes of, this Code may be proposed in writing to the Commission by any geoscientist at any time. If accepted for consideration by a majority vote of the Commission, they may be adopted by a two-thirds vote of the Commission at an annual meeting not less than a year after publication of the proposal.

FORMAL UNITS DISTINGUISHED BY CONTENT, PROPERTIES, OR PHYSICAL LIMITS

LITHOSTRATIGRAPHIC UNITS

Nature and Boundaries

Article 22.—**Nature of Lithostratigraphic Units.** A lithostratigraphic unit is a defined body of sedimentary, extrusive igneous, metasedimentary, or metavolcanic strata which is distinguished and delimited on the basis of lithic characteristics and stratigraphic position. A lithostratigraphic unit generally conforms to the Law of Superposition and commonly is stratified and tabular in form.

Remarks. (a) **Basic units.**—Lithostratigraphic units are the basic units of general geologic work and serve as the foundation for delineating strata, local and regional structure, economic resources, and geologic history in regions of stratified rocks. They are recognized and defined by observable rock characteristics; boundaries may be placed at clearly distinguished contacts or drawn arbitrarily within a zone of gradation. Lithification or cementation is not a necessary property; clay, gravel, till, and other unconsolidated deposits may constitute valid lithostratigraphic units.

(b) **Type section and locality.**—The definition of a lithostratigraphic unit should be based, if possible, on a stratotype consisting of readily accessible rocks in place, e.g., in outcrops, excavations, and mines, or of rocks accessible only to remote sampling devices, such as those in drill holes and underwater. Even where remote methods are used, definitions must be based on lithic criteria and not on the geophysical characteristics of the rocks, nor the implied age of their contained fossils. Definitions must be based on descriptions of actual rock material. Regional validity must be demonstrated for all such units. In regions where the stratigraphy has been established through studies of surface exposures, the naming of new units in the subsurface is justified only where the subsurface section differs materially from the surface section, or where there is doubt as to the equivalence of a subsurface and a surface unit. The establishment of subsurface reference sections for units originally defined in outcrop is encouraged.

(c) **Type section never changed.**—The definition and name of a lithostratigraphic unit are established at a type section (or locality) that, once specified, must not be changed. If the type section is poorly designated or delimited, it may be redefined subsequently. If the originally specified stratotype is incomplete, poorly exposed, structurally complicated, or unrepresentative of the unit, a principal reference section or several reference sections may be designated to supplement, but not to supplant, the type section (Article 8e).

(d) **Independence from inferred geologic history.**—Inferred geologic history, depositional environment, and biological sequence have no place in the definition of a lithostratigraphic unit, which must be based on composition and other lithic characteristics; nevertheless, considerations of well-documented geologic history properly may influence the choice of vertical and lateral boundaries of a new unit. Fossils may be valuable during mapping in distinguishing between two lithologically similar, noncontiguous lithostratigraphic units. The fossil content of a lithostratigraphic unit is a legitimate lithic characteristic; for example, oyster-rich sandstone, coquina, coral reef, or graptolitic shale. Moreover, otherwise similar units, such as the Formación Mendez and Formación Velasco mudstones, may be distinguished on the basis of coarseness of contained fossils (foraminifera).

(e) **Independence from time concepts.**—The boundaries of most lithostratigraphic units may transgress time horizons, but some may be approximately synchronous. Inferred time-spans, however measured, play no part in differentiating or determining the boundaries of any lithostratigraphic unit. Either relatively short or relatively long intervals of time may be represented by a single unit. The accumulation of material assigned to a particular unit may have begun or ended earlier in some localities than in others; also, removal of rock by erosion, either within the time-span of deposition of the unit or later, may reduce the time-span represented by the unit locally. The body in some places may be entirely younger than in other places. On the other hand, the establishment of formal units that straddle known, identifiable, regional disconformities is to be avoided, if at all possible. Although concepts of time or age play no part in defining lithostratigraphic units nor in determining their boundaries, evidence of age may aid recognition of similar lithostratigraphic units at localities far removed from the type sections or areas.

(f) **Surface form.**—Erosional morphology or secondary surface form may be a factor in the recognition of a lithostratigraphic unit, but properly should play a minor part at most in the definition of such units. Because the surface expression of lithostratigraphic units is an important aid in mapping, it is commonly advisable, where other factors do not countervail, to define lithostratigraphic boundaries so as to coincide with lithic changes that are expressed in topography.

(g) **Economically exploited units.**—Aquifers, oil sands, coal beds, and quarry layers are, in general, informal units even though named. Some such units, however, may be recognized formally as beds, members, or formations because they are important in the elucidation of regional stratigraphy.

(h) **Instrumentally defined units.**—In subsurface investigations, certain bodies of rock and their boundaries are widely recognized on borehole geophysical logs showing their electrical resistivity, radioactivity, density, or other physical properties. Such bodies and their boundaries may or may not correspond to formal lithostratigraphic units and their boundaries. Where other considerations do not countervail, the boundaries of subsurface units should be defined so as to correspond to useful geophysical markers; nevertheless, units defined exclusively on the basis of remotely sensed physical properties, although commonly useful in stratigraphic analysis, stand completely apart from the hierarchy of formal lithostratigraphic units and are considered informal.

(i) **Zone.**—As applied to the designation of lithostratigraphic units, the term "zone" is informal. Examples are "producing zone," "mineralized zone," "metamorphic zone," and "heavy-mineral zone." A zone may include all or parts of a bed, a member, a formation, or even a group.

(j) **Cyclothems.**—Cyclic or rhythmic sequences of sedimentary rocks, whose repetitive divisions have been named cyclothems, have been recognized in sedimentary basins around the world. Some cyclothems have

been identified by geographic names, but such names are considered informal. A clear distinction must be maintained between the division of a stratigraphic column into cyclothems and its division into groups, formations, and members. Where a cyclothem is identified by a geographic name, the word *cyclothem* should be part of the name, and the geographic term should not be the same as that of any formal unit embraced by the cyclothem.

(k) **Soils and paleosols.**—Soils and paleosols are layers composed of the in-situ products of weathering of older rocks which may be of diverse composition and age. Soils and paleosols differ in several respects from lithostratigraphic units, and should not be treated as such (see "Pedostratigraphic Units," Articles 55 et seq).

(l) **Depositional facies.**—Depositional facies are informal units, whether objective (conglomeratic, black shale, graptolitic) or genetic and environmental (platform, turbiditic, fluvial), even when a geographic term has been applied, e.g., Lantz Mills facies. Descriptive designations convey more information than geographic terms and are preferable.

Article 23.—Boundaries. Boundaries of lithostratigraphic units are placed at positions of lithic change. Boundaries are placed at distinct contacts or may be fixed arbitrarily within zones of gradation (Fig. 2a). Both vertical and lateral boundaries are based on the lithic criteria that provide the greatest unity and utility.

Remarks. (a) **Boundary in a vertically gradational sequence.**—A named lithostratigraphic unit is preferably bounded by a single lower and a single upper surface so that the name does not recur in a normal stratigraphic succession (see Remark b). Where a rock unit passes vertically into another by intergrading or interfingering of two or more kinds of rock, unless the gradational strata are sufficiently thick to warrant designation of a third, independent unit, the boundary is necessarily arbitrary and should be selected on the basis of practicality (Fig. 2b). For example, where a shale unit overlies a unit of interbedded limestone and shale, the boundary commonly is placed at the top of the highest readily traceable limestone bed. Where a sandstone unit grades upward into shale, the boundary may be so gradational as to be difficult to place even arbitrarily; ideally it should be drawn at the level where the rock is composed of one-half of each component. Because of creep in outcrops and caving in boreholes, it is generally best to define such arbitrary boundaries by the highest occurrence of a particular rock type, rather than the lowest.

(b) **Boundaries in lateral lithologic change.**—Where a unit changes laterally through abrupt gradation into, or intertongues with, a markedly different kind of rock, a new unit should be proposed for the different rock type. An arbitrary lateral boundary may be placed between the two equivalent units. Where the area of lateral intergradation or intertonguing is sufficiently extensive, a transitional interval of interbedded rocks may constitute a third independent unit (Fig. 2c). Where tongues (Article 25b) of formations are mapped separately or otherwise set apart without being formally named, the unmodified formation name should not be repeated in a normal stratigraphic sequence, although the modified name may be repeated in such phrases as "lower tongue of Mancos Shale" and "upper tongue of Mancos Shale." To show the order of superposition on maps and cross sections, the unnamed tongues may be distinguished informally (Fig. 2d) by number, letter, or other means. Such relationships may also be dealt with informally through the recognition of depositional facies (Article 22-l).

(c) **Key beds used for boundaries.**—Key beds (Article 26b) may be used as boundaries for a formal lithostratigraphic unit where the internal lithic characteristics of the unit remain relatively constant. Even though bounding key beds may be traceable beyond the area of the diagnostic overall rock type, geographic extension of the lithostratigraphic unit bounded thereby is not necessarily justified. Where the rock between key beds becomes drastically different from that of the type locality, a new name should be applied (Fig. 2e), even though the key beds are continuous (Article 26b). Stratigraphic and sedimentologic studies of stratigraphic units (usually informal) bounded by key beds may be very informative and useful, especially in subsurface work where the key beds may be recognized by their geophysical signatures. Such units, however, may be a kind of chronostratigraphic, rather than lithostratigraphic, unit (Article 75, 75c), although others are diachronous because one, or both, of the key beds are also diachronous.

(d) **Unconformities as boundaries.**—Unconformities, where recognizable objectively on lithic criteria, are ideal boundaries for lithostratigraphic units. However, a sequence of similar rocks may include an

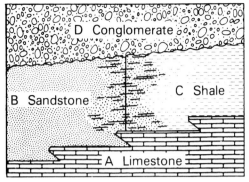

A.--Boundaries at sharp lithologic contacts and in laterally gradational sequence.

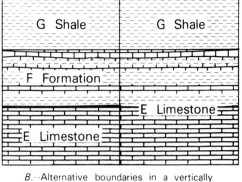

B.--Alternative boundaries in a vertically gradational or interlayered sequence.

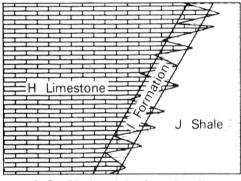

C.--Possible boundaries for a laterally intertonguing sequence.

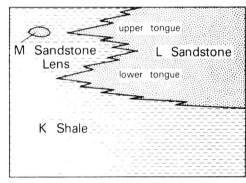

D. Possible classification of parts of an intertonguing sequence.

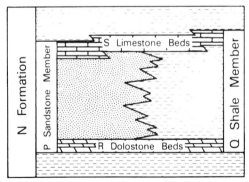

E.--Key beds, here designated the R Dolostone Beds and the S Limestone Beds, are used as boundaries to distinguish the Q Shale Member from the other parts of the N Formation. A lateral change in composition between the key beds requires that another name, P Sandstone Member, be applied. The key beds are part of each member.

EXPLANATION

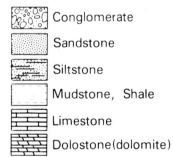

Conglomerate

Sandstone

Siltstone

Mudstone, Shale

Limestone

Dolostone(dolomite)

FIG. 2.—Diagrammatic examples of lithostratigraphic boundaries and classification.

obscure unconformity so that separation into two units may be desirable but impracticable. If no lithic distinction adequate to define a widely recognizable boundary can be made, only one unit should be recognized, even though it may include rock that accumulated in different epochs, periods, or eras.

(e) **Correspondence with genetic units.**—The boundaries of lithostratigraphic units should be chosen on the basis of lithic changes and, where feasible, to correspond with the boundaries of genetic units, so that subsequent studies of genesis will not have to deal with units that straddle formal boundaries.

Ranks of Lithostratigraphic Units

Article 24.—Formation. The formation is the fundamental unit in lithostratigraphic classification. A formation is a body of rock identified by lithic characteristics and stratigraphic position; it is prevailingly but not necessarily tabular and is mappable at the Earth's surface or traceable in the subsurface.

Remarks. (a) **Fundamental unit.**—Formations are the basic lithostratigraphic units used in describing and interpreting the geology of a region. The limits of a formation normally are those surfaces of lithic change that give it the greatest practicable unity of constitution. A formation may represent a long or short time interval, may be composed of materials from one or several sources, and may include breaks in deposition (see Article 23d).

(b) **Content.**—A formation should possess some degree of internal lithic homogeneity or distinctive lithic features. It may contain between its upper and lower limits (i) rock of one lithic type, (ii) repetitions of two or more lithic types, or (iii) extreme lithic heterogeneity which in itself may constitute a form of unity when compared to the adjacent rock units.

(c) **Lithic characteristics.**—Distinctive lithic characteristics include chemical and mineralogical composition, texture, and such supplementary features as color, primary sedimentary or volcanic structures, fossils (viewed as rock-forming particles), or other organic content (coal, oilshale). A unit distinguishable only by the taxonomy of its fossils is not a lithostratigraphic but a biostratigraphic unit (Article 48). Rock type may be distinctively represented by electrical, radioactive, seismic, or other properties (Article 22h), but these properties by themselves do not describe adequately the lithic character of the unit.

(d) **Mappability and thickness.**—The proposal of a new formation must be based on tested mappability. Well-established formations commonly are divisible into several widely recognizable lithostratigraphic units; where formal recognition of these smaller units serves a useful purpose, they may be established as members and beds, for which the requirement of mappability is not mandatory. A unit formally recognized as a formation in one area may be treated elsewhere as a group, or as a member of another formation, without change of name. Example: the Niobrara is mapped at different places as a member of the Mancos Shale, of the Cody Shale, or of the Colorado Shale, and also as the Niobrara Formation, as the Niobrara Limestone, and as the Niobrara Shale.

Thickness is not a determining parameter in dividing a rock succession into formations; the thickness of a formation may range from a feather edge at its depositional or erosional limit to thousands of meters elsewhere. No formation is considered valid that cannot be delineated at the scale of geologic mapping practiced in the region when the formation is proposed. Although representation of a formation on maps and cross sections by a labeled line may be justified, proliferation of such exceptionally thin units is undesirable. The methods of subsurface mapping permit delineation of units much thinner than those usually practicable for surface studies; before such thin units are formalized, consideration should be given to the effect on subsequent surface and subsurface studies.

(e) **Organic reefs and carbonate mounds.**—Organic reefs and carbonate mounds ("buildups") may be distinguished formally, if desirable, as formations distinct from their surrounding, thinner, temporal equivalents. For the requirements of formalization, see Article 30f.

(f) **Interbedded volcanic and sedimentary rock.**—Sedimentary rock and volcanic rock that are interbedded may be assembled into a formation under one name which should indicate the predominant or distinguishing lithology, such as Mindego Basalt.

(g) **Volcanic rock.**—Mappable distinguishable sequences of stratified volcanic rock should be treated as formations or lithostratigraphic units of higher or lower rank. A small intrusive component of a dominantly stratiform volcanic assemblage may be treated informally.

(h) **Metamorphic rock.**—Formations composed of low-grade metamorphic rock (defined for this purpose as rock in which primary structures are clearly recognizable) are, like sedimentary formations, distinguished mainly by lithic characteristics. The mineral facies may differ from place to place, but these variations do not require definition of a new formation. High-grade metamorphic rocks whose relation to established formations is uncertain are treated as lithodemic units (see Articles 31 et seq).

Article 25.—Member. A member is the formal lithostratigraphic unit next in rank below a formation and is always a part of some formation. It is recognized as a named entity within a formation because it possesses characteristics distinguishing it from adjacent parts of the formation. A formation need not be divided into members unless a useful purpose is served by doing so. Some formations may be divided completely into members; others may have only certain parts designated as members; still others may have no members. A member may extend laterally from one formation to another.

Remarks. (a) **Mapping of members.**—A member is established when it is advantageous to recognize a particular part of a heterogeneous formation. A member, whether formally or informally designated, need not be mappable at the scale required for formations. Even if all members of a formation are locally mappable, it does not follow that they should be raised to formational rank, because proliferation of formation names may obscure rather than clarify relations with other areas.

(b) **Lens and tongue.**—A geographically restricted member that terminates on all sides within a formation may be called a lens (lentil). A wedging member that extends outward beyond a formation or wedges ("pinches") out within another formation may be called a tongue.

(c) **Organic reefs and carbonate mounds.**—Organic reefs and carbonate mounds may be distinguished formally, if desirable, as members within a formation. For the requirements of formalization, see Article 30f.

(d) **Division of members.**—A formally or informally recognized division of a member is called a bed or beds, except for volcanic flow-rocks, for which the smallest formal unit is a flow. Members may contain beds or flows, but may never contain other members.

(e) **Laterally equivalent members.**—Although members normally are in vertical sequence, laterally equivalent parts of a formation that differ recognizably may also be considered members.

Article 26.—Bed(s). A bed, or beds, is the smallest formal lithostratigraphic unit of sedimentary rocks.

Remarks. (a) **Limitations.**—The designation of a bed or a unit of beds as a formally named lithostratigraphic unit generally should be limited to certain distinctive beds whose recognition is particularly useful. Coal beds, oil sands, and other beds of economic importance commonly are named, but such units and their names usually are not a part of formal stratigraphic nomenclature (Articles 22g and 30g).

(b) **Key or marker beds.**—A key or marker bed is a thin bed of distinctive rock that is widely distributed. Such beds may be named, but usually are considered informal units. Individual key beds may be traced beyond the lateral limits of a particular formal unit (Article 23c).

Article 27.—Flow. A flow is the smallest formal lithostratigraphic unit of volcanic flow rocks. A flow is a discrete, extrusive, volcanic body distinguishable by texture, composition, order of superposition, paleomagnetism, or other objective criteria. It is part of a member and thus is equivalent in rank to a bed or beds of sedimentary-rock classification. Many flows are informal units. The designation and naming of flows as formal rock-stratigraphic units should be limited to those that are distinctive and widespread.

Article 28.—Group. A group is the lithostratigraphic unit next higher in rank to formation; a group may consist entirely of named formations, or alternatively, need not be composed entirely of named formations.

Remarks. (a) **Use and content.**—Groups are defined to express the natural relationships of associated formations. They are useful in small-scale mapping and regional stratigraphic analysis. In some reconnaissance work, the term "group" has been applied to lithostratigraphic units that appear to be divisible into formations, but have not yet been so divided. In such cases, formations may be erected subsequently for one or all of the practical divisions of the group.

(b) **Change in component formations.**—The formations making up a group need not necessarily be everywhere the same. The Rundle Group, for example, is widespread in western Canada and undergoes several changes in formational content. In southwestern Alberta, it comprises the Livingstone, Mount Head, and Etherington Formations in the Front Ranges, whereas in the foothills and subsurface of the adjacent plains, it comprises the Pekisko, Shunda, Turner Valley, and Mount Head Formations. However, a formation or its parts may not be assigned to two vertically adjacent groups.

(c) **Change in rank.**—The wedge-out of a component formation or formations may justify the reduction of a group to formation rank, retaining the same name. When a group is extended laterally beyond where it is divided into formations, it becomes in effect a formation, even if it is still called a group. When a previously established formation is divided into two or more component units that are given formal formation rank, the old formation, with its old geographic name, should be raised to group status. Raising the rank of the unit is preferable to restricting the old name to a part of its former content, because a change in rank leaves the sense of a well-established unit unchanged (Articles 19b, 19g).

Article 29.—**Supergroup.** A supergroup is a formal assemblage of related or superposed groups, or of groups and formations. Such units have proved useful in regional and provincial syntheses. Supergroups should be named only where their recognition serves a clear purpose.

Remark. (a) **Misuse of "series" for group or supergroup.**—Although "series" is a useful general term, it is applied formally only to a chronostratigraphic unit and should not be used for a lithostratigraphic unit. The term "series" should no longer be employed for an assemblage of formations or an assemblage of formations and groups, as it has been, especially in studies of the Precambrian. These assemblages are groups or supergroups.

Lithostratigraphic Nomenclature

Article 30.—**Compound Character.** The formal name of a lithostratigraphic unit is compound. It consists of a geographic name combined with a descriptive lithic term or with the appropriate rank term, or both. Initial letters of all words used in forming the names of formal rock-stratigraphic units are capitalized.

Remarks. (a) **Omission of part of a name.**—Where frequent repetition would be cumbersome, the geographic name, the lithic term, or the rank term may be used alone, once the full name has been introduced; as "the Burlington," "the limestone," or "the formation," for the Burlington Limestone.

(b) **Use of simple lithic terms.**—The lithic part of the name should indicate the predominant or diagnostic lithology, even if subordinate lithologies are included. Where a lithic term is used in the name of a lithostratigraphic unit, the simplest generally acceptable term is recommended (for example, limestone, sandstone, shale, tuff, quartzite). Compound terms (for example, clay shale) and terms that are not in common usage (for example, calcirudite, orthoquartzite) should be avoided. Combined terms, such as "sand and clay," should not be used for the lithic part of the names of lithostratigraphic units, nor should an adjective be used between the geographic and the lithic terms, as "Chattanooga Black Shale" and "Biwabik Iron-Bearing Formation."

(c) **Group names.**—A group name combines a geographic name with the term "group," and no lithic designation is included; for example, San Rafael Group.

(d) **Formation names.**—A formation name consists of a geographic name followed by a lithic designation or by the word "formation." Examples: Dakota Sandstone, Mitchell Mesa Rhyolite, Monmouth Formation, Halton Till.

(e) **Member names.**—All member names include a geographic term and the word "member;" some have an intervening lithic designation, if useful; for example, Wedington Sandstone Member of the Fayetteville Shale. Members designated solely by lithic character (for example, siliceous shale member), by position (upper, lower), or by letter or number, are informal.

(f) **Names of reefs.**—Organic reefs identified as formations or members are formal units only where the name combines a geographic name with the appropriate rank term, e.g., Leduc Formation (a name applied to the several reefs enveloped by the Ireton Formation), Rainbow Reef Member.

(g) **Bed and flow names.**—The names of beds or flows combine a geographic term, a lithic term, and the term "bed" or "flow;" for example, Knee Hills Tuff Bed, Ardmore Bentonite Beds, Negus Variolitic Flows.

(h) **Informal units.**—When geographic names are applied to such informal units as oil sands, coal beds, mineralized zones, and informal members (see Articles 22g and 26a), the unit term should not be capitalized. A name is not necessarily formal because it is capitalized, nor does failure to capitalize a name render it informal. Geographic names should be combined with the terms "formation" or "group" only in formal nomenclature.

(i) **Informal usage of identical geographic names.**—The application of identical geographic names to several minor units in one vertical sequence is considered informal nomenclature (lower Mount Savage coal, Mount Savage fireclay, upper Mount Savage coal, Mount Savage rider coal, and Mount Savage sandstone). The application of identical geographic names to the several lithologic units constituting a cyclothem likewise is considered informal.

(j) **Metamorphic rock.**—Metamorphic rock recognized as a normal stratified sequence, commonly low-grade metavolcanic or metasedimentary rocks, should be assigned to named groups, formations, and members, such as the Deception Rhyolite, a formation of the Ash Creek Group, or the Bonner Quartzite, a formation of the Missoula Group. High-grade metamorphic and metasomatic rocks are treated as lithodemes and suites (see Articles 31, 33, 35).

(k) **Misuse of well-known name.**—A name that suggests some well-known locality, region, or political division should not be applied to a unit typically developed in another less well-known locality of the same name. For example, it would be inadvisable to use the name "Chicago Formation" for a unit in California.

LITHODEMIC UNITS

Nature and Boundaries

Article 31.—**Nature of Lithodemic Units.** A lithodemic[6] unit is a defined body of predominantly intrusive, highly deformed, and/or highly metamorphosed rock, distinguished and delimited on the basis of rock characteristics. In contrast to lithostratigraphic units, a lithodemic unit generally does not conform to the Law of Superposition. Its contacts with other rock units may be sedimentary, extrusive, intrusive, tectonic, or metamorphic (Fig. 3).

Remarks. (a) **Recognition and definition.**—Lithodemic units are defined and recognized by observable rock characteristics. They are the practical units of general geological work in terranes in which rocks generally lack primary stratification; in such terranes they serve as the foundation for studying, describing, and delineating lithology, local and regional structure, economic resources, and geologic history.

(b) **Type and reference localities.**—The definition of a lithodemic unit should be based on as full a knowledge as possible of its lateral and vertical variations and its contact relationships. For purposes of nomenclatural stability, a type locality and, wherever appropriate, reference localities should be designated.

(c) **Independence from inferred geologic history.**—Concepts based on inferred geologic history properly play no part in the definition of a lithodemic unit. Nevertheless, where two rock masses are lithically similar but display objective structural relations that preclude the possibility of their being even broadly of the same age, they should be assigned to different lithodemic units.

(d) **Use of "zone."**—As applied to the designation of lithodemic units, the term "zone" is informal. Examples are: "mineralized zone," "contact zone," and "pegmatitic zone."

---

[6]From the Greek *demas, -os:* "living body, frame".

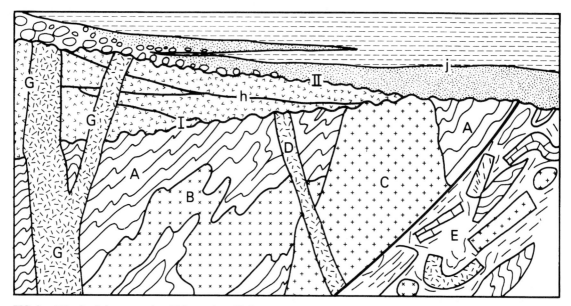

FIG. 3.—Lithodemic (upper case) and lithostratigraphic (lower case) units. A *lithodeme* of *gneiss* (**A**) contains an *intrusion* of diorite (**B**) that was deformed with the gneiss. A and B may be treated jointly as a *complex*. A younger *granite* (**C**) is cut by a dike of *syenite* (**D**), that is cut in turn by unconformity I. All the foregoing are in fault contact with a *structural complex* (**E**). A *volcanic complex* (**G**) is built upon unconformity I, and its feeder dikes cut the unconformity. Laterally equivalent volcanic strata in orderly, mappable succession (**h**) are treated as lithostratigraphic units. A *gabbro* feeder (**G'**), to the volcanic complex, where surrounded by gneiss is readily distinguished as a separate lithodeme and named as a *gabbro* or an *intrusion*. All the foregoing are overlain, at unconformity II, by sedimentary rocks (**j**) divided into formations and members.

Article 32.—**Boundaries.** Boundaries of lithodemic units are placed at positions of lithic change. They may be placed at clearly distinguished contacts or within zones of gradation. Boundaries, both vertical and lateral, are based on the lithic criteria that provide the greatest unity and practical utility. Contacts with other lithodemic and lithostratigraphic units may be depositional, intrusive, metamorphic, or tectonic.

Remark. (a) **Boundaries within gradational zones.**—Where a lithodemic unit changes through gradation into, or intertongues with, a rockmass with markedly different characteristics, it is usually desirable to propose a new unit. It may be necessary to draw an arbitrary boundary within the zone of gradation. Where the area of intergradation or intertonguing is sufficiently extensive, the rocks of mixed character may constitute a third unit.

Ranks of Lithodemic Units

Article 33.—**Lithodeme.** The lithodeme is the fundamental unit in lithodemic classification. A lithodeme is a body of intrusive, pervasively deformed, or highly metamorphosed rock, generally non-tabular and lacking primary depositional structures, and characterized by lithic homogeneity. It is mappable at the Earth's surface and traceable in the subsurface. For cartographic and hierarchical purposes, it is comparable to a formation (see Table 2).

Remarks. (a) **Content.**—A lithodeme should possess distinctive lithic features and some degree of internal lithic homogeneity. It may consist of (i) rock of one type, (ii) a mixture of rocks of two or more types, or (iii) extreme heterogeneity of composition, which may constitute in itself a form of unity when compared to adjoining rock-masses (see also "complex," Article 37).

(b) **Lithic characteristics.**—Distinctive lithic characteristics may include mineralogy, textural features such as grain size, and structural features such as schistose or gneissic structure. A unit distinguishable

from its neighbors only by means of chemical analysis is informal.

(c) **Mappability.**—Practicability of surface or subsurface mapping is an essential characteristic of a lithodeme (see Article 24d).

Article 34.—**Division of Lithodemes.** Units below the rank of lithodeme are informal.

Article 35.—**Suite.** A *suite* (metamorphic suite, intrusive suite, plutonic suite) is the lithodemic unit next higher in rank to lithodeme. It comprises two or more associated lithodemes of the same class (e.g., plutonic, metamorphic). For cartographic and hierarchical purposes, suite is comparable to group (see Table 2).

Remarks. (a) **Purpose.**—Suites are recognized for the purpose of expressing the natural relations of associated lithodemes having significant lithic features in common, and of depicting geology at compilation scales too small to allow delineation of individual lithodemes. Ideally, a suite consists entirely of named lithodemes, but may contain both named and unnamed units.

(b) **Change in component units.**—The named and unnamed units constituting a suite may change from place to place, so long as the original sense of natural relations and of common lithic features is not violated.

(c) **Change in rank.**—Traced laterally, a suite may lose all of its formally named divisions but remain a recognizable, mappable entity. Under such circumstances, it may be treated as a lithodeme but retain the same name. Conversely, when a previously established lithodeme is divided into two or more mappable divisions, it may be desirable to raise its rank to suite, retaining the original geographic component of the name. To avoid confusion, the original name should not be retained for one of the divisions of the original unit (see Article 19g).

Article 36.—**Supersuite.** A supersuite is the unit next higher in rank to a suite. It comprises two or more suites or complexes having a degree of natural relationship to one another, either in the vertical or the lateral sense. For cartographic and hierarchical purposes, supersuite is similar in rank to supergroup.

714

Article 37.—**Complex.** An assemblage or mixture of rocks of *two or more genetic classes*, i.e., igneous, sedimentary, or metamorphic, with or without highly complicated structure, may be named a *complex*. The term "complex" takes the place of the lithic or rank term (for example, Boil Mountain Complex, Franciscan Complex) and, although unranked, commonly is comparable to suite or supersuite and is named in the same manner (Articles 41, 42).

Remarks (a) **Use of "complex."**—Identification of an assemblage of diverse rocks as a complex is useful where the mapping of each separate lithic component is impractical at ordinary mapping scales. "Complex" is unranked but commonly comparable to suite or supersuite; therefore, the term may be retained if subsequent, detailed mapping distinguishes some or all of the component lithodemes or lithostratigraphic units.

(b) **Volcanic complex.**—Sites of persistent volcanic activity commonly are characterized by a diverse assemblage of extrusive volcanic rocks, related intrusions, and their weathering products. Such an assemblage may be designated a *volcanic complex*.

(c) **Structural complex.**—In some terranes, tectonic processes (e.g., shearing, faulting) have produced heterogeneous mixtures or disrupted bodies of rock in which some individual components are too small to be mapped. *Where there is no doubt that the mixing or disruption is due to tectonic processes*, such a mixture may be designated as a structural complex, whether it consists of two or more classes of rock, or a single class only. A simpler solution for some mapping purposes is to indicate intense deformation by an overprinted pattern.

(d) **Misuse of "complex".**—Where the rock assemblage to be united under a single, formal name consists of diverse types of a *single class* of rock, as in many terranes that expose a variety of either intrusive igneous or high-grade metamorphic rocks, the term "intrusive suite," "plutonic suite," or "metamorphic suite" should be used, rather than the unmodified term "complex." Exceptions to this rule are the terms *structural complex* and *volcanic complex* (see Remarks c and b, above).

Article 38.—**Misuse of "Series" for Suite, Complex, or Supersuite.** The term "series" has been employed for an assemblage of lithodemes or an assemblage of lithodemes and suites, especially in studies of the Precambrian. This practice now is regarded as improper; these assemblages are suites, complexes, or supersuites. The term "series" also has been applied to a sequence of rocks resulting from a succession of eruptions or intrusions. In these cases a different term should be used; "group" should replace "series" for volcanic and low-grade metamorphic rocks, and "intrusive suite" or "plutonic suite" should replace "series" for intrusive rocks of group rank.

Lithodemic Nomenclature

Article 39.—**General Provisions.** The formal name of a lithodemic unit is compound. It consists of a geographic name combined with a descriptive or appropriate rank term. The principles for the selection of the geographic term, concerning suitability, availability, priority, etc, follow those established in Article 7, where the rules for capitalization are also specified.

Article 40.—**Lithodeme Names.** The name of a lithodeme combines a geographic term with a lithic or descriptive term, e.g., Killarney Granite, Adamant Pluton, Manhattan Schist, Skaergaard Intrusion, Duluth Gabbro. The term *formation* should not be used.

Remarks. (a) **Lithic term.**—The lithic term should be a common and familiar term, such as schist, gneiss, gabbro. Specialized terms and terms not widely used, such as websterite and jacupirangite, and compound terms, such as graphitic schist and augen gneiss, should be avoided.

(b) **Intrusive and plutonic rocks.**—Because many bodies of intrusive rock range in composition from place to place and are difficult to characterize with a single lithic term, and because many bodies of plutonic rock

---

[7]Pluton—a mappable body of plutonic rock.

are considered not to be intrusions, latitude is allowed in the choice of a lithic or descriptive term. Thus, the descriptive term should preferably be compositional (e.g., gabbro, granodiorite), but may, if necessary, denote form (e.g., dike, sill), or be neutral (e.g., intrusion, pluton[7]). In any event, specialized compositional terms not widely used are to be avoided, as are form terms that are not widely used, such as bysmalith and chonolith. Terms implying genesis should be avoided as much as possible, because interpretations of genesis may change.

Article 41.—**Suite Names.** The name of a suite combines a geographic term, the term "suite," and an adjective denoting the fundamental character of the suite; for example, Idaho Springs Metamorphic Suite, Tuolumne Intrusive Suite, Cassiar Plutonic Suite. The geographic name of a suite may not be the same as that of a component lithodeme (see Article 19f). Intrusive assemblages, however, may share the same geographic name if an intrusive lithodeme is representative of the suite.

Article 42.—**Supersuite Names.** The name of a supersuite combines a geographic term with the term "supersuite."

MAGNETOSTRATIGRAPHIC UNITS

Nature and Boundaries

Article 43.—**Nature of Magnetostratigraphic Units.** A magnetostratigraphic unit is a body of rock unified by specified remanent-magnetic properties and is distinct from underlying and overlying magnetostratigraphic units having different magnetic properties.

Remarks. (a) **Definition.**—Magnetostratigraphy is defined here as all aspects of stratigraphy based on remanent magnetism (paleomagnetic signatures). Four basic paleomagnetic phenomena can be determined or inferred from remanent magnetism: polarity, dipole-field-pole position (including apparent polar wander), the non-dipole component (secular variation), and field intensity.

(b) **Contemporaneity of rock and remanent magnetism.**—Many paleomagnetic signatures reflect earth magnetism at the time the rock formed. Nevertheless, some rocks have been subjected subsequently to physical and/or chemical processes which altered the magnetic properties. For example, a body of rock may be heated above the blocking temperature or Curie point for one or more minerals, or a ferromagnetic mineral may be produced by low-temperature alteration long after the enclosing rock formed, thus acquiring a component of remanent magnetism reflecting the field at the time of alteration, rather than the time of original rock deposition or crystallization.

(c) **Designations and scope.**—The prefix *magneto* is used with an appropriate term to designate the aspect of remanent magnetism used to define a unit. The terms "magnetointensity" or "magnetosecular-variation" are possible examples. This Code considers only polarity reversals, which now are recognized widely as a stratigraphic tool. However, apparent-polar-wander paths offer increasing promise for correlations within Precambrian rocks.

Article 44.—**Definition of Magnetopolarity Unit.** A magnetopolarity unit is a body of rock unified by its remanent magnetic polarity and distinguished from adjacent rock that has different polarity.

Remarks. (a) **Nature.**—Magnetopolarity is the record in rocks of the polarity history of the Earth's magnetic-dipole field. Frequent past reversals of the polarity of the Earth's magnetic field provide a basis for magnetopolarity stratigraphy.

(b) **Stratotype.**—A stratotype for a magnetopolarity unit should be designated and the boundaries defined in terms of recognized lithostratigraphic and/or biostratigraphic units in the stratotype. The formal definition of a magnetopolarity unit should meet the applicable specific requirements of Articles 3 to 16.

(c) **Independence from inferred history.**—Definition of a magnetopolarity unit does not require knowledge of the time at which the unit acquired its remanent magnetism; its magnetism may be primary or secondary. Nevertheless, the unit's present polarity is a property that may be ascertained and confirmed by others.

(d) **Relation to lithostratigraphic and biostratigraphic units.**—Magnetopolarity units resemble lithostratigraphic and biostratigraphic units in that they are defined on the basis of an objective recognizable property, but differ fundamentally in that most magnetopolarity unit boundaries are thought not to be time transgressive. Their boundaries may coincide with those of lithostratigraphic or biostratigraphic units, or be parallel to but displaced from those of such units, or be crossed by them.

(e) **Relation of magnetopolarity units to chronostratigraphic units.**—Although transitions between polarity reversals are of global extent, a magnetopolarity unit does not contain within itself evidence that the polarity is primary, or criteria that permit its unequivocal recognition in chronocorrelative strata of other areas. Other criteria, such as paleontologic or numerical age, are required for both correlation and dating. Although polarity reversals are useful in recognizing chronostratigraphic units, magnetopolarity alone is insufficient for their definition.

Article 45.—**Boundaries.** The upper and lower limits of a magnetopolarity unit are defined by boundaries marking a change of polarity. Such boundaries may represent either a depositional discontinuity or a magnetic-field transition. The boundaries are either polarity-reversal horizons or polarity transition-zones, respectively.

Remark. (a) **Polarity-reversal horizons and transition-zones.**—A polarity-reversal horizon is either a single, clearly definable surface or a thin body of strata constituting a transitional interval across which a change in magnetic polarity is recorded. Polarity-reversal horizons describe transitional intervals of 1 m or less; where the change in polarity takes place over a stratigraphic interval greater than 1 m, the term "polarity transition-zone" should be used. Polarity-reversal horizons and polarity transition-zones provide the boundaries for polarity zones, although they may also be contained within a polarity zone where they mark an internal change subsidiary in rank to those at its boundaries.

Ranks of Magnetopolarity Units

Article 46.—**Fundamental Unit.** A polarity zone is the fundamental unit of magnetopolarity classification. A polarity zone is a unit of rock characterized by the polarity of its magnetic signature. Magnetopolarity zone, rather than polarity zone, should be used where there is risk of confusion with other kinds of polarity.

Remarks. (a) **Content.**—A polarity zone should possess some degree of internal homogeneity. It may contain rocks of (1) entirely or predominantly one polarity, or (2) mixed polarity.

(b) **Thickness and duration.**—The thickness of rock of a polarity zone or the amount of time represented should play no part in the definition of the zone. The polarity signature is the essential property for definition.

(c) **Ranks.**—When continued work at the stratotype for a polarity zone, or new work in correlative rocks elsewhere, reveals smaller polarity units, these may be recognized formally as polarity subzones. If it should prove necessary or desirable to group polarity zones, these should be termed polarity superzones. The rank of a polarity unit may be changed when deemed appropriate.

Magnetopolarity Nomenclature

Article 47.—**Compound Name.** The formal name of a magnetopolarity zone should consist of a geographic name and the term *Polarity Zone*. The term may be modified by *Normal, Reversed,* or *Mixed* (example: Deer Park Reversed Polarity Zone). In naming or revising magnetopolarity units, appropriate parts of Articles 7 and 19 apply. The use of informal designations, e.g., numbers or letters, is not precluded.

BIOSTRATIGRAPHIC UNITS

Nature and Boundaries

Article 48.—**Nature of Biostratigraphic Units.** A biostratigraphic unit is a body of rock defined or characterized by its fossil content. The basic unit in biostratigraphic classification is the biozone, of which there are several kinds.

Remarks. (a) **Enclosing strata.**—Fossils that define or characterize a biostratigraphic unit commonly are contemporaneous with the body of rock that contains them. Some biostratigraphic units, however, may be represented only by their fossils, preserved in normal stratigraphic succession (e.g., on hardgrounds, in lag deposits, in certain types of remanié accumulations), which alone represent the rock of the biostratigraphic unit. In addition, some strata contain fossils derived from older or younger rocks or from essentially coeval materials of different facies; such fossils should not be used to define a biostratigraphic unit.

(b) **Independence from lithostratigraphic units.**—Biostratigraphic units are based on criteria which differ fundamentally from those for lithostratigraphic units. Their boundaries may or may not coincide with the boundaries of lithostratigraphic units, but they bear no inherent relation to them.

(c) **Independence from chronostratigraphic units.**—The boundaries of most biostratigraphic units, unlike the boundaries of chronostratigraphic units, are both characteristically and conceptually diachronous. An exception is an abundance biozone boundary that reflects a mass-mortality event. The vertical and lateral limits of the rock body that constitutes the biostratigraphic unit represent the limits in distribution of the defining biotic elements. The lateral limits never represent, and the vertical limits rarely represent, regionally synchronous events. Nevertheless, biostratigraphic units are effective for interpreting chronostratigraphic relations.

Article 49.—**Kinds of Biostratigraphic Units.** Three principal kinds of biostratigraphic units are recognized: *interval, assemblage,* and *abundance* biozones.

Remark: (a) **Boundary definitions.**—Boundaries of interval zones are defined by lowest and/or highest occurrences of single taxa; boundaries of some kinds of assemblage zones (Oppel or concurrent range zones) are defined by lowest and/or highest occurrences of more than one taxon; and boundaries of abundance zones are defined by marked changes in relative abundances of preserved taxa.

Article 50.—**Definition of Interval Zone.** An interval zone (or subzone) is the body of strata between two specified, documented lowest and/or highest occurrences of single taxa.

Remarks. (a) **Interval zone types.**—Three basic types of interval zones are recognized (Fig. 4). These include the range zones and interval zones of the International Stratigraphic Guide (ISSC, 1976, p. 53, 60) and are:

1. The interval between the documented lowest and highest occurrences of a single taxon (Fig. 4A). This is the *taxon range zone* of ISSC (1976, p. 53).

2. The interval included between the documented lowest occurrence of one taxon and the documented highest occurrence of another taxon (Fig. 4B). When such occurrences result in stratigraphic overlap of the taxa (Fig. 4B-1), the interval zone is the *concurrent range zone* of ISSC (1976, p. 55), that involves only two taxa. When such occurrences do not result in stratigraphic overlap (Fig. 4B-2), but are used to partition the range of a third taxon, the interval is the *partial range zone* of George and others (1969).

3. The interval between documented successive lowest occurrences or successive highest occurrences of two taxa (Fig. 4C). When the interval is between successive documented lowest occurrences within an evolutionary lineage (Fig. 4C-1), it is the *lineage zone* of ISSC (1976, p. 58). When the interval is between successive lowest occurrences of unrelated taxa or between successive highest occurrences of either related or unrelated taxa (Fig. 4C-2), it is a kind of *interval zone* of ISSC (1976, p. 60).

(b) **Unfossiliferous intervals.**—Unfossiliferous intervals between or within biozones are the *barren interzones* and *intrazones* of ISSC (1976, p. 49).

Article 51.—**Definition of Assemblage Zone.** An assemblage zone is a biozone characterized by the association of three or more taxa. It may be based on all kinds of fossils present, or restricted to only certain kinds of fossils.

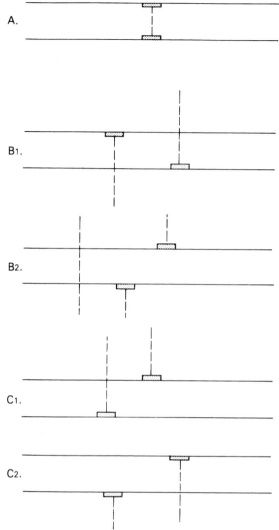

FIG. 4.—Examples of biostratigraphic interval zones.
Vertical broken lines indicate ranges of taxa; bars indicate lowest or highest documented occurrences.

Remarks. (a) **Assemblage zone contents.**—An assemblage zone may consist of a geographically or stratigraphically restricted assemblage, or may incorporate two or more contemporaneous assemblages with shared characterizing taxa (*composite assemblage zones* of Kauffman, 1969) (Fig. 5c).

(b) **Assemblage zone types.**—In practice, two assemblage zone concepts are used:

1. The *assemblage zone* (or cenozone) of ISSC (1976, p. 50), which is characterized by taxa without regard to their range limits (Fig. 5a). Recognition of this type of assemblage zone can be aided by using techniques of multivariate analysis. Careful designation of the characterizing taxa is especially important.

2. The *Oppel zone*, or the *concurrent range zone* of ISSC (1976, p. 55, 57), a type of zone characterized by more than two taxa and having boundaries based on two or more documented first and/or last occurrences of the included characterizing taxa (Fig. 5b).

Article 52.—Definition of Abundance Zone. An abundance zone is a biozone characterized by quantitatively distinctive maxima of relative abundance of one or more taxa. This is the *acme zone* of ISSC (1976, p. 59).

Remark. (a) **Ecologic controls.**—The distribution of biotic assemblages used to characterize some assemblage and abundance biozones may reflect strong local ecological control. Biozones based on such assemblages are included within the concept of ecozones (Vella, 1964), and are informal.

Ranks of Biostratigraphic Units

Article 53.—Fundamental Unit. The fundamental unit of biostratigraphic classification is a biozone.

Remarks. (a) **Scope.**—A single body of rock may be divided into various kinds and scales of biozones or subzones, as discussed in the International Stratigraphic Guide (ISSC, 1976, p. 62). Such usage is recommended if it will promote clarity, but only the unmodified term *biozone* is accorded formal status.

(b) **Divisions.**—A biozone may be completely or partly divided into formally designated sub-biozones (subzones), if such divisions serve a useful purpose.

Biostratigraphic Nomenclature

Article 54.—Establishing Formal Units. Formal establishment of a biozone or subzone must meet the requirements of Article 3 and requires a unique name, a description of its content and its boundaries, reference to a stratigraphic sequence in which the zone is characteristically developed, and a discussion of its spatial extent.

Remarks. (a) **Name.**—The name, which is compound and designates the kind of biozone, may be based on:

1. One or two characteristic and common taxa that are restricted to the biozone, reach peak relative abundance within the biozone, or have their total stratigraphic overlap within the biozone. These names most commonly are those of genera or subgenera, binomial designations of species, or trinomial designations of subspecies. If names of the nominate taxa change, names of the zones should be changed accordingly. Generic or subgeneric names may be abbreviated. Trivial species or subspecies names should not be used alone because they may not be unique.

2. Combinations of letters derived from taxa which characterize the biozone. However, alpha-numeric code designations (e.g., N1, N2, N3...) are informal and not recommended because they do not lend themselves readily to subsequent insertions, combinations, or eliminations. Biozonal systems based *only* on simple progressions of letters or numbers (e.g., A, B, C, or 1, 2, 3) are also not recommended.

(b) **Revision.**—Biozones and subzones are established empirically and may be modified on the basis of new evidence. Positions of established biozone or subzone boundaries may be stratigraphically refined, new characterizing taxa may be recognized, or original characterizing taxa may be superseded. If the concept of a particular biozone or subzone is substantially modified, a new unique designation is required to avoid ambiguity in subsequent citations.

(c) **Specifying kind of zone.**—Initial designation of a formally proposed biozone or subzone as an abundance zone, or as one of the types of interval zones, or assemblage zones (Articles 49-52), is strongly recommended. Once the type of biozone is clearly identified, the designation may be dropped in the remainder of a text (e.g., *Exus albus* taxon range zone to *Exus albus* biozone).

(d) **Defining taxa.**—Initial description or subsequent emendation of a biozone or subzone requires designation of the defining and characteristic taxa, and/or the documented first and last occurrences which mark the biozone or subzone boundaries.

(e) **Stratotypes.**—The geographic and stratigraphic position and boundaries of a formally proposed biozone or subzone should be defined precisely or characterized in one or more designated reference sections. Designation of a stratotype for each new biostratigraphic unit and of reference sections for emended biostratigraphic units is required.

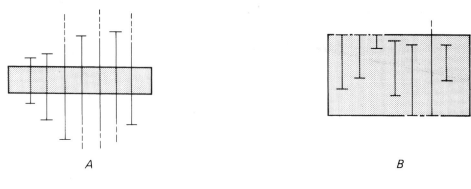

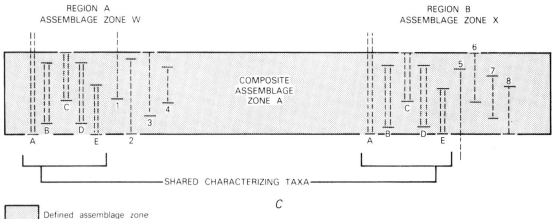

FIG. 5.—Examples of assemblage zone concepts.

PEDOSTRATIGRAPHIC UNITS

Nature and Boundaries

Article 55.—**Nature of Pedostratigraphic Units.** A pedostratigraphic unit is a body of rock that consists of one or more pedologic horizons developed in one or more lithostratigraphic, allostratigraphic, or lithodemic units (Fig. 6) and is overlain by one or more formally defined lithostratigraphic or allostratigraphic units.

Remarks. (a) **Definition.**—A pedostratigraphic[8] unit is a buried, traceable, three-dimensional body of rock that consists of one or more differentiated pedologic horizons.

(b) **Recognition.**—The distinguishing property of a pedostratigraphic unit is the presence of one or more distinct, differentiated, pedologic horizons. Pedologic horizons are products of soil development (pedogenesis) which occurred subsequent to formation of the lithostratigraphic, allostratigraphic, or lithodemic unit or units on which the buried soil was formed; these units are the parent materials in which pedogenesis occurred. Pedologic horizons are recognized in the field by diagnostic features such as color, soil structure, organic-matter accumulation, texture, clay coatings, stains, or concretions. Micromorphology, particle size, clay mineralogy, and other properties determined in the laboratory also may be used to identify and distinguish pedostratigraphic units.

(c) **Boundaries and stratigraphic position.**—The upper boundary of a pedostratigraphic unit is the top of the uppermost pedologic horizon formed by pedogenesis in a buried soil profile. The lower boundary of a pedostratigraphic unit is the lowest *definite* physical boundary of a pedologic horizon within a buried soil profile. The stratigraphic position of a pedostratigraphic unit is determined by its relation to overlying and underlying stratigraphic units (see Remark d).

(d) **Traceability.**—Practicability of subsurface tracing of the upper boundary of a buried soil is essential in establishing a pedostratigraphic unit because (1) few buried soils are exposed continuously for great distances, (2) the physical and chemical properties of a specific pedostratigraphic unit may vary greatly, both vertically and laterally, from place to place, and (3) pedostratigraphic units of different stratigraphic significance in the same region generally do not have unique identifying physical and chemical characteristics. Consequently, extension of a pedostratigraphic unit is accomplished by lateral tracing of the contact between a buried soil and an overlying, formally defined lithostratigraphic or allostratigraphic unit, or between a soil and two or more demonstrably correlative stratigraphic units.

(e) **Distinction from pedologic soils.**—Pedologic soils may include organic deposits (e.g., litter zones, peat deposits, or swamp deposits) that overlie or grade laterally into differentiated buried soils. The organic deposits are not products of pedogenesis, and O horizons are not included in a pedostratigraphic unit (Fig. 6); they may be classified as biostratigraphic or lithostratigraphic units. Pedologic soils also include the entire C horizon of a soil. The C horizon in pedology is not rigidly defined; it is merely the part of a soil profile that underlies the B horizon. The base of the C horizon in many soil profiles is gradational or unidentifiable; commonly it is placed arbitrarily. The need for clearly defined and easily recognized physical boundaries for a stratigraphic unit requires that the lower boundary of a pedostratigraphic unit be defined as the lowest *definite* physical boundary of a pedologic horizon in a buried soil profile, and part or all of the C horizon may be excluded from a pedostratigraphic unit.

---

[8]Terminology related to pedostratigraphic classification is summarized on page 850.

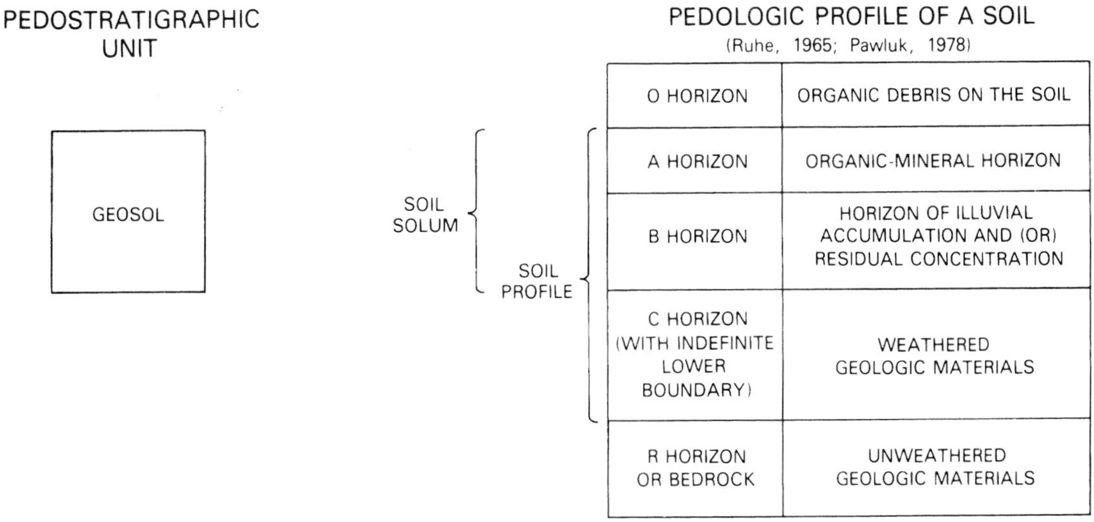

PEDOSTRATIGRAPHIC
UNIT

PEDOLOGIC PROFILE OF A SOIL
(Ruhe, 1965; Pawluk, 1978)

GEOSOL	SOIL SOLUM	SOIL PROFILE	O HORIZON	ORGANIC DEBRIS ON THE SOIL
			A HORIZON	ORGANIC-MINERAL HORIZON
			B HORIZON	HORIZON OF ILLUVIAL ACCUMULATION AND (OR) RESIDUAL CONCENTRATION
			C HORIZON (WITH INDEFINITE LOWER BOUNDARY)	WEATHERED GEOLOGIC MATERIALS
			R HORIZON OR BEDROCK	UNWEATHERED GEOLOGIC MATERIALS

FIG. 6.—Relationship between pedostratigraphic units and pedologic profiles.
The base of a geosol is the lowest clearly defined physical boundary of a pedologic horizon in a buried soil profile. In this example it is the lower boundary of the B horizon because the base of the C horizon is not a clearly defined physical boundary. In other profiles the base may be the lower boundary of a C horizon.

(f) **Relation to saprolite and other weathered materials.**—A material derived by in situ weathering of lithostratigraphic, allostratigraphic, and(or) lithodemic units (e.g., saprolite, bauxite, residuum) may be the parent material in which pedologic horizons form, but is not a pedologic soil. A pedostratigraphic unit may be based on the pedologic horizons of a buried soil developed in the product of in-situ weathering, such as saprolite. The parents of such a pedostratigraphic unit are both the saprolite and, indirectly, the rock from which it formed.

(g) **Distinction from other stratigraphic units.**—A pedostratigraphic unit differs from other stratigraphic units in that (1) it is a product of surface alteration of one or more older material units by specific processes (pedogenesis), (2) its lithology and other properties differ markedly from those of the parent material(s), and (3) a single pedostratigraphic unit may be formed in situ in parent material units of diverse compositions and ages.

(h) **Independence from time concepts.**—The boundaries of a pedostratigraphic unit are time-transgressive. Concepts of time spans, however measured, play no part in defining the boundaries of a pedostratigraphic unit. Nonetheless, evidence of age, whether based on fossils, numerical ages, or geometrical or other relationships, may play an important role in distinguishing and identifying non-contiguous pedostratigraphic units at localities away from the type areas. The name of a pedostratigraphic unit should be chosen from a geographic feature in the type area, and not from a time span.

Pedostratigraphic Nomenclature and Unit

Article 56.—**Fundamental Unit.** The fundamental and only unit in pedostratigraphic classification is a geosol.

Article 57.—**Nomenclature.**—The formal name of a pedostratigraphic unit consists of a geographic name combined with the term "geosol." Capitalization of the initial letter in each word serves to identify formal usage. The geographic name should be selected in accordance with recommendations in Article 7 and should not duplicate the name of another formal geologic unit. Names based on subjacent and superjacent rock units, for example the super-Wilcox–sub-Claiborne soil, are informal, as are

those with time connotations (post-Wilcox–pre-Claiborne soil).

Remarks. (a) **Composite geosols.**—Where the horizons of two or more merged or "welded" buried soils can be distinguished, formal names of pedostratigraphic units based on the horizon boundaries can be retained. Where the horizon boundaries of the respective merged or "welded" soils cannot be distinguished, formal pedostratigraphic classification is abandoned and a combined name such as Hallettville-Jamesville geosol may be used informally.

(b) **Characterization.**—The physical and chemical properties of a pedostratigraphic unit commonly vary vertically and laterally throughout the geographic extent of the unit. A pedostratigraphic unit is characterized by the *range* of physical and chemical properties of the unit in the type area, rather than by "typical" properties exhibited in a type section. Consequently, a pedostratigraphic unit is characterized on the basis of a composite stratotype (Article 8d).

(c) **Procedures for establishing formal pedostratigraphic units.**—A formal pedostratigraphic unit may be established in accordance with the applicable requirements of Article 3, and additionally by describing major soil horizons in each soil facies.

ALLOSTRATIGRAPHIC UNITS

Nature and Boundaries

Article 58.—**Nature of Allostratigraphic Units.** An allostratigraphic[9] unit is a mappable stratiform body of sedimentary rock that is defined and identified on the basis of its bounding discontinuities.

Remarks. (a) **Purpose.**—Formal allostratigraphic units may be defined to distinguish between different (1) superposed discontinuity-bounded deposits of similar lithology (Figs. 7, 9), (2) contiguous discontinuity-bounded deposits of similar lithology (Fig. 8), or (3) geographically separated discontinuity-bounded units of similar lithology (Fig. 9), or to distinguish as single units discontinuity-bounded deposits characterized by lithic heterogeneity (Fig. 8).

(b) **Internal characteristics.**—Internal characteristics (physical, chemical, and paleontological) may vary laterally and vertically throughout the unit.

(c) **Boundaries.**—Boundaries of allostratigraphic units are laterally traceable discontinuities (Figs. 7, 8, and 9).

(d) **Mappability.**—A formal allostratigraphic unit must be mappable

---

[9]From the Greek *allo*: "other, different."

719

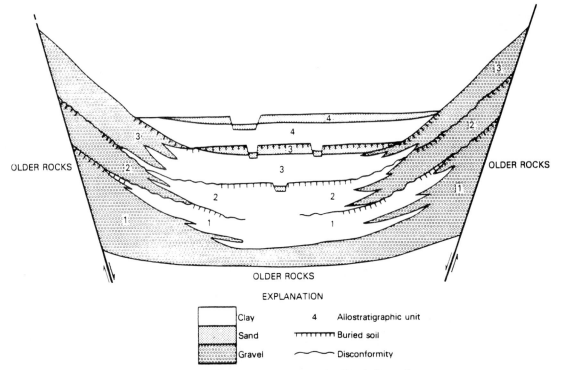

FIG. 7.—Example of allostratigraphic classification of alluvial and lacustrine deposits in a graben.

The alluvial and lacustrine deposits may be included in a single formation, or may be separated laterally into formations distinguished on the basis of contrasting texture (gravel, clay). Textural changes are abrupt and sharp, both vertically and laterally. The gravel deposits and clay deposits, respectively, are lithologically similar and thus cannot be distinguished as members of a formation. Four allostratigraphic units, each including two or three textural facies, may be defined on the basis of laterally traceable discontinuities (buried soils and disconformities).

EXPLANATION

Clay		4	Allostratigraphic unit
Sand		⊓⊓⊓⊓⊓	Buried soil
Gravel		∿∿∿	Disconformity

at the scale practiced in the region where the unit is defined.

(e) **Type locality and extent.**—A type locality and type area must be designated; a composite stratotype or a type section and several reference sections are desirable. An allostratigraphic unit may be laterally contiguous with a formally defined lithostratigraphic unit; a vertical cut-off between such units is placed where the units meet.

(f) **Relation to genesis.**—Genetic interpretation is an inappropriate basis for defining an allostratigraphic unit. However, genetic interpretation may influence the choice of its boundaries.

(g) **Relation to geomorphic surfaces.**—A geomorphic surface may be used as a boundary of an allostratigraphic unit, but the unit should not be given the geographic name of the surface.

(h) **Relation to soils and paleosols.**—Soils and paleosols are composed of products of weathering and pedogenesis and differ in many respects from allostratigraphic units, which are depositional units (see "Pedostratigraphic Units," Article 55). The upper boundary of a surface or buried soil may be used as a boundary of an allostratigraphic unit.

(i) **Relation to inferred geologic history.**—Inferred geologic history is not used to define an allostratigraphic unit. However, well-documented geologic history may influence the choice of the unit's boundaries.

(j) **Relation to time concepts.**—Inferred time spans, however measured, are not used to define an allostratigraphic unit. However, age relationships may influence the choice of the unit's boundaries.

(k) **Extension of allostratigraphic units.**—An allostratigraphic unit is extended from its type area by tracing the boundary discontinuities or by tracing or matching the deposits between the discontinuities.

Ranks of Allostratigraphic Units

Article 59.—**Hierarchy.** The hierarchy of allostratigraphic units, in order of decreasing rank, is allogroup, alloformation, and allomember.

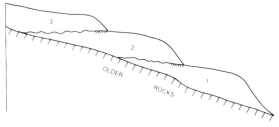

FIG. 8.—Example of allostratigraphic classification of contiguous deposits of similar lithology.

Allostratigraphic units 1, 2, and 3 are physical records of three glaciations. They are lithologically similar, reflecting derivation from the same bedrock, and constitute a single lithostratigraphic unit.

Remarks. (a) **Alloformation.**—The alloformation is the fundamental unit in allostratigraphic classification. An alloformation may be completely or only partly divided into allomembers, if some useful purpose is served, or it may have no allomembers.

(b) **Allomember.**—An allomember is the formal allostratigraphic unit next in rank below an alloformation.

(c) **Allogroup.**—An allogroup is the allostratigraphic unit next in rank above an alloformation. An allogroup is established only if a unit of that rank is essential to elucidation of geologic history. An allogroup may consist entirely of named alloformations or, alternatively, may contain one or more named alloformations which jointly do not comprise the entire allogroup.

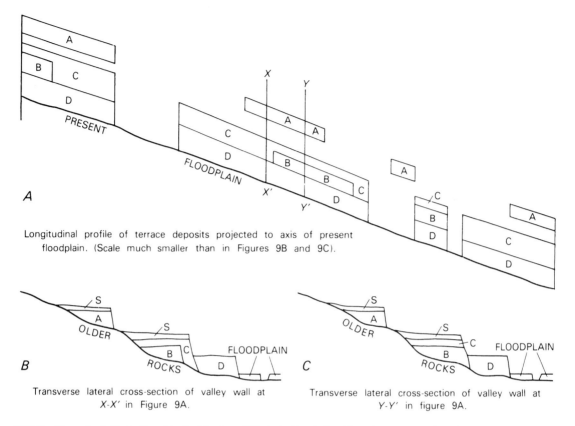

Longitudinal profile of terrace deposits projected to axis of present floodplain. (Scale much smaller than in Figures 9B and 9C).

Transverse lateral cross-section of valley wall at X-X' in Figure 9A.

Transverse lateral cross-section of valley wall at Y-Y' in figure 9A.

FIG. 9.—Example of allostratigraphic classification of lithologically similar, discontinuous terrace deposits.

A, B, C, and D are terrace gravel units of similar lithology at different topographic positions on a valley wall. The deposits may be defined as separate formal allostratigraphic units if such units are useful and if bounding discontinuities can be traced laterally. Terrace gravels of the same age commonly are separated geographically by exposures of older rocks. Where the bounding discontinuities cannot be traced continuously, they may be extended geographically on the basis of objective correlation of internal properties of the deposits other than lithology (e.g., fossil content, included tephras), topographic position, numerical ages, or relative-age criteria (e.g., soils or other weathering phenomena). The criteria for such extension should be documented. Slope deposits and eolian deposits (S) that mantle terrace surfaces may be of diverse ages and are not included in a terrace-gravel allostratigraphic unit. A single terrace surface may be underlain by more than one allostratigraphic unit (units B and C in sections b and c).

(d) **Changes in rank.**—The principles and procedures for elevation and reduction in rank of formal allostratigraphic units are the same as those in Articles 19b, 19g, and 28.

Allostratigraphic Nomenclature

Article 60.—**Nomenclature.** The principles and procedures for naming allostratigraphic units are the same as those for naming of lithostratigraphic units (see Articles 7, 30).

Remark. (a) **Revision.**—Allostratigraphic units may be revised or otherwise modified in accordance with the recommendations in Articles 17 to 20.

FORMAL UNITS DISTINGUISHED BY AGE

GEOLOGIC-TIME UNITS

Nature and Types

Article 61.—**Types.** Geologic-time units are conceptual, rather than material, in nature. Two types are recognized: those based

on material standards or referents (specific rock sequences or bodies), and those independent of material referents (Fig. 1).

Units Based on Material Referents

Article 62.—**Types Based on Referents.** Two types of formal geologic-time units based on material referents are recognized: they are isochronous and diachronous units.

Article 63.—**Isochronous Categories.** Isochronous time units and the material bodies from which they are derived are twofold: geochronologic units (Article 80), which are based on corresponding material chronostratigraphic units (Article 66), and polarity-geochronologic units (Article 88), based on corresponding material polarity-chronostratigraphic units (Article 83).

Remark. (a) **Extent.**—Isochronous units are applicable worldwide; they may be referred to even in areas lacking a material record of the named span of time. The duration of the time may be represented by a unit-stratotype referent. The beginning and end of the time are represented by point-boundary-stratotypes either in a single stratigraphic sequence or in separate stratotype sections (Articles 8b, 10b).

Article 64.—**Diachronous Categories.** Diachronic units (Article 91) are time units corresponding to diachronous material allostratigraphic units (Article 58), pedostratigraphic units (Article 55), and most lithostratigraphic (Article 22) and biostratigraphic (Article 48) units.

Remarks. (a) **Diachroneity.**—Some lithostratigraphic and biostratigraphic units are clearly diachronous, whereas others have boundaries which are not demonstrably diachronous within the resolving power of available dating methods. The latter commonly are treated as isochronous and are used for purposes of chronocorrelation (see biochronozone, Article 75). However, the assumption of isochroneity must be tested continually.

(b) **Extent.**—Diachronic units are coextensive with the diachronous material stratigraphic units on which they are based and are not used beyond the extent of their material referents.

Units Independent of Material Referents

Article 65.—**Numerical Divisions of Time.** Isochronous geologic-time units based on numerical divisions of time in years are geochronometric units (Article 96) and have no material referents.

CHRONOSTRATIGRAPHIC UNITS

Nature and Boundaries

Article 66.—**Definition.** A chronostratigraphic unit is a body of rock established to serve as the material reference for all rocks formed during the same span of time. Each of its boundaries is synchronous. The body also serves as the basis for defining the specific interval of time, or geochronologic unit (Article 80), represented by the referent.

Remarks. (a) **Purposes.**—Chronostratigraphic classification provides a means of establishing the temporally sequential order of rock bodies. Principal purposes are to provide a framework for (1) temporal correlation of the rocks in one area with those in another, (2) placing the rocks of the Earth's crust in a systematic sequence and indicating their relative position and age with respect to earth history as a whole, and (3) constructing an internationally recognized Standard Global Chronostratigraphic Scale.

(b) **Nature.**—A chronostratigraphic unit is a material unit and consists of a body of strata formed during a specific time span. Such a unit represents all rocks, and only those rocks, formed during that time span.

(c) **Content.**—A chronostratigraphic unit may be based upon the time span of a biostratigraphic unit, a lithic unit, a magnetopolarity unit, or any other feature of the rock record that has a time range. Or it may be any arbitrary but specified sequence of rocks, provided it has properties allowing chronocorrelation with rock sequences elsewhere.

Article 67.—**Boundaries.** Boundaries of chronostratigraphic units should be defined in a designated stratotype on the basis of observable paleontological or physical features of the rocks.

Remark. (a) **Emphasis on lower boundaries of chronostratigraphic units.**—Designation of point boundaries for both base and top of chronostratigraphic units is not recommended, because subsequent information on relations between successive units may identify overlaps or gaps. One means of minimizing or eliminating problems of duplication or gaps in chronostratigraphic successions is to define formally as a point-boundary stratotype only the base of the unit. Thus, a chronostratigraphic unit with its base defined at one locality, will have its top defined by the base of an overlying unit at the same, but more commonly another, locality (Article 8b).

Article 68.—**Correlation.** Demonstration of time equivalence is required for geographic extension of a chronostratigraphic unit from its type section or area. Boundaries of chronostratigraphic units can be extended only within the limits of resolution of available means of chronocorrelation, which currently include paleontology, numerical dating, remanent magnetism, thermoluminescence, relative-age criteria (examples are superposition and cross-cutting relations), and such indirect and inferential physical criteria as climatic changes, degree of weathering, and relations to unconformities. Ideally, the boundaries of chronostratigraphic units are independent of lithology, fossil content, or other material bases of stratigraphic division, but, in practice, the correlation or geographic extension of these boundaries relies at least in part on such features. Boundaries of chronostratigraphic units commonly are intersected by boundaries of most other kinds of material units.

Ranks of Chronostratigraphic Units

Article 69.—**Hierarchy.** The hierarchy of chronostratigraphic units, in order of decreasing rank, is eonothem, erathem, system, series, and stage. Of these, system is the primary unit of worldwide major rank; its primacy derives from the history of development of stratigraphic classification. All systems and units of higher rank are divided completely into units of the next lower rank. Chronozones are non-hierarchical and commonly lower-rank chronostratigraphic units. Stages and chronozones in sum do not necessarily equal the units of next higher rank and need not be contiguous. The rank and magnitude of chronostratigraphic units are related to the time interval represented by the units, rather than to the thickness or areal extent of the rocks on which the units are based.

Article 70.—**Eonothem.** The unit highest in rank is eonothem. The Phanerozoic Eonothem encompasses the Paleozoic, Mesozoic, and Cenozoic Erathems. Although older rocks have been assigned heretofore to the Precambrian Eonothem, they also have been assigned recently to other (Archean and Proterozoic) eonothems by the IUGS Precambrian Subcommission. The span of time corresponding to an eonothem is an *eon*.

Article 71.—**Erathem.** An erathem is the formal chronostratigraphic unit of rank next lower to eonothem and consists of several adjacent systems. The span of time corresponding to an erathem is an *era*.

Remark. (a) **Names.**—Names given to traditional Phanerozoic erathems were based upon major stages in the development of life on Earth: Paleozoic (old), Mesozoic (intermediate), and Cenozoic (recent) life. Although somewhat comparable terms have been applied to Precambrian units, the names and ranks of Precambrian divisions are not yet universally agreed upon and are under consideration by the IUGS Subcommission on Precambrian Stratigraphy.

Article 72.—**System.** The unit of rank next lower to erathem is the system. Rocks encompassed by a system represent a time-span and an episode of Earth history sufficiently great to serve as a worldwide chronostratigraphic reference unit. The temporal equivalent of a system is a *period*.

Remark. (a) **Subsystem and supersystem.**—Some systems initially established in Europe later were divided or grouped elsewhere into units ranked as systems. *Subsystems* (Mississippian Subsystem of the Carboniferous System) and *supersystems* (Karoo Supersystem) are more appropriate.

Article 73.—**Series.** Series is a conventional chronostratigraphic unit that ranks below a system and always is a division of a system. A series commonly constitutes a major unit of chronostratigraphic correlation within a province, between provinces, or between continents. Although many European series are being adopted increasingly for dividing systems on other continents, provincial series of regional scope continue to be useful. The temporal equivalent of a series is an *epoch*.

Article 74.—**Stage.** A stage is a chronostratigraphic unit of smaller scope and rank than a series. It is most commonly of greatest use in intra-continental classification and correlation, although it has the potential for worldwide recognition. The geochronologic equivalent of stage is *age*.

Remark. (a) **Substage.**—Stages may be, but need not be, divided completely into substages.

Article 75.—**Chronozone.** A chronozone is a non-hierarchical, but commonly small, formal chronostratigraphic unit, and its boundaries may be independent of those of ranked units. Although a chronozone is an isochronous unit, it may be based on a biostratigraphic unit (example: *Cardioceras cordatum* Biochronozone), a lithostratigraphic unit (Woodbend Lithochronozone), or a magnetopolarity unit (Gilbert Reversed-Polarity Chronozone). Modifiers (litho-, bio-, polarity) used in formal names of the units need not be repeated in general discussions where the meaning is evident from the context, e.g., *Exus albus* Chronozone.

Remarks. (a) **Boundaries of chronozones.**—The base and top of a *chronozone* correspond in the unit's stratotype to the observed, defining, physical and paleontological features, but they are extended to other areas by any means available for recognition of synchroneity. The temporal equivalent of a chronozone is a chron.

(b) **Scope.**—The scope of the non-hierarchical chronozone may range markedly, depending upon the purpose for which it is defined either formally or informally. The informal "biochronozone of the ammonites," for example, represents a duration of time which is enormous and exceeds that of a system. In contrast, a biochronozone defined by a species of limited range, such as the *Exus albus* Chronozone, may represent a duration equal to or briefer than that of a stage.

(c) **Practical utility.**—Chronozones, especially thin and informal biochronozones and lithochronozones bounded by key beds or other "markers," are the units used most commonly in industry investigations of selected parts of the stratigraphy of economically favorable basins. Such units are useful to define geographic distributions of lithofacies or biofacies, which provide a basis for genetic interpretations and the selection of targets to drill.

Chronostratigraphic Nomenclature

Article 76.—**Requirements.** Requirements for establishing a formal chronostratigraphic unit include: (i) statement of intention to designate such a unit; (ii) selection of name; (iii) statement of kind and rank of unit; (iv) statement of general concept of unit including historical background, synonymy, previous treatment, and reasons for proposed establishment; (v) description of characterizing physical and/or biological features; (vi) designation and description of boundary type sections, stratotypes, or other kinds of units on which it is based; (vii) correlation and age relations; and (viii) publication in a recognized scientific medium as specified in Article 4.

Article 77.—**Nomenclature.** A formal chronostratigraphic unit is given a compound name, and the initial letter of all words, except for trivial taxonomic terms, is capitalized. Except for chronozones (Article 75), names proposed for new chronostratigraphic units should not duplicate those for other stratigraphic units. For example, naming a new chronostratigraphic unit simply by adding "-an" or "-ian" to the name of a lithostratigraphic unit is improper.

Remarks. (a) **Systems and units of higher rank.**—Names that are generally accepted for systems and units of higher rank have diverse origins, and they also have different kinds of endings (Paleozoic, Cambrian, Cretaceous, Jurassic, Quaternary).

(b) **Series and units of lower rank.**—Series and units of lower rank are commonly known either by geographic names (Virgilian Series, Ochoan Series) or by names of their encompassing units modified by the capitalized adjectives Upper, Middle, and Lower (Lower Ordovician). Names of chronozones are derived from the unit on which they are based (Article 75). For series and stage, a geographic name is preferable because it may be related to a type area. For geographic names, the adjectival endings -an or -ian are recommended (Cincinnatian Series), but it is permissible to use the geographic name without any special ending, if more euphonious. Many series and stage names already in use have been based on lithic units (groups, formations, and members) and bear the names of these units

(Wolfcampian Series, Claibornian Stage). Nevertheless, a stage preferably should have a geographic name not previously used in stratigraphic nomenclature. Use of internationally accepted (mainly European) stage names is preferable to the proliferation of others.

Article 78.—**Stratotypes.** An ideal stratotype for a chronostratigraphic unit is a completely exposed unbroken and continuous sequence of fossiliferous stratified rocks extending from a well-defined lower boundary to the base of the next higher unit. Unfortunately, few available sequences are sufficiently complete to define stages and units of higher rank, which therefore are best defined by boundary-stratotypes (Article 8b).

Boundary-stratotypes for major chronostratigraphic units ideally should be based on complete sequences of either fossiliferous monofacial marine strata or rocks with other criteria for chronocorrelation to permit widespread tracing of synchronous horizons. Extension of synchronous surfaces should be based on as many indicators of age as possible.

Article 79.—**Revision of units.** Revision of a chronostratigraphic unit without changing its name is allowable but requires as much justification as the establishment of a new unit (Articles 17, 19, and 76). Revision or redefinition of a unit of system or higher rank requires international agreement. If the definition of a chronostratigraphic unit is inadequate, it may be clarified by establishment of boundary stratotypes in a principal reference section.

GEOCHRONOLOGIC UNITS

Nature and Boundaries

Article 80.—**Definition and Basis.** Geochronologic units are divisions of time traditionally distinguished on the basis of the rock record as expressed by chronostratigraphic units. A geochronologic unit is not a stratigraphic unit (i.e., it is not a material unit), but it corresponds to the time span of an established chronostratigraphic unit (Articles 65 and 66), and its beginning and ending corresponds to the base and top of the referent.

Ranks and Nomenclature of Geochronologic Units

Article 81.—**Hierarchy.** The hierarchy of geochronologic units in order of decreasing rank is *eon, era, period, epoch,* and *age.* Chron is a non-hierarchical, but commonly brief, geochronologic unit. Ages in sum do not necessarily equal epochs and need not form a continuum. An eon is the time represented by the rocks constituting an eonothem; era by an erathem; period by a system; epoch by a series; age by a stage; and chron by a chronozone.

Article 82.—**Nomenclature.** Names for periods and units of lower rank are identical with those of the corresponding chronostratigraphic units; the names of some eras and eons are independently formed. Rules of capitalization for chronostratigraphic units (Article 77) apply to geochronologic units. The adjectives Early, Middle, and Late are used for the geochronologic epochs equivalent to the corresponding chronostratigraphic Lower, Middle, and Upper series, where these are formally established.

POLARITY-CHRONOSTRATIGRAPHIC UNITS

Nature and Boundaries

Article 83.—**Definition.** A polarity-chronostratigraphic unit is a body of rock that contains the primary magnetic-polarity record imposed when the rock was deposited, or crystallized, during a specific interval of geologic time.

Remarks. (a) **Nature.**—Polarity-chronostratigraphic units depend fundamentally for definition on actual sections or sequences, or measurements on individual rock units, and without these standards they are meaningless. They are based on material units, the polarity zones of magnetopolarity classification. Each polarity-chronostratigraphic unit is the record of the time during which the rock formed and the Earth's magnetic field had a designated polarity. Care should be taken to define polarity-chronologic units in terms of polarity-chronostratigraphic units, and not vice versa.

(b) **Principal purposes.**—Two principal purposes are served by polarity-chronostratigraphic classification: (1) correlation of rocks at one place with those of the same age and polarity at other places; and (2) delineation of the polarity history of the Earth's magnetic field.

(c) **Recognition.**—A polarity-chronostratigraphic unit may be extended geographically from its type locality only with the support of physical and/or paleontologic criteria used to confirm its age.

Article 84.—**Boundaries.** The boundaries of a polarity chronozone are placed at polarity-reversal horizons or polarity transition-zones (see Article 45).

Ranks and Nomenclature of Polarity-Chronostratigraphic Units

Article 85.—**Fundamental Unit.** The polarity chronozone consists of rocks of a specified primary polarity and is the fundamental unit of worldwide polarity-chronostratigraphic classification.

Remarks. (a) **Meaning of term.**—A polarity chronozone is the worldwide body of rock strata that is collectively defined as a polarity-chronostratigraphic unit.

(b) **Scope.**—Individual polarity zones are the basic building blocks of polarity chronozones. Recognition and definition of polarity chronozones may thus involve step-by-step assembly of carefully dated or correlated individual polarity zones, especially in work with rocks older than the oldest ocean-floor magnetic anomalies. This procedure is the method by which the Brunhes, Matuyama, Gauss, and Gilbert Chronozones were recognized (Cox, Doell, and Dalrymple, 1963) and defined originally (Cox, Doell, and Dalrymple, 1964).

(c) **Ranks.**—Divisions of polarity chronozones are designated polarity subchronozones. Assemblages of polarity chronozones may be termed polarity superchronozones.

Article 86.—**Establishing Formal Units.** Requirements for establishing a polarity-chronostratigraphic unit include those specified in Articles 3 and 4, and also (1) definition of boundaries of the unit, with specific references to designated sections and data; (2) distinguishing polarity characteristics, lithologic descriptions, and included fossils; and (3) correlation and age relations.

Article 87.—**Name.** A formal polarity-chronostratigraphic unit is given a compound name beginning with that for a named geographic feature; the second component indicates the normal, reversed, or mixed polarity of the unit, and the third component is *chronozone*. The initial letter of each term is capitalized. If the same geographic name is used for both a magnetopolarity zone and a polarity-chronostratigraphic unit, the latter should be distinguished by an -an or -ian ending. Example: Tetonian Reversed-Polarity Chronozone.

Remarks: (a) **Preservation of established name.**—A particularly well-established name should not be displaced, either on the basis of priority, as described in Article 7c, or because it was not taken from a geographic feature. Continued use of Brunhes, Matuyama, Gauss, and Gilbert, for example, is endorsed so long as they remain valid units.

(b) **Expression of doubt.**—Doubt in the assignment of polarity zones to polarity-chronostratigraphic units should be made explicit if criteria of time equivalence are inconclusive.

POLARITY-CHRONOLOGIC UNITS

Nature and Boundaries

Article 88.—**Definition.** Polarity-chronologic units are divi-

sions of geologic time distinguished on the basis of the record of magnetopolarity as embodied in polarity-chronostratigraphic units. No special kind of magnetic time is implied; the designations used are meant to convey the parts of geologic time during which the Earth's magnetic field had a characteristic polarity or sequence of polarities. These units correspond to the time spans represented by polarity chronozones, e.g., Gauss Normal Polarity Chronozone. They are not material units.

Ranks and Nomenclature of Polarity-Chronologic Units

Article 89.—**Fundamental Unit.** The polarity chron is the fundamental unit of geologic time designating the time span of a polarity chronozone.

Remark. (a) **Hierarchy.**—Polarity-chronologic units of decreasing hierarchical ranks are polarity superchron, polarity chron, and polarity subchron.

Article 90.—**Nomenclature.** Names for polarity chronologic units are identical with those of corresponding polarity-chronostratigraphic units, except that the term chron (or superchron, etc) is substituted for chronozone (or superchronozone, etc).

DIACHRONIC UNITS

Nature and Boundaries

Article 91.—**Definition.** A diachronic unit comprises the unequal spans of time represented either by a specific lithostratigraphic, allostratigraphic, biostratigraphic, or pedostratigraphic unit, or by an assemblage of such units.

Remarks. (a) **Purposes.**—Diachronic classification provides (1) a means of comparing the spans of time represented by stratigraphic units with diachronous boundaries at different localities, (2) a basis for broadly establishing in time the beginning and ending of deposition of diachronous stratigraphic units at different sites, (3) a basis for inferring the rate of change in areal extent of depositional processes, (4) a means of determining and comparing rates and durations of deposition at different localities, and (5) a means of comparing temporal and spatial relations of diachronous stratigraphic units (Watson and Wright, 1980).

(b) **Scope.**—The scope of a diachronic unit is related to (1) the relative magnitude of the transgressive division of time represented by the stratigraphic unit or units on which it is based and (2) the areal extent of those units. A diachronic unit is not extended beyond the geographic limits of the stratigraphic unit or units on which it is based.

(c) **Basis.**—The basis for a diachronic unit is the diachronous referent.

(d) **Duration.**—A diachronic unit may be of equal duration at different places despite differences in the times at which it began and ended at those places.

Article 92.—**Boundaries.** The boundaries of a diachronic unit are the times recorded by the beginning and end of deposition of the material referent at the point under consideration (Figs. 10, 11).

Remark. (a) **Temporal relations.**—One or both of the boundaries of a diachronic unit are demonstrably time-transgressive. The varying time significance of the boundaries is defined by a series of boundary reference sections (Article 8b, 8e). The duration and age of a diachronic unit differ from place to place (Figs. 10, 11).

Ranks and Nomenclature of Diachronic Units

Article 93.—**Ranks.** A diachron is the fundamental and non-hierarchical diachronic unit. If a hierarchy of diachronic units is needed, the terms episode, phase, span, and cline, in order of decreasing rank, are recommended. The rank of a hierarchical

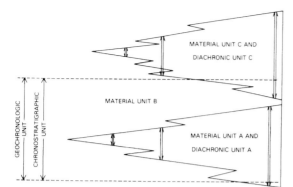

FIG. 10.—Comparison of geochronologic, chronostratigraphic, and diachronic units.

unit is determined by the scope of the unit (Article 91 b), and not by the time span represented by the unit at a particular place.

Remarks. (a) **Diachron.**—Diachrons may differ greatly in magnitude because they are the spans of time represented by individual or grouped lithostratigraphic, allostratigraphic, biostratigraphic, and(or) pedostratigraphic units.

(b) **Hierarchical ordering permissible.**—A hierarchy of diachronic units may be defined if the resolution of spatial and temporal relations of diachronous stratigraphic units is sufficiently precise to make the hierarchy useful (Watson and Wright, 1980). Although all hierarchical units of rank lower than episode are part of a unit next higher in rank, not all parts of an episode, phase, or span need be represented by a unit of lower rank.

(c) **Episode.**—An episode is the unit of highest rank and greatest scope in hierarchical classification. If the "Wisconsinan Age" were to be redefined as a diachronic unit, it would have the rank of episode.

Article 94.—**Name.** The name for a diachronic unit should be compound, consisting of a geographic name followed by the term diachron or a hierarchical rank term. Both parts of the compound name are capitalized to indicate formal status. If the diachronic unit is defined by a single stratigraphic unit, the geographic name of the unit may be applied to the diachronic unit. Otherwise, the geographic name of the diachronic unit should not duplicate that of another formal stratigraphic unit. Genetic terms (e.g., alluvial, marine) or climatic terms (e.g., gla-

cial, interglacial) are not included in the names of diachronic units.

Remarks. (a) **Formal designation of units.**—Diachronic units should be formally defined and named only if such definition is useful.

(b) **Inter-regional extension of geographic names.**—The geographic name of a diachronic unit may be extended from one region to another if the stratigraphic units on which the diachronic unit is based extend across the regions. If different diachronic units in contiguous regions eventually prove to be based on laterally continuous stratigraphic units, one name should be applied to the unit in both regions. If two names have been applied, one name should be abandoned and the other formally extended. Rules of priority (Article 7d) apply. Priority in publication is to be respected, but priority alone does not justify displacing a well-established name by one not well-known or commonly used.

(c) **Change from geochronologic to diachronic classification.**—Lithostratigraphic units have served as the material basis for widely accepted chronostratigraphic and geochronologic classifications of Quaternary nonmarine deposits, such as the classifications of Frye et al (1968), Willman and Frye (1970), and Dreimanis and Karrow (1972). In practice, time-parallel horizons have been extended from the stratotypes on the basis of markedly time-transgressive lithostratigraphic and pedostratigraphic unit boundaries. The time ("geochronologic"), defined on the basis of the stratotype sections but extended on the basis of diachronous stratigraphic boundaries, are diachronic units. Geographic names established for such "geochronologic" units may be used in diachronic classification if (1) the chronostratigraphic and geochronologic classifications are formally abandoned and diachronic classifications are proposed to replace the former "geochronologic" classifications, and (2) the units are redefined as formal diachronic units. Preservation of well-established names in these specific circumstances retains the intent and purpose of the names and the units, retains the practical significance of the units, enhances communication, and avoids proliferation of nomenclature.

Article 95.—**Establishing Formal Units.** Requirements for establishing a formal diachronic unit, in addition to those in Article 3, include (1) specification of the nature, stratigraphic relations, and geographic or areal relations of the stratigraphic unit or units that serve as a basis for definition of the unit, and (2) specific designation and description of multiple reference sections that illustrate the temporal and spatial relations of the defining stratigraphic unit or units and the boundaries of the unit or units.

Remark. (a) **Revision or abandonment.**—Revision or abandonment of the stratigraphic unit or units that serve as the material basis for defini-

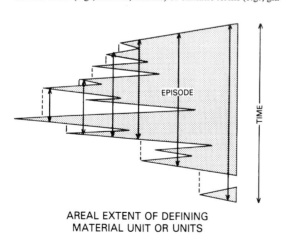

| AREAL EXTENT OF DEFINING MATERIAL UNIT OR UNITS | AREAL EXTENT OF DEFINING MATERIAL UNIT OR UNITS |

FIG. 11.—Schematic relation of phases to an episode.
Parts of a phase similarly may be divided into spans, and spans into clines. Formal definition of spans and clines is unnecessary in most diachronic unit hierarchies.

tion of a diachronic unit may require revision or abandonment of the diachronic unit. Procedure for revision must follow the requirements for establishing a new diachronic unit.

GEOCHRONOMETRIC UNITS

Nature and Boundaries

Article 96.—**Definition.** Geochronometric units are units established through the direct division of geologic time, expressed in years. Like geochronologic units (Article 80), geochronometric units are abstractions, i.e., they are not material units. Unlike geochronologic units, geochronometric units are not based on the time span of designated chronostratigraphic units (stratotypes), but are simply time divisions of convenient magnitude for the purpose for which they are established, such as the development of a time scale for the Precambrian. Their boundaries are arbitrarily chosen or agreed-upon ages in years.

Ranks and Nomenclature of Geochronometric Units

Article 97.—**Nomenclature.** Geochronologic rank terms (eon, era, period, epoch, age, and chron) may be used for geochronometric units when such terms are formalized. For example, Archean Eon and Proterozoic Eon, as recognized by the IUGS Subcommission on Precambrian Stratigraphy, are formal geochronometric units in the sense of Article 96, distinguished on the basis of an arbitrarily chosen boundary at 2.5 Ga. Geochronometric units are not defined by, but may have, corresponding chronostratigraphic units (eonothem, erathem, system, series, stage, and chronozone).

PART III: REFERENCES[10]

American Commission on Stratigraphic Nomenclature, 1947, Note 1—Organization and objectives of the Stratigraphic Commission: American Association of Petroleum Geologists Bulletin, v. 31, no. 3, p. 513-518.
——— ,1961, Code of Stratigraphic Nomenclature: American Association of Petroleum Geologists Bulletin, v. 45, no. 5, p. 645-665.
——— ,1970, Code of Stratigraphic Nomenclature (2d ed.): American Association of Petroleum Geologists, Tulsa, Okla., 45 p.
——— ,1976, Note 44—Application for addition to code concerning magnetostratigraphic units: American Association of Petroleum Geologists Bulletin, v. 60, no. 2, p. 273-277.
Caster, K. E., 1934, The stratigraphy and paleontology of northwestern Pennsylvania, Part 1, Stratigraphy: Bulletins of American Paleontology, v. 21, 185 p.
Chang, K. H., 1975, Unconformity-bounded stratigraphic units: Geological Society of America Bulletin, v. 86, no. 11, p. 1544-1552.
Committee on Stratigraphic Nomenclature, 1933, Classification and nomenclature of rock units: Geological Society of America Bulletin, v. 44, no. 2, p. 423-459, and American Association of Petroleum Geologists Bulletin, v. 17, no. 7, p. 843-868.
Cox, A. V., R. R. Doell, and G. B. Dalrymple, 1963, Geomagnetic polarity epochs and Pleistocene geochronometry: Nature, v. 198, p. 1049-1051.
——— ,1964, Reversals of the Earth's magnetic field: Science, v. 144, no. 3626, p. 1537-1543.
Cross, C. W., 1898, Geology of the Telluride area: U.S. Geological Survey 18th Annual Report, pt. 3, p. 759.
Cumming, A. D., J. G. C. M. Fuller, and J. W. Porter, 1959, Separation of strata: Paleozoic limestones of the Williston basin: American Journal of Science, v. 257, no. 10, p. 722-733.

Dreimanis, Aleksis, and P. F. Karrow, 1972, Glacial history of the Great Lakes–St. Lawrence region, the classification of the Wisconsin(an) Stage, and its correlatives: International Geologic Congress, 24th Session, Montreal, 1972, Section 12, Quaternary Geology, p. 5-15.
Dunbar, C. O., and John Rodgers, 1957, Principles of stratigraphy: Wiley, New York, 356 p.
Forgotson, J. M., Jr., 1957, Nature, usage and definition of marker-defined vertically segregated rock units: American Association of Petroleum Geologists Bulletin, v. 41, no. 9, p. 2108-2113.
Frye, J. C., H. B. Willman, Meyer Rubin, and R. F. Black, 1968, Definition of Wisconsinan Stage: U.S. Geological Survey Bulletin 1274-E, 22 p.
George, T. N., and others, 1969, Recommendations on stratigraphical usage: Geological Society of London, Proceedings no. 1656, p. 139-166.
Harland, W. B., 1977, Essay review [of] International Stratigraphic Guide, 1976: Geology Magazine, v. 114, no. 3, p. 229-235.
——— ,1978, Geochronologic scales, in G. V. Cohee et al, eds., Contributions to the Geologic Time Scale: American Association of Petroleum Geologists, Studies in Geology, no. 6, p. 9-32.
Harrison, J. E., and Z. E. Peterman, 1980, North American Commission on Stratigraphic Nomenclature Note 52—A preliminary proposal for a chronometric time scale for the Precambrian of the United States and Mexico: Geological Society of America Bulletin, v. 91, no. 6, p. 377-380.
Henbest, L. G., 1952, Significance of evolutionary explosions for diastrophic division of Earth history: Journal of Paleontology, v. 26, p. 299-318.
Henderson, J. B., W. G. E. Caldwell, and J. E. Harrison, 1980, North American Commission on Stratigraphic Nomenclature, Report 8—Amendment of code concerning terminology for igneous and high-grade metamorphic rocks: Geological Society of America Bulletin, v. 91, no. 6, p. 374-376.
Holland, C. H., and others, 1978, A guide to stratigraphical procedure: Geological Society of London, Special Report 10, p. 1-18.
Huxley, T. H., 1862, The anniversary address: Geological Society of London, Quarterly Journal, v. 18, p. xl-liv.
International Commission on Zoological Nomenclature, 1964: International Code of Zoological Nomenclature adopted by the XV International Congress of Zoology: International Trust for Zoological Nomenclature, London, 176 p.
International Subcommission on Stratigraphic Classification (ISSC), 1976, International Stratigraphic Guide (H. D. Hedberg, ed.): John Wiley and Sons, New York, 200 p.
International Subcommission on Stratigraphic Classification, 1979, Magnetostratigraphy polarity units—a supplementary chapter of the ISSC International Stratigraphic Guide: Geology, v. 7, p. 578-583.
Izett, G. A., and R. E. Wilcox, 1981, Map showing the distribution of the Huckleberry Ridge, Mesa Falls, and Lava Creek volcanic ash beds (Pearlette family ash beds) of Pliocene and Pleistocene age in the western United States and southern Canada: U. S. Geological Survey Miscellaneous Geological Investigations Map I-1325.
Kauffman, E. G., 1969, Cretaceous marine cycles of the Western Interior: Mountain Geologist: Rocky Mountain Association of Geologists, v. 6, no. 4, p. 227-245.
Matthews, R. K., 1974, Dynamic stratigraphy—an introduction to sedimentation and stratigraphy: Prentice-Hall, New Jersey, 370 p.
McDougall, Ian, 1977, The present status of the geomagnetic polarity time scale: Research School of Earth Sciences, Australian National University, Publication no. 1288, 34 p.
McElhinny, M. W., 1978, The magnetic polarity time scale; prospects and possibilities in magnetostratigraphy, in G. V. Cohee et al, eds., Contributions to the Geologic Time Scale, American Association of Petroleum Geologists, Studies in Geology, no. 6, p. 57-65.
McIver, N. L., 1972, Cenozoic and Mesozoic stratigraphy of the Nova Scotia shelf: Canadian Journal of Earth Science, v. 9, p. 54-70.
McLaren, D. J., 1977, The Silurian-Devonian Boundary Committee. A final report, in A. Martinsson, ed., The Silurian-Devonian boundary: IUGS Series A, no. 5, p. 1-34.
Morrison, R. B., 1967, Principles of Quaternary soil stratigraphy, in R. B. Morrison and H. E. Wright, Jr., eds., Quaternary soils: Reno, Nevada, Center for Water Resources Research, Desert Research Institute, Univ. Nevada, p. 1-69.
North American Commission on Stratigraphic Nomenclature, 1981, Draft North American Stratigraphic Code: Canadian Society of

---

[10]Readers are reminded of the extensive and noteworthy bibliography of contributions to stratigraphic principles, classification, and terminology cited by the International Stratigraphic Guide (ISSC, 1976, p. 111-187).

Petroleum Geologists, Calgary, 63 p.

Palmer, A. R., 1965, Biomere-a new kind of biostratigraphic unit: Journal of Paleontology, v. 39, no. 1, p. 149-153.

Parsons, R. B., 1981, Proposed soil-stratigraphic guide, *in* International Union for Quaternary Research and International Society of Soil Science: INQUA Commission 6 and ISSS Commission 5 Working Group, Pedology, Report, p. 6-12.

Pawluk, S., 1978, The pedogenic profile in the stratigraphic section, *in* W. C. Mahaney, ed., Quaternary soils: Norwich, England, GeoAbstracts, Ltd., p. 61-75.

Ruhe, R. V., 1965, Quaternary paleopedology, *in* H. E. Wright, Jr., and D. G. Frey, eds., The Quaternary of the United States: Princeton, N.J., Princeton University Press, p. 755-764.

Schultz, E. H., 1982, The chronosome and supersome--terms proposed for low-rank chronostratigraphic units: Canadian Petroleum Geology, v. 30, no. 1, p. 29-33.

Shaw, A. B., 1964, Time in stratigraphy: McGraw-Hill, New York, 365 p.

Sims, P. K., 1979, Precambrian subdivided: Geotimes, v. 24, no. 12, p. 15.

Sloss, L. L., 1963, Sequences in the cratonic interior of North America: Geological Society of America Bulletin, v. 74, no. 2, p. 94-114.

Tracey, J. I., Jr., and others, 1971, Initial reports of the Deep Sea Drilling Project, v. 8: U.S. Government Printing Office, Washington, 1037 p.

Valentine, K. W. G., and J. B. Dalrymple, 1976, Quaternary buried paleosols: A critical review: Quaternary Research, v. 6, p. 209-222.

Vella, P., 1964, Biostratigraphic units: New Zealand Journal of Geology and Geophysics, v. 7, no. 3, p. 615-625.

Watson, R. A., and H. E. Wright, Jr., 1980, The end of the Pleistocene: A general critique of chronostratigraphic classification: Boreas, v. 9, p. 153-163.

Weiss, M. P., 1979a, Comments and suggestions invited for revision of American Stratigraphic Code: Geological Society of America, News and Information, v. 1, no. 7, p. 97-99.

———, 1979b, Stratigraphic Commission Note 50--Proposal to change name of Commission: American Association of Petroleum Geologists Bulletin, v. 63, no. 10, p. 1986.

Weller, J. M., 1960, Stratigraphic principles and practice: Harper and Brothers, New York, 725 p.

Willman, H. B., and J. C. Frye, 1970, Pleistocene stratigraphy of Illinois: Illinois State Geological Survey Bulletin 94, 204 p.

REFERENCES

Abrahão, D., and J. E. Warme, 1990, Lacustrine and associated deposits in a rifted continental margin—Lower Cretaceous Lagoa Feia Formation, Campos Basin, offshore Brazil, *in* B. J. Katz (ed.), Lacustrine basin exploration: Am. Assoc. Petroleum Geologists Mem. 50, p. 287–305.

Adams, A. E., W. S. Mackenzie, and G. Guilford, 1984, Atlas of sedimentary rocks under the microscope: John Wiley & Sons, New York, 104 p.

Ager, D. V., 1981, The nature of the stratigraphical record, 2nd ed.: John Wiley & Sons, New York, 122 p.

Agterberg, F. P., 1990, Automated stratigraphic correlation: Elsevier, Amsterdam, 424 p.

Ahlbrandt, T. S., and S. G. Fryberger, 1981, Sedimentary features and significance of interdune deposits, *in* F. G. Ethridge and R. O. Flores (eds.), Recent and ancient nonmarine depositional environments: Models for exploration: Soc. Econ. Paleontologists and Mineralogists Spec. Pub. 31, p. 293–314.

_____ 1982, Introduction to eolian deposits, *in* P. A. Scholle and D. Spearing (eds.), Sandstone depositional environments: Am. Assoc. Petroleum Geologists Mem. 31, p. 11–47.

Aigner, T., 1985, Storm depositional systems: Springer-Verlag, Berlin, 174 p.

Aitken, J. D., 1967, Classification and environmental significance of cryptalgal limestones and dolomites, with illustrations from the Cambrian and Ordovician of southwestern Alberta: Jour. Sed. Petrology, v. 37, p. 1163–1179.

Allen, D. R., and G. V. Chilingarian, 1975, Mechanics of sand compaction, *in* G. V. Chilingarian and K. H. Wolf (eds.), Compaction of coarse-grained sediments, I: Developments in Sedimentology 18A, Elsevier, Amsterdam, p. 43–77.

Allen, J. R. L., 1963, The classification of cross-stratified units with notes on their origin: Sedimentology, v. 2, p. 93–114.

_____ 1964, Studies in fluviatile sedimentation: Six cyclothems from the Lower Old Red Sandstone, Anglo-Welsh Basin: Sedimentology, v. 3, p. 163–198.

_____ 1965a, A review of the origin and characteristics of recent alluvial sediments: Sedimentology, v. 5, p. 89–191.

_____ 1965b, The sedimentation and paleogeography of the Old Red Sandstone of Anglesey, North Wales: Yorkshire Geol. Soc. Proc., v. 35, p. 139–185.

_____ 1968, Current ripples: Their relation to patterns of water and sediment motion: North Holland Publishing, Amsterdam, 433 p.

_____ 1970a, Physical processes of sedimentation: George Allen & Unwin, London, 248 p.

_____ 1970b, Studies in fluviatile sedimentation: A comparison of fining-upward cyclothems, with special reference to coarse-member composition and interpretation: Jour. Sed. Petrology, v. 40, p. 298–323.

_____ 1982, Sedimentary structures: Their character and physical basis, v. 1–2: Elsevier, Amsterdam, 664 p.

_____ 1984, Laminations developed from upper-stage plane beds: A model based on the larger coherent structures of the turbulent boundary layer: Sed. Geology, v. 39, p. 227–242.

_____ 1985, Loose-boundary hydraulics and fluid mechanics: Selected advances since 1961, *in* P. J. Brenchley and B. P. J. Williams (eds.), Sedimentology: Recent developments and applied aspects: Geol. Soc., Blackwell, Oxford, p. 7–30.

Allen, P. A., and J. R. Allen, 1990, Basin analysis—Principles and applications: Blackwell, Oxford, 451 p.

Allen, P. A., and P. Homewood (eds.), 1986, Foreland basins: Blackwell, Oxford, 453 p.

Aller, R. C., 1982, The effects of macrobenthos on chemical properties of marine sediments and overlying water, *in* P. L. McCall and M. J. S. Tevesz (eds.), Animal-sediment relations: The biogenic alteration of sediments: Plenum, New York, p. 53–102.

Alvarez, W., M. A. Arthur, A. G. Fischer, W. Lowrey, G. Napoleone, I. Premoli Silva, and M. W. Roggenthen, 1977, Upper Cretaceous–Paleocene magnetic stratigraphy at Gubbio, Italy: V. Type section for the Late Cretaceous–Paleocene geomagnetic reversal time scale: Geol. Soc. America Bull., v. 88, p. 383–389.

Anadón, P., Ll. Cabrera, and K. Kelts (eds.), 1991, Lacustrine facies analysis: Internat. Assoc. Sedimentologists Spec. Pub. 13, Blackwell, Oxford, 318 p.

Anderson, J. B., and G. M. Ashley (eds.), 1991, Glacial marine sedimentation; Paleoclimatic significance: Geol. Soc. America Spec. Paper 261, 232 p.

Anderson, R. S., M. Sorensen, and B. B. Willets, 1991, A review of recent progress in our understanding of aeolian sediment transport: Acta Mechanica Supplementum 1, Springer-Verlag, Wien, New York, p. 1–19.

Anderson, R. Y., and D. W. Kirkland, 1966, Intrabasin varve correlation: Geol. Soc. America Bull., v. 77, p. 241–256.

Anderton, R., 1985, Clastic facies models and facies analysis, *in* P. J. Brenchley and B. P. J. Williams (eds.), Sedimentology, recent developments and applied aspects: Geol. Soc., Blackwell, Oxford, p. 31–48.

Anstey, N. A., 1982, Simple seismics: International Human Resources Development Corporation, Boston, 168 p.

Appel, P. W. U., and G. L. LaBerge, 1987, Precambrian iron-formations: Theophrastus, S. A., Athens, Greece, 674 p.

Arthur, M. A., T. F. Anderson, I. R. Kaplan, J. Veizer, and L. S. Land, 1983, Stable isotopes in geology: Soc. Econ. Paleontologists and Mineralogists Short Course No. 10.

Arthur, M. A., and A. G. Fischer, 1977, Upper Cretaceous–Paleocene magnetic stratigraphy at Gubbio, Italy: I. Lithostratigraphy and sedimentology: Geol. Soc. America Bull., v. 88, p. 367–371.

Ashley, G. A., 1990, Classification of large-scale subaqueous bedforms: A new look at an old problem: Jour. Sed. Petrology, v. 60, p. 160–172.

Ashley, G. M. (ed.), 1988, The hydrodynamics and sedimentology of a back-barrier lagoon-salt marsh system, Great Sound, New Jersey: Marine Geology, v. 82, p. 1–132.

Ashley, G. M., N. D. Smith, and J. D. Shaw, 1985, Glacial sedimentary environments: Soc. Econ. Paleontologists and Mineralogists Short Course Notes No. 16, 246 p.

Asquith, G., 1982, Basic well log analysis for geologists, Methods in exploration No. 3: Am. Assoc. Petroleum Geologists, Tulsa, Okla., 216 p.

Aston, S. R. (ed.), 1983, Silicon geochemistry and biochemistry: Academic Press, London, 248 p.

Badiozamani, K., 1973, The Dorag dolomitization model—Application to the Middle Ordovician of Wisconsin: Jour. Sed. Petrology, v. 43, p. 965–984.

Bagnold, R. A., 1941, 1954, The physics of blown sand and desert dunes: Methuen, London, 265 p.

_____ 1956, The flow of cohesionless grains in fluids: Royal Soc. London, Philos. Trans., Ser. A., v. 249, p. 235–297.

_____ 1962, Auto-suspension of transported sediment: Turbidity currents: Royal Soc. London Proc. (A), v. 265, p. 315–319.

Bagnold, R. A., and O. Barndorff-Nielsen, 1980, The pattern of natural size distribution: Sedimentology, v. 27, p. 199–207.

Baker, E. T., and G. J. Massoth, 1987, Characteristics of hydrothermal plumes from two vent fields on the Juan de Fuca Ridge, northeast Pacific Ocean: Earth and Planetary Science Letters, v. 85, p. 59–73.

Baker, P. A., and M. Kastner, 1981, Constraints on the formation of sedimentary dolomite: Science, v. 213, p. 214–216.

Baker, P. A., M. Kastner, J. D. Byerlee, and D. A. Lockner, 1980, Pressure solution and hydrothermal recrystallization of carbonate sediments—An experimental study: Marine Geology, v. 38, p. 185–203.

Bally, A. W. (comp.), 1981, Geology of passive continental margins: Am. Assoc. Petroleum Geologists Short Course Notes No. 19.

Bally, A. W. (ed.), 1987, 1988, 1989, Atlas of seismic stratigraphy: Am. Assoc. Petroleum Geologists Studies in Geology 27, 3 volumes (v. 1, 1987, 125 p; v. 2, 1988, 277 p.; v. 3, 1989, 244 p).

Bally, A. W., and S. Snelson, 1980, Realms of subsidence, in A. D. Miall (ed.), Facts and principles of world petroleum occurrence: Canadian Soc. Petroleum Geologists Mem. 6, p. 9–44.

Bandy, O. L. (ed.), 1970, Radiometric dating and paleontologic zonation: Geol. Soc. America Spec. Paper 124, 247 p.

Banerjee, I., 1991, Tidal sand sheets of the Late Albian Joi Fou–Kiowa–Skull Creek marine transgression, Western Interior Seaway of North America, in D. G. Smith, G. E. Reinson, G. A. Zaitlin, and R. A. Rahmani (eds.), Clastic tidal sedimentology: Canadian Soc. Petroleum Geologists Mem. 16, p. 335–348.

Barndorff-Nielsen, O., 1977, Exponentially decreasing distributions for the logarithm of particle size: Proc. Royal Society London A., v. 353, p. 401–419.

Barndorff-Nielsen, O., K. Dalsgaard, C. Halgreen, H. Kuhlman, J. T. Moller, and G. Schou, 1982, Variation in particle size distribution over a small dune: Sedimentology, v. 29, p. 53–65.

Barndorff-Nielsen, O. E., and B. B. Willets (eds.), 1991, Aeolian grain transport 1—Mechanics: Springer-Verlag, Wien, New York, 181 p.

Barnes, R. S. K., 1980, Coastal lagoons: Cambridge University Press, Cambridge, 106 p.

Barrett, P. J., 1980, The shape of rock particles, a critical review: Sedimentology, v. 27, p. 291–303.

Barron, J. A., 1987, Diatomite: Environmental and geologic factors affecting its distribution, in J. R. Hein (ed.), Siliceous sedimentary rock-hosted ores and petroleum: Van Nostrand Reinhold, New York, p. 164–178.

Basan, P. B. (ed.), 1978, Trace fossil concepts: Soc. Econ. Paleontologists and Mineralogists Short Course No. 5, 181 p.

Basu, A., S. W. Young, L. J. Suttner, W. C. James, and G. H. Mack, 1975, Reevaluation of the use of undulatory extinction and polycrystallinity in detrital quartz for provenance interpretation: Jour. Sed. Petrology, v. 45, p. 873–882.

Bates, C. C., 1953, Rational theory of delta formation: Am. Assoc. Petroleum Geologists Bull., v. 37, p. 2119–2161.

Bates, R. L., and J. A. Jackson (comps.), 1980, Glossary of geology, 2nd ed.: American Geological Institute, Falls Church, Va., 749 p.

Bathurst, R. G. C., 1975, Carbonate sediments and their diagenesis, 2nd ed.: Developments in Sedimentology 12, Elsevier, Amsterdam, 658 p.

Baturin, G. N., 1982, Phosphorites on the sea floor: Origin, composition, and distribution: Developments in Sedimentology 33, Elsevier, Amsterdam, 343 p.

Belderson, R. H., M. A. Johnson, and N. H. Kenyon, 1982, Bedforms, in A. H. Stride (ed.), Offshore tidal sands: Chapman and Hall, London, p. 27–57.

Belderson, R. H., N. H. Kenyon, A. H. Stride, and A. R. Stubbs, 1972, Sonographs of the sea floor: Elsevier, Amsterdam, 185 p.

Bentor, Y. K., 1980a, Phosphorites—The unsolved problem, in Y. K. Bentor (ed.), Marine phosphorites—Geochemistry, occurrence, genesis: Soc. Econ. Paleontologists and Mineralogists Spec. Pub. 29, p. 3–18.

Bentor, Y. K. (ed.), 1980b, Marine phosphorites—Geochemistry, occurrence, genesis: Soc. Econ. Paleontologists and Mineralogists Spec. Pub. 29, 249 p.

Berg, O. R., and D. B. Wolverton (eds.), 1985, Seismic stratigraphy II, an integrated approach: Am. Assoc. Petroleum Geologists Mem. 39, 276 p.

Berger, R., and H. E. Suess (eds.), 1979, Radiocarbon dating: University of California Press, Berkeley, 787 p.

Berger, W. H., 1974, Deep-sea sedimentation, in C. A. Burk and C. L. Drake (eds.), The geology of continental margins: Springer-Verlag, New York, p. 213–241.

Berggren, W. A., and J. A. Van Couvering, 1978, Biochronology, in G. V. Cohee, M. F. Glaessner, and H. D. Hedberg (eds.), Contributions to the geologic time scale: Am. Assoc. Petroleum Geologists, Tulsa, Okla., p. 39–55.

Berner, R. A., 1971, Principles of chemical sedimentology: McGraw-Hill, New York, 240 p.

_____ 1975, The role of magnesium in crystal growth of aragonite from sea water: Geochim. et Cosmochim. Acta, v. 39, p. 489–505.

_____ 1980, Early diagenesis: A theoretical approach: Princeton University Press, Princeton, N.J., 224 p.

Berner, R. A., T. Baldwin, and G. R. Holdren, Jr., 1979, Authigenic iron sulfides as paleosalinity indicators: Jour. Sed. Petrology, v. 49, p. 1345–1350.

Berner, R. A., J. T. Westrich, R. Graber, J. Smith, and C. S. Martens, 1978, Inhibition of aragonite precipitation from supersaturated seawater: A laboratory and field study: Am. Jour. Sci., v. 278, p. 816–837.

Berry, W. B. N., 1968, Growth of a prehistoric time scale: W. H. Freeman, San Francisco, 158 p.

_____ 1987, Growth of a prehistoric time scale: Based on organic evolution, 2nd ed.: Blackwell, Palo Alto, Calif., 202 p.

Bhattacharyya, A., and G. M. Friedman (eds.), 1983, Modern carbonate environments: Benchmark Papers in Geology, v. 74, Dowden, Hutchinson and Ross, Stroudsburg, Pa., 376 p.

Bhattacharya, J. P., and R. G. Walker, 1991, River- and wave-dominated depositional systems of the Upper Cretaceous Dunvegan Formation, northwestern Alberta: Bull. Canadian Petroleum Geol., v. 39, p. 165–191.

_____ 1992, Deltas, in R. G. Walker and N. P. James (eds.), Facies models—Response to sea level changes: Geol. Assoc. Canada, p. 157–178.

Biddle, K. T. (ed.), 1991, Active margin basins: Am. Assoc. Petroleum Geologists Mem. 52, Tulsa, Okla., 324 p.

Bien, G. S., D. E. Contois, and W. H. Thomas, 1959, The removal of silica from fresh water entering the sea, in H. A. Ireland (ed.), Silica in sediments: Soc. Econ. Paleontologists and Mineralogists Spec. Pub. 7, p. 20–35.

Bigarella, J. J., 1972, Eolian environments: Their characteristics, recognition and importance, in J. K. Rigby and W. K. Hamblin (eds.), Recognition of ancient sedimentary environments: Soc. Econ. Paleontologists and Mineralogists Spec. Pub. 16, p. 12–62.

Biggs, R. B., 1978, Coastal bays, in R. A. Davis, Jr. (ed.), Coastal sedimentary environments: Springer-Verlag, New York, p. 69–99.

Birkland, P. W., 1974, Pedology, weathering, and geomorphological research: Oxford University Press, New York, 285 p.

Black, M., 1933, The precipitation of calcium carbonate on the Great Bahama Bank: Geol. Mag., v. 70, p. 455–466.

Blake, G. H., 1991, Review of the Neogene biostratigraphy and stratigraphy of the Los Angeles Basin and implications for basin evolution, in K. T. Biddle (ed.), Active margin basins: (Am. Assoc. Petroleum Geologists Mem. 52, Tulsa, Okla., p. 135–184.

Blatt, H., 1967, Original characteristics of quartz grains: Jour. Sed. Petrology, v. 37, p. 401–424.

_____ 1970, Determination of mean sediment thickness in the crust: A sedimentological model: Geol. Soc. America Bull., v. 81, p. 255–262.

_____ 1979, Diagenetic processes in sandstones, in P. A. Scholle and P. R. Schluger (eds.), Aspects of diagenesis: Soc. Econ. Paleontologists and Mineralogists Spec. Pub. 26, p. 141–157.

_____ 1982, Sedimentary petrology: W. H. Freeman, San Francisco, 564 p.

Blatt, H., G. V. Middleton, and R. Murray, 1980, Origin of sedimentary rocks, 2nd ed.: Prentice-Hall, Englewood Cliffs, N.J., 782 p.

Blatt, H., and M. W. Totten, 1981, Detrital quartz as an indicator of distance from shore in marine mudrocks: Jour. Sed. Petrology, v. 51, p. 1259–1266.

Boggs, S., Jr., 1967a, Measurement of roundness and sphericity parameters using an electronic particle size analyzer: Jour. Sed. Petrology, v. 37, p. 908–913.

_____ 1967b, A numerical method for sandstone classification: Jour. Sed. Petrology, v. 37, p. 548–555.

_____ 1968, Experimental study of rock fragments: Jour. Sed. Petrology, v. 38, p. 1326–1339.

_____ 1969, Relationship of size and composition in pebble counts: Jour. Sed. Petrology, v. 39, p. 1243–1247.

_____ 1974, Sand-wave fields in Taiwan Strait: Geology, v. 2, p. 251–253.

_____ 1975, Seabed resources of the Taiwan continental shelf: Acta Oceanographica Taiwanica, v. 5, p. 1–18.

_____ 1984, Quaternary sedimentation in the Japan arc-trench system: Geol. Soc. America Bull., v. 95, p. 669–685.

_____ 1992, Petrology of sedimentary rocks: Macmillan, New York, 707 p.

Boggs, S., Jr., and C. A. Jones, 1976, Seasonal reversal of flood-tide dominant sediment transport in a small Oregon estuary: Geol. Soc. America Bull., v. 87, p. 419–426.

Boggs, S., Jr., D. G. Livermore, and M. G. Seitz, 1985, Humic macromolecules in natural waters: Jour, Macromolecular Science, v. C25 (4), p. 599–657.

Boggs, S., Jr., W. C. Wang, and F. S. Lewis, 1979, Sediment properties and water characteristics of the Taiwan shelf and slope: Acta Oceanographica Taiwanica, v. 10, p. 10–49.

Boguchwal, L. A., and J. B. Southard, 1990, Bed configurations in steady unidirectional water flows. Pt. 1.

Scale model study using fine sands: Jour. Sed. Petrology, v. 60, p. 649–657.

Bohn, H. L., B. L. McNeal, and G. A. O'Connor, 1979, Soil chemistry: John Wiley & Sons, New York, 329 p.

Bohor, B. F., and H. J. Gluskoter, 1973, Boron in illites as an indicator of paleosalinity of Illinois coals: Jour. Sed. Petrology, v. 43, p. 945–956.

Boothroyd, J. C., 1978, Mesotidal inlets and estuaries, *in* R. A. Davis, Jr. (ed.), Coastal sedimentary environments: Springer-Verlag, New York, p. 287–360.

_____ 1985, Tidal inlets and tidal deltas, *in* R. A. Davis, Jr. (ed.), Coastal sedimentary environments, 2nd ed.: Springer-Verlag, New York, p. 445–532.

Borchert, H., and R. O. Muir, 1964, Salt deposits: The origin, metamorphism, and deformation of evaporites: Van Nostrand, London, 338 p.

Bostock, N. H., 1979, Microscopic measurement of level of catagenesis of solid organic matter in sedimentary rocks to aid exploration for petroleum and to determine former burial temperatures—A review, *in* P. A. Scholle and P. R. Schluger (eds.), Aspects of diagenesis: Soc. Econ. Paleontologists and Mineralogists Spec. Pub. 26, p. 17–44.

Boucot, A. J., 1975, Evolution and extinction rate controls: Developments in Paleontology and Stratigraphy, I, Elsevier, New York, 427 p.

Boucot, A. J., and R. S. Carney, 1981, Principles of marine benthic paleoecology: Academic Press, New York, 463 p.

Bouisset, P. M., and A. M. Augustin, 1993, Borehole magnetostratigraphy, absolute age dating, and correlation of sedimentary rocks, with examples from the Paris Basin, France: Am. Assoc. Petroleum Geologists Bull., v. 77, p. 569–587.

Bouma, A., 1962, Sedimentology of some flysch deposits: Elsevier, Amsterdam, 168 p.

Bouma, A. H., 1969, Methods for the study of sedimentary structures: John Wiley & Sons, New York, 457 p.

_____ 1979, Continental slopes, *in* L. J. Doyle and O. H. Pilkey (eds.), Geology of continental slopes: Soc. Econ. Paleontologists and Mineralogists Spec. Pub. 27, p. 1–15.

Bouma, A. H., H. L. Berryhill, R. L. Brenner, and H. J. Knebel, 1982, Continental shelf and epicontinental seaways, *in* P. A. Scholle and D. Spearing (eds.), Sandstone depositional environments: Am. Assoc. Petroleum Geologists Mem. 31, p. 281–327.

Bouma, A. H., J. M. Coleman et al., 1986, Initial reports of Deep Sea Drilling Project 96: U.S. Government Printing Office, Washington, D.C.

Bouma, A. H., G. T. Moore, and J. M. Coleman (eds.), 1978, Framework, facies, and oil-trapping characteristics of the upper continental margin: Am. Assoc. Petroleum Geologists Studies in Geology 7, 326 p.

Bouma, A. H., W. R. Normark, and N. E. Barnes (eds.), 1985, Submarine fans and related turbidite systems: Springer-Verlag, New York, 351 p.

Bowen, R., 1988, Isotopes in the earth sciences: Elsevier, London, 647 p.

Braitsch, O., 1971, Salt deposits: Their origin and composition: Springer-Verlag, Berlin, 279 p.

Bremner, J. M., and J. Rogers, 1990, Phosphorite deposits on the Namibian continental shelf, *in* W. C. Burnett and S. R. Riggs (eds.), Phosphate deposits of the world: v. 3: Neogene to Modern phosphorites: Cambridge University Press, Cambridge, p. 143–152.

Brenchley, P. J., and B. P. J. Williams (eds.), 1985, Sedimentology: Recent developments and applied aspects: Geol. Soc. Spec. Pub. 12, Blackwell, Oxford, 320 p.

Bricker, O. P., 1971, Carbonate cements: Johns Hopkins University Press, Baltimore, Md., 376 p.

Briggs, D. E. G., E. N. K. Clarkson, and R. J. Aldridge, 1983, The conodont animal: Lethaea, v. 16, p. 1–14.

Brodzikowski, K., and A. J. van Loon, 1991, Glacigenic sediments: Elsevier, Amsterdam, 674 p.

Bromley, R. G., 1990, Trace fossils, biology and taphonomy: Special topics in paleontology 3: Unwin Hyman, London, 280 p.

Brookfield, M. E., 1984, Eolian sands, *in* R. G. Walker (ed.), Facies models, 2nd ed.: Geoscience Canada Reprint Ser. 1, p. 91–104.

Brookfield, M. E., 1992, Eolian systems, *in* R. G. Walker and N. P. James (eds.), Facies models: Response to sea level changes, Geol. Assoc. Canada, p. 143–156.

Brookfield, M. E., and T. S. Ahlbrandt (eds.), 1983, Eolian sediments and processes: Elsevier, Amsterdam, 660 p.

Broussard, M. L. (ed.), 1975, Deltas: Models for exploration: Houston Geological Society, 555 p.

Brower, J. C., 1981, Quantitative biostratigraphy, 1830–1980, *in* D. F. Merriam (ed.), Computer applications in the earth sciences: An update of the 70s: Plenum, New York, p. 63–103.

Brown, L. F., Jr., and W. L. Fisher, 1980, Seismic stratigraphic interpretation and petroleum exploration: Geophysical principles and techniques: Am. Assoc. Petroleum Geologists Continuing Education Course Notes Ser. 16, 56 p.

Brownridge, S., and T. F. Moslow, 1991, Tidal estuary and marine facies of the Glauconitic Member, Drayton Valley, central Alberta, *in* D. G. Smith, G. E. Reinson, B. A. Zaitlin, and R. A. Rahmani (eds.), Clastic tidal sedimentology: Canadian Soc. Petroleum Geologists, p. 107–122.

Bruce, C. H., 1984, Smectite dehydration—Its relation to structural development and hydrocarbon accumulation in northern Gulf of Mexico Basin: Am. Assoc. Petroleum Geologists Bull., v. 68, p. 673–683.

Bubb, J. N., and W. G. Hatelid, 1977, Recognition of carbonate buildups, *in* C. E. Payton (ed.), Seismic stratigraphy—Application to hydrocarbon exploration: Am. Assoc. Petroleum Geologists Mem. 26, p. 185–204.

Budd, D. A., and P. M. Harris (eds.), 1990, Carbonate-siliciclastic mixtures: Soc. Econ. Paleontologists and Mineralogists Reprint Ser. 14, 272 p.

Bull, P. A., 1986, Procedures in environmental reconstruction by SEM analysis, *in* G. De C. Sieveking and M. B. Hart (eds.), The scientific study of flint and chert: Cambridge University Press, Cambridge, p. 221–226.

Bull, W. B., 1972, Recognition of alluvial-fan deposits in the stratigraphic record, *in* J. K. Rigby and W. K. Hamblin (eds.), Recognition of ancient sedimentary environments: Soc. Econ. Paleontologists and Mineralogists Spec. Pub. 16, p. 63–83.

Buol, W. W., F. D. Hole, and R. J. McCracken, 1989, Soil genesis and classification, 3rd ed.: Iowa State University Press, Ames, 446 p.

Burger, H., and W. Skala, 1976, Comparison of sieve and thin-section techniques by a Monte-Carlo-model: Computer Geoscience, v. 2, p. 123–139.

Burk, C. A., and C. L. Drake (eds.), 1974, The geology of continental margins: Springer-Verlag, New York, 1009 p.

Burnett, W. C., and P. N. Froelich (eds.), 1988, The origin of marine phosphorites: The results of the *R. V. Robert D. Conrad* Cruise 23–06 to the Peru Shelf: Special issue of Marine Geology, v. 80, p. 181–346.

Burnett, W. C., and S. R. Riggs (eds.), 1990, Phosphate deposits of the world, v. 3: Neogene to Modern phosphorites: Cambridge University Press, Cambridge, 464 p.

Burst, J. F., 1965, Subaqueously formed shrinkage cracks in clay: Jour. Sed. Petrology, v. 35, p. 348–353.

_____ 1976, Argillaceous sediment dewatering: Ann. Rev. Earth and Planetary Sci., v. 4, p. 293–318.

Bushman, J. R., 1983, Twelve fallacies of uniformitarianism: Comment: Geology, v. 11, p. 312–313.

Bustin, R. M., A. R. Cameron, D. A. Grieve, and W. D. Kalkreuth, 1985, Coal petrology, its principles, methods, and applications: Geol. Assoc. Canada Short Course Notes, v. 3, 230 p.

Butler, R. F., 1992, Paleomagnetism: Blackwell, Boston, 319 p.

Button, A., T. D. Brock, P. J. Cook, H. P. Eugster, A. M. Goodwin, H. L. James, L. Margulis, K. H. Nealson, J. O. Nriagu, A. F. Trendall, and M. R. Walter, 1982, Sedimentary iron deposits, evaporites, and phosphorites, *in* H. D. Holland and M. Schidlowski (eds.), Mineral deposits and evolution of the biosphere: Springer-Verlag, New York, p. 259–273.

Calvert, S. E., 1974, Deposition and diagenesis of silica in marine sediments, *in* K. J. Hsü and H. C. Jenkyns (eds.), Pelagic sediments: On land and under the sea: Internat. Assoc. Sedimentologists Spec. Pub. 1, p. 273–300.

_____ 1983, Sedimentary geochemistry of silicon, *in* R. R. Aston (ed.), Silicon geochemistry and biogeochemistry: Academic Press, London, p. 143–186.

Cameron, E. M., and R. M. Garrels, 1980, Geochemical composition of some Precambrian shales from the Canadian Shield: Chem. Geology, v. 28, p. 181–197.

Campbell, C. V., 1967, Lamina, laminaset, bed and bedset: Sedimentology, v. 8, p. 7–26.

Cande, S. C., and D. V. Kent, 1992, A new geomagnetic polarity time scale for the Late Cretaceous and Cenozoic: Jour. Geophys. Research, v. 97, p. 13,917–13,951.

Cann, J. R., and M. R. Strens, 1989, Modeling periodic megaplume emission by black smoker systems: Jour. Geophys. Research, v. 94, p. 12,227–12,237.

Cant, D. J., 1982, Fluvial facies models and their application, *in* P. A. Scholle and D. Spearing (eds.), Sandstone depositional environments: Am. Assoc. Petroleum Geologists Mem. 31, p. 115–138.

Cant, D. J., and Walker, R. G., 1976, Development of a braided fluvial model for the Devonian Battery Point Sandstone, Quebec: Canadian Jour. Earth Sci., v. 13, p. 102–119.

Carballo, J. D., L. S. Land, and D. E. Miser, 1987, Holocene dolomitization of supratidal sediments by active tidal pumping, Sugarloaf Key, Florida: Jour. Sed. Petrology, v. 57, p. 153–165.

Carlson, W. D., 1983, The polymorphs of $CaCO_3$ and the aragonite-calcite transformation, *in* R. J. Reeder (ed.), Carbonates: Mineralogy and chemistry: Reviews in Mineralogy, v. 11, Mineralog. Soc. America, p. 191–225.

Carr, T. R., 1982, Log-linear models, Markov chains and cyclic sedimentation: Jour. Sed. Petrology, v. 52, p. 905–912.

Carroll, D., 1970, Rock weathering: Plenum, New York, 203 p.

Carver, R. E. (ed.), 1971, Procedures in sedimentary petrology: John Wiley & Sons, New York, 653 p.

Castanares, A. A., and F. B. Phleger (eds.), 1969, Coastal lagoons—A symposium: Universidad Nacional Autonoma de Mexico/UNESCO, Mexico City, 686 p.

Catacosinos, P. A., W. B. Harrison III, and P. A. Daniels, Jr., 1990, Structure, stratigraphy and petroleum potential of the Michigan Basin, *in* M. W. Leighton, D. R. Kolata, D. F. Oltz, and J. J. Eidel (eds.), Interior cratonic basins, Am. Assoc. Petroleum Geologists Mem. 51 p. 561–601.

Cathcart, J. B., 1989, The phosphate deposits of Florida with a note on the deposits in Georgia and South Carolina, USA, *in* A. J. G. Nothold R. P. Sheldon, and D. F. Davidson (eds.), Phosphate deposits of the world, v. 2: Phosphate rock resources: Cambridge University Press, Cambridge, p. 62–70.

Catt, J. A., 1986, Soils and Quaternary geology: A handbook for field scientists: Clarendon Press, Oxford, 267 p.

Cayeux, L., 1931, Introduction à l'étude pétrographique des roches sédimentaires: Imp. Nat., Paris, 524 p.

_____ 1935, Les roches sédimentaires de France: Roches carbonatées: Masson et Cie, Paris, 447 p.

Chambers, R. L., and S. B. Upchurch, 1979, Multivariate analysis of sedimentary environments using grain size frequency distributions: Jour. Math. Geology, v. 11, p. 27–43.

Chambre Syndical de la Recherche et de la Production due Pétrole et du Gaz Naturel (eds.), 1981, Evaporite deposits: Illustration and interpretation of some environmental sequences: Editions Technip, Paris, and Gulf Publishing, Houston, 266 p.

Chan, M. A., 1989, Erg margin of the Permian White Rim Sandstone, SE Utah: Sedimentology, v. 36, p. 235–252.

Channell, J. E. T., 1982, Palaeomagnetic stratigraphy as a correlation technique, *in* G. S. Odin (ed.), Numerical dating in stratigraphy: John Wiley & Sons, New York, p. 81–106.

Cheel, R. J., and D. A. Leckie, 1992, Coarse-grained storm beds of the Upper Cretaceous Chungo Member (Wapiabi Formation), Southern Alberta, Canada: Jour. Sed. Petrology, v. 62, p. 933–945.

Childs, O. E., 1983, Correlation of stratigraphic units of North America, COSUNA, 1977–83: Am. Assoc. Petroleum Geologists and U.S. Geol. Survey, 49 p.

Chilingarian, G. V., H. J. Bissell, and R. W. Fairbridge (eds.), 1967, Carbonate rocks. Part A: Origin, occurrence, and classification, 471 p.: Part B: Physical and chemical aspects, 413 p.: Elsevier, New York.

Chilingarian, G. V., and K. H. Wolf (eds.), 1976a, Compaction of coarse-grained sediments, I: Elsevier, New York, 552 p.

_____ 1976b, Compaction of coarse-grained sediments, II: Elsevier, New York, 808 p.

Chilingarian, G. V., and T. F. Yen, 1978, Bitumens, asphalts and tar sands: Elsevier, New York, 331 p.

Choquette, P. W., and L. C. Pray, 1970, Geologic nomenclature and classification of porosity in sedimentary carbonates: Am. Assoc. Petroleum Geologists Bull., v. 54, p. 207–250.

Chowns, T. M., and J. E. Elkins, 1974, The origin of quartz geodes and cauliflower cherts through the silicification of anhydrite nodules: Jour. Sed. Petrology, v. 44, p. 885–903.

Christie-Blick, N., G. S. Mountain, and K. G. Miller, 1990, Seismic stratigraphic record of sea-level change, *in* R. R. Reville et al. (eds.), Sea level change: National Research Council Studies in Geophysics, National Academy Press, Washington, D.C., p. 116–140.

Clarke, F. W., 1924, The data of geochemistry, 5th ed.: U.S. Geol. Survey Bull. 770, 841 p.

Clauer, N., 1982, The rubidium-strontium method applied to sediments: Certitudes and uncertainties, *in* G. S. Odin (ed.), Numerical dating in stratigraphy: John Wiley & Sons, New York, p. 245–276.

Claypool, G. E., W. T. Holser, I. R. Kaplan, H. Sakai, and I. Zak, 1980, The age curves of sulfur and oxygen isotopes in marine sulfate and their mutual interpretation: Chem. Geology, v. 28, p. 199–260.

Clayton, R. N., and E. T. Degens, 1959, Use of carbon isotope analyses of carbonates in differentiating freshwater and marine sediments: Am. Assoc. Petroleum Geologists Bull., v. 43, p. 890–897.

Clement, B. M., D. V. Kent, and N. D. Opdyke, 1982, Brunhes-Matuyama polarity transition in three deep-sea cores: Philos. Trans. Royal Soc. London, v. A306, p. 113–119.

Clemmensen, L. B., and J. Hegner, 1991, Eolian sequence and erg dynamics: The Permian Corrie Sandstone, Scotland: Jour. Sed. Petrology, v. 61, p. 768–774.

Clifton, H. E., 1969, Beach laminations: Nature and origin: Marine Geology, v. 7, p. 553–559.

_____ 1976, Wave-formed sedimentary structures: A conceptual model, *in* R. A. Davis and R. L. Ethington (eds.), Beach and nearshore sedimentation: Soc. Econ. Paleontologists and Mineralogists Spec. Pub. 24, p. 126–148.

_____ 1982, Estuarine deposits, *in* P. A. Scholle and D. Spearing (eds.), Sandstone depositional environments: Am. Assoc. Petroleum Geologists Mem. 31, p. 179–189.

_____ 1988, Sedimentologic relevance of convulsive geologic events, *in* H. E. Clifton (ed.), Sedimentological consequences of convulsive geologic events: Geol. Soc. America Spec. Paper 229, p. 1–5.

Clifton, H. E. (ed.), 1988, Sedimentologic consequences of convulsive geologic events: Geol. Soc. America Spec. Paper 229, 157 p.

Clifton, H. E., and J. R. Dingler, 1984, Wave-formed structures and paleoenvironmental reconstruction: Marine Geology, v. 60, p. 165–198.

Clifton, H. E., R. E. Hunter, and J. V. Gardner, 1988, Analysis of eustatic, tectonic, and sedimentologic influences on transgressive and regressive cycles in the Upper Cenozoic Merced Formation, San Francisco, California, *in* K. L. Kleinspehn and C. Paola (eds.), 1988, New perspectives in basin analysis: Springer-Verlag, New York, p. 109–128.

Cloud, P. E., 1973, Paleoecological significance of banded iron formations: Econ. Geology, v. 68, p. 1135–1143.

Clube, S. V. M. (ed.), 1989, Catastrophes and evolution: Astronomical foundations: Cambridge University Press, Cambridge, 239 p.

Cohee, G. V., M. F. Glaessner, and H. D. Hedberg, 1978, Contributions to the geologic time scale: Am. Assoc. Petroleum Geologists, Tulsa, Okla., 388 p.

Cole, G. A., 1975, Textbook of limnology: C. V. Mosby, St. Louis, Mo., 283 p.

Colella, A., and D. B. Prior (eds.), 1990, Coarse-grained deltas: Internat. Assoc. Sedimentologists Spec. Pub. 10, Blackwell, Oxford, 357 p.

Coleman, J. M., 1976, Deltas: Processes of deposition and models for exploration: Continuing Education Publication Co., Champaign, Ill., 102 p.

_____ 1981, Deltas: Processes of deposition and models for exploration, 2nd ed.: Burgess, 124 p.

Coleman, J. M., and D. B. Prior, 1980, Deltaic sand bodies: Am. Assoc. Petroleum Geologists Education Short Course Notes No. 15, 171 p.

_____ 1982, Deltaic environments of deposition, *in* P. A. Scholle and D. Spearing (eds.), Sandstone depositional environments: Am. Assoc. Petroleum Geologists Mem. 31, p. 139–178.

_____ 1983, Deltaic influences on shelfedge instability processes, *in* D. J. Stanley and G. T. Moore (eds.), The shelfbreak: Critical interface on continental margins: Soc. Econ. Paleontologists and Mineralogists Spec. Pub. 33, p. 121–127.

Coleman, J. M., and L. D. Wright, 1975, Modern river deltas: Variability of process and sand bodies, *in* M. L. Broussard (ed.), Deltas: Models for exploration: Houston Geological Society, p. 99–149.

Collins, A. G., 1975, Geochemistry of oilfield waters: Developments in petroleum science 1, Elsevier, Amsterdam, 496 p.

_____ 1980, Oilfield brines, *in* G. D. Hobson (ed.), Developments in petroleum geology 2: Applied Science Publishers, London, p. 139–187.

Collinson, J. D., 1978a, Alluvial sediments, *in* H. G. Reading (ed.), Sedimentary environments and facies: Elsevier, New York, p. 15–60.

_____ 1978b, Deserts, *in* H. G. Reading (ed.), Sedimentary environments and facies: Elsevier, New York, p. 80–96.

_____ 1978c, Lakes, *in* H. G. Reading (ed.), Sedimentary environments and facies: Elsevier, New York, p. 61–79.

_____ 1986a, Alluvial sediments, *in* H. G. Reading (ed.), Sedimentary environments and facies: Blackwell, Oxford, p. 20–62.

_____ 1986b, Deserts, *in* H. G. Reading (ed.), Sedimentary environments and facies: Blackwell, Oxford, p. 95–112.

Collinson, J. D., and J. Lewin (eds.), 1983, Modern and ancient fluvial systems: Internat. Assoc. Sedimentologists Spec. Pub. 6, Blackwell, Oxford, 575 p.

Collinson, J. D., and D. B. Thompson, 1982, Sedimentary structures: George Allen & Unwin, London, 194 p.

Collinson, J. D., and D. B. Thompson, 1989, Sedimentary structures, 2nd ed.: HarperCollins Academic, New York, 208 p.

Colman, S. M., and D. P. Dethier, 1986, Rates of chemical weathering of rocks and minerals: Academic Press, Orlando, 603 p.

Colombo, G., 1977, Lagoons, *in* R. S. K. Barnes (ed.), The coastline: John Wiley & Sons, New York, p. 63–81.

Conde, K. C., 1982, Plate tectonics and crustal evolution, 2nd ed.: Pergamon, New York, 310 p.

Conglio, M., and G. R. Dix, 1992, Carbonate slopes, *in* R. G. Walker and N. P. James (eds.), Facies models: Response to sea level change: Geol. Assoc. Canada, p. 349–373.

Conklin, B. A., and J. E. Conklin (eds.), 1984, Stratigraphy: Foundations and concepts: Benchmark Papers in Geology, v. 82, Van Nostrand Reinhold, New York, 365 p.

Conybeare, C. E. B., and K. A. W. Crook, 1968, Manual of sedimentary structures: Department of National Development, Bureau of Mineral Resources, Geology and Geophysics Bull. 102, 327 p.

Cook, H. E., 1983, Introductory perspectives, basic carbonate principles, and stratigraphic and depositional models, *in* H. E. Cook, A. C. Hine, and H. T. Mullins (eds.), Platform margin and deep water carbonates: Soc. Econ. Paleontologists and Mineralogists Short Course Notes No. 12, p. 1-1–1-89.

Cook, H. E., and P. Enos (eds.), 1977, Deep-water carbonate environments: Soc. Econ. Paleontologists and Mineralogists Spec. Pub. 25, 336 p.

Cook, H. E., M. E. Field, and J. V. Gardner, 1982, Characteristics of sediments on modern and ancient continental slopes, *in* P. A. Scholle and D. Spearing (eds.), Sandstone depositional environments: Am. Assoc. Petroleum Geologists Mem. 31, p. 329–364.

Cook, H. E., A. C. Hine, and H. T. Mullins (eds.), 1983, Platform margin and deep water carbonates: Soc. Econ. Paleontologists and Mineralogists Short Course Notes No. 12.

Cook, H. E., and H. T. Mullins, 1983, Basin margin, *in* P. A. Scholle, D. G. Bebout, and C. H. Moore (eds.), Carbonate depositional environments: Am. Assoc. Petroleum Geologists Mem. 33, p. 539–618.

Cook, P. J., 1976, Sedimentary phosphate deposits, *in* K. H. Wolf (ed.), Handbook of strata-bound and stratiform ore deposits, v. 7: Elsevier, New York, p. 505–536.

Cook, P. J., and S. A. Elgueta, 1986, Proterozoic and Cambrian phosphorites—deposits: Lady Annie, Queensland, Australia, *in* P. J. Cook and J. H. Shergold (eds.), Phosphate deposits of the world, v. 1: Proterozoic and Cambrian phosphorites: Cambridge University Press, Cambridge, p. 132–148.

Cook, P. J., and J. H. Shergold, 1986a, Proterozoic and Cambrian phosphorites—An introduction, *in* P. J. Cook and J. H. Shergold (eds.), Phosphate deposits of the world, v. 1: Proterozoic and Cambrian phosphorites: Cambridge University Press, Cambridge, p. 1–8.

Cook, P. J. K., and J. H. Shergold, 1986b, Proterozoic and Cambrian phosphorites—Nature and origin, *in* P. J. Cook and J. H. Shergold (eds.), Phosphate deposits of the world, v. 1: Proterozoic and Cambrian phosphorites: Cambridge University Press, Cambridge, p. 369–386.

Cook, P. J., and J. H. Shergold (eds.), 1986c, Phosphate deposits of the world, v. 1: Proterozoic and Cambrian phosphorites: Cambridge University Press, Cambridge, 386 p.

Coombs, D. S., 1971, Present status of the zeolite facies: Advances in Chemistry Ser. 101 (Molecular sieve zeolites, Am. Chem. Soc.—I), p. 317–327.

Coplen, T. B., C. Kendall, and J. Hopple, 1983, Comparison of stable isotope reference samples: Nature, v. 302, p. 236–238.

Corliss, J. B., J. Dymond, L. I. Gordon, J. M. Edmond, R. P. von Herzen, R. D. Ballard, K. Green, D. Williams, A. Bainbridge, K. Crane, and T. H. van Andel, 1979, Submarine thermal springs on the Galápagos Rift: Science, v. 203, p. 1073–1083.

Couch, E. L., 1971, Calculation of paleosalinities from boron and clay mineral data: Am. Assoc. Petroleum Geologists Bull., v. 55, p. 1829–1837.

Courtney, F. M., and S. T. Trudgill, 1984, The soil, 2nd ed.: Edward Arnold, London, 123 p.

Cowen, R., 1988, The role of algal symbiosis in reefs through time. Palaios, v. 3, p. 221–227.

Cox, A., 1969, Geomagnetic reversals: Science, v. 163, p. 237–245.

_____ 1973a, Plate tectonics and geomagnetic reversals: Introduction and reading list, *in* A. Cox (ed.), Plate tectonics and geomagnetic reversals: W. H. Freeman, San Francisco, p. 138–153.

Cox, A. (ed.), 1973b, Plate tectonics and geomagnetic reversals: W. H. Freeman, San Francisco, 702 p.

Cox, A. V., 1982, Magnetostratigraphic time scale, *in* W. B. Harland, A. V. Cox, P. G. Llewellyn, C. A. G. Pickton, A. G. Smith, and R. A. Walters (eds.), Geologic time scale: Cambridge University Press, Cambridge, p. 63–84.

Crelling, J. C., and R. R. Dutcher, 1980, Principles and applications of coal petrology: Soc. Econ. Paleontologists and Mineralogists Short Course Notes No. 8, 127 p.

Cressman, E. R., 1962, Nondetrital siliceous sediments: U.S. Geol. Survey Prof. Paper 440–T, 22 p.

Crevello, P. D., and P. M. Harris (eds.), 1985, Deepwater carbonates: Buildups, turbidites, debris flows and chalks: Soc. Econ. Paleontologists and Mineralogists Core Workshop No. 6, 527 p.

Crevello, P. D., J. L. Wilson, J. F. Sarg, and J. F. Read (eds.), 1989, Controls on carbonate platform and basin development: Soc. Econ. Paleontologists and Mineralogists Spec. Pub. 44, 405 p.

Crimes, T. P., 1975, The stratigraphical significance of trace fossils, *in* R. W. Frey (ed.), The study of trace fossils: Springer-Verlag, New York, p. 109–130.

Crimes, T. P., and J. C. Harper (eds.), 1970, Trace fossils: Seel House Press, Liverpool, 547 p.

_____ 1977, Trace fossils 2: Seel House Press, Liverpool, 351 p.

Cronin, L. E. (ed.), 1975, Estuarine research, v. 2: Geology and engineering: Academic Press, New York, 587 p.

Crosby, E. J., 1972, Classification of sedimentary environments, *in* J. K. Rigby and W. K. Hamblin (eds.), Recognition of ancient sedimentary environments: Soc. Econ. Paleontologists and Mineralogists Spec. Pub. 16, p. 4–11.

Cross, T. A., and M. A. Lessenger, 1988, Seismic stratigraphy: Ann. Rev. Earth and Planetary Sciences, v. 16, p. 319–354.

Cubitt, J. M., and R. A. Reyment (eds.), 1982, Quantitative stratigraphic correlation: John Wiley & Sons, New York, 301 p.

Curran, H. A. (ed.), 1985, Biogenic structures: Their use in interpreting depositional environments: Soc. Econ. Paleontologists and Mineralogists Spec. Pub. 35, 347 p.

Curray, J. R., W. R. Dickinson, W. G. Dow, K. O. Emery, D. R. Seeley, P. R. Vail, and H. Yarbough, 1977, Geology of continental margins: Am. Assoc. Petroleum Geologists Short Course Notes No. 5, variously paginated.

Curray, R. R., 1966, Observations of alpine mudflows in the Tenmile Range, central Colorado: Geol. Soc. America Bull., v. 77, p. 771–776.

Dalrymple, R. W., 1992, Tidal depositional systems, *in* R. G. Walker and N. P. James (eds.), Facies models: Geol. Assoc. Canada, p. 195–238.

Dalrymple, R. W., B. A. Zaitlin, and R. Boyd, 1992, Estuarine facies models: Conceptual basin and stratigraphic implications: Jour. Sed. Petrology, v. 62, p. 1130–1146.

Daniels, E. J., S. P. Altaner, S. Marshak, and J. R. Eggleson, 1990, Hydrothermal alteration in anthracite in eastern Pennsylvania: Implications for the mechanisms of anthracite formation: Geology, v. 18, p. 247–250.

Dapples, E. C., 1979, Diagenesis of sandstones, *in* G. Larsen and G. V. Chilingarian (eds.), Diagenesis in sediments and sedimentary rocks: Developments in Sedimentology 25A, Elsevier, Amsterdam, p. 31–141.

Darby, D. A., and Y. W. Tsang, 1987, Variations in ilmenite element composition within and among drainage basins: Implications for provenance. Jour. Sed. Petrology, v. 57, p. 831–838.

Darby, D. A., G. R. Whittecar, R. A. Barringer, and J. R. Garrett, 1990, Alluvial lithofacies recognition in a humid-tropical setting: Sed. Geology, v. 67, p. 161–174.

Davies, D. K., and F. G. Ethridge, 1975, Sandstone composition and depositional environments: Am. Assoc. Petroleum Geologists Bull., v. 59, p. 239–264.

Davies, D. K., F. G. Ethridge, and R. R. Berg, 1971, Recognition of barrier environments: Am. Assoc. Petroleum Geologists Bull., v. 55, p. 550–565.

Davies, T. A., and D. S. Gorsline, 1976, Oceanic sediments and sedimentary processes, *in* J. P. Riley and R. Chester (eds.), Chemical oceanography, v. 5, 2nd ed.: Academic Press, New York, p. 1–80.

Davis, R. A., Jr., 1968, Algal stromatolites composed of quartz sandstone: Jour. Sed. Petrology, v. 38, p. 953–955.

_____ 1978, Beach and nearshore zone, *in* R. A. Davis, Jr. (ed.), Coastal sedimentary environments: Springer-Verlag, New York, p. 237–285.

_____ 1983, Depositional systems: A genetic approach to sedimentary geology: Prentice-Hall, Englewood Cliffs, N.J., 669 p.

_____ 1992, Depositional systems, 2nd ed.: Prentice-Hall, Englewood Cliffs, N.J., 604 p.

Davis, R. A., Jr. (ed.), 1985, Coastal sedimentary environments, 2nd ed.: Springer-Verlag, New York, 716 p.

Davis, R. A., Jr., and R. L. Ethington (eds.), 1976, Beach and nearshore sedimentation: Soc. Econ. Paleontologists and Mineralogists Spec. Pub. 24, 187 p.

Davis, R. A., Jr., and W. T. Fox, 1972, Four-dimensional model for beach and inner nearshore sedimentation: Jour. Geology, v. 80, p. 484–493.

Davis, T. L., 1984, Seismic-stratigraphic facies model, *in* R. G. Walker (ed.), Facies models, 2nd ed.: Geoscience Canada Reprint Ser. 1, p. 311–317.

Dean, W. E., G. R. Davies, and R. Y. Anderson, 1975, Sedimentological significance of nodular and laminated anhydrite: Geology, v. 3, p. 367–372.

Dean, W. E., M. Leinen, and D. A. V. Stow, 1985, Classification of deep-sea, fine-grained sediments: Jour. Sed. Petrology, v. 55, p. 250–256.

Dean, W. E., and B. C. Schreiber, 1978, Marine evaporites: Soc. Econ. Paleontologists and Mineralogists Short Course Notes No. 4, 193 p.

de Boer, P. L., 1991, Pelagic black shale–carbonate rhythms: Orbital forcing and oceanographic response, *in* G. Einsele, W. Ricken, and A. Seilacher (eds.), Cycles and events in stratigraphy: Springer-Verlag, Berlin, p. 63–78.

de Boer, P. L., A. van Gelder, and S. D. Nio (eds.), 1988, Tide-influenced sedimentary environments and facies: D. Reidel, Dordrecht, 530 p.

Dec, T., 1992, Textural characteristics and interpretation of second-cycle, debris-flow dominated alluvial fans (Devonian of Northern Scotland): Sed. Geology, v. 77, p. 269–296.

Degens, E. T., 1965, Geochemistry of sediments: Prentice-Hall, Englewood Cliffs, N.J., 342 p.

Degens, E. T., and S. Epstein, 1964, Oxygen and carbon isotope ratios in coexisting calcites and dolomites from recent and ancient sediments: Geochim. et Cosmochim. Acta, v. 28, p. 23–44.

Depaolo, D. J., and K. L. Finger, 1988, Applications of strontium isotopes in correlating Monterey Formation, California: Am. Assoc. Petroleum Geologists Bull., v. 72, p. 379.

Dickey, P. A., 1969, Increasing concentration of subsurface brines with depth: Chem. Geology, v. 4, p. 361–370.

Dickinson, W. R., 1974, Plate tectonics and sedimentation, *in* W. R. Dickinson (ed.), Tectonics and sedimentation: Soc. Econ. Paleontologists and Mineralogists Spec. Pub. 22, p. 1–27.

Dickinson, W. R., 1982, Composition of sandstones in cirum-Pacific subduction complexes and fore-arc basins: Am. Assoc. Petroleum Geologists Bull., v. 66, p. 121–137.

Dickinson, W. R., L. S. Beard, G. R. Brakenridge, J. L. Erjavec, R. C. Ferguson, K. F. Inman, R. A. Knepp, F. A. Lindberg, and P. T. Ryberg, 1983, Provenance of North American Phanerozoic sandstones in relation to tectonic setting: Geol. Soc. America Bull., v. 94, p. 222–235.

Dickinson, W. R., and C. A. Suczek, 1979, Plate tectonics and sandstone composition: Am. Assoc. Petroleum Geologists Bull., v. 63, p. 2164–2182.

Dickinson, W. R., and H. Yarborough, 1976, Plate tectonics and hydrocarbon accumulation: Am. Assoc. Petroleum Geologists, Ed. Course Note Ser. No. 1, Tulsa, Okla., 34 p.

Didyk, B. M., B. T. R. Simoreit, S. C. Brassel, and G. Eglinton, 1978, Organic geochemical indicators of paleoenvironmental conditions of sedimentation: Nature, v. 272, p. 216–222.

Diemer, J. A., E. S. Belt 1991, Sedimentology and paleohydrology of the meandering river systems of the Fort Union Formation southeastern Montana: Sed. Geology, v. 75, p. 85-108.

Diessel, C. F. K., and R. Offler, 1975, Change in physical properties of coalified and graphitized phytoclasts with grade of metamorphism: Neues Jahrg. Mineralogie Monatsh., Jahrg. 1975, p. 11–26.

Dill, D. F., 1966, Sand flows and sand falls, *in* R. W. Fairbridge (ed.), Encyclopedia of oceanography: Reinhold, New York, p. 763–765.

Dimroth, E., 1976, Aspects of the sedimentary petrology of cherty iron-formation, *in* K. H. Wolf (ed.), Handbook of strata-bound and stratiform ore deposits, v. 7: Elsevier, New York, p. 203–254.

_____ 1979, Models of physical sedimentation of iron formations, *in* R. G. Walker (ed.), Facies models: Geoscience Canada Reprint Ser. 1, p. 159–174.

Dimroth, E., and M. M. Kimberley, 1976, Precambrian atmospheric oxygen: Evidence in the sedimentary distribution of carbon, sulfur, uranium and iron: Canadian Jour. Earth Sci., v. 13, p. 1161–1185.

Dobkins, J. E., and R. L. Folk, 1970, Shape development on Tahati-Nui: Jour. Sed. Petrology, v. 40, p. 1167–1203.

Dobrin, M. B., 1976, Introduction to geophysical prospecting: McGraw-Hill, New York, 630 p.

Dodd, J. R., and R. J. Stanton, 1975, Paleosalinities within a Pliocene bay, Kettleman Hills, California: A study of the resolving power of isotope and faunal techniques: Geol. Soc. America Bull., v. 86, p. 51–64.

_____ 1981, Paleoecology, concepts and applications: John Wiley & Sons, New York, 559 p.

Dominguez, J. M. L., L. Martin, and A. C. S. P. Bittencourt, 1987, Sea-level history and Quaternary evolution of river mouth–associated beach-ridge plains along the east-southeast Brazilian Coast: A summary, *in* D. Nummedal, O. H. Pilkey, and J. D. Howard (eds.), Sea-level fluctuations and coastal evolution: Soc. Econ. Paleontologists and Mineralogists Spec. Pub. 41, p. 115–127.

Donovan, D. T., 1966, Stratigraphy—An introduction to principles: John Wiley & Sons, New York, 199 p.

Donovan, S. K. (ed.), 1989, Mass extinctions: Processes and evidence: Columbia University Press, New York, 266 p.

Donselaar, M. E., 1989, The Cliff House Sandstone, San Juan Basin, New Mexico: Model for the stacking of "transgressive" barrier complexes: Jour. Sed. Petrology, v. 59, p. 13–27.

Dott, R. H., 1983, SEPM presidential address: Episodic sedimentation—How normal is average? How rare is rare? Does it matter?: Jour. Sed. Petrology, v. 53, p. 5–23.

Dott, R. H., Jr., and J. Bourgeois, 1982, Hummocky stratification: Significance of its variable bedding sequences: Geol. Soc. America Bull., v. 93, p. 663–680.

Dott, R. H., and R. H. Shaver (eds.), 1974, Modern and ancient geosynclinal sedimentation: Soc. Econ. Paleontologists and Mineralogists Spec. Pub. 19, 380 p.

Dowdeswell, J. A., 1982, Scanning electron micrographs of quartz sand grains from cold environments examined using Fourier shape analysis: Jour. Sed. Petrology, v. 52, p. 1315–1326.

Dowdeswell, J. D., and J. D. Scourse (eds.), 1990, Glaciomarine environments: Processes and sediments: Geol. Soc. London Spec. Pub. 53, 423 p.

Doyle, L. J., W. J. Cleary, and D. U. Pilkey, 1968, Mica: Its use in determining shelf-depositional regime: Marine Geology, v. 6, p. 381–389.

Doyle, L. J., and O. H. Pilkey (eds.), 1979, Geology of continental slopes: Soc. Econ. Paleontologists and Mineralogists Spec. Pub. 27, 374 p.

Doyle, L. J., and H. H. Roberts (eds.), 1988, Carbonate-clastic transitions: Elsevier, Amsterdam, 304 p.

Drever, J. I., 1971, Magnesium iron replacements in clay minerals in anoxic marine sediments: Science, v. 172, p. 1334–1336.

_____ 1974, Geochemical model for the origin of Precambrian banded iron formations: Geol. Soc. America Bull., v. 85, p. 1099–1106.

Drever, J. I. (ed.), 1985, The chemistry of weathering: D. Reidel, Hingham, Mass., 336 p.

Drewry, D., 1986, Glacial geologic processes: Edward Arnold, London, 276 p.

Droz, L., and G. Bellaiche, 1985, Rhone deep-sea fan: Morphostructure and growth pattern: Am. Assoc. Petroleum Geologists Bull., v. 69, p. 460–479.

Duff, P. McL. D., A. Hallam, and E. K. Walton, 1967, Cyclic sedimentation: Developments in Sedimentology 10, Elsevier, Amsterdam, 280 p.

Duke, W. L., 1985, Hummocky cross-stratification, tropical hurricanes and intense winter storms: Sedimentology, v. 32, p. 167–194.

Duke, W. L., R. W. C. Arnott, and R. J. Cheel, 1991, Shelf sandstones and hummocky cross-stratification: New insights on a stormy debate: Geology, v. 19, p. 625–628.

Dunbar, C. O., and J. Rodgers, 1957, Principles of stratigraphy: John Wiley & Sons, New York, 356 p.

Duncan, D. C., 1976, Geologic setting of oil shale deposits and world prospects, in T. F. Yen and G. V. Chilingarian (eds.), Oil shale: Elsevier, New York, p. 13–26.

Dunham, R. J., 1962, Classification of carbonate rocks according to depositional textures, in W. E. Ham (ed.), Classification of carbonate rocks. Am. Assoc. Petroleum Geologists Mem. 1, p. 108–121.

_____ 1970, Stratigraphic reef versus ecologic reefs: Am. Assoc. Petroleum Geologists Bull., v. 54, p. 1931–1932.

Dunoyer de Segonzac, G., 1970, The transformation of clay minerals during diagenesis and low-grade metamorphism: A review: Sedimentology, v. 15, p. 281–346.

Dyer, K. R., 1973, Estuaries: A physical introduction: John Wiley & Sons, New York, 140 p.

Dyer, K. R. (ed.), 1979, Estuarine hydrography and sedimentation: A handbook: Cambridge University Press, Cambridge, England, 230 p.

Dzulynski, S., and E. K. Walton, 1965, Sedimentary features of flysch and greywackes: Developments in Sedimentology 7, Elsevier, Amsterdam, 274 p.

Easterbrook, D. J., 1982, Characteristic features of glacial sediments, in P. A. Scholle and D. Spearing (eds.), Sandstone depositional environments: Am. Assoc. Petroleum Geologists Mem. 31, p. 1–10.

Easterbrook, D. J. (ed.), 1988, Dating Quaternary sediments: Geol. Soc. America Spec. Paper 227, 165 p.

Eckel, E. C., 1904, On the chemical composition of American shales and roofing slates: Jour. Geology, v. 12, p. 25–29.

Edmond, J. M., 1980, Ridge crest hot springs: The story so far: EOS, v. 61, p. 129–131.

Edmond, J. M., K. L. Von Damm, R. E. McDuff, and C. I. Measures, 1982, Chemistry of hot springs on the East Pacific Rise and their effluent dispersal: Nature, v. 297, p. 187–191.

Edwards, J. D., and P. A. Santogrossi (eds.), 1990, Divergent/passive margin basins: Am. Assoc. Petroleum Geologists Mem. 48, Tulsa, Okla., 252 p.

Edwards, M., 1986, Glacial environments, in H. G. Reading (ed.), Sedimentary environments and facies, 2nd ed.: Blackwell, Oxford, p. 445–470.

Edwards, M. B., 1978, Glacial environments, in H. G. Reading (ed.), Sedimentary environments and facies: Elsevier, New York, p. 416–438.

Ehrlich, R., and B. Weinberg, 1970, An exact method for characterization of grain shape: Jour. Sed. Petrology, v. 40, p. 205–212.

Eicher, D. L., 1976, Geologic time, 2nd ed.: Prentice-Hall, Englewood, Cliffs, N.J., 152 p.

Eichler, J., 1976, Origin of the Precambrian banded iron-formations, in K. H. Wolf (ed.), Handbook of strata-bound and stratiform ore deposits, v. 7: Elsevier, New York, p. 157–201.

Einsele, G., 1992, Sedimentary basins: Springer-Verlag, Berlin, 628 p.

Einsele, G., W. Ricken, and A. Seilacher, 1991a, Cycles and events in stratigraphy—Basic concepts and terms, in G. Einsele, W. Ricken, and A. Seilacher (eds.), Cycles and events in stratigraphy: Springer-Verlag, Berlin, p. 1–19.

Einsele, G., W. Ricken, and A. Seilacher (eds.), 1991b, Cycles and events in stratigraphy: Springer-Verlag, Berlin, 955 p.

Einsele, G., and A. Seilacher, 1982, Cyclic and event stratification: Springer-Verlag, Berlin, 536 p.

Ekdale, A. A., R. G. Bromley, and S. G. Pemberton, 1984, Ichnology, trace fossils in sedimentology and stratigraphy: Soc. Econ. Paleontologists and Mineralogists Short Course No. 15, 317 p.

Eldredge, N., and S. J. Gould, 1972, Punctuated equilibria: An alternative to phyletic gradualism, in T. J. M. Schopf (ed.), Models in paleobiology: Freeman, Cooper, San Francisco, p. 82–115.

_____ 1977, Evolutionary models and biostratigraphic strategies, in E. G. Kauffman and J. E. Hazel (eds.), Concepts and methods of biostratigraphy: Dowden, Hutchinson and Ross, Stroudsburg, Pa., p. 25–40.

Eliott, T., 1978a, Deltas, in H. G. Reading (ed.), Sedimentary environments and facies: Blackwell, Oxford, p. 97–142.

_____ 1978b, Clastic shorelines, in H. G. Reading (ed.), Sedimentary environments and facies: Elsevier, New York, p. 143–177.

_____ 1986a, Deltas, in H. G. Reading (ed.), Sedimentary environments and facies, 2nd ed.: Blackwell, Oxford, p. 113–154.

_____ 1986b, Siliciclastic shorelines, *in* H. G. Reading (ed.), Sedimentary environments and facies, 2nd ed.: Blackwell, Oxford, p. 155–188.

Embry, A. F., and J. E. Klovan, 1971, A late Devonian reef tract on the northeastern Banks Island, N.W.T.: Canadian Petroleum Geology Bull., v. 19, p. 730–781.

_____ 1972, Absolute water depth limits of late Devonian paleoecological zones: Geol. Rundschau, v. 61, p. 672–686.

Emery, K. O., 1968, Relict sediments on continental shelves of the world: Am. Assoc. Petroleum Geologists Bull., v. 52, p. 445–464.

_____ 1977, Structure and stratigraphy of divergent continental margins: Am. Assoc. Petroleum Geologists Continuing Education Course Notes Ser. 5, Geology of continental margins, p. B1–B20.

Emiliani, C., 1955, Pleistocene temperatures: Jour. Geology, v. 63, p. 538–578.

Enos, P., 1983, Shelf environment, *in* P. A. Scholle, D. G. Bebout, and C. H. Moore (eds.), Carbonate depositional environments: Am. Assoc. Petroleum Geologists Mem. 33, p. 267–296.

Enos, P., and C. H. Moore, 1983, Fore-reef slope, *in* P. A. Scholle, D. G. Bebout, and C. H. Moore (eds.), Carbonate depositional environments: Am. Assoc. Petroleum Geologists Mem. 33, p. 507–538.

Epstein, A. G., J. B. Epstein, and L. D. Harris, 1977, Conodont color alteration—An index to organic metamorphism: U.S. Geol. Survey Prof. Paper 995, 27 p.

Eriksson, K. A., 1979, Marginal marine processes from the Archaen Moodies Group, Barberton Mountain Land, South Africa: Evidence and significance: Precambrian Research, v. 8, p. 153–182.

Ernst, W., 1970, Geochemical facies analysis: Elsevier, Amsterdam, 152 p.

Eslinger, E. V., S. M. Savin, and H-W. Yeh, 1979, Oxygen isotope geothermometry of diagenetically altered shales, *in* P. A. Scholle and P. R. Schluger (eds.), Aspects of diagenesis: Soc. Econ. Paleontologists and Mineralogists Spec. Pub. 26, p. 113–124.

Ethridge, F. G., and R. M. Flores (eds.), 1981, Recent and ancient nonmarine depositional environments: Models for exploration: Soc. Econ. Paleontologists and Mineralogists Spec. Pub. 31, 349 p.

Eugster, H. P., and L. A. Hardie, 1978, Saline lakes, *in* A. Lerman (ed.), Lakes: Chemistry, geology, physics: Springer-Verlag, New York, p. 237–294.

Eugster, H. P., and K. Kelts, 1983, Lacustrine chemical sediments, *in* A. S. Goudie and K. Pye (eds.), Chemical sediments and geomorphology: Academic Press, London and New York, p. 321–368.

Evans, G., 1965, Intertidal flat sediments and their environments of deposition in the Wash: Geol. Soc. London Quart. Jour., v. 121, p. 209–241.

Evans, I., and C. G. St. C. Kendall, 1977, An interpretation of the depositional setting of some deep-water Jurassic carbonates of the central High Atlas Mountains, Morocco, *in* H. E. Cook and P. Enos (eds.), Deep-water carbonate environments: Soc. Econ. Paleontologists and Mineralogists Spec. Pub. 25, p. 249–261.

Evenson, E. B., Ch. Schlüchter, and J. Rabassa, 1983, Till and related deposits: A. A. Balkema, 454 p.

Ewers, W. E., 1983, Chemical factors in the deposition and diagenesis of banded iron-formation, *in* A. F. Trendall and R. C. Morris (eds.), Developments in Precambrian Geology 6: Elsevier, Amsterdam, p. 491–512.

Ewing, M., and E. M. Thorndike, 1965, Suspended matter in deep-ocean water: Science, v. 147, p. 1291–1294.

Eyles, N. (ed.), 1983, Glacial geology: An introduction for engineers and earth scientists: Pergamon, Oxford, 409 p.

Eyles, N., and C. H. Eyles, 1992, Glacial depositional systems, *in* R. G. Walker and N. P. James (eds.), Facies models: Response to sea level changes: Geol. Assoc. Canada, p. 73–100.

Eyles, N., and A. D. Miall, 1984, Glacial facies, *in* R. G. Walker (ed.), Facies models, 2nd ed.: Geoscience Canada Reprint Ser. 1, p. 15–38.

Fabricus, F. H., 1977, Origin of marine ooids and grapestones: Contrib. Sedimentology 7, 113 p.

Fagerstrom, J. A., 1987, The evolution of reef communities: John Wiley & Sons, New York, 600 p.

Fairbridge, R. W., 1980, The estuary: Its definition and geodynamic cycle, *in* E. Olausson and I. Cato (eds.), Chemistry and biochemistry of estuaries: John Wiley & Sons, New York, p. 1–35.

Fanning, D. S., and M. C. B. Fanning, 1989, Soil: Morphology, genesis, classification: John Wiley & Sons, New York, 395 p.

Farrow, G. E., N. H. Allen, and E. B. Akpan, 1984, Bioclastic carbonate sedimentation on a high-latitude, tide-dominated shelf: Northeast Orkney Islands, Scotland: Jour. Sed. Petrology, v. 54, p. 373–393.

Faul, H., 1966, Ages of rocks, planets, and stars: McGraw-Hill, New York, 109 p.

Faure, G., 1977, Principles of isotope geology: John Wiley & Sons, New York, 464 p.

_____ 1982, The marine-strontium geochronometer, *in* G. S. Odin (ed.), Numerical dating in stratigraphy, Part I: John Wiley & Sons, p. 73–80.

_____ 1986, Principles of isotope geology, 2nd ed.: John Wiley & Sons, New York, 589 p.

Fedo, C. M., and J. D. Cooper, 1990, Braided fluvial to marine transition: The basal Lower Cambrian Wood Canyon Formation, southern Marble Mountains, Mojave Desert, California: Jour. Sed. Petrology, v. 60, p. 220–234.

Fenwick, I., 1985, Paleosols, problems of recognition and interpretation, *in* J. Boardman (ed.), Soils and Quaternary evolution: Wiley Interscience, New York, p. 3–21.

Fieller, N. R. J., D. D. Gilbertson, and W. Olbricht, 1984, A new method for environmental analysis of particle size distribution data from shoreline sediments: Nature, v. 311, p. 648–651.

Fischer, A. G., 1961, Stratigraphic record of transgressing seas in the light of sedimentation on the Atlantic coast of New Jersey: Am. Assoc. Petroleum Geologists Bull., v. 45, p. 1656–1666.

Fisher, J. S., and R. Dolan (eds.), 1977, Beach processes and coastal hydrodynamics: Dowden, Hutchinson and Ross, Stroudsburg, Pa., 382 p.

Fisher, W. L., and J. H. McGowen, 1967, Depositional systems in the Wilcox Group of Texas and their relationship to occurrences of oil and gas: Gulf Coast Assoc. Geol. Soc. Trans., v. 17, p. 105–125.

Fitzpatrick, E. A., 1980, Soils: Longman, London, 353 p.

Fleming, B. W., 1980, Sand transport and bedform patterns on the continental shelf between Durban and Port Elizabeth (southeast Africa continental margin): Sed. Geology, v. 26, p. 179–205.

Flessa, K. W., 1990, The "facts" of mass extinctions, in V. L. Sharpton and P. D. Ward (eds.), Global catastrophes in Earth history: Geol. Soc. America Spec. Paper 247, p. 1–7.

Flint, R. F., 1971, Glacial and Quaternary geology: John Wiley & Sons, New York, 892 p.

Folk, R. L., 1955, Student operator error in determination of roundness, sphericity, and grain size: Jour. Sed. Petrology, v. 25, p. 297–301.

_____ 1959, Practical petrographic classification of limestones: Am. Assoc. Petroleum Geologists Bull., v. 43, p. 1–38.

_____ 1962, Spectral subdivision of limestone types, in W. E. Ham (ed.), Classification of carbonate rocks: Am. Assoc. Petroleum Geologists Mem. 1, p. 62–84.

_____ 1965, Some aspects of recrystallization in ancient limestones, in L. C. Pray and R. C. Murray (eds.), Dolomitization and limestone diagenesis: Soc. Econ. Paleontologists and Mineralogists Spec. Pub. 13, p. 14–48.

_____ 1974, Petrology of sedimentary rocks: Hemphill, Austin, Tex., 182 p.

Folk, R. L., P. B. Andrews, and D. W. Lewis, 1970, Detrital sedimentary rock classification and nomenclature for use in New Zealand: New Zealand Jour. Geol. and Geophysics, v. 13, p. 937–968.

Folk, R. L., and L. S. Land, 1975, Mg/Ca ratio and salinity: Two controls over crystallization of dolomite: Am. Assoc. Petroleum Geologists Bull., v. 59, p. 60–68.

Forrest, J., and N. R. Clark, 1989, Characterizing grain size distributions: Evaluation of a new approach using multivariate extension of entropy analysis: Sedimentology, v. 36, p. 711–722.

Fouch, T. D., and W. E. Dean, 1982, Lacustrine and associated clastic depositional environments, in P. A. Scholle and D. Spearing (eds.), Sandstone depositional environments: Am. Assoc. Petroleum Geologists Mem. 31, p. 87–114.

Fox, W. T., 1983, At the sea's edge, Chapter 4, Tides, p. 93–124: Prentice-Hall, Englewood Cliffs, N.J., 317 p.

Fraser, G. S., and L. Suttner, 1986, Alluvial fans and fan deltas: International Human Resources Development Corporation, Boston, 199 p.

Freeman, W. E., and G. S. Visher, 1975, Stratigraphic analysis of the Navajo Sandstone: Jour. Sed. Petrology, v. 45, p. 651–668.

Frey, R. W. (ed.), 1975, The study of trace fossils: Springer-Verlag, New York, 562 p.

Frey, R. W., 1978, Behavioral and ecological implications of trace fossils, in P. B. Basan (ed.), Trace fossil concepts: Soc. Econ. Paleontologists and Mineralogists Short Course No. 5, p. 43–66.

Frey, R. W., and P. B. Basan, 1978, Coastal salt marshes, in R. A. Davis, Jr. (ed.), Coastal sedimentary environments: Springer-Verlag, New York, p. 101–170.

Frey, R. W., and S. G. Pemberton, 1984, Trace fossil facies models, in R. G. Walker (ed.), Facies models: Geoscience Canada Reprint Ser. 1, p. 189–207.

_____ 1985, Biogenic structures in outcrops and cores. 1. Approaches to ichnology: Canadian Petroleum Geology Bull., v. 33, p. 72–115.

Frey, R. W., S. G. Pemberton, and J. A. Fagerstrom, 1984, Morphological, ethological and environmental significance of the ichnogenera Scoyenia and Ancorichnus: Jour. Paleontology, v. 58, p. 511–528.

Frey, R. W., and A. Seilacher, 1980, Uniformity in marine invertebrate ichnology: Lethaea, v. 13, p. 183–207.

Frey, R. W., and R. A. Wheatcroft, 1989, Organism-substrate relations and their impact on sedimentary petrology: Jour. Geological Education, v. 37, p. 261–279.

Friedman, G. M., 1961, Distinction between dune, beach, and river sands from their textural characteristics: Jour. Sed. Petrology, v. 31, p. 514–529.

_____ 1967, Dynamic processes and statistical parameters compared for size frequency distribution of beach and river sands: Jour. Sed. Petrology, v. 37, p. 327–354.

Friedman, G. M. (ed.), 1969, Depositional environments in carbonate rocks: Soc. Econ. Paleontologists and Mineralogists Spec. Pub. 14, 209 p.

Friedman, G. M., 1979, Address of the retiring president of the International Association of Sedimentologists: Differences in size distributions of populations of particles among sands of various origins: Sedimentology, v. 26, p. 3–32.

Friedman, G. M., and J. E. Sanders, 1978, Principles of sedimentology: John Wiley & Sons, New York, 792 p.

Froelich, P. N., M. A. Arthur, W. C. Burnett, M. Deakin, V. Hensley, R. Jahnke, L. Kaul, K.-H Kim, K. Roe, A. Soutar, and C. Vathakanon, 1988, Early diagenesis of organic matter in Peru continental margin sediments: Phosphorite precipitation: Marine Geology, v. 80, p. 309–343.

Frost, S. H., M. P. Weiss, and J. B. Saunders (eds.), 1977, Reefs and related carbonates—Ecology and sedimentology: Am. Assoc. Petroleum Geologists Studies in Geology 4, 421 p.

Frostick, L. E., and I. Reid, 1987, Desert sediments: Ancient and Modern: Geol. Soc. Spec. Pub. 35, Blackwell, Oxford, 401 p.

Fryberger, S. G., and G. Dean, 1979, Dune forms and wind regime, in E. D. McKee (ed.), 1979, A study of global sand seas: U.S. Geol. Survey Prof. Paper 1052, p. 137–169.

Faüchtbauer, H., 1974, Zur diagenese fluviatiler sandsteine: Geol. Rundschau, v. 63, p. 904–925.

Gagan, M. K., D. P. Johnston, and R. M. Carter, 1988, The Cyclone Winifred storm beds, central Great Barrier Reef shelf, Australia: Jour. Sed. Petrology, v. 58, p. 845–856.

Gains, A. M., 1980, Dolomitization kinetics: Recent experimental studies, *in* D. H. Zenger, J. B. Dunham, and R. L. Ethington (eds.), Concepts and models of dolomitization: Soc. Econ. Paleontologists and Mineralogists Spec. Pub. 28, p. 81–86.

Galehouse, J. S., 1971, Sedimentation analysis, *in* R. E. Carver (ed.), Procedures in sedimentary petrology: John Wiley & Sons, New York, p. 69–94.

Gall, J. C., 1983, Ancient sedimentary environments and the habitats of living organisms: Springer-Verlag, Berlin, 219 p.

Galloway, W. E., 1975, Process framework for describing the morphologic and stratigraphic evolution of deltaic depositional systems, *in* M. L. Broussard (ed.), Deltas: Models for exploration: Houston Geological Society, p. 87–98.

_____ 1989, Clastic facies models, depositional systems, sequences and correlation: A sedimentologist's view of the dimensional and temporal resolution of lithostratigraphy, *in* T. A. Cross (ed.), Quantitative dynamic stratigraphy: Prentice-Hall, Englewood Cliffs, N.J., p. 459–477.

Galloway, W. E., and D. K. Hobday, 1983, Terrigenous clastic depositional systems: Springer-Verlag, New York, 423 p.

Garde, R. J., and K. G. Ranga Raju, 1978, Mechanics of sediment transport and alluvial stream problems: Halsted, New York, 483 p.

Garrels, R. M., 1987, A model for the deposition of the microbanded Precambrian iron formations: Am. Jour. Science, v. 287, p. 81–106.

Garrels, R. M., and C. L. Christ, 1965, Solutions, minerals, and equilibria: Harper & Row, New York, 450 p.

Garrels, R. M., and F. T. McKenzie, 1971, Evolution of sedimentary rocks: W. W. Norton, New York, 397 p.

Garrison, R. E., 1974, Radiolarian cherts, pelagic limestone, and igneous rocks in eugeosynclinal settings, *in* K. J. Hsü and H. C. Jenkyns (eds.), Pelagic sediments on land and under the sea: Internat. Assoc. Sedimentologists Spec. Pub. 1, p. 367–399.

Garrison, R. E., R. B. Douglas, K. E. Pisciotto, C. M. Isaacs, and J. C. Ingle (eds.), 1981, The Monterey Formation and related siliceous rocks of California: Soc. Econ. Paleontologists and Mineralogists, Pacific Section, Los Angeles, 327 p.

Garven, G., and R. A. Freeze, 1984, Theoretical analysis of the role of groundwater flow in the genesis of stratabound ore deposits: Am. Jour. Science, v. 284, p. 1085–1174.

Geblein, C. D., 1974, Guidebook for modern Bahamian environments: Geol. Soc. American Ann. Mtg. Field Trip Guide, 93 p.

Geldsetzer, H. H. J., N. P. James, and G. E. Tebbutt (eds.), 1988, Reefs, Canada and adjacent areas: Canadian Soc. Petroleum Geologists Mem. 13, 775 p.

Geyh, M. A., and H. Schleicher, 1990, Absolute age determination: Springer-Verlag, Berlin, 503 p.

Ghent, E. D., 1979, Problems in zeolite facies geothermometry, geobarometry and fluid compositions, *in* P. S. Scholle and P. R. Schluger (eds.), Aspects of diagenesis, Soc. Econ. Paleontologists and Mineralogists Spec. Pub. 26, p. 81–87.

Gill, D., and D. F. Merriam (eds.), 1979, Geomathematical and petrophysical studies in sedimentology: Pergamon, New York, 267 p.

Gingerich, P. D., 1985, Species in the fossil record: Concepts, trends, and transitions: Paleobiology, v. 11, p. 27–41.

Ginsburg, R. N., 1956, Environmental relationships of grain size and constituent particles in some south Florida carbonate environments: Am. Assoc. Petroleum Geologists Bull., v. 40, p. 2384–2427.

_____ 1971, Landward movement of carbonate mud: New model for regressive cycles in carbonates (abs.): Am. Assoc. Petroleum Geologists Bull., v. 55, p. 340.

Ginsburg, R. N. (ed.), 1975, Tidal deposits, a casebook of Recent examples and fossil counterparts: Springer-Verlag, New York, 428 p.

Ginsburg, R. N., and N. P. James, 1974, Holocene carbonate sediments of continental shelves, *in* C. A. Burk and C. L. Drake (eds.), The geology of continental margins: Springer-Verlag, New York, p. 137–155.

Glaister, R. P., and H. W. Nelson, 1974, Grain-size distributions, an aid to facies identifications: Canadian Petroleum Geology Bull., v. 22, p. 203–240.

Glass, B. P., and M. J. Zwart, 1979, North American microtektites in Deep Sea Drilling Project cores from the Caribbean Sea and Gulf of Mexico: Geol. Soc. America Bull., v. 90, p. 595–602.

Glen, W., 1982, The road to Jaramillo: Critical years of the revolution in the earth sciences: Stanford University Press, Stanford, 459 p.

Glenn, G. R., and M. A. Arthur, 1988, Petrology and major element geochemistry of Peru margin phosphorites and associated diagenetic minerals: Authigenesis in modern organic-rich sediments: Marine Geology, v. 80, p. 231–267.

Glennie, K. W., 1970, Desert sedimentary environments: Developments in Sedimentology 14: Elsevier, Amsterdam, 222 p.

_____ 1972, Permian Rotliegendes of northwest Europe interpreted in light of modern desert sedimentation studies: Am. Assoc. Petroleum Geologists Bull., v. 56, p. 1048–1071.

_____ 1987, Desert sedimentary environments, present and past—A summary: Sed. Geology, v. 50, p. 135–166.

Godwin, P. D., 1991, Fining-upward cycles in the sandy braided-river deposits of the Westwater Canyon Member (Upper Jurassic), Morrison Formation, New Mexico: Sed. Geology, v. 70, p. 61–82.

Goldich, S. S., 1938, A study of rock weathering: Jour. Geology, v. 46, p. 17–58.

Goldthwait, R. P. (ed.), 1971, Till: A symposium: The Ohio State University Press, Columbus, 402 p.

_____ 1975, Glacial deposits: Dowden, Hutchinson and Ross, Stroudsburg, Pa., 464 p.

Gole, M. J., and C. Klein, 1981, Banded iron-formations through much of Precambrian time: Jour. Geology, v. 89, p. 169–183.

Goodwin, P. W., and E. J. Anderson, 1985, Punctuated aggradational cycles: A general hypothesis of episodic stratigraphic accumulation: Jour. Geology, v. 93, p. 515–533.

Gould, H. R., 1972, Environmental indicators—A key to the stratigraphic record, *in* J. K. Rigby and W. K. Hamblin (eds.), Recognition of ancient sedimentary environments: Soc. Econ. Paleontologists and Mineralogists Spec. Pub. 16, p. 1–3.

Gould, S. J., and N. Eldredge, 1986, Punctuated equilibrium and the third stage: Systematic Zoology, v. 35, p. 143–148.

Grabeau, A. W., 1904, On the classification of sedimentary rocks: Amer. Geol., v. 33, p. 228–247.

_____ 1913, Principles of stratigraphy: A. G. Seiler, 1185 p.

Gradstein, F. M., F. P. Atterberg, J. C. Brower, and W. S. Schwarzacher, 1985, Quantitative stratigraphy: D. Reidel, Dordrecht, 598 p.

Gray, J., and A. J. Boucot (eds.), 1979, Historical biogeography, plate tectonics, and the changing environment: Oregon State University Press, Corvallis, 500 p.

Green, H. G., S. H. Clarke, Jr., and M. P. Kennedy, 1991, Tectonic evolution of submarine canyons along the California continental margin, *in* R. H. Osborne (ed.), From shoreline to abyss: Soc. Economic Paleontologists and Mineralogists Spec. Pub. 46, p. 231–248.

Greensmith, J. T., 1989, Petrology of the sedimentary rocks, 7th ed.: Unwin Hyman, London, 262 p.

Greenwood, B., and R. A. Davis, Jr. (eds.), 1984, Hydrodynamics and sedimentation in wave-dominated coastal environments: Elsevier, New York, 473 p.

Griffith, J. C. 1967, Scientific methods in analysis of sediments: McGraw-Hill, New York, 508 p.

Griffiths, C. M., 1982, A proposed geologically significant segmentation and reassignment algorithm for petrophysical borehole logs, *in* J. M. Cubitt and R. A. Reyment (eds.), Quantitative stratigraphic correlation: John Wiley & Sons, New York, p. 287–298.

Griggs, D. T., 1936, The factor of fatigue in rock exfoliation: Jour. Geology, v. 44, p. 781–796.

Gromet, L. P., R. F. Dymek, L. A. Haskin, and R. L. Korotev, 1984, The "North American shale composite": Its compilation and major and trace element characteristics: Geochim. et Cosmochim. Acta, v. 48, p. 2469–2482.

Gross, G. A., 1980, A classification of iron formations based on depositional environments: Canadian Mineralogist, v. 18, p. 215–222.

Gross, M. G., 1982, Oceanography, 3rd ed.: Prentice-Hall, Englewood Cliffs, N.J., 498 p.

Guex, J., 1991, Biochronological correlations: Springer-Verlag, Berlin, 252 p.

Gulbrandsen, R. A., and C. E. Roberson, 1973, Inorganic phosphorites in seawater: Environmental phosphorus handbook: John Wiley & Sons, New York, Chapter 5, p. 117–140.

Guy, H. P., D. B. Simons, and E. V. Richardson, 1966, Summary of alluvial channel data from flume experiments, 1956–1961: U.S. Geol. Survey Prof. Paper 462-I, 96 p.

Hails, J., and A. Carr (eds.), 1975, Nearshore sediment dynamics and sedimentation: John Wiley & Sons, New York, 361 p.

Hails, J. R., 1976, Placer deposits, *in* K. H. Wolf (ed.), Handbook of strata-bound and stratiform ore deposits, v. 3: Elsevier, New York, p. 213–244.

Hailwood, E. A., 1989, Magnetostratigraphy: Geol. Soc. Spec. Report 19, Blackwell, Oxford, 84 p.

Håkanson, L., and M. Jansson, 1983, Lake sedimentation: Springer-Verlag, Berlin, 320 p.

Hallam, A., 1973a, A revolution in the earth sciences: From continental drift to plate tectonics: Oxford University Press, London, 127 p.

Hallam, A. (ed.), 1973b, Atlas of paleobiogeography: Elsevier, Amsterdam, 531 p.

Hallam, A., 1973c, Distributional patterns in contemporary terrestrial and marine animals, *in* N. F. Hughes (ed.), Organisms and continents through time: The Paleontological Association, London, p. 93–105.

_____ 1975, Evolutionary size increase and longevity in Jurassic bivalves and ammonites: Nature, v. 258, p. 493–496.

Hallam, A. (ed.), 1977, Patterns of evolution as illustrated by the fossil record: Elsevier, New York, 591 p.

Hallam, A., 1981, Facies interpretation and the stratigraphic record: W. H. Freeman, San Francisco, 291 p.

_____ 1984, Pre-Quaternary sea-level changes: Ann. Rev. Earth and Planetary Sciences, v. 12, p. 205–243.

_____ 1989a, Catastrophism in geology, *in* S. V. M. Clube (ed.), Catastrophes and evolution: Astronomical foundations: Cambridge University Press, p. 25–55.

_____ 1989b, Great geological controversies, 2nd ed.: Oxford University Press, Oxford, 244 p.

Ham, W. E. (ed.), 1962, Classification of carbonate rocks: Am. Assoc. Petroleum Geologists Mem. 1, 279 p.

Ham, W. E., and L. C. Pray, 1962, Modern concepts and classifications of carbonate rocks, *in* W. E. Ham (ed.), Classification of carbonate rocks: Am. Assoc. Petroleum Geologists Mem. 1, p. 2–19.

Hamblin, W. K., 1965, Internal structures of "homogeneous" sandstones: Kansas Geol. Survey Bull. 175, pt. 1, p. 1–37.

Hamilton, E. I., and R. M. Farquhar (eds.), 1968, Radiometric dating for geologists: John Wiley & Sons, New York, 506 p.

Hancock, J. M., 1977, The historic development of biostratigraphic correlation, *in* E. G. Kauffman and J. E. Hazel (eds.), Concepts and methods of biostratigraphy: Dowden, Hutchinson and Ross, Stroudsburg, Pa., p. 3–22.

Handschumacher, D. W., W. W. Sager, T. W. C. Hilde, and D. R. Bracey, 1988, Pre-Cretaceous tectonic evolution of the Pacific plate and extension of the geomagnetic polarity reversal time scale with implications for the origin of the Jurassic "Quiet Zone": Tectonophysics, v. 155, p. 365–380.

Haney, W. D., and L. I. Briggs, 1964, Cyclicity of textures in evaporite rocks of the Lucas Formation, *in* D. F. Merriam (ed.), Symposium on cyclic sedimentation: Kansas Geol. Survey, p. 191–197.

Hanor, J. S., 1979, The sedimentary genesis of hydrothermal fluids, *in* H. L. Barnes (ed.), Geochemistry of hydrothermal ore deposits: John Wiley & Sons, New York, p. 137–172.

Hanshaw, B. B., W. Back, and R. G. Deike, 1971, A geochemical hypothesis for dolomitization by ground water: Econ. Geology, v. 66, p. 710–724.

Häntzschel, W., 1975, Trace fossils and problematica, 2nd ed.: Treatise on invertebrate paleontology, pt. W, Misc., Suppl. 1, C. Teichert (ed.): Geol. Soc. America and Univ. Kansas, Boulder, Colo., and Lawrence, Kan., v. XXi +, 269 p.

Haq, B. U., J. Hardenbol, and P. R. Vail, 1987, The chronology of fluctuating sea level since the Triassic: Science, v. 235, p. 1156–1167.

_____ 1988, Mesozoic and Cenozoic chronostratigraphy and eustatic cycles, *in* C. K. Wilgus, B. S. Hastings, C. G. St. C. Kendall, H. W. Posamentier, C. A. Ross, and J. C. Van Wagoner (eds.), Sea-level changes: An integrated approach: Soc. Econ. Paleontologists and Mineralogists Spec. Pub. 42, p. 71–108.

Haq, B. U., and T. R. Worsley, 1982, Biochronology—Biological events in time resolution, their potential and limitations, *in* G. S. Odin (ed.), Numerical dating in stratigraphy, pt. I: John Wiley & Sons, New York, p. 19–36.

Harbaugh, J. W., and D. F. Merriam, 1968, Computer applications in stratigraphic analysis: John Wiley & Sons, New York, 282 p.

Hardage, B. A. (ed.), 1987, Seismic stratigraphy: Handbook of geophysical exploration, v. 9: Geophysical Press, London, 432 p.

Harder, H., 1970, Boron content of sediments as a tool in facies analysis: Sed. Geology, v. 4, p. 153–175.

Hardie, L. A. (ed.), 1977, Sedimentation on the modern carbonate tidal flats of northwest Andros Island, Bahamas: Johns Hopkins University Press, Baltimore, Md., 202 p.

Hardie, L. A., 1987, Dolomitization: A critical view of some current views: Jour. Sed. Petrology, v. 57, p. 166–183.

_____ 1991, On the significance of evaporites: Ann. Rev. Earth and Planetary Sciences, v. 19, p. 131–168.

Hardisty, J., 1990, Beaches—Form and process: Unwin Hyman, London, 324 p.

Harland, W. B., 1978, Geochronologic scales, *in* G. V. Cohee, M. F. Glaessner, and H. D. Hedberg (eds.), Contributions to the geologic time scale: Am. Assoc. Petroleum Geologists, Tulsa, Okla., p. 9–32.

Harland, W. B., R. L. Armstrong, A. V. Cox, L. E. Craig, A. G. Smith, and D. G. Smith, 1990a, The magnetostratigraphic time scale, *in* W. B. Harland, R. L. Armstrong, A. V. Cox, L. E. Craig, A. G. Smith, and D. G. Smith (eds.), A geologic time scale 1989: Cambridge University Press, Cambridge, p. 140–167.

Harland, W. B., R. L. Armstrong, A. V. Cox, L. E. Craig, A. G. Smith, and D. G. Smith, 1990b, A geologic time scale 1989, 2nd ed.: Cambridge University Press, Cambridge, 263 p.

Harland, W. B., A. V. Cox, P. G. Llewellyn, C. A. G. Pickton, A. G. Smith, and R. Walters, 1982, A geologic time scale: Cambridge University Press, Cambridge, 131 p.

Harms, J. C., and R. K. Fahnestock, 1965, Stratification, bed forms and flow phenomena (with examples from the Rio Grande), *in* G. V. Middleton (ed.), Primary sedimentary structures and their hydrodynamic interpretation: Soc. Econ. Paleontologists and Mineralogists Spec. Pub. 12, p. 84–155.

Harms, J. C., J. B. Southard, D. R. Spearing, and R. G. Walker, 1975, Depositional environments as interpreted from primary sedimentary structures and stratification sequences: Soc. Econ. Paleontologists and Mineralogists Short Course No. 2, 161 p.

Harms, J. C., J. B. Southard, and R. G. Walker, 1982, Structures and sequences in clastic rocks: Soc. Econ. Paleontologists and Mineralogists Lecture Notes for Short Course No. 9.

Harper, C. W., Jr., 1984, Improved methods of facies sequence analysis, *in* R. G. Walker (ed.), Facies models, 2nd ed.: Geoscience Canada Reprint Ser. 1, p. 11–13.

Harris, A. G., 1979, Conodont color alteration, an organo-mineral metamorphic index, and its application to Appalachian Basin geology, *in* P. A. Scholle and P. A. Schluger (eds.), Aspects of diagenesis: Soc. Econ. Paleontologists and Mineralogists Spec. Pub. 26, p. 3–16.

Harris, P. M., C. H. Moore, and J. L. Wilson, 1985, Carbonate depositional environments, modern and ancient: Pt. 2: Carbonate platforms: Colorado School of Mines Quarterly, v. 80, no. 4, p. 1–60.

Harrison, J. E., and Z. E. Peterman, 1980, North American Commission on Stratigraphic Nomenclature Note 52—A preliminary proposal for a chronometric time scale for the Precambrian of the United States and Mexico: Geol. Soc. America Bull., v. 91, p. 377–380.

Harrison, W. E., 1979, Levels of graphitization of kerogen as a potentially useful method of assessing paleotemperatures, *in* P. A. Scholle and P. R. Schluger (eds.), Aspects of diagenesis: Soc. Econ. Paleontologists and Mineralogists Spec. Pub. 26, p. 45–53.

Haworth, E. Y., and J. W. G. Lund (eds.), 1984, Lake sediments and environmental history: University of Minnesota Press, Minneapolis, 411 p.

Hay, W. W. (ed.), 1974, Studies in paleooceanography: Soc. Econ. Paleontologists and Mineralogists Spec. Pub. 20, 218 p.

Hay, W. W., and M. A. Leslie, 1990, Could possible changes in global groundwater reservoir cause eustatic sea level fluctuations? *in* R. R. Reville (ed.), Sea level change: National Research Council, Studies in Geophysics, National Academy Press, Washington, D.C., p. 161–170.

Hayes, M. O., 1967, Hurricanes as geological agents: Case studies of Hurricanes Carla, 1961, and Cindy, 1963: Texas Bur. Econ. Geology Rep. Inv. 61, 54 p.

――― 1975, Morphology of sand accumulations in estuaries, *in* L. E. Cronin (ed.), Estuarine research, v. 2: Geology and engineering: Academic Press, New York, p. 3–22.

Hayes, M. O., and T. W. Kana (eds.), 1976, Terrigenous clastic depositional environments: Some modern examples. Univ. South Carolina Tech. Rept. 11–CRD, 315 p.

Haywick, D. W., R. M. Carter, and R. A. Henderson, 1992, Sedimentology of 40,000 year Milankovich-controlled cyclothems from central Hawke's Bay, New Zealand: Sedimentology, v. 39, p. 675–696.

Hazel, J. E., 1977, Use of certain multivariate and other techniques in assemblage zonal biostratigraphy: Examples utilizing Cambrian, Cretaceous, and Tertiary benthic invertebrates, *in* E. G. Kauffman and J. E. Hazel (eds.), Concepts and methods of biostratigraphy: Dowden, Hutchinson and Ross, Stroudsburg, Pa., p. 187–212.

Heath, G. R., 1974, Dissolved silica and deep-sea sediments, *in* W. W. Hay (ed.), Studies in paleo-oceanography: Soc. Econ. Paleontologists and Mineralogists Spec. Pub. 20, p. 77–94.

Heckel, P. H., 1972, Recognition of ancient shallow marine environments, *in* J. K. Rigby and W. K. Hamblin (eds.), Recognition of ancient sedimentary environments: Soc. Econ. Paleontologists and Mineralogists Spec. Pub. 16, p. 226–286.

――― 1974, Carbonate buildups in the geologic record: A review, *in* L. F. Laporte (ed.), Reefs in time and space: Soc. Econ. Paleontologists and Mineralogists Spec. Pub. 18, p. 90–154.

――― 1986, Sea-level curve for Pennsylvanian eustatic marine transgressive-regressive depositional cycles along midcontinent outcrop belt, North America: Geology, v. 14, p. 330–334.

Hedberg, H. D. (ed.), 1976, International Stratigraphic Guide: A guide to stratigraphical classification, terminology, and procedure: International Subcommission on Stratigraphic Classification of IUGS Commission on Stratigraphy: John Wiley & Sons, New York, 200 p.

Heezen, B. C., and C. D. Hollister, 1971, The face of the deep: Oxford University Press, New York, 659 p.

Hein, J. R., and S. M. Karl, 1983, Comparisons between open-ocean and continental margin chert sequences, *in* A. Iijima, J. R. Hein, and R. Siever (eds.), Siliceous deposits in the Pacific region: Elsevier, Amsterdam, p. 25–43.

Hein, J. R., and J. T. Parrish, 1987, Distribution of siliceous deposits in space and time, *in* J. R. Hein (ed.), Siliceous sedimentary rock–hosted ores and petroleum: Van Nostrand Reinhold, New York, p. 10–57.

Hein, J. R., H.-W. Yeh, and J. A. Barron, 1990, Eocene diatom chert from Adak Island, Alaska: Jour. Sed. Petrology, v. 60, p. 250–257.

Heirtzler, J. R., G. O. Dickson, E. M. Herron, W. C. Pitman, III, and X. Le Pichon, 1968, Marine magnetic anomalies, geomagnetic field reversals, and motions of the ocean floor and continents: Jour. Geophys. Research, v. 73, p. 2119–2136.

Heller, F., and L. Tsungsheng, 1984, Magnetism of Chinese loess deposits: Geophys. J. R. Astron. Society, v. 77, p. 125–141.

Hesp, P., and S. G. Fryberger (eds.), 1988, Eolian sediments: Sed. Geology, v. 55, p. 1–163 (special issue devoted to eolian sediments).

Heward, A. P., 1981, A review of wave-dominated clastic shoreline deposits: Earth Science Rev., v. 17, p. 223–276.

Hine, A. C., 1983, Modern shallow water carbonate platform margins, *in* H. E. Cook, A. C. Hine, and H. T. Mullins (eds.), Platform margin and deep water carbonates: Soc. Econ. Paleontologists and Mineralogists Short Course Notes No. 12, p. 3-1–3-100.

Hiscott, R. N., and G. V. Middleton, 1980, Fabric of coarse deep-water sandstones, Tourelle Formation, Quebec, Canada: Jour. Sed. Petrology, v. 50, p. 703–722.

Hobday, D. K., and R. A. Morton, 1984, Lower Cretaceous shelf storm deposits, northeast Texas, *in* R. W. Tillman and C. T. Siemers (eds.), Siliciclastic shelf sediments: Soc. Econ. Paleontologists and Mineralogists Spec. Pub. 34, p. 205–213.

Hoefs, J., 1987, Stable isotope geochemistry, 3rd ed.: Springer-Verlag, Berlin, 241 p.

Hoffman, J., and J. Hower, 1979, Clay mineral assemblages as low grade metamorphic geothermometers: Applications to the thrust faulted disturbed belt of Montana, U.S.A., *in* P. A. Scholle and P. R. Schluger (eds.), Aspects of diagenesis: Soc. Econ. Paleontologists and Mineralogists Spec. Pub. 26, p. 55–79.

Holland, H. D., 1984, The chemical evolution of the atmosphere and oceans: Princeton University Press, Princeton, N.J., 582 p.

Holland, H. D., and F. F. Trendall (eds.), 1984, Patterns of change in earth evolution: Springer-Verlag, Berlin, 432 p.

Hollister, D. D., and A. R. M. Nowell, 1991, HEBBLE epilogue: Marine Geology, v. 99, p. 445–460.

Holmes, A., 1965, Principles of physical geology, 2nd ed., Thomas Nelson, London, 1288 p.

Holser, W. T., 1977, Catastrophic chemical events in the history of the ocean: Nature, v. 267, p. 403–408.

――― 1984, Gradual and abrupt shifts in ocean chemistry during Phanerozoic time, *in* H. D. Holland and A. F. Trendall (eds.), Patterns of change in Earth evolution: Springer-Verlag, Berlin, p. 123–143.

Holser, W. T., M. Mordeckai, and D. L. Clark, 1986, Carbon-isotope stratigraphic correlations in the Late Permian: Am. Jour. Sci., v. 286, p. 390–402.

Horne, J. C., J. C. Ferm, F. T. Caruccio, and B. P. Bagnaz, 1978, Depositional models in coal exploration and mine planning in Appalachian Region: Amer. Assoc. Petroleum Geol. Bull., v. 62, p. 2379–2411.

Horowitz, A. S., and P. E. Potter, 1971, Introductory petrography of fossils: Springer-Verlag, New York, 302 p.

Howard, J. D., 1978, Sedimentology and trace fossils, *in* P. B. Basan (ed.), Trace fossil concepts: Soc. Econ. Paleontologists and Mineralogists Short Course No. 5, p. 13–47.

Howarth, M. J., 1982, Tidal currents of the continental shelf, *in* A. H. Stride (ed.), Offshore tidal sands: Processes and deposits: Chapman and Hall, London, p. 10–26.

Hsü, K. J., 1989, Physical principles of sedimentology: Springer-Verlag, Berlin, 233 p.

Hsü, K. J., and H. C. Jenkyns (eds.), 1974, Pelagic sediments on land and under the sea: Internat. Assoc. Sedimentologists Spec. Pub. 1, Blackwell, London, 447 p.

Hubbard, D. K., 1992, Hurricane-induced sediment transport in open-shelf tropical systems—An example from St. Croix, U.S. Virgin Islands: Jour. Sed. Petrology, v. 62, p. 946–960.

Hubbard, R. J., J. Pape, and D. G. Roberts, 1985a, Depositional sequence mapping as a technique to establish tectonic and stratigraphic framework and evaluate hydrocarbon potential on a passive continental margin, *in* O. R. Berg and D. G. Wolverton (eds.), Seismic stratigraphy II: An integrated approach: Am. Assoc. Petroleum Geologists Mem. 39, p. 79–91.

———— 1985b, Depositional sequence mapping to illustrate the evolution of a passive continental margin, *in* O. R. Berg and D. G. Wolverton (eds.), Seismic stratigraphy II: An integrated approach: Am. Assoc. Petroleum Geologists Mem. 39, p. 93–115.

Hughes, N. F. (ed.), 1973, Organisms and continents through time: The Palaeontological Association, London, 334 p.

Hunt, J. M., 1979, Petroleum, geochemistry and geology: W. H. Freeman, San Francisco, 617 p.

Hunter, R. E., 1977, Basic types of stratification in small eolian dunes: Sedimentology, v. 24, p. 361–387.

Hutton, J., 1888, Theory of the earth, or an investigation of the laws observable in the composition, dissolution and restoration of land upon the globe: Royal Soc. Edinburgh Trans. v. 1, p. 109–304.

Iijima, A., J. R. Hein, and R. Siever (eds.), 1983, Siliceous deposits in the Pacific region: Elsevier, Amsterdam, 472 p.

Iijima, A., H. Inagaki, and Y. Kakuwa, 1979, Nature and origin of the Paleogene cherts in the Setogawa Terrain, Shizuoka, Central Japan: Jour. Fac. Sci., University of Tokyo, v. 20, p. 1–30.

Iller, R. K., 1979, Chemistry of silica: Wiley-Interscience, New York, 866 p.

Illing, L. V., 1954, Bahamian calcareous sands: Am. Assoc. Petroleum Geologists Bull., v. 38, p. 1–95.

Inden, R. F., and C. H. Moore, 1983, Beach, *in* P. A. Scholle, D. G. Bebout, and C. H. Moore (eds.), Carbonate depositional environments: Am. Assoc. Petroleum Geologists Mem. 33, p. 211–266.

Ingersoll, R. V., 1982, Initiation and evolution of the Great Valley forearc basin of northern and central California, U.S.A., *in* J. K. Leggett (ed.), Trench-forearc geology: Modern and ancient active plate margins: Geol. Soc. Spec. Pub. 10, p. 458–467.

———— 1988, Tectonics of sedimentary basins: Geol. Soc. America Bull., v. 100, p. 1704–1719.

Inman, D. L., and Nordstrom, C. E., 1971, On the tectonic and morphologic classification of coasts: Jour. Geology, v. 79, p. 1–21.

International Committee for Coal Petrology, 1963, International handbook of coal petrology, 2nd ed.: Centre National de la Recherche Scientifique, Paris. Supplements published in 1971, 1975.

Ireland, H. A., 1959, Silica in sediments: Soc. Econ. Paleontologists and Mineralogists Spec. Pub. 7, 185 p.

Isaacs, C. M., 1984, Hemipelagic deposits in a Miocene basin, California: Toward a model of lithologic variation and sequence, *in* D. A. V. Stow and D. J. W. Piper, 1984, Fine-grained sediments: Deep-water processes and facies: The Geological Society, Blackwell, Oxford, p. 481–496.

IUGS International Subcommission on Stratigraphic Classification and IUGS/IAGA Subcommission on a Magnetic Polarity Time Scale, 1979, Magnetostratigraphic polarity units—A supplementary chapter of the ISSC International Stratigraphic Guide: Geology, v. 7, p. 578–583.

Jackson, M. L., 1968, Weathering of primary and secondary minerals in soils: Transactions, 9th International Congress Soil Science, v. 4, p. 281–292.

James, H. L., 1966, Chemistry of the iron-rich sedimentary rocks: Data of geochemistry, 6th ed.: U.S. Geol. Survey Prof. Paper 440–W, 61 p.

James, H. L., and P. K. Sims (eds.), 1973, Precambrian iron formations of the world: Econ. Geology, v. 68, p. 913–1179.

James, H. L., and A. F. Trendall, 1982, Banded iron formation: Distribution in time and paleoenvironmental significance, *in* H. D. Holland and M. Schidlowski (eds.), Mineral deposits and the evolution of the biosphere: Springer-Verlag, Berlin, p. 199–218.

James, N. P., 1983, Reef environment, *in* P. A. Scholle, D. G. Bebout, and C. H. Moore (eds.), Carbonate depositional environments: Am. Assoc. Petroleum Geologists Mem. 33, p. 345–440.

———— 1984a, Introduction to carbonate facies models, *in* R. G. Walker (ed.), Facies models: Geoscience Canada Reprint Ser. 1, p. 209–212.

———— 1984b, Shallowing-upward sequences in carbonates, *in* R. G. Walker (ed.), Facies models: Geoscience Canada Reprint Ser. 1, p. 213–228.

———— 1984c, Reefs, *in* R. G. Walker (ed.), Facies models: Geoscience Canada Reprint Ser. 1, p. 229–244.

James, N. P., and P-A. Bourque, 1992, Reefs and mounds, *in* R. G. Walker and N. P. James (eds.), Facies models—Response to sea level change: Geol. Assoc. Canada, p. 323–348.

James, N. P., and A. C. Kendall, 1992, Introduction to carbonate and evaporite facies models, *in* R. G. Walker and N. P. James (eds.), Facies models—Response to sea level change: Geol. Assoc. Canada, p. 265–276.

James, N. P., and I. G. Macintyre, 1985, Carbonate depositional environments, modern and ancient: Pt. 1: Reefs: Colorado School of Mines Quarterly, v. 80, no. 3, p. 1–70.

Jenkyns, H. C., 1978, Pelagic environments, *in* H. G. Reading (ed.), Sedimentary environments and facies: Elsevier, New York, p. 314–371.

Jenkyns, H. C., 1986, Pelagic environments, *in* H. G. Reading (ed.), Sedimentary environments and facies, 2nd ed.: Blackwell, Oxford, p. 343–397.

Johansson, E. E., 1976, Structural studies of frictional sediments: Geograf. Annaler, v. 58A, p. 200–300.

Johnson, D. W., 1919, Shore processes and shoreline development: John Wiley & Sons, New York, 584 p.

Johnson, H. D., 1978, Shallow siliciclastic seas, *in* H. G. Reading (ed.), Sedimentary environments and facies: Elsevier, New York, p. 207–258.

Johnsson, M. J., R. F. Stallard, and R. H. Meade, 1988, First-cycle quartz arenites in the Orinoco River basin, Venezuela and Colombia: Jour. Geology, v. 96, p. 263–277.

Jones, B., and A. Desrochers, 1992, Shallow platform carbonates, *in* R. G. Walker and N. P. James (eds.), Facies models: Response to sea level change: Geol. Assoc. Canada, p. 277–301.

Jones, B. F., and C. J. Bowser, 1978, The mineralogy and related chemistry of lake sediments, *in* A. Lerman (ed.), Lakes: Chemistry, geology, physics: Springer-Verlag, New York, p. 179–235.

Jones, D. L., and B. Murchey, 1986, Geologic significance of Paleozoic and Mesozoic radiolarian chert: Ann. Rev. Earth and Planetary Science Letters, v. 14, p. 455–492.

Jones, K. P. N., I. N. McCave, and P. D. Patel, 1988, A computer-interfaced sedigraph for modal size analysis of fine-grained sediment: Sedimentology, v. 35, p. 163–172.

Jones, T. A., D. E. Hamilton, and C. R. Johnson, 1986, Contouring geological surfaces with the computer: Van Nostrand Reinhold, New York, 314 p.

Jopling, A. V., and B. C. McDonald, 1975, Glaciofluvial and glaciolacustrine sedimentation: Soc. Econ. Paleontologists and Mineralogists Spec. Pub. 23, 320 p.

Jopling, A. V., and R. G. Walker, 1968, Morphology and origin of ripple-drift cross-lamination with examples from the Pleistocene of Massachusetts: Jour. Sed. Petrology, v. 38, p. 971–984.

Jordan, C. F., G. E. Freyer, and E. H. Hemmen, 1971, Size analysis of silt and clay by hydrophotometer: Jour. Sed. Petrology, v. 41, p. 489–496.

Journal Geological Society (London), 1980, v. 136, pt. 6 (an issue devoted to phosphatic and glauconitic sediments), p. 657–805.

Karl, D. M., G. M. McMurtry, A. Malahoff, and M. O. Garcia, 1988, Loihi Seamount, Hawaii: A mid-plate volcano with a distinctive hydrothermal system: Nature, v. 335, p. 532–535.

Kastner, M., and J. M. Gieskes, 1983, Opal-A to opal-CT transformation: A kinetic study, *in* A. Iijima, J. R. Hein, and R. Siever (eds.), Siliceous deposits in the Pacific region: Developments in Sedimentology 36, Elsevier, Amsterdam, p. 211–228.

Katz, B. J. (ed.), 1990, Lacustrine basin exploration: Amer. Assoc. Petroleum Geologists Mem. 50, 340 p.

Kauffman, E. G., 1977, Evolutionary rates and biostratigraphy, *in* E. G. Kauffman and J. E. Hazel (eds.), Concepts and methods of biostratigraphy: Dowden, Hutchinson and Ross, Stroudsburg, Pa., p. 109–142.

Kauffman, E. G., and J. E. Hazel (eds.), 1977, Concepts and methods of biostratigraphy: Dowden, Hutchinson and Ross, Stroudsburg, Pa., 658 p.

Kauffman, E. G., and O. H. Walliser (eds.), 1990, Extinction events in Earth history: Springer-Verlag, Berlin, 432 p.

Keith, M. L., G. M. Anderson and R. Eichler, 1964, Carbon and oxygen isotopic composition of mollusk shells from marine and fresh-water environments: Geochim. et Cosmochim. Acta, v. 28, p. 1757–1786.

Keith, M. L., and J. N. Weber, 1964, Carbon and oxygen isotope composition of selected limestones and fossils: Geochim. et Cosmochim. Acta, v. 28, p. 1787–1816.

Keller, W. D., 1955, The principles of chemical weathering: Lucas Brothers, Columbia, Mo., 88 p.

Kelts, K., and K. J. Hsü, 1978, Calcium carbonate sedimentation in freshwater lakes and the formation of non-glacial varves in Lake Zurich, *in* A. Lerman (ed.), Lakes: Chemistry, geology, physics: Springer-Verlag, New York, p. 295–324.

Kempema, E. W., E. Reimnitz, and P. W. Barnes, 1988, Sea ice sediment entrainment and rafting in the Arctic: Jour. Sed. Petrology, v. 59, p. 308–317.

Kendall, A. C., 1979, Subaqueous evaporites, *in* R. G. Walker (ed.), Facies models: Geoscience Canada Reprint Ser. 1, p. 159–174.

Kennedy, S. K., and J. Mazzullo, 1991, Image analysis method of grain size measurement, *in* J. P. M. Syvitski (ed.), Principles, methods, and applications of particle size analysis: Cambridge University Press, Cambridge, p. 76–87.

Kennedy, V. S., 1984, The estuary as a filter: Academic Press, Orlando, Fla., 511 p.

Kennett, J. P. (ed.), 1980, Magnetic stratigraphy of sediments: Benchmark Papers in Geology 54, Dowden, Hutchinson and Ross, Stroudsburg, Pa., 438 p.

Kennett, J. P., 1982, Marine geology: Prentice-Hall, Englewood Cliffs, N.J., 812 p.

Kerr, A., B. J. Smith, W. B. Whalley, and J. P. McGreevy, 1984, Rock temperatures from southeast Morocco and their significance for experimental rock-weathering studies: Geology, v. 12, p. 306–309.

Kerr, R. A., 1984, Vail's sea-level curves aren't going away: Science, v. 226, p. 677–678.

Kersey, D. G., and K. J. Hsü, 1976, Energy relations and density current flows: An experimental investigation: Sedimentology, v. 23, p. 761–790.

Ketchum, B. H. (ed.), 1983, Estuaries and enclosed seas: Ecosystems of the World 26: Elsevier, Amsterdam, 500 p.

Khramov, A. N., 1987, Paleomagnetology: Springer-Verlag, Berlin, 308 p.

Kingston, D. R., C. P. Dishroon, and P. A. Williams, 1983, Global basin classification system: Am. Assoc. Petroleum Geologists Bull., v. 67, p. 2175–2193.

Kirkland, D. W., and R. Evans (eds.), 1973, Marine evaporites: Origin, diagenesis and geochemistry: Dowden, Hutchinson and Ross, Stroudsburg, Pa., 426 p.

Kjerfve, B. (ed.), 1978, Estuarine transport processes: University of South Carolina Press, 331 p.

Kjerfve, B., and K. E. Magill, 1989, Geographic and hydrodynamic characteristics of shallow coastal lagoons: Marine Geology, v. 88, p. 187–199.

Klein, G. deV., 1963, Analysis and review of sandstone classifications in the North American geological literature, 1940–1960: Geol. Soc. America Bull. v. 74, p. 555–576.

_____ 1965, Diverse origins of graded bedding: Geol. Soc. America Spec. Paper 82, 109 p.

_____ 1970, Depositional and dispersal dynamics of intertidal sand bars: Jour. Sed. Petrology, v. 40, p. 1095–1127.

Klein, G. deV. (ed.), 1976, Holocene tidal sedimentation: Benchmark Papers in Geology 5: Dowden, Hutchinson and Ross, Stroudsburg, Pa., 423 p.

Klein, G. deV., 1977, Clastic tidal facies: Continuing Education Publication Co., Champaign, Ill., 149 p.

_____ 1985, Intertidal flats and intertidal sand bodies, *in* R. A. Davis, Jr. (ed.), Coastal sedimentary environments, 2nd ed.: Springer-Verlag, New York, p. 187–224.

_____ 1987, Current aspects of basin analysis: Sed. Geology, v. 50, p. 95–118.

Kleinspehn, K. L., and C. Paola (eds.), 1988, New perspectives in basin analysis: Springer-Verlag, New York, 453 p.

Kleinspehn, K. L., R. J. Steel, E. Johannessen, and A. Netland, 1984, Conglomeratic fan-delta sequences, Late Carboniferous–Early Permian, Western Spitsbergen, *in* E. H. Koster and R. J. Steel (eds.), Sedimentology of gravels and conglomerates: Canadian Soc. Petroleum Geologists Mem. 10, p. 279–294.

Knauth, L. P., 1979, A model for the origin of chert in limestone: Geology, v. 7, p. 274–277.

Kocurek, G., and J. Nielson, 1986, Conditions favorable for the formation of warm-climate aeolian sand sheets: Sedimentology, v. 33, p. 795–816.

Kocurek, G., M. Townsley, E. Yeh, K. Havholm, and M. L. Sweet, 1992, Dune and dune-field development on Padre, Island, Texas, with implications for interdune deposition and water-table–controlled accumulation: Jour. Sed. Petrology, v. 62, p. 622–635.

Kohout, F. A., H. R. Henry, and J. E. Banks, 1977, Hydrogeology related to geothermal conditions of the Floridan Plateau, *in* K. L. Smith and G. M. Griffin (eds.), The geothermal nature of the Floridian Plateau: Florida Dept. Nat. Resources Bur. Geology Spec. Pub. 21, p. 1–34.

Kolla, V., and F. Coumes, 1985, Indus Fan, Indian Ocean, *in* A. H. Bouma, W. R. Normark, and N. E. Barnes (eds.), Submarine fans and related turbidite systems: Springer-Verlag, New York, p. 129–136.

Kolodny, Y., 1980, The origin of phosphorite deposits in the light of occurrences of Recent sea-floor phosphorites, *in* Y. K. Bentor (ed.), Marine phosphorites: Soc. Econ. Paleontologists and Mineralogists Spec. Pub. 29, p. 249.

_____ 1981, Phosphorites, *in* C. Emiliani (ed.), The ocean lithosphere: The sea, v. 7: John Wiley & Sons, New York, p. 981–1023.

Kolodny, Y., and I. R. Kaplan, 1970, Uranium isotopes in sea floor phosphorites: Geochim. et Cosmochim. Acta, v. 34, p. 3–24.

Komar, P. D., 1976, Beach processes and sedimentation: Prentice-Hall, Englewood Cliffs, N.J., 429 p.

Komar, P. D., and M. C. Miller, 1975, Sediment threshold under oscillatory waves: 14th Conference on Coastal Engineering Proc., p. 756–775.

Koster, E. H., and R. J. Steel (eds.), 1984, Sedimentology of gravels and conglomerates: Canadian Soc. Petroleum Geologists Mem. 10, 441 p.

Krauskopf, K. B., 1959, The geochemistry of silica in sedimentary environments, *in* H. A. Ireland (ed.), Silica in sediments: Soc. Econ. Paleontologists and Mineralogists Spec. Pub. 7, p. 4–19.

_____ 1979, Introduction to geochemistry, 2nd ed.: McGraw-Hill, New York, 617 p.

Krinsley, D., 1962, Applications of electron microscopy to geology: New York Acad. Sci. Trans., v. 25, p. 3–22.

Krinsley, D., and J. Doornkamp, 1973, Atlas of quartz sand surface textures: Cambridge University Press, Cambridge, 91 p.

Krinsley, D. H., and P. Trusty, 1986, Sand grain surface textures, *in* G. De C. Sieveking and M. B. Hart (eds.), The scientific study of flint and chert: Cambridge University Press, Cambridge, p. 201–207.

Krumbein, W. C., 1934, Size frequency distribution of sediments: Jour. Sed. Petrology, v. 4, p. 65–77.

_____ 1941, Measurement and geological significance of shape and roundness of sedimentary particles: Jour. Sed. Petrology, v. 11, p. 64–72.

Krumbein, W. C., and F. A. Graybill, 1965, An introduction to statistical models in geology: McGraw-Hill, New York, 475 p.

Krumbein, W. C., and F. J. Pettijohn, 1938, Manual of sedimentary petrography: Appleton-Century Crofts, New York, 549 p.

Krumbein, W. C., and L. L. Sloss, 1963, Stratigraphy and sedimentation: W. H. Freeman, San Francisco, 660 p.

Kuenen, Ph. H., 1958, Experiments in geology: Geol. Soc. Glasgow Trans., v. 23, p. 1–28.

_____ 1959, Experimental abrasion, pt. 3: Fluviatile action on sand: Am. Jour. Sci., v. 257, p. 172–190.

_____ 1960, Experimental abrasion, part 4: Eolian action: Jour. Geology, v. 68, p. 427–449.

_____ 1964, Experimental abrasion, pt. 6: Surf action: Sedimentology, v. 3, p. 29–43.

LaBerge, G. L., E. I. Robbins, and T.-M. Han, 1987, A model for the biological precipitation of Precambrian iron-formation—A: Geologic evidence, *in* P. W. U. Appel and G. L. LaBerge (eds.), Precambrian iron-formations: Theophrastus, S. A., Athens, Greece, p. 69–96.

LaBrecque, J. L., D. V. Kent, and S. C. Cande, 1977, Revised magnetic polarity time scale for late Cretaceous and Cenozoic time: Geology, v. 5, p. 330–335.

Lancaster, N., and J. T. Teller, 1988, Interdune deposits of the Namib sand sea: Sed. Geology, v. 55, p. 91–107.

Landis, C. A., 1971, Graphitization of dispersed carbonaceous material in metamorphic rocks: Contr. Mineralogy and Petrology, v. 30, p. 34–45.

Laporte, L. F., 1968, Ancient environments: Prentice-Hall, Englewood Cliffs, N.J., 116 p.

Laporte, L. F. (ed.), 1974, Reefs in time and space: Soc. Econ. Paleontologists and Mineralogists Spec. Pub. 18, 256 p.

Larsen, G., and G. V. Chilingarian (eds.), 1967, Diagenesis in sediments: Developments in Sedimentology 8, Elsevier, Amsterdam, 551 p.

_____ 1979, Diagenesis in sediments and sedimentary rocks: Elsevier North Holland, New York, 579 p.

_____ 1983, Diagenesis in sediments and sedimentary rocks, 2: Elsevier, New York, 572 p.

Larsen, R. L., and T. W. C. Hilde, 1975, A revised time scale of magnetic reversals for the Early Cretaceous and Late Jurassic: Jour. Geophys. Research, v. 80, p. 2586–2594.

Larwood, G. P. (ed.), 1988, Extinctions and survival in the fossil record: Oxford University Press, Oxford, 365 p.

Lauff, G. H. (ed.), 1967, Estuaries: Am. Assoc. for the Advancement of Science Spec. Pub. 83, 757 p.

Leatherman, S. P. (ed.), 1979, Barrier islands from the Gulf of St. Lawrence to the Gulf of Mexico: Academic Press, New York, 325 p.

LeBlanc, R. J., 1975, Significant studies of modern and ancient deltaic sediments, *in* M. L. Broussard (ed.), Deltas: Models for exploration: Houston Geological Society, p. 13–85.

Leckie, D. A., and R. G. Walker, 1982, Storm- and tide-dominated shorelines in Cretaceous Moosebar–Lower Gates interval—Outcrop equivalents of deep basin gas trap in western Canada: Amer. Assoc. Petroleum Geologists Bull., v. 66, p. 138–157.

Leeder, M. R., 1982, Sedimentology: Process and product: George Allen & Unwin, London, 344 p.

Lees, A., and A. T. Buller, 1972, Possible influences of salinity and temperature on modern shelf carbonate sedimentation: Marine Geology, v. 13, p. 1767–1773.

Legget, R. F. (ed.), 1976, Glacial till: An inter-disciplinary study: Royal Soc. Canada Spec. Pub. 12, 412 p.

Leggett, J. K. (ed.), 1982, Trench-forearc geology: Sedimentation and tectonics on modern and ancient active plate margins: Geol. Soc. London Spec. Pub. 10, Blackwell, Oxford, 576 p.

Leighton, M. W., D. R. Kolata, D. F. Oltz, and J. J. Eidel (eds.), 1990, Interior cratonic basins: Am. Assoc. Petroleum Geologists Mem. 51, Tulsa, Okla., 819 p.

Leopold, L. B., and M. G. Wolman, 1957, River channel patterns, braided, meandering, and straight: U.S. Geol. Survey Prof. Paper 282–B, p. 39–85.

Lepp, H. (ed.), 1975, Geochemistry of iron: Benchmark Papers in Geology, v. 18, Dowden, Hutchinson and Ross, Stroudsburg, Pa., 464 p.

Lepp, H., 1987, Chemistry and origin of Precambrian iron formations, *in* P. W. U. Appel and G. L. LaBerge, 1987, Precambrian iron-formations: Theophrastus, S. A., Athens, Greece, p. 3–30.

Lepp, H., and S. S. Goldich, 1964, Origin of Precambrian iron formation: Econ. Geology, v. 58, p. 1025–1061.

Lerman, A. (ed.), 1978, Lakes: Chemistry, geology, physics: Springer-Verlag, New York, 363 p.

Lerman, A., and M. Meybeck (eds.), 1988, Physical and chemical weathering in geochemical cycles: Kluwer Academic Pub., Dordrecht, 375 p.

Levinton, J., 1988, Genetics, paleontology, and macroevolution: Cambridge University Press, Cambridge, 637 p.

Levinton, J. S., and C. M. Simon, 1980, A critique of the punctuated equilibria model and implications for the detection of speciation in the fossil record: Systematic Zoology, v. 29, p. 130–142.

Lewan, M. D., 1978, Laboratory classification at very fine-grained sedimentary rocks: Geology, v. 6, p. 745–748.

Lewis, D. W., 1984, Practical sedimentology: Dowden, Hutchinson and Ross, Stroudsburg, Pa., 229 p.

Lindholm, R., 1987, A practical approach to sedimentology: Allen and Unwin, London, 276 p.

Link, M. H., and R. H. Osborne, 1978, Lacustrine facies in the Pliocene Ridge Basin Group: Ridge Basin, California, *in* A. Matter and M. E. Tucker (eds.), Modern and ancient lake sediments: Internat. Assoc. Sedimentologists Spec. Pub. 2, Blackwell, Oxford, p. 169–187.

Lippman, F., 1973, Sedimentary carbonate minerals: Springer-Verlag, New York, 228 p.

Lisitzin, A. P., 1972, Sedimentation in the world ocean: Soc. Econ. Paleontologists and Mineralogists Spec. Pub. 17, 218 p.

Logan, B. W., G. F. Davies, J. F. Read, and D. Cebulski, 1970, Carbonate sedimentation and environments, Shark Bay, Western Australia: Am. Assoc. Petroleum Geologists Mem. 13, 223 p.

Logan, B. W., J. F. Read, G. M. Hagen, P. Hoffman, R. G. Brown, P. J. Woods, and C. D. Gebelien, 1974, Evolution and diagenesis of Quaternary carbonate sequences, Shark Bay, Western Australia: Am. Assoc. Petroleum Geologists Mem. 22, 358 p.

Logan, B. W., R. Rezak, and R. N. Ginsburg, 1964, Classification and environmental significance of algal stromatolites: Jour. Geology, v. 72, p. 68–83.

Lomando, A. J., and P. M., Harris, (eds.), 1991, Mixed carbonate-clastic siliciclastic sequences: Soc. Econ. Paleontologists and Mineralogists Core Workshop No. 15, 568 p.

Longman, M. W., 1981, A process approach to recognizing facies of reef complexes, *in* D. F. Toomey (ed.), European fossil reef models: Soc. Econ. Paleontologists and Mineralogists Spec. Pub. 30, p. 9–40.

———— 1982, Carbonate diagenesis as a control on stratigraphic traps: Am. Assoc. Petroleum Geologists Education Course Notes Ser. 21, 159 p.

Loughnan, F. C., 1969, Chemical weathering of silicate minerals: Elsevier, New York, 154 p.

Lowe, D. R., 1976, Subaqueous liquefied and fluidized sediment flows and their deposits: Sedimentology, v. 23, p. 285–308.

———— 1982, Sedimentary gravity flows: II. Depositional models with special reference to the deposits of high-density turbidity currents: Jour. Sed. Petrology, v. 52, p. 279–297.

Lowe, D. R., and R. D. LoPiccolo, 1974, The characteristics and origin of dish and pillar structures: Jour. Sed. Petrology, v. 44, p. 484–501.

Lowrie, W., and W. Alvarez, 1977, Upper Cretaceous–Paleocene magnetic stratigraphy at Gubbio, Italy: III. Upper Cretaceous magnetic stratigraphy: Geol. Soc. America Bull., v. 88, p. 374–377.

———— 1981, One hundred million years of geomagnetic polarity history: Geology, v. 9, p. 392–397.

Lowrie, W., W. Alvarez, G. Napeleone, K. Perch-Neilsen, I. P. Silva, and M. Toumarkine, 1982, Paleogene magnetic stratigraphy in Umbrian pelagic carbonate rocks: The Contessa sections, Gubbio: Geol. Soc. America Bull. v. 93, p. 414–432.

Lowrie, W., and J. G. Ogg, 1986, A magnetic polarity time scale for the Early Cretaceous and Late Jurassic: Earth and Planetary Science Letters, v. 76, p. 341–349.

Lundegard, P. D., and N. D. Samuels, 1980, Field classification of fine-grained rocks: Jour. Sed. Petrology, v. 50, p. 781–786.

Lyell, Sir Charles, 1833, Principles of geology, v. 1: John Murray, London, 511 p.

MacDonald, D. I. M., 1991, Sedimentation, tectonics and eustasy: Sea-level changes at active margins: Internat. Assoc. Sedimentologists Spec. Pub. 12, Blackwell, Oxford, 518 p.

Machel, G.-G., and E. W. Mountjoy, 1986, Chemistry and environments of dolomitization—A reappraisal: Earth Science Rev., v. 23, p. 175–22.

Macintyre, I. G., and R. P. Reid, 1992, A comment on the origin of aragonite needle mud: A picture is worth a thousand words: Jour. Sed. Petrology, v. 62, p. 1095–1097.

MacQueen, R. W., 1983, Carbonate sedimentology for the mineral explorationist: Geol. Assoc. Canada, Cordilleran Section, Short Course 1.

MacQueen, R. W., and D. A. Leckie, 1992, Foreland basins and fold belds: Am. Assoc. Petroleum Geologists Mem. 55, Tulsa, Okla., 460 p.

Mahaney, W. C. (ed.), 1984, Quaternary dating methods: Elsevier, New York, 431 p.

Maiklem, W. R., D. G. Bebolt, and R. P. Glaister, 1969, Classification of anhydrite—A practical approach: Canadian Petroleum Geology Bull., v. 17, p. 194–233.

Maliva, R. G., 1989, Nodular chert formation in carbonate rocks: Jour. Geology, v. 97, p. 421–433.

Maliva, R. G., and R. Siever, 1988a, Pre-Cenozoic nodular cherts: Evidence for opal-CT precursors and direct quartz replacement: Am. Jour. Science, v. 288, p. 798–809.

Maliva, R. G., and R. Siever, 1988b, Diagenetic replacement controlled by force of crystallization: Geology, v. 16, p. 688–691.

Maliva, R. G., and R. Siever, 1989, Chertification histories of some Late Mesozoic and Middle Paleozoic platform carbonates: Sedimentology, v. 36, p. 907–926.

Manley, P. L., and R. D. Flood, 1988, Cyclic sediment deposition within Amazon deep-sea fan: Am. Assoc. Petroleum Geologists Bull., v. 72, p. 912–925.

Mann, C. J., 1981, Stratigraphic analysis: Decades of revolution (1970–1979) and refinements (1980–1989), *in* D. F. Merriam (ed.), Computer applications in earth sciences—An update of the 70s: Plenum, New York, p. 211–242.

Manspeizer, W. (ed.), 1988, Triassic–Jurassic rifting, 2 vols.: Elsevier, New York, 998 p.

Markevich, V. P., 1960, The concept of facies: Internat. Geol. Rev., v. 2, p. 376–379, 498–507, 582–604.

Marshall, J. R. (ed.), 1987, Clastic particles: Scanning electon microscopy and shape analysis of sedimentary and volcanic clasts: Van Nostrand Reinhold, New York, 346 p.

Martens, C. S., and R. C. Harris, 1970, Inhibition of apatite precipitation in marine environment by magnesium ions: Geochim. et Cosmochim. Acta, v. 34, p. 621–625.

Matter, A., and M. E. Tucker (eds.), 1978, Modern and ancient lake sediments: Internat. Assoc. Sedimentologists Spec. Pub. 2, 290 p.

McBride, E. F., 1963, A classification of common sandstones: Jour. Sed. Petrology, v. 33, p. 664–669.

McBride, E. F. (ed.), 1979, Silica in sediments: Nodular and bedded chert: Soc. Econ. Paleontologists and Mineralogists Reprint Ser. 8, 184 p.

McBride, E. F., R. G. Shepard, and R. A. Crawley, 1975, Origin of parallel, near-horizontal laminae by migration of bed forms in a small flume: Jour. Sed. Petrology, v. 45, p. 132–139.

McCabe, P. J., 1984, Depositional environments of coal and coal-bearing strata, *in* R. A. Rahmani and R. M. Flores (eds.), Sedimentology of coal and coal-bearing sequences: Internat. Assoc. Sedimentologists Spec. Pub. 7, p. 13–42.

McCall, P. L., and M. J. S. Tevesz, 1982, Animal-sediment relations. The biogenic alteration of sediments: Plenum, New York, 336 p.

McCave, I. N., R. J. Bryant, H. F. Cook, and C. A. Coughanowr, 1986, Evaluation of a laser-diffraction-size analyzer for use with natural sediments: Jour. Sed. Petrology, v. 56, p. 561–564.

McConachy, T. F., R. D. Ballard, M. J. Mottl, and R. P. von Herzen, 1986, Geologic form and setting of a hydrothermal vent field at lat 10° 56′ N, East Pacific Rise: A detailed study using Angus and Alvin: Geology, v. 14, p. 295–298.

McCubbin, D. J., 1982, Barrier-island and strand-plain facies, *in* P. A. Scholle and D. Spearing (eds.), Sandstone depositional environments: Am. Assoc. Petroleum Geologists Mem. 31, p. 247–279.

McDonald, D. A., and R. C. Surdam, 1984, Clastic diagenesis: Am. Assoc. Petroleum Geologists Mem. 37, 434 p.

McDonald, K. C., K. B. F. N. Spiess, and R. D. Ballard, 1980, Hydrothermal heat flux of the "black smoker" vents on the East Pacific Rise: Earth and Planetary Sci. Letters, v. 48, p. 1–7.

McDougall, I., 1977, The present status of the geomagnetic polarity time scale, *in* M. W. McElhinney (ed.), The earth: Its origin, structure, and evolution (a volume in honor of J. C. Jaeger and A. L. Hales): Academic Press, New York, p. 543–566.

McDougall, I., K. Saemundsson, H. Johannesson, N. D. Watkins, and L. Kristjansson, 1977, Extension of the geomagnetic polarity time scale to 6.5 m.y.: K-Ar dating, geological and paleomagnetic study of a 3500-m lava succession in western Iceland: Geol. Soc. America Bull., v. 88, p. 1–15.

McDougall, I., and T. M. Harrison, 1988, Geochronology and thermochronology by the 40 Ar/39 Ar method: Oxford University Press, New York, 212 p.

McElhinney, M. W., 1978, The magnetic polarity time scale: Prospects and possibilities in magnetostratigraphy, *in* G. V. Cohee, M. F. Glaessner, and H. D. Hedberg (eds.), Contributions to the geologic time scale: Am. Assoc. Petroleum Geologists Studies in Geology 6, p. 57–66.

McGowen, J. H., and C. G. Groat, 1971, Van Horn Sandstone, West Texas: An alluvial fan model for mineral exploration: Texas Bur. Econ. Geology Rept. Inv. 72, Austin, Tex., 57 p.

McGrail, D. W., and M. Carnes, 1983, Shelfedge dynamics and the nepheloid layer in the northwest Gulf of Mexico, *in* D. J. Stanley and G. T. Moore (eds.), The shelfbreak: Critical interface on continental margins: Soc. Econ. Paleontologists and Mineralogists Spec. Pub. 33, p. 251–264.

McIlreath, I. A., and N. P. James, 1984, Carbonate slopes, *in* R. G. Walker (ed.), Facies models, 2nd ed.: Geoscience Canada Reprint Ser. 1, p. 245–257.

McKee, E. D., 1965, Experiments on ripple lamination, *in* G. V. Middleton (ed.), Primary sedimentary structures and their hydrodynamic interpretation: Soc. Econ. Paleontologists and Mineralogists Spec. Pub. 12, p. 66–83.

McKee, E. D. (ed.), 1979a, A study of global sand seas: U.S. Geol. Suvey Prof. Paper 1052, 429 p.

McKee, E. D., 1979b, Sedimentary structures in dunes, *in* E. D. McKee (ed.), A study of global sand seas: U.S. Geol. Survey Prof. Paper 1052, p. 83–134.

_____ 1979c, Introduction to a study of global sand seas, *in* E. D. McKee (ed.), A study of global sand seas: U.S. Geol. Survey Prof. Paper 1052, p. 1–19.

_____ 1982, Sedimentary structures in dunes of the Namib Desert, Southwest Africa: Geol. Soc. America Spec. Paper 188, 64 p.

McKee, E. D., J. R. Douglass, and S. Rittenhouse, 1971, Deformation of lee-side laminae in eolian dunes: Geol. Soc. America Bull., v. 82, p. 359–378.

McKee, E. D., and G. W. Weir, 1953, Terminology for stratification and cross-stratification in sedimentary rocks: Geol. Soc. America Bull., v. 64, p. 381–390.

McKelvey, V. E., 1973, Abundance and distribution of phosphorus in the lithosphere, *in* Environmental phosphorus handbook: John Wiley & Sons, New York, p. 13–31.

McKelvey, V. E., J. S. Williams, R. P. Sheldon, E. R. Cressman, and T. M. Channey, 1959, The Phosphoria, Park City and Shedhorn formations in the Western Phosphate Field: U.S. Geol. Survey Prof. Paper 313–A.

McNeill, D. F., R. N. Ginsburg, S-B. R. Chang, and J. L. Kirschvink, 1988, Magnetostratigraphic dating of shallow-water carbonates from San Salvador, Bahamas: Geology, v. 16, p. 8–12.

Melnik, Y. P., 1982, Precambrian banded iron-formations: Developments in Precambrian Geology 5, Elsevier, Amsterdam, 310 p.

Melvin, J. L. (ed.), 1991, Evaporites, petroleum and mineral resources: Elsevier, Amsterdam, 555 p.

Merino, E., 1975a, Diagenesis in Tertiary sandstones from Kettleman North Dome, California—I. Diagenetic mineralogy: Jour. Sed. Petrology, v. 45, p. 320–336.

_____ 1975b, Diagenesis in Tertiary sandstones from Kettleman North Dome, California—II. Interstitial solutions: Distribution of aqueous species at 100°C and chemical relation to the diagenetic mineralogy: Geochim. et Cosmochim. Acta, v. 39, p. 1629–1645.

Mero, J. L., 1965, The mineral resources of the sea: Elsevier, New York, 312 p.

Merriam, D. F., 1964, Symposium on cyclic sedimentation: Geol. Survey Kansas Bull. 169, v. 1 and 2, 636 p.

Merriam, D. F. (ed.), 1981, Computer applications in the earth sciences—An update of the 70s: Plenum, New York, 385 p.

Meybeck, M., 1981, River transport of organic carbon to the ocean, *in* Carbon dioxide effects research and assessment program: Flux of organic carbon by rivers to the ocean: Committee on Flux of Organic Carbon to the Ocean, G. E. Likens (chm.), Div. Biol. Sci., Natl. Research Council, U.S. Dept. Energy, Off. Energy Research, Washington, D.C., Reference: CONF-80009140 UC, 219 p.

Miall, A. D., 1973, Markov chain analysis applied to an ancient alluvial plain succession: Sedimentology, v. 20, p. 347–365.

_____ 1977, A review of the braided-river depositional environment: Earth Science Rev., v. 13, p. 1–62.

Miall, A. D. (ed.), 1978, Fluvial sedimentology: Canadian Soc. Petroleum Geologists Mem. 5, 589 p.

Miall, A. D., 1982, Analysis of fluvial depositional systems: Am. Assoc. Petroleum Geologists Education Course Notes Ser. 20, 75 p.

_____ 1984a, Principles of sedimentary basin analysis: Springer-Verlag, New York, 490 p.

_____ 1984b, Deltas, in R. G. Walker (ed.), Facies models: Geoscience Canada Reprint Ser. 1, 2nd ed., p. 105–118.

_____ 1986, Eustatic sea-level changes interpreted from seismic stratigraphy: A critique of the methodology with particular reference to the North Sea Jurassic record: Am. Assoc. Petroleum Geologists Bull., v. 70, p. 131–137.

_____ 1990, Principles of sedimentary basin analysis, 2nd ed.: Springer-Verlag, New York, 668 p.

_____ 1991a, Exxon global cycle chart: An event for every occasion? Geology, v. 20, p. 797–790.

_____ 1991b, Stratigraphic sequences and their chronostratigraphic correlation: Jour. Sed. Petrology, v. 61, p. 497–505.

_____ 1992, Alluvial deposits, in R. G. Walker and N. P. James (eds.), Facies models: Response to sea level changes: Geol. Assoc. Canada, p. 119–142.

Middlemass, F. A., P. F. Rawson, and G. Newall (eds.), 1971, Faunal provinces in space and time: Geology Jour. Spec. Issue 4, Seel House Press, Liverpool, 236 p.

Middleton, G. V. (ed.), 1965, Primary sedimentary structures and their hydrodynamic interpretation: Soc. Econ. Paleontologists and Mineralogists Spec. Pub. 12, 265 p.

Middleton, G. V., 1973, Johannes Walther's Law of the Correlation of Facies: Geol. Soc. America Bull., v. 84, p. 979–988.

_____ 1976, Hydraulic interpretation of sand size distributions: Jour. Geology, v. 84, p. 405–426.

_____ 1978, Facies, in R. W. Fairbridge and J. Bourgeois (eds.), Encyclopedia of sedimentology: Dowden, Hutchinson and Ross, Stroudsburg, Pa., p. 323–325.

_____ 1991, Mechanics of sediment gravity flows (unpublished manuscript).

_____ 1993, Sediment deposition from turbidity currents: Ann. Rev. Earth and Planetary Sciences, v. 21, p. 89–114.

Middleton, G. V., and A. H. Bouma (eds.), 1973, Turbidites and deep water sedimentation: Soc. Econ. Paleontologists and Mineralogists, Pacific Section, Short Course, Anaheim, Calif., 157 p.

Middleton, G. V., and M. A. Hampton, 1976, Subaqueous sediment transport and deposition by sediment gravity flows, in D. J. Stanley and D. J. P. Swift (eds.), Marine sediment transport and environmental management: John Wiley & Sons, New York, p. 197–218.

Middleton, G. V., and J. B. Southard, 1978, Mechanics of sediment movement: Soc. Econ. Paleontologists and Mineralogists Short Course Notes No. 3, variously paginated.

_____ 1984, Mechanics of sediment movement, 2nd ed.: Eastern Section, Soc. Econ. Paleontologists and Mineralogists Short Course No. 3, 401 p.

Miller, F. X., 1977, The graphic correlation method in biostratigraphy, in E. G. Kauffman and J. E. Hazel (eds.), Concepts and methods in biostratigraphy: Dowden, Hutchinson and Ross, Stroudsburg, Pa., p. 165–186.

Miller, M. F., A. A. Ekdale, and M. D. Picard (eds.), 1984, Trace fossils and paleoenvironments: Marine carbonate, marginal marine terrigenous and continental terrigenous settings: Jour. Paleontology, v. 58, p. 283–597.

Milliman, J. D., 1974, Marine carbonates: Springer-Verlag, New York, 375 p.

Milner, H. B., 1962, Sedimentary petrography, v. 2: Principles and applications: Macmillan, New York, 715 p.

Mitchell, A. H. C., and H. G. Reading, 1986, Sedimentation and tectonics, in H. G. Reading (ed.), Sedimentary environments and facies, 2nd ed.: Blackwell, p. 471–519.

Mitchum, R. M., Jr., and P. R. Vail, 1977, Seismic stratigraphic interpretation procedures, in C. E. Payton (ed.), Seismic stratigraphy—applications to hydrocarbon exploration: Am. Assoc. Petroleum Geologists Mem. 26, p. 135–143.

Mitchum, R. M., Jr., P. R. Vail, and J. B. Sangree, 1977, Seismic stratigraphy and global change of sea level. Part 6: Stratigraphic interpretation of seismic reflection patterns in depositional sequences, in C. E. Payton (ed.), Seismic stratigraphy—Applications to hydrocarbon exploration: Am. Assoc. Petroleum Geologists Mem. 26, p. 117–133.

Mitchum, R. M., Jr., P. R. Vail, and S. Thompson, III, 1977, Seismic stratigraphy and global change of sea level. Part 2: The depositional sequence as a basic unit for stratigraphic analysis, in C. E. Payton (ed.), Seismic stratigraphy—Applications to hydrocarbon exploration: Am. Assoc. Petroleum Geologists Mem. 26, p. 53–62.

Molnia, B. F., 1983, Glacial-marine sedimentation: Plenum, New York, 844 p.

Moore, C. H., 1989, Carbonate diagenesis and porosity: Elsevier, Amsterdam.

Moore, D. M., 1978, A sample of the Purington Shale prepared as a geochemical standard: Jour. Sed. Petrology, v. 48, p. 995–998.

Moore, R. C., 1949, Meaning of facies: Geol. Soc. America Mem. 39, p. 1–34.

Morey, G. W., R. O. Fournier, and J. J. Rowe, 1962, The solubility of quartz in water in the temperature interval from 25°C to 300°C: Geochim. et Cosmochim. Acta, v. 26, p. 1029–1043.

_____ 1964, The solubility of amorphous silica at 25°C: Jour. Geophys. Research, v. 69, p. 1995–2002.

Morgan, J. P. (ed.), 1970, Deltaic sedimentation—Modern and ancient: Soc. Econ. Paleontologists and Mineralogists Spec. Pub. 15, 312 p.

Morrow, D. W., and B. D. Ricketts, 1988, Experimental investigation of sulfate inhibition of dolomite and its mineral analogs, in V. Shukla and P. A. Baker (eds.), Sedimentology and geochemistry of dolostones: Soc. Econ. Paleontologists and Mineralogists Spec. Pub. 43, p. 25–38.

Morse, J. W., and S. He, 1993, Influences of T, S and PCO₂ on the pseudohomogeneous precipitation of CaCO₃ from seawater: Implications for whitings formation: Marine Chemistry, v. 41, p. 291–297.

Morse, J. W., and F. T. Mackenzie, 1990, Geochemistry of sedimentary carbonates: Elsevier, Amsterdam, 707 p.

Morton, A. C., S. P. Todd, and P. D. W. Haughton (eds.), 1991, Developments in sedimentary provenance studies: Geol. Soc. Spec. Pub. 57, London, 370 p.

Morton, R. A., 1988, Nearshore responses to great storms, in H. E. Clifton (ed.), Sedimentologic consequences of convulsive geologic events: Geol. Soc. America Spec. Paper 229, p. 7–22.

Moslow, T. F., 1985, Depositional models of shelf and shoreline sandstones: Am. Assoc. Petroleum Geologists Education Course Notes Ser. 17, 102 p.

Mount, J., 1985, Mixed siliciclastic and carbonate sediments: A proposed first-order textural and compositional classification: Sedimentology, v. 32, p. 435–442.

Mucci, A., and J. W. Morse, 1983, The incorporation of Mg² and Sr²⁺ into calcite overgrowths: Influence of growth rates and solution composition: Geochim. et Cosmochim. Acta, v. 47, p. 217–233.

Muerdter, D. R., J. P. Dauphin, and G. Steele, 1981, An interactive computerized system for grain size analysis of silt using electro-resistance: Jour. Sed. Petrology, v. 51, p. 647–650.

Mullins, H. T., and H. E. Cook, 1986, Carbonate apron models: Alternatives to the submarine fan model for paleoenvironmental analysis and hydrocarbon exploration: Sed. Geology, v. 48, p. 37–79.

Murray, R. W., D. L. Jones, and M. R. Bucholtz ten Brink, 1992, Diagenetic formation of bedded chert: Evidence from chemistry of chert-shale couplet: Geology, v. 20, p. 271–274.

Mutti, E., 1985, Turbidite systems and their relations to depositional sequences, in G. G. Zuffa (ed.), Provenance of arenites: D. Riedel, Dordrecht, p. 65–93.

Nahon, D. B., 1991, Introduction to the petrology of soils and chemical weathering: John Wiley & Sons, New York, 313 p.

Nanz, R. H., 1953, Chemical composition of Precambrian slates with notes on the geochemical evolution of lutites: Jour. Geology, v. 61, p. 51–64.

Nardin, T. R., B. D. Edwards, and D. S. Gorsline, 1979, Santa Cruz Basin, California borderland: Dominance of slope processes in basin sedimentation, in L. J. Doyle and O. H. Pilkey (eds.), Geology of continental slopes: Soc. Econ. Paleontologists and Mineralogists Spec. Pub. 27, p. 209–221.

Nardin, T. R., F. J. Hein, D. S. Gorsline, and B. D. Edwards, 1979, A review of mass movement processes, sediment and acoustic characteristics, and contrasts in slope and base-of-slope systems versus canyon-fan-basin floor system, in L. J. Doyle and O. R. Pilkey (eds.), Geology of continental slopes: Soc. Econ. Paleontologists and Mineralogists Spec. Pub. 27, p. 61–73.

Nathan, Y., 1984, The mineralogy and geochemistry of phosphorites, in J. O. Nriagu and P. B. Moore (eds.), Phosphate minerals: Springer-Verlag, Berlin, p. 275–291.

Neidell, N. S., 1979, Stratigraphic modeling and interpretation: Geophysical principles and techniques: Am. Assoc. Petroleum Geologists Education Course Notes No. 13, 141 p.

Nelson, B. W. (ed.), 1972, Environmental framework of coastal plain estuaries: Geol. Soc. America Mem. 133, 619 p.

Nelson, C. S., 1988, An introductory perspective on non-tropical carbonates: Sed. Geology, v. 60, p. 3–12.

Nelson, C. H., and T. H. Nilsen, 1984, Modern and ancient deep-sea fan sedimentation: Soc. Econ. Paleontologists and Mineralogists Short Course No. 14, 404 p.

Nemec, W., 1990, Deltas—Remarks on terminology and classification, in A. Colella and D. B. Prior (eds.), Coarse-grained deltas: Internat. Assoc. Sedimentologists Spec. Pub. 10, Blackwell, Scientific Publications, Oxford, p. 3–12.

Nemec, W., and R. J. Steel (eds.), 1988a, What is a fan delta and how do we recognize it?, in W. Nemec and R. J. Steel (eds.), Fan deltas: Sedimentology and tectonic settings: Blackie, Glasgow and London, p. 3–13.

Nemec, W., and R. J. Steel (eds.), 1988b, Fan deltas: Sedimentology and tectonic settings: Blackie, Glasgow and London, 444 p.

Nettleton, L. L., 1940, Geophysical prospecting for oil: McGraw-Hill, New York, 444 p.

Neumann, A. C., J. W. Kofoed, and G. H. Keller, 1977, Lithoherms in the Straits of Florida: Geology, v. 5, p. 4–11.

Neumann, A. C., and L. S. Land, 1975, Lime mud deposition and calcareous algae in the Bight of Abaco, Bahamas: A budget: Jour. Sed. Petrology, v. 45, p. 763–786.

Newell, N. D., J. K. Rigby, A. J. Whitman, and J. S. Bradley, 1951, Shoal-water geology and environments, eastern Andros Island, Bahamas: Am. Mus. Nat. History Bull., v. 97, p. 1–29.

Newton, R. S., 1968, Internal structure of wave-formed ripple marks in the nearshore zone: Sedimentology, v. 11, p. 275–292.

Nichols, M. M., and R. B. Biggs, 1985, Estuaries, in R. A. Davis, Jr. (ed.), Coastal sedimentary environments, 2nd ed.: Springer-Verlag, New York, p. 77–186.

Nilsen, T. H., 1980, Modern and ancient submarine fans: Discussion of papers by R. G. Walker and W. R. Normark: Am. Assoc. Petroleum Geologists Bull., v. 64, p. 1094–1112.

——— 1982, Alluvial fan deposits, in P. A. Scholle and D. Spearing (eds.), Sandstone depositional environments: Am. Assoc. Petroleum Geologists Mem. 31, p. 49–86.

Nilsen, T. H. (ed.), 1985, Modern and ancient alluvial fan deposits: Van Nostrand Reinhold, New York, 372 p.

Nilsen, T. H., and P. L. Abbott, 1981, Paleogeography and sedimentology of Upper Cretaceous turbidites, San Diego, California: Am. Assoc. Petroleum Geologists Bull., v. 65, p. 1256–1284.

Nio, S. D., and C. Yang, 1991, Diagnostic attributes of clastic tidal deposits: A review, *in* D. G. Smith, G. E. Reinson, B. A. Zaitlin, and R. A. Rahmani (eds.), Clastic tidal sedimentology: Canadian Soc. Petroleum Geologists, p. 3d–28.

Nisbet, E. G., and I. Price, 1974, Siliceous turbidites: Bedded cherts as redeposited, ocean ridge-derived sediments, *in* K. J. Hsü and H. C. Jenkyns (eds.), Pelagic sediments: On land and under the sea: Internat. Assoc. Sedimentologists Spec. Pub. 1, p. 351–366.

Nitecki, M. H. (ed.), 1984, Extinctions: University of Chicago Press, 354 p.

Nittrouer, C. A., 1981, Sedimentary dynamics of continental shelves: Elsevier, Amsterdam, 449 p.

Norholt, A. J. G., R. P. Sheldon, and D. F. Davidson (eds.), 1989, Phosphate deposits of the world, v. 2: Phosphate rock resources: Cambridge University Press, Cambridge, 566 p.

Normark, W. R., 1978, Fan valleys, channels, and depositional lobes on modern submarine fans: Characters for recognition of sandy turbidite environments: Am. Assoc. Petroleum Geologists Bull., v. 62, p 912–931.

Normark, W. R., and D. J. W. Piper, 1991, Initiation process and flow evolution of turbidity currents: Implications for the depositional record, *in* R. H. Osborne (ed.), From shoreline to abyss: Soc. Econ. Paleontologists and Mineralogists Spec. Pub. 46, p. 207–230.

North American Commission on Stratigraphic Nomenclature, 1983, North American Stratigraphic Code: Am. Assoc. Petroleum Geologists Bull., v. 67, p. 841–875.

Nriagu, J. O., and P. B. Moore (eds.), 1984, Phosphate minerals: Springer-Verlag, New York, 434 p.

Nummedal, D., O. H. Pilkey, and J. D. Howard (eds.), 1987, Sea-level fluctuation and coastal evolution: Soc. Econ. Paleontologists and Mineralogists Spec. Pub. 41, 267 p.

Odin, G. S. (ed.), 1982a, Numerical dating in stratigraphy: John Wiley & Sons, New York. Pt. 1, p. 1–630, pt. II, p. 631–1040.

Odin, G. S., 1982b, Introduction: Uncertainties in evaluating the numerical time scale, *in* G. S. Odin (ed.), Numerical dating in stratigraphy: John Wiley & Sons, New York, p. 3–16.

———— 1982c, Zero isotopic ages of glauconites, *in* G. S. Odin (ed.), Numerical dating in stratigraphy: John Wiley & Sons, New York, p. 277–305.

Odin, G. S., D. Curry, N. H. Gale, and W. J. Kennedy, 1982, The Phanerozoic time scale in 1981; *in* G. S. Odin (ed.), Numerical dating in stratigraphy: John Wiley & Sons, New York, p. 957–960.

Odin, G. S., and M. H. Dodson, 1982, Zero isotopic ages of glauconites, *in* G. S. Odin (ed.), Numerical dating in stratigraphy: John Wiley & Sons, New York, p. 277–306.

Odin, G. S., M. Renard, and C. V. Grazzini, 1982, Geochemical events as a means of correlation, *in* G. S. Odin (ed.), Numerical dating in stratigraphy: John Wiley & Sons, New York, p. 37–72.

Oertel, G. F., and S. P. Leatherman (eds.), 1985, Barrier islands: Marine Geology, v. 63, p. 1–396.

Officer, C. B. (chm.), 1977, Estuaries, geophysics, and the environment: Nat. Acad. Sci., Washington, D.C., 127 p.

Ogg, J. G., Magnetic polarity time scale of the Phanerozoic: Am. Geophys. Union handbook of physical constants (in press).

Okada, H., 1971, Classification of sandstone: Analysis and proposal: Jour. Geology, v. 79, p. 509–525.

Olausson, E., and I. Cato (eds.), 1980, Chemistry and biogeochemistry of estuaries: John Wiley & Sons, New York, 452 p.

Olea, R. A., 1988, Correlator—An interactive computer system for lithostratigraphic correlation of wireline logs: Kansas Geological Survey, Lawrence, Kansas, Petrophysical Series 4, 85 p.

Ollier, C., 1969, Weathering: American Elsevier, New York, 304 p.

Ondrick, C. W., and J. C. Griffiths, 1969, Frequency distribution of elements in Rensselaer graywacke, Troy, New York: Geol. Soc. Amer. Bull., v. 80, p. 509–518.

Orton, G. J., 1988, A spectrum of Middle Ordovician fan deltas and braid plain deltas, North Wales: A consequence of varying fluvial clastic input, *in* W. Nemic and R. J. Steel (eds.), Fan deltas: Sedimentology and tectonic setting: Blackie, Glasgow and London, p. 23–49.

Osborne, R. H. (ed.), 1991, From shoreline to abyss: Contributions in marine geology in honor of Francis Parker Shepard: SEPM (Society for Sedimentary Geology) Spec. Pub. 46, 320 p.

Osleger, D., and J. F. Read, 1991, Relation of eustasy to stacking patterns of meter-scale carbonate cycles, Late Cambrian, U.S.A.: Jour. Sed. Petrology, v. 61, p. 1225–1252.

Otto, G. H., 1938, The sedimentation unit and its use in field sampling: Jour. Geology, v. 46, p. 569–582.

Palmer, A. R. (comp.), 1983, The Decade of North American Geology 1983 Geologic Time Scale: Geology, v. 11, p. 503–504.

Pantin, H. M., 1979, Interaction between velocity and effective density in turbidity flow: Phase plane analysis with criteria for autosuspension: Marine Geology, v. 31, p 59–99.

Park, W. C., and E. H. Schot, 1968, Stylolitization in carbonate rocks, *in* C. Müller and G. M. Friedman (eds.), Recent developments in carbonate sedimentology in central Europe: Springer-Verlag, New York, p. 66–74.

Parkash, B., and G. V. Middleton, 1970, Downcurrent textural changes in Ordovician turbidite graywackes: Sedimentology, v. 14, p. 259–293.

Parker, A., and B. W. Sellwood (eds.), 1983, Sediment diagenesis: D. Reidel, Boston, 427 p.

Parker, G., 1982, Conditions for the ignition of catastrophically erosive turbidity currents: Marine Geology, v. 46, p. 307–327.

Parrish, J. T., 1990, Paleooceanographic and paleoclimatic setting of the Miocene phosphogenic episode, *in* W. C. Burnett and S. R. Riggs (eds.), Phosphate deposits of the world, v. 3: Neogene to Modern phosphorites: Cambridge University Press, Cambridge, p. 223–240.

Passega, R., 1957, Texture as a characteristic of clastic deposition: Am. Assoc. Petroleum Geologists Bull., v. 41, p. 1952–1984.

——— 1964, Grain size representation by CM patterns as a geological tool: Jour. Sed. Petrology, v. 34, p. 830–847.

——— 1977, Significance of CM diagrams of sediments deposited by suspensions: Sedimentology, v. 24, p. 723–733.

Payton, C. E. (ed.), 1977, Seismic stratigraphy—Applications to hydrocarbon exploration: Am. Assoc. Petroleum Geologists Mem. 26, 516 p.

Pemberton, S. G., J. A. MacEachern, and R. W. Frey, 1992, Trace fossil facies models: Environmental and allostratigraphic significance, *in* R. G. Walker and N. P. James (eds.), Facies models—Response to sea level change: Geol. Assoc. Canada, p. 47–72

Peterson, M. N., and C. C. von der Borch, 1965, Chert: Modern inorganic deposition in a carbonate-precipitating locality: Science, v. 149, p. 1501–1503.

Petrakis, L., and D. W. Grandy, 1980, Coal analysis, characterization and petrography: Jour. Chem. Education, v. 57, p. 689–694.

Pettijohn, F. J., 1941, Persistence of minerals and geologic age: Jour. Geology, v. 49, p. 610–625.

——— 1963, Chemical composition of sandstones— excluding carbonate and volcanic sands, *in* Data of geochemistry, 6th ed.: U.S. Geol. Survey Prof. Paper 440S, 19 p.

——— 1975, Sedimentary rocks, 3rd ed.: Harper & Row, New York, 628 p.

Pettijohn, F. J., and P. E. Potter, 1964, Atlas and glossary of primary sedimentary structures: Springer-Verlag, New York, 370 p.

Pettijohn, F. J., P. E. Potter, and R. Siever, 1973, Sand and sandstone: Springer-Verlag, New York, 618 p.

——— 1987, Sand and sandstone, 2nd ed.: Springer-Verlag, New York, 553 p.

Phleger, F. B., 1969, Some general characteristics of coastal lagoons, *in* A. A. Castanares and F. B. Phleger (eds.), Coastal lagoons—A symposium: Universidad Nacional Autonoma de Mexico/UNESCO, Mexico City, p. 5–26.

Piazzola, J., and V. V. Cavaroc, 1991, Comparison of grain-size-distribution statistics determined by sieving and thin-section analyses: Jour. Geological Education, v. 39, p. 364–367.

Picard, M. D., 1971, Classification of fine-grained sedimentary rocks: Jour. Sed. Petrology, v. 41, p. 179–195.

——— 1977, Stratigraphic analysis of the Navajo Sandstone: A discussion: Jour. Sed. Petrology, v. 47, p. 475–483.

Picard, M. D., and L. R. High, Jr., 1972, Criteria for recognizing lacustrine rocks, *in* J. K. Rigby and W. K. Hamblin (eds.), Recognition of ancient sedimentary environments: Soc. Econ. Paleontologists and Mineralogists Spec. Pub. 16, p. 108–145.

——— 1973, Sedimentary structures of ephemeral streams: Elsevier, New York, 223 p.

——— 1981, Physical stratigraphy of ancient lacustrine deposits, *in* F. G. Ethridge and R. M. Flores (eds.), Recent and nonmarine depositional environments: Models for exploration: Soc. Econ. Paleontologists and Mineralogists Spec. Pub. 31, p. 233–259.

Pickering, K. T., R. N. Hiscott, and F. J. Hein, 1989, Deep marine environments: Clastic sedimentation and tectonics: Unwin Hyman, London, 416 p.

Pilkey, O. H., 1983, The beaches are moving: Duke University Press, Durham, N.C., 336 p.

Piper, D. J. W., 1987, Paleomagnetism and the continental crust: The Open University Press, Milton Keynes, 434 p.

Piper, D. J. W., A. N. Shor, and J. E. H. Clarke, 1988, The 1929 "Grand Banks" earthquake, slump, and turbidity current, *in* H. E. Clifton (ed.), Sedimentologic consequences of convulsive geologic events: Geol. Soc. America Spec. Paper 229, p. 77–92.

Piper, D. J. W., D. A. V. Stow, and W. R. Normark, 1984, The Laurentian Fan; Sohm Abyssal Plain: Geo-Marine Letters, v. 3, p. 141–146.

Pisciotto, K. A., and R. E. Garrison, 1981, Lithofacies and depositional environments of the Monterey Formation, California, *in* R. E. Garrison, R. G. Douglas, K. E. Pischiotto, C. M. Isaacs, and J. C. Ingle (eds.), The Monterey Formation and related siliceous rocks of California: Pacific Section, Soc. Econ. Paleontologists and Mineralogists, Los Angeles, p. 97–122.

Pitman, W. C., III, 1978, Relationship between eustasy and stratigraphic sequences of passive margins: Geol. Soc. America Bull., v. 89, p. 1389–1403.

Plint, A. G., N. Eyles, C. H. Eyles, and R. G. Walker, 1992, Control of sea level change, *in* R. G. Walker and N. P. James (eds.), Facies models—Response to sea level change: Geol. Assoc. Canada, p. 15–25.

Plummer, P. S., and V. A. Gostin, 1981, Shrinkage cracks: Desiccation or synaeresis? Jour. Sed. Petrology, v. 51, p. 1147–1156.

Porrenga, D. H., 1967, Glauconite and chamosite as depth indicators in the marine environment, *in* A. Hallam (ed.), Depth indicators in marine sedimentary rocks: Marine Geology, Spec. Issue 5, no. 5/6, p. 495–502.

Posamentier, H. W., M. T. Jervey, and P. R. Vail, 1988, Eustatic controls on clastic deposition 1—Conceptual framework, *in* C. K. Wilgus, B. S. Hastings, C. G. St. C. Kendall, H. W. Posamentier, C. A. Ross, and J. C. Van Wagoner (eds.), Sea-level changes: An integrated approach: Soc. Econ. Paleontologists and Mineralogists Spec. Pub. 42, p. 109–124.

Postma, H., 1980, Sediment transport and sedimentation, *in* G. Olausson and I. Cato (eds.), Chemistry and biogeochemistry of estuaries: John Wiley & Sons, Chichester, p. 153–186.

Potter, P. E., 1962, Late Mississippian sandstones of Illinois Basin: Illinois Geol. Survey Circ. 340, 36 p.

_____ 1967, Sand bodies and sedimentary environments. A review: Am. Assoc. Petroleum Geologists Bull., v. 51, p. 337–365.

Potter, P. E., J. B. Maynard, and W. A. Pryor, 1980, Sedimentology of shale: Springer-Verlag, New York, 306 p.

Potter, P. E., and F. J. Pettijohn, 1977, Paleocurrents and basin analysis, 2nd ed.: Springer-Verlag, New York, 460 p.

Potter, P. E., N. F. Shimp, and J. Witters, 1963, Trace elements in marine and fresh-water argillaceous sediments: Geochim. et Cosmochim. Acta, v. 27, p. 669–694.

Powers, D. W., and R. G. Easterling, 1982, Improved methodology for using embedded Markov chains to describe cyclic sediments: Jour. Sed. Petrology, v. 56, p. 913–923.

Powers, M. C., 1953, A new roundness scale for sedimentary particles: Jour. Sed. Petrology, v. 23, p. 117–119.

Pratt, B. R., and N. P. James, 1992, Peritidal carbonates, *in* R. G. Walker and N. P. James, (eds.), Facies models: Response to sea level change: Geol. Assoc. Canada, p. 303–322.

Pray, L. C., and R. C. Murray (eds.), 1965, Dolomitization and limestone diagenesis: A symposium: Soc. Econ. Paleontologists and Mineralogists, Tulsa, Okla., 180 p.

Prospero, J. M., 1981, Eolian transport to the world ocean, *in* C. Emiliani (ed.), The oceanic lithosphere: The sea, v. 7, John Wiley & Sons, New York, p. 801–874.

Pulham, A. J., 1989, Controls on internal structure and architecture of sandstone bodies within Upper Carboniferous fluvial-dominated deltas, County Clare, western Ireland, *in* M. K. G. Whateley and K. T. Pickering (eds.), Deltas: Sites and traps for fossil fuels: Geol. Soc. Spec. Pub. 41, Blackwell, Oxford, p. 179–203.

Pye, K., and H. Tsoar, 1990, Aeolian sand and sand dunes: Unwin Hyman, London, 396 p.

Rachocki, A., 1981, Alluvial fans: John Wiley & Sons, New York, 161 p.

Rachocki, A. H., and M. Church (eds.), 1990, Alluvial fans: John Wiley & Sons, Chichester and New York, 391 p.

Rahmani, R. A., and R. M. Flores (eds.), 1985, Sedimentology of coal and coal-bearing sequences: Internat. Assoc. Sedimentologists Spec. Pub. 7, Blackwell, Oxford, 412 p.

Ramsbottom, W. H. C., 1979, Rates of transgression and regression in Carboniferous of NW Europe: Jour. Geol. Soc. London, v. 136, p. 147–153.

Raup, D. M., 1977, Stochastic models in evolutionary palaeontology, *in* A. Hallam (ed.), Patterns of evolution as illustrated by the fossil record: Elsevier, New York, p. 59–78.

Rautman, C. A., and R. H. Dott, Jr., 1977, Dish structures formed by fluid escape in Jurassic shallow marine sandstones: Jour. Sed. Petrology, v. 47, p. 101–106.

Raymo, M. E., W. F. Ruddiman, J. Backman, B. M. Clement, and D. G. Martinson, 1989, Late Pliocene variation in Northern Hemisphere ice sheets and North Atlantic deep water circulation: Paleoceanography, v. 4, p. 413–446.

Read, J. F., 1982, Carbonate platforms of passive (extensional) continental margins: Types, character and evolution: Tectonophysics, v. 81, p. 195–212.

_____ 1985, Carbonate platform facies models: Am. Assoc. Petroleum Geologists Bull., v. 69, p. 1–21.

Reading, H. G., 1978a, Facies, *in* H. G. Reading (ed.), Sedimentary environments and facies: Elsevier, New York, p. 4–14.

Reading, H. G. (ed.), 1978b, Sedimentary environments and facies: Elsevier, New York, 557 p.

Reddy, M. M., and K. K. Wang, 1980, Crystallization of calcium carbonate in the presence of metal ions. I. Inhibition by magnesium ions at pH 8.8 and 25°C: Jour. Crystal Growth, v. 50, p. 470–480.

Reed, W. R., R. LeFever, and G. J. Moir, 1975, Depositional environment interpretation from settling-velocity (psi) distributions: Geol. Soc. America Bull., v. 86, p. 1321–1328.

Reeder, R. J. (ed.), 1983, Carbonates: Mineralogy and chemistry: Rev. in Mineralogy, v. 11, 394 p.

Reeves, C. C., Jr., 1968, Introduction to paleolimnology: Elsevier, New York, 228 p.

Reineck, H. E., 1972, Tidal flats, *in* J. K. Rigby and W. K. Hamblin (eds.), Recognition of ancient depositional environments: Soc. Econ. Paleontologists and Mineralogists Spec. Pub. 16, p. 146–159.

Reineck, H. E., and I. B. Singh, 1980, Depositional sedimentary environments, 2nd ed.: Springer-Verlag, Berlin, 549 p.

Reinhardt, J., and W. R. Sigleo (eds.), 1988, Paleosols and weathering through geologic time: Principles and applications: Geol. Soc. America Spec. Paper 216, 181 p.

Reinson, G. E., 1984, Barrier-island and associated strand-plain systems, *in* R. G. Walker (ed.), Facies models: Geoscience Canada Reprint Ser. 1, 2nd ed., p. 119–140.

_____ 1992, Transgressive barrier island and estuarine systems, *in* R. G. Walker and N. P. James (eds.), Facies models: Geol. Assoc. Canada, p. 179–194.

Retallack, G. J., 1988, Field recognition of paleosols, *in* J. Reinhardt and W. R. Sigleo (eds.), Paleosols and weathering through geologic time: Geol. Soc. Amer. Spec. Paper 216, p. 1–20.

_____ 1990, Soils of the past: Unwin Hyman, Boston, 520 p.

Revelle, R. (ed.), 1990, Sea level change: National Research Council, Studies in Geophysics: National Academy Press, Washington, D.C., 234 p.

Rich, J. L., 1951, Three critical environments of deposition and criteria for recognition of rocks deposited in each of them: Geol. Soc. America Bull., v. 62, p. 1–20.

Rider, M. H., 1986, The geological interpretation of well logs: John Wiley & Sons, New York, 175 p.

Riecke, H. H., III, and G. V. Chilingarian (eds.), 1974, Compaction of argillaceous sediments: Elsevier, New York, 424 p.

Rigby, J. K., and W. K. Hamblin (eds.), 1972, Recognition of ancient sedimentary environments: Soc. Econ. Paleontologists and Mineralogists Spec. Pub. 16, 340 p.

Riley, J. P., and R. L. Chester (eds.), 1976, Chem. Oceanography, v. 5, 2nd ed.: Academic Press, New York, 401 p.

Robbin, L. L., and P. L. Blackwelder, 1992, Biochemical and ultrastructural evidence for the origin of whitings: A biologically induced calcium carbonate precipitation mechanism: Geology, v. 20, p. 464–468.

Roberts, H. H., and C. H. Moore, Jr., 1971, Recently cemented aggregates (grapestones), Grand Cayman Island, BWI, in O. P. Bricker (ed.), Carbonate cements: Johns Hopkins University Press, Baltimore, Md., p. 88–90.

Robinson, J. E., 1982, Computer applications in petroleum geology: Dowden, Hutchinson and Ross, New York, 164 p.

Roedder, E., 1976, Fluid-inclusion evidence on the genesis of ores in sedimentary and volcanic rocks, in K. H. Wolf (ed.), Handbook of strata-bound and stratiform ore deposits: Elsevier, Amsterdam, v. 4, no. 2, p. 67–110.

———— 1979, Fluid inclusion evidence on the environments of sedimentary diagenesis, a review, in P. A. Scholle and P. R. Schluger (eds.), Aspects of diagenesis: Soc. Econ. Paleontologists and Mineralogists Spec. Pub. 26, p. 89–107.

Rodgers, J., 1959, The meaning of correlation: Am. Jour. Sci., v. 257, p. 684–691.

Rona, P. A., G. Klinkhammer, T. A. Nelsen, J. H. Trefry, and H. Elderfield, 1986, Black smokers, massive sulphides and vent biota at the Mid-Atlantic Ridge: Nature, v. 321, p. 33–37.

Rona, P. A., and R. P. Lowell (eds.), 1980, Seafloor spreading centers: Hydrothermal systems: Benchmark Papers in Geology, v. 56, Dowden, Hutchinson and Ross, Stroudsburg, Pa., 424 p.

Ronov, A. B., V. E. Khain, A. N. Balukhovsky, and K. B. Seslavinsky, 1980, Quantitative analysis of Phanerozoic sedimentation: Sed. Geology, v. 25, p. 311–325.

Ronov, A. B., and A. A. Migdisov, 1971, Geochemical history of the crystalline basement and sedimentary cover of the Russian and North American platforms: Sedimentology, v. 16, p. 137–185.

Ross, C. A., and J. R. P. Ross (eds.), 1984, Geology of coal: Benchmark Papers in Geology 77: Dowden, Hutchinson and Ross, Stroudsburg, Pa., 349 p.

Ross, S., 1989, Soil processes: Routledge, London and New York, 444 p.

Rothe, P., J. Hoefs, and V. Sonne, 1974, The isotopic composition of Tertiary carbonates from the Mainz Basin: An example of isotopic fractionation in "closed basins": Sedimentology, v. 21, p. 373–395.

Rouse, H., and J. W. Howe, 1953, Basic mechanics of fluids: John Wiley & Sons, New York, 245 p.

Ruddiman, W. F., 1977, Late Quaternary deposition of ice-rafted sand in the subpolar North Atlantic (lat. 40° to 65° N): Geol. Soc. America Bull., v. 88, p. 1813–1827.

Ruddiman, W. F., M. E. Raymo, D. G. Martinson, B. M. Clement, and J. Backman, 1989, Pleistocene evolution: Northern Hemisphere ice sheets and North Atlantic Ocean: Paleooceanography, v. 4, p. 353–412.

Rupke, N. A., 1978, Deep clastic seas, in H. G. Reading (ed.), Sedimentary environments and facies: Elsevier, New York, p. 372–415.

Russell, P. L., 1990, Oil shales of the world: Their origin, occurrence and exploitation: Pergamon Press, Oxford, 736 p.

Russell, R. D., and R. E. Taylor, 1937, Roundness and shape of Mississippi River sands: Jour. Geology, v. 45, p. 225–267.

Rust, B. R., and B. G. Jones, 1987, The Hawkesbury Sandstone south of Sydney, Australia: Triassic analogue for the deposits of a large, braided river: Jour. Sed. Petrology, v. 57, p. 222–233.

Ryer, T. A., and A. W. Langer, 1980, Thickness change involved in peat-to-coal transformation for a bituminous coal of Cretaceous age in central Utah: Jour. Sed. Petrology, v. 50, p. 987–992.

Sagoe, K-M. O., and G. S. Visher, 1977, Population breaks in grain-size distributions of sand—A theoretical model: Jour. Sedimentary Petrology, v. 47, p. 285–310.

Sah, M. P., and R. A. K. Srivastava, 1992, Morphology and facies of the alluvial-fan sedimentation in the Kangra Valley, Himachal Himalaya: Sed. Geology, v. 76, p. 23–42.

Salvador, A., 1985, Chronostratigraphic and geochronometric scales in COSUNA Stratigraphic Correlation Charts of the United States: Am. Assoc. Petroleum Geologists Bull, v. 69, p. 181–189.

Sanders, J. E., and N. Kumar, 1975, Evidence of shoreface retreat and in-place "drowning" during Holocene submergence of barriers, shelf off Fire Island, New York: Geol. Soc. America Bull., v. 86, p. 65–76.

Sarjeant, A. A. S. (ed.), 1983, Terrestrial trace fossils: Benchmark Papers in Geology, v. 76. Dowden, Hutchinson and Ross, Stroudsburg, Pa., 415 p.

Saxov, S., and J. K. Nieuwenhuis (eds.), 1982, Marine slides and other mass movements: NATO Conference Series IV: Marine Sciences, v. 6, Plenum, New York, 353 p.

Scheltema, R. S., 1977, Dispersal of marine invertebrate organisms: Paleobiogeographic and biostratigraphic implications, in E. G. Kauffman and J. E. Hazel (eds.), Concepts and methods of biostratigraphy: Dowden, Hutchinson and Ross, Stroudsburg, Pa., p. 73–108.

Schidlowski, M., 1982, Content and isotopic composition of reduced carbon in sediments, *in* H. D. Holland and M. Schidloski (eds.), Mineral deposits and the evolution of the biosphere: Springer-Verlag, New York, p. 103–122.

Schieber, J., 1987, Small scale sedimentary iron deposits in a mid-Proterozoic basin: Viability of iron supply by rivers, *in* P. W. U. Appel and G. L. LaBerge (eds.), Precambrian iron-formations: Theophrastus, S. A., Athens, Greece, p. 267–295.

Schmidt, V., and D. A. McDonald, 1979, The role of secondary porosity in the course of sandstone diagenesis, *in* P. A. Scholle and P. R. Schluger (eds.), Aspects of diagenesis: Soc. Econ. Paleontologists and Mineralogists Spec. Pub. 26, p. 175–208.

Scholle, P. A., 1978, A color illustrated guide to carbonate rock constituents, textures, cements, and porosities: Am. Assoc. Petroleum Geologists Mem. 27, 241 p.

——— 1979, A color illustrated guide to constituents, textures, cements, and porosities of sandstones and associated rocks: Am. Assoc. Petroleum Geologists Mem. 28, 201 p.

Scholle, P. A., M. A. Arthur, and A. A. Ekdale, 1983, Pelagic environment, *in* P. A. Scholle, D. G. Bebout, and C. H. Moore (eds.), Carbonate depositional environments: Am. Assoc. Petroleum Geologists Mem. 33, p. 620–691.

Scholle, P. A., D. G. Bebout, and C. H. Moore (eds.), 1983, Carbonate depositional environments: Am. Assoc. Petroleum Geologists Mem. 33, 708 p.

Scholle, P. A., and P. R. Schluger (eds.), 1979, Aspects of diagenesis: Soc. Econ. Paleontologists and Mineralogists Spec. Pub. 26, 400 p.

Scholle, P. A., and D. Spearing (eds.), 1982, Sandstone depositional environments: Am. Assoc. Petroleum Geologists Mem. 31, 410 p.

Schopf, J. M., 1956, A definition of coal: Econ. Geology, v. 51, p. 521–527.

Schopf, T. J. M., 1977, Patterns of evolution: A summary and discussion, *in* A. Hallam (ed.), Patterns of evolution as illustrated by the fossil record: Elsevier, New York, p. 547–561.

——— 1980, Paleoceanography: Harvard University Press, Cambridge, Mass., 341 p.

Schreiber, B. C., 1988, Subaqueous evaporite deposition, *in* B. C. Schreiber (ed.), Evaporites and hydrocarbons: Columbia University Press, New York, p. 182–255.

Schreiber, B. C., M. E. Tucker, and R. Till, 1986, Arid shorelines and evaporites, *in* H. G. Reading (ed.), Sedimentary environments and facies: Blackwell, p. 189–228.

Schroeder, J. H., and B. H. Purser, 1986, Reef diagenesis: Springer-Verlag, Berlin, 455 p.

Schumm, S. A., 1977, The fluvial system: John Wiley & Sons, New York, 338 p.

Schwan, J., 1988, The structure and genesis of Weichselian to Early Holocene aeolian sheets in western Europe: Sed. Geology, v. 55, p. 197–232.

Schwartz, M. L. (ed.), 1972, Spits and bars: Benchmark Papers in Geology, v. 3, Dowden, Hutchinson and Ross, Stroudsburg, Pa., 452 p.

——— 1973, Barrier islands: Benchmark Papers in Geology, v. 9, Dowden, Hutchinson and Ross, Stroudsburg, Pa., 451 p.

Schwarzacher, W. J., 1975, Sedimentation models and quantitative stratigraphy: Developments in Sedimentology 19, Elsevier, Amsterdam, 382 p.

Scoffin, T. P., 1987, An introduction to carbonate sediments and rocks: Blackie, Glasgow, 274 p.

Sedimentation Seminar, 1981, Comparison of methods of size analysis for sands of the Amazon-Solimtões Rivers, Brazil and Peru: Sedimentology, v. 28, p. 123–128.

Seibold, E., and W. H. Berger, 1982, The seafloor: Springer-Verlag, Berlin, 288 p.

Seibold, E., and W. H. Berger, 1993, The seafloor: An introduction to marine geology, 2nd ed.: Springer-Verlag, Berlin, 356 p.

Seilacher, A., 1964, Biogenic sedimentary structures, *in* J. Imbrie and N. D. Newell (eds.), Approaches to paleoecology: John Wiley & Sons, New York, p. 296–315.

——— 1992, Event stratigraphy: A dynamic view of the sedimentary record, *in* G. C. Brown, C. J. Hawkesworth, and R. C. L. Wilson (eds.), Understanding the earth: Cambridge University Press, Cambridge, p. 375–385.

Selley, R. C., 1970, Studies of sequences in sediments using a simple mathematical device: Geol. Soc. London Quart. Jour., v. 125, p. 557–581.

——— 1978, Ancient sedimentary environments, 2nd ed.: Cornell University Press, Ithaca, N.Y., 287 p.

Sellwood, B. W., 1972, Tidal flat sedimentation in the Lower Jurassic of Bornholm, Denmark: Palaeogeography, Palaeoclimatology, and Palaeoecology, v. 11, p. 93–106.

——— 1975, Lower Jurassic tidal-flat deposits, Bornholm, Denmark, *in* R. N. Ginsburg (ed.), Tidal deposits: Springer-Verlag, New York, p. 93–101.

——— 1978, Shallow-water carbonate environments, *in* H. G. Reading (ed.), Sedimentary environments and facies: Elsevier, New York, p. 259–313.

——— 1986, Shallow-marine carbonate environments, *in* H. G. Reading (ed.), Sedimentary environments and facies, 2nd ed.: Blackwell, Oxford, p. 283–342.

Sestini, G., 1989, Nile delta: A review of depositional environments and geological history, *in* M. K. G. Whateley and K. T. Pickering (eds.), Deltas: Sites and traps for fossil fuels: Geol. Soc. Spec. Pub. 41, Blackwell, Oxford, p. 99–127.

Shackleton, N. J., 1967, Oxygen isotope analyses and paleotemperatures reassessed: Nature, v. 215, p. 15–17.

Sharp, R. F., and L. H. Nobles, 1953, Mudflow of 1941 at Wrightwood, southern California: Geol. Soc. America Bull., v. 64, p. 547–560.

Sharp, R. P., 1988, Living ice: Understanding glaciers and glaciation: Cambridge University Press, Cambridge, 225 p.

Sharpton, V. L., and P. D. Ward (eds.), 1990, Global catastrophes in earth history: Geol. Soc. America Spec. Paper 247, 631 p.

Shaw, A. B., 1964, Time in stratigraphy: McGraw-Hill, New York, 365 p.

Shaw, B. R., 1982, A short note on the correlation of geologic sequences, *in* J. M. Cubitt and R. A. Reyment (eds.), Quantitative stratigraphic correlation: John Wiley & Sons, New York, p. 7–12.

Shaw, B. R., and J. M. Cubitt, 1978, Stratigraphic correlation of well logs: An automated approach, *in* D. Gill and D. F. Merriam (eds.), Geomathematical and petrophysical studies in sedimentology: Pergamon, Oxford, p. 127–148.

Shaw, D. M., 1956, Geochemistry of pelitic rocks, Part III: Major elements and general geochemistry: Geol. Soc. America Bull., v. 67, p. 919–934.

Shaw, J., 1985, Subglacial and ice marginal environments, *in* G. M. Ashley, J. Shaw, and N. D. Smith (eds.), Glacial sedimentary environments: Soc. Econ. Paleontologists and Mineralogists Short Course No. 16, p. 7–84.

Shea, J. H., 1982, Twelve fallacies of uniformitarianism: Geology, v. 10, p. 455–460.

Shearman, D. J., 1978, Evaporites of coastal sabkhas, *in* W. E. Dean and B. C. Schreiber (eds.), Marine evaporites: Soc. Econ. Paleontologists and Mineralogists Short Course Notes No. 4, p. 6–42.

Sheldon, R. P., 1989, Phosphorite deposits of the Phosphoria Formation, western United States, *in* A. J. G. Nothold, R. P. Sheldon, and D. F. Davidson (eds.), Phosphate deposits of the world, v. 2: Phosphate rock resources: Cambridge University Press, Cambridge, p. 53–61.

Shepard, F. P., 1932, Sediments on the continental shelves: Geol. Soc. America Bull., v. 43, p. 1017–1039.

_____ 1961, Deep-sea sand: 21st Internat. Geol. Cong. Rept., p. 23, p. 26–42.

_____ 1973, Submarine geology, 3rd ed.: Harper & Row, New York, 551 p.

_____ 1977, Geological oceanography: Crane, Russak, New York, 214 p.

_____ 1979, Currents in submarine canyons and other types of seavalleys, *in* L. J. Doyle and O. H. Pilkey (eds.), Geology of continental slopes: Soc. Econ. Paleontologists and Mineralogists Spec. Pub. 27, p. 85–94.

Shepard, F. P., and R. F. Dill, 1966, Submarine canyons and other sea valleys: Rand McNally, Chicago, 381 p.

Shepard, F. P., and D. G. Moore, 1955, Central Texas coast sedimentation: Characteristics of sedimentary environment, recent history and diagenesis: Am. Assoc. Petroleum Geologists Bull., v. 39, p. 1463–1593.

Sheriff, R. E., 1980, Seismic stratigraphy: International Human Resources Development Corp., Boston, 227 p.

Sheriff, R. E., and L. P. Geldart, 1982, Exploration seismology: History, theory, and data acquisition: Cambridge University Press, Cambridge, 253 p.

Shields, A., 1936, Application of similarity principles and turbulence research to bed-load movements. Mitteilungen der Preussischen Versuchanstalt für Wasserbau und Schiffbau, Berlin, *in* W. P. Ott and J. C. Van Uehelen (trans.), Calif. Inst. Tech., W. M. Keck Laboratory of Hydraulics and Water Research, Rept. No. 167.

Shimp, N. F., J. Witters, P. E. Potter, and J. A. Schleicher, 1969, Distinguishing marine and freshwater muds: Jour. Geology, v. 77, p. 566–580.

Shinn, E. A., 1971, Holocene submarine sedimentation in the Persian Gulf, *in* O. P. Bricker (ed.), Carbonate cements: Johns Hopkins University Press, Baltimore, Md., p. 63–65.

_____ 1983, Tidal flat environment, *in* P. A. Scholle, D. G. Bebout, and C. H. Moore (eds.), Carbonate depositional environments: Am. Assoc. Petroleum Geologists Mem. 33, p. 171–210.

Shinn, E. A., R. B. Halley, H. H. Hudson, and B. H. Lidz, 1977, Limestone compaction: An enigma: Geology, v. 5, p. 21–24.

Shinn, E. A., and D. M. Robbin, 1983, Mechanical and chemical compaction in fine-grained shallow-water limestones: Jour. Sed. Petrology, v. 53, p. 595–618.

Shinn, E. A., R. P. Steinen, B. H. Lidz, and P. K. Swart, 1989, Whitings, a sedimentologic dilemma: Jour. Sed. Petrology, v. 59, p. 147–161.

Shirley, M. L., and J. A. Ragsdale (eds.), 1966, Deltas in their geologic framework: Houston Geological Society, 251 p.

Sibley, D. F., and H. Blatt, 1976, Intergranular pressure solution and cementation of the Tuscarora Orthoquartzite: Jour. Sed. Petrology, v. 46, p. 881–896.

Sibley, D. F., and J. M. Gregg, 1987, Classification of dolomite rock textures: Jour. Sed. Petrology, v. 57, p. 967–975.

Siemers, C. T., R. W. Tillman, and C. R. Williamson, 1981, Deep-water clastic sediments: A core workshop: Soc. Econ. Paleontologists and Mineralogists Core Workshop No. 2, 416 p.

Siever, R., 1979, Plate-tectonic controls on diagenesis: Jour. Geology, v. 87, p. 127–155.

_____ 1983, Evolution of chert at active and passive continental margins, *in* A. Iijima, J. R. Hein, and R. Siever (eds.), Siliceous deposits in the Pacific region: Developments in Sedimentology 36, Elsevier, Amsterdam, p. 7–24.

Simons, D. B., and E. V. Richardson, 1961, Forms of bed roughness in alluvial channels: Am. Soc. Civil Engineers Proc., Jour. Hydraulics Div., v. 87 (HY3), p. 87–105.

Simonson, B. M., 1985, Sedimentological constraints on the origins of Precambrian iron-formations: Geol. Soc. America Bull., v. 96, p. 244–252.

Simpson, E. L., 1991, An exhumed Lower Cambrian tidal-flat: The Antietam Formation, central Virginia, U.S.A., *in* D. G. Smith, G. E. Reinson, B. A. Zaitlin, and R. A. Rahmani (eds.), Clastic tidal sedimentology: Canadian Soc. Petroleum Geologists, p. 123–136.

Simpson, S., 1975, Classification of trace fossils, *in* R. W. Frey (ed.), The study of trace fossils: Springer-Verlag, New York, p. 39–54.

Singer, J. K., J. B. Anderson, M. T. Ledbetter, I. N. McCave, K. P. N. Jones, and R. Wright, 1988, An assessment of analytical techniques for size analysis of fine-grained sediments: Jour. Sed. Petrology, v. 58, p. 534–543.

Slansky, M., 1986, Geology of sedimentary phosphates: North Oxford Academic Pub., Essex, Great Britain, 210 p.

Sloss, L. L., 1963, Sequences in the cratonic interior of North America: Geol. Soc. America Bull., v. 74, p. 93–114.

Sly, P. G., 1978, Sedimentation processes in lakes, *in* A. Lerman (ed.), Lakes—Chemistry, geology, physics: Springer-Verlag, New York, p. 65–89.

Smith, D. G., G. E. Reinson, B. A. Zaitlin, and R. A. Rahmani (eds.), 1991, Clastic tidal sedimentology: Canadian Soc. Petroleum Geologists Mem. 16, 307 p.

Snedden, J. W., D. Nummedal, and A. F. Amos, 1988, Storm- and fair-weather combined flow on the central Texas continental shelf: Jour. Sed. Petrology, v. 58, p. 580–595.

Sneed, E. D., and R. L. Folk, 1958, Pebbles in the lower Colorado River, Texas, a study in particle morphogenesis: Jour. Geology, v. 66, p. 114–150.

Snelling, N. J. (ed.), 1985a, The chronology of the geological record: Geol. Soc. Mem. 10, Blackwell, Oxford, 343 p.

Snelling, N. J., 1985b, Geochronology and the geological record, *in* N. J. Snelling (ed.), The chronology of the geological record: Geol. Soc. Mem. 10, Blackwell, Oxford, p. 3–9.

Soil Survey Staff, 1975, Soil taxonomy: Handbook, U.S. Department of Agriculture No. 436, U.S. Government Printing Office, Washington, D. C.

Sonnenfeld, P., 1984, Brines and evaporites: Academic Press, London, 624 p.

Sonnenfeld, P., and G. C. St. C. Kendall (convenors), 1989, Marine evaporites: Genesis, alteration, and associated deposits: Penrose Conference Report, Geology, v. 17, p. 573–574.

Soudry, D., 1992, Primary bedded phosphorites in the Campanian Mishash Formation, Negev, Southern Israel: Sed. Geology, v. 80, p. 77–88.

Southard, J. B., and L. A. Boguchwal, 1990, Bed configurations in steady unidirectional water flows, Pt. 2: Synthesis of flume data: Jour. Sed. Petrology, v. 60, p. 658–679.

Spencer, A. M., 1975, Late Precambrian glaciation in the North Atlantic region, *in* A. E. Wright and F. Moseley (eds.), Ice Ages: Ancient and modern: Seel House Press, Liverpool, England, p. 7–42.

Sperling, C. H. B., and R. U. Cooke, 1980, Salt weathering in arid environments: Experimental investigations of the relative importance of hydration and crystallization processes. II. Laboratory studies: Bedford College, London, Papers in Geography 9, 53 p.

Stach, E., 1975, Handbook of coal petrology, 2nd ed.: Gebrüder Borntraeger, Berlin, 428 p.

Stanley, D. J. (ed.), 1969, New concepts of continental margin sedimentation: Am. Geol. Inst. Short Course Notes: Washington, D.C., 400 p.

Stanley, D. J., and G. T. Moore (eds.), 1983, The shelfbreak: Critical interface on continental margins: Soc. Econ. Paleontologists and Mineralogists Spec. Pub. 33, 467 p.

Stanley, D. J., and D. J. P. Swift (eds.), 1976, Marine sediment transport and environmental management: John Wiley & Sons, New York, 602 p.

Stanley, G. D., Jr., and S. D. Cairns, 1988, Constructional azooxanthellate coral communities: An overview with implications for the fossil record: Palaios, v. 3, p. 233–242.

Stanley, G. D., Jr., and J. A. Fagerstrom (eds.), 1988, Ancient reef ecosystems: Palaios, v. 3, p. 111–254.

Stanley, S. M., 1979, Macroevolution, pattern and process: W. H. Freeman, San Francisco, 332 p.

_____ 1985, Rates of evolution: Paleobiology, v. 11, p. 13–26.

_____ 1989, Adaptive radiation and macroevolution, *in* P. D. Taylor and G. P. Larwood (eds.), Major evolutionary radiations: Oxford University Press, Oxford, p. 1–16.

Staplin, F. L., W. G. Dow, C. W. D. Milner, D. I. O'Connor, S. A. J. Pocock, P. van Gijzel, D. H. Welte, and M. A. Yükler, 1982, How to assess maturation and paleotemperatures: Soc. Econ. Paleontologists and Mineralogists Short Course 7, 298 p.

Stein, R., 1985, Rapid grain-size analyses of clay and silt fraction by Sedigraph 5000D: Comparison with Coulter counter and Atterberg methods: Jour. Sed. Petrology, v. 55, p. 590–593.

Stewart, F. H., 1963, Marine evaporites, *in* M. Fleischer (ed.), Data of geochemistry: U.S. Geol. Survey Prof. Paper 440–Y, 54 p.

Stockman, K. W., R. N. Ginsburg, and E. A. Shinn, 1967, The production of lime mud by algae in south Florida: Jour. Sed. Petrology, v. 37, p. 633–648.

Stokes, S., C. S. Nelson, and T. R. Healy, 1989, Textural procedures for the environmental discrimination of late Neogene coastal sand deposits, southwest Auckland, New Zealand: Sed. Geology, v. 61, p. 135–150.

Stopes, M. C., 1919, On the four visible ingredients in banded bituminous coal. Studies in the composition of coal: Royal Soc. London Proc., Ser. B, v. 90, p. 470–487.

_____ 1935, On the petrology of banded bituminous coal: Fuel, London, v. 14, p. 4–13.

Stow, D. A. V., 1986, Deep clastic seas, *in* H. G. Reading (ed.), Sedimentary environments and facies, 2nd ed.: Blackwell, Oxford, p. 399–444.

Stow, D. A. V., and D. J. W. Piper, 1984a, Deep-water fine-grained sediments: Facies models, *in* D. A. V. Stow and D. J. W. Piper (eds.), Fine-grained sediments: Deep-water processes and facies: The Geological Society, Blackwell, Oxford, p. 611–646.

Stow, D. A. V., and Piper (eds.), 1984b, Fine-grained sediments: Deep-water processes and facies: Geol. Soc. Spec. Pub. 15, Blackwell, Oxford, 659 p.

Stride, A. H. (ed.), 1982, Offshore tidal sands: Processes and deposits: Chapman and Hall, London, 222 p.

Stride, A. H., R. H. Belderson, N. H. Kenyon, and M. A. Johnson, 1982, Offshore tidal deposits: Sand sheet and sand bank facies, *in* A. H. Stride (ed.), Offshore tidal sands: Processes and deposits: Chapman and Hall, London, p. 95–125.

Stuiver, M., S. W. Robinson, and I. C. Yang, 1979, ^{14}C dating to 60,000 years B.P. with proportional counters, *in* R. Berger and H. E. Suess (eds.), Radiocarbon dating: University of California Press, Berkeley, p. 202–215.

Stumm, W. (ed.), 1985, Chemical processes in lakes: John Wiley & Sons, New York, 435 p.

Stumm, W., and J. J. Morgan, 1981, Aquatic chemistry. An introduction emphasizing chemical equilibria in natural waters: John Wiley & Sons, New York, 583 p.

Sugden, D. E., and B. S. John, 1976, Glaciers and landscape: Edward Arnold, London, 376 p.

Sundborg, A., 1956, the River Klarälven, a study of fluvial processes: Geograf. Annaler, v. 38, p. 125–316.

Swift, D. J. P., 1975a, Barrier island genesis: Evidence from the Middle Atlantic Shelf of North America: Sed. Geology, v. 14, p. 1–43.

Swift, D. J. P., 1975b, Tidal sand ridges and shoal retreat massifs: Marine Geology, v. 18, p. 105–134.

Swift, D. J. P., D. B. Duane, and O. H. Pilkey (eds.), 1972, Shelf sediment transport: Process and pattern: Dowden, Hutchinson and Ross, Stroudsburg, Pa., 656 p.

Swift, D. J. P., J. R. Schubel, and R. E. Sheldon, 1972, Size analysis of fine-grained suspended sediments: A review: Jour. Sed. Petrology, v. 42, p. 122–134.

Swift, D. J. P., D. J. Stanley, and J. R. Curray, 1971, Relict sediments on continental shelves: A recommendation: Jour. Geology, v. 79, p. 322–346.

Sylvester-Bradley, P. C., 1977, Biostratigraphical tests of evolutionary theory, *in* E. G. Kauffman and J. E. Hazel (eds.), Concepts and methods of biostratigraphy: Dowden, Hutchinson and Ross, Stroudsburg, Pa., p. 41–64.

Syvitski, J. P. M., 1991, Principles, methods, and applications of particle size analysis: Cambridge University Press, Cambridge, 368 p.

Taira, A., and P. A. Scholle, 1979, Discrimination of depositional environments using setling tube data: Jour. Sed. Petrology, v. 49, p. 787–800.

Tankard, A. J., and J. H. Barwis, 1982, Wave-dominated deltaic sedimentation in the Devonian Bokkeveld Basin of South Africa: Jour. Sed. Petrology, v. 52, p. 959–974.

Tardy, Y., G. Bocquier, H. Paquet, and G. Millot, 1973, Formation of clay from granite and its distribution in relation to climate and topography: Geoderma, v. 10, p. 271–284.

Tarling, D. H., 1983, Paleomagnetism: Principles and applications in geology, geophysics and archaeology: Chapman and Hall, London, 379 p.

Taylor, J. M., 1950, Pore-space reduction in sandstones: Am. Assoc. Petroleum Geologists Bull., v. 34, p. 701–716.

Teichert, C., 1958, Concepts of facies: Am. Assoc. Petroleum Geologists Bull., v. 42, p. 2718–2744.

Terwindt, J. H. J., 1988, Paleo-tidal reconstructions of inshore tidal depositional environments, *in* P. L. de Boer, A. van Gelder, and S. D. Nio (eds.), Tide-influenced sedimentary environments and facies: D. Reidel, Dordrecht, p. 233–263.

Tetzlaff, D. M., and J. W. Harbaugh, 1989, Simulating clastic sedimentation: Van Nostrand Reinhold, New York, 202 p.

The Open University Team, 1989, Waves, tides and shallow-water processes: Pergamon Press, Oxford, 187 p.

Thode, H. G., and J. Monster, 1965, Sulfur-isotope geochemistry of petroleum evaporites and ancient seas: Am. Assoc. Petroleum Geologists Mem. 4, p. 367–377.

Thompson, R. W., 1968, Tidal flat sedimentation on the Colorado River delta, northwest Gulf of California: Geol. Soc. America Mem. 107, 133 p.

Tickell, F. G., 1965, The techniques of sedimentary mineralogy: Elsevier, New York, 220 p.

Tietz, G., and G. Müller, 1971, Recent beachrocks, Fuerteventura, Canary Islands, Spain, *in* O. P. Bricker (ed.), Carbonate cements: Johns Hopkins University Press, Baltimore, Md., p. 4–8.

Tillman, R. W., and C. T. Siemers (eds.), 1984, Siliciclastic shelf sediments: Soc. Econ. Paleontologists and Mineralogists Spec. Pub. 34, 268 p.

Tissot, B. P., and Welte, D. H., 1978, Petroleum formation and occurrence: Springer-Verlag, Berlin, 538 p.

Tissot, B. P. and D. H. Welte, 1984, Petroleum formation and occurrence, 2nd ed.: Springer-Verlag, Berlin, 699 p.

Toomey, D. F. (ed.), 1981, European fossil reef models: Soc. Econ. Paleontologists and Mineralogists Spec. Pub. 30, 546 p.

Tourtelot, H. A., 1960, Origin and use of the word "shale": Am. Jour. Sci., Bradley Volume, v. 258–A, p. 335–343.

Trask, P. D. (ed.), 1950, Applied sedimentation: John Wiley & Sons, New York, 707 p.

Trendall, A. F., 1983, Introduction, *in* A. F. Trendall and R. C. Morris (eds.), 1983, Iron-formation facts and problems: Developments in Precambrian Geology 6, Elsevier, Amsterdam, p. 1–12.

Trendall, A. F., and R. C. Morris (eds.), 1983, Iron-formations facts and problems: Developments in Precambrian Geology 6, Elsevier, Amsterdam, 558 p.

Tucker, M. E., J. L. Wilson, P. D. Crevello, J. R. Sarg, and J. F. Read, 1990, Carbonate platforms—Facies, sequences and evolution: Internat. Assoc. Sedimentologists Spec. Pub. 9, Blackwell, Oxford, 328 p.

Tucker, M. E., and V. P. Wright, 1990, Carbonate sedimentology: Blackwell, Oxford, 482 p.

Tucker, R. W., and H. L. Vacher, 1980, Effectiveness of discriminating beach, dune, and river sands by moments and the cumulative weight percentages: Jour. Sed. Petrology, v. 50, p. 165–172.

Turner, C. E., and N. S. Fishman, 1991, Jurassic Lake T'oo'dichi': A large alkaline, saline lake, Morrison Formation, eastern Colorado Plateau: Geol. Soc. America Bull., v. 103, p. 538–558.

Twenhofel, W. H., 1950, Principles of sedimentation, 2nd ed.: McGraw-Hill, New York, 673 p.

Twenhofel, W. H., and collaborators, 1926, Treatise on sedimentation: Williams and Wilkins, Baltimore, 661 p.

Udden, J. A., 1898, Mechanical composition of wind deposits: Augustana Library Pub. 1, 69 p.

Uffen, R. J., 1963, Influence of the earth's core on origin and evolution of life: Nature, v. 198, p. 143.

Vacquier, V., 1972, Geomagnetism in marine geology: Elsevier Oceanography Ser. 6, Elsevier, Amsterdam, 185 p.

Vail, P. R., 1987, Seismic stratigraphy interpretation procedure, in A. W. Bally, (ed.), Atlas of seismic stratigraphy, Am. Assoc. Petroleum Geologists Studies in Geology No. 27, v. 1, p. 1–10.

Vail, P. R., F. Audemard, S. A. Bowman, P. N. Eisner, and C. Perez-Cruz, 1991, The stratigraphic signatures of tectonics, eustasy, and sedimentology—An overview, in G. Einsele, W. Ricken, and A. Seilacher (eds.), Cycles and events in stratigraphy: Springer-Verlag, Berlin, p. 617–659.

Vail, P. R., J. Hardenbol, and R. G. Todd, 1984, Jurassic unconformities, chronostratigraphy, and sea-level changes from seismic stratigraphy and biostratigraphy, in J. S. Schlee (ed.), Interregional unconformities and hydrocarbon accumulation: Am. Assoc. Petroleum Geologists Mem. 36, Tulsa, Okla., p. 129–144.

Vail, P. R., and R. M. Mitchum, Jr., 1977, Seismic stratigraphy and global change of sea level, Part 1: Overview, in C. E. Payton (ed.), Seismic stratigraphy—Applications to hydrocarbon exploration: Am. Assoc. Petroleum Geologists Mem. 26, p. 51–52.

Vail, P. R., R. M. Mitchum, Jr., and S. Thompson, III, 1977a, Seismic stratigraphy and global change of sea level, Part 3: Relative changes of sea level from coastal onlap, in C. E. Payton (ed.), Seismic stratigraphy—Applications to hydrocarbon exploration: Am. Assoc. Petroleum Geologists Mem. 26, p. 63–81.

_____ 1977b, Seismic stratigraphy and global change of sea level, Part 4: Global cycles of relative changes of sea level, in C. E. Payton (ed.), Seismic stratigraphy—Applications to hydrocarbon exploration: Am. Assoc. Petroleum Geologists Mem. 26, p. 83–97.

Vail, P. R., and R. G. Todd, 1981, Northern North Sea Jurassic unconformities, chronostratigraphy, and sea-level changes from seismic stratigraphy, in Petroleum geology of the continental shelf of northwest Europe: Heyden, London, p. 216–235.

Valentine, J. W., 1971, Plate tectonics and shallow marine diversity and endemism, and actualistic model: Systematic Zoology, v. 20, p. 253–264.

_____ 1977a, General patterns of metazoan evolution, in A. Hallam (ed.), Patterns of evolution as illustrated by the fossil record: Elsevier, New York, p. 27–57.

_____ 1977b, Biogeography and biostratigraphy, in E. G. Kauffmann and J. E. Hazel (eds.), Concepts and methods of biostratigraphy: Dowden, Hutchinson and Ross, Stroudsburg, Pa., p. 143–162.

van de Kreeke, J. (ed.), 1986, Physics of shallow estuaries and bays: Lecture notes on coastal and estuarine studies: Springer-Verlag, Berlin, 280 p.

Vandenberghe, N., 1975, An evaluation of CM patterns for grain-size studies of fine grained sediments: Sedimentology, v. 22, p. 615–622.

Van der Leeden, F., 1975, Water resources of the world—Selected statistics: Water Information Centre, Point Washington, N. Y., 568 p.

Van der Linder, G. J. (ed.), 1977, Diagenesis of deep-sea biogenic sediments: Benchmark Papers in Geology, v. 40. Dowden, Hutchinson and Ross, Stroudsburg, Pa., 385 p.

Van der Voo, R., 1993, Paleomagnetism of the Atlantic, Tethys, and Iapetus oceans: Cambridge University Press, New York, 411 p.

Van der Voo, R., C. R. Scotese, and N. Bonhommet (eds.), 1984, Plate reconstruction from Paleozoic paleomagnetism: Geodynamics Series, v. 12, Am. Geophys. Union, Washington, D.C., 136 p.

Van Houten, F. B., 1964, Cyclic lacustrine sedimentation, Upper Triassic Lockatong Formation, central New Jersey and adjacent Pennsylvania, in D. F. Merriam (ed.), Symposium on cyclic sedimentation: Kansas Geol. Survey Bull., v. 169, no. 2, p. 497–531.

Van Houten, F. B. (ed.), 1977, Ancient continental deposits: Benchmark Papers in Geology 43, Dowden, Hutchinson and Ross, Stroudsburg, Pa., 367 p.

Van Houten, F. B., and D. P. Bhattacharyya, 1982, Phanerozoic oolitic ironstone: Geologic record and facies models: Ann. Rev. Earth and Planetary Sci., v. 10, p. 441–457.

Van Straaten, L. M. J. U., 1961, Sedimentation in tidal flat areas: Alberta Soc. Petroleum Geologists Jour., v. 9, p. 203–213, 216–226.

Van Straaten, L. M. J. U. (ed.), 1964, Deltaic and shallow marine deposits: Developments in Sedimentology, v. 1, Elsevier, New York, 464 p.

Van Valen, L., 1973, A new evolutionary law: Evolution Theory, v. 1, p. 1–30.

Van Wagoner, J. C., H. W. Posamentier, R. M. Mitchum, P. R. Vail, J. F. Sarg, T. S. Loutit, and J. Hardenbol, 1988, An overview of the fundamentals of sequence stratigraphy and key definitions, in C. K. Wilgus, B. S. Hastings, C. G. St. C. Kendall, H. W. Posamentier, C. A. Ross, and J. C. Van Wagoner (eds.), Sea-level changes: An integrated approach: Soc. Econ. Paleontologists and Mineralogists Spec. Pub. 42, p. 39–45.

Van Wagoner, J. C., R. M. Mitchum, K. M. Campion, and V. D. Rahmanian, 1990, Siliciclastic sequence stratigraphy in well logs, cores, and outcrops: Am. Assoc. Petroleum Geologists, Methods in Exploration Series No. 7, Tulsa, Okla., 55 p.

Varshal, G. M., I. Ya. Koshcheyeva, I. S. Sirotkina, T. K. Velyukhanova, L. N. Intskirveli, and N. S. Zamokina, 1979, Interactions of metal ions with organic matter in surface waters: Trans. from Geokhimiya No. 4, p. 598–607.

Veizer, J., and R. Demovic, 1974, Strontium as a tool in facies analysis: Jour. Sed. Petrology, v. 44, p. 93–115.

Veizer, J., W. T. Holser, and C. K. Wilgus, 1980, Correlation of $^{13}C/^{12}C$ and $^{34}S/^{32}S$ secular variations: Geochim. et Cosmochim. Acta, v. 44, p. 579–587.

Verdier, A. C., T. Oki, and A. Suardy, 1980, Geology of the Handil Field (East Kalimantan-Indonesia), in M. T. Halbouty (ed.), Giant oil and gas fields of the decade 1968–1978: Am. Assoc. Petroleum Geologists, Tulsa, Okla., p. 399–421.

Vincent, E., and W. H. Berger, 1985, Carbon dioxide and polar cooling in the Miocene: The Monterey hypothesis, in E. T. Sundquist and W. S. Broecker (eds.), The carbon cycle and atmospheric CO_2: Natural variations archean to present: Geophysics Monograph Ser., v. 32, Am. Geophys. Union, Washington, D.C., p. 455–468.

Vine, F. H., and D. H. Matthews, 1963, Magnetic anomalies over oceanic ridges: Nature, v. 199, p. 947–949.

Visher, G. S., 1965, Use of vertical profiles in environmental reconstruction: Am. Assoc. Petroleum Geologists Bull. v. 49, p. 41–61.

_____ 1969, Grain size distributions and depositional processes: Jour. Sed. Petrology, v. 39, p. 1074–1106.

_____ 1984, Exploration stratigraphy: Pennwell, Tulsa, Okla., 334 p.

Visser, J. N. J., 1991, The paleoclimatic setting of the late Paleozoic marine ice sheet in the Karoo Basin of southern Africa, in J. B. Anderson and G. M. Ashley (eds.), Glacial marine sedimentation; paleoclimatic significance: Geol. Soc. Amer. Spec. Paper 261, p. 181–189.

Von Damm, K. L., J. M. Edmond, B. Grant, and C. I. Measures, 1985, Chemistry of submarine hydrothermal solutions at 21° N, East Pacific Rise: Geochim. et Cosmochim. Acta, v. 49, p. 2197–2220.

Wadell, H., 1932, Volume, shape and roundness of rock particles: Jour. Geology, v. 40, p. 443–451.

Walker, R. G., 1978, Deep-water sandstone facies of ancient submarine fans: Models for exploration for stratigraphic traps: Am. Assoc. Petroleum Geologists Bull., v. 62, p. 932–966.

_____ 1979a, Facies and facies models. General introduction, in R. G. Walker (ed.), Facies models: Geoscience Canada Reprint Ser. 1, p. 1–8.

Walker, R. G. (ed.), 1979b, Facies models: Geoscience Canada Reprint Ser. 1, 211 p.

Walker, R. G., 1984a, Shelf and shallow marine sands, in R. G. Walker (ed.), Facies models, 2nd ed.: Geoscience Canada Reprint Ser. 1, p. 141–170.

Walker, R. G. (ed.), 1984b, Facies models, 2nd ed.: Geoscience Canada Reprint Ser. 1, 317 p.

Walker, R. G., 1984c, Turbidites and associated coarse clastic deposits, in R. G. Walker (ed.), Facies models: Geoscience Canada Reprint Ser. 1, p. 171–188.

_____ 1990, Facies modeling and sequence stratigraphy: Jour. Sed. Petrology, v. 60, p. 777–786.

_____ 1992, Facies, facies models and modern stratigraphic concepts, in R. G. Walker and N. P. James (eds.), Facies models—Response to sea level change: Geol. Assoc. Canada, p. 1–14.

Walker, R. G., and N. P. James (eds.), 1992, Facies models—Response to sea level changes: Geol. Assoc. Canada, 407 p.

Walker, R. G., and A. G. Plint, 1992, Wave- and storm-dominated shallow marine systems, in R. G. Walker and N. P. James (eds.), Facies models—Response to sea level changes: Geol. Assoc. Canada, p. 219–238.

Walker, R. G., and D. J. Cant, 1979, Facies models 3. Sandy fluvial systems, in R. G. Walker (ed.), Facies models: Geoscience Canada Reprint Ser. 1, p. 23–31.

_____ 1984, Sandy fluvial systems, in R. G. Walker (ed.), Facies models: Geoscience Canada Reprint Ser. 1, p. 71–89.

Walker, R. G., and G. V. Middleton, 1979, Facies models 4. Eolian sands, in R. G. Walker (ed.), Facies models: Geoscience Canada Reprint Ser. 1, p. 33–41.

Walker, T. R., 1962, Reversible nature of chert-carbonate replacement in sedimentary rocks: Geol. Soc. America Bull., v. 73, p. 237–242.

_____ 1967, Formation of red beds in modern and ancient deserts: Geol. Soc. America Bull., v. 78, p. 353–368.

_____ 1984, 1984 SEPM Presidential Address: Diagenetic albitization of potassium feldspars in arkosic sandstones: Jour. Sed. Petrology, v. 54, p. 3–16.

Walker, T. R., and J. C. Harms, 1972, Eolian origin of flagstone beds, Lyons Sandstone (Permian), type area, Boulder County, Colorado: Mountain Geologists, v. 9, p. 279–288.

Walter, M. R. (ed.), 1976, Stromatolites: Elsevier, New York, 790 p.

Wanless, H. R., and J. M. Weller, 1932, Correlation and extent of Pennsylvanian cyclothems: Geol. Soc. America Bull., v. 43, p. 1003–1016.

Ward, C. R. (ed.), 1984, Coal geology and coal technology: Blackwell, Oxford, 345 p.

Ward, L. G., and G. M. Ashley, 1989, Physical processes and sedimentology of siliciclastic-dominated lagoonal systems: Marine Geology, v. 88, p. 181–364.

Warme, J. E., R. G. Douglas, and E. L. Winterer (eds.), 1981, The Deep Sea Drilling Project: A decade of progress: Soc. Econ. Paleontologists and Mineralogists Spec. Pub. 32, 564 p.

Warren, J. K., 1989, Evaporite sedimentology: Prentice-Hall, Englewood Cliffs, N.J., 285 p.

Warren, J. K., and G. C. St. C. Kendall, 1985, Comparison of marine sabkhas (subaerial) and salina (subaqueous) evaporites: Modern and ancient: Am. Assoc. Petroleum Geologists Bull., v. 69, p. 843–858.

Watkins, N. D., 1972, A review of the development of the geomagnetic polarity time scale and discussion of prospects for its finer definition: Geol. Soc. America Bull., v. 83, p. 551–574.

Weaver, M., and S. W. Wise, Jr., 1974, Opaline sediments of the southeastern coastal plain and Horizon A: Biogenic origin: Science, v. 184, p. 899–901.

Wei, W., 1993, Calibration of Upper Pliocene–Lower Pleistocene nannofossil events with oxygen isotope stratigraphy: Paleooceanography, v. 8, p. 85–99.

Weimer, P., 1989, Sequence stratigraphy, facies geometries, and depositional history of the Mississippi Fan, Gulf of Mexico: Am. Assoc. Petroleum Geologists Bull., v. 74, p. 425–453.

Weimer, R. J., J. D. Howard, and D. R. Lindsay, 1982, Tidal flats and associated tidal channels, *in* P. A. Scholle and D. Spearing (eds.), Sandstone depositional environments: Am. Assoc. Petroleum Geologists Mem. 31, p. 191–245.

Weimer, P., and M. H. Link (eds.), 1991, Seismic facies and sedimentary processes of submarine fans and turbidite systems: Springer-Verlag, New York, 447 p.

Weise, B. R., 1980, Wave-dominated deltaic systems of the Upper Cretaceous San Miguel Formation, Maverick Basin, South Texas: Texas Bur. Econ. Geology Report of Investigations 107, 39 p.

Weller, J. M., 1958, Stratigraphic facies differentiation and nomenclature: Am. Assoc. Petroleum Geologists Bull., v. 42, p. 609–639.

———— 1960, Stratigraphic principles and practices: Harper and Brothers, New York, 725 p.

Wentworth, C. K., 1919, A laboratory and field study of cobble abrasion: Jour. Geology, v. 27, p. 507–521.

———— 1922, A scale of grade and class terms for clastic sediments: Jour. Geology, v. 30, p. 377–392.

Whateley, M. K. G., and K. T. Pickering (eds.), 1989, Deltas—Sites and traps for fossil fuels: Geol. Soc. Spec. Pub. 41, Blackwell, Oxford, 360 p.

Wheeler, H. E., and V. S. Mallory, 1956, Factors in lithostratigraphy: Am. Assoc. Petroleum Geologists Bull., v. 40, p. 2711–2723.

Whitaker, F. F., and P. L. Smart, 1990, Active circulation of saline ground water in carbonate platforms: Evidence from the Great Bahama Bank: Geology, v. 18, p. 200–203.

White, D. E., 1965, Saline waters of sedimentary rocks, *in* A. Young and J. E. Galley (eds.), Fluids in subsurface environments: Am. Assoc. Petroleum Geologists Mem. 4, p. 342–366.

Wiley, M. (ed.), 1976, Estuarine processes, v. II, Circulation, sediments, and transfer of material in the estuary: Academic Press, New York, 428 p.

Wilgus, C. K., B. S. Hastings, C. G. St. C. Kendall, H. W. Posamentier, C. A. Ross, and J. C. Van Wagoner (eds.), 1988, Sea-level changes: An integrated approach: Soc. Econ. Paleontologists and Mineralogists Spec. Pub. 42, 407 p.

Wilkinson, B. H., R. M. Owen, and A. R. Carroll, 1985, Submarine hydrothermal weathering, global eustasy, and carbonate polymorphism in Phanerozoic marine oolites: Jour. Sed. Petrology, v. 55, p. 171–183.

Wilkinson, B. R., 1982, Cyclic cratonic carbonates and Phanerozoic calcite seas: Jour. Geol. Education, v. 30, p. 189–203.

Williams, D. F., I. Lerche, and W. E. Full, 1988, Isotope chronostratigraphy: Theory and methods: Academic Press, San Diego, 345 p.

Williams, H., F. J. Turner, and C. M. Gilbert, 1982, Petrography, 2nd ed.: W. H. Freeman, San Francisco, 626 p.

Williams, L. A., and D. A. Crerar, 1985, Silica diagenesis, II. General mechanisms: Jour. Sed. Petrology, v. 55, p. 312–321.

Williams, L. A., G. A. Parks, and D. A. Crerar, 1985, Silica diagenesis, I. Solubility controls: Jour. Sed. Petrology, v. 55, p. 301–311.

Wilson, C., 1992, Sequence stratigraphy: An introduction, *in* G. C. Brown, C. J. Hawkesworth, and R. C. L. Wilson (eds.), Understanding the earth: Cambridge University Press, Cambridge, p. 388–414.

Wilson, I. G., 1972, Aeolian bedforms—Their development and origins: Sedimentology, v. 19, p. 173–210.

Wilson, J. L., 1975, Carbonate facies in geologic history: Springer-Verlag, Berlin, 471 p.

Wilson, J. L., and C. Jordan, 1983, Middle shelf, *in* P. A. Scholle, D. G. Bebout, and C. H. Moore (eds.), Carbonate depositional environments: Am. Assoc. Petroleum Geologists Mem. 33, p. 297–344.

Winland, H. D., and R. K. Matthews, 1974, Origin and significance of grapestones, Bahama Islands: Jour. Sed. Petrology, v. 44, p. 921–927.

Wolf, K. H., G. V. Chilingarian, and F. W. Beales, 1967, Elemental composition of carbonate skeletons, minerals, and sediments, *in* G. V. Chilingarian, H. J. Bissell, and R. W. Fairbridge (eds.), Carbonate rocks: Developments in Sedimentology 9B, Elsevier, New York, p. 23–150.

Woodruff, F., and S. M. Savin, 1991, Mid-Miocene isotope stratigraphy in the deep sea: High-resolution correlations, paleoclimatic cycles, and sediment preservation: Paleooceanography, v. 6, p. 755–806.

Wright, A. E., and F. Moseley, 1975, Ice ages: Ancient and modern: Geol. Jour., Spec. Issue 6, Seel House Press, Liverpool, 320 p.

Wright, L. D., 1977, Sediment transport and deposition at river mouths: A synthesis: Geol. Soc. America Bull., v. 88, p. 857–868.

———— 1978, River deltas, *in* R. A. Davis, Jr. (ed.), Coastal sedimentary environments: Springer-Verlag, New York, p. 5–68.

———— 1985, River deltas, *in* R. A. Davis, Jr. (ed.), Coastal sedimentary environments, 2nd ed.: Springer-Verlag, New York, p. 1–76.

Wright, V. P., 1992, A revised classification of limestones: Sed. Geology, v. 76, p. 177–185.

Wuellner, E. E., L. R. Lehtonen, and W. C. James, 1986, Sedimentary-tectonic development of the Marathon and Val Verde Basins, west Texas, U.S.A.: A Permo-Carboniferous migrating foredeep, *in* P. A. Allen and P. Homewood (eds.), Foreland basins: Internat. Assoc. Sedimentologists Spec. Pub. 8, Blackwell, Oxford, p. 347–368.

Wyllie, P. J., 1976, The way the earth works: John Wiley & Sons, New York, 296 p.

Wyrwoll, K.-H., and G. K. Smyth, 1985, On using the log-hyperbolic distribution to describe the textural characteristics of eolian sediments: Jour. Sed. Petrology, v. 55, p. 471–478.

Yalin, M. S., 1977, Mechanics of sediment transport, 2nd ed.: Pergamon, New York, 298 p.

Yanov, E. N., 1978, Classification of sandstones and siltstones by composition of grains: Lithology and Mineral Resources, v. 12, p. 466–472.

Yeh, H., and S. Savin, 1977, The mechanisms of burial diagenetic reactions in argillaceous sediments: 3. Oxygen isotope evidence: Geol. Soc. America Bull., v. 88, p. 1321–1330.

Yen, T. F., and G. V. Chilingarian, 1976a, Introduction to oil shales, *in* T. F. Yen and G. V. Chilingarian (eds.), Oil shales: Elsevier, New York, p. 1–12.

Yen, T. F., and G. V. Chilingarian (eds.), 1976b, Oil shales: Elsevier, New York, 292 p.

York, D., and R. M. Farquhar, 1972, The earth's age and geochronology: Pergamon, New York, 178 p.

Young, F. G., and G. E. Reinson, 1975, Sedimentology of Blood Reserve and adjacent formations (Upper Cretaceous), St. Mary River, southern Alberta, *in* M. S. Shawa (ed.), Guidebook to selected sedimen-tary environments in southwestern Alberta, Canada: Canadian Soc. Petroleum Geologists Field Conference, p. 10–20.

Young, T. P., 1989, Phanerozoic ironstones: An introduction and review, *in* T. P. Young and W. E. G. Taylor (eds.), Phanerozoic ironstones: Geol. Soc. Spec. Pub. 46, The Geological Society, London, p. ix–xxv.

Young, T. P., and W. E. G. Taylor (eds.), 1989, Phanerozoic ironstones: Geol. Soc. Spec. Pub. 46, The Geological Society, London, 251 p.

Zenger, D. H., and J. B. Dunham, 1980, Concepts and models of dolomitization—An introduction, *in* D. H. Zenger, J. B. Dunham, and R. L. Ethington (eds.), Concepts and models of dolomitization: Soc. Econ. Paleontologists and Mineralogists Spec. Pub. 28, p. 1–9.

Zenger, D. H., J. B. Dunham, and R. L. Ethington (eds.), 1980, Concepts and models of dolomitization: Soc. Econ. Paleontologists and Mineralogists Spec. Pub. 28, 320 p.

Zenger, D. H., and S. J. Mazzullo (eds.), 1982, Dolomitization: Benchmark Papers in Geology, v. 65. Dowden, Hutchinson and Ross, Stroudsburg, Pa., 426 p.

Zingg, Th., 1935, Beiträge zur Schotteranalyse: Schweiz. Mineralog. Petrog. Mitt., v. 15, p. 39–140.

Zuffa, G. G., 1980, Hybrid arenites: Their composition and classification: Jour. Sed. Petrology, v. 50, p. 21–29.

Zuffa, G. G. (ed.), 1985, Provenance of arenites: D. Reidel, Dordrecht, 408 p.

Index